Feedback Control System Analysis and Synthesis

McGraw-Hill Electrical and Electronic Engineering Series

FREDERICK EMMONS TERMAN, *Consulting Editor*
W. W. HARMAN AND J. G. TRUXAL, *Associate Consulting Editors*

AHRENDT AND SAVANT · Servomechanism Practice
ANGELO · Electronic Circuits
ASELTINE · Transform Method in Linear System Analysis
ATWATER · Introduction to Microwave Theory
BAILEY AND GAULT · Alternating-current Machinery
BERANEK · Acoustics
BRACEWELL · The Fourier Transform and Its Application
BRENNER AND JAVID · Analysis of Electric Circuits
BROWN · Analysis of Linear Time-invariant Systems
BRUNS AND SAUNDERS · Analysis of Feedback Control Systems
CAGE · Theory and Application of Industrial Electronics
CAUER · Synthesis of Linear Communication Networks
CHEN · The Analysis of Linear Systems
CHEN · Linear Network Design and Synthesis
CHIRLIAN · Analysis and Design of Electronic Circuits
CHIRLIAN AND ZEMANIAN · Electronics
CLEMENT AND JOHNSON · Electrical Engineering Science
COTE AND OAKES · Linear Vacuum-tube and Transistor Circuits
CUCCIA · Harmonics, Sidebands, and Transients in Communication Engineering
CUNNINGHAM · Introduction to Nonlinear Analysis
D'AZZO AND HOUPIS · Feedback Control System Analysis and Synthesis
EASTMAN · Fundamentals of Vacuum Tubes
FEINSTEIN · Foundations of Information Theory
FITZGERALD AND HIGGINBOTHAM · Basic Electrical Engineering
FITZGERALD AND KINGSLEY · Electric Machinery
FRANK · Electrical Measurement Analysis
FRIEDLAND, WING, AND ASH · Principles of Linear Networks
GHAUSI · Principles and Design of Linear Active Circuits
GHOSE · Microwave Circuit Theory and Analysis
GREINER · Semiconductor Devices and Applications
HAMMOND · Electrical Engineering
HANCOCK · An Introduction to the Principles of Communication Theory
HAPPELL AND HESSELBERTH · Engineering Electronics
HARMAN · Fundamentals of Electronic Motion
HARMAN · Principles of the Statistical Theory of Communication
HARMAN AND LYTLE · Electrical and Mechanical Networks
HARRINGTON · Introduction to Electromagnetic Engineering
HARRINGTON · Time-harmonic Electromagnetic Fields
HAYASHI · Nonlinear Oscillations in Physical Systems
HAYT · Engineering Electromagnetics
HAYT AND KEMMERLY · Engineering Circuit Analysis
HILL · Electronics in Engineering
JAVID AND BRENNER · Analysis, Transmission, and Filtering of Signals
JAVID AND BROWN · Field Analysis and Electromagnetics
JOHNSON · Transmission Lines and Networks
KOENIG AND BLACKWELL · Electromechanical System Theory
KRAUS · Antennas
KRAUS · Electromagnetics
KUH AND PEDERSON · Principles of Circuit Synthesis
LEDLEY · Digital Computer and Control Engineering

Feedback Control

System Analysis

and Synthesis

John J. D'Azzo
Constantine H. Houpis

ASSOCIATE PROFESSORS OF ELECTRICAL ENGINEERING
DEPARTMENT OF ELECTRICAL ENGINEERING
AIR FORCE INSTITUTE OF TECHNOLOGY

Second Edition

McGraw-Hill Book Company
NEW YORK, ST. LOUIS, SAN FRANCISCO, TORONTO, LONDON, SYDNEY

Feedback Control System Analysis and Synthesis

Library of Congress Catalog Card Number 65-17391

ISBN 07-016175-5

15 16 17 K P K P   7 9 8 7 6

Preface

The main objective in writing this book has been to produce a good textbook from the student's point of view. The authors have been teaching courses in automatic feedback controls, at both the undergraduate level and the graduate level, for a number of years. As a result of this experience, they felt the need for a textbook on feedback controls that would unify the diverse methods of analysis and would present the fundamentals explicitly and clearly. This enlarged second edition gives the material in a sequence that, in the authors' experience, enables the reader to understand and apply the basic principles. Terminology and definitions are introduced and explained as needed; emphasis is placed on the topics considered most important; and the material is arranged for ease in understanding.

As science and engineering proceed into the arena of knowns and unknowns, the past must continuously be reevaluated. This was borne in mind in the preparation of the second edition. Some topics were compacted and others amplified; some were relocated; and some were introduced in order to achieve a more cohesive and up-to-date text. An engineer having a thorough understanding of the material in this text and having available both analog and digital computers has both an up-to-date knowledge and powerful analytic means for the analysis and synthesis of control systems.

The first four chapters present the basic mathematics of the analysis of control systems. Account is given, also, of the physical laws of nature used in writing the dynamic equations describing the performance of various systems, followed by an outline of the classical and Laplace-transform methods of solving these differential equations. Chapter 4 emphasizes utilization of the Spirule for graphically determining the coefficients in partial-fraction expansions (residues) and for determining the frequency response from a pole-zero diagram. Next there is introduced the block-diagram method for simplifying the representation of complex control systems; each basic function in the control system is represented by a block, and the transfer function is developed to represent the block mathematically. The signal flow-graph technique is given as an alternative means for conveying the same information as the block diagram. Basic servomechanism characteristics are described, followed by details of the root-locus method of analysis. Important specialized topics are discussed in Chapter 8; if time permits, an instructor can include some or all of them in his automatic control system course.   Then the frequency-response method of analysis is presented, using both the log and the polar plots.   The chapters on compensation describe the possible improvements in system performance, along with examples of the techniques for applying compensators.   Next follows a chapter on a-c carrier systems, including a discussion of a-c compensators.   The remaining chapters cover some criteria for determining the optimum response that can be achieved, the principles of analysis of complex or multiple-loop systems with several inputs, several basic methods of treating systems with certain types of non-linearities, the use of analog computers for system simulation, and the principles of over-all system design.   Computers are used extensively by control systems engineers, especially for the design of systems that contain non-linearities.   Chapter 21 contains an introduction to adaptive and sampled-data systems.   The Appendix contains tables, root-solving methods, etc., that are useful to a feedback control system engineer.   Engineers who are required to plot many root-locus curves will find references 10 and 11 at the end of Chapter 7, describing methods of obtaining the root locus by means of a computer, particularly useful.

This book differs from the majority of existing textbooks in presenting the root-locus method of analysis before the frequency-response method.   It is believed that the root-locus method gives a better "feel" for the performance that can be achieved by a specified system.   Also, the log magnitude and phase angle plots are introduced before the polar plots, which appears to be the logical sequence.   In the realm of design criteria the authors have attempted to be as explicit as possible and to explain the philosophy of design in those portions based purely on experience.   The chapter on feedback compensation contains more detail and description of the basis for design than given in most other books on introductory control systems theory.   Numerous examples illustrate all major points and aspects of the theory developed.

The scope of the text is such that it can be used for both undergraduate and graduate courses in feedback control systems.   The starting point for a particular group depends upon the background of the students.   For graduate students with a previous introductory course the first several chapters may be omitted, utilized for quick review, or viewed as a good reference.   The material in Appendixes F and I can be included as part of a graduate course.

The text is arranged so that it can be used for self-study by the engineer in practice.   Included are as many examples of feedback control systems in various areas of practice (electrical, aeronautical, mechanical, etc.) as space permits while maintaining a strong basic feedback control text that can be used for study in any of the various branches of engineering.   To make the text meaningful and valuable to all engineers, the authors have attempted to unify the treatment of physical control systems through use of mathematical and block-diagram models common to all.   A large portion of the text has been class-tested, thus enhancing its value for classroom and self-study use.

The styling of the second edition has been changed from the first edition in a manner to provide more space around the figures. This necessarily has resulted in an increase in the number of pages over the first edition.   It is believed that this will make the book easier to read and more pleasing to the eye.

Many varied problems are included at the end of the text; and in this connection the authors wish to thank Professors L. A. Gould and G. C. Newton, Jr., of the Massachusetts Institute of Technology, who have generously permitted the inclusion of some of their problems.

The authors express their thanks to the students who have used this book and to the faculty who have reviewed it for their helpful comments and corrections.   Especial appreciation is expressed to Dr. C. M. Zieman, Head of the Electrical Engineering Department, Air Force Institute of Technology, for the encouragement he has given, and to Dr. John G. Truxal, Vice-President, Polytechnic Institute of Brooklyn, and Dr. T. J. Higgins, Professor of Electrical Engineering, University of Wisconsin, for their thorough review of the manuscript.

The authors thank all those who have used the first edition and were kind enough to submit their helpful comments.   In particular, appreciation is expressed to Professor T. A. Rogers, University of California, for his detailed review of the first edition, his suggestions for the second edition, and his permission to include some of his problems.

John J. D'Azzo

Constantine H. Houpis

Contents

1

Introduction

1-1   General Introduction

To a great extent, the art of automatic control systems dominates the way of life of modern America and can be considered as a catalyst in ensuring the peace or destruction of the world.   The automatic toaster, the alarm clock, the thermostat, and the control systems that have speeded up the production and quality of manufactured goods have all influenced the current way of life in America.   The control systems that are so necessary in the area of nuclear work and in guided missiles have resulted in a destructive capability that can be a powerful deterrent to war, resulting in peace, or can cause the destruction of mankind.   In the near future, technological development will make it possible to travel to outer space and other planets.   Space vehicles whose successful operation will depend on the proper functioning of a large number of control systems will be used in this venture.

1-2   Introduction to Control Systems

Assume that a toaster timer is set for the desired darkness of the toasted bread.   The setting of the "darkness," or timer, knob represents the input quantity, and the degree of darkness of the toast produced is the output quantity.   If the degree of darkness is not satisfactory, because of the condition of the bread or for some similar reason, this condition can in no way automatically alter the time that the heat is applied.   Thus it can be said that the output quantity has no influence on the input quantity.   The heater portion of the toaster, excluding the timer unit, represents the dynamic part of the over-all system.

Another example is a d-c shunt motor.   For a given value of field current, a certain value of voltage is applied to the armature to produce the desired value of motor speed.   In this case the motor is the dynamic part of the system, the applied armature voltage is the input quantity, and the speed is the output quantity.   A variation of the speed from the desired value, due to a change of mechanical load on the shaft, can in no way cause a change in the value of the applied armature voltage to maintain the desired speed.   In this example it can also be said that the output quantity has no influence on the input quantity.

Systems in which the output quantity has no effect upon the input quantity are called *open-loop control systems*.   The two examples just cited can be represented symbolically by a functional block diagram, as shown in Fig. 1-1.   The desired darkness of the toast or the desired speed of the motor is the command input; the selection of the value of time on the toaster timer or the value of voltage applied to the motor armature is represented by the reference-selector block; and the output of this block is identified as the reference input.   The reference input is applied to the dynamic-unit block that performs the desired control function, and the output of this block is the desired output.

A human being can be added to the systems above for the purpose of sensing the actual value of the output with respect to the reference input. If the output does not have the desired value, he can alter the reference-selector position to achieve this value.   Addition of the human being has provided a means through which the output is fed back and by which the

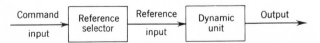

Fig. 1-1   *Functional block diagram of an open-loop control system.*

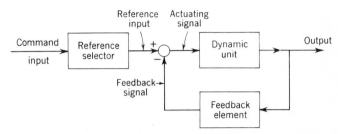

Fig. 1-2 *Functional block diagram of a closed-loop control system.*

output is compared with the input.   Any necessary change is then made in order to cause the output to equal the desired value.   The feedback action has controlled the input to the dynamic unit.   Thus, systems in which the output has an effect upon the input quantity are called *closed-loop control systems.*

To improve the performance of the closed-loop system with respect to maintaining the output quantity as close as possible to the desired quantity, the human being is replaced by a mechanical, electrical, or other form of comparison unit.   The functional block diagram of a closed-loop control system is illustrated in Fig. 1-2.   The comparison between the reference input and the feedback signals results in an actuating signal that is the difference between these two quantities.   The actuating signal acts in a manner to maintain the output at the desired value.

This system may now be properly labeled a *closed-loop control system.* The designation *closed-loop* implies the action resulting from the comparison between the output and input quantities in order to maintain the output at the desired value.   Thus, the output is controlled in order to maintain the desired value.

1-3   *Definitions*

From the preceding discussion the following definitions are evolved, based in part on the proposed standards of the IEEE* which are given in more detail in Appendix C.   The variations from the standards in the definitions below are intended to tie in with the introductory material.

1. *System.*   A combination of components that act together.   The word *systems* as used herein shall be interpreted to include physical,

*Formerly AIEE and IRE.

biological, organizational and other entities, and combinations thereof, which can be represented through a common mathematical symbolism. The formal name *systems engineering* may be assigned to this definition of the word *systems*. Thus the study of feedback control systems is essentially a study of an important aspect of systems engineering.

2. *Command Input.* The motivating input signal to the system, which is independent of the output of the system.

3. *Reference Selector* (reference input element). The unit that establishes the value of the reference input. The reference selector is calibrated in terms of the desired value of the system output.

4. *Reference Input.* The reference signal produced by the reference selector. It is the actual signal input to the control system.

5. *Control Unit* (dynamic element). The unit that reacts to an actuating signal to produce a desired output. This unit does the work of controlling the output and thus may be a power amplifier.

6. *Output* (controlled variable). The quantity that must be maintained at a prescribed value.

7. *Open-loop Control System.* A system in which the output has no effect upon the input signal.

8. *Feedback Element.* The unit that provides the means for feeding back the output quantity, or a function of the output, in order to compare it with the reference input.

9. *Actuating Signal.* The signal that is the difference between the reference input and the feedback signal. It actuates the control unit in order to maintain the output at the desired value.

10. *Closed-loop Control System.* A system in which the output has an effect upon the input quantity in such a manner as to maintain the desired output value.

Note that the fundamental difference between the open- and closed-loop systems is the *feedback action*, which can be continuous or discontinuous. Continuous control implies that the output is continuously being fed back, in time, and compared with the reference input. In one form of discontinuous control the input and output quantities are periodically sampled and compared; that is, the control action is discontinuous in time. This type is commonly called a pulsed-data or sampled-data feedback control system. In another form of discontinuous control system, the actuating signal must reach a prescribed value before the dynamic unit reacts to it; that is, the control action is discontinuous in amplitude rather than in time. This type of discontinuous control system is commonly called an on-off or relay feedback control system. Both forms may be present in a system. In this text only continuous control systems are considered since they lend themselves readily to a basic understanding of feedback control systems, without use of complicated mathematics.

With the above introductory material, it seems proper to state the proposed definition[1]* of a feedback system: "A feedback control system is a control system which tends to maintain a prescribed relationship of one system variable to another by comparing functions of these variables and using the difference as a means of control." In other books and papers on this subject the following terms may be used.

Servomechanism (often abbreviated as servo). This term is sometimes used to refer to a mechanical system. Occasionally it refers to a mechanical system in which the steady-state error is zero for a constant input signal.

Regulator. This term is used to refer to systems in which there is a constant steady-state value for a constant input signal. The name is derived from the early speed and voltage controls, called speed and voltage regulators.

The reader is cautioned to ascertain the meaning of a particular author. Throughout this text an attempt is made to conform to the proposed IEEE definitions.

1-4   *Historical Background*

The action of a human being in walking from a starting point to a destination point along a prescribed path satisfies the definition of a feedback control system. In Fig. 1-3, the prescribed path is the reference input. The eyes perform the function of comparing the actual path of movement with the prescribed path, the desired output. The eyes transmit a signal to the brain, which amplifies this signal and transmits a signal to the legs to correct the actual path of movement to bring it in line with the desired path. Thus, based upon the definition, from the beginning of the existence of the human race there has existed a feedback control system.

One of the earliest open-loop control systems was Hero's device for opening the doors of a temple in the first century.[2] The command input to the system (see Fig. 1-4) was the lighting of the fire upon the altar. The expanding hot air under the fire drove the water from the container into the bucket. As the bucket became heavier, it descended and turned the door spindles by means of a rope, causing the counterweight to rise. The door could be closed by dousing the fire. As the air in the container cooled and

---

* Superscript numbers refer to items in the Bibliography at the end of the chapter.

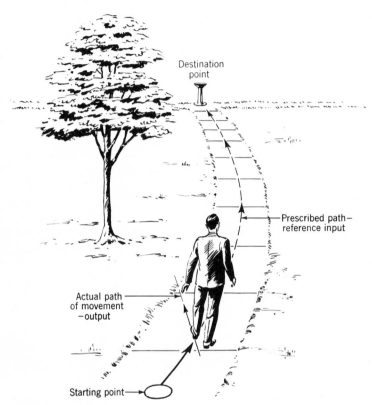

Fig. 1-3   *A pictorial demonstration of a human being acting as a feedback control system.*

the pressure was thereby reduced, the water from the bucket siphoned back into the storage container.   Thus the bucket became lighter and the counterweight, being heavier, moved down, thereby closing the door.   This occurs as long as the bucket is higher than the container.   The device was probably actuated when the ruler and his entourage started to ascend the temple steps. The system for opening the door was not visible or known to the masses. Thus it created an air of mysticism and demonstrated the power of the Olympian gods.

James Watts's flyball governor for controlling speed, developed in 1788, can be considered the first widely used feedback control system not involving a human being.   Maxwell, in 1868, made an analytic study of the stability of the flyball governor.   This was followed by a more detailed solution of the stability of a third-order flyball governor in 1876 by the Russian engineer Wischnegradsky.[3]   Minorsky made one of the early deliberate applications

Fig. 1-4   *Hero's device for opening temple doors.*

of nonlinear elements in closed-loop systems in his study of automatic ship steering about 1922.[4]

A significant date in the history of automatic feedback control systems is 1934, when Hazen's paper "Theory of Servomechanisms" was published in the *Journal of the Franklin Institute*, marking the beginning of the very intense modern interest in this new field.   It was in this paper that the word *servomechanism* originated, from the words *servant* (or slave) and *mechanism*. The word *servomechanism* thus implies a slave mechanism.   It is interesting to note that in the same year Black's important paper on feedback amplifiers appeared.[5]   During the following six years, further basic work was accomplished.   Owing to World War II security restrictions, the develop-

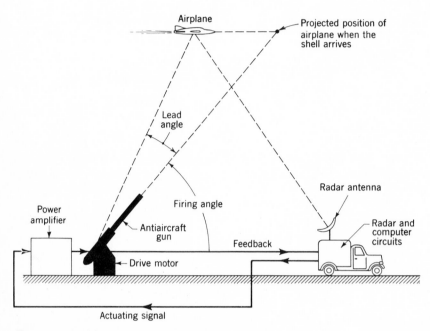

Fig. 1-5 *Antiaircraft radar tracking control system.*

ments in the period 1940 to 1945 were obscured, delaying rapid progress in this field.   During this time three important laboratories organized at the Massachusetts Institute of Technology—the Servomechanisms, Radiation, and Instrumentation Laboratories—contributed much to the advancement of the control field.   The research done by many companies during this period also helped to strengthen the foundation of this new science.   Since the lifting of wartime security restrictions in 1945, rapid progress has been made in the control field.   Since then, books and thousands of articles and technical papers have been written, and the application of control systems in the industrial and military fields has been extensive.   This rapid growth of feedback control systems was accelerated by the equally rapid development and widespread use of automatic computers.

   An example of a military application of a feedback control system is the antiaircraft radar tracking control system shown in Fig. 1-5.   The radar antenna detects the position and velocity of the target airplane, and the computer circuit takes this information and determines the correct firing angle for the gun.   This angle includes the necessary lead angle so that the shell reaches the projected position at the same time as the airplane.   The output signal of the computer, a function of the firing angle, is fed into a

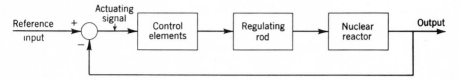

Fig. 1-6 *A nuclear-reactor power-level control system.*

power amplifier whose output voltage is applied to the drive motor. The motor then tilts the gun to the desired firing angle. A feedback signal proportional to the gun position ensures correct alignment with the position determined by the computer. The gun must be positioned both horizontally and vertically. Therefore this system has two drive motors which are parts of two separate feedback loops.

The advent of the nuclear reactor is a milestone in the advancement of science and technology. For proper operation the power level of the reactor must be maintained at a desired value or must vary in a prescribed manner. This must be accomplished automatically with minimum supervision by a human being. Figure 1-6 is a block diagram of a feedback control system for the power output level of a reactor. If the power output level differs from the reference input value, the actuating signal produces a signal at the output of the control elements. This, in turn, moves the regulating rod in the proper direction to achieve the desired power level of the nuclear reactor. The position of the regulating rod determines the rate of nuclear fission and, therefore, the total power generated. This output nuclear power, as an example, can be converted to steam power for generating electric energy.

Throughout the period of recorded history, man has striven to simplify the manner of doing things, to speed up the production and quality of manufactured goods, and to improve the performance of control devices. The advent of the steam engine, and the subsequent industrial revolution, provided man with larger quantities of controlled power than he had had at his command earlier. This created the need for controlling large amounts of power by means of low-power input signals and paved the way for the development of feedback control system analysis. As is frequently true of a science in its infant stage, the large number of individuals with many different backgrounds who have worked in this field has resulted in a large number of terms and definitions. Since about 1950 an effort has been made—through the creation of terminology and definition committees in several major engineering societies (IEEE, ASME, AIChE)—to eliminate this confusion and to establish a set of definitions that can serve as a standard.

In considering the historical background, one cannot overlook the fact that the development of the control concept in the engineering field brought with it the necessity for greater control in the realm of human engineering. The basic concept of feedback control has come to be utilized extensively in the field of management. Thus standards of optimum performance are established in all areas of endeavor, the actual performance is compared with the standard, and any difference between the two is used to bring them into closer agreement.

1-5   *Mathematical Background*

The early studies of control systems were based upon the solution of differential equations by classical means. Other than for simple systems, the analysis in this approach is tedious and does not readily indicate what changes should be made to improve system performance. Use of the Laplace transform simplifies this analysis somewhat. Nyquist's paper[6] published in 1932 dealt with the application of steady-state frequency-response calculations to feedback amplifier design. This work was extended by Black[5] and Bode.[7] Hall[8] and Harris[9] then applied frequency-response analysis in the study of feedback control systems. This latter development was of considerable aid in the development of control theory as a whole.

Another advance occurred in 1948 when Evans[10] presented his root-locus theory. This theory affords a graphical display of the stability properties of a system and permits the graphical evaluation of the frequency response. Laplace-transform theory and network theory are joined in the root-locus calculation. Further on in the text the reader will come to appreciate the simplicity and value of this phase of analysis.

In the various phases of linear analysis presented in this text, mathematical models are used. Once a physical system has been described by a set of mathematical equations, they are manipulated to achieve one of the mathematical models. Once this has been done, the manner of analysis is independent of whether the physical system is electrical, mechanical, etc., in nature. This technique helps the designer to spot similarities with his previous experience. The reader should keep in mind, as he covers the various phases of analysis presented here, that no one aspect is intended to be used to the exclusion of the others. Depending upon the known factors and the simplicity (or complexity) of a control-system problem, a designer may use one phase exclusively or a combination of various aspects of each phase. With experience in the field of feedback control system design comes the ability to utilize to a greater extent the advantages of each method. As mentioned earlier, the use of automatic computers greatly aids the designer in making a synthesis of his control problem.

1-6 *General Nature of the Engineering Control Problem*

In general, a control problem can be divided into the following steps:

1. A set of performance specifications must be established.
2. As a result of the performance specifications a control problem exists.
3. A set of differential equations that describe the physical system must be formulated.
4. The performance of the basic (or original) system is determined by application of one of the methods of analysis (or a combination of them).
5. If the performance of the original system does not meet the required specifications, equipment must be added to improve the response.
6. The achievement of optimal performance for the required system response specifications may be required. This step is referred to as system *optimization*. (There may be more than one "optimal" response since there are various definitions of *optimal*.)

To maintain the actual performance of a system close to the desired performance creates a control problem. The necessary basic equipment is then assembled into a system to perform the desired control function. To a varied extent, most systems are nonlinear. In many cases the nonlinearity is so small that it can be neglected, or the limits of operation are small enough that a linear analysis can be made. In this textbook linear systems or those that can be approximated as linear systems are considered. Because of the relative simplicity and straightforwardness of this approach, the reader can obtain a thorough understanding of linear systems. After mastering the terminology, definitions, and methods of analysis for linear control systems, the engineer will find it easier to undertake a study of nonlinear systems. Some of the principles of nonlinear operation are included in a separate chapter.

A basic system has the minimum amount of equipment necessary to accomplish the control function. The differential equations that describe the physical system are derived, and an analysis of the basic system is made. If the analysis indicates that the desired performance has not been achieved with this basic system, additional equipment must be inserted into the system. Generally this analysis also indicates the characteristics for the additional equipment that are necessary to achieve the desired performance. After the system is synthesized to achieve the desired performance, based upon a linear analysis, final adjustments can be made on the actual system to take into account the nonlinearities that were neglected. It should be noted that, in conjunction with the analysis, a computer is often utilized, depending upon the complexity of the system.

Step 5 above may yield many possible sets of design characteristics for the additional equipment that result in a satisfactory response.   One of these sets yields an optimum response obtainable for the basic system.   To achieve this response for control systems, special optimization techniques are usually applied; the optimum response *might* be obtainable in some instances *without* special techniques.

1-7   Outline of Text

The first few chapters deal with the mathematics underlying the analysis of control systems.   In conjunction with this presentation, various basic physical units are discussed.   Once the technique of writing the differential equations (and, in turn, the Laplace transforms) that describe the performance of a dynamic system has been mastered, the ideas of block diagrams and transfer functions are developed.   When physical systems are described in terms of block diagrams and transfer functions, they exhibit basic servo characteristics.   These characteristics are described and discussed.   Then follows a presentation of various phases of analysis that can be utilized in the study of feedback control systems.   These phases are based on root-locus and steady-state frequency-response analysis.   The manner in which a basic system can be improved by the use of compensators, if it does not meet the desired specifications, is then presented.   Up to this point simple systems have been considered, with intelligence-bearing signals that are d-c or low-frequency a-c, ranging from zero to a few cycles per second.   In a later chapter a-c carrier systems, those in which the intelligence-bearing signal is amplitude-modulated by a suppressed-carrier signal, are discussed. Within the restrictions advanced, a-c carrier systems can be analyzed by utilizing much the same techniques as for d-c systems.   For more precise a-c system design, special theory is needed.[11]

The remaining portion of the text deals with complex (or multiple-loop) control systems, methods of treating systems with certain types of non-linearities, and analog computers.   Chapter 21 is included so that the reader may be informed of the advanced areas in control theory.   In the Appendix are tables, methods for obtaining roots of polynomials, etc., that will be useful to a feedback control system engineer.

It seems appropriate to close this chapter by pointing out that a feedback control engineer is essentially a "system engineer," that is, a person whose primary concern is with the design and synthesis of the over-all system.   To a varied extent, depending on his own background and experience, he will rely on engineers in the various recognized branches of engineering (electrical, mechanical, aeronautical, etc.) to furnish him with the transfer functions of various portions of a control system.

Bibliography

1. Proposed Symbols and Terms for Feedback Control Systems, AIEE Committee Reports, *Elec. Eng.*, vol. 70, pp. 905–909, 1951.
2. Singer, C., E. J. Holmyard, A. R. Hall, T. I. Williams, et al.: "A History of Technology," vol. 2, Oxford University Press, Fair Lawn, N.J., 1956.
3. Trinks, W.: "Governors and the Governing of Prime Movers," D. Van Nostrand Company, Inc., Princeton, N.J., 1919.
4. Minorsky, N.: Directional Stability and Automatically Steered Bodies, *J. Am. Soc. Naval Engrs.*, vol. 34, p. 280, 1922.
5. Black, H. S.: Stabilized Feedback Amplifiers, *Bell System Tech. J.*, 1934.
6. Nyquist, H.: Regeneration Theory, *Bell System Tech. J.*, 1932.
7. Bode, H. W.: "Network Analysis and Feedback Amplifier Design," D. Van Nostrand Company, Inc., Princeton, N.J., 1945.
8. Hall, A. C.: Application of Circuit Theory to the Design of Servomechanisms, *J. Franklin Inst.*, 1946.
9. Harris, H.: The Frequency Response of Automatic Control Systems, *Trans. AIEE*, vol. 65, pp. 539–546, 1946.
10. Evans, W. R.: Graphical Analysis of Control Systems, *Trans. AIEE*, vol. 67, Part II, pp. 547–551, 1948.
11. Ivey, K. A.: "A-C Carrier Control Systems," John Wiley & Sons, Inc., New York, 1964.

2

Methods of
Writing
Differential
Equations

2-1   *Introduction*

Analysis of a dynamic system requires the ability to predict its performance. This ability and the precision of the results depend on how well the characteristics of each component can be expressed mathematically. Techniques for solving linear equations with constant parameters are very comprehensive. When the equations are time-varying or nonlinear, it is more difficult to solve them. Except for low-order equations, the solution of a time-varying or nonlinear equation may require numerical procedures.

The purpose of this chapter is to present methods of writing the differential equations for a variety of electrical, mechanical, thermal, and hydraulic systems.[1] This is the first step that must be mastered by the would-be control systems engineer. The basic laws are given for each system and the parameters are defined. Examples are included to show the application of the basic laws to physical equipment. The result is a differential equation that describes the system. The equations derived are limited to linear systems or to those systems that can be represented by linear equations over their useful operating range. The solution of the differential equations is covered in the next chapter.

To simplify the writing of differential equations, the so-called classical D operator notation is used.[2] The symbols D and $1/D$ are defined by

$$Dy \equiv \frac{dy}{dt} \qquad D^2y \equiv \frac{d^2y}{dt^2} \qquad \cdots \qquad (2\text{-}1)$$

$$D^{-1}y \equiv \frac{1}{D}y \equiv \int_0^t y\,dt + Y_0 \qquad (2\text{-}2)$$

where Y_0 is the value of the integral at time $t = 0$, that is,

$$Y_0 = \int_{-\infty}^0 y\,dt$$

It is therefore the initial value of the integral. As a consequence of these definitions

$$\frac{1}{D}Dy = \int_0^t \frac{dy}{dt}\,dt + Y_0 = y(t) - Y_0 + Y_0 = y$$

$$D\left(\frac{1}{D}y\right) = \frac{d}{dt}\left(\int_0^t y\,dt + Y_0\right) = y$$

Thus the voltage equation for an RLC series circuit is

$$L\frac{di}{dt} + Ri + \frac{1}{C}\int_0^t i\,dt + \frac{Q_0}{C} = e$$

where each term in the equation is voltage. The initial voltage across the capacitor is $V_0 = Q_0/C$, that is, the value of the integral at $t = 0$ is the charge Q_0. The circuit equation may be written in operator notation as

$$\left(LD + R + \frac{1}{CD}\right)i = e$$

Operating on each side of this equation with the operator D, that is, differentiating both sides,

$$\left(LD^2 + RD + \frac{1}{C}\right)i = De$$

As will be seen later, this operator notation not only facilitates the writing of differential equations but also is of considerable use in finding their solutions.

In general, time-varying quantities are indicated by small letters. These are sometimes indicated by the form $x(t)$, but more often this is written just as x.   There are some exceptions, because of established convention in the use of certain symbols.

2-2   *Electric Circuits and Components*[3]

The equations for an electric circuit obey Kirchhoff's laws, which state:

1. The algebraic sum of the potential differences around a closed circuit must equal zero.   This may be restated as follows: In any closed loop the sum of the voltage rises must equal the sum of the voltage drops.
2. The algebraic sum of the currents at a junction, or node, must equal zero.   In other words, the sum of the currents flowing toward the junction must equal the sum of the currents flowing away from the junction.

Both of these laws are used in examples in this chapter.

The voltage sources are generators.   The usual direct-current (d-c) voltage source is a battery or d-c generator.   The voltage drops appear across the three basic electrical elements: resistors, inductors, and capacitors.

The voltage drop across a resistor follows Ohm's law, which states that the voltage across a resistor is equal to the product of the current flowing through the resistor and its resistance.   Resistors absorb energy from the system.   Symbolically this is written

$$v_R = Ri \tag{2-3}$$

The voltage drop across a pure inductor is given by Faraday's law, which is written in the form

$$v_L = L\frac{di}{dt} = L\,Di \tag{2-4}$$

This equation states that the voltage drop across an inductor is equal to the product of the inductance and the time rate of change of current.   An increasing current means that the voltage is positive, and a decreasing current means that it is negative.

The voltage drop across a capacitor is defined as the ratio of the magnitude of the positive electric charge on its positive plate to the value of its capacitance.   The charge on the capacitor plate is equal to the time integral of the current plus the initial value of the charge.   The capacitor voltage is

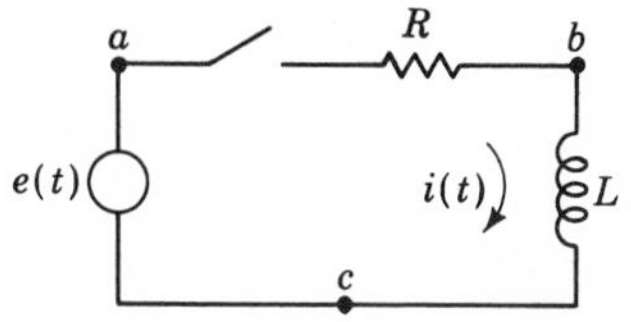

Fig. 2-1 *Series resistor-inductor circuit.*

Table 2-1. **Electrical symbols and units**

Symbol	Quantity	Units
e or v	Voltage	Volts
i	Current	Amperes
L	Inductance	Henrys
C	Capacitance	Farads
R	Resistance	Ohms

written in the form

$$v_C = \frac{q}{C} = \frac{1}{C} \int_0^t i\, dt + \frac{Q_0}{C} = \frac{i}{CD} \tag{2-5}$$

The mks units for these quantities in the practical system are given in Table 2-1.

Series Resistor-Inductor Circuit

The voltage e (see Fig. 2-1) is a function of time. Setting the voltage rise equal to the sum of the voltage drops produces

$$v_R + v_L = e$$
$$Ri + L\frac{di}{dt} = Ri + L\,Di = e \tag{2-6}$$

The voltage across the inductor, v_L, may be desired and may be obtained in the following manner. The voltage across the inductor is

$$v_L = L\,Di$$

The current through the inductor is therefore

$$i = \frac{1}{LD} v_L$$

Substituting these values into the original equation gives

$$\frac{R}{LD} v_L + v_L = e \tag{2-7}$$

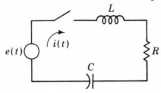

Fig. 2-2   *Series RLC circuit.*

The node method is also convenient for writing the system equations directly in terms of the voltages.   The junctions of any two elements are called *nodes*.   This circuit has three nodes, labeled a, b, and c (see Fig. 2-1).   One node is used as a reference point; in this circuit it will be node c.   The voltages at the other nodes are all considered with respect to the reference node.   Thus v_{ac} is the voltage drop from node a to node c, and v_{bc} is the voltage drop from node b to the reference node c.   For simplicity, these voltages are written just as v_a and v_b.

The voltage $v_a = e$ is known; therefore there is only one unknown voltage, v_b, and only one node equation is necessary.   Kirchhoff's second law, that the algebraic sum of the currents at a node must equal zero, will be applied to node b.

The current flowing away from node b through the resistor R is $(v_b - v_a)/R$.   The current flowing away from node b through the inductor L is v_b/LD.   The sum of these currents must equal zero:

$$\frac{v_b - v_a}{R} + \frac{1}{LD}\, v_b = 0 \tag{2-8}$$

By rearranging terms, this equation becomes

$$\left(\frac{1}{R} + \frac{1}{LD}\right) v_b - \left(\frac{1}{R}\right) v_a = 0 \tag{2-9}$$

Except for the use of different symbols, this is the same as Eq. (2-7).   Note that the node method required the writing of only one equation.

Series Resistor-Inductor-Capacitor Circuit

For the series RLC circuit shown in Fig. 2-2, the applied voltage is equal to the sum of the voltage drops when the switch is closed:

$$v_L + v_R + v_C = e$$
$$L\, Di + Ri + \frac{1}{CD}\, i = e \tag{2-10}$$

The circuit equation can be written in terms of the voltage drop across any circuit element.   For example, in terms of the voltage across the resistor, $v_R = Ri$, Eqs. (2-10) become

$$v_R + \frac{L}{R} Dv_R + \frac{1}{RCD} v_R = e \tag{2-11}$$

Multiloop Electric Circuits

Multiloop electric circuits (see Fig. 2-3) may be solved by use of either loop or nodal equations. The following example illustrates both methods. The problem is to solve for the output voltage v_o.

1. *Loop Method.* A loop current is drawn in each closed loop; then Kirchhoff's voltage equation is written for each loop:

$$\left(R_1 + \frac{1}{CD}\right) i_1 - R_1 i_2 - \frac{1}{CD} i_3 = e \tag{2-12}$$

$$- R_1 i_1 + (R_1 + R_2 + LD)i_2 - R_2 i_3 = 0 \tag{2-13}$$

$$-\frac{1}{CD} i_1 - R_2 i_2 + \left(R_2 + R_3 + \frac{1}{CD}\right) i_3 = 0 \tag{2-14}$$

The output voltage is

$$v_o = R_3 i_3 \tag{2-15}$$

These four equations must be solved simultaneously to obtain $v_o(t)$ in terms of the input voltage $e(t)$ and the circuit parameters.

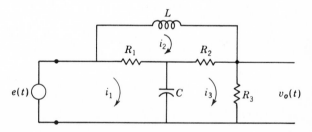

Fig. 2-3 *Multiloop network.*

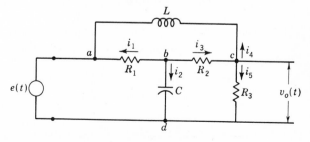

Fig. 2-4 *Multinode network.*

2. *Node Method.*   The junctions, or nodes, are labeled by letters in Fig. 2-4.   Kirchhoff's current equations are written for each node in terms of the node voltages, where node d is taken as reference. The voltage v_{bd} is the voltage of node b with reference to node d. For simplicity, the voltage v_{bd} will be written just as v_b.   There are two unknown voltages v_b and v_o and therefore two equations are required:

For Node b: $$i_1 + i_2 + i_3 = 0 \tag{2-16}$$
For Node c: $$-i_3 + i_4 + i_5 = 0 \tag{2-17}$$

In terms of the node voltages, these equations are

$$\frac{v_b - v_a}{R_1} + C\,Dv_b + \frac{v_b - v_o}{R_2} = 0 \tag{2-18}$$

$$\frac{v_o - v_b}{R_2} + \frac{v_o}{R_3} + \frac{1}{LD}\,(v_o - e) = 0 \tag{2-19}$$

Rearranging the terms to systematize the form of the equations gives

$$\left(\frac{1}{R_1} + CD + \frac{1}{R_2}\right)v_b - \frac{1}{R_2}v_o = \frac{e}{R_1} \tag{2-20}$$

$$-\frac{1}{R_2}v_b + \left(\frac{1}{R_2} + \frac{1}{R_3} + \frac{1}{LD}\right)v_o = \frac{e}{LD} \tag{2-21}$$

For this example, only two nodal equations are needed to solve for the potential at node c.   An additional equation must be used if the current in R_3 is required.   With the loop method, three equations must be solved simultaneously; an additional equation must be used if the voltage across R_3 is required.   The method that requires the solution of the smaller number of equations should be used.   This varies with the circuit.

The rules for writing the node equations can be summarized as follows:

1. The number of equations required is equal to the number of unknown node voltages.
2. An equation is written for each node.
3. The equation includes the following terms:
   a. The node voltage multiplied by the sum of all the admittances that are connected to this node.   This term is positive.
   b. The node voltage at the other end of each branch multiplied by the admittance connected between the two nodes.   This term is negative.

The reader should learn to apply these rules so that Eqs. (2-20) and (2-21) can be written directly for the circuit of Fig. 2-4.

2-3   *Mechanical Translation Systems*

Mechanical systems obey the basic law that the sum of the forces must equal zero. This is known as Newton's law and may be restated as follows: The sum of the applied forces must be equal to the sum of the reactive forces. The following analysis includes only linear functions. Static friction, coulomb friction, and other nonlinear friction terms are not included. The three qualities characterizing elements in a mechanical translation* system are mass, elastance, and damping. Basic elements entailing these qualities are represented as network elements,[4] and a mechanical network is drawn for each mechanical system to facilitate writing the differential equations.

The mass M is the inertial element. A force applied to a mass produces an acceleration of the mass. The reaction force f_M is equal to the product of mass and acceleration and is opposite in direction to the applied force. In terms of displacement x, velocity v, and acceleration a, the force equation is

$$f_M = Ma = M \, Dv = M \, D^2x \qquad (2\text{-}22)$$

The network representation of mass is shown in Fig. 2-5a. One terminal, a, has the motion of the mass; and the other terminal, b, is considered to have the motion of the reference. The reaction force f_M is a function of time and acts "through" M.

The elastance, or stiffness, K provides a restoring force as represented by a spring. Thus, if stretched, the spring tries to contract; if compressed, it tries to expand to its normal length. The reaction force f_K on each end of the spring is the same and is equal to the product of the stiffness K and the amount of deformation of the spring.

The network representation of a spring is shown in Fig. 2-5b. The displacement of each end of the spring is measured from the original or equilibrium position. End c has a position x_c and end d has a position x_d measured from the respective equilibrium positions. The force equation, in accordance with Hooke's law, is

$$f_K = K(x_c - x_d) \qquad (2\text{-}23)$$

If the end d is stationary, the above equation reduces to

$$f_K = Kx_c \qquad (2\text{-}24)$$

The damping, or viscous friction, B characterizes the element that absorbs energy. The damping force is proportional to the difference in velocity of two bodies, and the assumption is made that the viscous friction is linear. This assumption simplifies the solution of the dynamic equation.

---

* Translation means motion in a straight line.

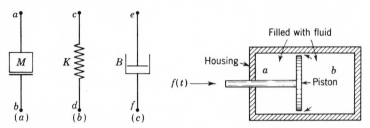

Fig. 2-5 *Network elements of mechanical translation systems.*

Fig. 2-6 *Dashpot construction.*

The network representation of damping action is a dashpot, as shown in Fig. 2-5c.   It should be realized that damping either may be intentional or may occur unintentionally and is present because of physical construction.

The reaction damping force f_B is equal to the product of damping B and the relative velocity of the two ends of the dashpot.   The direction of this force, given by Eq. (2-25), depends on the relative magnitudes and directions of the velocities Dx_e and Dx_f:

$$f_B = B(v_e - v_f) = B(Dx_e - Dx_f) \qquad (2\text{-}25)$$

Damping may be added to a system by use of a dashpot.   The basic operation of a dashpot in which the housing is filled with a fluid is shown in Fig. 2-6.   If a force f is applied to the shaft, the piston presses against the fluid, increasing the pressure on side b and decreasing the pressure on side a. As a result, the fluid flows around the piston from side b to side a.   If necessary, a small hole can be drilled through the piston to provide a positive path for the flow of fluid.   The force required to move the piston inside the housing is given by Eq. (2-25), where the damping B depends on the dimensions and the fluid used.

Before writing the differential equations of a complete system, the first step is to draw the mechanical network.   This is done by connecting the terminals of those elements that have the same displacement.   Then the force equation is written for each node or position by equating the sum of the forces at each position to zero.   The equations are similar to the node equations in an electric circuit, with force analogous to current, velocity analogous to voltage, and the mechanical elements with their appropriate operators analogous to admittance.   Several examples are shown in the following pages.   The reference positions in all cases should be taken from the static equilibrium positions.   The force of gravity therefore does not appear in the system equations.   The English and mks systems of units are shown in Table 2-2.

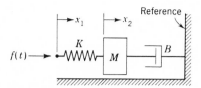

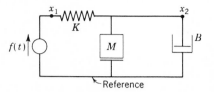

Fig. 2-7 *Simple mass-spring-damper mechanical system.*

Fig. 2-8 *Mechanical network corresponding to Fig. 2-7.*

Table 2-2. *Mechanical translation symbols and units*

Symbol	Quantity	English units	mks units
f	Force	Pounds	Newtons
x	Distance	Feet	Meters
v	Velocity	Feet/second	Meters/second
a	Acceleration	Feet/second^2	Meters/second^2
M*	Mass	Slugs $= \dfrac{\text{pound-second}^2}{\text{foot}}$	Kilograms
K	Stiffness coefficient	Pounds/foot	Newtons/meter
B	Damping coefficient	Pounds/(foot/second)	Newtons/(meter/second)

* Mass M in the English system above has the dimensions of slugs. Sometimes it is given in units of pounds. If so, then in order to use the consistent set of units above, the mass must be expressed in slugs by using the conversion factor 1 slug = 32 lb.

Simple Mechanical Translation System

The system shown in Fig. 2-7 is initially at rest. The end of the spring and the mass have positions denoted as the reference positions. Any displacement from these reference positions is labeled x_1 or x_2, respectively. A force f applied at the end of the spring must be balanced by a compression of the spring. The same force is also transmitted through the spring and acts at point x_2.

To draw the mechanical network, the points x_1 and x_2 and the reference are located. The network elements are then connected between these points. For example, one end of the spring has the position x_1 and the other end has the position x_2. Therefore the spring is connected between these points. The complete mechanical network is drawn in Fig. 2-8.

The displacements x_1 and x_2 are nodes of the circuit. At each node the sum of the forces must add to zero. Accordingly, the equations may be written

$$f = f_K = K(x_1 - x_2)$$
$$f_K = f_M + f_B = M\, D^2 x_2 + B\, D x_2 \tag{2-26}$$

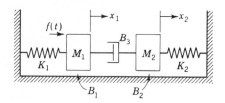

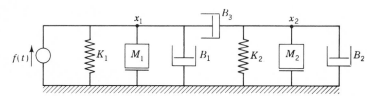

Fig. 2-9 *Multiple-element mechanical system.*

Fig. 2-10 *Mechanical network for the mechanical system of Fig. 2-9.*

These two equations can be solved for the two displacements x_1 and x_2 and their respective velocities Dx_1 and Dx_2.

It is possible to obtain one equation relating x_1 to x_2 by combining Eqs. (2-26):

$$(MD^2 + BD + K)x_2 = Kx_1 \qquad (2\text{-}27)$$

The solution of Eq. (2-27) would show the motion x_2 resulting from a given motion x_1. It would also be possible from Eqs. (2-26) to solve for the force f required in order for x_1 to have the desired motion.

Multiple-element Mechanical Translation System

A force $f(t)$ is applied to the mass M_1 of Fig. 2-9. The system equations can be written in terms of the two displacements x_1 and x_2. The mechanical network is drawn by connecting the terminals of the elements that have the same displacement (see Fig. 2-10). Since the forces at each node must add to zero, the equations are written according to the rules for node equations:

$$\begin{aligned} For\,Node\,1\colon &\quad (M_1D^2 + B_1D + B_3D + K_1)x_1 - (B_3D)x_2 = f &\quad (2\text{-}28) \\ For\,Node\,2\colon &\quad -(B_3D)x_1 + (M_2D^2 + B_2D + B_3D + K_2)x_2 = 0 &\quad (2\text{-}29) \end{aligned}$$

A definite pattern to these equations can be detected. Observe that K_1, M_1, B_1, and B_3 are connected to node 1 and that the equation for node 1 contains all four of these terms as coefficients of x_1. Notice also that element B_3 is also connected to node 2 and that the term $-B_3$ appears as a coefficient of x_2. By using this pattern, Eq. (2-29) can be written directly. Thus, since K_2, M_2, B_2, and B_3 are connected to node 2, they appear as coefficients of x_2. B_3 is also connected to node 1, and $-B_3$ appears as the coefficient of x_1.

The node equations for a mechanical system follow directly from the

Table 2-3. *Electrical and mechanical analogs*

Mechanical translation element		Electrical element	
Symbol	Quantity	Symbol	Quantity
f	Force	i	Current
$v = Dx$	Velocity	e or v	Voltage
M	Mass	C	Capacitance
K	Stiffness coefficient	$1/L$	Reciprocal inductance
B	Damping coefficient	$G = 1/R$	Conductance

mechanical network.   They are similar in form to the node equations for an electric circuit and follow the same rules.

2-4   *Analogous Circuits*

Analogous circuits represent systems for which the differential equations have the same form.   The corresponding variables and parameters in two circuits represented by equations of the same form are called analogs.   An electric circuit can be drawn that looks like the mechanical circuit and is represented by node equations that have the same mathematical form as the mechanical equations.   The analogs are listed in Table 2-3.

In this table the force f and the current i are analogs and are classified as "through" variables.   There is a physical similarity since a measuring instrument must be placed in series in both cases; that is, an ammeter and a force indicator must be placed in series with the system.   Also, the velocity "across" a mechanical element is analogous to voltage across an electrical element.   Again, the physical similarity is present since a measuring instrument must be placed across the system in both cases.   A voltmeter must be placed across a circuit to measure voltage; it must have a point of reference. A velocity indicator must also have a point of reference.   Nodes in the mechanical network are analogous to nodes in the electric network.

By using the analogs of Table 2-3, the mechanical network of Fig. 2-9 is drawn in Fig. 2-11.   Note that the circuit diagram of the electrical analog in Fig. 2-11 and the mechanical network in Fig. 2-10 are similar.   The node equations for Fig. 2-11 are written by inspection as

$$\left(C_1 D + G_1 + G_3 + \frac{1}{L_1 D} \right) v_1 - (G_3) v_2 = i \qquad (2\text{-}30)$$

$$-(G_3) v_1 + \left(C_2 D + G_2 + G_3 + \frac{1}{L_2 D} \right) v_2 = 0 \qquad (2\text{-}31)$$

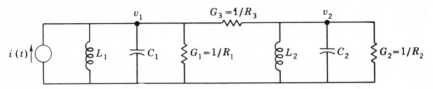

Fig. 2-11   *Electrical analog of Figs. 2-9 and 2-10.*

Equations (2-28) and (2-29) can be written in terms of the velocities instead of the displacements for a better comparison with the electric-circuit equations:

$$\left(M_1 D + B_1 + B_3 + \frac{K_1}{D}\right) Dx_1 - (B_3) Dx_2 = f \qquad (2\text{-}32)$$

$$-(B_3) Dx_1 + \left(M_2 D + B_2 + B_3 + \frac{K_2}{D}\right) Dx_2 = 0 \qquad (2\text{-}33)$$

A comparison of Eqs. (2-30) and (2-31) with Eqs. (2-32) and (2-33) shows that they have the same mathematical form. The solutions of the dependent variables in either set of equations must therefore also have the same mathematical form.

The advantage of the electrical analog is that it can be set up very easily in the laboratory. Also, a change in any parameter is accomplished more readily in the electric circuit. The electric circuit can readily be adjusted to produce a desired response. Then the parameters in the mechanical system must be changed by a corresponding amount to achieve the same desired response.

2-5   Mechanical Rotational Systems

Rotational systems are similar to translation systems except for the basic differences that torque equations are written in place of force equations and that the displacement, velocity, and acceleration terms are angular quantities. The applied torque is equal to the sum of the reaction torques. The three elements in a rotation system are inertia, the spring, and the dashpot.

The torque applied to a body having a moment of inertia J produces an angular acceleration. The reaction torque T_J is equal to the product of moment of inertia and acceleration. In terms of angular displacement θ, angular velocity ω, or angular acceleration α, the torque equation is

$$T_J = J\alpha = J D\omega = J D^2\theta \qquad (2\text{-}34)$$

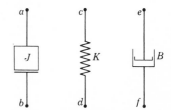

Fig. 2-12   *Network elements of mechanical rotational systems.*

When a torque is applied to a spring, the spring is twisted by an angle θ. The applied torque is transmitted through the spring and appears at the other end. The reaction spring torque T_K that is produced is equal to the product of the stiffness, or elastance, K of the spring and the angle of twist. By denoting the positions of the two ends of the spring, measured from the neutral position, as θ_c and θ_d, the reaction torque is

$$T_K = K(\theta_c - \theta_d) \tag{2-35}$$

Damping occurs whenever a body moves through a fluid, which may be either a liquid or a gas such as air. To produce motion of the body a torque must be applied to overcome the reaction damping torque. The damping is represented as a dashpot with a viscous friction coefficient B. The damping torque T_B is equal to the product of damping B and the relative angular velocity of the ends of the dashpot.

The reaction torque of a damper is

$$T_B = B(\omega_e - \omega_f) = B(D\theta_e - D\theta_f) \tag{2-36}$$

The mechanical-network representation of the rotational elements is shown in Fig. 2-12.

By first drawing the mechanical network for the system, the work of writing the differential equations can be simplified. This is done by first designating nodes that correspond to each angular displacement. Then each element is connected between the nodes that correspond to the two motions at each end of that element. The inertia elements are connected from the reference node to the node representing its position. The spring and dashpot elements are connected to the two nodes that represent the position of each end of the element. Then the torque equation is written for each node by equating the sum of the torques at each node to zero. These equations are similar to those for the case of mechanical translation and are analogous to those for electric circuits. Several examples are shown in the following pages. The units for these elements are given in Table 2-4.

An electrical analog can be obtained for a mechanical rotational system in the manner described in Sec. 2-4. The changes in Table 2-3 are due only to the fact that rotational quantities are involved. Thus torque is the analog of current, and the moment of inertia is the analog of capacitance.

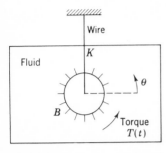

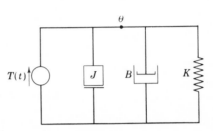

Fig. 2-13 *Simple rotational system.*

Fig. 2-14 *Mechanical network for the system of Fig. 2-13.*

Table 2-4. *Mechanical rotational symbols and units*

Symbol	Quantity	English units	mks units
T	Torque	Pound-feet	Newton-meters
θ	Angle	Radians	Radians
ω	Angular velocity	Radians/second	Radians/second
α	Angular acceleration	Radians/second^2	Radians/second^2
$J*$	Moment of inertia	Slug-feet^2 (or pound-feet-second^2)	Kilogram-meter^2
K	Stiffness coefficient	Pound-feet/radian	Newton-meters/radian
B	Damping coefficient	Pound-feet/(radian/second)	Newton-meters/(radian/second)

* The moment of inertia J has the dimensions of mass-distance^2 which have the units slug-feet^2 in the English system. Sometimes the units are given as pound-feet^2. To use the consistent set of units above, the moment of inertia must be expressed in slug-feet^2 by using the conversion factor 1 slug = 32 lb.

Simple Mechanical Rotational System

The system shown in Fig. 2-13 has a mass with a moment of inertia J immersed in a fluid. A torque T is applied to the mass. The wire produces a reactive torque proportional to a stiffness K and to the angle of twist. The fins moving through the fluid have a damping B which requires a torque proportional to the rate at which they are moving. The mechanical network is drawn in Fig. 2-14. There is one node having a displacement θ; therefore, only one equation is necessary:

$$J\,D^2\theta + B\,D\theta + K\theta = T(t) \tag{2-37}$$

Multiple-element Mechanical Rotational System

The system represented by Fig. 2-15a has two disks which have damping between them and also between each of them and the frame. The mechan-

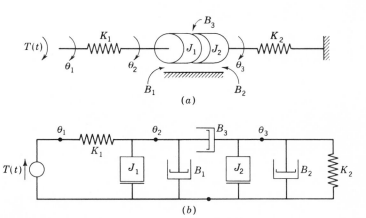

Fig. 2-15(a) *Rotational system.* (b) *Corresponding mechanical network.*

ical network is drawn in Fig. 2-15b by connecting the terminals of those elements that have the same angular displacement. The torques at each node must add to zero.

The equations are written directly in systematized form:

For Node 1: $\qquad [K_1]\theta_1 - [K_1]\theta_2 = T(t)$ (2-38)

For Node 2:

$$-[K_1]\theta_1 + [J_1D^2 + (B_1 + B_3)D + K_1]\theta_2 - [B_3D]\theta_3 = 0 \quad (2\text{-}39)$$

For Node 3:

$$-[B_3D]\theta_2 + [J_2D^2 + (B_2 + B_3)D + K_2]\theta_3 = 0 \qquad (2\text{-}40)$$

These three equations can be solved simultaneously for θ_1, θ_2, and θ_3 as a function of the applied torque.

Effective Inertia and Damping of a Gear Train

When the load is coupled to a drive motor through a gear train, the inertia and damping relative to the motor are important. Since the shaft length between gears is very short, the stiffness may be considered infinite. A simple representation of a gear train is shown in Fig. 2-16. The following definitions are used:

N = number of teeth on each gear
$\omega = D\theta$ = velocity of each gear
n_a = ratio of $\dfrac{\text{speed of driving shaft}}{\text{speed of driven shaft}}$
θ = angular position

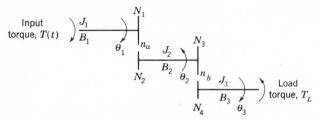

Fig. 2-16 *Representation of a gear train.*

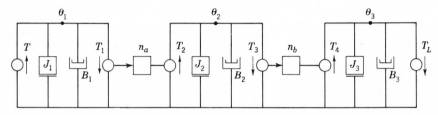

Fig. 2-17 *Mechanical network for the gear train of Fig. 2-16.*

The mechanical network for the gear train is shown in Fig. 2-17.   At each gear pair two torques are produced.   For example, a restraining torque T_1 is produced on gear 1 by the rest of the gear train.   There is also produced a driving torque T_2 on gear 2. T_1 is the load on gear 1 produced by the rest of the gear train. T_2 is the torque transmitted to gear 2 to drive the rest of the gear train.   These torques are inversely proportional to the speeds of the respective gears.   The block labeled n_a between T_1 and T_2 is used to show the relationship between them, that is, $T_2 = n_a T_1$.   The equations describing the system are listed below:

$$J_1 D^2 \theta_1 + B_1 D\theta_1 + T_1 = T$$
$$J_2 D^2 \theta_2 + B_2 D\theta_2 + T_3 = T_2 \qquad \theta_2 = \frac{\theta_1}{n_a}$$
$$J_3 D^2 \theta_3 + B_3 D\theta_3 + T_L = T_4 \qquad \theta_3 = \frac{\theta_2}{n_b} = \frac{\theta_1}{n_a n_b}$$
$$n_a = \frac{\omega_1}{\omega_2} = \frac{\theta_1}{\theta_2} = \frac{N_2}{N_1} \tag{2-41}$$
$$n_b = \frac{\omega_2}{\omega_3} = \frac{\theta_2}{\theta_3} = \frac{N_4}{N_3}$$
$$T_2 = n_a T_1$$
$$T_4 = n_b T_3$$

The equations can be combined to produce

$$J_1 D^2\theta_1 + B_1 D\theta_1 + \frac{1}{n_a}\bigg[J_2 D^2\theta_2 + B_2 D\theta_2$$
$$+ \frac{1}{n_b}(J_3 D^2\theta_3 + B_3 D\theta_3 + T_L)\bigg] = T \tag{2-42}$$

This equation can be expressed in terms of the input position θ_1 only:

$$\left(J_1 + \frac{J_2}{n_a^2} + \frac{J_3}{n_a^2 n_b^2}\right) D^2\theta_1 + \left(B_1 + \frac{B_2}{n_a^2} + \frac{B_3}{n_a^2 n_b^2}\right) D\theta_1 + \frac{T_L}{n_a n_b} = T \quad (2\text{-}43)$$

Equation (2-43) represents the system performance as a function of a single dependent variable. An equivalent system is one having an equivalent moment of inertia and damping equal to

$$J_{eq} = J_1 + \frac{J_2}{n_a^2} + \frac{J_3}{n_a^2 n_b^2} \quad (2\text{-}44)$$

$$B_{eq} = B_1 + \frac{B_2}{n_a^2} + \frac{B_3}{n_a^2 n_b^2} \quad (2\text{-}45)$$

Should the solution for θ_3 be desired, the equation can be altered by the substitution of $\theta_1 = n_a n_b \theta_3$. This system can be generalized for any number of gear stages.

For the case where the gear reduction ratio is large, the load inertia may contribute a negligible value to the equivalent moment of inertia.

2-6   *Thermal Systems*[5,6]

A limited number of thermal systems can be represented by linear differential equations. The basic requirement is that the temperature of a body be considered uniform. When the body is small, this approximation is valid. Also, when the region consists of a body of air or liquid, the temperature can be considered uniform if there is perfect mixing of the fluid.

The necessary condition of equilibrium requires that the heat added to the system equal the heat stored plus the heat carried away. This can also be expressed in terms of rate of heat flow.

The symbols shown in Table 2-5 are used for thermal systems.

Table 2-5.   Thermal symbols and units

Symbol	Quantity	English units
q	Rate of heat flow	Btu/minute
M	Mass	Pounds
S	Specific heat	(Btu/pound)/°F
C	Thermal capacitance $C = MS$	Btu/degree
K	Thermal conductance	(Btu/minute)/degree
R	Thermal resistance	Degree/(Btu/minute)
θ	Temperature	°F
h	Heat energy	Btu

(a)   (b)   **Fig. 2-18**   *Network elements of thermal systems.*

A thermal-system network will be drawn for each system in which the thermal capacitance and thermal resistance are represented by network elements.   From the thermal network the differential equations can be written.

The additional heat stored in a body whose temperature is raised from θ_1 to θ_2 is given by

$$h = \frac{q}{D} = C(\theta_2 - \theta_1)$$

In terms of rate of heat flow, this equation can be written as

$$q = C\,D(\theta_2 - \theta_1) \tag{2-46}$$

The thermal capacitance determines the amount of heat stored in a body. It is analogous to the electric capacitance of a capacitor in an electric circuit, which determines the amount of charge stored.   The network representation of thermal capacitance is shown in Fig. 2-18a.

Rate of heat flow through a body in terms of the two boundary temperatures θ_3 and θ_4 is

$$q = \frac{\theta_3 - \theta_4}{R} \tag{2-47}$$

The thermal resistance determines the rate of heat flow through the body. This is analogous to the resistance of a resistor in an electric circuit, which determines the current flow.   The network representation of thermal resistance is shown in Fig. 2-18b.

In the thermal network the temperature is analogous to potential.

Simple Mercury Thermometer

Consider a thin glass-wall thermometer filled with mercury which has stabilized at a temperature θ_1.   It is plunged into a bath of temperature θ_0 at $t = 0$.   In its simplest form, the thermometer can be considered to have a capacitance C which stores heat and a resistance R which limits the heat flow.   The flow of heat into the thermometer is

$$q = \frac{\theta_0 - \theta}{R}$$

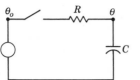
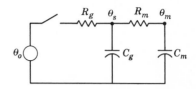

Fig. 2-19 *Simple net-work representation of a thermometer.*

Fig. 2-20 *More exact network representation of a thermometer.*

The heat entering the thermometer is stored in the thermal capacitance and is given by

$$h = \frac{q}{D} = C(\theta - \theta_1)$$

These equations can be combined to form

$$\frac{\theta_0 - \theta}{RD} = C(\theta - \theta_1) \qquad (2\text{-}48)$$

Differentiating Eq. (2-48) and rearranging terms,

$$RC\,D\theta + \theta = \theta_0 \qquad (2\text{-}49)$$

The thermal network is drawn in Fig. 2-19. The node equation for this circuit, with the temperature considered as a voltage, gives Eq. (2-49) directly.

More Exact Analysis of a Mercury Thermometer

A more exact analysis of the thermometer should take into account both the resistance of the glass, R_g, and the resistance of the mercury, R_m. Also, the glass has a capacitance C_g and the mercury has a capacitance C_m. The thermal network can be represented by Fig. 2-20, where θ_s represents the temperature at the inner surface between the glass case and the mercury. The equations can be written from the thermal network:

$$\text{For Node } s: \quad \left(\frac{1}{R_g} + \frac{1}{R_m} + C_g D\right)\theta_s - \frac{1}{R_m}\theta_m = \frac{\theta_0}{R_g} \qquad (2\text{-}50)$$

$$\text{For Node } m: \quad -\frac{1}{R_m}\theta_s + \left(\frac{1}{R_m} + C_m D\right)\theta_m = 0 \qquad (2\text{-}51)$$

The two equations can be combined to eliminate θ_s:

$$\left[C_g C_m D^2 + \left(\frac{C_g}{R_m} + \frac{C_m}{R_g} + \frac{C_m}{R_m}\right)D + \frac{1}{R_g R_m}\right]\theta_m = \frac{1}{R_m R_g}\theta_0 \qquad (2\text{-}52)$$

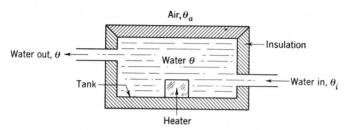

Fig. 2-21   *Electric water heater.*

This equation is of the form

$$(A_2 D^2 + A_1 D + A_0)\theta_m = k\theta_0$$

Simple Heat-transfer System

The electric water heater used in many homes to supply hot water is a good example of a simple heat-transfer problem.   A sketch of such a heater is shown in Fig. 2-21.   The tank is insulated to reduce heat loss to the surrounding air.   The electrical heating element is turned on and off by a thermostatic switch to maintain a reference temperature.   A demand from any faucet in the house causes hot water to leave and cold water to enter the tank.   The necessary simplifying assumptions are as follows:

1. There is no heat storage in the insulation.   This is valid since the specific heat of the insulation is small and the water temperature variation is small.
2. All the water in the tank is at a uniform temperature.   This requires perfect mixing of the water.

Definitions of the system parameters and variations are as follows:

q = rate of heat-flow of heating element
q_t = rate of heat-flow into water in tank
q_o = rate of heat-flow carried out by hot water leaving tank
q_i = rate of heat-flow carried in by cold water entering tank
q_e = rate of heat-flow through tank insulation
θ = temperature of water in tank
θ_i = temperature of water entering tank
θ_a = temperature of air surrounding tank
C = thermal capacitance of water in tank
R = thermal resistance of insulation
n = water flow from tank
S = specific heat of water

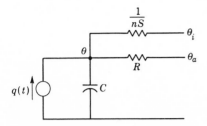

Fig. 2-22 *Thermal network of an electric water heater.*

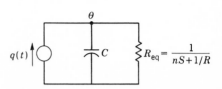

Fig. 2-23 *Simplified network of a water heater.*

There are three temperatures: θ, θ_i, and θ_a; therefore the thermal network must have three nodes. The network can be drawn first or after the equations are written. The equilibrium equation for rate of heat flow is

$$q_t + q_o - q_i + q_e = q \tag{2-53}$$

where
$$q_t = C\,D\theta$$
$$q_o = nS\theta$$
$$q_i = nS\theta_i$$
$$q_e = \frac{\theta - \theta_a}{R}$$

Combining these equations gives

$$C\,D\theta + nS(\theta - \theta_i) + \frac{\theta - \theta_a}{R} = q \tag{2-54}$$

The thermal network is shown in Fig. 2-22. The heat loss, due to the difference $\theta - \theta_i$, can be represented as occurring through the resistor of value $1/nS$. In this thermal network the rate of heat-flow from the heater is analogous to the current from a constant-current source in an electric circuit.

There are four variables in this system: θ, θ_i, θ_a, and n. Three of them must be specified in order to solve the problem. For the special case in which n is a constant and $\theta_a = \theta_i$, Eq. (2-54) can be simplified. In terms of θ, which now is the temperature above the reference θ_a, the equation is

$$C\,D\theta + \left(nS + \frac{1}{R}\right)\theta = q \tag{2-55}$$

The thermal network is drawn in Fig. 2-23.

2-7   *Hydraulic Linear Actuator*

The valve-controlled hydraulic actuator is used in many applications as a power amplifier. Very little power is required to position the valve, but a

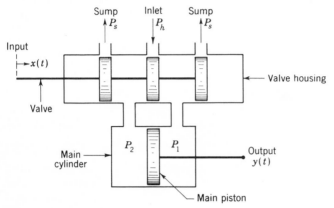

Fig. 2-24   *Hydraulic actuator.*

large power output is controlled.   The hydraulic unit is relatively small, which makes its use very attractive.   Figure 2-24 shows a simple hydraulic actuator in which motion of the valve regulates the flow of oil to either side of the main cylinder.   An input motion of a few thousandths of an inch results in a large change of oil flow.   The resulting difference in pressure on the main piston causes motion of the output shaft.   The oil flowing in is at a high pressure P_h, and the oil on the opposite side of the piston flows into the drain at low pressure P_s.   The load-induced pressure P_L is the difference between the pressures on each side of the main piston:

$$P_L = P_1 - P_2 \tag{2-56}$$

The flow of fluid through an inlet orifice is given by[7]

$$q = ca \sqrt{2g \frac{\Delta p}{w}} \tag{2-57}$$

where c = orifice coefficient
a = orifice area
w = specific weight of fluid
Δp = pressure drop across orifice
g = gravitational acceleration constant
q = rate of flow of fluid

Simplified Analysis

As a first-order approximation, it can be assumed that the orifice coefficient and the pressure drop across the orifice are constant and independent of valve position.   Also, the orifice area can be expressed in terms of the valve displacement. Equation (2-57), which gives the rate of flow of hydraulic fluid through the valve, can then be rewritten as

$$q = C_x x \tag{2-58}$$

where x is the displacement of the valve. The displacement of the main piston is directly proportional to the flow of fluid into the main cylinder. By neglecting the compressibility of the fluid and the leakage around the valve and main piston, the equation of motion of the main piston becomes

$$q = C_b \, Dy \tag{2-59}$$

Combining the two equations gives

$$Dy = \frac{C_x}{C_b} x = C_1 x \tag{2-60}$$

This analysis is essentially correct when the load reaction is small.

More Complete Analysis

When the load reaction is not negligible, a more complete analysis should take into account the pressure drop across the orifice, the leakage of oil around the piston, and the compressibility of the oil.

The pressure drop across the orifice, Δp, is a function of the source pressure P_h and the load pressure P_L. Since P_h is assumed constant, the flow equation is a function of valve displacement x and load pressure P_L:

$$q = f(x, P_L) \tag{2-61}$$

The differential of q, expressed in terms of partial derivatives, is

$$dq = \frac{\partial q}{\partial x} \, dx + \frac{\partial q}{\partial P_L} \, dP_L \tag{2-62}$$

If q, x, and P_L are measured from zero values as reference points and if the partial derivatives are constant at the values they have at zero, the integration of Eq. (2-62) gives

$$q = \left(\frac{\partial q}{\partial x}\right)_0 x + \left(\frac{\partial q}{\partial P_L}\right)_0 P_L \tag{2-63}$$

By defining

$$C_x \equiv \left(\frac{\partial q}{\partial x}\right)_0$$

and

$$C_p \equiv \left(\frac{-\partial q}{\partial P_L}\right)_0$$

the flow equation becomes

$$q = C_x x - C_p P_L \tag{2-64}$$

A comparison with Eq. (2-58) shows that the load pressure reduces the flow into the main cylinder.

The flow of fluid into the cylinder must satisfy the conditions of equilibrium. This flow is equal to the following components:

$$q = q_o + q_l + q_c \tag{2-65}$$

where q_o = incompressible component (causes motion of piston)
q_l = leakage component
q_c = compressible component

The component q_o produces a motion y of the main piston which is given by:

$$q_o = C_b\, Dy \tag{2-66}$$

The compressible component is derived in terms of the bulk modulus of elasticity, which is defined as the ratio of incremental stress to incremental strain. Thus

$$K_B = \frac{\Delta P_L}{\Delta V/V}$$

Solving for ΔV and dividing both sides of the equation by Δt gives

$$\frac{\Delta V}{\Delta t} = \frac{V}{K_B}\frac{\Delta P_L}{\Delta t}$$

Taking the limit as Δ approaches zero and letting $q_c = dV/dt$ results in

$$q_c = \frac{V}{K_B}\, DP_L \tag{2-67}$$

where V is the effective volume of fluid under compression and K_B is the bulk modulus of the hydraulic oil. The volume V at the middle position of the piston stroke is often used in order to linearize the differential equation.

The leakage component is

$$q_l = LP_L \tag{2-68}$$

where L is the leakage coefficient of the whole system.

Combining these equations gives

$$q = C_x x - C_p P_L = C_b\, Dy + \frac{V}{K_B}\, DP_L + LP_L \tag{2-69}$$

Rearranging terms,

$$C_b\, Dy + \frac{V}{K_B}\, DP_L + (L + C_p)P_L = C_x x \tag{2-70}$$

The force developed by the main piston is

$$F = n_F A P_L = C P_L \tag{2-71}$$

where n_F is the force conversion efficiency of the unit and A is the area of the main actuator piston.

An example of a specific type of load consisting of a mass and a dashpot is shown in Fig. 2-25. The equation for this system is obtained by equating

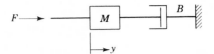

Fig. 2-25   *Load on a hydraulic piston.*

the force produced by the piston, which is given by Eq. (2-71), to the reactive load forces:

$$F = M D^2y + B Dy = CP_L \tag{2-72}$$

Substituting the value of P_L from Eq. (2-72) into Eq. (2-70) gives the equation relating the input motion x to the response y:

$$\frac{MV}{CK_B} D^3y + \left[\frac{BV}{CK_B} + \frac{M}{C}(L + C_p)\right] D^2y + \left[C_b + \frac{B}{C}(L + C_p)\right] Dy = C_x x \tag{2-73}$$

The analysis above is based on perturbations about the reference set of values $x = 0$, $q = 0$, $P_L = 0$. For the entire range of motion x of the valve, the quantities $\partial q/\partial x$ and $-\partial q/\partial P_L$ can be determined experimentally. While they are not constant at values equal to the values C_x and C_p at the zero reference point, average values can be assumed in order to simulate the system by linear equations. For conservative design the volume V is determined for the main piston at the midpoint.

2-8   *Positive-displacement Rotational Hydraulic Transmission*[8,9]

When a large torque is required in a control device, it is possible to use a hydraulic transmission. The transmission contains a variable displacement pump driven at constant speed. It pumps a quantity of oil that is proportional to a control stroke and independent of back pressure. The direction of fluid flow is determined by the direction of displacement of the control stroke. The hydraulic motor has an angular velocity proportional to the volumetric flow rate and in the direction of the oil flow from the pump.

The assumption is made that over a limited range of operation the hydraulic transmission is linear. A schematic picture of the system is shown in Fig. 2-26.

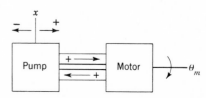

Fig. 2-26   *Hydraulic transmission.*

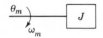

Fig. 2-27   *Inertia load on a hydraulic transmission.*

The following symbols are used:

q_p = total volumetric flow rate from pump
q_m = volumetric flow rate through motor
q_l = volumetric leakage flow rate of both pump and motor
q_c = compressibility flow rate
x = control stroke (x varies from zero to ± 1)
ω_p = angular velocity of pump shaft (constant)
ω_m = angular velocity of motor shaft (variable)
θ_m = angular position of motor shaft
d_p = volumetric displacement (at $x = 1$) per unit angular displacement
d_m = volumetric motor displacement per unit angular displacement
L = leakage coefficient of complete system, $(\text{ft}^3/\text{sec})/(\text{lb}/\text{ft}^2)$
V = total volume of liquid under compression, ft^3
K_B = bulk modulus of oil, lb/ft^2
P_L = load-induced pressure drop across motor, lb/ft^2
C = motor torque constant, ft^3
n_T = torque conversion efficiency of motor

The basic equation is based on the fact that the fluid flow rate from the pump must equal the sum of the flow rates in the system.   This is given by

$$q_p = q_m + q_l + q_c \tag{2-74}$$

where
$$q_p = x\, d_p \omega_p$$
$$q_m = d_m \omega_m$$
$$q_l = LP_L$$
$$q_c = DV = \frac{V}{K_B} DP_L$$

Combining these flow rates into the original equation produces

$$d_m \omega_m + LP_L + \frac{V}{K_B} DP_L = d_p \omega_p x \tag{2-75}$$

The torque produced at the motor shaft is

$$T = n_T\, d_m P_L = CP_L \tag{2-76}$$

Since the torque required depends on the load, two cases are considered.

Case 1.   Inertia load

Figure 2-27 shows an inertia load on a hydraulic transmission.   Equating the generated torque to the load reaction torque,

$$T = J\, D^2 \theta_m = CP_L \tag{2-77}$$

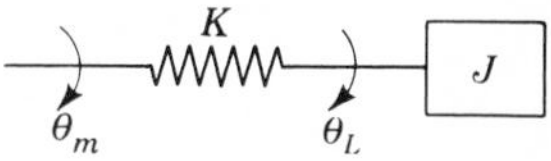

Fig. 2-28 Spring and inertia load on a hydraulic transmission.

When P_L from Eq. (2-77) is inserted in Eq. (2-75), the result is

$$d_m \, D\theta_m + \frac{LJ}{C} \, D^2\theta_m + \frac{VJ}{K_B C} \, D^3\theta_m = d_p \omega_p x \qquad (2\text{-}78)$$

This equation can be solved for the motor position θ_m in terms of the stroke position x.

Case 2. Inertia load coupled through a spring

Figure 2-28 shows a spring and inertia load on a hydraulic transmission. Equating the torque generated by the hydraulic motor to the load torque and then solving for P_L in terms of θ_m gives

$$T = K(\theta_m - \theta_L) = J \, D^2\theta_L = CP_L \qquad (2\text{-}79)$$

$$P_L = \frac{KJ \, D^2\theta_m}{(JD^2 + K)C} \qquad (2\text{-}80)$$

Using the value of P_L from Eq. (2-80) in the system equation (2-75) relates the motor output position θ_m to the stroke input x by

$$\left[\left(\frac{J}{K} + \frac{VJ}{K_B C d_m} \right) D^3 + \frac{LJ}{C d_m} \, D^2 + D \right] \theta_m = \frac{d_p \omega_p}{d_m} \left(\frac{J}{K} D^2 + 1 \right) x \quad (2\text{-}81)$$

The solution of these differential equations is considered in the next chapter. It will be found that the leakage L is essential for stable response. Without leakage the system would have a sustained steady-state oscillation. Therefore, rather than rely on accidental leakage, the designer provides for a positive and finite leakage of hydraulic fluid around the pistons of the motor. This leakage can also serve to lubricate the piston joints.

2-9   *Rotating Power Amplifiers*[10,11]

A d-c generator can be used as a power amplifier in which the power required to excite the field circuit is lower than the power output rating of the armature circuit. The voltage e_g induced in the armature circuit is directly proportional to the product of the flux ϕ set up by the field and the speed of rotation ω of the armature. This can be expressed by

$$e_g = K_1 \phi \omega \qquad (2\text{-}82)$$

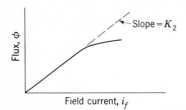

Fig. 2-29   *Magnetization curve.*

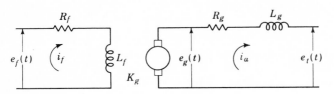

Fig. 2-30   *Schematic diagram of a generator.*

The flux is a function of field current and the type of iron used in the field. A typical magnetization curve showing flux as a function of field current is given in Fig. 2-29.   Up to saturation the relation is approximately linear and the flux is directly proportional to field current:

$$\phi = K_2 i_f \tag{2-83}$$

Combining these equations gives

$$e_g = K_1 K_2 \omega i_f \tag{2-84}$$

For the special case where the armature is driven at constant speed, this equation becomes

$$e_g = K_g i_f \tag{2-85}$$

Single-stage Rotating Amplifier

A generator is represented schematically in Fig. 2-30, in which L_f and R_f and L_g and R_g are the inductance and resistance of the field and armature circuits, respectively.   The equations for the generator are

$$(L_f D + R_f) i_f = e_f$$
$$e_g = K_g i_f \tag{2-86}$$
$$e_t = e_g - (L_g D + R_g) i_a$$

The armature current depends on the load connected to the generator terminals.   Combining the first two equations gives

$$(L_f D + R_f) e_g = K_g e_f \tag{2-87}$$

Equation (2-87) relates the generated voltage e_g to the input field voltage e_f.

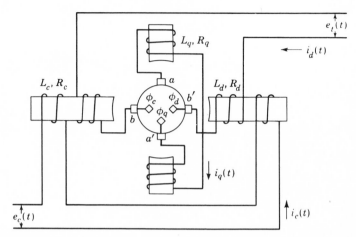

Fig. 2-31   *Schematic diagram of an amplidyne.*

Two-stage Rotating Amplifier: Amplidyne[12,13]

Two single stages of rotating amplifiers can be cascaded to obtain a large power gain. This function can be performed in one machine. A schematic drawing of the amplidyne, which is a two-stage amplifier, is shown in Fig. 2-31.

The armature is wound as a two-pole machine, and it is rotated at a constant speed. The input voltage e_c is applied to the control field, which has a large number of turns and a high impedance so that the power input is small. The current i_c in the control field sets up the small flux ϕ_c. The flux ϕ_c induces a voltage e_q in the armature between the quadrature brushes aa'. The voltage e_q is applied to the field coils located on poles which are perpendicular to the control-field axis. The impedance of this quadrature field winding and the armature is small, so a large current i_q flows. The current i_q flowing in the armature and quadrature winding produces a large flux ϕ_q. Since this flux ϕ_q is in quadrature with the control-field flux, it is termed the quadrature flux. The flux ϕ_q induces a voltage e_d in the armature between the direct axis brushes bb'. The load current i_d flowing through the armature would set up a flux ϕ_d that would buck the flux ϕ_c and cancel the action of the machine. To offset this effect, the current i_d is passed through series compensating windings located on the control-field poles. The net flux due to i_d flowing in the armature and the direct axis windings is zero. Therefore, as long as the iron is not saturated, the induced output voltage e_d is proportional to the input voltage. The terminal voltage e_t is reduced by the internal impedance drop in proportion to the load current.

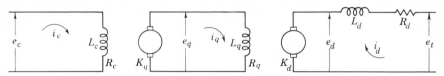

Fig. 2-32   *Equivalent two-stage representation of an amplidyne.*

To simplify the understanding of the amplidyne, an equivalent two-stage rotating amplifier is shown in Fig. 2-32. The equations can be written from either the actual or the equivalent schematic diagrams:

$$(L_cD + R_c)i_c = e_c$$
$$e_q = K_q i_c$$
$$(L_qD + R_q)i_q = e_q \tag{2-88}$$
$$e_d = K_d i_q$$
$$e_t = e_d - (L_dD + R_d)i_d$$

Combining the first four equations results in one equation that relates the output induced voltage e_d to the input control voltage e_c:

$$(L_cD + R_c)(L_qD + R_q)e_d = K_q K_d e_c \tag{2-89}$$
$$[L_cL_qD^2 + (L_qR_c + L_cR_q)D + R_cR_q]e_d = K_q K_d e_c \tag{2-90}$$

2-10   D-C Servomotor

A current-carrying conductor located in a magnetic field experiences a force proportional to the magnitude of the flux, the current, the length of the conductor, and the sine of the angle between the conductor and the direction of the flux. When the conductor is a fixed distance from an axis about which it can rotate, a torque is produced that is proportional to the product of the force and the radius. In a motor the resultant torque is the sum of the torques produced by each conductor. For any given motor the only two adjustable quantities are the flux and armature current. The torque can be expressed as

$$T(t) = K_3 \phi i_m \tag{2-91}$$

In this case, to avoid confusion with t for time, the capital letter T is used to indicate torque. It may mean either a constant or a function that varies with time. There are two modes of operation of a servomotor. In one mode the field current is held constant and an adjustable voltage is applied to the armature. In the second mode the armature current is held constant and an

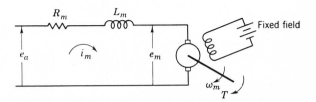

Fig. 2-33 *Circuit diagram of a d-c motor.*

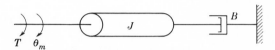

Fig. 2-34 *Inertia and friction as a motor load.*

adjustable voltage is applied to the field. These methods of operation are considered separately.

Armature Control

A constant field current is obtained by separately exciting the field from a fixed d-c source. The flux is produced by the field current and is therefore constant. Thus the torque is proportional only to the armature current and is given as

$$T(t) = K_T i_m \qquad (2\text{-}92)$$

When the motor armature is rotating, there is induced a voltage proportional to the product of flux and speed. This voltage is called the back emf. Since the flux is held constant, the induced voltage e_m is directly proportional to the speed ω_m:

$$e_m = K_1 \phi \omega_m = K_b \omega_m = K_b \, D\theta_m \qquad (2\text{-}93)$$

Control of the motor speed is obtained by adjusting the voltage applied to the armature. Its polarity determines the direction of the armature current and therefore the direction of the torque generated. This, in turn, determines the direction of rotation of the motor. A circuit diagram of the armature-controlled d-c motor is shown in Fig. 2-33. The armature inductance and resistance are labeled L_m and R_m. The voltage equation of the armature circuit is

$$L_m \, Di_m + R_m i_m + e_m = e_a \qquad (2\text{-}94)$$

The current that flows in the armature produces the required torque according to Eq. (2-92). The required torque depends on the load connected to the motor shaft. If the load consists only of an inertia and damper (friction), as shown in Fig. 2-34, the torque equation can be written as

$$J \, D\omega_m + B\omega_m = T(t) \qquad (2\text{-}95)$$

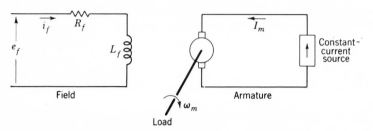

Fig. 2-35 *Circuit diagram of a field-controlled d-c motor.*

The required armature current i_m can be obtained by equating the generated torque of Eq. (2-92) to the required load torque of Eq. (2-95). Inserting this current and the back emf from Eq. (2-93) into Eq. (2-94) produces the system equation in terms of the velocity ω_m:

$$\frac{L_m J}{K_T} D^2 \omega_m + \frac{L_m B + R_m J}{K_T} D \omega_m + \left(\frac{R_m B}{K_T} + K_b \right) \omega_m = e_a \qquad (2\text{-}96)$$

This equation can also be written in terms of motor position θ_m:

$$\frac{L_m J}{K_T} D^3 \theta_m + \frac{L_m B + R_m J}{K_T} D^2 \theta_m + \frac{R_m B + K_b K_T}{K_T} D \theta_m = e_a \qquad (2\text{-}97)$$

This equation is of the form

$$(A_3 D^3 + A_2 D^2 + A_1 D)\theta_m = e_a \qquad (2\text{-}98)$$

The armature inductance is small and can usually be neglected. Equation (2-97) is thus reduced to a second-order equation.

Field Control

If the armature current i_m is constant, the torque $T(t)$ is proportional only to the flux ϕ. In the unsaturated region (see Fig. 2-29) the flux is directly proportional to the field current [see Eq. (2-83)]. Therefore the torque equation can be written as

$$T(t) = K_3 \phi i_m = K_3 K_2 i_m i_f = K_f i_f \qquad (2\text{-}99)$$

Control of the motor speed is obtained by adjusting the voltage applied to the field. Its magnitude and polarity determine the magnitude of the torque and the direction of rotation. A circuit diagram of the field-controlled d-c motor is shown in Fig. 2-35. The field winding inductance and resistance are labeled L_f and R_f. The voltage equation of the field circuit is

$$L_f D i_f + R_f i_f = e_f \qquad (2\text{-}100)$$

If the load consists of inertia and viscous friction, as shown in Fig. 2-34, the torque required to drive the load is given by Eq. (2-95). By combining

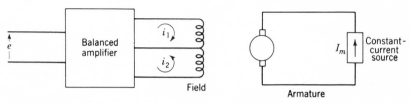

Fig. 2-36 *Circuit diagram of a split-field-controlled d-c motor.*

Eqs. (2-95), (2-99), and (2-100), the system equation in terms of the motor velocity is

$$(L_f D + R_f)(J D + B)\omega_m = K_f e_f \tag{2-101}$$

An advantage of field control compared with armature control is that the power required by the field is much smaller than that required by the armature. Since this power is usually supplied by an amplifier, smaller power capacity is required in the amplifier.

Although a constant armature current i_m is specified in the derivation above, this is not easily achieved. If the armature is connected to a fixed d-c voltage source E_a, the armature current depends on the back emf, as shown by

$$i_m = \frac{E_a - e_m}{R_m} = \frac{E_a - K_2\phi\omega_m}{R_m} \tag{2-102}$$

For the range of speed that the motor undergoes, the armature circuit may be designed so that $E_a \gg e_m$. Thus

$$i_m \approx \frac{E_a}{R_m} \tag{2-103}$$

A practical circuit for field control of a d-c motor with a split field is shown in Fig. 2-36. The field is energized by a balanced amplifier which has the input e. When this input is zero, the currents i_1 and i_2 are equal. Since they flow in opposite directions the net flux is zero, so the torque is zero and the motor is stationary. If e is not zero, one of the currents increases and the other decreases in proportion to the magnitude of e. The resulting flux is proportional to the magnitude of e, and its direction depends on the polarity of e. The size and direction of the generated torque and the resulting speed therefore respond to the magnitude and polarity of the input e.

2-11   A-C Servomotor[14,15]

An a-c servomotor is basically a two-phase induction motor that has two stator field coils placed 90 electrical degrees apart, as shown in Fig. 2-37.

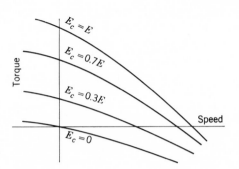

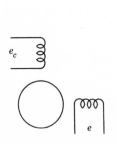

Fig. 2-37 *Schematic diagram of a two-phase induction motor.*

Fig. 2-38 *Servomotor characteristics.*

In a two-phase motor the a-c voltages e and e_c are equal in magnitude and separated by a phase angle of 90°.   A two-phase induction motor runs at a speed slightly below the synchronous speed and is essentially a constant-speed motor.   The synchronous speed is determined by the number of poles produced by the stator windings and the frequency of the voltage applied to the stator windings.

When the unit is used as a servomotor, the speed must be proportional to an input voltage.   The two-phase motor can be used as a servomotor by applying an a-c voltage e of fixed amplitude to one of the motor windings. When the other voltage e_c is varied, the torque and speed are a function of this voltage.   Figure 2-38 shows a set of torque-speed curves for various control voltages.

It is to be noted that the curve for zero control-field voltage goes through the origin and that the slope is negative.   This means that when the control-field voltage becomes zero the motor develops a decelerating torque until it stops.   The curves show a large torque at zero speed.   This is a requirement for a servomotor in order to provide rapid acceleration.   It is accomplished in an induction motor by building the rotor with a high resistance.

The torque-speed curves are not straight lines.   Therefore, a linear differential equation cannot be used to represent the exact motor characteristics.   Sufficient accuracy may be obtained by approximating the characteristics by straight lines.   The following analysis is based on this approximation.

The torque generated is a function of both the speed ω and the control-field voltage e_c.   In terms of partial derivatives, the torque equation is

$$\frac{\partial T}{\partial e_c} e_c + \frac{\partial T}{\partial \omega} \omega = T(e_c,\omega) \tag{2-104}$$

By approximating the torque-speed motor curves by parallel straight lines the partial-derivative coefficients of Eq. (2-104) are constants which can be evaluated from the graph.   Let

$$\frac{\partial T}{\partial e_c} = K_c$$
$$\frac{\partial T}{\partial \omega} = K_\omega$$

(2-105)

For a load consisting of inertia and damping, the load torque required is

$$T_L = J\,D\omega + B\omega$$

(2-106)

Since the generated and load torques must be equal, Eqs. (2-104) and (2-106) are equated:

$$K_c e_c + K_\omega \omega = J\,D\omega + B\omega$$

Rearranging terms,

$$J\,D\omega + (B - K_\omega)\omega = K_c e_c$$

(2-107)

In terms of position θ, this equation can be written as

$$J\,D^2\theta + (B - K_\omega)\,D\theta = K_c e_c$$

(2-108)

In order for the system to be stable (see Chap. 3) the coefficient $B - K_\omega$ must be positive.   Observation of the motor characteristics shows that $K_\omega = \partial T/\partial \omega$ is negative; therefore the stability requirement is satisfied.

2-12   *Lagrange's Equation*

Section 2-2 shows the application of Kirchhoff's laws for writing the differential equations of electric networks, and Secs. 2-3 and 2-5 show the application of Newton's laws for writing the equations of motion of mechanical systems.   These laws can be applied, depending on the complexity of the system, with relative ease.   In many instances there are systems that contain both electric and mechanical components.   The use of Lagrange's equation provides a systematic unified approach for handling a broad class of physical systems, no matter how complex in structure they may be.[16]

Lagrange's equation is given by

$$\frac{d}{dt}\left(\frac{\partial T}{\partial \dot{q}_n}\right) - \frac{\partial T}{\partial q_n} + \frac{\partial D}{\partial \dot{q}_n} + \frac{\partial V}{\partial q_n} = Q_n$$

(2-109)

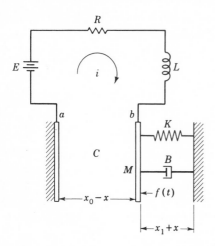

Fig. 2-39 *Electromechanical system with capacitive coupling.*

where $n = 1, 2, 3, \ldots$ are the independent coordinates or degrees of freedom which exist in the system and

$\quad T$ = total kinetic energy of system
$\quad D$ = dissipation function of system
$\quad V$ = total potential energy of system
$\quad Q_n$ = generalized applied force at the coordinate n
$\quad q_n$ = generalized coordinate
$\quad \dot{q}_n = dq_n/dt$ (generalized velocity)

The total kinetic energy T includes all energy terms, regardless of whether they are electrical or mechanical in nature. The dissipation function D represents one-half the rate at which energy is dissipated as heat; dissipation is produced by friction in mechanical systems and by resistance in electric circuits. The total potential energy stored in the system is designated by V. The forcing functions applied to a system are designated by Q_n; they take the form of externally applied forces or torques in mechanical systems and appear as voltage or current sources in electric circuits. Tables 2-6 to 2-10 list the energy terms associated with electrical and mechanical parameters. With the aid of these tables the writing of the Lagrange equation is facilitated. The application of Lagrange's equation is illustrated by the following example.

Example: Electromechanical System with Capacitive Coupling

Figure 2-39 shows a system in which mechanical motion is converted into electric energy. This represents the action which takes place in a capacitor microphone. Plate a of the capacitor is fastened rigidly to the frame. Sound waves impinge upon and exert a force on plate b of mass M, which is

Table 2-6. Identification of energy functions for electric circuits based on the loop or mesh analysis

Definitions	Element and symbol	Kinetic energy T	Potential energy V	Dissipation function D	Forcing function Q
q = charge (position coordinate) $i = dq/dt = \dot{q}$ = current e = voltage	Voltage source, e	—	—	—	e
	Inductance, L	$\tfrac{1}{2}Li^2 = \tfrac{1}{2}L\dot{q}^2$	—	—	—
	Capacitance, C	—	$\dfrac{(\int i\,dt)^2}{2C} = \dfrac{q^2}{2C}$	—	—
	Resistance, R	—	—	$\tfrac{1}{2}Ri^2 = \tfrac{1}{2}R\dot{q}^2$	—
	Mutual inductance, L_M	$\pm L_M i_1 i_2$ $= \pm L_M \dot{q}_1 \dot{q}_2$	—	—	—

Use the $+$ sign if both currents enter (or both leave) the dot ends of the coils. Otherwise use the $-$ sign.

Table 2-7. Identification of energy functions for translational mechanical elements

Definitions	Element and symbol	Kinetic energy T	Potential energy V	Dissipation factor D	Forcing function Q
x = displacement $v = dx/dt = \dot{x}$ = velocity f = force	Force, f	—	—	—	f
	Mass, M	$\tfrac{1}{2}Mv^2 = \tfrac{1}{2}M\dot{x}^2$	—	—	—
	Elastance, K	—	$\dfrac{K}{2}\left[\displaystyle\int (v_1 - v_2)\,dt\right]^2$ $= \tfrac{1}{2}K(x_1 - x_2)^2$	—	—
	Viscous damping, B	—	—	$\tfrac{1}{2}B(v_1 - v_2)^2$ $= \tfrac{1}{2}B(\dot{x}_1 - \dot{x}_2)^2$	—

Table 2-8. *Identification of energy functions for rotational mechanical elements*

Definitions	Element and symbol	Kinetic energy T	Potential energy V	Dissipation factor D	Forcing function Q
θ = angular displacement	Torque, τ	—	—	—	τ
$\omega = d\theta/dt = \dot{\theta}$ = angular velocity τ = torque	Inertia, J	$\tfrac{1}{2}J\omega^2 = \tfrac{1}{2}J\dot{\theta}^2$	—	—	—
	Elastance, K	—	$\tfrac{1}{2}K(\theta_1 - \theta_2)^2$	—	—
	Viscous damping, B	—	—	$\tfrac{1}{2}B(\omega_1 - \omega_2)^2 = \tfrac{1}{2}B(\dot{\theta}_1 - \dot{\theta}_2)^2$	—

Table 2-9. Identification of energy functions for electric circuits based on the nodal analysis

Definitions	Element and symbol	Kinetic energy T	Potential energy V	Dissipation function D	Forcing function Q
v = voltage drop $\int v\,dt = \phi$ is the position coordinate. $v = d\phi/dt = \dot{\phi}$ is the velocity coordinate.	Current source, i	—	—	—	i
	Inductance, L v_L	—	$\dfrac{1}{2L}\left(\displaystyle\int v\,dt\right)^2$ $= \dfrac{1}{2L}\phi^2$	—	—
	Capacitance, C v_C	$\tfrac{1}{2}Cv^2 = \tfrac{1}{2}C\dot{\phi}^2$	—	—	—
	Conductance $G = 1/R$ v_R	—	—	$\tfrac{1}{2}Gv^2$	—

Table 2-10. *Identification of energy functions for electromechanical elements*

Definitions	Element and symbol	Kinetic energy T
x = relative motion i = current in coil $U = l\beta n$ (electro-mechanical coupling constant) β = flux density produced by the permanent magnet l = length of coil n = no. of turns in coil		Uxi

suspended from the frame by a spring K and which has damping B. The output voltage which appears across the resistor R is intended to reproduce electrically the sound-wave patterns which strike the plate b.

At equilibrium, with no external force exerted on plate b, there is a charge q_0 on the capacitor. This produces a force of attraction between the plates so that the spring is stretched by an amount x_1 and the space between the plates is x_0. When sound waves exert a force on plate b there will be a resulting motion x which is measured from the equilibrium position. The distance between the plates will then be $x_0 - x$, and the charge on the plates will be $q_0 + q$.

The capacitance is approximated by

$$C = \frac{\epsilon A}{x_0 - x} \quad \text{and} \quad C_0 = \frac{\epsilon A}{x_0}$$

where ϵ is the dielectric constant for air and A is the area of the plate.

The energy factors for this system are

$$T = \tfrac{1}{2}L\dot{q}^2 + \tfrac{1}{2}M\dot{x}^2$$
$$D = \tfrac{1}{2}R\dot{q}^2 + \tfrac{1}{2}B\dot{x}^2$$
$$V = \frac{1}{2C}(q_0 + q)^2 + \tfrac{1}{2}K(x_1 + x)^2$$
$$= \frac{1}{2\epsilon A}(x_0 - x)(q_0 + q)^2 + \tfrac{1}{2}K(x_1 + x)^2$$

The method is simple and direct. It is merely necessary to include all the energy terms, whether electrical or mechanical. The electromechanical coupling in this example appears in the potential energy. Here, the presence of charge on the plates of the capacitor exerts a force on the mechanical system. Also, motion of the mechanical system produces an equivalent emf in the electric circuit.

The two degrees of freedom are the motion x of plate b and the current $i = \dot{q}$. Applying Lagrange's equation twice gives

$$M\ddot{x} + B\dot{x} - \frac{1}{2\epsilon A}(q_0 + q)^2 + K(x_1 + x) = f(t) \qquad (2\text{-}110)$$

$$L\ddot{q} + R\dot{q} + \frac{1}{\epsilon A}(x_0 - x)(q_0 + q) = E \qquad (2\text{-}111)$$

These equations are nonlinear. However, a good linear approximation can be obtained when it is realized that x and q are very small quantities and therefore the x^2, q^2, and xq terms can be neglected. This gives

$$(q_0 + q)^2 \approx q_0^2 + 2q_0 q$$
$$(x_0 - x)(q_0 + q) \approx x_0 q_0 - q_0 x + x_0 q$$

With these approximations the system equations become

$$M\ddot{x} + Kx_1 + Kx - \frac{q_0^2}{2\epsilon A} - \frac{2q_0 q}{2\epsilon A} + B\dot{x} = f(t) \qquad (2\text{-}112)$$

$$L\ddot{q} + \frac{x_0 q_0}{\epsilon A} - \frac{q_0 x}{\epsilon A} + \frac{x_0 q}{\epsilon A} + R\dot{q} = E \qquad (2\text{-}113)$$

From the first equation, by setting $f(t) = 0$ and taking steady-state conditions, the result is

$$Kx_1 - \frac{q_0^2}{2\epsilon A} = 0$$

This simply equates the force on the spring and the force due to the charges at the equilibrium condition. Similarly, in the second equation at equilibrium

$$\frac{x_0 q_0}{\epsilon A} = \frac{q_0}{C_0} = E$$

Therefore the two system equations can be written in linearized form as

$$M\ddot{x} + B\dot{x} + Kx - \frac{q_0}{\epsilon A} q = f(t) \qquad (2\text{-}114)$$

$$L\ddot{q} + R\dot{q} + \frac{q}{C_0} - \frac{q_0}{\epsilon A} x = 0 \qquad (2\text{-}115)$$

These equations show that $q_0/\epsilon A$ is the coupling factor between the electrical and mechanical portions of the system.

Another form of electromechanical coupling exists when current in a coil produces a force which is exerted on a mechanical system and, simultaneously, motion of a mass induces an emf in an electric circuit. In that case the kinetic energy includes a term*

$$T = l\beta nxi = Uxi \qquad (2\text{-}116)$$

where l = length of coil

β = flux density produced by a permanent magnet which links the coil

n = number of turns in coil

x = displacement of coil

i = current through coil

The main advantage of Lagrange's equation is the use of a single systematic procedure. This replaces the need to consider separately Kirchhoff's laws for the electrical aspects and Newton's law for the mechanical aspects of the system in formulating the statements of equilibrium. Once this procedure is mastered, the differential equations which describe the system are readily obtained.

2-13   *Conclusions*

The examples in this chapter cover many of the basic aspects encountered in control systems. In order to write the differential equations, the basic laws governing performance are first stated for electric, mechanical, thermal, and hydraulic systems. These basic laws are then applied to specific devices and their differential equations of performance obtained. Lagrange's equation has been introduced to provide a systematized method for writing the differential equations of electric, mechanical, and electromechanical systems. Use of the tables provided makes the application of Lagrange's equation a straightforward procedure. This chapter should be a future reference for the engineer who would refresh himself on the fundamental concepts involved in writing differential equations of performance.

Bibliography

1. Blackburn, J. F. (ed.): "Components Handbook," McGraw-Hill Book Company, New York, 1948.
2. Wylie, C. R., Jr.: "Advanced Engineering Mathematics," 2d ed., McGraw-Hill Book Company, New York, 1960.
3. Corcoran, G. F., and R. M. Kerchner: "Alternating-current Circuits," 4th ed., John Wiley & Sons, Inc., New York, 1960.

* The energy for this system, which is shown in Table 2-10, may also be considered potential energy. This influences the sign on the corresponding term in the differential equation.

4. Gardner, M. F., and J. L. Barnes: "Transients in Linear Systems," chap. 2, John Wiley & Sons, Inc., New York, 1942.
5. Hornfeck, A. J.: Response Characteristics of Thermometer Elements, *Trans. ASME*, vol. 71, pp. 121–132, 1949.
6. Trimmer, J. D.: "Response of Physical Systems," p. 17, John Wiley & Sons, Inc., New York, 1950.
7. "Flow Meters: Their Theory and Application," American Society of Mechanical Engineers, New York, 1937.
8. Newton, G. C., Jr.: Hydraulic Variable Speed Transmissions as Servomotors, *J. Franklin Inst.*, vol. 243, no. 6, pp. 439–469, June, 1947.
9. Bruns, R. A., and R. M. Saunders: "Analysis of Feedback Control Systems," chap. 4, McGraw-Hill Book Company, New York, 1955.
10. Saunders, R. M.: The Dynamo Electric Amplifier—Class A Operation, *Trans. AIEE*, vol. 68, pp. 1368–1373, 1949.
11. Litman, B.: An Analysis of Rotating Amplifiers, *Trans. AIEE*, vol. 68, pt. II, pp. 1111–1117, 1949.
12. Alexanderson, E. F.: The Amplidyne System of Control, *Proc. IRE*, vol. 32, pp. 513–520, 1944.
13. Bowers, J. L.: Fundamentals of the Amplidyne Generator, *Trans. AIEE*, vol. 64, pp. 873–881, 1945.
14. Koopman, R. J. W.: Operating Characteristics of Two-phase Servomotors, *Trans. AIEE*, vol. 68, pp. 319–328, 1949.
15. Hopkin, A. M.: Transient Response of Small Two-phase Servomotors, *Trans. AIEE*, vol. 70, pp. 881–886, 1951.
16. Ogar, G. W., and J. J. D'Azzo: A Unified Procedure for Deriving the Differential Equations of Electrical and Mechanical Systems, *IRE Trans. Educ.*, vol. E-5, no. 1, pp. 18–26, March, 1962.

Solution of
Differential Equations

The preceding chapter has covered the writing of differential equations for dynamic components and systems. The accurate representation of a physical system by the correct dynamic equations is, of course, very necessary. The next requirement is to determine a solution of the differential equations.[1-4]

The general solution of a differential equation is the sum of the particular integral and the complementary function. Those terms of the general solution that disappear as t approaches infinity can be considered as comprising the transient component of the solution. If the ultimate component is periodic or constant in nature, it can be termed the steady-state component. Often the particular integral is the steady-state component of the solution of the differential equation; and the complementary function, which is the solution of the corresponding homogeneous equation, is the transient component of the solution. Often the steady-state component of the response has the same form as the driving function. In this book, the particular integral is called the steady-state solution even when it is not periodic. The form of the transient component of the response depends only on the roots of the characteristic equation. The instantaneous amplitude of the transient component depends on both the initial conditions and the instantaneous magnitude of the steady-state component.

For a particular system either the steady-state or the transient component of response may be the more important component. For some systems the two are equally important. This is determined entirely by the objective of the system.

This chapter covers methods of solving for the steady-state and the transient components of the solution. These components are first determined separately and then added to form the complete solution. The transient component of the solution is determined by the classical method, which should give the student a "feel" for the solution to be expected.

3-2   Input to Control Systems

For some control systems the input has a specific form which may be represented either by an analytical expression or as a specific curve. An example of the latter is the pattern used in a machining operation where the cutting tool is required to follow the path indicated by the pattern outline.

For other control systems the input may be random in shape. In this case it cannot be expressed analytically and it is not repetitive. An example is the camera platform used in a photographic airplane. The airplane flies at a fixed altitude and speed, and the camera takes a series of pictures of the terrain below it, which are then fitted together to form one large picture of the area. This requires that the camera platform remain level regardless of the motion of the airplane. Since the attitude of the airplane varies with the wind gusts present and depends on the stability of the airplane itself, it can be realized that the input to the camera platform is a random function.

It is important to have a basis of comparison for various systems. One way of doing this is comparing the response with a standardized input. The input or inputs used as a basis of comparison must be determined from the required response of the system and the actual form of its input.

3-3   Standardized Inputs

The following standard inputs are often used in checking the response of a system:

1. *Sinusoidal Function:* $r = \cos \omega t$
2. *Power-series Function:* $r = a_0 + a_1 t + a_2 t^2$
$$+ \cdots$$
3. *Unit Step Function:* $r = u(t)$
4. *Unit Ramp (Step Velocity) Function:* $r = tu(t)$
5. *Unit Parabolic (Step Acceleration) Function:* $r = t^2 u(t)$
6. *Unit Impulse Function:* $r = \delta(t)$

For each of these inputs a complete solution of the differential equation is determined in this chapter. First, generalized methods are developed to determine the steady-state output for each type of input. These methods are applicable to differential equations of any order. Next, the method of evaluating the transient component of response is determined, and it is shown that the form of the transient component of response depends on the characteristic equation. The coefficients of the transient terms are determined by the steady-state component and the initial conditions. Addition of the steady-state component and the transient component gives the complete solution. Several examples are used to illustrate these principles.

The solution of the differential equation with a pulse input is postponed until the next chapter, where the Laplace transform is used.

3-4   *Steady-state Response: Sinusoidal Input*

The input quantity r is assumed to be a sinusoidal function of the form

$$r(t) = R \cos (\omega t + \alpha) \tag{3-1}$$

The general integrodifferential equation to be solved is of the form

$$A_v D^v c + A_{v-1} D^{v-1} c + \cdots + A_0 D^0 c + A_{-1} D^{-1} c + \cdots$$
$$A_{-w} D^{-w} c = r \tag{3-2}$$

The steady-state solution can be obtained directly by use of Euler's identity,

$$e^{j\omega t} = \cos \omega t + j \sin \omega t$$

The input can then be written

$$r = R \cos (\omega t + \alpha) = \text{real part of } (Re^{j(\omega t + \alpha)}) = \text{Re} \ (Re^{j(\omega t + \alpha)})$$
$$= \text{Re} \ (Re^{j\alpha} e^{j\omega t}) = \text{Re} \ (\mathbf{R} e^{j\omega t}) \tag{3-3}$$

For simplicity, the phrase *real part of* or its symbolic equivalent Re will often be omitted, but it is to be remembered that the real part is intended. The quantity $R = Re^{j\alpha}$ is the phasor* representation of the input; i.e., it has both a magnitude R and an angle α. The magnitude R represents the maximum value of the input quantity $r(t)$. For simplicity the angle $\alpha = 0°$ is chosen for R. The input r from Eq. (3-3) is inserted in Eq. (3-2). Then, in order for the expression to be an equality, the response c must be of the form

$$c(t) = C \cos(\omega t + \phi) = \text{Re}\ (Ce^{j\phi}e^{j\omega t}) = \text{Re}\ (Ce^{j\omega t}) \qquad (3-4)$$

where $C = Ce^{j\phi}$ is a phasor quantity having the magnitude C and the angle ϕ. The nth derivative of c with respect to time is

$$D^n c = \text{Re}\ [(j\omega)^n Ce^{j\omega t}] \qquad (3-5)$$

Inserting c and its derivatives from Eqs. (3-4) and (3-5) into Eq. (3-2) gives

$$\text{Re}\ [A_v(j\omega)^v Ce^{j\omega t} + A_{v-1}(j\omega)^{v-1}Ce^{j\omega t} + \cdots + A_{-w}(j\omega)^{-w}Ce^{j\omega t}]$$
$$= \text{Re}\ (Re^{j\omega t}) \qquad (3-6)$$

Canceling $e^{j\omega t}$ from both sides of the equation and solving for C gives

$$C =$$
$$\frac{R}{A_v(j\omega)^v + A_{v-1}(j\omega)^{v-1} + \cdots + A_0 + A_{-1}(j\omega)^{-1} + \cdots + A_{-w}(j\omega)^{-w}}$$
$$(3-7)$$

where C is the phasor representation of the output; i.e., it has a magnitude C and an angle ϕ. Since the values of C and ϕ are functions of the frequency ω, it may be written as $C(j\omega)$ to show this relationship. Similarly, $R(j\omega)$ denotes the fact that the input is sinusoidal and may be a function of frequency.

When Eqs. (3-2) and (3-6) are compared, it can be seen that one equation can be determined easily from the other. Substituting $j\omega$ for D, $C(j\omega)$ for c, and $R(j\omega)$ for r in Eq. (3-2) results in Eq. (3-6). The reverse is also true and is independent of the order of the equation. It should be realized that this is simply a *rule of thumb* which yields the desired expression.

The time response can be obtained directly from the phasor response. The output is

$$c(t) = \text{Re}\ (Ce^{j\omega t}) = |C| \cos(\omega t + \phi) \qquad (3-8)$$

Determination of the steady-state sinusoidal response for two systems follows.

---

* The term *phasor* has replaced the term *vector* when applied to sinusoidal quantities.

RLC Circuit

For the *RLC* circuit described in Sec. 2-2 the equation of the circuit is

$$L\,Di + Ri + \frac{1}{CD}\,i = e \tag{3-9}$$

The sinusoidal voltage e is given by

$$e = \sqrt{2}\,\mathbf{E}(j\omega)\,\cos \omega t = \mathrm{Re}\,[\sqrt{2}\,\mathbf{E}(j\omega)e^{j\omega t}]$$

The use of capital letters without subscripts to designate voltages and currents denotes effective values; therefore the maximum value of e is equal to $\sqrt{2}\,|\mathbf{E}(j\omega)|$.

Solving for the phasor current $\mathbf{I}(j\omega)$,

$$\mathbf{I}(j\omega) = \frac{\mathbf{E}(j\omega)}{j\omega L + R + 1/j\omega C} \tag{3-10}$$

The impedance of the circuit is

$$\mathbf{Z}(j\omega) = j\omega L + R + \frac{1}{j\omega C} \tag{3-11}$$

Both phasors $\mathbf{E}(j\omega)$ and $\mathbf{I}(j\omega)$ have a magnitude and an angle. If the voltage e is taken as zero at the reference time $t = 0$, its angle is $0°$. The current i leads or lags the voltage e by an angle ϕ which is determined by the impedance. The impedance and phasor diagrams are shown in Fig. 3-1.

For the same *RLC* circuit the equation relating the input voltage to the voltage across the resistor is

$$\frac{L}{R}\,Dv_R + v_R + \frac{1}{RCD}\,v_R = e \tag{3-12}$$

Solving for the output voltage in phasor form,

$$\mathbf{V}_R(j\omega) = \frac{\mathbf{E}(j\omega)}{j\omega L/R + 1 + 1/j\omega RC} = \frac{R\mathbf{E}(j\omega)}{j\omega L + R + 1/j\omega C} \tag{3-13}$$

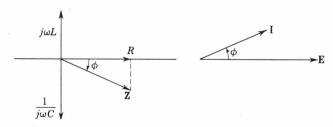

Fig. 3-1 *Impedance and phasor diagrams for an RLC circuit.*

The ratio of phasor output to phasor input is termed the *frequency transfer function* and is often designated by the symbol $\mathbf{G}(j\omega)$. This ratio, in terms of the steady-state sinusoidal phasors, is

$$\mathbf{G}(j\omega) = \frac{\mathbf{V}_R(j\omega)}{\mathbf{E}(j\omega)} = \frac{R}{j\omega L + R + 1/j\omega C} \tag{3-14}$$

Hydraulic Transmission

The hydraulic transmission described in Sec. 2-8 has an input x and an output angular position θ_m. Equation (2-78), which relates input to output in terms of system parameters, is repeated below:

$$d_m D\theta_m + \frac{LJ}{C} D^2\theta_m + \frac{VJ}{K_B C} D^3\theta_m = d_p\omega_p x \tag{3-15}$$

The ratio of steady-state sinusoidal output to input is

$$\mathbf{G}(j\omega) = \frac{\theta_m(j\omega)}{\mathbf{X}(j\omega)} = \frac{d_p\omega_p/d_m}{j\omega[(VJ/K_B Cd_m)(j\omega)^2 + (LJ/Cd_m)(j\omega) + 1]} \tag{3-16}$$

This ratio of the phasor output to the phasor input is the frequency transfer function.

3-5   Steady-state Response: Power-series Input

A general development of the steady-state solution of a differential equation with a general power-series input is first given. Particular cases of the power series are then covered in detail.

The general differential equation is repeated here:

$$A_v D^v c + A_{v-1} D^{v-1} c + \cdots + A_0 c + A_{-1} D^{-1} c + \cdots \\ + A_{-w} D^{-w} c = r \tag{3-17}$$

The power-series input is of the form

$$r(t) = a_0 + a_1 t + a_2 t^2 + \cdots + a_k t^k \tag{3-18}$$

where the highest-order term in the input is $a_k t^k$. The problem is to find the steady-state or particular solution of the dependent variable c. The method used is to assume a power-series solution of the form

$$c(t) = b_0 + b_1 t + b_2 t^2 + \cdots + b_y t^y \tag{3-19}$$

The assumed solution is then substituted into the differential equation. The coefficients b_0, b_1, b_2, etc., of the power-series solution are evaluated by equating the coefficients of like powers of t on both sides of the equation. This condition must be satisfied if the solution is correct. The highest power of t on the right side of Eq. (3-17) is k; therefore t^k must also appear on the left side of this equation. If a higher power of t appears on the left side of the equation, then the coefficient of this term must equal zero. The highest power of t on the left side of the equation is produced by the lowest-order derivative term and is equal to y minus the order of the lowest derivative. With this information, the value of the highest-order exponent of t to use in the assumed solution of Eq. (3-17) is

$$y = k + X \qquad\qquad y \geq 0 \qquad\qquad (3\text{-}20)$$

where k is the highest exponent appearing in the input and X is the order of the lowest derivative appearing in the differential equation. When integral terms are present, X is a negative number. X is the lowest-order exponent of the differential operator D appearing in the differential equation. For the general differential equation (3-17), the value of X is equal to $-w$. Equation (3-20) is valid only for positive values of y.

Examples of power-series input to various components are shown in the following pages. For each of the examples the response is a power series since the input is of that form. However, the highest power in the response may not be the same as that of the input.

Step-function Input

A convenient input $r(t)$ to a system is an abrupt change represented by a unit step function, as shown in Fig. 3-2. This type of input cannot always be put into a system since it takes a definite length of time to make the change in input, but it represents a good mathematical input for checking system response.

Provided that the system is stable and the input is held constant, the output must eventually reach a constant value. This means that in the steady state the derivatives and integrals reach a constant value. In determining the solution, care must be taken to consider the correct form of response. The following example using a servomotor shows that with a step input the velocity of the output reaches a constant value (a step response) and the output position is therefore a ramp that increases linearly with time. The unit step function can be considered a power series where the highest

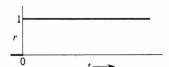

Fig. 3-2   Unit step function.

power of t is $k = 0$. The method of solution for a power series can therefore be used.

The servomotor described in Sec. 2-10 is used as an example. The response of motor velocity ω_m in terms of the voltage e applied to the armature, as given by Eq. (2-96), is of the form

$$A_2 D^2\omega_m + A_1 D\omega_m + A_0\omega_m = e \qquad (3\text{-}21)$$

The input is the unit step function

$$e = u(t)$$

For the step function the highest exponent of t is $k = 0$. The lowest-order differential in Eq. (3-21) is $X = 0$; therefore the response is of the form

$$\omega_m = b_0 \qquad (3\text{-}22)$$

The derivatives are

$$D\omega_m = 0$$
$$D^2\omega_m = 0$$

Inserting these values into Eq. (3-21),

$$(A_2 \times 0) + (A_1 \times 0) + A_0 b_0 = 1$$

Solving for b_0 gives

$$b_0 = \frac{1}{A_0}$$

With the value of A_0 for the original system as described by Eq. (2-96), the steady-state output velocity is

$$\omega_m = \frac{K_T}{R_m B + K_b K_T} \qquad (3\text{-}23)$$

Note that a constant voltage applied to the motor armature produces a constant speed. If the position of the motor shaft is desired, it can be obtained by integrating Eq. (3-23):

$$\theta_m = \frac{1}{D}\,\omega_m = \frac{K_T}{R_m B + K_m K_T}\,t + C_0 \qquad (3\text{-}24)$$

C_0 is an arbitrary constant which must be evaluated in a specific case from knowledge of the initial position of the motor shaft.

Ramp-function Input (Step-function of Velocity)

The ramp function is a fixed rate of change of a variable as a function of time.

This input and its rate of change are shown in Fig. 3-3. The input is expressed mathematically as

$$r = tu(t) \qquad Dr = u(t)$$

A ramp input is a power-series input where the highest power of t is $k = 1$. Therefore, the method for the steady-state solution of a power-series input can be used.

For a stable system with a ramp input there must be some function of the output that varies at the same rate.

The electric water heater of Sec. 2-6 is an example of a system that operates with a ramp input. Consider the case of a constant rate of flow of water, n, through the tank. When the heater is turned on, the heat h entering the water tank is a ramp and the rate q at which heat enters the water tank is a constant. In terms of the temperature of the water in the tank, θ, and the heat entry into the tank, h, the system response is obtained by integrating Eq. (2-55):

$$C\theta + \left(nS + \frac{1}{R}\right) D^{-1}\theta = h \qquad (3\text{-}25)$$

The highest exponent of t in the input, $h = t$, is $k = 1$, and the order of the lowest derivative in the system is $X = -1$. Therefore, the response, which is the temperature, is of the form

$$\theta = b_0 \qquad (3\text{-}26)$$

and its integral is

$$D^{-1}\theta = b_0 t + C_0$$

When these values are inserted in Eq. (3-25), the result is

$$Cb_0 + \left(nS + \frac{1}{R}\right)(b_0 t + C_0) = t$$

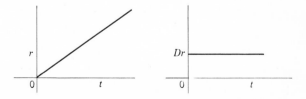

Fig. 3-3 *A ramp function and its derivative.*

Equating coefficients of t raised to the same power,

t^0:
$$Cb_0 + \left(nS + \frac{1}{R}\right)C_0 = 0$$

t^1:
$$\left(nS + \frac{1}{R}\right)b_0 = 1$$

from which

$$b_0 = \frac{R}{SRn + 1}$$

Thus, from Eq. (3-26), the steady-state temperature is

$$\theta = \frac{R}{SRn + 1} \tag{3-27}$$

This indicates that the rate of water flow, n, determines the steady-state temperature. If the flow is too large, the steady-state temperature will be lower than desired. If the flow is too small, then the temperature will be too high. Of course, in an actual electric water heater a thermostat shuts off the heater when the temperature reaches a preset upper limit.

The same results could be obtained from Eq. (2-54) expressed in terms of the rate of heat flow, q, from the heater. In this case the rate of heat flow, q, is a step function.

Parabolic-function Input (Step-function of Acceleration)

A parabolic-function input represents a constant second derivative, which means that the first derivative is a ramp. The input, the first derivative, and the second derivative are shown in Fig. 3-4.

The input is expressed mathematically as

$$r = t^2u(t) \qquad Dr = 2tu(t) \qquad D^2r = 2u(t)$$

A parabolic input is a power-series input where the highest power of t is $k = 2$. A steady-state solution can be found in the conventional manner for a power-series input.

Fig. 3-4 *Parabolic function and its derivatives.*

The simple mechanical system of Sec. 2-3 is used as an example. Equation (2-27), repeated below, relates the position x_2 to the position x_1:

$$M\,D^2x_2 + B\,Dx_2 + Kx_2 = Kx_1 \tag{3-28}$$

Assume that the input x_1 moves with constant acceleration, as illustrated in Fig. 3-4. The highest exponent of t in the input is $k = 2$, and the order of the lowest derivative in the system equation is $X = 0$. The value of y is therefore equal to 2, and the steady-state response is of the form

$$x_2(t) = b_0 + b_1 t + b_2 t^2 \tag{3-29}$$

The derivatives of x_2 are

$$Dx_2 = b_1 + 2b_2 t$$
$$D^2x_2 = 2b_2$$

Inserting these values into Eq. (3-28) gives

$$M(2b_2) + B(b_1 + 2b_2 t) + K(b_0 + b_1 t + b_2 t^2) = Kt^2$$

Equating the coefficients of t raised to the same power,

$$t^0: \qquad 2Mb_2 + Bb_1 + Kb_0 = 0$$
$$t^1: \qquad 2Bb_2 + Kb_1 = 0$$
$$t^2: \qquad Kb_2 = K$$

From these equations, $b_2 = 1$, $b_1 = -2B/K$, and $b_0 = -2M/K + 2B^2/K^2$. The resulting steady-state solution of x_2 is

$$x_2(t) = -2\frac{M}{K} + 2\frac{B^2}{K^2} - 2\frac{B}{K}t + t^2 \tag{3-30}$$

In control systems it is desired that the output follow the input with a minimum of error and time lag. The error (equal to the input minus the output) is used to evaluate the suitability of the system performance. In the example above, the solution is significant because it shows that the error between x_2 and x_1 has two components. One component is constant and equals $2(M/K - B^2/K^2)$, and the other component $(2Bt/K)$ increases in magnitude with time. The acceleration of x_2 is equal to the acceleration of x_1.

3-6   *Transient Response: Classical Method*

The classical method of solving for the complementary function or transient response of a differential equation requires, first, the writing of the homogeneous equation. The general differential equation has the form

$$b_v D^v c + b_{v-1} D^{v-1} c + \cdots + b_0 D^0 c + b_{-1} D^{-1} c + \cdots$$
$$+ b_{-w} D^{-w} c = r \quad (3\text{-}31)$$

where r is the forcing function and c is the response.

The homogeneous equation is formed by letting the right-hand side of the differential equation equal zero:

$$b_v D^v c_t + b_{v-1} D^{v-1} c_t + \cdots + b_0 c_t + b_{-1} D^{-1} c_t + \cdots$$
$$+ b_{-w} D^{-w} c_t = 0 \quad (3\text{-}32)$$

where c_t is the transient component of the general solution.

The general expression for the transient response, which is the solution of the homogeneous equation, is obtained by assuming a solution of the form

$$c_t = e^{mt} \quad (3\text{-}33)$$

where m is a constant yet to be determined. Substituting this value of c_t into Eq. (3-32) and factoring e^{mt} from all terms gives

$$e^{mt}(b_v m^v + b_{v-1} m^{v-1} + \cdots + b_0 + \cdots + b_{-w} m^{-w}) = 0 \quad (3\text{-}34)$$

Equation (3-34) must be satisfied for e^{mt} to be a solution. Since e^{mt} can not be zero for all values of time t, it is necessary that

$$b_v m^v + b_{v-1} m^{v-1} + \cdots + b_0 + \cdots + b_{-w} m^{-w} = 0 \quad (3\text{-}35)$$

This is purely an algebraic equation and is termed the *characteristic equation.* There are $v + w$ roots of the characteristic equation; therefore the complete transient solution will contain the same number of terms of the form e^{mt} if all the roots are simple. Thus the transient component, where there are no multiple roots, is

$$c_t = A_1 e^{m_1 t} + A_2 e^{m_2 t} + \cdots + A_k e^{m_k t} + \cdots + A_{v+w} e^{m_{v+w} t} \quad (3\text{-}36)$$

If there is a root m_q of multiplicity r, the transient will include corresponding terms of the form

$$A_{q1} e^{m_q t} + A_{q2} t e^{m_q t} + \cdots + A_{qr} t^{r-1} e^{m_q t} \quad (3\text{-}37)$$

Instead of using the detailed procedure outlined above, the characteristic equation is usually obtained directly from the homogeneous equation by substituting m for Dc_t, m^2 for $D^2 c_t$, etc. Since the coefficients of the transient solution must be determined from the initial conditions, there must be $v + w$ known initial conditions. These conditions are values of the variable c and its derivatives which are known at specific times. The $v + w$ initial conditions are used to set up $v + w$ simultaneous equations of c and its derivatives. The value of c includes both the steady-state and transient components. Since determination of the coefficients includes consideration of the steady-state component, it is seen that the input affects the magnitude of the coefficient of each exponential term.

Complex Roots

If all values of m_k are real, the transient terms can be evaluated as indicated above. Frequently, some values of m_k are complex. When this happens, they always occur in pairs that are complex conjugates and are of the form

$$
\begin{aligned}
m_k &= \sigma + j\omega_d \\
m_{k+1} &= \sigma - j\omega_d
\end{aligned}
$$
(3-38)

The transient terms corresponding to these values of m are

$$
A_k e^{(\sigma+j\omega_d)t} + A_{k+1} e^{(\sigma-j\omega_d)t}
$$
(3-39)

These terms are combined to a more useful form by factoring the term $e^{\sigma t}$:

$$
e^{\sigma t}(A_k e^{j\omega_d t} + A_{k+1} e^{-j\omega_d t})
$$
(3-40)

By using the Euler formula expansion $e^{\pm j\omega_d t} = \cos \omega_d t \pm j \sin \omega_d t$, expression (3-40) can be written as

$$
e^{\sigma t}(A_k \cos \omega_d t + jA_k \sin \omega_d t + A_{k+1} \cos \omega_d t - jA_{k+1} \sin \omega_d t)
$$

By collecting terms, this becomes

$$
e^{\sigma t}[(A_k + A_{k+1}) \cos \omega_d t + j(A_k - A_{k+1}) \sin \omega_d t]
$$
(3-41)

Remember that this is a physical system, which means that each term of expression (3-41) must be real. This imposes limitations on the values of A_k and A_{k+1} which require that

$$
\begin{aligned}
A_k + A_{k+1} &= \text{real number} \\
j(A_k - A_{k+1}) &= \text{real number}
\end{aligned}
$$
(3-42)

To satisfy simultaneously Eqs. (3-42) the necessary condition is that A_k and A_{k+1} be complex conjugates and of the form

$$
A_k = A_0 e^{j\psi} \qquad A_{k+1} = A_0 e^{-j\psi}
$$

Inserting these values into expression (3-40) gives

$$
e^{\sigma t}(A_0 e^{j(\omega_d t+\psi)} + A_0 e^{-j(\omega_d t+\psi)})
$$
(3-43)

The cosine expressed in terms of exponentials is

$$
\cos \beta = \frac{e^{j\beta} + e^{-j\beta}}{2}
$$

Therefore, by using this form and putting it in terms of the sine, where $\phi = \psi + \pi/2$, expression (3-43) is written as

$$2A_0 e^{\sigma t} \sin (\omega_d t + \phi) = A e^{\sigma t} \sin (\omega_d t + \phi) \tag{3-44}$$

The two constants A and ϕ must be determined from two initial conditions. ω_d is defined as the damped natural frequency.

When two roots of the characteristic equation are found to be complex conjugates, the transient term can be written in the form

$$A e^{\sigma t} \sin (\omega_d t + \phi) \tag{3-45}$$

This is a very convenient form for plotting the transient response. The student must learn to use this equation directly without deriving it each time. Often it will be found that the constants in the transient term can be evaluated more readily from the initial conditions by using the form

$$e^{\sigma t}(B_1 \cos \omega_d t + B_2 \sin \omega_d t) \tag{3-46}*$$

This term is called an exponentially damped sinusoid; it consists of a sine wave of frequency ω_d whose magnitude is $A e^{\sigma t}$, that is, it is decreasing exponentially with time if σ is negative. It has the form shown in Fig. 3-5. The curves $\pm A e^{\sigma t}$ comprise the *envelope*. The plot of the time solution always remains between the two branches of the envelope.

In the analysis above, the complex roots were given in the form

$$m_{1,2} = \sigma \pm j\omega_d$$

When σ is negative, the transient decays with time and eventually dies out. This represents a stable system. When σ is positive, the transient increases with time and will destroy the equipment unless otherwise restrained. This

---

* Expression (3-46) can be converted to expression (3-45) by use of the relationships $A = \sqrt{B_1^2 + B_2^2}$ and $\phi = \tan^{-1} B_1/B_2$.

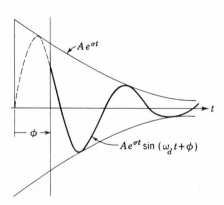

Fig. 3-5 *Sketch of an exponentially damped sinusoid.*

represents the undesirable case of an unstable system. Control systems must be designed so that they are always stable.

Damping Ratio and Undamped Natural Frequency

When the characteristic equation has a pair of complex conjugate roots, it has a quadratic factor of the form $b_2 m^2 + b_1 m + b_0$. The roots of this factor are

$$m_{1,2} = -\frac{b_1}{2b_2} \pm j \sqrt{\frac{4b_2 b_0 - b_1^2}{4b_2^2}} = \sigma \pm j\omega_d \qquad (3\text{-}47)$$

The real part σ is recognized as the exponent of e, and ω_d is the damped natural frequency of the oscillatory portion of the component stemming from this pair of roots, as given by expression (3-45).

The quantity b_1 represents the effective damping constant of the system. If b_1 has the value $2\sqrt{b_2 b_0}$, the two roots m are equal. This represents the critical value of damping constant and is written $b_1' = 2\sqrt{b_2 b_0}$.

The damping ratio ζ is defined as the ratio of the actual damping constant to the critical value of damping constant:

$$\zeta = \frac{\text{actual damping coefficient}}{\text{critical damping coefficient}} = \frac{b_1}{b_1'} = \frac{b_1}{2\sqrt{b_2 b_0}} \qquad (3\text{-}48)$$

When ζ is positive and less than unity, the roots are complex and the transient is a damped sinusoid of the form of expression (3-45). When ζ is less than unity, the response is said to be underdamped. When ζ is greater than unity, the roots are real and the response is overdamped; i.e., the transient solution consists of two exponential terms.

The undamped natural frequency ω_n is defined as the frequency of oscillation of the transient if the damping is zero:

$$\omega_n = \sqrt{\frac{b_0}{b_2}} \qquad (3\text{-}49)$$

The case of zero damping constant means that the transient response does not die out; it is a sine wave of constant amplitude.

The quadratic factors are frequently written in terms of the damping ratio and the undamped natural frequency. Underdamped systems are generally analyzed in terms of these two parameters. By factoring b_0, the quadratic factor of the characteristic equation is

$$\frac{b_2}{b_0} m^2 + \frac{b_1}{b_0} m + 1 = \frac{1}{\omega_n^2} m^2 + \frac{2\zeta}{\omega_n} m + 1 \qquad (3\text{-}50)$$

When it is multiplied through by ω_n^2, the quadratic appears in the form

$$m^2 + 2\zeta\omega_n m + \omega_n^2 \qquad (3\text{-}51)$$

The forms given by Eqs. (3-50) and (3-51) are called the standard forms of the quadratic factor, and the corresponding roots are

$$m_{1,2} = \sigma \pm j\omega_d = -\zeta\omega_n \pm j\omega_n \sqrt{1 - \zeta^2} \qquad (3\text{-}52)$$

The transient response of Eq. (3-45) for the underdamped case, written in terms of ζ and ω_n, is

$$Ae^{-\zeta\omega_n t} \sin (\omega_n \sqrt{1 - \zeta^2}\, t + \phi) \qquad (3\text{-}53)$$

From this expression the effect on the transient of the terms ζ and ω_n can be readily seen. The larger the product $\zeta\omega_n$, the faster the transient will decay. These terms also affect the damped natural frequency of oscillation, $\omega_d = \omega_n \sqrt{1 - \zeta^2}$, of the transient, which varies directly as the undamped natural frequency and decreases with an increase in the damping ratio.

3-7   Definition of Time Constant

The transient terms have the exponential form Ae^{mt}. When $m = -a$ is real and negative, the plot of Ae^{-at} has the form shown in Fig. 3-6. The value of time that makes the exponent of e equal to -1 is called the time constant T. Thus

$$-aT = -1$$
$$T = \frac{1}{a} \qquad (3\text{-}54)$$

In a duration of time equal to one time constant the exponential e^{-at} decreases from the value 1 to the value 0.368. Geometrically the tangent drawn to the curve Ae^{-at} at $t = 0$ intersects the time axis at the value of time equal to the time constant T.

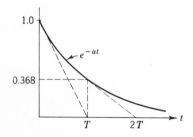

Fig 3-6 *Plot of the exponential e^{-at}.*

When $m = \sigma \pm j\omega_d$ is a complex quantity, the transient has the form $Ae^{\sigma t} \sin(\omega_d t + \phi)$. A plot of this function is shown in Fig. 3-5. In the case of the damped sinusoid the time constant is defined in terms of the parameter σ that characterizes the envelope $Ae^{\sigma t}$. Thus the time constant T is equal to

$$T = \frac{1}{|\sigma|} \tag{3-55}$$

In terms of the damping ratio and undamped natural frequency, the time constant is $T = 1/\zeta\omega_n$. Therefore, the larger the product $\zeta\omega_n$, the greater the instantaneous rate of decay of the transient.

3-8 Example: First-order System

The series RL circuit of Fig. 3-7 will be used as an example. The equation relating output voltage v_R to the generator voltage e is

$$e = \frac{L}{R} D v_R + v_R \tag{3-56}$$

The output voltage is $v_R = v_{R,\text{ss}} + v_{R,t}$, where $v_{R,\text{ss}}$ is the steady-state solution and $v_{R,t}$ is the transient solution. The steady-state solution is found first by using the method of Sec. 3-4. The phasor voltage $\mathbf{V}_{R,\text{ss}}(j\omega)$ is

$$\mathbf{V}_{R,\text{ss}}(j\omega) = \frac{\mathbf{E}(j\omega)}{1 + (L/R)j\omega} = \frac{\mathbf{E}(j\omega)}{\sqrt{1 + [(L/R)\omega]^2}} \Big/ -\tan^{-1}\frac{\omega L}{R}$$

The steady-state voltage as a function of time is therefore

$$v_{R,\text{ss}} = \frac{E\sqrt{2}}{\sqrt{1 + (\omega L/R)^2}} \sin\left(\omega t - \tan^{-1}\frac{\omega L}{R}\right) \tag{3-57}$$

The characteristic equation obtained from the differential equation is

$$\frac{L}{R}m + 1 = 0$$

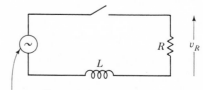

Fig. 3-7 *RL circuit.* $e(t) = E\sqrt{2}\sin\omega t$

and the root of this characteristic equation is

$$m = -\frac{R}{L}$$

The transient solution therefore has only one exponential term:

$$v_{R,t} = Ae^{-(R/L)t} \tag{3-58}$$

The complete solution is the sum of the transient and steady-state solutions:

$$v_R = \frac{E\sqrt{2}}{\sqrt{1 + (\omega L/R)^2}} \sin\left(\omega t - \tan^{-1}\frac{\omega L}{R}\right) + Ae^{-(R/L)t} \tag{3-59}$$

Evaluation of the constant A must be made by using the initial conditions. For example, suppose that the switch is closed at the time when the generator voltage is zero. Since the current cannot change instantaneously in an inductor, the current at the time the switch is closed is zero. Therefore, the voltage v_R is zero at $t = 0$. Inserting $v_R = 0$ and $t = 0$ into Eq. (3-59) and solving for A gives

$$A = \frac{-E\sqrt{2}}{\sqrt{1 + (\omega L/R)^2}} \sin\left(-\tan^{-1}\frac{\omega L}{R}\right)$$

$$= \frac{-E\sqrt{2}}{\sqrt{1 + (\omega L/R)^2}} \frac{-\omega L}{\sqrt{R^2 + (\omega L)^2}}$$

$$= \frac{\omega RLE\sqrt{2}}{R^2 + (\omega L)^2}$$

Notice that the differential equation is of the first order, there is one constant to be evaluated, and one initial condition is used to evaluate the constant.

3-9   *Example: Second-order System— Mechanical*

The simple mechanical system of Sec. 2-3 is used as an example and is shown in Fig. 3-8. Equation (2-27) relates the displacement x_2 to x_1:

$$M\,D^2x_2 + B\,Dx_2 + Kx_2 = Kx_1 \tag{3-60}$$

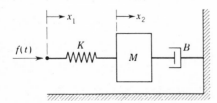

Fig. 3-8 *Simple mechanical system.*

The system is considered to be originally at rest.   The function x_1 moves one unit at time $t = 0$; that is, the input is a unit step function.   The problem is to find the motion $x_2(t)$.

The displacement of the mass is given by

$$x_2(t) = x_{2,\text{ss}} + x_{2,t}$$

where $x_{2,\text{ss}}$ is the steady-state solution and $x_{2,t}$ is the transient solution.

The steady-state solution is found first by using the method of Sec. 3-5. In this example it may be easier to consider the following:

1. x_2 must reach a fixed steady-state position.
2. When x_2 reaches a constant value, the velocity and acceleration become zero.

By putting $D^2 x_2 = D x_2 = 0$ into Eq. (3-60), the final or steady-state value of x_2 is $x_2(\infty) = x_1$.   When the steady-state solution has been found, the transient solution is determined.   The characteristic equation is

$$Mm^2 + Bm + K = K\left(\frac{M}{K}m^2 + \frac{B}{K}m + 1\right) = 0$$

Putting this in terms of ζ and ω_n gives

$$\frac{1}{\omega_n^2}m^2 + \frac{2\zeta}{\omega_n}m + 1 = 0 \tag{3-61}$$

for which the roots are $m = -\zeta\omega_n \pm \omega_n\sqrt{\zeta^2 - 1}$.   The transient solution depends on whether the damping ratio ζ is (1) greater than unity, (2) equal to unity, or (3) smaller than unity.   For ζ greater than unity the roots are real and the transient response is

$$x_{2,t} = A_1 e^{(-\zeta + \sqrt{\zeta^2-1})\omega_n t} + A_2 e^{(-\zeta - \sqrt{\zeta^2-1})\omega_n t} \tag{3-62}$$

For ζ equal to unity, the roots are real and equal; that is, $m_1 = m_2 = -\zeta\omega_n$. Since there are multiple roots, the transient response is

$$x_{2,t} = A_1 e^{-\zeta\omega_n t} + A_2 t e^{-\zeta\omega_n t} \tag{3-63}$$

For ζ less than unity, the roots are complex,

$$m_{1,2} = -\zeta\omega_n \pm j\omega_n\sqrt{1 - \zeta^2}$$

and the transient solution, as outlined in Sec. 3-6, is

$$x_{2,t} = A e^{-\zeta\omega_n t}\sin\left(\omega_n\sqrt{1 - \zeta^2}\,t + \phi\right) \tag{3-64}$$

The complete solution is the sum of the steady-state and transient solutions. For the underdamped case, $\zeta < 1$, the complete solution to Eq. (3-60) is

$$x_2(t) = 1 + A e^{-\zeta\omega_n t}\sin\left(\omega_n\sqrt{1 - \zeta^2}\,t + \phi\right) \tag{3-65}$$

The two constants A and ϕ must next be determined from the initial conditions. In this example the system was initially at rest; therefore $x_2(0) = 0$. Since *the velocity of a system with mass cannot change instantaneously*, $Dx_2(0) = 0$. Two equations are necessary, one for $x_2(t)$ and one for $Dx_2(t)$. Differentiating Eq. (3-65) yields

$$Dx_2(t) = -\zeta\omega_n A e^{-\zeta\omega_n t}\sin(\omega_n\sqrt{1-\zeta^2}\,t + \phi)$$
$$+ \omega_n\sqrt{1-\zeta^2}\,A e^{-\zeta\omega_n t}\cos(\omega_n\sqrt{1-\zeta^2}\,t + \phi) \quad (3\text{-}66)$$

Inserting in Eqs. (3-65) and (3-66) the initial conditions

$$x_2(0) = 0 \qquad Dx_2(0) = 0 \qquad t = 0$$

yields
$$0 = 1 + A\sin\phi$$
$$0 = -\zeta\omega_n A\sin\phi + \omega_n\sqrt{1-\zeta^2}\,A\cos\phi$$

These equations are then solved for A and ϕ:

$$A = \frac{-1}{\sqrt{1-\zeta^2}} \tag{3-67}$$

$$\phi = \tan^{-1}\frac{\sqrt{1-\zeta^2}}{\zeta} = \cos^{-1}\zeta \tag{3-68}$$

Thus the complete solution is

$$x_2(t) = 1 - \frac{e^{-\zeta\omega_n t}}{\sqrt{1-\zeta^2}}\sin(\omega_n\sqrt{1-\zeta^2}\,t + \cos^{-1}\zeta) \tag{3-69}$$

When the complete solution has been obtained, it should be checked to see that it satisfies the known conditions. For example, putting $t = 0$ into Eq. (3-69) gives $x_2(0) = 0$; therefore the solution checks. In a like manner, the constants can be evaluated for the other two cases of damping.

3-10   Example: Second-order System— Electrical

The electric circuit of Fig. 3-9 is used to further illustrate the determination of initial conditions. The circuit, as shown, is in the steady state. At time $t = 0$ the switch is closed. The problem is to solve for the current flowing through the inductance for $t \geq 0$.

Two loop equations are written for this circuit:

$$10 = 20i_1 - 10i_2 \tag{3-70}$$

$$0 = -10i_1 + \left(D + 25 + \frac{100}{D}\right)i_2 \tag{3-71}$$

Eliminating i_1 from these equations yields

$$10 = \left(2D + 40 + \frac{200}{D}\right) i_2 \qquad (3\text{-}72)$$

After differentiating Eq. (3-72), the steady-state solution is found by using the method of Sec. 3-5. Since the input is a step function, the steady-state output is

$$i_{2,ss} = 0 \qquad (3\text{-}73)$$

This can also be deduced from an inspection of the circuit. When a branch contains a capacitor, the steady-state current is always zero for a d-c source.
Next the transient solution is determined. The characteristic equation is

$$m + 20 + \frac{100}{m} = 0 \qquad (3\text{-}74)$$

for which the roots are $m_{1,2} = -10$. Thus the circuit is critically damped and the current through the coil can be expressed by

$$i_2(t) = i_{2,t}(t) = A_1 e^{-10t} + A_2 t e^{-10t} \qquad (3\text{-}75)$$

Two equations and two initial conditions are necessary to evaluate the two constants A_1 and A_2. Equation (3-75) and its derivative, given below, are utilized in conjunction with the initial conditions $i_2(0+)$ and $D i_2(0+)$.

$$D i_2(t) = -10 A_1 e^{-10t} + A_2(1 - 10t)e^{-10t} \qquad (3\text{-}76)$$

In this example the currents just before the switch is closed are

$$i_1(0-) = i_2(0-) = 0$$

The energy stored in the magnetic field of a single inductor is given by

$$W = \tfrac{1}{2} L i^2 \qquad (3\text{-}77)$$

Since this energy cannot change instantly, the current through an inductor also cannot change instantly. Therefore

$$i_2(0+) = i_2(0-) = 0$$

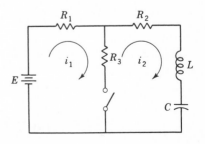

Fig. 3-9 *Electric circuit. E = 10 volts, R_1 = 10 ohms, R_2 = 15 ohms, R_3 = 10 ohms, L = 1 h, C = 0.01 farad.*

$Di_2(t)$ is found from the original circuit equation, Eq. (3-71):

$$Di_2(t) = 10i_1(t) - 25i_2(t) - \frac{100i_2(t)}{D}$$
$$= 10i_1(t) - 25i_2(t) - v_c(t) \tag{3-78}$$

To determine $Di_2(0+)$, it is necessary to first determine $i_1(0+)$, $i_2(0+)$, and $v_c(0+)$. Since $i_2(0+) = 0$, Eq. (3-70) yields

$$i_1(0+) = 0.5 \text{ amp} \tag{3-79}$$

Since the energy $W = \tfrac{1}{2}Cv^2$ stored in a capacitor cannot change instantly, the voltage across the capacitor also cannot change instantly. The steady-state value of capacitor voltage for $t < 0$ is 10 volts, thus

$$v_c(0-) = v_c(0+) = 10 \text{ volts} \tag{3-80}$$

Inserting these values into Eq. (3-78) yields

$$Di_2(0+) = 10i_1(0+) - 25i_2(0+) - v_c(0+)$$
$$= 5 - 0 - 10 = -5 \tag{3-81}$$

Substituting the initial conditions into Eqs. (3-75) and (3-76) results in

$$0 = A_1 \tag{3-82}$$
$$-5 = A_2 \tag{3-83}$$

Therefore the current flowing through the inductor for $t \geq 0$ is given by

$$i_2(t) = -5te^{-10t} \tag{3-84}$$

3-11   *Second-order Transients*[5]

The response to a unit-step-function input is, as previously stated, usually used as a means of evaluating the response of a system. This is done in this section for a second-order system. With the example of the preceding section used as an illustrative system, the differential equation is

$$\frac{D^2c}{\omega_n^2} + \frac{2\zeta}{\omega_n} Dc + c = r$$

This is defined as a *simple* second-order equation because there are no derivatives of r on the right side of the equation. The response to a unit step input, subject to zero initial conditions, is derived in Sec. 3-9 and is given by

$$c(t) = 1 - \frac{e^{-\zeta\omega_n t}}{\sqrt{1 - \zeta^2}} \sin(\omega_n \sqrt{1 - \zeta^2}\, t + \cos^{-1} \zeta) \tag{3-85}$$

A family of curves representing this equation is shown in Fig. 3-10, where the abscissa is the dimensionless variable $\omega_n t$. The curves are thus a function only of the damping ratio ζ.

These curves show that the amount of overshoot depends on the damping ratio ζ. For the overdamped and critically damped case, $\zeta \geq 1$, there is no overshoot. For the underdamped case, $\zeta < 1$, the system oscillates around the final value. The oscillations decrease with time, and the system response approaches the final value. The peak overshoot for the underdamped system is the first overshoot. The time at which the peak overshoot occurs, t_p, can be found by differentiating $c(t)$ from Eq. (3-85) with respect to time and setting this derivative equal to zero:

$$\frac{dc}{dt} = \zeta \omega_n e^{-\zeta \omega_n t} \sin\left(\omega_n \sqrt{1 - \zeta^2}\, t + \cos^{-1} \zeta\right)$$
$$- \omega_n \sqrt{1 - \zeta^2}\, e^{-\zeta \omega_n t} \cos\left(\omega_n \sqrt{1 - \zeta^2}\, t + \cos^{-1} \zeta\right) = 0$$

This derivative is zero at $\omega_n \sqrt{1 - \zeta^2}\, t = 0, \pi, 2\pi, \ldots$. The peak overshoot occurs at the first value after zero, provided there are zero initial conditions; therefore

$$t_p = \frac{\pi}{\omega_n \sqrt{1 - \zeta^2}} \qquad (3\text{-}86)$$

Inserting this value of time in Eq. (3-85) gives the peak overshoot as

$$M_p = c_p = 1 + e^{-\zeta \pi / \sqrt{1 - \zeta^2}} \qquad (3\text{-}87)$$

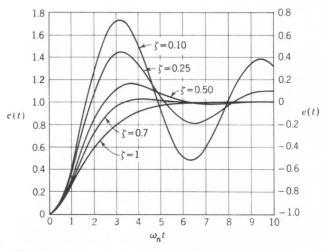

Fig. 3-10 *Simple second-order transients.*

The per unit overshoot M_o as a function of damping ratio is shown in Fig. 3-11, where

$$M_o = \frac{c_p - c_{ss}}{c_{ss}} \qquad (3\text{-}88)$$

The variation of the frequency of oscillation of the transient with variation of damping ratio is also of interest. In order to represent this variation by one curve, the quantity ω_d/ω_n is plotted against ζ in Fig. 3-12. If the scales of ordinate and abcissa are equal, the curve is an arc of a circle. Note that this curve has been plotted for $\zeta \leq 1$. Values of damped natural frequency for $\zeta > 1$ are mathematical only, not physical.

The error in the system is the difference between the input and output; thus the error equation is

$$e = r - c = \frac{e^{-\zeta \omega_n t}}{\sqrt{1 - \zeta^2}} \sin \left(\omega_n \sqrt{1 - \zeta^2}\, t + \cos^{-1} \zeta \right) \qquad (3\text{-}89)$$

The variation of error with time is sometimes plotted. These curves can be obtained from Fig. 3-10 by realizing that the curves start at $e(0) = +1$ and have the final value $e(\infty) = 0$.

Response Characteristics

The transient-response curves for the second-order system show a number of significant characteristics.

The overdamped system is slow-acting and does not oscillate about the final position. For some applications the absence of oscillations may be

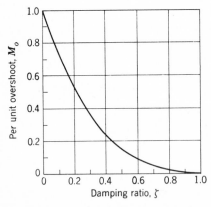

Fig. 3-11 *Peak overshoot versus damping ratio for a simple second-order equation* $D^2 c/\omega_n^2 + (2\zeta/\omega_n)Dc + 1 = r(t)$.

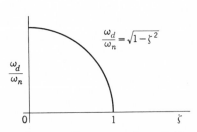

Fig. 3-12 *Frequency of oscillation versus damping ratio.*

Table 3-1. Exponential
values

t	$e^{-\zeta\omega_n t}$	Error, %
$1T$	0.368	36.8
$2T$	0.135	13.5
$3T$	0.050	5.0
$4T$	0.018	1.8
$5T$	0.007	0.7

necessary.   For example, an elevator cannot be allowed to oscillate at each stop.   But for systems where a fast response is necessary, the slow response of an overdamped system cannot be tolerated.

The underdamped system reaches the final value faster than the over-damped system, but the response oscillates about this final value.   If this oscillation can be tolerated, the underdamped system is faster-acting.   The amount of permissible overshoot determines the desirable value of damping ratio.   A damping ratio $\zeta = 0.4$ has an overshoot of 25.4 per cent, and a damping ratio $\zeta = 0.8$ has an overshoot of 1.6 per cent.

The settling time is the time required for the oscillations to decrease to a specified absolute percentage of the final value and thereafter remain less than this value.   Errors of 2 or 5 per cent are common values used to determine settling time.   For second-order systems the size of the transient at any time is equal to or less than the exponential $e^{-\zeta\omega_n t}$.   The value of this term is given in Table 3-1 for several values of t expressed in a number of time constants T.

The settling time for a 2 per cent error criterion is approximately four time constants; for a 5 per cent error criterion, it is three time constants. The per cent error criterion used must be determined from the response desired for the system.   The time for the envelope of the transient to die out is

$$T_s = \frac{\text{number of time constants}}{\zeta\omega_n} \qquad (3\text{-}90)$$

Since ζ must be determined and adjusted for the permissible overshoot, the undamped natural frequency determines the settling time.

3-12   Time-response Specifications[6]

The desired performance characteristics of a system of any order may be specified in terms of the transient response to a unit-step-function input.

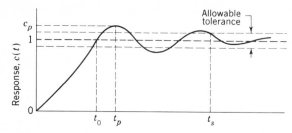

Fig. 3-13   *Typical underdamped response to a step function.*

The performance of a system may be evaluated in terms of the following quantities, as shown in Fig. 3-13.

1. Maximum overshoot c_p is the magnitude of the first overshoot. This may also be expressed in per cent of the final value.
2. Time to maximum overshoot, t_p, is the time required to reach the maximum overshoot.
3. Time to first zero error, t_0, is the time required to reach the final value the first time.   It is often referred to as duplicating time.
4. Settling time t_s is the time required for the output response first to reach and thereafter remain within a prescribed percentage of the final value.   This percentage must be specified in the individual case.   Common values used for settling time are 2 and 5 per cent. As commonly used, the 2 or 5 per cent is applied to the envelope which yields T_s.   The actual t_s may be smaller than T_s.
5. Frequency of oscillation of the transient, ω_d.

The time response differs for each set of initial conditions.   Therefore, to compare the time response of various systems it is necessary to start with standard initial conditions.   The most practical standard is to start with the system at rest.   Then the response characteristics, such as maximum overshoot and settling time, can be compared significantly.

For some systems these specifications are also applied for a ramp input. In such cases the plot of error with time is used with the definitions.   For systems subject to shock inputs the response due to an impulse is used as a criterion of performance.

3-13   Conclusions

This chapter has established the manner of solving differential equations. The steady-state and transient solutions are determined separately, and then the constants of the transient solution are evaluated to satisfy the initial conditions.   The ability to anticipate the form of the response is very important.   When a solution is obtained, it should be reasonable.   Checking the solution at known times is a method of checking its reasonableness.

The next chapter is concerned with the solution of differential equations by use of the Laplace transform.   That method has both advantages and disadvantages as compared with the classical method.

Bibliography

1. Wylie, C. R., Jr.: "Advanced Engineering Mathematics," 2d ed., McGraw-Hill Book Company, New York, 1960.
2. Doherty, R. E., and E. G. Keller: "Mathematics of Modern Engineering," vol. I, John Wiley & Sons, Inc., New York, 1936.
3. Weber, E.: "Linear Transient Analysis," vol. I, John Wiley & Sons, Inc., New York, 1954.
4. Electrical Engineering Staff, Massachusetts Institute of Technology: "Electric Circuits," The Technology Press of the Massachusetts Institute of Technology, Cambridge, Mass., and John Wiley & Sons, Inc., New York, 1940.
5. Trimmer, J. D.: "Response of Physical Systems," p. 17, John Wiley & Sons, Inc., New York, 1950.
6. Truxal, J. G.: "Automatic Feedback Control System Synthesis," p. 76, McGraw-Hill Book Company, New York, 1955.

4

Laplace
Transforms

4-1   Introduction

Solution of differential equations with discontinuous inputs or of higher than second order is laborious by the classical method. Also, inserting the initial conditions to evaluate the constants of integration requires the simultaneous solution of a number of algebraic equations equal to the order of the differential equation. To facilitate and systematize the solution of ordinary constant-coefficient differential equations, the Laplace-transform method is used extensively.[1-4]   This operational method has found wide acceptance and is used in the technical literature and books on feedback systems.   The advantages of an operational method are as follows:

1. It automatically includes the boundary or initial conditions.
2. The work involved in the solution is simple algebra.
3. The work is systematized.
4. The use of a table of transforms reduces the labor required.
5. Discontinuous inputs are easily treated.
6. The transient and steady-state components of the solution are obtained simultaneously.

The disadvantage of operational methods is that, if used mechanically, without knowledge of the actual theory involved, they sometimes yield erroneous results. Also, a particular equation may sometimes be solved more simply and with less work by the classical method.

4-2 Definition of the Laplace Transform

The direct Laplace transformation of a function of time, $f(t)$, is given by the equation

$$\mathcal{L}[f(t)] = \int_0^\infty f(t)e^{-st}\,dt = F(s) \qquad (4\text{-}1)$$

where the symbol $\mathcal{L}[f(t)]$ is a shorthand notation for the Laplace integral. Evaluation of the integral results in a function $F(s)$ which has s as the variable. This variable s is a complex quantity of the form $\sigma + j\omega$. It is to be noted that, as the limits of integration are zero and infinity, it is immaterial what value $f(t)$ has for negative or zero time.

There are limitations on the functions $f(t)$ that are Laplace-transformable. Basically the requirement is that the Laplace integral converge, which means that this integral has a definite functional value. To meet this requirement[5] the function $f(t)$ must be (1) piecewise continuous over every finite interval $0 \le t_1 \le t \le t_2$ and (2) of exponential order. A function is piecewise continuous in a finite interval if that interval can be divided into a finite number of subintervals over each of which the function is continuous and at the ends of each of which $f(t)$ possesses finite right- and left-hand limits. A function $f(t)$ is of exponential order if there exists a constant a such that the product $e^{-at}|f(t)|$ is bounded for all values of t greater than some finite value T. This imposes the restriction that σ, the real part of s, must be greater than a lower bound σ_a for which the product $e^{-\sigma_a t}|f(t)|$ is of exponential order. A linear differential equation with constant coefficients and with a finite number of terms is Laplace-transformable if the driving function is Laplace-transformable.

All cases covered in this book are Laplace-transformable. The basic purpose of using the Laplace transform is to obtain a method of solution which involves only simple algebraic operations in the complex plane (s plane).

4-3   *Derivation of Laplace Transforms of Simple Functions*

A number of examples are worked out showing the derivation of the Laplace transforms.   A list of transform pairs frequently encountered is contained in Appendix A.

Step Function $u(t)$

The Laplace transform of the unit step function $u(t)$ is

$$\mathcal{L}[u(t)] = \int_0^\infty u(t)e^{-st}\,dt = U(s) \tag{4-2}$$

Since $u(t)$ has the value 1 over the limits of integration,

$$U(s) = \int_0^\infty e^{-st}\,dt = -\frac{e^{-st}}{s}\bigg|_0^\infty = \frac{1}{s} \qquad \text{if } \sigma > 0 \tag{4-3}$$

The step function may be undefined at $t = 0$; but this is immaterial, for it must be remembered that the integral is defined by a limit process,

$$\int_0^\infty f(t)e^{-st}\,dt = \lim_{\substack{T \to \infty \\ \epsilon \to 0}} \int_\epsilon^T f(t)e^{-st}\,dt \tag{4-4}$$

The value of the integral as obtained by taking limits is implied in each case but is not written out explicitly.

Decaying Exponential $e^{-\alpha t}$

The exponent α is a positive real number:

$$\begin{aligned}
\mathcal{L}[e^{-\alpha t}] &= \int_0^\infty e^{-\alpha t}e^{-st}\,dt = \int_0^\infty e^{-(s+\alpha)t}\,dt \\
&= -\frac{e^{-(s+\alpha)t}}{s+\alpha}\bigg|_0^\infty = \frac{1}{s+\alpha} \qquad \sigma > -\alpha
\end{aligned} \tag{4-5}$$

Sinusoid $\cos \omega t$

Here ω is a positive real number:

$$\mathcal{L}[\cos \omega t] = \int_0^\infty \cos \omega t\, e^{-st}\,dt \tag{4-6}$$

Expressing cos ωt in exponential form,

$$\cos \omega t = \frac{e^{j\omega t} + e^{-j\omega t}}{2}$$

Then $\quad \mathcal{L}[\cos \omega t] = \frac{1}{2}\left(\int_0^\infty e^{(j\omega - s)t}\, dt + \int_0^\infty e^{(-j\omega - s)t}\, dt\right)$

$$= \frac{1}{2}\left[\frac{e^{(j\omega - s)t}}{j\omega - s} + \frac{e^{(-j\omega - s)t}}{-j\omega - s}\right]_0^\infty$$

$$= \frac{1}{2}\left(-\frac{1}{j\omega - s} - \frac{1}{-j\omega - s}\right) = \frac{s}{s^2 + \omega^2} \qquad \sigma > 0 \qquad (4\text{-}7)$$

Ramp Function $f(t) = t$

$$\mathcal{L}[t] = \int_0^\infty t e^{-st}\, dt \qquad \sigma > 0 \qquad (4\text{-}8)$$

This is integrated by parts by using

$$\int u\, dv = uv - \int v\, du$$

Let $u = t$ and $dv = e^{-st}\, dt$. Then $du = dt$ and $v = -e^{-st}/s$. Thus

$$\int_0^\infty t e^{-st}\, dt = -\left.\frac{t e^{-st}}{s}\right|_0^\infty - \int_0^\infty \left(-\frac{e^{-st}}{s}\right) dt$$

$$= 0 - \left.\frac{e^{-st}}{s^2}\right|_0^\infty = \frac{1}{s^2} \qquad \sigma > 0 \qquad (4\text{-}9)$$

4-4   Laplace-transform Theorems

A number of theorems that are useful in applying the Laplace transform are presented in this section. In general, they are helpful in evaluating transforms.

Theorem 1.   Linearity

If a is a constant or is independent of s and t and if $f(t)$ is transformable, then

$$\mathcal{L}[af(t)] = a\mathcal{L}[f(t)] = aF(s) \qquad (4\text{-}10)$$

Theorem 2.   Superposition

If $f_1(t)$ and $f_2(t)$ are both Laplace-transformable, the principle of superposition holds:

$$\mathcal{L}[f_1(t) \pm f_2(t)] = \mathcal{L}[f_1(t)] \pm \mathcal{L}[f_2(t)] = F_1(s) \pm F_2(s) \qquad (4\text{-}11)$$

Theorem 3.   Translation in time

If the Laplace transform of $f(t)$ is $F(s)$, a is a positive real number, and $f(t - a) = 0$ for $0 < t < a$, then

$$\mathcal{L}[f(t - a)] = e^{-as}F(s) \tag{4-12}$$

Translation in the positive t direction in the real domain becomes multiplication by the exponential e^{-as} in the s domain.

Theorem 4.   Complex differentiation

If the Laplace transform of $f(t)$ is $F(s)$, then

$$\mathcal{L}[tf(t)] = -\frac{d}{ds}F(s) \tag{4-13}$$

Multiplication by time in the real domain entails differentiation with respect to s in the s domain.

Example:

$$\mathcal{L}[te^{-\alpha t}] = -\frac{d}{ds}\mathcal{L}[e^{-\alpha t}]$$

$$= -\frac{d}{ds}\left(\frac{1}{s + \alpha}\right) = \frac{1}{(s + \alpha)^2}$$

Theorem 5.   Translation in the s domain

If the Laplace transform of $f(t)$ is $F(s)$ and a is either real or complex, then

$$\mathcal{L}[e^{at}f(t)] = F(s - a) \tag{4-14}$$

Multiplication by e^{at} in the real domain is equivalent to translation in the s domain.

Example:
Starting with $\mathcal{L}[\sin \omega t] = \omega/(s^2 + \omega^2)$ and applying Theorem 5 gives

$$\mathcal{L}[e^{-\alpha t} \sin \omega t] = \frac{\omega}{(s + \alpha)^2 + \omega^2}$$

Theorem 6.   Real differentiation

If the Laplace transform of $f(t)$ is $F(s)$ and if the first derivative of $f(t)$ with respect to time, $Df(t)$, is transformable, then

$$\mathcal{L}[Df(t)] = sF(s) - f(0+) \tag{4-15}$$

The term $f(0+)$ is the value of the limit of the function $f(t)$ as the origin,

$t = 0$, is approached from the right side (thus through positive values of time). This enfolds functions, such as the step function, that may be undefined at $t = 0$. For simplicity, the plus sign following the zero is omitted in the future although its presence is implied.

The transform of the second derivative $D^2 f(t)$ is

$$\mathcal{L}[D^2 f(t)] = s^2 F(s) - sf(0) - Df(0) \tag{4-16}$$

where $Df(0)$ is the value of the limit of the derivative of $f(t)$ as the origin, $t = 0$, is approached from the right side.

The transform of the nth derivative $D^n f(t)$ is

$$\mathcal{L}[D^n f(t)] = s^n F(s) - s^{n-1} f(0) - \cdots - D^{n-1} f(0) \tag{4-17}$$

Note that the transform includes the initial conditions, whereas in the classical method of solution the initial conditions are introduced separately to evaluate the coefficients of the solution of the differential equation.

Theorem 7.   Real integration

If the Laplace transform of $f(t)$ is $F(s)$, its integral

$$D^{-1} f(t) = \int_0^t f(t) \, dt + D^{-1} f(0+)$$

is transformable and the value of its transform is

$$\mathcal{L}[D^{-1} f(t)] = \frac{F(s)}{s} + \frac{D^{-1} f(0+)}{s} \tag{4-18}$$

The term $D^{-1} f(0+)$ is the constant of integration and is equal to the value of the integral as the origin is approached from the positive, or right, side. The plus sign will be omitted in the remainder of this text.

The transform of the double integral $D^{-2} f(t)$ is

$$\mathcal{L}[D^{-2} f(t)] = \frac{F(s)}{s^2} + \frac{D^{-1} f(0)}{s^2} + \frac{D^{-2} f(0)}{s} \tag{4-19}$$

The transform of the higher-order integral $D^{-n} f(t)$ is

$$\mathcal{L}[D^{-n} f(t)] = \frac{F(s)}{s^n} + \frac{D^{-1} f(0)}{s^n} + \cdots + \frac{D^{-n} f(0)}{s} \tag{4-20}$$

Theorem 8.   Final value

If $f(t)$ and $Df(t)$ are Laplace-transformable, if the Laplace transform of $f(t)$ is $F(s)$, and if the limit $f(t)$ as $t \to \infty$ exists, then

$$\lim_{s \to 0} sF(s) = \lim_{t \to \infty} f(t) \tag{4-21}$$

This theorem states that the behavior of $f(t)$ in the neighborhood of $t = \infty$ is related to the behavior of $sF(s)$ in the neighborhood of $s = 0$. If $sF(s)$ has poles [values of s for which $sF(s)$ becomes infinite] on the imaginary axis (excluding the origin) or in the right-half s plane, there is no finite final value of $f(t)$ and the theorem cannot be used. If the driving function is sinusoidal, the theorem is invalid, since $\mathcal{L}[\sin \omega t]$ has poles at $s = \pm j\omega$; also $\lim\limits_{t \to \infty} \sin \omega t$ does not exist. However, for poles of $sF(s)$ at the origin, $s = 0$, this theorem gives the final value of $f(\infty) = \infty$. This correctly describes the behavior of $f(t)$ as $t \to \infty$.

Theorem 9.   Initial value

If the function $f(t)$ and its first derivative are Laplace-transformable, if the Laplace transform of $f(t)$ is $F(s)$, and if $\lim\limits_{s \to \infty} sF(s)$ exists, then

$$\lim_{s \to \infty} sF(s) = \lim_{t \to 0} f(t) \tag{4-22}$$

This theorem states that the behavior of $f(t)$ in the neighborhood of $t = 0$ is related to the behavior of $sF(s)$ in the neighborhood of $s = \infty$. There are no limitations on the location of the poles of $sF(s)$.

Theorem 10.   Complex integration

If the Laplace transform of $f(t)$ is $F(s)$ and if $f(t)/t$ has a limit as $t \to 0+$, then

$$\mathcal{L}\left[\frac{f(t)}{t}\right] = \int_s^\infty F(s)\,ds \tag{4-23}$$

This theorem states that division by the variable in the real domain entails integration with respect to s in the s domain.

4-5   Application of the Laplace Transform to Differential Equations

The Laplace transform is now applied to the solution of the differential equation for the simple mechanical system of Sec. 2-3, which was solved by the classical method in Sec. 3-9. The differential equation of the system was given by Eq. (2-27) and is repeated below:

$$M\,D^2 x_2 + B\,D x_2 + K x_2 = K x_1 \tag{4-24}$$

The position $x_1(t)$ undergoes a unit step displacement. This is the input and is called the *driving* function. The unknown quantity for which the equation is to be solved is the output $x_2(t)$ and is called the *response* function. The Laplace transform of Eq. (4-24) is

$$\mathcal{L}[Kx_1] = \mathcal{L}[M\,D^2x_2 + B\,Dx_2 + Kx_2] \tag{4-25}$$

The transform of each term is listed below:

$$\mathcal{L}[Kx_1] = KX_1(s)$$
$$\mathcal{L}[Kx_2] = KX_2(s)$$
$$\mathcal{L}[B\,Dx_2] = B[sX_2(s) - x_2(0)]$$
$$\mathcal{L}[M\,D^2x_2] = M[s^2X_2(s) - sx_2(0) - Dx_2(0)]$$

Inserting these terms into Eq. (4-25) and collecting terms gives

$$KX_1(s) = (Ms^2 + Bs + K)X_2(s) - [Msx_2(0) + MDx_2(0) + Bx_2(0)] \tag{4-26}$$

Equation (4-26) is the transform equation and shows the way in which the initial conditions—the initial position $x_2(0)$ and the initial velocity $Dx_2(0)$—are incorporated into the equation. The function $X_1(s)$ is called the driving transform; the function $X_2(s)$ is called the response transform; and the coefficient of $X_2(s)$, which is $Ms^2 + Bs + K$, is called the *characteristic* function. The equation formed by setting the characteristic function equal to zero is called the characteristic equation of the system. Solving for $X_2(s)$,

$$X_2(s) = \frac{K}{Ms^2 + Bs + K}\,X_1(s) + \frac{Msx_2(0) + Bx_2(0) + MDx_2(0)}{Ms^2 + Bs + K} \tag{4-27}$$

The coefficient of $X_1(s)$ is defined in Chap. 5 as the *system transfer function*. The second term on the right side of the equation is called the *initial-condition operator*. Combining the terms of Eq. (4-27),

$$X_2(s) = \frac{KX_1(s) + Msx_2(0) + Bx_2(0) + MDx_2(0)}{Ms^2 + Bs + K} \tag{4-28}$$

The problem is to find the function $x_2(t)$ whose transform is given by Eq. (4-28). The inverse transform is indicated by $\mathcal{L}^{-1}$; thus

$$x_2(t) = \mathcal{L}^{-1}[X_2(s)] = \mathcal{L}^{-1}\left[\frac{KX_1(s) + Msx_2(0) + Bx_2(0) + MDx_2(0)}{Ms^2 + Bs + K}\right] \tag{4-29}$$

The function $x_2(t)$ might be found in the table of Appendix A after inserting numerical values into Eq. (4-29). For this example the initial conditions are $x_2(0) = Dx_2(0) = 0$. Assume, as was done in Sec. 3-9, that the damping ratio ζ is less than unity. Since $x_1(t)$ is a step function, $X_1(s) = 1/s$ and

$$x_2(t) = \mathcal{L}^{-1}\left[\frac{1}{s\left(\dfrac{M}{K}s^2 + \dfrac{B}{K}s + 1\right)}\right] = \mathcal{L}^{-1}\left[\frac{1}{s\left(\dfrac{1}{\omega_n^2}s^2 + \dfrac{2\zeta}{\omega_n}s + 1\right)}\right] \tag{4-30}$$

Reference to transform pair 27a in Appendix A provides the solution directly as

$$x_2(t) = 1 - \frac{e^{-\zeta\omega_n t}}{\sqrt{1 - \zeta^2}} \sin\left(\omega_n \sqrt{1 - \zeta^2}\, t + \cos^{-1}\zeta\right)$$

For this example the inverse transform is available directly in the table of Appendix A. When this is not the case, a more general method of obtaining the inverse transform must be used, as described in the following section.

4-6   *Inverse Transformation*

The application of the Laplace transforms to a differential equation yields an algebraic equation. From the algebraic equation the transform of the response function is readily found. To complete the solution the inverse transform must be found. In some cases the inverse-transform operation

$$f(t) = \mathcal{L}^{-1}[F(s)] \tag{4-31}$$

can be performed by direct reference to transform tables. The more complete the table, the better will be the possibility of finding an $F(s)$ that will give its corresponding $f(t)$ directly. The linearity and translation theorems are useful in extending the tables.

When the response transform cannot be found in the tables, the general procedure is to express $F(s)$ as the sum of partial fractions with constant coefficients. The partial fractions have a first-order or quadratic factor in the denominator and are readily found in the table of transforms. The complete inverse transform is the sum of the inverse transforms of each fraction.

The response transform $F(s)$ can be expressed in general, as the ratio of two polynomials $P(s)$ and $Q(s)$. Consider that these polynomials are of degree w and v, respectively, and are arranged in descending order of the powers of the variable s; thus,

$$F(s) = \frac{P(s)}{Q(s)} = \frac{a_w s^w + a_{w-1}s^{w-1} + \cdots + a_1 s + a_0}{s^v + b_{v-1}s^{v-1} + \cdots + b_1 s + b_0} \tag{4-32}$$

The a's and b's are real constants, and the coefficient of the highest power of s in the denominator has been made equal to unity. Only those $F(s)$ which are proper fractions—that is, those in which v is equal to or greater than w—are considered.*

The first step is to factor $Q(s)$ into linear and quadratic factors with real coefficients:

$$F(s) = \frac{P(s)}{Q(s)} = \frac{P(s)}{(s - s_1)(s - s_2) \cdots (s - s_k) \cdots (s - s_v)} \tag{4-33}$$

* If $v = w$, first divide $P(s)$ by $Q(s)$ to obtain

$$F(s) = 1 + \frac{P'(s)}{Q'(s)} = 1 + F'(s)$$

Then express $F'(s)$ as the sum of partial fractions with constant coefficients.

The values s_1, s_2, . . . , s_v in the finite plane that make the denominator equal to zero are called the *zeros* of the denominator. These values of s, which may be either real or complex, also make $F(s)$ equal to infinity, so they are called *poles* of $F(s)$. Therefore, the values s_1, s_2, . . . , s_v are referred to as zeros of the denominator or poles of the complete function in the finite plane; i.e., there are v poles of $F(s)$. Several methods of factoring polynomials are shown in Appendix B and in Sec. 8-5.

The transform $F(s)$ can be expressed as a series of fractions, where the number of fractions is equal to v, the number of poles of $F(s)$. If there are no repeated factors, the function $F(s)$ can be expanded as

$$F(s) = \frac{P(s)}{Q(s)} = \frac{A_1}{s - s_1} + \frac{A_2}{s - s_2} + \cdots + \frac{A_k}{s - s_k} + \cdots + \frac{A_v}{s - s_v} \quad (4\text{-}34)$$

The problem is to evaluate the constants A_1, A_2, . . . , A_v corresponding to the poles s_1, s_2, . . . , s_v. Since the partial fractions contain only non-repeated factors of $Q(s)$, the coefficients A_1, A_2, . . . are termed the *residues* of $F(s)$ at the corresponding poles. Cases of repeated factors and complex factors are treated separately. Several ways of evaluating the constants are taken up in the following section.

4-7   Heaviside Partial-fraction Expansion Theorems

The technique of partial-fraction expansion is set up to take care of all cases systematically. There are four classes of problems, depending on the denominator $Q(s)$.

Case 1. $F(s)$ has first-order real poles.
Case 2. $F(s)$ has repeated first-order real poles.
Case 3. $F(s)$ has a pair of complex conjugate poles (a quadratic factor in the denominator).
Case 4. $F(s)$ has repeated pairs of complex conjugate poles (a repeated quadratic factor in the denominator).

Each of these cases is worked out separately below.

Case 1.   First-order real poles

The position of three real poles of $F(s)$ in the s plane is shown in Fig. 4-1. The poles may be positive, zero, or negative, and they lie on the real axis in the s plane. In this example, s_1 is positive, s_0 is zero, and s_2 is negative.

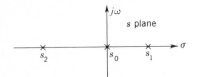

Fig. 4-1   *Location of real poles in the s plane.*

For the poles shown in Fig. 4-1 the transform $F(s)$ and its partial fractions are

$$F(s) = \frac{P(s)}{Q(s)} = \frac{P(s)}{s(s - s_1)(s - s_2)} = \frac{A_0}{s} + \frac{A_1}{s - s_1} + \frac{A_2}{s - s_2} \quad (4\text{-}35)$$

There are as many fractions as there are factors in the denominator of $F(s)$. Since $s_0 = 0$, the factor $s - s_0$ is written simply as s. The inverse transform of $F(s)$ is

$$f(t) = A_0 + A_1 e^{s_1 t} + A_2 e^{s_2 t} \quad (4\text{-}36)$$

The pole s_1 is positive; therefore the term $A_1 e^{s_1 t}$ is an increasing exponential and the system is unstable. The pole s_2 is negative, and the term $A_2 e^{s_2 t}$ is a decaying exponential with a final value of zero. Therefore, for a system to be stable, all real poles that contribute to the complementary solution must be in the left-half s plane.

To evaluate a typical coefficient A_k, multiply both sides of Eq. (4-34) by the factor $s - s_k$. The result is

$$(s - s_k)F(s) = (s - s_k)\frac{P(s)}{Q(s)} = A_1 \frac{s - s_k}{s - s_1} + A_2 \frac{s - s_k}{s - s_2} + \cdots + A_k$$

$$+ \cdots + A_v \frac{s - s_k}{s - s_v} \quad (4\text{-}37)$$

The multiplying factor $s - s_k$ on the left side of the equation and the same factor of $Q(s)$ should be divided out. By letting $s = s_k$, all terms on the right side of the equation are zero except A_k. Thus, a general rule for evaluating the constants for single-order real poles is

$$A_k = \left[(s - s_k)\frac{P(s)}{Q(s)} \right]_{s=s_k} = \left[\frac{P(s)}{Q'(s)} \right]_{s=s_k} \quad (4\text{-}38)$$

where $Q'(s)$ is $dQ(s)/ds = Q(s)/(s - s_k)$. The coefficients A_k are called the *residues* of $F(s)$ at the corresponding poles. For the case of

$$F(s) = \frac{s + 2}{s(s + 1)(s + 3)} = \frac{A_0}{s} + \frac{A_1}{s + 1} + \frac{A_2}{s + 3}$$

the constants are

$$A_0 = [sF(s)]_{s=0} = \left[\frac{s+2}{(s+1)(s+3)} \right]_{s=0} = \frac{2}{3}$$

$$A_1 = [(s+1)F(s)]_{s=-1} = \left[\frac{s+2}{s(s+3)} \right]_{s=-1} = -\frac{1}{2}$$

$$A_2 = [(s+3)F(s)]_{s=-3} = \left[\frac{s+2}{s(s+1)} \right]_{s=-3} = -\frac{1}{6}$$

The solution as a function of time is

$$f(t) = \frac{2}{3} - \frac{e^{-t}}{2} - \frac{1}{6} e^{-3t} \tag{4-39}$$

Case 2. Multiple-order real poles

The position of real poles of $F(s)$, some of which are repeated, is shown in Fig. 4-2. The symbol $\rceil_r$ is intended to indicate a pole of order r. All real poles lie on the real axis of the s plane. For the poles shown in Fig. 4-2 the transform $F(s)$ and its partial fractions are

$$F(s) = \frac{P(s)}{Q(s)} = \frac{P(s)}{(s-s_1)^3(s-s_2)}$$

$$= \frac{A_{13}}{(s-s_1)^3} + \frac{A_{12}}{(s-s_1)^2} + \frac{A_{11}}{s-s_1} + \frac{A_2}{s-s_2} \tag{4-40}$$

The order of $Q(s)$ in this case is four, and there are four fractions. Note that the multiple pole s, which is of order three, has resulted in three fractions on the right side of Eq. (4-40). To designate the constants in the partial fractions, a single subscript is used for a first-order pole. For multiple-order poles a double-subscript notation is used. The first subscript designates the pole, and the second subscript designates the order of the pole in the partial fraction. The constants associated with first-order denominators in the partial-fraction expansion are called residues; therefore only the constants A_{11} and A_2 are residues of Eq. (4-40).

Fig. 4-2 *Location of real poles in the s plane.*

The inverse transform of $F(s)$ is

$$f(t) = A_{13}\frac{t^2}{2}e^{s_1t} + A_{12}te^{s_1t} + A_{11}e^{s_1t} + A_2e^{s_2t} \tag{4-41}$$

For the general transform with repeated real roots,

$$
\begin{aligned}
F(s) &= \frac{P(s)}{Q(s)} = \frac{P(s)}{(s - s_q)^r(s - s_1) \cdots} \\
&= \frac{A_{qr}}{(s - s_q)^r} + \frac{A_{q(r-1)}}{(s - s_q)^{r-1}} + \cdots + \frac{A_{q(r-k)}}{(s - s_q)^{r-k}} + \cdots \\
&\qquad\qquad + \frac{A_{q1}}{s - s_q} + \frac{A_1}{s - s_1} + \cdots \tag{4-42}
\end{aligned}
$$

The constant A_{qr} can be evaluated simply. Multiplying both sides of Eq. (4-42) by $(s - s_q)^r$ gives

$$
\begin{aligned}
(s - s_q)^r F(s) &= \frac{(s - s_q)^r P(s)}{Q(s)} = \frac{P(s)}{(s - s_1) \cdots} = A_{qr} + A_{q(r-1)}(s - s_q) \\
&\quad + \cdots + A_{q1}(s - s_q)^{r-1} + A_1\frac{(s - s_q)^r}{s - s_1} + \cdots \tag{4-43}
\end{aligned}
$$

Notice that the factor $(s - s_q)^r$ is divided out of the left side of the equation.

For $s = s_q$, all terms on the right side of the equation are zero except for A_{qr}:

$$A_{qr} = \left[(s - s_q)^r\frac{P(s)}{Q(s)}\right]_{s=s_q} \tag{4-44}$$

Evaluation of $A_{q(r-1)}$ cannot be performed in a similar manner. Multiplying both sides of Eq. (4-42) by $(s - s_q)^{r-1}$ and letting $s = s_q$ would result in both sides being infinite, which leaves $A_{q(r-1)}$ indeterminate. If the term A_{qr} were eliminated from Eq. (4-43), $A_{q(r-1)}$ could be evaluated. This can be done by differentiating Eq. (4-43) with respect to s:

$$\frac{d}{ds}\left[(s - s_q)^r\frac{P(s)}{Q(s)}\right] = A_{q(r-1)} + 2A_{q(r-2)}(s - s_q) + \cdots \tag{4-45}$$

Letting $s = s_q$,

$$A_{q(r-1)} = \frac{d}{ds}\left[(s - s_q)^r\frac{P(s)}{Q(s)}\right]_{s=s_q} \tag{4-46}$$

Repeating the differentiation gives the coefficient $A_{q(r-2)}$ as

$$A_{q(r-2)} = \frac{1}{2}\frac{d^2}{ds}\left[(s - s_q)^r\frac{P(s)}{Q(s)}\right]_{s=s_q} \tag{4-47}$$

This process can be repeated until each constant is determined. A general formula for finding these coefficients associated with the repeated real pole of order r is

$$A_{q(r-k)} = \frac{1}{k!} \frac{d^k}{ds^k} \left[(s - s_q)^r \frac{P(s)}{Q(s)} \right]_{s=s_q} \tag{4-48}$$

For the case of

$$F(s) = \frac{1}{(s + 2)^3 (s + 3)} = \frac{A_{13}}{(s + 2)^3} + \frac{A_{12}}{(s + 2)^2} + \frac{A_{11}}{s + 2} + \frac{A_2}{s + 3} \tag{4-49}$$

the constants are

$$A_{13} = [(s + 2)^3 F(s)]_{s=-2} = 1$$

$$A_{12} = \frac{d}{ds} [(s + 2)^3 F(s)]_{s=-2} = -1$$

$$A_{11} = \frac{d^2}{2ds^2} [(s + 2)^3 F(s)]_{s=-2} = 1$$

$$A_2 = [(s + 3)F(s)]_{s=-3} = -1$$

and the solution as a function of time is

$$f(t) = \frac{t^2}{2} e^{-2t} - te^{-2t} + e^{-2t} - e^{-3t} \tag{4-50}$$

Case 3. Complex conjugate poles

The position of complex poles of $F(s)$ in the s plane is shown in Fig. 4-3. Complex poles always are present in complex conjugate pairs; their real part may be either positive or negative. For the poles shown in Fig. 4-3 the transform $F(s)$ and its partial fractions are

$$F(s) = \frac{P(s)}{Q(s)} = \frac{P(s)}{(s^2 + 2\zeta\omega_n s + \omega_n^2)(s - s_3)} = \frac{A_1}{s - s_1} + \frac{A_2}{s - s_2} + \frac{A_3}{s - s_3}$$

$$= \frac{A_1}{s + \zeta\omega_n - j\omega_n \sqrt{1 - \zeta^2}} + \frac{A_2}{s + \zeta\omega_n + j\omega_n \sqrt{1 - \zeta^2}} + \frac{A_3}{s - s_3} \tag{4-51}$$

The inverse transform of $F(s)$ is

$$f(t) = A_1 e^{(-\zeta\omega_n + j\omega_n\sqrt{1-\zeta^2})t} + A_2 e^{(-\zeta\omega_n - j\omega_n\sqrt{1-\zeta^2})t} + A_3 e^{s_3 t} \tag{4-52}$$

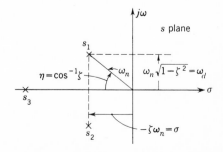

Fig. 4-3 *Location of complex conjugate poles in the s plane.*

Since the poles s_1 and s_2 are complex conjugates and since $f(t)$ is a real quantity, the coefficients A_1 and A_2 must also be complex conjugates, as stated in Sec. 3-6. Equation (4-52) can be written with the first two terms combined to a more useful damped sinusoidal form:

$$\begin{aligned} f(t) &= 2|A_1|e^{-\zeta\omega_n t}\sin\left(\omega_n\sqrt{1-\zeta^2}\,t+\phi\right)+A_3 e^{s_3 t}\\ &= 2|A_1|e^{\sigma t}\sin\left(\omega_d t+\phi\right)+A_3 e^{s_3 t} \end{aligned} \qquad (4\text{-}53)$$

where the angle

$$\phi = \text{angle of } A_1 + 90°$$

The values of A_1 and A_3, as found in the manner shown previously, are

$$A_1 = [(s-s_1)F(s)]_{s=s_1}$$
$$A_3 = [(s-s_3)F(s)]_{s=s_3}$$

Since s_1 is complex, the constant A_1 is also complex. *Remember that A_1 is associated with the complex pole with the positive imaginary part.*

In Fig. 4-3 the complex poles have a negative real part, $\sigma = -\zeta\omega_n$, where the damping ratio ζ is positive. For this case the corresponding transient response is known as a damped sinusoid and is shown in Fig. 3-5. Its final value is zero. The angle η shown in Fig. 4-3 is measured from the negative real axis and is related to the damping ratio by

$$\cos\eta = \zeta$$

If the complex pole has a positive real part, the time response increases exponentially with time and the system is unstable. If the complex roots are in the right half of the s plane, the damping ratio ζ is negative. The angle η for this case is measured from the positive real axis and is given by

$$\cos\eta = |\zeta|$$

For the case of

$$F(s) = \frac{1}{(s^2+6s+25)(s+2)} = \frac{A_1}{s+3-j4}+\frac{A_2}{s+3+j4}+\frac{A_3}{s+2} \qquad (4\text{-}54)$$

the constants are

$$A_1 = \left[(s+3-j4)\,\frac{1}{(s^2+6s+25)(s+2)}\right]_{s=-3+j4}$$

$$= \left[\frac{1}{(s+3+j4)(s+2)}\right]_{s=-3+j4} = 0.0303\underline{/-194°}$$

$$A_3 = \left[(s+2)\,\frac{1}{(s^2+6s+25)(s+2)}\right]_{s=-2}$$

$$= \left[\frac{1}{s^2+6s+25}\right]_{s=-2} = 0.059$$

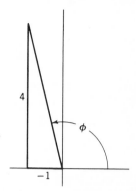

Fig. 4-4 Angle of Eq. (4-56).

The solution is

$$f(t) = 0.06e^{-3t} \sin (4t - 104°) + 0.059e^{-2t} \qquad (4\text{-}55)$$

This example is used to illustrate techniques in partial-fraction expansion. However, the particular function $F(s)$ of Eq. (4-54) does appear in Appendix A as transform pair 29. With the symbols of Appendix A, the phase angle in the damped sinusoidal term is

$$\phi = \tan^{-1} \frac{b}{c-a} = \tan^{-1} \frac{4}{2-3} = \tan^{-1} \frac{4}{-1} = 104° \qquad (4\text{-}56)$$

It is important to note that

$$\tan^{-1} \frac{4}{-1} \neq \tan^{-1} \frac{-4}{1}$$

To get the correct value for the angle ϕ, it is useful to draw a sketch, as shown in Fig. 4-4. This avoids ambiguity and ensures that ϕ is evaluated correctly.

Imaginary Poles

The position of imaginary poles of $F(s)$ in the s plane is shown in Fig. 4-5. As the real part of the poles is zero, the poles lie on the imaginary axis. This situation is a special case of complex poles, i.e., the damping ratio $\zeta = 0$. For the poles shown in Fig. 4-5 the transform $F(s)$ and its partial fractions are

$$F(s) = \frac{P(s)}{Q(s)} = \frac{P(s)}{(s^2 + \omega_n^2)(s - s_3)} = \frac{A_1}{s - s_1} + \frac{A_2}{s - s_2} + \frac{A_3}{s - s_3} \qquad (4\text{-}57)$$

The quadratic can be factored in terms of the poles s_1 and s_2; thus,

$$s^2 + \omega_n^2 = (s - j\omega_n)(s + j\omega_n) = (s - s_1)(s - s_2)$$

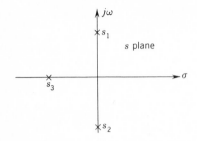

Fig. 4-5 *Poles of $F(s)$ containing imaginary conjugate poles in the s plane.*

The inverse transform of Eq. (4-57) is

$$f(t) = A_1 e^{j\omega_n t} + A_2 e^{-j\omega_n t} + A_3 e^{s_3 t} \qquad (4\text{-}58)$$

As $f(t)$ is a real quantity, the coefficients A_1 and A_2 are complex conjugates, as was shown in Sec. 3-6. The terms in Eq. (4-58) can be combined to the more useful form

$$f(t) = 2|A_1| \sin(\omega_n t + \phi) + A_3 e^{s_3 t} \qquad (4\text{-}59)$$

Note that, since there is no damping term multiplying the sinusoid, this term represents a steady-state value. The angle

$$\phi = \text{angle of } A_1 + 90°$$

The values of A_1 and A_3, found in the conventional manner, are

$$A_1 = [(s - s_1)F(s)]_{s=s_1}$$
$$A_3 = [(s - s_3)F(s)]_{s=s_3}$$

For the case where

$$F(s) = \frac{100}{(s^2 + 25)(s + 2)} = \frac{A_1}{s - j5} + \frac{A_2}{s + j5} + \frac{A_3}{s + 2} \qquad (4\text{-}60)$$

the values of the coefficients are

$$A_1 = [(s - j5)F(s)]_{s=j5} = \left[\frac{100}{(s + j5)(s + 2)}\right]_{s=j5} = 1.86\underline{/-158.2°}$$

$$A_3 = [(s + 2)F(s)]_{s=-2} = \left[\frac{100}{s^2 + 25}\right]_{s=-2} = 3.45$$

The solution is

$$f(t) = 3.72 \sin(5t - 68.2°) + 3.45 e^{-2t} \qquad (4\text{-}61)$$

Case 4.   Multiple-order complex poles

Although multiple-order complex conjugate poles do not occur very often, it is important to know how to proceed if they do. They can be treated in the same fashion as repeated real poles. The partial-fraction expansion of the transform $F(s)$ with multiple complex poles is

$$F(s) = \frac{P(s)}{(s^2 + 2\zeta\omega_n s + \omega_n^2)^r Q_1(s)}$$

$$= \frac{A_r}{(s + \zeta\omega_n - j\omega_n \sqrt{1 - \zeta^2})^r} + \frac{A_r'}{(s + \zeta\omega_n + j\omega_n \sqrt{1 - \zeta^2})^r}$$

$$+ \frac{A_{r-1}}{(s + \zeta\omega_n - j\omega_n \sqrt{1 - \zeta^2})^{r-1}} + \frac{A_{r-1}'}{(s + \zeta\omega_n + j\omega_n \sqrt{1 - \zeta^2})^{r-1}}$$

$$+ \cdots$$

$$+ \frac{A_1}{(s + \zeta\omega_n - j\omega_n \sqrt{1 - \zeta^2})} + \frac{A_1'}{(s + \zeta\omega_n + j\omega_n \sqrt{1 - \zeta^2})} + \cdots$$

$$(4\text{-}62)$$

The constants are evaluated in the same manner as for repeated real roots, as shown previously.

The inverse transform is of the form

$$f(t) = 2|A_r| \frac{t^{r-1}}{(r-1)!} e^{-\zeta\omega_n t} \sin(\omega_n \sqrt{1 - \zeta^2}\, t + \phi_r)$$

$$+ 2|A_{r-1}| \frac{t^{r-2}}{(r-2)!} e^{-\zeta\omega_n t} \sin(\omega_n \sqrt{1 - \zeta^2}\, t + \phi_{r-1}) + \cdots \quad (4\text{-}63)$$

4-8   *Example of Inverse Transformation*

This example assumes that the equations for a system have been written and the Laplace transform of the response has been found. This is given by

$$X(s) = \frac{100}{s(s+2)(s+3)^2(s^2 + 8s + 25)} \tag{4-64}$$

Since this expression does not appear in the table of Appendix A, the solution $x(t)$ is obtained by expressing the transform as the sum of simpler fractions. The function $X(s)$ is equal to the sum of the following partial fractions:

$$X(s) = \frac{A_0}{s} + \frac{A_1}{s+2} + \frac{A_{22}}{(s+3)^2} + \frac{A_{21}}{s+3} + \frac{A_3}{s+4-j3} + \frac{A_4}{s+4+j3} \tag{4-65}$$

Note that the repeated factor $(s + 3)^2$ results in a series of two fractions, the same number of fractions as the order of the factor. The inverse transform of each fraction in Eq. (4-65) can be obtained from Appendix A, so the problem now is to evaluate the constants, which are:

$$A_0 = [sX(s)]_{s=0} = 0.22$$
$$A_1 = [(s+2)X(s)]_{s=-2} = -3.85$$
$$A_{22} = [(s+3)^2X(s)]_{s=-3} = 3.33$$
$$A_{21} = \frac{d}{ds}[(s+3)^2X(s)]_{x=-3} = 3.78$$
$$A_3 = [(s+4-j3)X(s)]_{s=-4+j3} = 0.0925\underline{/-214°}$$

The inverse transform of Eq. (4-65) is

$$x(t) = 0.22 - 3.85e^{-2t} + 3.33te^{-3t} + 3.78e^{-3t}$$
$$+ 0.185e^{-4t}\sin(3t - 124°) \quad (4\text{-}66)$$

4-9   *Partial-fraction Shortcuts*

Some shortcuts can be used to simplify the partial-fraction expansion procedures previously described. These new procedures are generally most useful for transform functions that have multiple poles or complex poles. With multiple poles, the evaluation of the constants by the process of repeated differentiation given by Eq. (4-48) can be very tedious. This is particularly true when a factor containing s appears in the numerator of $F(s)$. With complex poles the residues are complex numbers, with the result that algebraic errors are easily made. The first step in the procedure is to evaluate those residues at the simple real poles of the function $F(s)$. The corresponding partial fractions are then subtracted from the original function. The resultant function contains only the multiple or complex poles, and it is easier to work with this to get the additional partial fractions than to work with the original function $F(s)$.

As a first example, consider the function

$$X(s) = \frac{12(s+1)}{s(s+2)^2(s+3)}$$
$$= \frac{A_0}{s} + \frac{A_{12}}{(s+2)^2} + \frac{A_{11}}{s+2} + \frac{A_2}{s+3} \quad (4\text{-}67)$$

The residues at $s = 0$ and $s = -3$ are found readily as

$$A_0 = [sX(s)]_{s=0} = 1$$
$$A_2 = [(s+3)X(s)]_{s=-3} = 8$$

These two fractions are subtracted from $X(s)$, and the result is put over a common denominator:

$$\frac{12(s + 1)}{s(s + 2)^2(s + 3)} - \frac{1}{s} - \frac{8}{s + 3} = -\frac{9s^3 + 39s^2 + 36s}{s(s + 2)^2(s + 3)}$$

Since the partial fractions containing the poles at $s = 0$ and $s = -3$ have been subtracted from $X(s)$, the remainder does not contain these poles. This means that the numerator must be divisible by $s(s + 3)$. After this division, the remainder is equal to the remaining terms in the partial-fraction expansion. In this example, the remainder is

$$-\frac{9s + 12}{(s + 2)^2} = \frac{A_{12}}{(s + 2)^2} + \frac{A_{11}}{s + 2}$$

The coefficient A_{12} is evaluated from

$$A_{12} = (s + 2)^2 \left[-\frac{9s + 12}{(s + 2)^2} \right]_{s=-2} = 6$$

In a similar manner, as used earlier, the corresponding fraction can be subtracted from the left side of the equation. The result, obtained by putting terms over a common denominator and simplifying, is

$$-\frac{9s + 12}{(s + 2)^2} - \frac{6}{(s + 2)^2} = -\frac{9s + 18}{(s + 2)^2} = \frac{-9}{s + 2}$$

This last remainder must be equal to the last term in the partial-fraction expansion: thus $A_{11} = -9$. The total effort in evaluating the constants is less than that required by the usual procedures.

As a second example, consider the function

$$F(s) = \frac{20}{(s^2 + 6s + 25)(s + 1)} = \frac{As + B}{s^2 + 6s + 25} + \frac{C}{s + 1} \tag{4-68}$$

Instead of partial fractions for each pole, leave the quadratic factor in one fraction. Note that the numerator of the fraction containing the quadratic contains s and is one degree lower than the denominator. The residue C is

$$C = [(s + 1)F(s)]_{s=-1} = 1$$

Subtracting this fraction from $F(s)$ gives

$$\frac{20}{(s^2 + 6s + 25)(s + 1)} - \frac{1}{s + 1} = \frac{-s^2 - 6s - 5}{(s^2 + 6s + 25)(s + 1)}$$

Since the pole at $s = -1$ has been removed, the numerator of the remainder must be divisible by $s + 1$. The simplified function must be equal to the remaining partial fraction; i.e.,

$$-\frac{s + 5}{s^2 + 6s + 25} = \frac{As + B}{s^2 + 6s + 25} = \frac{-(s + 5)}{(s + 3)^2 + (4)^2}$$

The inverse transform for this fraction can be obtained directly from Appendix A, pair 26. The complete inverse transform is given by

$$f(t) = e^{-t} - 1.12e^{-3t} \sin (4t + 63.4°) \tag{4-69}$$

Leaving the complete quadratic in one fraction has resulted in real constants for the numerator. The chances for error are probably less than evaluating the residues, which are complex numbers, at the complex poles.

A rule presented by Hazony and Riley[11] is very useful for evaluating the coefficients of partial-fraction expansions.

For a normalized ratio of polynomials:

1. *If the denominator is one degree higher than the numerator, the sum of the residues is one.*
2. *If the denominator is two or more degrees higher than the numerator, the sum of the residues is zero.*

Equation (4-32) with a_w factored from the numerator is a normalized ratio of polynomials. These rules are applied to the ratio $F(s)/a_w$. It should be noted that *residue* refers only to the coefficients of terms in a partial-fraction expansion with first-degree denominators. Coefficients of terms with higher-degree denominators are referred to only as coefficients.

These rules can be used to simplify the work involved in evaluating the coefficients of partial-fraction expansions. This is particularly true when the original function has a multiple-order pole. For example, in Eq. (4-49) it can immediately be written that $A_{11} + A_2 = 0$. Since A_2 is evaluated easily and equals -1, the value of $A_{11} = 1$ is obtained directly. The evaluation of derivatives associated with the residues for multiple-order poles is therefore eliminated. The functions in Eqs. (4-65) and (4-67) can be treated in a similar fashion.

4-10   Graphical Determination of Partial-fraction Coefficients[8,9]

The preceding sections describe the analytical evaluation of the partial-fraction coefficients. These constants are directly related to the pole-zero pattern of the function $F(s)$ and can be determined graphically, whether the poles and zeros are real or in complex conjugate pairs. In fact, as long as $P(s)$ and $Q(s)$ are in factored form, the coefficients may be determined graphically by inspection, and a table of Laplace transforms is unnecessary.

Referring to Eq. (4-32) and rewriting with both the numerator and denominator factored,

$$F(s) = \frac{P(s)}{Q(s)} = \frac{K(s - z_1)(s - z_2) \cdots (s - z_m) \cdots (s - z_w)}{(s - p_1)(s - p_2) \cdots (s - p_k) \cdots (s - p_v)}$$

$$= K \frac{\displaystyle\prod_{m=1}^{w} (s - z_m)}{\displaystyle\prod_{k=1}^{v} (s - p_k)} \tag{4-70}$$

The zeros of this function, $s = z_m$, are those values of s for which the function is zero, that is, $F(z_m) = 0$. Zeros are indicated by a small circle on the s plane. The poles of this function, $s = p_k$, are those values of s for which function is infinite, that is, $F(p_k) = \infty$. Poles are indicated by a small x on the s plane. Poles and zeros are also known as singularities of the function.

By assuming that $F(s)$ has only simple poles (first-order poles), it can be expanded into partial fractions of the form

$$F(s) = \frac{A_1}{s - p_1} + \frac{A_2}{s - p_2} + \cdots + \frac{A_k}{s - p_k} + \cdots + \frac{A_v}{s - p_v} \tag{4-71}$$

The coefficients are obtained from Eq. (4-38) and are given by

$$A_k = [(s - p_k)F(s)]_{s=pk} \tag{4-72}$$

The first coefficient is

$$A_1 = \frac{K(p_1 - z_1)(p_1 - z_2) \cdots (p_1 - z_w)}{(p_1 - p_2)(p_1 - p_3) \cdots (p_1 - p_v)} \tag{4-73}$$

Figure 4-6 shows the poles and zeros of a function $F(s)$. The quantity $s = p_1$ is drawn as an arrow from the origin of the s plane to the pole p_1.

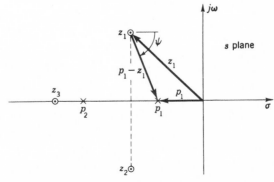

Fig. 4-6 *Directed line segments drawn on a pole-zero diagram of a function $F(s)$.*

This arrow is called a *directed line segment*. It has a magnitude equal to $|p_1|$ and an angle of 180°. Similarly, the directed line segment $s = z_1$ is drawn as an arrow from the origin to the zero z_1 and has a corresponding magnitude and angle. A directed line segment is also drawn from the zero z_1 to the pole p_1. By the rules of vector addition, it can be shown that this directed line segment is equal to $p_1 - z_1$; it has a magnitude equal to its length and an angle ψ, as shown. By referring to Eq. (4-73), it is seen that $p_1 - z_1$ appears in the numerator of A_1. While this quantity can be evaluated analytically, it can also be measured from the pole-zero diagram by means of a ruler and a protractor. The measured directed line segment is obtained in polar form, a magnitude and an angle, which is most useful in evaluating Eq. (4-73). In a similar fashion, each factor in Eq. (4-73) can be obtained graphically; the directed line segments are drawn *from* the zeros and each of the other poles *to* the pole p_1. The angle from a zero is indicated by the symbol ψ, and the angle from a pole is indicated by the symbol θ.

The general rule for evaluating the coefficients in the partial-fraction expansion is quite simple. The value of A_k is the product of K and the directed distances from each zero to the pole p_k divided by the product of the directed distances from each of the other poles to the pole p_k. Each of these directed distances is characterized by a magnitude and an angle. This statement can be written in equation form as

$$A_k = K \; \frac{\text{product of directed distances from each zero to the pole } p_k}{\text{product of directed distances from all other poles to the pole } p_k}$$

$$= K \; \frac{\displaystyle\prod_{m=1}^{w} (p_k - z_m) \Big/ \sum_{m=1}^{w} \psi(p_k - z_m)}{\displaystyle\prod_{\substack{c=1 \\ c \neq k}}^{v} (p_k - p_c) \Big/ \sum_{c=1}^{v} \theta(p_k - p_c)} \qquad (4\text{-}74)$$

For real poles the values of A_k must be real but can be either positive or negative. A_k is positive if the total number of real poles and real zeros to the right of p_k is even, and it is negative if the total number of real poles and real zeros to the right of p_k is odd. For complex poles the values of A_k are complex. If $F(s)$ has no finite zeros, the numerator of Eq. (4-74) is equal to K. For a repeated factor of the form $(s + a)^r$ in either the numerator or the denominator, its contribution to Eq. (4-74) is given by

$$|(s + a)^r| \; \underline{/(s + a)^r} = |s + a|^r \; \underline{/r/(s + a)}$$

where the values of $|s + a|$ and $\underline{/(s + a)}$ are obtained graphically.

The application of the graphical technique is illustrated by the following examples.

Example 1. A function $F(s)$ with real poles

$$F(s) = \frac{K(s - z_1)}{s(s - p_1)(s - p_2)} \tag{4-75}$$

The pole-zero diagram for $F(s)$ is shown in Fig. 4-7a. The pole at the origin is labeled p_0 in the figure. The partial-fraction expansion is

$$F(s) = \frac{A_0}{s} + \frac{A_1}{s - p_1} + \frac{A_2}{s - p_2} \tag{4-76}$$

The value of A_0 is given by

$$A_0 = [sF(s)]_{s=0} = \left[\frac{K(s - z_1)}{(s - p_1)(s - p_2)}\right]_{s=0}$$

$$= \frac{K(-z_1)}{(-p_1)(-p_2)} \tag{4-77}$$

By referring to Fig. 4-7b, the quantities are obtained graphically. The quantities $-z_1$, $-p_1$, and $-p_2$ each have an angle of $0°$; thus A_0 is a positive real number. Note that the constant K is not obtainable from the pole-zero diagram; it must be obtained from the original equation of $F(s)$.

Similarly, the value of A_1 is given by

$$A_1 = \frac{K(p_1 - z_1)}{p_1(p_1 - p_2)} \tag{4-78}$$

and the directed line segments are shown in Fig. 4-7c. The quantities

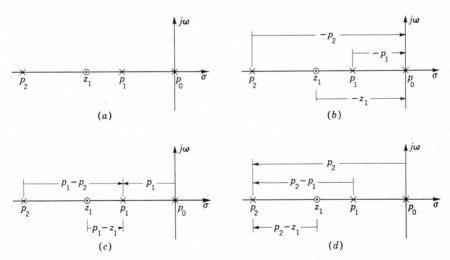

Fig. 4-7 (a) *Pole-zero diagram for* $F(s)$ *of Eq.* (4-75). (b) *Directed line segments for evaluating* A_0 *by Eq.* (4-77). (c) *Directed line segments for evaluating* A_1 *by Eq.* (4-78). (d) *Directed line segments for evaluating* A_2 *by Eq.* (4-79).

$p_1 - z_1$ and $p_1 - p_2$ have an angle of $0°$, and p_1 has an angle of $180°$. The coefficient A_1 is therefore a negative real number. Note that this agrees with the rule that A_1 is negative because the number of real poles and zeros to the right of p_1 is an odd number.

In like manner the value of A_2 is given by

$$A_2 = \frac{K(p_2 - z_1)}{p_2(p_2 - p_1)} \tag{4-79}$$

Figure 4-7d shows the directed line segments for Eq. (4-79). A_2 is a negative real number; this is evident from the fact that the number of real poles and zeros to the right of p_2 is an odd number.

The complete inverse transform of $F(s)$ is

$$f(t) = A_0 + A_1 e^{p_1 t} + A_2 e^{p_2 t} \tag{4-80}$$

It should be noted that zeros in $F(s)$ affect both the size and the sign of the coefficients in the time function.

Example 2.   A function $F(s)$ with complex poles

$$
\begin{aligned}
F(s) &= \frac{K(s + \alpha)}{s[(s + a)^2 + b^2]} = \frac{K(s + \alpha)}{s(s + a - jb)(s + a + jb)} = \frac{K(s - z_1)}{s(s - p_1)(s - p_2)} \\
&= \frac{A_0}{s} + \frac{A_1}{s + a - jb} + \frac{A_2}{s + a + jb}
\end{aligned} \tag{4-81}
$$

The coefficient A_0 is obtained by use of the directed line segments shown in Fig. 4-8a:

$$A_0 = \frac{K(\alpha)}{(a - jb)(a + jb)} = \frac{K\alpha}{\sqrt{a^2 + b^2}\, e^{j\psi_1} \sqrt{a^2 + b^2}\, e^{j\psi_2}} = \frac{K\alpha}{a^2 + b^2} \tag{4-82}$$

Note that the angles ψ_1 and ψ_2 are equal in magnitude but opposite in sign. The coefficient for a real pole is therefore always a real number.

The coefficient A_1 is obtained by use of the directed line segments shown in Fig. 4-8b:

$$
\begin{aligned}
A_1 &= \frac{K[(\alpha - a) + jb]}{(-a + jb)(j2b)} = \frac{K\sqrt{(\alpha - a)^2 + b^2}\, e^{j\psi}}{\sqrt{a^2 + b^2}\, e^{j\theta} 2b e^{j\pi/2}} \\
&= \frac{K}{2b}\sqrt{\frac{(\alpha - a)^2 + b^2}{a^2 + b^2}}\, e^{j(\psi - \theta - \pi/2)}
\end{aligned} \tag{4-83}
$$

where $\psi = \tan^{-1}[b/(\alpha - a)]$ and $\theta = \tan^{-1}(b/-a)$.

The constants A_1 and A_2 are complex conjugates; therefore it is not necessary to evaluate A_2.

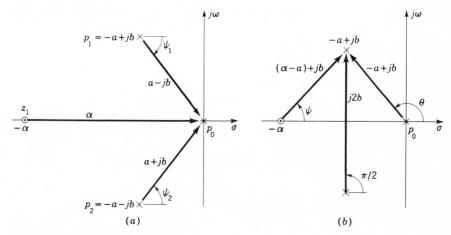

Fig. 4-8 (*a*) *Directed line segments for evaluating* A_0 *of Eq.* (*4-82*). (*b*) *Directed line segments for evaluating* A_1 *of Eq.* (*4-83*).

The response as a function of time is

$$f(t) = A_0 + 2|A_1|e^{-at} \cos\left(bt + \psi - \theta - \frac{\pi}{2}\right) \qquad (4\text{-}84)$$

The trigonometric function in the damped sinusoidal term can be changed from a cosine to a sine by a shift of $\pi/2$ in the phase angle. This gives

$$f(t) = A_0 + 2|A_1|e^{-at} \sin\ (bt + \underline{/A_1} + 90°)$$

$$= \frac{K\alpha}{a^2 + b^2} + \frac{K}{b} \sqrt{\frac{(\alpha - a) + b^2}{a^2 + b^2}}\ e^{-at} \sin\ (bt + \phi) \qquad (4\text{-}85)$$

where

$$\phi = \psi - \theta = \tan^{-1} \frac{b}{\alpha - a} - \tan^{-1} \frac{b}{-a}$$

$$= \text{angle of } A_1 + 90° \qquad (4\text{-}86)$$

This agrees with transform pair 28 in Appendix A. The evaluation of A_0 and A_1 is intended to be performed by measuring the directed line segments directly from the pole-zero diagram.

Example 3. A function $F(s)$ with repeated real poles

$$F(s) = \frac{K(s - z_1)(s - z_2)}{s(s - p_1)^2} = \frac{A_0}{s} + \frac{A_{12}}{(s - p_1)^2} + \frac{A_{11}}{s - p_1} \qquad (4\text{-}87)$$

The coefficients A_0 and A_{12} are found in the conventional manner from

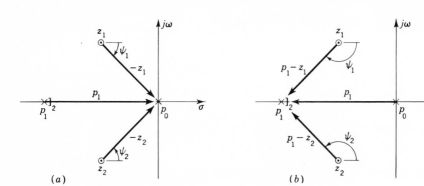

Fig. 4-9 (a) *Directed line segments for evaluating A_0 of Eq. (4-88).* (b) *Directed line segment for evaluating A_{12} of Eq. (4-89).*

the pole-zero diagram.   The value A_0 is given by

$$A_0 = \left[\frac{K(s - z_1)(s - z_2)}{(s - p_1)^2} \right]_{s=0} = \frac{K(-z_1)(-z_2)}{(-p_1)^2} \qquad (4\text{-}88)$$

The factors used to evaluate A_0 are shown on the pole-zero diagram of Fig. 4-9a.   The zeros in this example are complex conjugates; therefore the angles ψ_1 and ψ_2 are equal in magnitude but opposite in sign.   Thus, A_0 is a real number.   Note that in the evaluation of A_0 the double pole at p_1 is treated as two superimposed first-order poles.   The denominator therefore contains the magnitude $|-p_1|^2$ and the angle $2/\!-p_1$.   The value of A_{12} is given by

$$A_{12} = \left[\frac{K(s - z_1)(s - z_2)}{s} \right]_{s=p_1} = \frac{K(p_1 - z_1)(p_1 - z_2)}{p_1} \qquad (4\text{-}89)$$

Figure 4-9b shows the pole-zero diagram with these factors indicated. A_{12} is a negative real number.

The coefficient A_{11} must be treated in a different fashion.   One approach is the method in Sec. 4-9.   The denominator of $F(s)$ is one degree higher than the numerator; therefore the sum of the residues is K:

$$A_0 + A_{11} = K \qquad (4\text{-}90)$$

Since A_0 is known from Eq. (4-88), the value $A_{11} = K - A_0$ is readily determined.   The second approach is to use Eq. (4-48), which gives

$$A_{11} = \frac{d}{ds}\, [(s - p_1)^2 F(s)]_{s=p_1} \qquad (4\text{-}91)$$

After the differentiation, the value of A_{11} can be determined by obtaining the factors in the expression from the pole-zero diagram.   It can be shown also that Eq. (4-91) can be put in the general form

$$A_{11} = A_{12}\left[\sum_{m=1}^{w}\frac{1}{s-z_m} - \sum_{k=2}^{v}\frac{1}{s-p_k}\right]_{s=p_1} \tag{4-92}$$

where $s = p_1$ is the double pole. The coefficient A_{11} can now be computed graphically. If the function $F(s)$ does not contain finite zeros, the term $1/(s - z_m)$ will be taken as zero.

4-11   Use of the Spirule

The preceding section shows the procedure for graphically evaluating the coefficients in the partial-fraction expansion of a function $F(s)$. The directed line segments used to compute these coefficients must be measured on the pole-zero diagram. Since they have both a magnitude and an angle, these measurements can be made with a ruler and a protractor. A Spirule* is a convenient device for facilitating the computation of the coefficients A_k. It is essentially a protractor with a pivoted arm, as shown in Fig. 4-10. Its function is to measure and multiply or divide the lengths of the directed line segments to obtain the magnitude of A_k. It also algebraically adds the angles of the directed line segments in order to obtain the angle of A_k. Outlined below are steps of multiplication and division and the algebraic addition of angles.

For a given function

$$F(s) = \frac{K(s-z_1)(s-z_2)\cdots(s-z_w)}{(s-p_1)(s-p_2)\cdots(s-p_v)} = \frac{A_1}{s-p_1} + \frac{A_2}{s-p_2}$$
$$+ \cdots + \frac{A_k}{s-p_k} + \cdots \tag{4-93}$$

the steps in evaluating a coefficient A_k are listed below.

* Available from The Spirule Company, 9728 El Venado, Whittier, California.

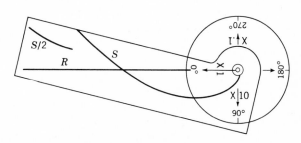

Fig. 4-10 *Outline of a Spirule.*

Magnitude of A_k

1. Plot the poles and zeros on rectangular coordinate graph paper. Determine the *scale factor* between the scale used on the graph paper and the unit length on the edge of the arm of the Spirule.
2. Line up the reference line R on the Spirule arm with the X—1 arrow on the disk. Place the pivot of the Spirule over the pole p_k.
3. Rotate the arm and disk together to align the reference line R with the first zero z_1. Hold the disk stationary and turn the arm until the S curve crosses over the zero z_1. This procedure sets the magnitude of $p_k - z_1$ on the Spirule.
4. Release the disk so that the arm and disk rotate together. Repeat step 3 for all zeros of $F(s)$. This sets the product of the zero factors on the Spirule.
5. Rotate the arm and disk together so that the S curve crosses over the pole p_1. Hold the disk stationary and rotate the arm until the reference line R coincides with the pole p_1. This procedure performs the division by $p_k - p_1$.
6. Repeat step 5 for all poles of $F(s)$.
7. The final *Spirule reading* is obtained from the *index arrow* on the disk and the scale on the arm. Three of the index arrows are marked X—.1, X—1, and X—10.
8. The magnitude of A_k is given by

$$A_k = K(\text{Spirule reading})(\text{index arrow marking})(\text{scale factor})^x$$
$$(4\text{-}94)$$

The exponent x is equal to the number of zeros minus the number of poles of $(s - p_k)F(s)$.

Angle of A_k

1. Line up the reference line R on the Spirule arm with the 0° marking on the disk. Place the pivot of the Spirule over the pole p_k.
2. Rotate the arm and disk together so that the line R is pointing horizontally to the left. Hold the disk stationary and turn the arm until the R line crosses the first zero z_1. The Spirule reading is now the angle of $p_k - z_1$.
3. Release the disk so that the arm and disk rotate together, and repeat step 2 for all zeros of $F(s)$. This procedure adds the angles of the zero factors on the Spirule.
4. Rotate the arm and disk together so that the line R crosses over the pole p_1. Hold the disk stationary and turn the arm until line R is pointing horizontally to the left. This procedure subtracts the angle of $p_k - p_1$.

5. Release the disk so that the arm and disk rotate together, and repeat step 4 for all poles of $F(s)$.

6. The final reading after step 5 is the angle of A_k. The scale on the Spirule disk is from 0 to 360° in the positive or counterclockwise direction. If the angle is greater than 180°, it may be more convenient to convert it to its equivalent negative angle.

It is recommended that facility with the Spirule be developed by the student by working a number of problems; with practice, it is possible to obtain good accuracy. Until confidence in its use is developed, the results should be checked by analytically calculating the coefficients. The Spirule is also used in Chap. 7 for plotting the root locus.

4-12   *Frequency Response from the Pole-Zero Diagram*

The frequency response of a system is described as the steady-state response with a sine-wave forcing function for all values of frequency. This information is often presented graphically. One form of presentation requires the use of two curves. One curve shows the ratio of output amplitude to input amplitude, M, and the other curve the phase angle of the output, α, where both are plotted as a function of frequency.

Consider the input to a system as sinusoidal and given by

$$x_1(t) = X_1 \sin \omega t \tag{4-95}$$

Thus the magnitude of the frequency response is given by

$$M = \left| \frac{\mathbf{X}_2(j\omega)}{\mathbf{X}_1(j\omega)} \right| = \left| \frac{\mathbf{P}(j\omega)}{\mathbf{Q}(j\omega)} \right| = \left| \frac{K(j\omega - z_1)(j\omega - z_2) \cdots}{(j\omega - p_1)(j\omega - p_2) \cdots} \right| \tag{4-96}$$

and the angle is given by

$$\alpha = \underline{/\mathbf{P}(j\omega)} - \underline{/\mathbf{Q}(j\omega)}$$
$$= \underline{/K} + \underline{/j\omega - z_1} + \underline{/j\omega - z_2} + \cdots - \underline{/j\omega - p_1} - \underline{/j\omega - p_2} - \cdots \tag{4-97}$$

Figure 4-11 shows a possible pole-zero diagram and the directed line segments for evaluating M and α corresponding to a frequency ω_1. As ω increases, each of the directed line segments changes in both magnitude and angle. Several significant characteristics can be noted as the driving frequency ω increases from 0 to ∞. The poles and zeros referred to below are shown in Fig. 4-11.

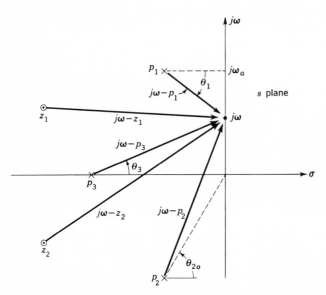

Fig. 4-11 *Pole-zero diagram of Eq. (4-96) showing the directed line segments for evaluating the frequency response* $M\underline{/\alpha}$.

1. *Real Pole p_3.* The directed line segment from a real pole in the left half of the s plane has the form $j\omega - p_3$. Its magnitude starts at $-p_3$ for $\omega = 0$ and increases to infinity as $\omega \to \infty$. Therefore, it causes a decrease in the magnitude of M as ω increases, because the term $j\omega - p_3$ appears in the denominator of Eq. (4-96). Also, the angle θ_3 of $j\omega - p_3$ starts at $0°$ for $\omega = 0$ and increases to $90°$ as $\omega \to \infty$. The frequency response therefore experiences an angle varying from 0 to $-90°$ due to $j\omega - p_3$ as ω increases from 0 to ∞. The angle of $j\omega - p_3$ is exactly $45°$ at $\omega = |p_3|$.

2. *Complex Pole in the Third Quadrant, p_2.* The directed line segment from a complex pole in the third quadrant is equal to $j\omega - p_2$. At $\omega = 0$ it has a finite magnitude and a positive angle θ_{20}. As ω increases and approaches infinity, the magnitude of $j\omega - p_2$ also increases and approaches infinity. The angle of $j\omega - p_2$ increases and approaches $90°$. The effect on the frequency response is to decrease the magnitude M and vary the angle α from $-\theta_{20}$ to $-90°$ as ω increases from 0 to ∞.

3. *Complex Pole in the Second Quadrant, p_1.* The directed line segment from a complex pole in the second quadrant starts with a finite magnitude and a negative angle θ_{10} at $\omega = 0$. As ω increases, the

magnitude of $j\omega - p_1$ first decreases. When $\omega = \omega_a$ is on a horizontal line with p_1, the magnitude $j\omega - p_1$ reaches a minimum value. As ω increases further, the length of $j\omega - p_1$ increases and approaches infinity. Since this term is in the denominator, the magnitude M shows a peak in the vicinity of ω_a. The angle of $j\omega - p_1$ increases from its initial negative angle to zero degrees at ω_a and reaches an eventual limit of 90°. The rate of change of this angle is greatest at ω_a.

For zeros the effects are reciprocal to the effects for poles because the terms $j\omega - z_m$ appear in the numerator of Eq. (4-96). The magnitude M and the angle α are a composite of the effects of all the poles and zeros. In particular, for a system function which has all poles and zeros in the left half of the s plane, the following characteristics of the frequency response are noted:

1. At $\omega = 0$ the magnitude is a finite value and the angle is 0°.
2. If there are more poles than zeros, as $\omega \to \infty$, the magnitude approaches zero and the angle is $-90°$ times the difference between the number of poles and zeros.
3. The magnitude can have a peak value M_m only if there are complex poles fairly close to the imaginary axis. It is shown later that the damping ratio ζ of the complex poles must be less than 0.707 to obtain a peak value for M. Of course, the presence of zeros could counteract the effect of the complex poles, so it is possible that no peak would be present even if ζ of the poles were less than 0.707.

The frequency-response characteristics of several system transforms are illustrated next. As a first example,

$$\frac{X_2(s)}{X_1(s)} = \frac{P(s)}{Q(s)} = \frac{a}{s + a} \tag{4-98}$$

The frequency response starts with a magnitude of unity and an angle of 0°. As ω increases, the magnitude decreases and approaches zero while the angle approaches $-90°$. At $\omega = a$ the magnitude is 0.707 and the angle is $-45°$. The frequency $\omega = a$ is called the corner frequency ω_{cf} or the break frequency. Figure 4-12 shows the pole location, the frequency-response magnitude M, and the frequency-response angle α.

The second example is a system transform with a pair of conjugate poles:

$$\frac{X_2(s)}{X_1(s)} = \frac{\omega_n^2}{s^2 + 2\zeta\omega_n s + \omega_n^2}$$

$$= \frac{\omega_n^2}{(s + \zeta\omega_n - j\omega_n \sqrt{1 - \zeta^2})(s + \zeta\omega_n + j\omega_n \sqrt{1 - \zeta^2})} = \frac{\omega_n^2}{(s - p_1)(s - p_2)} \tag{4-99}$$

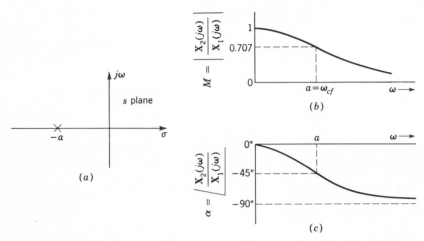

Fig. 4-12   (a) Pole location of Eq. (4-98).   (b) M versus ω for Eq. (4-98).   (c) α versus ω for Eq. (4-98).

The frequency response is given by

$$M/\underline{\alpha} = \frac{X_2(j\omega)}{X_1(j\omega)} = \frac{\omega_n^2}{(j\omega + \zeta\omega_n - j\omega_n\sqrt{1 - \zeta^2})(j\omega + \zeta\omega_n + j\omega_n\sqrt{1 - \zeta^2})}$$
$$(4\text{-}100)$$

The location of the poles of Eq. (4-99) and the directed line segments for $s = j\omega$ are shown in Fig. 4-13a.   At $\omega = 0$ the values are $M = 1$ and $\alpha = 0°$. As ω increases, the magnitude M first increases because $j\omega - p_1$ is decreasing faster than $j\omega - p_2$ is increasing.   By differentiating the magnitude of M, obtained from Eq. (4-100), with respect to ω, it can be shown (see Sec. 11-3) that the maximum value M_m and the frequency ω_m at which it occurs are given by

$$M_m = \frac{1}{2\zeta\sqrt{1 - \zeta^2}} \qquad (4\text{-}101)$$

$$\omega_m = \omega_n\sqrt{1 - 2\zeta^2} \qquad (4\text{-}102)$$

Above ω_m the value of M decreases and approaches zero as ω approaches infinity.   A circle drawn on the s plane with the poles p_1 and p_2 as the diameter intersects the imaginary axis at the value of ω given by Eq. (4-102). Both Eq. (4-102) and the geometrical construction of the circle show that the M curve has a maximum value, other than at $\omega = 0$, only if $\zeta < 0.707$.   The angle α becomes more negative as ω increases.   At $\omega = \omega_n$ the angle is equal to 90°.   This is the corner frequency for a quadratic factor with complex roots.   As $\omega \to \infty$, the angle α approaches $-180°$.   The magnitude and angle of the frequency response for $\zeta < 0.707$ are shown in Fig. 4-13.   Note that the M and α curves are continuous for all linear systems.

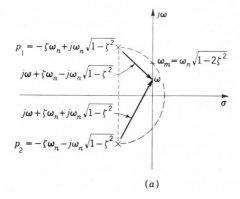

(a)

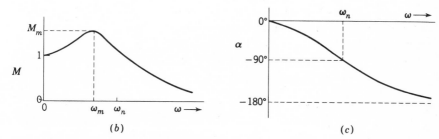

(b) (c)

Fig. 4-13 (a) *Pole-zero diagram, $\zeta < 0.707$.* (b) *Magnitude of frequency response.* (c) *Angle of frequency response.*

The pole-zero diagram of the system transform is used above to show the effect of poles and zeros in the left half of the s plane on the frequency response. In a similar fashion it is easy to see the effect of poles and zeros in the right half of the s plane. Figure 4-14 contains a positive real zero $z_1 = a$ and a negative real zero $z_2 = -a$. It can be seen from the figure that the directed

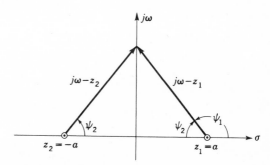

Fig. 4-14 *Comparison of directed line segments due to zeros in the right- and left-half s planes.*

lines $j\omega - z_1$ and $j\omega - z_2$ have the same magnitude for a given frequency, but the angles are supplements of each other.   The angle of $j\omega - z_2$ increases from 0 to 90°, while the angle of $j\omega - z_1$ decreases from 180 to 90° as ω varies from zero to infinity.

When the system transform contains poles and zeros confined to the left half of the s plane, it is a *minimum-phase system*.[13]   When there are poles or zeros in the right half of the s plane, it is a *non-minimum-phase system*.   It should be realized that poles in the right half of the s plane mean that the system is unstable.   Zeros of the system transform in the right-half s plane do not affect stability, but they can affect the size of each of the terms in the time response.   Some authors have incorrectly defined non-minimum systems as those which allow only zeros in the right-half s plane, i.e., stable systems.   But systems with poles in the right-half s plane must be included in the definition of non-minimum-phase systems.   Some high-performance aircraft have inherently non-minimum-phase characteristics.   Feedback, as described in later chapters, can be used to stabilize such systems.

4-13   Location of Poles and Stability

The stability and the corresponding response of a system can be determined from the location of the poles of the response transform $F(s)$ in the s plane. The possible positions of the poles are shown in Fig. 4-15, and the responses are given in Table 4-1.   These poles are the roots of the characteristic equation.

Poles of the response transform at the origin or on the imaginary axis that are not contributed by the forcing function result in a continuous output. These outputs are undesirable in a control system.   Poles in the right-half s

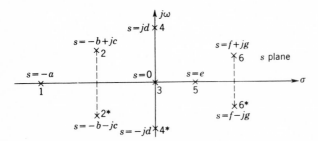

Fig. 4-15   *Location of poles in the s plane.   (Numbers are used to identify the poles.)*

Table 4-1.   Relation of response to location of poles

Position of pole	Form of response	Characteristics
1	Ae^{-at}	Damped exponential
2–2*	$Ae^{-bt} \sin (ct + \phi)$	Exponentially damped sinusoid
3	A	Constant
4–4*	$A \sin (dt + \phi)$	Constant sinusoid
5	Ae^{et}	Increasing exponential (unstable)
6–6*	$Ae^{ft} \sin (gt + \phi)$	Exponentially increasing sinusoid (unstable)

plane result in transient terms that increase with time.   Such performance characterizes an unstable system; therefore poles in the right-half s plane are not desirable, in general.

4-14   Routh's Stability Criterion

The response transform $X_2(s)$ has the general form given by Eq. (4-32), which is repeated here in slightly modified form. $X_1(s)$ is the driving transform.

$$X_2(s) = \frac{P(s)}{Q(s)} X_1(s) = \frac{P(s)X_1(s)}{b_v s^v + b_{v-1}s^{v-1} + b_{v-2}s^{v-2} + \cdots + b_1 s + b_0}$$

$$(4\text{-}103)$$

Sections 4-6 and 4-7 entail the methods used to evaluate the inverse transform $\mathcal{L}^{-1}[F(s)] = f(t)$.   However, before the inverse transformation can be performed, the polynomial $Q(s)$ must be factored.   Appendix B contains several techniques for factoring a polynomial.   The higher the degree of $Q(s)$, the more laborious is the process of finding its factors.   Section 4-13 has shown that stability of the response $x_2(t)$ requires that all zeros of $Q(s)$ have negative real parts.   Since it is usually not necessary to find the exact solution when the response is unstable, a simple procedure to determine the existence of zeros with positive real parts is desired.   If such zeros of $Q(s)$ with positive real parts are found, the system must be modified.   Routh's criterion is a simple method of determining the number of zeros with positive real parts without actually solving for the zeros of $Q(s)$.   Another method of determining stability is Hurwitz's criterion, which establishes the necessary conditions in terms of the system determinants.   Numerous accounts of this criterion can be found in the literature.[6]

The characteristic equation is

$$Q(s) = b_v s^v + b_{v-1} s^{v-1} + b_{v-2} s^{v-2} + \cdots + b_1 s + b_0 = 0 \quad (4\text{-}104)$$

The b's are real coefficients. All powers of s from s^v to s^0 must be present in the characteristic equation. If any coefficients other than b_0 are zero or if not all the coefficients are positive, there are imaginary roots or roots with positive real parts and the system is unstable. It is therefore unnecessary to continue if only stability or instability is to be determined. In special situations it may be necessary to determine the actual number of roots in the right-half s plane. For these situations the procedure described in this section can be used.

The coefficients of the characteristic equation are arranged in the pattern shown in the following Routhian array. These coefficients are then used to evaluate the rest of the constants to complete the array.

$$
\begin{array}{c|cccccc}
s^v & b_v & b_{v-2} & b_{v-4} & b_{v-6} & \cdots \\
s^{v-1} & b_{v-1} & b_{v-3} & b_{v-5} & b_{v-7} & \cdots \\
s^{v-2} & c_1 & c_2 & c_3 & \cdots \\
s^{v-3} & d_1 & d_2 & \cdots \\
\cdot \\
\cdot \\
\cdot \\
s^1 & j_1 \\
s^0 & k_1
\end{array}
$$

The constants c_1, c_2, c_3, etc., in the third row are evaluated as follows:

$$c_1 = \frac{b_{v-1} b_{v-2} - b_v b_{v-3}}{b_{v-1}} \quad (4\text{-}105)$$

$$c_2 = \frac{b_{v-1} b_{v-4} - b_v b_{v-5}}{b_{v-1}} \quad (4\text{-}106)$$

$$c_3 = \frac{b_{v-1} b_{v-6} - b_v b_{v-7}}{b_{v-1}} \quad (4\text{-}107)$$

This pattern is continued until the rest of the c's are all equal to zero. Then the d row is formed by using the s^{v-1} and s^{v-2} row. The constants are

$$d_1 = \frac{c_1 b_{v-3} - b_{v-1} c_2}{c_1} \quad (4\text{-}108)$$

$$d_2 = \frac{c_1 b_{v-5} - b_{v-1} c_3}{c_1} \quad (4\text{-}109)$$

$$d_3 = \frac{c_1 b_{v-7} - b_{v-1} c_4}{c_1} \quad (4\text{-}110)$$

This is continued until no more d terms are present. The rest of the rows are formed in this way down to the s^0 row. The complete array is triangular, ending with the s^0 row. Notice that s^1 and s^0 rows contain only one term each. Once the array has been found, *Routh's criterion states:*

The number of roots of the characteristic equation with positive real parts is equal to the number of changes of sign of the coefficients in the first column. Therefore the system is stable if all terms in the first column have the same sign.

The following example illustrates this criterion:

$$Q(s) = s^5 + s^4 + 10s^3 + 72s^2 + 152s + 240 \qquad (4\text{-}111)$$

Routh's array is formed by using the procedure described above:

$$
\begin{array}{c|ccc}
s^5 & 1 & 10 & 152 \\
s^4 & 1 & 72 & 240 \\
s^3 & -62 & -88 & \\
s^2 & 70.6 & 240 & \\
s^1 & 122.6 & & \\
s^0 & 240 & &
\end{array}
$$

In the first column there are two changes of sign, from 1 to -62 and from -62 to 70.6; therefore $Q(s)$ has two roots in the right-half s plane. Note that this criterion gives the number of roots with positive real parts but does not tell the values of the roots. If Eq. (4-111) is factored, the roots are

$$
\begin{aligned}
s &= -3 \\
s &= -1 \pm j\sqrt{3} \\
s &= +2 \pm j4
\end{aligned}
$$

This confirms the fact that there are two roots with positive real parts. The Routh criterion does not distinguish between real and complex roots.

Theorem 1.   Division of a row

The coefficients of any row may be multiplied or divided by a positive number without changing the signs of the first column. The labor of evaluating the coefficients in Routh's array may be reduced by multiplying or dividing any row by a constant. This may result, for example, in reducing the size of the coefficients and therefore simplifying the evaluation of the remaining coefficients. The following example illustrates this theorem:

$$Q(s) = s^6 + 3s^5 + 2s^4 + 9s^3 + 5s^2 + 12s + 20 \qquad (4\text{-}112)$$

The Routhian array is

$$
\begin{array}{c|cccc}
s^6 & 1 & 2 & 5 & 20 \\
s^5 & \cancel{3} & \cancel{9} & \cancel{12} & \\
 & 1 & 3 & 4 & \text{(after dividing by 3)} \\
s^4 & -1 & 1 & 20 & \\
s^3 & \cancel{4} & \cancel{24} & & \\
 & 1 & 6 & & \text{(after dividing by 4)} \\
s^2 & 7 & 20 & & \\
s^1 & 22\tfrac{2}{7} & & & \\
s^0 & 20 & & &
\end{array}
$$

Notice that the size of the numbers has been reduced by dividing the s^5 row by 3 and the s^3 by 4. The result is unchanged; i.e., there are two changes of signs in the first column and therefore there are two roots with positive real parts.

Theorem 2.   A zero coefficient in the first column

When the first term in a row is zero but not all the other terms are zero, the following methods can be used:

1. Substitute a small positive number δ for the zero and proceed to evaluate the rest of the array.
2. Substitute in the original equation $s = 1/x$; then solve for the roots of x with positive real parts. The number of roots x with positive real parts will be the same as the number of s roots with positive real parts.

Both of these methods are illustrated in the following example:

$$Q(s) = s^5 + s^4 + 2s^3 + 2s^2 + 3s + 15 = 0 \qquad (4\text{-}113)$$

Method 1. The number of roots with positive real parts is found from the Routhian array:

s^5	1	2	3
s^4	1	2	15
s^3	$\cancel{0}$	$\cancel{-12}$	
	δ	-12	(after replacing 0 by δ)
s^2	$\dfrac{(\delta \times 2) - [1 \times (-12)]}{\delta}$	15	
s^1	A_1		
s^0	15		

The value of A_1 is

$$A_1 = \frac{\{(\delta \times 2) - [1 \times (-12)]\}(-12) - (\delta^2 \times 15)}{(\delta \times 2) - [1 \times (-12)]}$$

The first term in the s^2 row is seen to be large but positive, and A_1 has a limiting value of -12. Therefore there are two changes of sign and two roots in the right-half s plane.

Method 2. Letting $s = 1/x$ and rearranging the polynomial gives

$$15x^5 + 3x^4 + 2x^3 + 2x^2 + x + 1 \qquad (4\text{-}114)$$

The new Routhian array is

$$
\begin{array}{c|ccc}
x^5 & 15 & 2 & 1 \\
x^4 & 3 & 2 & 1 \\
x^3 & -8 & -4 & \\
x^2 & 0.5 & 1 & \\
x^1 & 12 & & \\
x^0 & 1 & &
\end{array}
$$

There are two changes of sign; therefore there are two roots of x in the right-half s plane. The number of roots s with positive real parts is also two.

Theorem 3. A zero row

When all the coefficients of one row are zero, the procedure is as follows:

1. The auxiliary equation can be formed from the preceding row, as shown below.
2. The Routhian array can be completed by replacing the all-zero row by the coefficients obtained by differentiating the auxiliary equation.
3. The roots of the auxiliary equation are also roots of the original equation. These roots occur in pairs and are the negatives of each other. Therefore these roots may be imaginary (complex conjugates) or real (one positive and one negative), may lie in quadruplets (two pairs of complex conjugate roots), etc.

Consider the system which has the characteristic equation

$$Q(s) = s^4 + 2s^3 + 11s^2 + 18s + 18 = 0 \qquad \text{(4-115)}$$

The Routhian array is

$$
\begin{array}{c|ccl}
s^4 & 1 & 11 & 18 \\
s^3 & \cancel{2} & \cancel{18} & \\
 & 1 & 9 & \text{(after dividing by 2)} \\
s^2 & \cancel{2} & \cancel{18} & \\
 & 1 & 9 & \text{(after dividing by 2)} \\
s^1 & 0 & 0 &
\end{array}
$$

The presence of a zero row indicates that there are roots that are the negatives of each other. The next step is to form the auxiliary equation from the preceding row. The auxiliary equation is formed from the s^2 row. The highest power of s is s^2, and only even powers of s appear. Therefore the auxiliary equation is

$$s^2 + 9 = 0 \qquad \text{(4-116)}$$

The roots of this equation are

$$s = \pm j3$$

These are also roots of the original equation. The presence of imaginary roots indicates that the output includes a sinusoidally oscillating component.

To complete the Routhian array, the auxiliary equation is differentiated and is

$$2s + 0 = 0 \tag{4-117}$$

The coefficients of this equation are inserted in the s^1 row and the array is then completed:

$$\begin{array}{c|c} s^1 & 2 \\ s^0 & 9 \end{array}$$

Since there are no changes of sign in the first column, there are no roots with positive real parts.

In feedback systems, covered in detail in the following chapters, the ratio of the output to the input does not have an explicitly factored denominator. An example of such a function is

$$\frac{X_2(s)}{X_1(s)} = \frac{P(s)}{Q(s)} = \frac{K(s+2)}{s(s+5)(s^2+2s+5) + K(s+2)} \tag{4-118}$$

The value K is an adjustable parameter in the system and may be positive or negative. The value of K determines the location of the poles and therefore the stability of the system. It is important to know the range of values of K for which the system is stable. This information must be obtained from the characteristic equation, which is

$$Q(s) = s^4 + 7s^3 + 15s^2 + (25 + K)s + 2K = 0 \tag{4-119}$$

The coefficients must all be positive in order for the roots of $Q(s)$ to lie in the left half of the s plane, but this is not a sufficient condition for stability. The Routhian array permits evaluation of precise boundaries for K:

$$\begin{array}{c|ccc} s^4 & 1 & 15 & 2K \\ s^3 & 7 & 25 + K \\ s^2 & 80 - K & 14K \\ s^1 & \dfrac{(80 - K)(25 + K) - 98K}{80 - K} \\ s^0 & 14K \end{array}$$

The term $80 - K$ from the s^2 row imposes the restriction $K < 80$, and the s^0 row requires $K > 0$. The numerator of the first term in the s^1 row is equal to $-K^2 - 43K + 2{,}000$. By use of the quadratic formula the zeros of this function are $K = 28.1$ and $K = -71.1$. The combined restrictions on K for stability of the system are therefore $0 < K < 28.1$. For the value $K = 28.1$ the characteristic equation has imaginary roots which can be evaluated by applying Theorem 3. Also, for $K = 0$, it can be seen from

Eq. (4-118) that there is no output present. The methods for selecting the "best" value of K in the range between 0 and 28.1 are contained in later chapters. At this point it is important to note that the Routh criterion has provided useful but restricted information.

4-15   Regional Location of Roots

To evaluate the inverse transform of a function $F(s) = P(s)/Q(s)$ it is necessary that $Q(s)$ be factored. Some methods of factoring polynomials are described in Appendix B. In general, these methods require an iterative procedure. For example, Lin's method applies a trial divisor. Based on the size of the remainder, a second trial divisor is used. The procedure is repeated until the remainder is sufficiently small. The number of repetitions can be reduced by starting with roots that are approximately correct.

The approximate location of the roots of the characteristic equation $Q(s) = 0$ can be found by shifting the origin of the s plane and applying the Routh criterion. For example, the poles of $F(s)$ are shown in Fig. 4-16. Construction of the Routhian array would show that there are no roots of the characteristic equation in the right half of the s plane. If the origin is moved to the left by the amount $-\alpha$, there are two poles to the right of the new origin. This information is available from the Routh criterion when it is applied after shifting the origin.

Shifting the origin to the left by the amount $-\alpha$ requires a change of variable

$$s = s_1 - \alpha \qquad \text{or} \qquad s_1 = s + \alpha$$

The change of variable can be accomplished by substituting $s = s_1 - \alpha$ into $Q(s)$, but this procedure becomes tedious as the degree of $Q(s)$ increases. A simple procedure is illustrated below.

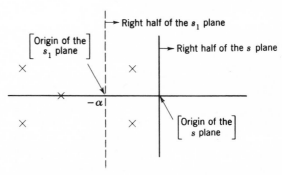

Fig. 4-16   *Roots of $Q(s) = 0$ in the s and s_1 planes.*

Start with the polynomial

$$Q(s) = s^v + b_{v-1}s^{v-1} + \cdots + b_1 s + b_0 \qquad (4\text{-}120)$$

If the origin is to be shifted α units to the left, first divide $Q(s)$ by $s + \alpha$, giving

$$Q(s) = (s + \alpha)\{Q_1(s)\} + R_0 \qquad (4\text{-}121)$$

Equation (4-121) contains a remainder R_0. Continue the process by dividing $Q_1(s)$ by $s + \alpha$, giving

$$Q(s) = (s + \alpha)\{(s + \alpha)Q_2(s) + R_1\} + R_0 \qquad (4\text{-}122)$$

Again there is a remainder, R_1. Continue the process by dividing $Q_2(s)$ by $s + \alpha$, giving

$$Q(s) = (s + \alpha)\{(s + \alpha)[(s + \alpha)Q_3(s) + R_2] + R_1\} + R_0 \quad (4\text{-}123)$$

The procedure is repeated until $Q_v(s) = 1$ is obtained. The division by $s + \alpha$ must be performed v times since this is the degree of the original polynomial. Synthetic division, as described in Appendix B, simplifies the work. Expanding Eq. (4-123) by inspection gives

$$Q(s) = (s + \alpha)^v + R_{v-1}(s + \alpha)^{v-1} + \cdots + R_2(s + \alpha)^2$$
$$+ R_1(s + \alpha) + R_0 \quad (4\text{-}124)$$

The change of variable $s + \alpha = s_1$ makes this equation

$$s_1^v + R_{v-1}s_1^{v-1} + \cdots + R_2 s_1^2 + R_1 s_1 + R_0 \qquad (4\text{-}125)$$

Example

$$s^4 + 10s^3 + 36s^2 + 70s + 75 = 0 \qquad (4\text{-}126)$$

The roots of this equation are $s = -1 \pm j2, -3, -5$.
The Routhian array is

$$
\begin{array}{c|ccc}
s^4 & 1 & 36 & 75 \\
s^3 & 10 & 70 & \\
s^2 & 29 & 75 & \\
s^1 & 1{,}280 & & \\
s^0 & 75 & &
\end{array}
$$

This shows that there are no roots in the right-half s plane and the system is stable. Now shift the axis to $s = -2$. The results of the successive division by $(s + 2)$ are

$$(s + 2)\{s^3 + 8s^2 + 20s + 30\} + 15$$
$$(s + 2)\{(s + 2)[s^2 + 6s + 8] + 14\} + 15$$
$$(s + 2)\{(s + 2)[(s + 2)(s + 4) + 0] + 14\} + 15$$
$$(s + 2)\{(s + 2)[(s + 2)((s + 2) + 2) + 0] + 14\} + 15$$

The polynomial in terms of the new variable $s_1 = s + 2$ is

$$s_1^4 + 2s_1^3 + 0s_1^2 + 14s_1 + 15 \qquad (4\text{-}127)$$

Applying the Routh criterion to Eq. (4-127) gives

$$
\begin{array}{c|ccc}
s_1^4 & 1 & 0 & 15 \\
s_1^3 & 2 & 14 & \\
s_1^2 & -7 & 15 & \\
s_1^1 & 128 & & \\
s_1^0 & 15 & &
\end{array}
$$

Since there are two changes of sign in the first column of this array, there are two roots in the right half of the s_1 plane. But $s_1 = s + 2$; therefore there are two roots of $Q(s) = 0$ to the right of $s = -2$. Since there are no roots to the right of $s = 0$, the roots must be in the region $-2 < s < 0$.

Shifting the axis to the right of $s = -2$ will locate the roots more closely. The successive division required can be performed by synthetic division. For example, shifting the axis to $s = -1$ requires division by $s + 1$:

$$
\begin{array}{rrrrrl}
1 & 10 & 36 & 70 & 75 & /-1 \\
 & -1 & -9 & -27 & -43 & \\
\hline
1 & 9 & 27 & 43 & 32 & = R_0 \\
 & -1 & -8 & -19 & & \\
\hline
1 & 8 & 19 & 24 = R_1 & & \\
 & -1 & -7 & & & \\
\hline
1 & 7 & 12 = R_2 & & & \\
 & -1 & & & & \\
\hline
1 & 6 = R_3 & & & &
\end{array}
$$

The polynomial in terms of the new variable $s_2 = s + 1$ is

$$s_2^4 + 6s_2^3 + 12s_2^2 + 24s_2 + 32$$

The new Routhian array is

$$
\begin{array}{c|cc}
s_2^4 & 1 & 12 & 32 \\
s_2^3 & \cancel{6} & \cancel{24} & \\
 & 1 & 4 & \\
s_2^2 & \cancel{8} & \cancel{32} & \\
 & 1 & 4 & \\
s_2^1 & 0 & &
\end{array}
$$

Roots are therefore located at $s_2 = \pm j2$, or $s = -1 \pm j2$. The procedure can be continued to locate the remaining roots.

The procedure outlined for shifting the origin, coupled with Routh's criterion, permits the determination of the regions in which the roots are located. With this information it is possible to determine whether any of the roots exist in desired regions for a desired time response and to minimize the work involved in the use of Lin's method for determining the exact location of the roots.

4-16   Laplace Transform of the Impulse Function

Figure 4-17 shows a rectangular pulse of amplitude $1/a$ and of duration a. The analytical expression for this pulse is

$$f(t) = \frac{u(t) - u(t - a)}{a} \tag{4-128}$$

and its Laplace transform is

$$F(s) = \frac{1 - e^{-as}}{as} \tag{4-129}$$

If a is decreased, the amplitude increases and the duration of the pulse decreases but the area under the pulse, and thus its strength, remains unity. The limit of $f(t)$ as $a \to 0$ is termed a unit impulse and is designated by $\delta(t)$:

$$\delta(t) = \lim_{a \to 0} f(t) = \lim_{a \to 0} \frac{u(t) - u(t - a)}{a} \tag{4-130}$$

The Laplace transform of the unit impulse, as evaluated by use of Lhopital's theorem, is

$$\Delta(s) = \lim_{a \to 0} \frac{1 - e^{-as}}{as} = \lim_{a \to 0} \frac{0 + se^{-as}}{s} = \lim_{a \to 0} e^{-as} = 1 \tag{4-131}$$

This transform pair is used in an example in a later section. The unit impulse is related to the derivative of the step function. The Laplace transform of the derivative of the unit step function obtained by the use of the

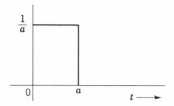

Fig. 4-17   Rectangular pulse.

differentiation theorem is

$$\mathcal{L}[Du(t)] = sU(s) - u(0+) = s\frac{1}{s} - 1 = 0 \qquad (4\text{-}132)$$

Since $\mathcal{L}[\delta(t)] = 1$, Eq. (4-132) can be rewritten as

$$\mathcal{L}[Du(t)] = \mathcal{L}[\delta(t)] - 1 \qquad (4\text{-}133)$$

When any function has a jump discontinuity, the derivative of the function has an impulse at the discontinuity.

4-17   Response to an Impulse Function

Some systems are subjected to shock inputs. For example, an airplane in flight may be subjected to a gust of wind of short duration that jolts the airplane. When a gun is fired there is a reaction force that has a large magnitude and very short duration. Also, a steel ball bouncing off a steel plate has a reaction force of large magnitude and very short duration.

When the duration of a disturbance or input to a system is very short compared with the natural periods of the system, often the response can be well approximated by considering the input to be an impulse of proper strength. An impulse of infinite magnitude and zero duration does not occur in nature; but if the pulse duration is much smaller than the time constants of the system, a representation of the input by an impulse is a good approximation. The shape of the impulse is unimportant as long as the strength of the equivalent impulse is equal to the strength of the actual pulse.

Since the Laplace transform of an impulse is defined, the approximate response of a system to a pulse is often found more easily by this method than by the classical method.

For the RLC circuit shown in Fig. 4-18, consider that the input voltage is an impulse. The problem is to find the voltage e_c across the capacitor as a function of time. The differential equation relating e_c to the input voltage e and the impedances is written by use of the node method:

$$e_c\left(CD + \frac{1}{R+LD}\right) - e\frac{1}{R+LD} = 0 \qquad (4\text{-}134)$$

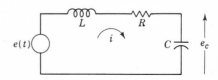

Fig. 4-18 *RLC series circuit.*

By rationalizing, the equation becomes

$$e_c(LCD^2 + RCD + 1) = e \tag{4-135}$$

Taking the Laplace transform of this equation,

$$E_c(s)(s^2LC + sRC + 1) - sLCe_c(0) - LCDe_c(0) - RCe_c(0) = E(s)$$

Consider an initial condition $e_c(0-) = 0$. The voltage across the capacitor is related to the current flowing in the series circuit by the expression $De_c = i/C$. Since the current cannot change instantaneously through an inductor, $De_c(0-) = 0$. These initial conditions which exist at $t = 0-$ are used in the Laplace-transformed equation when the forcing function is an impulse. The solution for the current will actually show a step change at $t = 0$.

The input voltage e is an impulse; therefore $E(s) = 1$. Solving for $E_c(s)$ gives

$$E_c(s) = \frac{1/LC}{s^2 + (R/L)s + 1/LC} = \frac{\omega_n^2}{s^2 + 2\zeta\omega_n s + \omega_n^2} \tag{4-136}$$

Depending on the relative sizes of the parameters, the circuit may be overdamped or underdamped. In the overdamped case the voltage $e_c(t)$ rises to a peak value and then decays to zero. In the underdamped case the voltage oscillates with a frequency $\omega_d = \omega_n \sqrt{1 - \zeta^2}$ and eventually decays to zero. The form of the voltage $e_c(t)$ for several values of damping is shown later in this section.

The stability of a system is revealed by the impulse response. With an impulse input the response transform contains only poles contributed by the system parameters. The impulse response is therefore the inverse transform of the system function. The impulse response provides one method of evaluating the system function; although the method is not simple, it is possible to take the impulse response which is a function of time and convert it to a function of s.

The system response $g(t)$ to an impulse input $\delta(t)$ can also be utilized to determine the response $c(t)$ to any input $r(t)$.[12] The response $c(t)$ is determined by the use of the convolution integral

$$c(t) = \int_0^t r(\tau)g(t - \tau) \, d\tau \tag{4-137}$$

This method is normally used for inputs which are not sinusoidal or a power series.

4-18 *Second-order System with Impulse Excitation*

The Laplace transform for a second-order system of the type developed in Sec. 4-17 with an impulse input is

$$F(s) = \frac{\omega_n^2}{s^2 + 2\zeta\omega_n s + \omega_n^2} \qquad (4\text{-}138)$$

The response of a second-order system to a step-function input was shown in Fig. 3-10. The impulse function is the derivative of the step function; therefore the response to an impulse is the derivative of the response to a step function. Figure 4-19 shows the response for several values of damping ratio ζ.

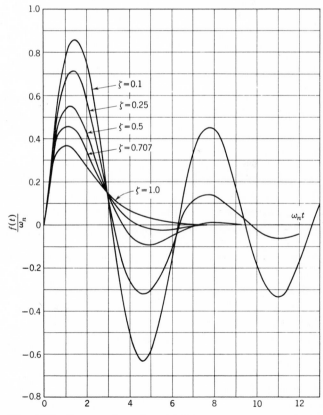

Fig. 4-19 *Response to an impulse for*
$$F(s) = \omega_n^2/(s^2 + 2\zeta\omega_n s + \omega_n^2)$$

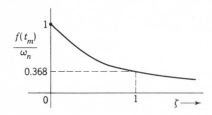

Fig. 4-20 *Maximum overshoot with an impulse input.*

The first zero of the impulse response occurs at t_p, which is the time at which the maximum value of the step-function response occurs.

For the underdamped case the response is

$$f(t) = \frac{\omega_n}{\sqrt{1 - \zeta^2}}\, e^{-\zeta\omega_n t} \sin \omega_n \sqrt{1 - \zeta^2}\, t \qquad (4\text{-}139)$$

By setting the derivative of $f(t)$ with respect to time equal to zero, it can be shown that the maximum overshoot occurs at

$$t_m = \frac{\cos^{-1} \zeta}{\omega_n \sqrt{1 - \zeta^2}} \qquad (4\text{-}140)$$

The maximum value of $f(t)$ is

$$f(t_m) = \omega_n e^{-(\zeta \cos^{-1}\zeta)/\sqrt{1-\zeta^2}} \qquad (4\text{-}141)$$

The maximum overshoot as a function of ζ is plotted in Fig. 4-20.

4-19   Conclusions

This chapter discusses the important characteristics and the use of the Laplace transform, which is employed extensively with differential equations, because it systematizes their solution. It is also used extensively in feedback-system synthesis. The pole-zero diagram has been introduced to represent a system function. The pole-zero pattern is significant because it determines the amplitudes of all the time-response terms. The frequency response has also been shown to be a function of the pole-zero pattern. Later chapters cover feedback system analysis and synthesis by two methods. The first of these is the root-locus method, which locates the poles and zeros of the system in the s plane. Knowing the poles and zeros permits an exact determination of the time response. The second method is based on the frequency response. Since the frequency response is a function of the pole-zero pattern, the two methods are complementary and give equivalent information in different forms.

System stability requires that all poles be located in the left half of

Laplace transforms   135

the s plane.   Routh's criterion is a straightforward method for finding the regional location of the poles and is therefore useful in determining stability. It also identifies the necessary restrictions on an adjustable parameter in order to maintain stability.   The question of stability is considered further in the chapters on the root locus and the Nyquist stability criterion.

Bibliography

1. Gardner, M. F., and J. L. Barnes: "Transients in Linear Systems," chap. 2, John Wiley & Sons, Inc., New York, 1942.
2. Churchill, R. V.: "Operational Mathematics," 2d ed., McGraw-Hill Book Company, New York, 1958.
3. Thomson, W. T.: "Laplace Transformation," Prentice-Hall, Inc., Englewood Cliffs, N.J., 1950.
4. Weber, E.: "Linear Transient Analysis," vol. 1, John Wiley & Sons, Inc., New York, 1954.
5. Wylie, C. R., Jr.: "Advanced Engineering Mathematics," 2d ed., p. 150, McGraw-Hill Book Company, New York, 1960.
6. Guillemin, E. A.: "The Mathematics of Circuit Analysis," pp. 395–409, John Wiley & Sons, Inc., New York, 1949.
7. Seshu, S., and N. Balabanian: "Linear Network Analysis," chap. 2, John Wiley & Sons, Inc., New York, 1959.
8. Aseltine, J. A.: "Transform Method in Linear System Analysis," chap. 7, McGraw-Hill Book Company, New York, 1958.
9. Murphy, G. J.: "Control Engineering," chap. 4, D. Van Nostrand Company, Inc., Princeton, N.J., 1959.
10. Kunz, K. S.: "Numerical Analysis," p. 19, McGraw-Hill Book Company, New York, 1957.
11. Hazony, D., and I. Riley: "Simplified Technique for Evaluating Residues in the Presence of High Order Poles," paper presented at The Western Electronic Show and Convention, August, 1959.
12. Cheng, D. K.: "Analysis of Linear Systems," chap. 8, Addison-Wesley Publishing Company, Inc., Reading, Mass., 1959.
13. Balabanian, N., and W. R. LePage: What Is a Minimum-phase Network?, *Trans. AIEE*, vol. 74, pt. II, pp. 785–788, January, 1956.

Block Diagrams;
Transfer Functions; Flow Graphs

5-1   Introduction: Block Diagrams[1-3]

Closed-loop systems generally have many components. Complete drawings of all the detail parts are frequently too congested to show the specific functions that are performed. To simplify the picture of the complete system, it is common to use a block diagram in which each element in the system is represented by a block. Each block is labeled with the name of the component, and the blocks are appropriately interconnected by line segments. This type of diagram removes excess detail from the picture and shows the functional operation of the system.

A block diagram represents the flow of information and the functions performed by each component in the system. The primary concern is the dynamic behavior of a complete feedback control system. The use of a block diagram provides a simple means by which the functional relationship of the various components may be shown and reveals the operation of the system more readily than observation of the physical system itself. The simple functional block diagram shows the similarities between different types of systems; it also shows clearly that apparently different physical systems may be analyzed by the same techniques. A block diagram is involved not with the physical characteristics of the system but only with the functional relationship between various points in the system and therefore reveals the similarity of apparently unrelated systems.

A further step taken to increase the information supplied by the block diagram is to label the input quantity into each block and the output quantity from each block. Arrows are used to show the direction of the flow of information.

The block represents the function or dynamic characteristics of the component. The ratio of the transform of the output to the transform of the input is called the *transfer function*. Commonly, the transfer function is written inside the block.

The complete block diagram shows the manner in which the functional components are connected and the mathematical equations that determine the response of each component.

5-2   Examples of Block Diagrams

The following examples show the formation of a block diagram to represent a physical system.

Example 1.   Stabilized platform

The first example is the control system of Fig. 5-1a, which shows a single-degree-of-freedom stabilized platform.[4] The base on which the motor is mounted is subject to a disturbance of angular velocity ω_b about its axis. A table is mounted on bearings which are supported from the base. This stabilized table is required to maintain its position and must not follow the displacement of the base. It may be used to hold an inertial space orientation very closely against arbitrarily great rotations of the base about the input axis. The command input signal, shown in the figure, is a second input to the table. Its effect is discussed later. The operation of this platform is as follows:

Case 1. Disturbance Input ω_b. Consider that the table is in the proper position and $\omega_c = 0$. The base is then subjected to a disturbance velocity ω_b. Mechanical coupling of base and table is made through the gear train, so the table follows the base. The integrating gyro, which is mounted on

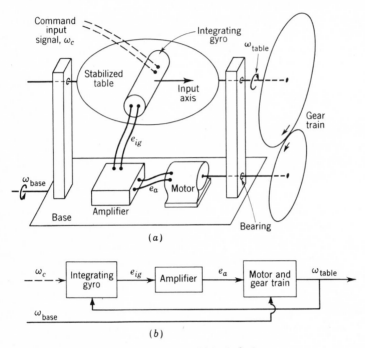

Fig. 5-1   *Single-degree-of-freedom stabilized platform.*

the table, provides the reference for stabilizing the table.   It senses the rotation ω_t of the table and produces a proportional electric signal e_{ig}.   This voltage is amplified and applied to the motor, which turns in a direction to return the table to its original position.

Case 2.   Command Input ω_c.   Consider that the table has a given orientation and that $\omega_b = 0$.   The table position θ_t can be changed by applying a command input signal ω_c (a voltage or a current) to the integrating gyro. The command input signal has the effect of rotating the reference orientation of the gyro about the input axis.   As the reference orientation changes with respect to the actual position of the table, the integrating gyro produces a signal e_{ig} which is amplified and causes the motor to drive the table toward the new reference.   As long as the signal ω_c is present, the reference position of the gyro is displaced from the table position and the signal e_{ig} exists.   The result is that the table rotates with respect to inertial space at an angular velocity proportional to the command input ω_c.   Therefore the angular displacement θ_t of the table is given by

$$\theta_t = \int_0^t \omega_c \, dt$$

When both ω_b and ω_c are present, the platform follows the signal ω_c

regardless of the disturbance ω_b.   When this unit is used with ω_c as the input and θ_t as the output, it is called an *inertial space integrator*.   The block diagram for this system is shown in Fig. 5-1b.   Mechanical coupling is shown by means of a solid arrow ($\longrightarrow$).   The angular velocity of the base (a disturbance) is a mechanical input, so this input to the motor is shown by the solid arrow.   Also, the angular velocity of the table is coupled mechanically to the integrating gyro.   As the outputs of the integrating gyro and the amplifier are electrical quantities, they are shown by thin arrows ($\rightarrow$).   The angular velocity of the table, ω_{table}, is shown by a thin arrow unless a mechanical coupling is intended.   The command input signal ω_c is shown by dashed lines.

Example 2.   Vertical Indicator

The inertial space integrator can be converted into a vertical indicator by adding an accelerometer to the table, with its sensitive axis horizontal and perpendicular to the input axis.   A vertical indicator shows the true vertical position with respect to the earth (the direction of gravity), independent of the location and attitude of the base.   It is used as a navigational instrument.   The arrangement of equipment for the vertical indicator is shown in Fig. 5-2a.   When the stabilized platform is displaced from the true horizontal by an angle θ_{table}, there is a component of gravitational acceleration, $g \sin \theta_t$, along the sensitive axis of the accelerometer.   The accelerometer produces a signal e_{ac} that is proportional to the acceleration along its sensitive axis.   The signal e_{ac} is integrated, producing a signal ω_c which is applied to the integrating gyro on the stabilized platform.   This signal changes the reference axis of the integrating gyro, resulting in a rotation of the table. When the table reaches a horizontal position, the signal e_{ac} becomes zero and the table stops.   The table therefore maintains a horizontal orientation (perpendicular to the direction of gravity) even though the orientation of the base may change.   Additional refinements are omitted from this description for the purpose of simplicity.   The block diagram representing this vertical indicator is shown in Fig. 5-2b.   One block is labeled "inertial space integrator" and represents the entire stabilized platform shown in Fig. 5-1. This is done in order to simplify the block diagram.

Example 3.   Command Guidance System

A more complex system is a command guidance system which directs the flight of a missile in space in order to intercept a moving target.   The target may be an enemy bomber whose aim is to drop bombs at some position. The defense uses the missile with the objective of intercepting and destroying the bomber before it launches its bombs.   A sketch of a generalized com-

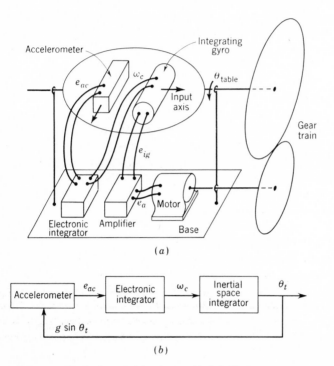

(a)

(b)

Fig. 5-2   *Single-degree-of-freedom vertical indicator.*

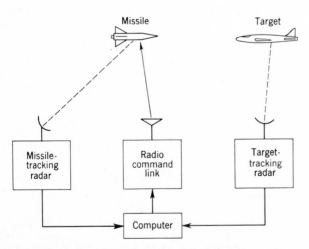

Fig. 5-3   *Command guidance interceptor system.*

mand guidance interceptor system is shown in Fig. 5-3.   The target-tracking radar is used first for detection and then for tracking of the target.   It supplies information on target range and angle and their rates of change (time derivatives).   This information is continuously fed into the computer, which calculates a predicted course for the target.   The missile-tracking radar supplies similar information which is used by the computer to determine its flight path.   The computer compares the two flight paths and determines the necessary change in missile flight path to produce a collision course.   The necessary flight-path changes are supplied to the radio command link, which transmits this information to the missile.   This electrical information containing corrections in flight path is used by a control system in the missile. The missile control system converts the error signals to mechanical displacements of the missile airframe control surfaces by means of motor drives. The missile responds to the positions of the aerodynamic control surfaces to follow the prescribed flight path, which is intended to produce a collision with the target.   Monitoring of the target is continuous so that changes in the missile course can be corrected up to the point of impact.   A block diagram depicting the functions of this command guidance system is shown in Fig. 5-4.   Within each of the blocks shown there are many individual functions performed.   Some of the components of the missile control system are shown within the block representing the missile.

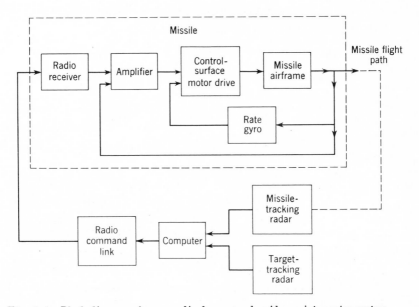

Fig. 5-4   *Block diagram of a generalized command guidance interceptor system.*

5-3   *Summation Points in Block Diagrams*

Control systems require the arithmetic manipulation of addition or subtraction. In a block diagram a circle is used as a junction point, as shown in the examples of Fig. 5-5. The quantities being added or subtracted must be expressed in the same units. Voltages can be added or subtracted, but a voltage and a force cannot be combined.

The summing point may include the addition of more than two variables. If necessary, this junction point can be replaced for purposes of analysis by several junction points. An example is shown in Fig. 5-6.

Feedback systems involve a comparison between the reference input $r(t)$ and the controlled variable $c(t)$. In systems with direct feedback these two quantities are compared and the difference between them is used to actuate the system. This difference is called the *actuating signal* $e(t)$. Various devices are used to measure this quantity. Reference to the literature and textbooks in the automatic control field will reveal many kinds of error-measuring devices. For example, when synchros or potentiometers measure the difference between two mechanical positions, the difference appears explicitly as a voltage and is used to actuate the rest of the system. In other devices the error does not appear explicitly. As an example, the reference input and controlled variable of a mechanical system may be connected to a control valve and its housing, respectively. A difference between the two causes the valve to open, and an actuating fluid flows to the main actuating piston. A block diagram showing direct feedback is given in Fig. 5-7.

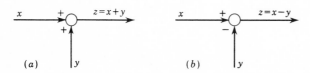

Fig. 5-5   *Representation of summation points.*

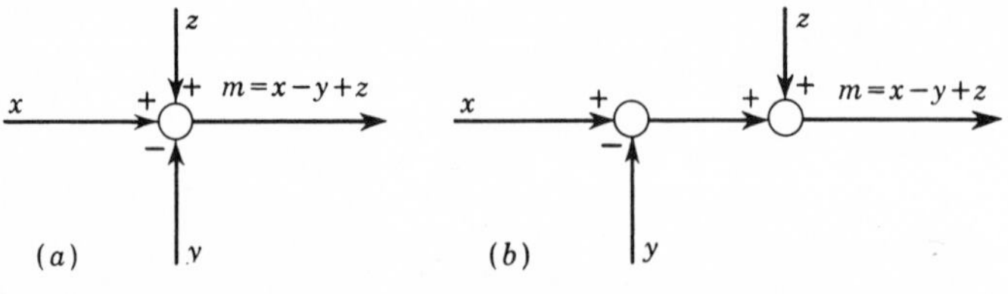

Fig. 5-6   *Equivalent summation representation.*

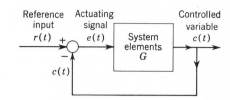

Fig. 5-7 *Block diagram of a feedback control system.*

The block diagram does not indicate the type of device used for comparison of the input and output of the system but does indicate that a comparison is made.

5-4 Definition of the Transfer Function

The RC circuit shown in Fig. 5-8 is used as a simple example in deriving a transfer function.   A voltage e_1 is applied at the input terminals and a voltage e_2 appears at the output terminals.   The nodal equation for this circuit is

$$\left(CD + \frac{1}{R}\right) e_2 - (CD)e_1 = 0 \tag{5-1}$$

The ratio of output to input voltage is the transfer function and is represented by the symbol G.   In terms of the derivative operator D, the transfer function for the circuit in operational form is

$$G(D) = \frac{e_2}{e_1} = \frac{RCD}{1 + RCD} \tag{5-2}$$

The transfer function is frequently expressed in terms of the Laplace-transform variable s.   Taking the Laplace transform of Eq. (5-1) results in

$$E_2(s)sC - Ce_2(0) + E_2(s)\frac{1}{R} - E_1(s)sC + Ce_1(0) = 0 \tag{5-3}$$

By rearranging terms, this equation becomes

$$E_2(s) = \underbrace{\frac{RCs}{1 + RCs}}_{\substack{\text{Transfer} \\ \text{function}}} E_1(s) + \underbrace{\frac{RC[e_2(0) - e_1(0)]}{1 + RCs}}_{\substack{\text{Initial-condition} \\ \text{operator}}} \tag{5-4}$$

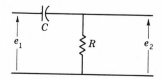

Fig. 5-8 *RC network.*

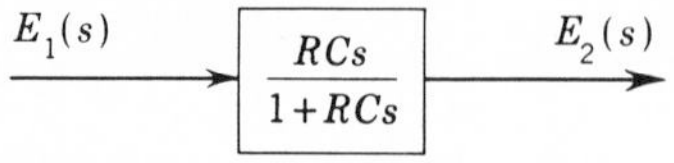

$$E_1(s) \longrightarrow \boxed{\dfrac{RCs}{1+RCs}} \xrightarrow{\;E_2(s)\;}$$

Fig. 5-9   *Block diagram of the RC network.*

If the initial conditions are zero, the ratio of the transform of the output to the transform of the input is defined as the transfer function:

$$G(s) = \frac{E_2(s)}{E_1(s)} = \frac{RCs}{1 + RCs} \tag{5-5}$$

Note that Eqs. (5-2) and (5-5) differ only in the use of the operator symbol D and the transform parameter s.

The frequency transfer function, used in the steady-state sinusoidal analysis, is defined as the ratio of the phasor $\mathbf{E}_2(j\omega)$ representing the sinusoidal output to the phasor $\mathbf{E}_1(j\omega)$ representing the sinusoidal input. The frequency transfer function is

$$\mathbf{G}(j\omega) = \frac{\mathbf{E}_2(j\omega)}{\mathbf{E}_1(j\omega)} = \frac{j\omega RC}{1 + j\omega RC} \tag{5-6}$$

Inasmuch as Eqs. (5-5) and (5-6) differ only in the respective arguments s and $j\omega$, it is obvious that one can be *formally* obtained from the other by replacement of s by $j\omega$, and vice versa. But it must be kept in mind that $E_1(s)$ and $\mathbf{E}_2(j\omega)$, and similarly $E_2(s)$ and $\mathbf{E}_2(j\omega)$, are different in nature, the former being a transform and the latter a phasor.

The transfer functions express the dynamic relationship between output and input, and it should be noted that the denominator of the transform transfer function $G(s)$ is used to form the characteristic equation

$$1 + RCs = 0 \tag{5-7}$$

The block diagram for this circuit is shown in Fig. 5-9. The arrows show the direction of the signal flow.

5-5   *Blocks in Cascade*

When a component is connected into a system so that its operation can be described independently of the rest of the system, a block can be used to represent that component. In the example of the preceding section the transfer function was derived for the case of no loading of the output. In other words, any device connected to the output terminals of the *RC* circuit of Fig. 5-8 must have a sufficiently high input impedance so that the output

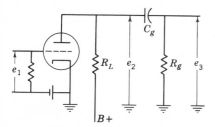

Fig. 5-10 *A-C amplifier with RC coupling.*

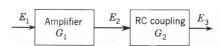

Fig. 5-11 *Block diagram of an amplifier with RC coupling*

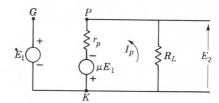

Fig. 5-12 *Equivalent circuit of a vacuum-tube amplifier.*

voltage is essentially unchanged. This is true if the output voltage is fed to the input of a vacuum-tube amplifier.

A simple example is an a-c vacuum-tube voltage amplifier with RC coupling of the output which is shown in Fig. 5-10. The load resistor R_L is selected for maximum output voltage and minimum distortion, and the coupling resistor R_g is much larger than R_L. Therefore, at midband frequencies the amplifier circuit with load resistance R_L and the R_gC_g coupling circuit are approximately independent. The block representation under these conditions is shown in Fig. 5-11. The transfer function of each block is now derived by use of the equivalent a-c circuit of the amplifier (see Fig. 5-12).

The nodal equation is

$$\left(\frac{1}{r_p} + \frac{1}{R_L}\right) E_2 + \left(\frac{1}{r_p}\right) \mu E_1 = 0 \tag{5-8}$$

The transfer function of the amplifier is

$$G_1(s) = \frac{E_2(s)}{E_1(s)} = \frac{-\mu R_L}{r_p + R_L} \tag{5-9}$$

The transfer function of the R_gC_g circuit, as previously derived, is

$$G_2(s) = \frac{E_3(s)}{E_2(s)} = \frac{R_gC_gs}{1 + R_gC_gs} \tag{5-10}$$

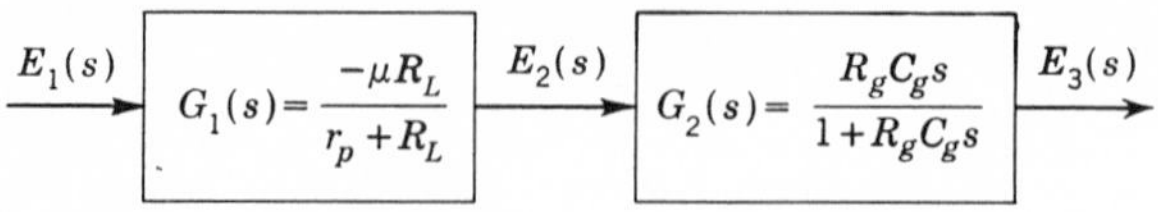

$E_1(s)$ → $G_1(s) = \dfrac{-\mu R_L}{r_p + R_L}$ → $E_2(s)$ → $G_2(s) = \dfrac{R_g C_g s}{1 + R_g C_g s}$ → $E_3(s)$

Fig. 5-13   *Block diagram of RC-coupled amplifier*

The block diagram is redrawn, with the transfer functions included, in Fig. 5-13.

It is important to reemphasize that the separation of this circuit into two independent components is based on the fact that the amplifier output voltage is independent of the loading by the $R_g C_g$ coupling circuit. When this condition does not apply, the circuit cannot be separated into two parts as is done here; an over-all transfer function must then be derived for the combined circuit.

5-6   Combination of Cascade Blocks

In the preceding section the amplifier and the coupling circuit are represented as two separate blocks, as shown in Fig. 5-13. The voltage $E_1(s)$ feeds into the transfer function $G_1(s)$, and $E_2(s)$ appears at its output. The voltage $E_2(s)$ then feeds into the transfer function $G_2(s)$, and $E_3(s)$ appears at the output. The following equations apply to this system:

$$\begin{aligned} E_2(s) &= G_1(s)E_1(s) \\ E_3(s) &= G_2(s)E_2(s) \end{aligned} \tag{5-11}$$

These equations can be combined so that only the voltages $E_1(s)$ and $E_3(s)$ appear and the internal voltage $E_2(s)$ does not appear explicitly:

$$E_3(s) = G_1(s)G_2(s)E_1(s) \tag{5-12}$$

The over-all transfer function from input to output of the blocks in cascade is

$$G(s) = \frac{E_3(s)}{E_1(s)} = G_1(s)G_2(s) \tag{5-13}$$

A simplified block diagram can now be drawn, as shown in Fig. 5-14.

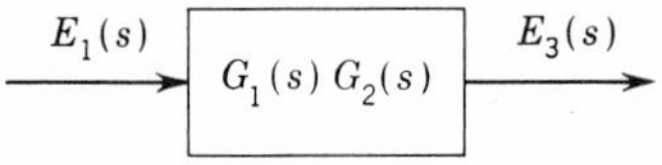

$E_1(s)$ → $G_1(s)\,G_2(s)$ → $E_3(s)$

Fig. 5-14   *Equivalent cascade block diagram.*

The over-all transfer function is the product of the transfer functions of each element in cascade. This is general and applies to any number of elements in cascade.

5-7 Determination of Control Ratio

Control ratio is defined as the ratio of the controlled variable C to the reference input R. This ratio may be expressed in operational, Laplace-transform, or phasor form; thus $C(D)$, $C(s)$, $\mathbf{C}(j\omega)$, and $R(D)$, $R(s)$, $\mathbf{R}(j\omega)$ are used in appropriate combinations.

Figure 5-15 shows a block diagram of a system with a negative feedback. The control elements and controlled system G respond to the actuating signal E to produce the controlled variable C. The feedback elements H comprise the portion of the system that responds to the controlled variable C to produce the primary feedback B. The actuating signal E is equal to the reference input R minus the primary feedback B. G and H are the transfer functions of the forward and feedback components, respectively, of the system.

The equations describing this system in terms of the transform variable are

$$C(s) = G(s)E(s)$$
$$B(s) = H(s)C(s) \qquad (5\text{-}14)$$
$$E(s) = R(s) - B(s)$$

Combining these equations produces the control ratio or system function:

$$\frac{C(s)}{R(s)} = \frac{G(s)}{1 + G(s)H(s)} \qquad (5\text{-}15)$$

The characteristic equation of the closed-loop system is obtained from the

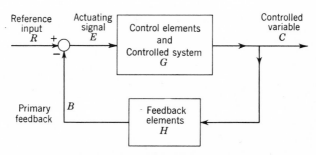

Fig. 5-15 *Block diagram of a feedback system.*

Fig. 5-16 *Control system with unity feedback.*

denominator of the control ratio:

$$1 + G(s)H(s) = 0 \qquad (5\text{-}16)$$

The stability and response of the closed-loop system are determined by the characteristic equation and are discussed more fully in later chapters.

In electronics books the gain of an amplifier with feedback, A', is given in terms of the gain without feedback, A, and a feedback factor β:

$$A' = \frac{A}{1 - \beta A} \qquad (5\text{-}17)$$

Equations (5-15) and (5-17) are similar, with $H = -\beta$ and $G = A$.

It is often useful to express the actuating signal E in terms of the input R. Solving from Eqs. (5-14) gives

$$\frac{E(s)}{R(s)} = \frac{1}{1 + G(s)H(s)}$$

For simplified systems where the feedback is unity, that is, $H(s) = 1$, the actuating signal is the actual error—reference input minus the controlled variable.   This is expressed by the equation

$$E(s) = R(s) - C(s) \qquad (5\text{-}18)$$

For this case the block diagram is shown in Fig. 5-16.

The *control ratio* with unity feedback is

$$\frac{C(s)}{R(s)} = \frac{G(s)}{1 + G(s)} \qquad (5\text{-}19)$$

The *open-loop transfer function* is defined as the ratio of the output of the feedback path $B(s)$ to the actuating signal $E(s)$ for any given feedback loop. In terms of Fig. 5-15, the open-loop transfer function is

$$\frac{B(s)}{E(s)} = G(s)H(s)$$

The *forward transfer function* is defined as the ratio of the controlled variable $C(s)$ to the actuating signal $E(s)$.   For the system shown in Fig. 5-15 the forward transfer function is

$$\frac{C(s)}{E(s)} = G(s)$$

Naturally, in the case of unity feedback where $H(s) = 1$ the open-loop and the forward transfer functions are the same.

The forward transfer function $G(s)$ may not be made up just of elements in cascade but may contain internal feedback loops. The algebra of combining these internal feedback loops is similar to that used above. Examples of controlled systems that have an internal feedback loop are shown in later sections.

Complex block diagrams can be manipulated and simplified by the rules described in Sec. 15-4, Intercoupled Multiple-loop Control Systems. Appendix J also lists a number of straightforward manipulations that can be used to revise the appearance of a block diagram.

5-8 Terminology for Feedback Control Systems

Previous sections have introduced some symbols used in feedback control systems. The terminology used in this book is that contained in a proposed standard issued by the IEEE Subcommittee on Terminology and Nomenclature of the Feedback Control Systems Committee. Their work in establishing these proposed standards was carried on in cooperation with the ASA Committee on Letter Symbols for Feedback Control Systems. Appendix C describes the standardized block diagram representing a feedback control system and gives the definitions of the variables and components used.

In using the block diagram to represent a linear feedback control system where the transfer functions of the elements are known, the letter symbol is capitalized, indicating that it is a transformed quantity; i.e., it is a function of the operator D, the complex variable s, or the frequency parameter $j\omega$. This applies for the transfer function, where G is used to represent $G(D)$, $G(s)$, or $\mathbf{G}(j\omega)$. It also applies to all variable quantities such as C, which represents $C(D)$, $C(s)$, or $\mathbf{C}(j\omega)$.

Lowercase symbols are used to represent any function in the time domain and indicate simply that there is a functional relationship between the input to any given element and the resulting output. In some cases this relationship cannot be written explicitly, but the symbols indicate that the functional relationship exists. Where the letter symbol is associated with the variable quantity represented by lines in the block diagram, the lowercase symbols represent a function of time. For example, the symbol c represents $c(t)$.

The concept of system error, y_e, is an important one. In Appendix C, it is defined as the deviation of the system output from an ideal or desired value. The ideal value establishes the desired performance of the system.

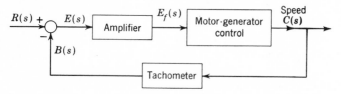

Fig. 5-17   *Speed control system.*

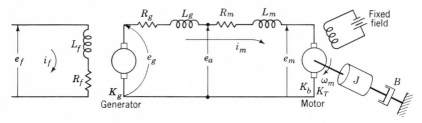

Fig. 5-18   *Motor-generator control.*

For unity feedback systems the actuating signal is an actual measure of the error and is directly proportional to the system error.

5-9   Speed Control Using Motor-Generator Control Unit

Figure 5-17 shows a simplified block diagram of a speed control system.   A voltage source is used to supply the reference input $R(s)$, and a tachometer is used to provide a voltage $B(s)$ proportional to the speed $C(s)$.   These two voltages are subtracted to produce the actuating signal $E(s)$.

The transfer function of the motor-generator control is derived in the next section.   This control system is discussed in more detail and its performance is analyzed in Sec. 6-5.

5-10   Transfer Function and Block Diagram of a D-C Motor-Generator Control

Figure 5-18 shows a d-c motor which has a constant field excitation and drives an inertia and friction load.   The armature voltage e_m for the motor is furnished by the generator, which is driven at constant speed by a prime mover. The generator voltage e_g is determined by the voltage e_f applied to the generator field. The generator acts as a power amplifier for the signal voltage e_f.

The equations for this system are

$$e_f = L_f \, Di_f + R_f i_f \qquad (5\text{-}20)$$

$$e_g = K_g i_f \qquad (5\text{-}21)$$

$$e_g - e_m = (L_g + L_m) \, Di_m + (R_g + R_m) i_m \qquad (5\text{-}22)$$

$$e_m = K_b \omega_m \qquad (5\text{-}23)$$

$$T = K_T i_m = J \, D\omega_m + B\omega_m \qquad (5\text{-}24)$$

The complete block diagram in Fig. 5-19 is based on the performance represented by these equations. Starting with the input quantity e_f, Eq. (5-20) shows that a current i_f is produced. Therefore a block is drawn with e_f as the input and i_f as the output. Equation (5-21) shows that a voltage e_g is generated as a function of the current i_f. Therefore a second block is drawn with the current i_f as the input and e_g as the output. Equation (5-22) relates the current i_m which flows through the motor to a difference of two voltages, $e_g - e_m$. To obtain this difference a summation point is introduced. The quantity e_g from the previous block enters this summation point. To obtain the quantity $e_g - e_m$ there must be added the quantity e_m entering the summation point with a minus sign. Up to this point the manner in which e_m is obtained has not yet been determined. The output of this summation point is used as the input to the next block, from which the current i_m is the output. In the same manner the block with current as the input and the generated torque as the output and the block with the torque input and the resultant motor speed as the output are drawn. There must be no "loose ends" in the complete diagram; i.e., every dependent variable must be connected through a block or blocks into the system. Therefore it is necessary to obtain e_m from the system. Equation (5-23) shows that e_m is derived from the motor speed, and a block representing this relationship is drawn with ω_m as the input and e_m as the output. When this procedure to set up the block diagram is completed, the functional relationships in the system are described. Note that the generator and the motor are no longer separately distinguishable. The input and the output of each block are not in the same units, as both electrical and mechanical quantities are included.

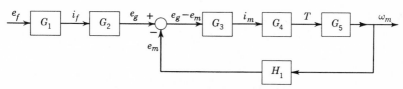

Fig. 5-19   *Block diagram of the motor-generator control.*

$E_f(s)$
K_g/R_f
$1+T_f s$
$E_g(s)$
$E_g(s)-E_m(s)$
+
−
$K_T/R_{gm}B$
$(1+T_{gm}s)(1+T_n s)$
$\Omega(s)$
$E_m(s)$
K_b

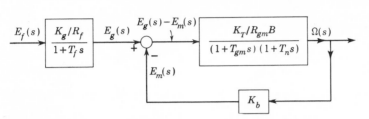

Fig. 5-20 Simplified block diagram.

$E_f(s)$
K_g/R_f
$1+T_f s$
$E_g(s)$
$K_T/R_{gm}B$
$(1+T_{gm}s)(1+T_n s)+K_T K_b/R_{gm}B$
$\Omega(s)$

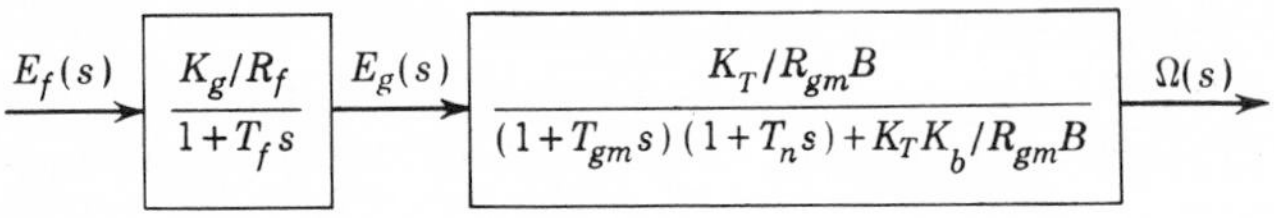

Fig. 5-21 Simplified block diagram.

The transfer functions of each block are determined in terms of the Laplace transform:

$$G_1(s) = \frac{I_f(s)}{E_f(s)} = \frac{1/R_f}{1 + (L_f/R_f)s} = \frac{1/R_f}{1 + T_f s} \tag{5-25}$$

$$G_2(s) = \frac{E_g(s)}{I_f(s)} = K_g \tag{5-26}$$

$$G_3(s) = \frac{I_m(s)}{E_g(s) - E_m(s)} = \frac{1/(R_g + R_m)}{1 + [(L_g + L_m)/(R_g + R_m)]s}$$

$$= \frac{1/R_{gm}}{1 + (L_{gm}/R_{gm})s} = \frac{1/R_{gm}}{1 + T_{gm}s} \tag{5-27}$$

$$G_4(s) = \frac{T(s)}{I_m(s)} = K_T \tag{5-28}$$

$$G_5(s) = \frac{\Omega(s)}{T(s)} = \frac{1/B}{1 + (J/B)s} = \frac{1/B}{1 + T_n s} \tag{5-29}$$

$$H_1(s) = \frac{E_m(s)}{\Omega(s)} = K_b \tag{5-30}$$

The block diagram can be simplified, as shown in Fig. 5-20, by combining the blocks in cascade. It is further simplified by evaluating an equivalent block from $E_g(s)$ to $\Omega(s)$, using the principle of Eq. (5-15). This block diagram is shown in Fig. 5-21.

The final simplification results in Fig. 5-22. The over-all transfer function $G(s)$ is

$$G(s) = \frac{K_g K_T/R_f B R_{gm}}{(1 + T_f s)[(1 + K_T K_b/BR_{gm}) + (T_{gm} + T_n)s + T_{gm}T_n s^2]} \tag{5-31}$$

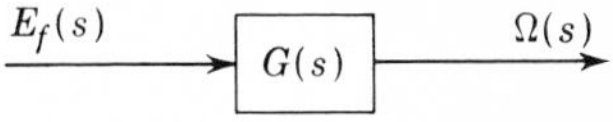

Fig. 5-22 Simplified block diagram.

$$E_f(s) \longrightarrow \boxed{G(s)} \longrightarrow \Omega(s)$$

This expression is the exact transfer function for the entire motor-generator combination. Certain approximations, if valid, may be made to simplify this expression. The first approximation is that the inductance of the generator and motor armatures is very small ($T_{gm} \approx 0$). With this approximation, the transfer function reduces to

$$\frac{\Omega(s)}{E_f(s)} = G(s) = \frac{K_x}{(1 + T_f s)(1 + T_m s)} \tag{5-32}$$

where

$$K_x = \frac{K_g K_T}{R_f(BR_{gm} + K_T K_b)} \tag{5-33}$$

$$T_f = \frac{L_f}{R_f} \tag{5-34}$$

$$T_m = \frac{JR_{gm}}{BR_{gm} + K_T K_b} \tag{5-35}$$

If the frictional effect of the load is very small, the approximation can be made that $B \approx 0$. With this additional approximation the transfer function remains of the same form as Eq. (5-32), but the constants are now given by

$$K_x = \frac{K_g}{R_f K_b} \tag{5-36}$$

$$T_f = \frac{L_f}{R_f} \tag{5-37}$$

$$T_m = \frac{JR_{gm}}{K_T K_b} \tag{5-38}$$

For simple components, as in this case, an over-all transfer function can often be derived more easily by combining the original system equations. This can be done without the intermediate steps shown in this example. However, the purpose of this example was to show the representation of dynamic components by individual blocks and the combination of blocks.

5-11 Aircraft Control System[6,7]

Consider the feedback control system used to keep an airplane on a predetermined course or heading. This is the fundamental navigation function for commercial airliners. Despite poor weather conditions and lack of visibility the airplane must maintain a specified heading and altitude in order to reach its destination safely. In addition, in spite of rough air the trip must be made as smooth and comfortable as possible for the passengers and crew. The problem is considerably complicated by the fact that the airplane moves in three-dimensional space. This makes control more difficult than the control of a ship whose motion is limited to the surface of the water.

An automatic pilot, or autopilot, is used to control the aircraft motion. Its function in the control and guidance system is described below. The input to the system is the correct flight path set by the pilot and the level position of the airplane. The ultimately controlled variable is the actual course and position of the airplane and is the indirectly controlled variable. The output of the control system, the controlled variable, is the aircraft heading.

The motion and flight path of an airplane can be described by a translation of the center of gravity and by a rotation about this point. A set of cartesian axes is affixed to the airplane. These are the principal airplane axes and are shown in Fig. 5-23. The three perpendicular axes are labeled X, Y, and Z, and the rotations of the airplane around these axes are labeled p, q, and r, respectively.

There are three primary control surfaces used to control the attitude (level position) of the airplane. These are the elevators, rudder, and ailerons and are shown in Fig. 5-24a, b, and c, respectively, together with the principal rotation produced by each. It is important to keep in mind that these motions are interdependent. For example, aileron deflection has a primary effect on the roll of the airplane, but it also affects yaw.

Equations can be written relating the response of the airplane with respect to each axis caused by the deflection of each control surface.

The directional gyroscope is used as the error-measuring device. Two gyros must be used to provide control of both heading and attitude of the

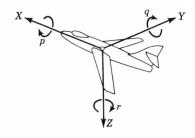

Axis	Angular velocity around axis
X (longitudinal)	p (rolling velocity)
Y (lateral)	q (pitching velocity)
Z (vertical)	r (yawing velocity)

Fig. 5-23   Principal airplane axes.

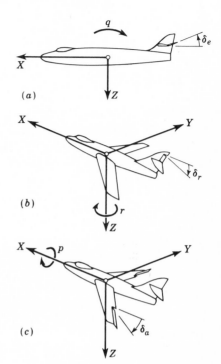

Fig. 5-24 *Airplane control surfaces. (a) Elevator deflection produces pitching velocity q. (b) Rudder deflection produces yawing velocity r. (c) Aileron deflection produces rolling velocity p.*

airplane.   The error that appears in the gyro as an angular displacement between the rotor and case is translated into a voltage by various methods, including use of potentiometers, synchros, transformers, or microsyns.   The selection of the method used depends on the personal preference of the gyro manufacturer and the sensitivity required.   The microsyn gives probably the best sensitivity but is also the most expensive.

Additional stabilization for the aircraft can be provided in the control system.   This is commonly done by means of rate feedback.   In other words, in addition to the primary feedback, which is the position of the airplane, another signal proportional to the angular rate of rotation is fed back for stability.   A rate gyro is used to supply this signal.   This additional stabilization may be absolutely necessary for some of the newer high-speed aircraft.

A typical block diagram of the aircraft control system is shown in Fig. 5-25.   This block diagram illustrates control of the airplane heading.

Another control included in the complete airplane control system keeps the airplane in level flight and operates the ailerons and elevators.

Notice that in the system of Fig. 5-25 the airplane heading is controlled and is the direction the airplane would travel in still air.   The pilot must

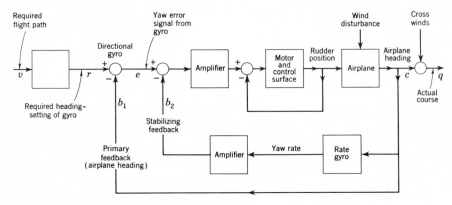

Fig. 5-25 *Airplane directional control system.*

correct this heading, depending on the cross winds present, so that the airplane will follow the desired path.

The factors affecting the design of an aircraft control system are covered in more detail in Chap. 15, which deals with complex control systems.

5-12   Hydraulic Position Control System

The block diagram for a rotational position control system is shown in Fig. 5-26. The components are hydraulic, electric, and mechanical. The hydraulic components are the power amplifiers, whereas the electric components are the sensing elements.

The input r to the system is the setting of a synchro-generator position. The actual output position c is determined by a synchro control transformer. The synchro system compares the two positions and delivers a voltage proportional to the difference. The error, $e = r - c$, in radians, does not appear explicitly in the system. However, it is common to show this error in radians followed by the synchro sensitivity. The sensitivity is expressed in volts per radian of error. The most common sensitivity used is 1 volt/deg or 57.3 volts/radian. The synchro transfer function is

$$G_1(s) = K_s \qquad (5\text{-}39)$$

The synchro-control-transformer voltage is a phase-sensitive a-c voltage. Either 60 or 400 cps is generally used. The magnitude of the voltage is proportional to the size of the error, and the phase of the voltage determines whether the error is in the clockwise or counterclockwise sense.

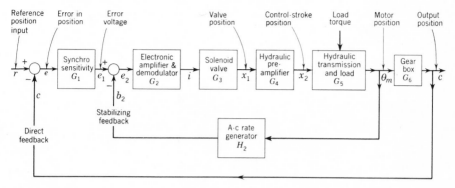

Fig. 5-26   *Block diagram of a rotational position control system.*

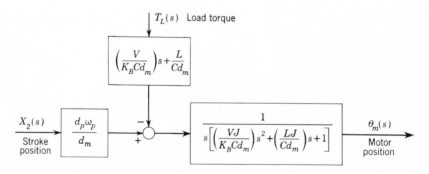

Fig. 5-27   *Block diagram of a hydraulic power actuator.*

The a-c rate generator delivers an a-c voltage proportional to the rate of change of motor position and has the same frequency as the synchro reference voltage.   The a-c generator transfer function is

$$H_2(s) = K_r s \tag{5-40}$$

The electronic amplifier and demodulator supplies a d-c current i to the coil of a solenoid.   The solenoid coil has both resistance and inductance; therefore the transfer function of the combination is of the form

$$G_2(s) = \frac{K_a}{1 + T_2 s} \tag{5-41}$$

The solenoid G_3 is polarized so that it can move the valve in either direction from the reference position.

The hydraulic preamplifier G_4 supplies enough power to move the control stroke of the hydraulic transmission.   The hydraulic transmission G_5 is the main power unit, which is described in Sec. 2-8.   By using Eq. (2-75), the block G_5 can be represented by the block diagram of Fig. 5-27.   This block diagram represents the case where the load can be assumed to be both a pure inertia and a load torque T_L. The equivalent inertia J_{eq} reflected by the gearbox to the motor shaft is used as derived in Sec. 2-5. For the case where T_L is zero the transfer function for this entire unit, as obtained directly from Eq. (2-78), is

$$G_5(s) = \frac{\theta_m(s)}{X_2(s)} = \frac{d_p\omega_p/d_m}{s[1 + (LJ_{eq}/Cd_m)s + (VJ_{eq}/K_BCd_m)s^2]} \tag{5-42}$$

This expression can also be obtained from the block diagram of Fig. 5-27, with $T_L = 0$.

The gearbox provides a ratio of output position to motor position, and its transfer function is

$$G_6(s) = K_g \tag{5-43}$$

The system represented in Fig. 5-26 is a typical closed-loop or feedback control system. The transfer functions for blocks G_3 and G_4 have been omitted. They are left to the reader to derive as problems, which are given at the end of the text.

5-13   *Transfer Functions of Compensating Networks*

Compensating networks are used in closed-loop systems to improve performance. The compensators shown are made up of electric resistors and capacitors, which are passive elements. Mechanical and hydraulic systems that have similar characteristics can be built for use as compensators. In this section the transfer function of each compensator is developed, but the use and application of these networks are treated in later chapters.

The transfer functions developed are based on no loading effect upon the output. All the transfer functions are expressed in nondimensional form.

Lag Compensator

The circuit shown in Fig. 5-28 is a typical lag or integral compensator. With a suitable choice of constants it approximates a proportional plus

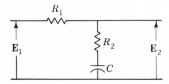

Fig. 5-28   Lag compensator.

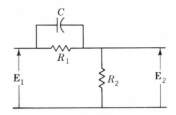

Fig. 5-29 *Lead compensator.*

integral compensator, that is, the output signal is proportional to the sum of the input signal and its integral. This characteristic is discussed in Chapter 12 on root-locus compensation. The designation *lag* applied to this network is based on the steady-state sinusoidal response. The sinusoidal response E_2 with a sinusoidal input E_1 can be determined by the method described in Sec. 3-4. These results show that the output E_2 lags the input E_1 by an angle that is a function of the frequency.

The output voltage E_2 can be expressed in terms of the impedances and the input voltage E_1. This expression will be written directly in terms of the Laplace-transform operator s. Initial conditions are considered to be zero.

$$E_2(s) = \frac{1/Cs + R_2}{1/Cs + R_1 + R_2} E_1(s) \tag{5-44}$$

By rationalizing this equation, it becomes

$$\frac{E_2(s)}{E_1(s)} = \frac{1 + R_2Cs}{1 + (R_1 + R_2)Cs} \tag{5-45}$$

Let

$$T = R_2C$$

and

$$\alpha = \frac{R_1 + R_2}{R_2}$$

As used in Eq. (5-45), the quantity α is always greater than unity. The transfer function, as expressed in terms of T and α, is

$$G(s) = \frac{E_2(s)}{E_1(s)} = \frac{1 + Ts}{1 + \alpha Ts} \tag{5-46}$$

In another useful form of this transfer function, the coefficients of s are equal to unity. In this form the transfer function is

$$G(s) = \frac{1}{\alpha} \frac{s + 1/T}{s + 1/\alpha T} \tag{5-47}$$

Lead Compensator

The circuit of Fig. 5-29 is a lead compensator. With suitable choice of parameters, it approximates a proportional plus derivative compensator, that is, the output signal is proportional to the sum of the input signal and its derivative. This characteristic is discussed in Chapter 12 on root-locus compensation. The *lead* designation of this network is based on the steady-state sinusoidal response. The sinusoidal output E_2 leads the sinusoidal input E_1. The angle of lead is a function of the frequency.

The output voltage $E_2(s)$ is

$$E_2(s) = \frac{R_2 E_1(s)}{R_2 + \dfrac{R_1(1/Cs)}{R_1 + 1/Cs}} \tag{5-48}$$

Rationalizing this equation and taking the ratio of voltages,

$$\frac{E_2(s)}{E_1(s)} = \frac{R_2}{R_1 + R_2} \frac{1 + R_1 Cs}{1 + [R_2/(R_1 + R_2)]R_1 Cs} \tag{5-49}$$

Let
$$T = R_1 C$$

and
$$\alpha = \frac{R_2}{R_1 + R_2}$$

As used in Eq. (5-48), the quantity α is always less than unity. The transfer function expressed in terms of T and α is

$$G(s) = \frac{E_2(s)}{E_1(s)} = \alpha \frac{1 + Ts}{1 + \alpha Ts} \tag{5-50}$$

In the second form of the transfer function, the coefficients of s are equal to unity:

$$G(s) = \frac{s + 1/T}{s + 1/\alpha T} \tag{5-51}$$

Lag-Lead Compensator

The circuit of Fig. 5-30 is a lag-lead compensator, which combines the characteristics of the lag and the lead compensators. This is called a *lag-lead* network because the phase of the sinusoidal response E_2, compared with the sinusoidal input E_1, varies from a lag to a lead angle as the frequency is increased from zero to infinity. The phase angle can be determined from the steady-state solution of the differential equation. For frequencies from

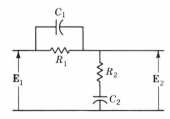

Fig. 5-30 *Lag-lead compensator.*

zero to a value ω_x the output voltage lags the input voltage.   For frequencies above ω_x the output voltage leads the sinusoidal input voltage.

The equation of the output voltage is

$$E_2(s) = \frac{R_2 + 1/C_2s}{R_2 + \dfrac{1}{C_2s} + \dfrac{R_1(1/C_1s)}{R_1 + 1/C_1s}} E_1(s)$$

Rationalizing and taking the ratio of voltages,

$$\frac{E_2(s)}{E_1(s)} = \frac{(1 + R_1C_1s)(1 + R_2C_2s)}{(1 + R_1C_1s)(1 + R_2C_2s) + R_1C_2s} \tag{5-52}$$

Let $R_1C_1 = T_1$, $R_2C_2 = T_2$, $R_1C_2 = T_{12}$.   The transfer function in terms of these time constants is

$$G(s) = \frac{E_2(s)}{E_1(s)} = \frac{1 + (T_1^{\cdot} + T_2)s + T_1T_2s^2}{1 + (T_1 + T_2 + T_{12})s + T_1T_2s^2} \tag{5-53}$$

Both numerator and denominator can be factored as follows:

$$G(s) = \frac{E_2(s)}{E_1(s)} = \frac{(1 + T_1s)(1 + T_2s)}{(1 + T_3s)(1 + T_4s)} = \frac{(1 + T_1s)(1 + T_2s)}{(1 + \alpha T_1s)[1 + (T_2/\alpha)s]} \tag{5-54}$$

The value of α can be determined by factoring the denominator of $G(s)$. The form of the transfer function with the coefficients of s equal to unity is

$$G(s) = \frac{(s + 1/T_1)(s + 1/T_2)}{(s + 1/\alpha T_1)(s + \alpha/T_2)} \tag{5-55}$$

5-14   *Signal Flow Graphs*[8-11]

The block diagram is a useful tool for simplifying the representation of a system.   The block diagrams of Figs. 5-15 and 5-17 have only one feedback loop and may be categorized as simple block diagrams.   The system represented in Fig. 5-25 has a total of three feedback loops and is no longer a simple system.   When intercoupling exists between feedback loops and when a system has more than one input and one output, the control system and block diagram are more complex.   Such systems are covered in Chap. 15. Having the block diagram simplifies the analysis of a complex system.   Such an analysis can be further simplified by using a signal flow graph, which has the appearance of a simplified block diagram.

A signal flow graph is a diagram which represents a set of simultaneous equations.   It consists of a *network* in which *nodes* are connected by directed *branches*.   The nodes represent each of the system variables.   A branch

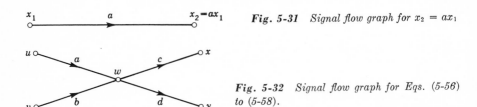

Fig. 5-31 *Signal flow graph for $x_2 = ax_1$*

Fig. 5-32 *Signal flow graph for Eqs. (5-56) to (5-58).*

connected between two nodes acts as a one-way signal multiplier: the direction of signal flow is indicated by an arrow placed on the branch, and the multiplication factor (transmittance or transfer function) is indicated by a letter placed near the arrow. Thus, in Fig. 5-31, the branch transmits the signal x_1 from left to right and multiplies it by the quantity a in the process. The quantity a is the transmittance, or transfer function. It may also be indicated by $a = t_{12}$, where the subscripts show that the signal flow is from node 1 to node 2.

Flow-graph Definitions

A *node* performs two functions:

1. *Addition* of the signals on all incoming branches.
2. *Transmission* of the total node signal (the sum of all incoming signals) to all outgoing branches. These functions are illustrated in the graph of Fig. 5-32, which represents the equations

$$w = au + bv \tag{5-56}$$
$$x = cw \tag{5-57}$$
$$y = dw \tag{5-58}$$

There are three types of nodes that are of particular interest:

Source Nodes (independent nodes). These represent independent variables and have only outgoing branches. In Fig. 5-32, nodes u and v are source nodes.

Sink Nodes (dependent nodes). These represent dependent variables and have only incoming branches. In Fig. 5-32, nodes x and y are sink nodes.

Mixed Nodes (general nodes). These have both incoming and outgoing branches. In Fig. 5-32, node w is a mixed node. A mixed node may be treated as a sink node by adding an outgoing branch of unity transmittance, as shown in Fig. 5-33.

A *path* is any connected sequence of branches whose arrows are in the same direction.

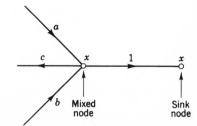

Fig. 5-33 *Mixed and sink nodes for a variable.*

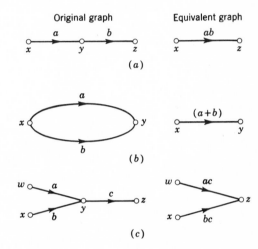

Fig. 5-34 *Flow-graph simplifications.*

A *forward path* between two nodes is one which follows the arrows of successive branches and in which a node appears only once. In Fig. 5-32 the path uwx is a forward path between the nodes u and x.

Flow-graph Algebra

The following rules are useful for simplifying a signal flow graph.

1. *Series Paths (cascade nodes).* Series paths may be combined into a single path by multiplying the transmittances as shown in Fig. 5-34a.
2. *Parallel Paths.* Parallel paths may be combined by adding the transmittances as shown in Fig. 5-34b.
3. *Node Absorption.* A node representing a variable other than a source or sink may be eliminated as shown in Fig. 5-34c.
4. *Feedback Paths.* The equations for the feedback system of Fig. 5-15 are

$$C = GE \tag{5-59}$$
$$B = HC \tag{5-60}$$
$$E = R - B \tag{5-61}$$

Note that an equation is written for each dependent variable. The corresponding signal flow graph is shown in Fig. 5-35a. The node B can be eliminated to produce Fig. 5-35b. The node E can be eliminated to produce Fig. 5-35c, which has a *self-loop* of value $-GH$. The final simplification is to eliminate the self-loop to produce the over-all transmittance shown in Fig. 5-35d.

General Flow-graph Analysis

If all the source nodes are brought to the left and all the sink nodes are brought to the right, the signal flow graph for an arbitrarily complex system may be represented by Fig. 5-36a. The effect of the *internal* nodes may be factored out by ordinary algebraic processes to yield the equivalent graph represented by Fig. 5-36b. This simplified graph is represented by

$$y_1 = T_a x_1 + T_d x_2 \qquad (5\text{-}62)$$
$$y_2 = T_b x_1 + T_e x_2 \qquad (5\text{-}63)$$
$$y_3 = T_c x_1 + T_f x_2 \qquad (5\text{-}64)$$

The T's, called over-all graph transmittances, are the over-all transmittances from a specified source node to a specified dependent node. For linear systems the principle of superposition may be used to "solve" the graph. This means that the sources can be considered one at a time. Then the output signal is equal to the sum of the contributions produced by each input.

The over-all transmittances can be found by the ordinary processes of linear algebra, that is, by the solution of the set of simultaneous equations representing the system. However, the same results can be obtained directly from the signal flow graph! The fact that they can produce answers to

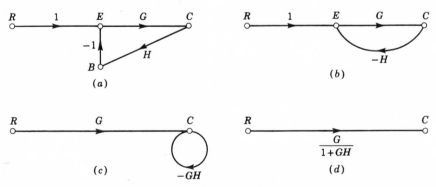

Fig. 5-35 *Successive reduction of the flow graph for the feedback system of Fig. 5-15.*

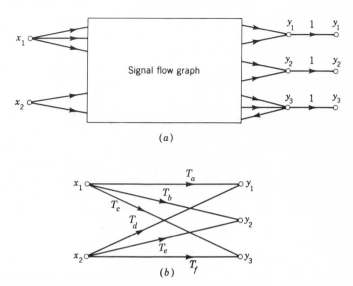

(a)

(b)

Fig. 5-36 *Equivalent signal flow graphs.*

large sets of linear equations *by inspection* gives the signal flow graphs their power and utility.

The Mason Rule

The over-all transmittance can be obtained from the formula developed by S. J. Mason. The formula and definitions are followed by an example to show its application. The over-all transmittance is given by

$$T = \frac{\Sigma T_n \, \Delta_n}{\Delta} \tag{5-65}$$

where

1. T_n is the transmittance of each forward path between a source and a sink node.
2. Δ is the graph determinant and is found from

$$\Delta = 1 - \Sigma L_1 + \Sigma L_2 - \Sigma L_3 + \cdots \tag{5-66}$$

in which

a. L_1 is the transmittance of each closed path and ΣL_1 is the sum of the transmittances of all closed paths in the graph.

b. L_2 is the product of the transmittances of two nontouching loops. Loops are nontouching if they do not have any common nodes. ΣL_2 is the sum of the product of transmittances of all possible combinations of nontouching loops taken two at a time.

    c. L_3 is the product of the transmittances of three nontouching
         loops. ΣL_3 is the sum of the product of transmittances of all
         possible combinations of nontouching loops taken three at a
         time.
3. Δ_n is the cofactor of T_n. It is the determinant of the remaining
    subgraph when the path which produces T_n is removed.

Example of Mason's Rule

Figure 5-37 shows a block diagram and its signal flow graph. Note that not
all the variables are shown. Since $E_1 = M_1 - B_1 = G_1 E - H_1 M_2$, it is not
necessary to show M_1 and B_1 explicitly. This is a fairly complex system, so
it is expected that the resulting equation is also complex. However, the
application of Mason's rule produces the resulting over-all transmittance in a
systematic manner. This system has four loops whose transmittances are
$-G_2 H_1$, $-G_5 H_2$, $-G_1 G_2 G_3 G_5$, and $-G_1 G_2 G_4 G_5$. Therefore

$$\Sigma L_1 = -G_2 H_1 - G_5 H_2 - G_1 G_2 G_3 G_5 - G_1 G_2 G_4 G_5 \tag{5-67}$$

Only two loops are nontouching; therefore

$$\Sigma L_2 = (-G_2 H_1)(-G_5 H_2) \tag{5-68}$$

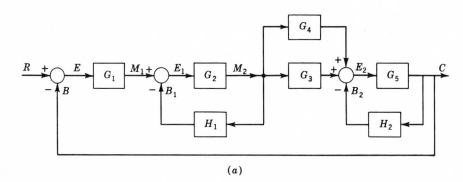

(a)

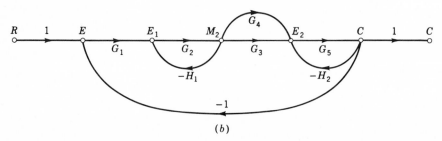

(b)

Fig. 5-37 *A block diagram and its signal flow graph.*

Although there are four loops, there is no set of three loops which are non-touching; therefore

$$\Sigma L_3 = 0 \qquad (5\text{-}69)$$

The system determinant can therefore be obtained from Eq. (5-66).

There are two forward paths between R and C. The corresponding forward transmittances are $G_1G_2G_3G_5$ and $G_1G_2G_4G_5$. If either path, with its corresponding nodes, is removed from the graph, the remaining subgraphs have no loops. The cofactors Δ_n are therefore both equal to unity. The complete over-all transmittance from R to C, obtained from Eq. (5-65), is

$$T = \frac{G_1G_2G_3G_5 + G_1G_2G_4G_5}{1 + G_2H_1 + G_5H_2 + G_1G_2G_3G_5 + G_1G_2G_4G_5 + G_2G_5H_1H_2} \qquad (5\text{-}70)$$

The over-all transmittance has been obtained by inspection from the signal flow graph. This is much simpler than solving the five simultaneous equations which represent this system. Further applications are shown in Chap. 15.

5-15   Conclusions

The purpose of this chapter is to develop the block-diagram representation for components and for complete systems. The transfer functions of individual blocks have been derived, and the methods of combining blocks have been shown. The block diagram represents functions performed in the system, and the transfer functions relate the dynamic response of each block. This is independent of the physical form that each component takes. The analysis of complex systems can be further simplified by using signal flow graphs. This way of representing systems aids in analyzing the stability and performance of the system, which are the ultimate concern of a designer.

Bibliography

1. Brown, G. S., and D. P. Campbell: "Principles of Servomechanisms," chap. 5, John Wiley & Sons, Inc., New York, 1948.
2. Stout, T. M.: A Block Diagram Approach to Network Analysis, *Trans. AIEE*, vol. 71, pp. 255–260, 1952.
3. "Methods of Analysis and Synthesis of Piloted Aircraft Flight Control Systems," BuAer Report AE-61-4, vol. 1, chap. 2, U.S. Navy, Bureau of Aeronautics, 1952.
4. Draper, C. S., W. Wrigley, and L. R. Grohe: "The Floating Integrating Gyro and Its Application to Geometrical Stabilizer Problems on Moving Bases,"

pp. 1–9, presented at the 23rd Annual Meeting of the Institute of Aeronautical Sciences, January, 1955. (Published as Sherman M. Fairchild Fund Paper FF-13.)

5. Locke, A. S.: "Guidance," chap. 16, D. Van Nostrand Company, Inc., Princeton, N.J., 1955.

6. Bollay, W.: Aerodynamic Stability and Automatic Control, *J. Aeron. Sci.*, vol. 18, no. 9, pp. 569–624, September, 1951.

7. Monroe, W. R.: Improving the Dynamic Response of Airplanes by Means of Electric Equipment, *Trans. AIEE*, pt. II, vol. 71, pp. 441–449, January, 1953.

8. Mason, S. J.: Feedback Theory: Further Properties of Signal Flow Graphs, *Proc. IRE*, vol. 44, no. 7, pp. 920–926, July, 1956.

9. Brown, F. M.: Signal-flow Graphs, unpublished notes, Air Force Institute of Technology, November, 1961.

10. Chow, Y., and E. Cassignol: "Linear Signal-flow Graphs and Applications," John Wiley & Sons, Inc., New York, 1962.

11. Robichaud, L. P. A., M. Boisvert, and J. Robert: "Signal Flow Graphs and Applications," Prentice-Hall, Inc., Englewood Cliffs, N.J., 1962.

6

Basic
Servo
Characteristics

6-1   Introduction

In Chap. 5, the transfer functions of both open and closed loops are developed for various types of controlled quantities.   One can readily determine that these transfer functions have certain basic characteristics that permit a steady-state analysis of the feedback-controlled system.   Two factors of prime importance in feedback-controlled systems are *the existence of a steady-state error* and *the magnitude of this steady-state error*.   For unity feedback systems both of these factors are obtainable from the open-loop transfer function and yield figures of merit and a ready means for classifying systems.[1-3]

6-2   *Mathematical and Physical Forms*

In the various systems that are devised the controlled variable C shown in Fig. 6-1 may have the physical form of position or speed of a rotor shaft, temperature, rate of change of temperature, voltage, rate of flow, pressure, etc. To generalize the study of feedback systems the controlled variable has been labeled C. Once the blocks in the diagram are related to transfer functions, it is immaterial to the analysis of the system what the physical form of the controlled variable may be.

Generally, the important quantities are the controlled quantity c, its rate of change Dc, and its second derivative D^2c, that is, the first several derivatives of c, including the zeroth derivative. For any specific control system each of these "mathematical" functions has a definite "physical" meaning. For example, if the controlled variable c is position, Dc is velocity and D^2c is acceleration. As a second example, if the controlled variable c is velocity, then Dc is acceleration and D^2c is the rate of change of acceleration.

As was pointed out, in the analysis of a given system the mathematical manipulations of c (or r) are of importance. For example, the input signal to a system may have the irregular form shown in Fig. 6-2 which cannot be expressed by any simple equation. This prevents a straightforward analysis of system response. One notes, though, that the signal form shown in Fig. 6-2 may be considered to be composed of three basic forms of known types of input signals, i.e., a step in the region cde, a ramp in the region $0b$, and a parabola in the region ef. This leads one to the conclusion that, if the given linear system is analyzed separately for each of these types of input signals, there is then established a good measure of performance with the irregular input. Also, use of these standard inputs provides a means for comparing the performance of different systems.

Consider that the system shown in Fig. 6-1 is a position control, position correspondence (pc, pc) system (position correspondence: output and input are compared by relative position) and that the input position signal is of the form shown in Fig. 6-2. In general, since feedback control systems are analyzed on the basis of a unit-step-input signal (see Fig. 6-3), this system can be analyzed on the basis that the unit-step-input signal $r(t)$ represents position. Since this gives only a limited idea of how the system responds

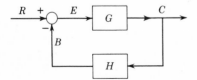

Fig. 6-1   *Simple block diagram.*

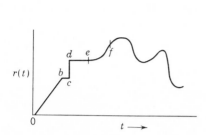

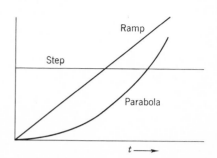

Fig. 6-2 *Input signal to a system.*

Fig. 6-3 *Graphical forms of step, ramp, and parabolic input functions.*

to the actual input signal, one can then analyze the system on the basis that the unit step signal represents a constant velocity $Dr = u(t)$. This in reality gives an input position signal of the form of a ramp (Fig. 6-3) and thus a closer idea of how the system responds to the actual input signal.

In the same manner one can consider that the unit-step-input signal represents a constant acceleration, $D^2r = u(t)$, to obtain the system's performance to a parabolic position input signal. The curves shown in Fig. 6-3 then represent acceleration, velocity, and position. In this case the mathematical and physical forms are identical.

6-3   Types of Feedback Systems

A simple closed-loop feedback system with unity feedback is shown in Fig. 6-4. The open-loop transfer function for this system is $G(s) = C(s)/E(s)$, which is determined by the components of the actual control system. Several examples were derived in the preceding chapter. Generally the transfer function has one of the following mathematical forms:

$$G(s) = \frac{K_0(1 + T_1s)(1 + T_2s) \cdots}{(1 + T_as)(1 + T_bs) \cdots} \tag{6-1}$$

$$G(s) = \frac{K_1(1 + T_1s)(1 + T_2s) \cdots}{s(1 + T_as)(1 + T_bs) \cdots} \tag{6-2}$$

$$G(s) = \frac{K_2(1 + T_1s)(1 + T_2s) \cdots}{s^2(1 + T_as)(1 + T_bs) \cdots} \tag{6-3}$$

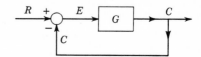

Fig. 6-4 *Unity feedback control system.*

The above equations are expressed in a more generalized manner by defining the standard form of the transfer function as

$$G(s) = \frac{K_n(1 + a_1s + a_2s^2 + \cdots + a_ws^w)}{s^n(1 + b_1s + b_2s^2 + \cdots + b_us^u)} = K_nG'(s) \qquad (6\text{-}4)$$

where $a_1, a_2, \ldots, b_1, b_2, \ldots$ = constant coefficients

   K_n = over-all gain of transfer function $G(s)$

   $n = 0, 1, 2, \ldots$ and denotes the *type* of the transfer function

The degree of the denominator is $v = n + u$. For a unity feedback system, E and C have the same units. Therefore, K_0 is nondimensional, K_1 has the units seconds^{-1}, and K_2 has the units seconds^{-2}.

Once a physical system has been expressed mathematically, the analysis that must follow is independent of the nature of the physical system. It is immaterial whether the system is electric, mechanical, hydraulic, thermal, or a combination of these. The most common types of feedback control systems fall into the three categories expressed by Eqs. (6-1) to (6-3). It is important to analyze each category thoroughly and to relate it as closely as possible to its transient and steady-state solution. In order to analyze best each control system, a "type" designation is introduced. The designation is based upon the order of the exponent n of s in Eq. (6-4). Thus, when $n = 0$ the system represented by this equation is called a Type 0 system; when $n = 1$ it is called a Type 1 system; when $n = 2$ it is called a Type 2 system; etc.

The various types exhibit the following properties:

Type 0.   A constant actuating signal results in a constant value for the controlled variable.

Type 1.   A constant actuating signal results in a constant rate of change (constant velocity) of the controlled variable.

Type 2.   A constant actuating signal results in a constant second derivative (constant acceleration) of the controlled variable.

Type 3.   A constant actuating signal results in a constant rate of change of acceleration of the controlled variable.

These classifications lend themselves to definition in terms of the differential equations of the system and to identification in terms of the forward transfer function.

As is shown later, the system classification by type can be readily determined from graphical plots in the $\mathbf{G}(j\omega)$, log $|\mathbf{G}(j\omega)|$ versus ω, and s planes. These plots also reveal many other important qualities of the system. For all classifications the degree of the denominator of the $G(s)H(s)$ function usually is either equal to or greater than the degree of the numerator. This is so because of the physical nature of feedback control systems. That is, in every physical system there are energy-storage and dissipative elements such that usually there can be no instantaneous transfer of energy from the input to the output. However, exceptions do occur.

6-4   *Analysis of System Types*

The properties presented in the preceding section are now examined in detail for each type of $G(s)$ appearing in stable systems. First, remember that

Final-value Theorem: $\quad \lim_{t \to \infty} f(t) = \lim_{s \to 0} sF(s)$ $\qquad$ (6-5)

Differentiation Theorem:

$\qquad \mathcal{L}[D^n c(t)] = s^n C(s) \qquad$ when all initial conditions are zero $\qquad$ (6-6)

It should also be remembered that the steady-state output of a stable closed-loop system that has *unity feedback* has the same form as the input when the input is a power series. Therefore, if the input is a ramp function, the steady-state output must also be a ramp function, and so forth.

From the preceding section it is seen that the forward transfer function defines the system type. In deriving the transfer function it is generally in the factored form

$$G(s) = \frac{C(s)}{E(s)} = \frac{K_n(1 + T_1 s)(1 + T_2 s) \cdots}{s^n(1 + T_a s)(1 + T_b s)(1 + T_c s) \cdots} \qquad (6\text{-}7)$$

Rearranging this equation yields

$$E(s) = \frac{(1 + T_a s)(1 + T_b s)(1 + T_c s) \cdots}{K_n(1 + T_1 s)(1 + T_2 s) \cdots} s^n C(s) \qquad (6\text{-}8)$$

Thus

$$e(t)_{ss} = \lim_{s \to 0} [sE(s)] = \lim_{s \to 0} \left[\frac{s(1 + T_a s)(1 + T_b s)(1 + T_c s) \cdots}{K_n(1 + T_1 s)(1 + T_2 s) \cdots} s^n C(s) \right]$$

$$= \lim_{s \to 0} \frac{s[s^n C(s)]}{K_n} \qquad (6\text{-}9)$$

But applying the final-value theorem to Eq. (6-6) gives

$$\lim_{s \to 0} s[s^n C(s)] = D^n c(t)_{ss} \qquad (6\text{-}10)$$

Therefore, Eq. (6-9) may be written as

$$e(t)_{ss} = \frac{D^n c(t)_{ss}}{K_n} \qquad (6\text{-}11)$$

or $\qquad\qquad\qquad K_n e(t)_{ss} = D^n c(t)_{ss} \qquad (6\text{-}12)$

This equation is most useful for the case where $D^n c(t)_{ss} = $ constant. Then $e(t)_{ss}$ must also equal a constant, that is, $e(t)_{ss} = E_0$. For this case Eq. (6-12) may be expressed as

$$K_n E_0 = D^n c(t)_{ss} = \text{constant} = C_n \qquad (6\text{-}13)$$

Note that $C(s)$ is given by

$$C(s) = \frac{G(s)}{1 + G(s)} R(s)$$

$$= \frac{K_n[(1 + T_1 s)(1 + T_2 s) \cdots]}{s^n(1 + T_a s)(1 + T_b s) \cdots + K_n(1 + T_1 s)(1 + T_2 s) \cdots} R(s) \quad (6\text{-}14)$$

The expression for $E(s)$ in terms of the input $R(s)$ is obtained as follows:

$$E(s) = \frac{C(s)}{G(s)} = \frac{1}{G(s)} \frac{G(s)R(s)}{1 + G(s)H(s)} = \frac{R(s)}{1 + G(s)H(s)} \qquad (6\text{-}15)$$

For the case of unity feedback $[H(s) = 1]$ and with $G(s)$ given by Eq. (6-7), the expression for $E(s)$ is

$$E(s) = \frac{s^n(1 + T_a s)(1 + T_b s) \cdots R(s)}{s^n(1 + T_a s)(1 + T_b s) \cdots + K_n(1 + T_1 s)(1 + T_2 s) \cdots} \qquad (6\text{-}16)$$

Applying the final-value theorem to Eq. (6-16) yields

$$e(t)_{ss} = \lim_{s \to 0} s \left[\frac{s^n(1 + T_a s)(1 + T_b s) \cdots R(s)}{s^n(1 + T_a s)(1 + T_b s) \cdots + K_n(1 + T_1 s)(1 + T_2 s) \cdots} \right]$$
$$(6\text{-}17)$$

Equation (6-17) is now analyzed for various system types and for step, ramp, and parabolic inputs.

Case 1. $n = 0$ (Type 0 system)

1. For a step input $r(t) = R_0 u(t)$, $R(s) = R_0/s$.   From Eq. (6-17)

$$e(t)_{ss} = \frac{R_0}{1 + K_0} = \text{constant} = E_0 \neq 0 \qquad (6\text{-}18)$$

From Eq. (6-18) it is seen that a Type 0 system with a constant input produces a constant value of the output with a constant actuating signal. The same results can be obtained by applying Eq. (6-13).   This means that in a Type 0 system a fixed error E_0 is required to produce a desired constant output C_0, that is, $K_0 E_0 = c(t)_{ss} = \text{constant} = C_0$.   For   steady-state conditions

$$e(t)_{ss} = r(t)_{ss} - c(t)_{ss} = R_0 - C_0 = E_0 \qquad (6\text{-}19)$$

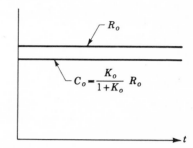

Fig. 6-5 *Steady-state response of a Type 0 system with a step input.*

Differentiating the above equation yields $Dr(t)_{ss} = Dc(t)_{ss} = 0$. Figure 6-5 illustrates the results obtained above.

2. For a ramp input $r(t) = R_1 tu(t)$, $R(s) = R_1/s^2$. From Eq. (6-17)

$$e(t)_{ss} = \infty \tag{6-20}$$

Also, by the use of the Heaviside partial-fraction expansion, the particular solution of $e(t)$ can be shown, by use of Eq. (6-16), to contain the term $[R_1/(1 + K_0)]t$. Therefore the conclusion is that a Type 0 system with a ramp-function input produces a ramp output with a smaller slope, thus there is an error which increases with time and approaches a value of infinity. This means that a Type 0 system cannot follow a ramp input.

3. In the same manner, it can be shown that a Type 0 system cannot follow a parabolic input $r(t) = R_2 t^2 u(t)$, that is, $e(t)_{ss} = r(t)_{ss} - c(t)_{ss}$ approaches a value of infinity.

Case 2. $n = 1$ **(Type 1 system)**

1. For a step input $R(s) = R_0/s$. From Eq. (6-17)

$$e(t)_{ss} = 0 \tag{6-21}$$

Therefore it can be stated that a Type 1 system with a constant input produces a steady-state constant output of value identical with the input. This means that for a Type 1 system there is zero steady-state error between the output and input for a step input, i.e.,

$$e(t)_{ss} = r(t)_{ss} - c(t)_{ss} = 0 \tag{6-22}$$

The above analysis is in agreement with Eq. (6-13).   That is, for a constant input $r(t) = R_0 u(t)$, the steady-state output must be a constant $c(t) = C_0 u(t)$ so that

$$Dc(t)_{ss} = 0 = K_1 E_0 \tag{6-23}$$

or

$$E_0 = 0 \tag{6-24}$$

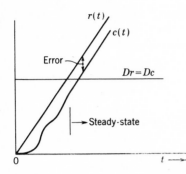

Fig. 6-6 *Steady-state response of a Type 1 system with a ramp input.*

2. For a ramp input $R(s) = R_1/s^2$.   From Eq. (6-17)

$$e(t)_{ss} = \frac{R_1}{K_1} = \text{constant} = E_0 \neq 0 \tag{6-25}$$

From Eq. (6-25) it is seen that a Type 1 system with a ramp input produces a ramp output with a constant actuating signal.   That is, in a Type 1 system a fixed error E_0 is required to produce a ramp output.   This result can also be obtained from Eq. (6-13), that is, $K_1 E_0 = Dc(t)_{ss} = \text{constant} = C_1$.   For steady-state conditions

$$e(t)_{ss} = r(t)_{ss} - c(t)_{ss} = E_0 \tag{6-26}$$

For

$$r(t) = R_1 t u(t) \tag{6-27}$$

the output has the form

$$c(t)_{ss} = C_0 + C_1 t \tag{6-28}$$

Substituting Eqs. (6-27) and (6-28) into Eq. (6-26) yields

$$E_0 = R_1 t - C_0 - C_1 t \tag{6-29}$$

Differentiating the above equation results in

$$0 = R_1 - C_1 \tag{6-30}$$

or

$$R_1 = C_1 \tag{6-31}$$

This result signifies that the slopes of the ramp input and the ramp output are equal.   This, of course, is a necessary condition if the difference between input and output is a constant.   Figure 6-6 illustrates the results obtained above.

3. For a parabolic input $r(t) = R_2 t^2 u(t)$, $R(s) = 2R_2/s^3$.   From Eq. (6-17), $e(t)_{ss} = \infty$.   Also, by use of the partial-fraction expansion, the particular solution of $e(t)$ obtained from Eq. (6-16) can be shown to contain the term

$$e(t) = 2\frac{R_2 t}{K_1}$$

Therefore, it can be stated that a Type 1 system with a parabolic input produces a parabolic output, but with an error which increases with time and approaches a value of infinity. This means that a Type 1 system cannot follow a parabolic input.

Case 3. $n = 2$ (Type 2 system)

1. For a step input $R(s) = R_0/s$.   From Eq. (6-17)

$$e(t)_{ss} = 0 \qquad\qquad (6\text{-}32)$$

Therefore, it can be stated that a Type 2 system with a constant input produces a constant output of value identical with the input. This means that for a Type 2 system, zero error exists in the steady state between the output and input for a step input, i.e.,

$$e(t)_{ss} = r(t)_{ss} - c(t)_{ss} = 0 \qquad\qquad (6\text{-}33)$$

The above analysis is in agreement with Eq. (6-13).   That is, for a constant input the output is also a constant and

$$D^2 c(t)_{ss} = 0 = K_2 E_0 \qquad\qquad (6\text{-}34)$$
or
$$E_0 = 0 \qquad\qquad (6\text{-}35)$$

2. For a ramp input $R(s) = R_1/s^2$.   From Eq. (6-17)

$$e(t)_{ss} = 0 \qquad\qquad (6\text{-}36)$$

Therefore, it can be stated that a Type 2 system with a ramp input produces a ramp output of identical slope with the input after steady-state conditions have been achieved.   This means that for a Type 2 system, zero error exists between the steady-state output and input for a ramp input, i.e.,

$$e(t)_{ss} = r(t)_{ss} - c(t)_{ss} = 0 \qquad\qquad (6\text{-}37)$$

Substituting Eqs. (6-27) and (6-28) into Eq. (6-37) yields

$$R_1 t - C_0 - C_1 t = 0 \qquad\qquad (6\text{-}38)$$

Differentiating the above gives

$$R_1 = C_1 \qquad\qquad (6\text{-}39)$$

Substituting this result into Eq. (6-38) yields

$$C_0 = 0 \qquad\qquad (6\text{-}40)$$
and
$$R_1 t = C_1 t \qquad\qquad (6\text{-}41)$$

Thus, the steady-state slopes are identical.   The above analysis is in agreement with Eq. (6-13); that is, for a ramp input the output has the form

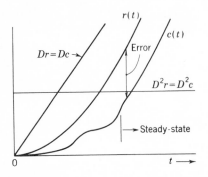

Fig. 6-7 *Steady-state response of a Type 2 system with a parabolic input.*

$C_1t + C_0$ and

$$D^2c(t)_{ss} = 0 = K_2E_0 \qquad (6\text{-}42)$$

or $\qquad\qquad\qquad E_0 = 0 \qquad\qquad\qquad (6\text{-}43)$

3. For a parabolic input $R(s) = 2R_2/s^3$. From Eq. (6-17)

$$e(t)_{ss} = \frac{2R_2}{K_2} = \text{constant} = E_0 \neq 0 \qquad (6\text{-}44)$$

From Eq. (6-44) it is seen that a Type 2 system with a parabolic input produces a parabolic output with a constant actuating signal.   That is, in a Type 2 system a fixed error E_0 is required to produce a parabolic output. This result is further confirmed by applying Eq. (6-13), that is,

$$K_2E_0 = D^2c(t)_{ss} = \text{constant} = C_2$$

Thus, for steady-state conditions

$$e(t)_{ss} = r(t)_{ss} - c(t)_{ss} = E_0 \qquad (6\text{-}45)$$

For a parabolic input given by

$$r(t) = R_2t^2 \qquad (6\text{-}46)$$

the particular solution of the output must be given by

$$c(t)_{ss} = C_2t^2 + C_1t + C_0 \qquad (6\text{-}47)$$

Substituting Eqs. (6-46) and (6-47) into Eq. (6-45) yields

$$E_0 = R_2t^2 - C_2t^2 - C_1t - C_0 \qquad (6\text{-}48)$$

Differentiating the above equation twice results in

$$2R_2t - 2C_2t - C_1 = 0 \qquad (6\text{-}49)$$
$$2R_2 - 2C_2 = 0 \qquad (6\text{-}50)$$

Therefore, $C_1 = 0$, $R_2 = C_2$, and $E_0 = -C_0$. Thus the input and output curves have the same shape but are displaced by a constant error.   Figure 6-7 illustrates the results obtained above.

Table 6-1. *Steady-state response characteristics for stable unity feedback systems*

System type n	$r(t)_{ss}$	$c(t)_{ss}$	$e(t)_{ss}$	$e(\infty)$	Derivatives
0	$R_0 u(t)$	$\dfrac{K_0}{1+K_0} R_0$	$\dfrac{R_0}{1+K_0}$	$\dfrac{R_0}{1+K_0}$	$Dr = Dc = 0$
	$R_1 t u(t)$	$\dfrac{K_0 R_1}{1+K_0} t + C_0$	$\dfrac{R_1}{1+K_0} t - C_0$	∞	$Dr \neq Dc$
	$R_2 t^2 u(t)$	$\dfrac{K_0 R_2}{1+K_0} t^2 + C_1 t + C_0$	$\dfrac{R_2}{1+K_0} t^2 - C_1 t - C_0$	∞	$Dr \neq Dc$
1	$R_0 u(t)$	R_0	0	0	$Dr = Dc = 0$
	$R_1 t u(t)$	$R_1 t - \dfrac{R_1}{K_1}$	$\dfrac{R_1}{K_1}$	$\dfrac{R_1}{K_1}$	$Dr = Dc = R_1$
	$R_2 t^2 u(t)$	$R_2 t^2 + C_1 t + C_0$	$- C_1 t - C_0$	∞	$Dr \neq Dc$
2	$R_0 u(t)$	R_0	0	0	$Dr = Dc = 0$
	$R_1 t u(t)$	$R_1 t$	0	0	$Dr = Dc = R_1$
	$R_2 t^2 u(t)$	$R_2 t^2 - \dfrac{2R_2}{K_2}$	$\dfrac{2R_2}{K_2}$	$\dfrac{2R_2}{K_2}$	$D^2r = D^2c = 2R_2$ $Dr = Dc = 2R_2 t$

The results determined in this section verify the properties stated in the previous section for the system types. The steady-state response characteristics for Type 0, 1, and 2 unity feedback systems are given in Table 6-1 and apply only to stable systems.

6-5 Examples of Types of Systems

The various types of feedback control systems illustrated in this section are in common use in industrial and military applications.

Type 0 Feedback Control System

A Type 0 system is one in which a constant actuating signal maintains a constant value of the output. In Fig. 6-8 is shown a system for speed control of a sheet-metal processing unit.[4] Figure 6-8a shows the sheet metal being moved along the production line by means of rollers. It is essential that the roller speeds ω_1, ω_2, ω_3, and ω_4 be equal to each other at all times so that a prescribed uniform tension is maintained on the sheet metal. To accomplish this, Fig. 6-8b shows a means by which the speeds of all the rollers

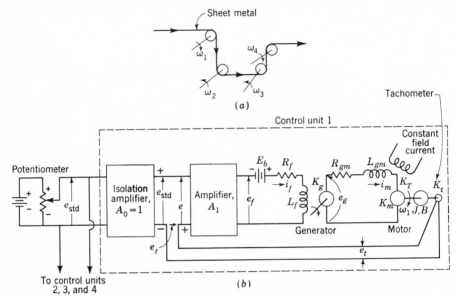

Fig. 6-8 *Speed control for sheet-metal processing unit (Type 0 control system).*

can be controlled and maintained at a standard speed ω_{std}.   When $\omega_1 = \omega_{std}$, the tachometer has a given output voltage $e_t = e_{std}$.   Therefore, the potentiometer is set for this value of voltage.   If $\omega_1 = \omega_{std}$, e_t is equal to e_{std} and the error signal e is zero.   The amplifier is designed so that its normal or quiescent output voltage supplies the nominal value of generator field current.   This is shown in Fig. 6-8b by a separate d-c source E_b.   Thus the motor, with constant field excitation, runs at the desired speed with $e = 0$. This is the reference condition.   The control units 2, 3, and 4 are identical to unit 1, so Fig. 6-8 applies to any of the drive units.

The motor-generator control unit is described in Sec. 5-10.   If the inductance L_{gm} of the generator and motor armatures is neglected, the transfer function is given by Eq. (5-32), and the constants in this transfer function are given by Eqs. (5-33) to (5-35).

A block diagram of the system is shown in Fig. 6-9.   *It should be realized that the values of $C(s)$ and $R(s)$ represent changes from the reference values.* The forward transfer function, as seen from the block diagram, is

$$G(s) = \frac{A_1 K_x}{(1 + T_f s)(1 + T_m s)} = \frac{K_0}{(1 + T_f s)(1 + T_m s)} \qquad (6\text{-}51)$$

where K_0 has the units of radians per second per volt.   The tachometer constant K_t has the units of volts per radian per second.   Since this a non-unity feedback system, K_0 is not dimensionless.

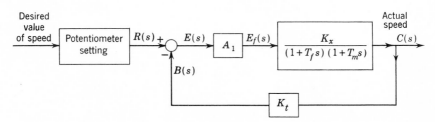

Fig. 6-9 *Block-diagram representation of Fig. 6-8.*

Applying the definitions of Sec. 6-3 to Eq. (6-51), it is seen that this is a Type 0 control system. Thus from Sec. 6-4 a constant actuating voltage E_0 yields a constant speed, or vice versa, i.e., a constant-speed output requires a constant actuating voltage for $H(s) = 1$ or a constant. Therefore, from Eq. (6-13),

$$K_0 E_0 = c(t)_{ss} = \text{constant}$$

The control ratio is

$$\frac{C(s)}{R(s)} = \frac{G(s)}{1 + G(s)H(s)} = \frac{K_0}{(1 + T_f s)(1 + T_m s) + K_0 K_t} \qquad (6\text{-}52)$$

$$C(s) = \frac{K_0}{(1 + T_f s)(1 + T_m s) + K_0 K_t} R(s) \qquad (6\text{-}53)$$

For a constant input $r(t) = R_0 u(t)$ the steady-state value of $c(t)_{ss}$ is found by applying the final-value theorem:

$$c(t)_{ss} = \lim_{s \to 0} sC(s) = \lim_{s \to 0} \frac{K_0 R_0}{(1 + T_f s)(1 + T_m s) + K_0 K_t}$$

$$= \frac{K_0}{1 + K_0 K_t} R_0 \qquad (6\text{-}54)$$

When the tachometer constant $K_t = 1$, the steady-state value $c(t)_{ss}$ is

$$c(t)_{ss} = \frac{K_0}{1 + K_0} R_0 \qquad (6\text{-}55)$$

Thus

$$E_0 = r(t)_{ss} - c(t)_{ss} = R_0 - \frac{K_0}{1 + K_0} R_0 = \frac{R_0}{1 + K_0} \qquad (6\text{-}56)$$

which agrees with the value given in Table 6-1.

The transfer function for the feedback loop is

$$H(s) = \frac{B(s)}{C(s)} = K_t \qquad (6\text{-}57)$$

By combining Eqs. (6-51) and (6-57), the over-all open-loop transfer function is

$$G(s)H(s) = \frac{K_0 K_t}{(1 + T_f s)(1 + T_m s)} \qquad (6\text{-}58)$$

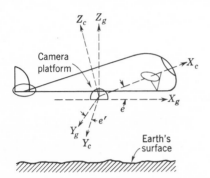

Fig. 6-10   *Camera platform mounted in an airplane.*

Fig. 6-11   *Vertical gyroscopes.*

The stability of the closed-loop system depends on the characteristic polynomial which is the denominator of the control ratio as given by Eq. (6-52). As shown in a later chapter, the stability can also be determined from the open-loop transfer function.

Type 1 Feedback Control System

A Type 1 system is one in which a constant actuating signal maintains a constant rate of change of the output.   In Fig. 6-10 is shown a camera platform mounted in an airplane.   The camera must be maintained parallel to the earth's surface irrespective, within given limits, of the airplane's movement.

By means of a vertical gyroscope, several of which are shown in Fig. 6-11, actuating signals are produced for any variation of the airplane's movement in either or both the XZ and YZ planes.[5,6]   The feedback occurs since the case of the gyro is fastened to the frame of the airplane and moves with it.   Consider two sets of axes: one for the camera and one for the gyro, as defined below.

Y_c and X_c = camera axes parallel to face of camera
Z_c = camera axis perpendicular to face of camera
Y_g and X_g = gyro axes parallel to earth's surface
Z_g = gyro axis perpendicular to earth's surface

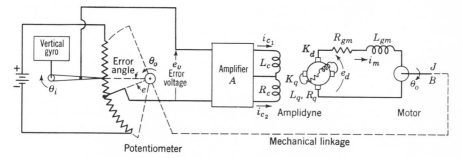

Fig. 6-12 *Position control for aircraft camera (Type 1 system).*

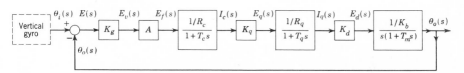

Fig. 6-13 *Block-diagram representation of system of Fig. 6-12.*

When the airplane is in level flight, both sets of axes coincide. In the following discussion only the movement in the XZ plane is considered since the control action is identical for the movement in the YZ plane.

Consider that an updraft has caused the plane, as shown in Fig. 6-10, to go from level to an ascending attitude. The input signal θ_i to the control system corresponds to the reference position X_g; the output signal θ_o corresponds to the actual camera position X_c; and the actuating signal is $e = \theta_i - \theta_o$. The actuating signal, as shown in Fig. 6-12, actuates the control system to cause the face of the camera to return to a position parallel to the earth's surface.

The parameters for the system are as follows:

K_g = potentiometer characteristic, volts/radian of error angle
A = amplifier gain, volts output/volts of input signal
K_q = quadrature volts/amp of amplidyne field current, $i_c = i_{c1} - i_{c2}$
T_q = time constant of quadrature circuit, sec
K_d = direct-axis volts/amp of quadrature axis
T_c = control-field time constant, sec
T_{gm} = electric time constant of generator-motor armature, sec (L_{gm}/R_{gm})
T_m = mechanical time constant, sec (B $\approx$ 0)
K_b = motor back emf constant, volts/(radian/sec)
K_T = motor torque constant, lb-ft/amp
R_q = total resistance in quadrature circuit, ohms
R_c = resistance of control-field winding, ohms

Figure 6-12 can be converted to a block-diagram representation, as shown in Fig. 6-13. In this representation the time constant T_{gm} is neglected.

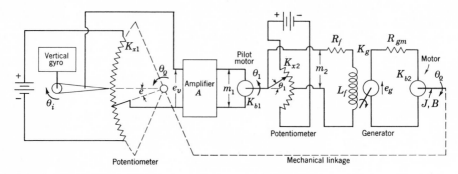

Fig. 6-14   *Position control of aircraft camera (Type 2 system).*

The forward transfer function is

$$G(s) = \frac{\theta_0(s)}{E(s)} = \frac{K_g A K_a K_d}{R_c R_q K_b} \frac{1}{s(1 + T_m s)(1 + T_c s)(1 + T_q s)}$$

$$= \frac{K_1}{s(1 + T_m s)(1 + T_c s)(1 + T_q s)} \tag{6-59}$$

where K_1 has the units of seconds^{-1}.

When the results of Sec. 6-3 are applied to Eq. (6-59), it is seen that this is a Type 1 control system.

For a constant input $r(t) = R_0 u(t)$ the steady-state value of $c(t)_{ss}$ is

$$c(t)_{ss} = \lim_{s \to 0} \left(s \frac{C}{R} \frac{R_0}{s} \right)$$

$$= \lim_{s \to 0} \left[\frac{K_1}{s(1 + T_m s)(1 + T_c s)(1 + T_q s) + K_1} R_0 \right] = R_0 \tag{6-60}$$

Thus
$$E_0 = r(t)_{ss} - c(t)_{ss} = R_0 - R_0 = 0 \tag{6-61}$$

Therefore the rate of change of output is zero. Since this is a unity feedback system, the forward transfer function and the open-loop transfer function are the same.

Type 2 Feedback Control System

A Type 2 system is one in which the second derivative of the output is maintained constant by a constant actuating signal. In Fig. 6-14 is shown a positioning system which is similar to the system in Fig. 6-12 but has been altered to make a Type 2 system.

The inertia and damping factor of the potentiometer are very small; therefore the pilot motor is very lightly loaded. The motor armature current is small, and the pilot motor armature IR drop can be neglected. Thus,

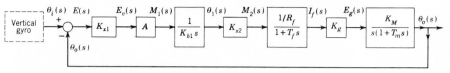

Fig. 6-15   *Block-diagram representation of the system of Fig. 6-14.*

the following equation can be written:

$$M_1(s) = K_{b1}s\theta_1(s) \tag{6-62}$$

where K_{b1} and K_{b2} = motor back emf constants, volts/(radian/sec)
K_{x1} and K_{x2} = potentiometer constants, volts/radian
K_T = motor torque constants, lb-ft/amp
A = amplifier gain
K_g = generator constant, volts/amp
T_m = motor mechanical constant = $JR_{gm}/(K_{b2}K_T + R_{gm}B)$, sec
T_f = generator field constant, sec
K_M = over-all motor constant = $K_T/(BR_{gm} + K_TK_{b2})$, radians/volt-sec

Figure 6-15 is the block-diagram representation of the Type 2 position control system shown in Fig. 6-14.   It is assumed that the inductance of the motor-generator circuit is negligible.   Thus, the open-loop transfer function is

$$G(s) = \frac{\theta_0(s)}{E(s)} = \frac{K_{x1}AK_{x2}K_gK_M/K_{b1}R_f}{s^2(1 + T_fs)(1 + T_ms)} \tag{6-63}$$

or

$$G(s) = \frac{K_2}{s^2(1 + T_fs)(1 + T_ms)} \tag{6-64}$$

where K_2 has the units of seconds^{-2}.

Again, when the results of Sec. 6-3 are applied to Eq. (6-63), it is seen that this is a Type 2 control system.   As the transfer function now stands, it represents an unstable system.   The truth of this statement is proved in Chap. 10.   By the addition of an appropriate compensator in cascade, the forward transfer function can be modified to produce a stable system.   This new transfer function has the form

$$G_0(s) = G_c(s)G(s) = \frac{K_2'(1 + T_1s)}{s^2(1 + T_fs)(1 + T_ms)(1 + T_2s)} \tag{6-65}$$

For a constant input $r(t) = R_0u(t)$ the steady-state value of $c(t)_{ss}$ is

$$c(t)_{ss} = \lim_{s \to 0}\left[s\frac{C(s)}{R(s)}\frac{R_0}{s}\right] = \lim_{s \to 0}\left[\frac{G_0(s)}{1 + G_0(s)}R_0\right] = R_0 \tag{6-66}$$

Thus

$$E_0 = r(t)_{ss} - c(t)_{ss} = R_0 - R_0 = 0 \tag{6-67}$$

Therefore, as expected, a Type 2 system follows a step-function input with no error.

6-6   Static Error Coefficients[3]

In the preceding sections of this chapter, system types are defined.   This is the first step toward establishing a set of standard characteristics that permit the engineer to obtain as much information as possible about a given system with a minimum amount of calculation.   Also, these standard characteristics must point the direction in which a given system must be modified to meet a given set of performance specifications.

Another item of importance is the ability of a system to maintain the output at the desired value with a minimum error.   Thus, in this section are defined static error coefficients that are a measure of a unity feedback control system's steady-state accuracy for a given desired output that is relatively constant or slowly varying.

In Eq. (6-13) it is shown that when the derivative of the output is constant, there exists a constant actuating signal.   This derivative is proportional to the actuating signal E_0 and to a constant K_n, which is the gain of the forward transfer function.   The conventional names for these constants for the Type 0, 1, and 2 systems are *position, velocity*, and *acceleration error coefficients*, respectively.   As a result of the analysis in Sec. 6-2 and of their own experiences, the authors believe that some other names should be assigned to these error coefficients.   When position, velocity, and acceleration are mentioned, the student, because of prior associations, immediately considers these terms to mean that the physical form of the controlled variable is position, velocity, or acceleration.

The conventional names *position, velocity*, and *acceleration error coefficients* were originally selected for application to mechanical position control systems (servomechanisms).   Thus, these names referred to the actual physical form of $c(t)$ or $r(t)$, which is position, as well as to the mathematical

Table 6-2.   Correspondence between the conventional and the authors' designation of static error coefficients

Conventional symbol	Conventional designation of error coefficients	Authors' symbol	Authors' designation of error coefficients
K_p	position	K_0	step
K_v	velocity	K_1	ramp
K_a	acceleration	K_2	parabolic

Table 6-3. *Definitions of static error coefficients for stable unity feedback systems*

Error coefficient	Definition of error coefficient	Value of error coefficient	Form of input signal $r(t)$
Step	$c(t)_{ss}/e(t)_{ss}$	$\lim_{s \to 0} G(s)$	$R_0 u(t)$
Ramp	$(Dc)_{ss}/e(t)_{ss}$	$\lim_{s \to 0} sG(s)$	$R_1 t u(t)$
Parabolic	$(D^2 c)_{ss}/e(t)_{ss}$	$\lim_{s \to 0} s^2 G(s)$	$R_2 t^2 u(t)$

forms $c(t)$, that is, c, Dc, and D^2c, which are involved in the properties exhibited by the Type 0, 1, and 2 systems. It is the belief of the authors that these names are ambiguous when this type of analysis is extended to cover control of temperature, velocity, etc. To avoid this ambiguity of terminology and to define general terms that are universally applicable, the authors have selected the terminology *step, ramp,* and *parabolic error coefficients.* Table 6-2 shows the correspondence between the conventional and the authors' designation of the error coefficients.

The *following derivations of the error coefficients are independent of the system type. They apply to any system type and are defined for specific forms of the input, i.e., for a step, ramp, or parabolic input. These error coefficients are useful only for stable unity feedback systems.* The results are summarized in Table 6-3.

Static Step Error Coefficient

The step error coefficient is defined as

$$\text{Step error coefficient} = \frac{\text{steady-state value of output, } c(t)_{ss}}{\text{steady-state actuating signal, } e(t)_{ss}} \quad (6\text{-}68)$$

and applies only for a step input $r(t) = R_0 u(t)$. The steady-state value of the output is obtained by applying the final-value theorem to Eq. (6-14):

$$c(t)_{ss} = \lim_{s \to 0} sC(s) = \lim_{s \to 0} \left[\frac{sG(s)}{1 + G(s)} \frac{R_0}{s} \right] = \lim_{s \to 0} \left[\frac{G(s)}{1 + G(s)} R_0 \right] \quad (6\text{-}69)$$

Similarly, from Eq. (6-15), for a unity feedback system

$$e(t)_{ss} = \lim_{s \to 0} \left[s \frac{1}{1 + G(s)} \frac{R_0}{s} \right] = \lim_{s \to 0} \frac{R_0}{1 + G(s)} \quad (6\text{-}70)$$

Substituting Eqs. (6-69) and (6-70) into Eq. (6-68) yields

$$\text{Step error coefficient} = \frac{\lim_{s \to 0} \left[\dfrac{G(s)}{1 + G(s)} R_0 \right]}{\lim_{s \to 0} \left[\dfrac{1}{1 + G(s)} R_0 \right]} \tag{6-71}$$

Since both the numerator and the denominator of Eq. (6-71) in the limit can never be zero or infinity simultaneously, where $K_n \neq 0$, the indeterminate forms $0/0$ and ∞/∞ never occur. Thus, this equation reduces to

$$\text{Step error coefficient} = \lim_{s \to 0} G(s) \tag{6-72}$$

Therefore, for a

Type 0 System:

$$\text{Step error coefficient} = \lim_{s \to 0} \frac{K_0(1 + T_1 s)(1 + T_2 s) \cdots}{(1 + T_a s)(1 + T_b s)(1 + T_c s) \cdots} = K_0 \tag{6-73}$$

Type 1 System: Step error coefficient $= \infty$ $\qquad$ (6-74)
Type 2 System: Step error coefficient $= \infty$ $\qquad$ (6-75)

Static Ramp Error Coefficient

The ramp error coefficient is defined as

$$\text{Ramp error coefficient} = \frac{\text{steady-state value of derivative of output, } (Dc)_{ss}}{\text{steady-state actuating signal, } e(t)_{ss}} \tag{6-76}$$

and applies only for a ramp input $r(t) = R_1 t u(t)$. The first derivative of the output is given by

$$\mathcal{L}[Dc] = sC(s) = \frac{sG(s)}{1 + G(s)} R(s) \tag{6-77}$$

The steady-state value of the derivative of the output is obtained by using the final-value theorem:

$$(Dc)_{ss} = \lim_{s \to 0} s[sC(s)] = \lim_{s \to 0} \left[\frac{s^2 G(s)}{1 + G(s)} \frac{R_1}{s^2} \right] = \lim_{s \to 0} \left[\frac{G(s)}{1 + G(s)} R_1 \right] \tag{6-78}$$

Similarly, from Eq. (6-15), for a unity feedback system

$$e(t)_{ss} = \lim_{s \to 0} \left[s \frac{1}{1 + G(s)} \frac{R_1}{s^2} \right] = \lim_{s \to 0} \left[\frac{1}{1 + G(s)} \frac{R_1}{s} \right] \tag{6-79}$$

Substituting Eqs. (6-78) and (6-79) into Eq. (6-76) yields

$$\text{Ramp error coefficient} = \frac{\lim_{s \to 0} \left[\dfrac{G(s)}{1 + G(s)} R_1 \right]}{\lim_{s \to 0} \left[\dfrac{1}{1 + G(s)} \dfrac{R_1}{s} \right]} \tag{6-80}$$

Since the above equation never has the indeterminate form $0/0$ or ∞/∞, it can be simplified to

$$\text{Ramp error coefficient} = \lim_{s \to 0} sG(s) \tag{6-81}$$

Therefore, for a

Type 0 System:

$$\text{Ramp error coefficient} = \lim_{s \to 0} \frac{sK_0(1 + T_1 s)(1 + T_2 s) \cdots}{(1 + T_a s)(1 + T_b s)(1 + T_c s) \cdots} = 0 \tag{6-82}$$

Type 1 System: Ramp error coefficient $= K_1$ (6-83)

Type 2 System: Ramp error coefficient $= \infty$ (6-84)

Static Parabolic Error Coefficient

The parabolic error coefficient is defined as

Parabolic error coefficient

$$= \frac{\text{steady-state value of second derivative of output, } (D^2 c)_{ss}}{\text{steady-state actuating signal, } e(t)_{ss}} \tag{6-85}$$

and applies only for a parabolic input $r(t) = R_2 t^2 u(t)$. The second derivative of the output is given by

$$\mathcal{L}[D^2 c] = s^2 C(s) = \frac{s^2 G(s)}{1 + G(s)} R(s) \tag{6-86}$$

The steady-state value of the second derivative of the output is obtained by using the final-value theorem:

$$(D^2 c)_{ss} = \lim_{s \to 0} s[s^2 C(s)] = \lim_{s \to 0} \left[\frac{s^3 G(s)}{1 + G(s)} \frac{2R_2}{s^3} \right] = \lim_{s \to 0} \left[\frac{G(s)}{1 + G(s)} 2R_2 \right] \tag{6-87}$$

Similarly, from Eq. (6-15), for a unity feedback system

$$e(t)_{ss} = \lim_{s \to 0} \left[s \frac{1}{1 + G(s)} \frac{2R_2}{s^3} \right] = \lim_{s \to 0} \left[\frac{1}{1 + G(s)} \frac{2R_2}{s^2} \right] \tag{6-88}$$

Substituting Eqs. (6-87) and (6-88) into Eq. (6-85) yields

$$\text{Parabolic error coefficient} = \frac{\lim_{s \to 0} \left[\dfrac{G(s)}{1 + G(s)} 2R_2 \right]}{\lim_{s \to 0} \left[\dfrac{1}{1 + G(s)} \dfrac{2R_2}{s^2} \right]} \tag{6-89}$$

Since the above equation never has the indeterminate form $0/0$ or ∞/∞, it can be simplified to

$$\text{Parabolic error coefficient} = \lim_{s \to 0} s^2 G(s) \tag{6-90}$$

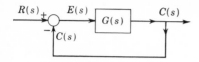

Fig. 6-16   *Simple block diagram.*

Therefore, for a

Type 0 System:

Parabolic error coefficient

$$= \lim_{s \to 0} \frac{s^2 K_0 (1 + T_1 s)(1 + T_2 s) \cdots}{(1 + T_a s)(1 + T_b s)(1 + T_c s) \cdots} = 0 \quad (6\text{-}91)$$

$$\text{\textit{Type 1 System:}} \quad \text{Parabolic error coefficient} = 0 \quad (6\text{-}92)$$

$$\text{\textit{Type 2 System:}} \quad \text{Parabolic error coefficient} = K_2 \quad (6\text{-}93)$$

6-7   Use of Static Error Coefficients

The use of static error coefficients is discussed for each type of unity feedback system.

Type 0 System

If the simple system shown in Fig. 6-16 is a Type 0 system, then

$$c(t)_{ss} = K_0 E_0 \quad (6\text{-}94)$$

Thus the higher K_c is made, the smaller the actuating signal $e(t)_{ss}$ has to be to maintain a fixed desired value of the output.   The stability considerations discussed in the following chapters limit the maximum value K_0 can have for a given system.   Thus K_0 is another standard characteristic that indicates a system's degree of steady-state accuracy.   With a step-function input $r(t) = R_0 u(t)$ there is an error present in the output.   A larger K_0 results in a smaller error.   This is shown by solving for $c(t)_{ss}$ for the case where $G(s)$ represents a Type 0 system and $R(s) = R_0/s$.   For simplicity, $G(s)$ in the following example has no factors in the numerator.

$$C(s) = \frac{G(s)}{1 + G(s)} R(s) = \frac{K_0 R(s)}{(1 + T_a s)(1 + T_b s) \cdots (1 + T_u s) + K_0} \quad (6\text{-}95)$$

$$c(t)_{ss} = \lim_{s \to 0} sC(s) = \frac{K_0}{1 + K_0} R_0 \quad (6\text{-}96)$$

From Eq. (6-81),

$$\text{Ramp error coefficient} = \lim_{s \to 0} sG(s) = \lim_{s \to 0} \frac{sC(s)}{E(s)} = 0 \quad (6\text{-}97)$$

For a ramp input $r(t) = R_1 t u(t)$ the steady-state error is

$$e(t)_{ss} = \frac{(Dc)_{ss}}{\text{Ramp error coefficient}} = \infty \qquad (6\text{-}98)$$

Similarly, the parabolic error coefficient is zero, so the steady-state error with a parabolic input is infinite.

For a Type 0 system the following can be concluded:

1. The output follows a step input with a steady-state error inversely proportional to $1 + K_0$.
2. The output cannot follow a ramp or parabolic input without eventually resulting in an indefinitely large error.

Type 1 System

For a Type 1 system considered at steady state, the value of $(Dc)_{ss}$ is

$$(Dc)_{ss} = K_1 E_0 \qquad (6\text{-}99)$$

Thus, for a larger K_1, the smaller is the size of the actuating signal necessary to maintain a constant rate of change of the output. From the standpoint of trying to maintain $c(t) = r(t)$ at all times, a larger K_1 results in a more sensitive system. In other words, a larger K_1 results in a greater speed of response of the system to a given actuating signal $e(t)$. Therefore, K_1 is another standard characteristic of a system's performance. The maximum value of K_1 is limited by stability considerations and is discussed in later chapters.

For the Type 1 system the step error coefficient is equal to infinity, and the steady-state error is zero for a step input. Therefore, the steady-state output $c(t)_{ss}$ for a Type 1 system is equal to the input when $r(t) = \text{constant}$.

Consider now a ramp input $r(t) = R_1 t u(t)$. The steady-state value of $(Dc)_{ss}$ is found by using the final-value theorem:

$$(Dc)_{ss} = \lim_{s \to 0} s[sC(s)] = \lim_{s \to 0} \left[s \frac{sG(s)}{1 + G(s)} R(s) \right] \qquad (6\text{-}100)$$

where $R(s) = R_1/s^2$ and

$$G(s) = \frac{K_1(1 + T_1 s)(1 + T_2 s) \cdots (1 + T_w s)}{s(1 + T_a s)(1 + T_b s) \cdots (1 + T_u s)}$$

Inserting these values in Eq. (6-100),

$(Dc)_{ss}$
$$= \lim_{s \to 0} \left[s \frac{sK_1(1 + T_1 s)(1 + T_2 s) \cdots (1 + T_w s)}{s(1 + T_a s)(1 + T_b s) \cdots (1 + T_u s) + K_1(1 + T_1 s)(1 + T_2 s) \cdots (1 + T_w s)} \frac{R_1}{s^2} \right] = R_1 \quad (6\text{-}101)$$

Therefore,
$$(Dc)_{ss} = (Dr)_{ss} \qquad (6\text{-}102)$$

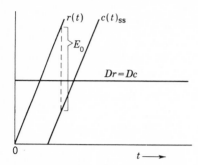

Fig. 6-17   *Steady-state response of a Type 1 system for Dr = constant.*

The magnitude of the steady-state error is found by using the ramp error coefficient.   From Eq. (6-81),

$$\text{Ramp error coefficient} = \lim_{s \to 0} sG(s) = K_1$$

From the definition of ramp error coefficient, the steady-state error is

$$e(t)_{ss} = \frac{(Dc)_{ss}}{K_1} \tag{6-103}$$

Since $(Dc)_{ss} = Dr$,

$$e(t)_{ss} = \frac{Dr}{K_1} = \frac{R_1}{K_1} \tag{6-104}$$

Therefore a Type 1 system follows a ramp input with a constant error E_0. Figure 6-17 illustrates these conditions graphically.

Table of Static Error Coefficients

Table 6-4 gives the values of the error coefficients for the Type 0, 1, and 2 systems.   These values are determined from Table 6-3.   The reader should be able to make ready use of Table 6-4 for evaluating the appropriate error coefficient.   The error coefficient is then used with the definitions given in Table 6-3 to evaluate the magnitude of the steady-state error.

Power-series Input

Assume that the control system shown in Fig. 6-16 is a Type 1 system having an open-loop transfer function

$$G(s) = \frac{10}{s(1 + 2s)} \tag{6-105}$$

and an input signal of the form

$$r(t) = 1 + t \tag{6-106}$$

Table 6-4. Static error coefficients for stable systems

System type	Step error coefficient	Ramp error coefficient	Parabolic error coefficient
0	K_0	0	0
1	∞	K_1	0
2	∞	∞	K_2

The steady-state actuating signal as a function of time can be determined by the use of linear superposition.   First determine $e_a(t)_{ss}$ due to $r_a = 1$ and then determine $e_b(t)_{ss}$ due to $r_b = t$.   The total actuating signal is equal to the sum of the two:

$$e(t)_{ss} = e_a(t)_{ss} + e_b(t)_{ss} \tag{6-107}$$

From Table 6-4,

$$\text{Step error coefficient} = \infty \tag{6-108}$$
$$\text{Ramp error coefficient } K_1 = 10$$

Thus for $\qquad\qquad\qquad r_a(t) = 1$

$$e_a(t)_{ss} = \frac{c_a(t)_{ss}}{\text{Step error coefficient}} = 0 \tag{6-109}$$

For $\qquad\qquad\qquad r_b(t) = t$

$$e_b(t)_{ss} = \frac{(Dc_b)_{ss}}{K_1} = \frac{Dr_b}{K_1} = \frac{1}{10} \tag{6-110}$$

In a Type 1 system with a ramp input the value $(Dc)_{ss} = (Dr)_{ss}$ is a constant. Therefore $(e_b)_{ss}$ must also be a constant.   The total actuating signal is

$$e(t)_{ss} = e_a(t)_{ss} + e_b(t)_{ss} = 0 + \frac{1}{10} = \frac{1}{10} \tag{6-111}$$

By a similar analysis the steady-state error can be determined for a Type 0 or Type 2 system with a power-series input.

t^{n+1} *Input*

Note that a Type n system can follow an input of the form t^{n-1} with zero steady-state error.   It can follow an input t^n but there is a constant steady-state error.   It cannot follow an input t^{n+1} because the steady-state error approaches infinity.   However, for this case the input may be present only for a finite length of time.   Thus the error is also finite.   Then the error may be evaluated by taking the inverse Laplace transform of Eq. (6-16) and inserting the value of time.   Thus the maximum permissible error limits the time $0 < t < t_1$) that an input t^{n+1} may be applied to a control system.

6-8   Conclusions

In this chapter an attempt is made to clarify the distinction between the physical forms that the reference input and the controlled variable may have and the mathematical manipulations of these quantities. It is seen that for all three cases c, Dc, and D^2c of a position correspondence system the error always represents position. For a system that controls some other physical quantity, such as temperature, the error represents temperature or whatever is being controlled.

Since, in general, the forward transfer functions of most feedback control systems fall into three categories, they can be identified as Type 0, 1, and 2 systems, with the corresponding definitions of the static error coefficients. It was seen that these coefficients are indicative of a system's steady-state performance. Thus, a start has been made in developing a set of standard characteristics.

Bibliography

1. James, H. M., N. B. Nichols, and R. S. Phillips: "Theory of Servomechanisms," chap. 4, McGraw-Hill Book Company, New York, 1947.
2. Brown, G. S., and D. P. Campbell: "Principles of Servomechanisms," chap. 6, John Wiley & Sons, Inc., New York, 1948.
3. Chestnut, H., and R. W. Mayer: "Servomechanisms and Regulating System Design," 2d ed., vol. 1, chap. 8, John Wiley & Sons, Inc., New York, 1959.
4. Graves, W. L. O., and E. H. Dinger: Three Control Systems for D-C Adjustable Speed Drives, *Elec. Mfg.*, vol. 44, pp. 82–87, July, 1949.
5. Oplinger, K. A.: Gyroscopes and Their Applications, *Westinghouse Engr.*, vol. 8, pp. 75–79, May, 1948.
6. Konet, H.: Electrically Operated Gyroscopic Instruments, *Trans. AIEE*, vol. 63, pp. 735–738, 1944.

7

Root

Locus

7-1   *Introduction*

A designer can determine whether his design of a control system meets the specifications if he knows the desired time response of the controlled variable. By deriving the differential equations for the control system and solving them, an accurate solution of the system's performance can be obtained, but this approach is not feasible for other than simple systems. If the response does not meet the specifications, it is not easy to determine from this solution just what physical parameters in the system should be changed to improve the response.

It is the desire of a designer to be able to predict a system's performance by an analysis that does not require the actual solution of the differential equations. Also, he would like this analysis to indicate readily the manner or method by which this system must be adjusted or compensated to produce the desired performance characteristics.

The first thing that a designer wants to know about a given system is whether or not it is stable. This can be determined by obtaining the roots of the characteristic equation $1 + G(s)H(s) = 0$. The work involved in determining the roots can become tedious (see Appendix B). Thus a simpler approach is desirable. By applying Routh's criterion to the characteristic equation it is possible in short order to determine whether the system is stable or unstable. Yet this does not satisfy the designer because it does not indicate the degree of stability of the system, that is, the amount of overshoot and the settling time of the controlled variable. Not only must the system be stable, but the overshoot must be maintained within prescribed limits and transients must die out in a sufficiently short time. The graphical methods to be described in this text not only indicate whether a system is stable or unstable but, for a stable system, also show the degree of stability.

There are two basic methods available to a designer. He can choose to analyze and interpret the steady-state sinusoidal response of the system's transfer function to obtain an idea of the system's response. This method is based upon the interpretation of the system's Nyquist plot, which is discussed in more detail in Chaps. 9 to 11. Although this frequency-response approach does not yield an exact quantitative prediction of the system's performance, i.e., the poles of the control ratio $C(s)/R(s)$ cannot be determined, enough information can be obtained to indicate whether the system needs to be adjusted or compensated. Also, the analysis indicates the manner in which the system should be compensated.

This chapter deals with the second method, *the root-locus method,*[1-5] which incorporates the more desirable features of both the classical method and the frequency-response method. *The root locus is a plot of the roots of the characteristic equation of the closed-loop system as a function of the gain.* Devised by Evans, this is a graphical approach which yields a clear indication of the effect of gain adjustment with relatively small effort compared with other methods. The underlying principle is based upon the fact that the poles of $C(s)/R(s)$ (transient-response modes) are related to the zeros and poles of the open-loop transfer function $G(s)H(s)$ and to the gain. An important advantage of the root-locus method is that the roots of the characteristic equation of the system can be obtained directly; this results in a complete and accurate solution of the transient and steady-state response of the controlled variable. Another important feature is that an approximate solution may be obtained with a reduction of the work required. As with any other design technique, when an individual has obtained sufficient experience with this method he is able to apply it and to synthesize a compensating network, if one is required, with relative ease.

7-2   *Plotting Roots of a Characteristic Equation*

To give a better insight into the root-locus plots, consider the position control system shown in Fig. 7-1. Its forward transfer function is

$$G(s) = \frac{\theta_0(s)}{E(s)} = \frac{A/J}{s(s + B/J)} = \frac{K}{s(s + a)} \tag{7-1}$$

where $K = A/J$ and $a = B/J$. Assume that $a = 2$. Thus

$$G(s) = \frac{C(s)}{E(s)} = \frac{K}{s(s + 2)} \tag{7-2}$$

When the transfer function is expressed with the coefficients of the highest powers of s in both the numerator and the denominator equal to unity, the value of K is defined as the *static loop sensitivity*. The control ratio (closed-loop transfer function) is

$$\frac{C(s)}{R(s)} = \frac{K}{s(s + 2) + K} = \frac{K}{s^2 + 2s + K} = \frac{K}{s^2 + 2\zeta\omega_n s + \omega_n^2} \tag{7-3}$$

where $\omega_n = \sqrt{K}$, $\zeta = 1/\sqrt{K}$, and K is considered to be adjustable from zero to an infinite value.

The problem is to determine the roots of the characteristic equation for all values of K and to plot these roots in the s plane. The roots of the characteristic equation are given by

$$s_{1,2} = -1 \pm \sqrt{1 - K} = -\zeta\omega_n \pm \omega_n \sqrt{\zeta^2 - 1} \tag{7-4}$$

For $K = 0$, the roots are $s_1 = 0$ and $s_2 = -2$, which also are the poles of the open-loop transfer function given by Eq. (7-2). When $K = 1$, then $s_{1,2} = -1$. Thus when $0 < K < 1$, the roots $s_{1,2}$ are real and lie on the negative real axis of the s plane between -2 to -1 and -1 to 0, respectively. For the case where $K > 1$, the roots are complex and are given by

$$s_{1,2} = \sigma \pm j\omega_d = -\zeta\omega_n \pm j\omega_n \sqrt{1 - \zeta^2} = -1 \pm j\sqrt{K-1} \tag{7-5}$$

Note that the real part of all the roots is constant for values of $K > 1$.

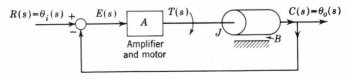

Fig. 7-1   *A position control system.*

Table 7-1.   Location of roots for the
characteristic equation $s^2 + 2s + K = 0$

K	s_1	s_2
0	$-0 + j0$	$-2.0 - j0$
0.5	$-0.293 + j0$	$-1.707 - j0$
1.0	$-1.0 + j0$	$-1.0 - j0$
2.0	$-1.0 + j1.0$	$-1.0 - j1.0$
3.0	$-1.0 + j1.414$	$-1.0 - j1.414$

The roots of the characteristic equation $s^2 + 2s + K = 0$ are determined for a number of values of K (see Table 7-1) and are plotted in Fig. 7-2. Curves are drawn through these plotted points.   On these curves lie all possible roots of the characteristic equation for all values of K from zero to infinity.   Note that these curves are calibrated with K as a parameter and the values of K at points on the locus are underlined; the arrows show the direction of increasing values of K.   *These curves are defined as the root-locus plot of Eq. (7-3).*   Once this plot is obtained, the roots that best fit the system performance specifications can be selected.   Corresponding to the selected roots there is a value of K which can be determined from the plot. When the roots have been selected, the time response can then be obtained. This process of finding the root locus by calculating the roots for various values of K becomes tedious for characteristic equations of higher than second

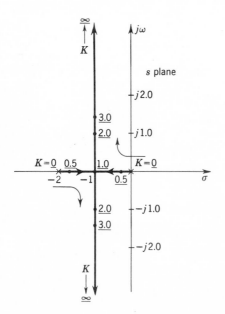

Fig. 7-2 *A plot of all roots of the characteristic equation $s^2 + 2s + K = 0$ for $0 \le K < \infty$. Values of K are underlined.*

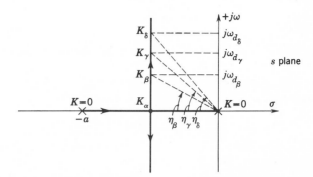

Fig. 7-3 *Root-locus plot of the sample position control system of Fig. 7-1.*

order. Thus a simpler method of obtaining the root locus is desired. The graphical methods for determining the root-locus plot are the subject of the rest of this chapter.

The value of K is normally considered to be positive. However, it is possible for K to be negative. This occurs in electronic amplifiers with an odd number of stages. For the example in this section, if the value of K is negative, Eq. (7-4) gives only real roots. Thus the entire locus lies on the real axis, that is, $0 \leq \sigma_1 < + \infty$ and $-2 \geq \sigma_2 > - \infty$ for $0 \geq K > - \infty$. For any negative value of K there is a root in the right half of the s plane and the system is unstable.

Once the root locus has been obtained for a control system, it is possible to determine the variation in system performance with respect to a variation in sensitivity, K. For the example of Fig. 7-1, the control ratio is written in terms of its roots, for $K > 1$, as

$$\frac{C(s)}{R(s)} = \frac{K}{(s + \sigma - j\omega_d)(s + \sigma + j\omega_d)} \tag{7-6}$$

Note, as defined in Sec. 4-7, that a root with a damping ratio ζ lies on a line making the angle $\eta = \cos^{-1} \zeta$ with the negative real axis. The damping ratio of several roots is indicated in Fig. 7-3. From the root locus it can be seen that an increase in the gain of the system results in the following:

1. A decrease in the damping ratio ζ. This increases the overshoot of the time response.
2. An increase in the undamped natural frequency ω_n. The value of ω_n is the distance from the origin to the complex root.
3. An increase in the damped natural frequency ω_d. The value of ω_d is the imaginary component of the complex root.

4. No effect on the rate of decay σ; that is, it remains constant for all values of gain equal to or greater than K_α. For more complex systems this will not be the case.

5. The root locus is a vertical line for $K \geq K_\alpha$, and $\zeta\omega_n = \sigma$ is constant. This means that no matter how much the gain is increased in a linear *simple* second-order system, the system can never become unstable. The time response of this system with a step-function input, for $\zeta < 1$, is of the form

$$c(t) = A_0 + A_1 e^{-\zeta\omega_n t} \sin{(\omega_d t + \phi)}$$

The root locus of each control system may be analyzed in a similar manner to obtain an idea of the variation in its time response which results from a variation in its sensitivity.

7-3   Qualitative Analysis of the Root Locus

A zero is added to the simple second-order system of the preceding section so that the transfer function is

$$G(s) = \frac{K(s + 1/T_2)}{s(s + 1/T_1)} \tag{7-7}$$

The root locus of the control system having this transfer function is shown in Fig. 7-4b. When this root locus is compared with that of the original system, shown in Fig. 7-4a, it is seen that the branches have been pulled to the left, or farther from the imaginary axis. For values of static loop sensitivity greater than K_α the roots are farther to the left than for the original system. Therefore the transients will decay faster, yielding a more stable system.

If a pole, instead of a zero, is added to the original system, the resulting transfer function is

$$G(s) = \frac{K}{s(s + 1/T_1)(s + 1/T_3)} \tag{7-8}$$

Figure 7-4c shows the root locus of the control system having this transfer function. Note that the addition of a pole has pulled the locus to the right so that two branches cross the imaginary axis. For values of static loop sensitivity greater than K_β the roots are closer to the imaginary axis than for the original system. Therefore the transients will decay more slowly, yielding a less stable system. Also for values of $K > K_\gamma$ two of the three roots lie in the right half of the s plane, resulting in an unstable system. The

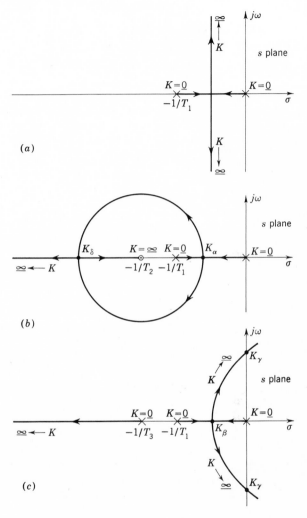

Fig. 7-4 *Various root-locus configurations.* (a) *Root locus of basic transfer function:*

$$G(s) = \frac{K}{s(s + 1/T_1)} \qquad H(s) = 1$$

(b) *Root locus with additional zero:*

$$G(s) = \frac{K(s + 1/T_2)}{s(s + 1/T_1)} \qquad H(s) = 1$$

(c) *Root locus with additional pole:*

$$G(s) = \frac{K}{s(s + 1/T_1)(s + 1/T_3)} \qquad H(s) = 1$$

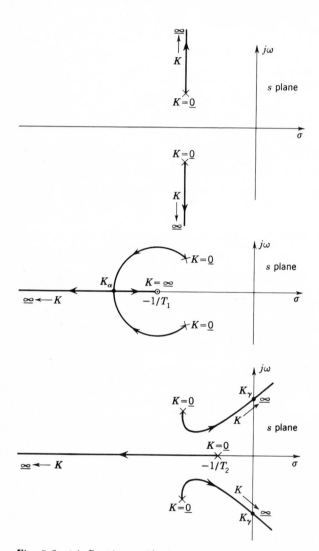

Fig. 7-5 (a) *Root locus of basic transfer function:*

$$G(s) = \frac{K}{s^2 + 2\zeta\omega_n s + \omega_n^2} \qquad \zeta < 1 \qquad H(s) = 1$$

(b) *Root locus of altered transfer function:*

$$G(s) = \frac{K(s + 1/T_1)}{s^2 + 2\zeta\omega_n s + \omega_n^2} \qquad H(s) = 1$$

(c) *Root locus of altered transfer function:*

$$G(s) = \frac{K}{(s^2 + 2\zeta\omega_n s + \omega_n^2)(s + 1/T_2)} \qquad H(s) = 1$$

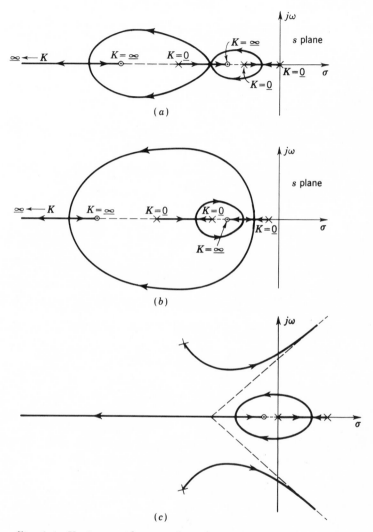

Fig. 7-6   *Various root-locus configurations.*

$$(a)\ \ G(s)H(s) = \frac{K(s + 1/T_2)(s + 1/T_4)}{s(s + 1/T_1)(s + 1/T_3)}$$

$$(b)\ \ G(s)H(s) = \frac{K(s + 1/T_2)(s + 1/T_4)}{(s + 1/T_1)(s + 1/T_3)(s + 1/T_5)}$$

$$(c)\ \ G(s)H(s) = \frac{K(s + 1/T_2)}{s(s - 1/T)(s^2 + 2\zeta\omega_n s + \omega_n^2)}$$

NOTE: *These figures are not drawn to scale. Several other root-locus shapes are possible, for a given pole-zero arrangement, depending on the specific values of the poles and zeros. Interesting variations of the possible root-locus plots for a given pole-zero arrangement are shown in V. C. M. Yeh, The Study of Transients in Linear Feedback Systems by Conformal Mapping and Root-locus Method, Trans. ASME, vol. 76, pp. 349–361, 1954.*

addition of the pole has resulted in a less stable system, compared with the original system.

Another example, shown in Fig. 7-5, illustrates the effect of adding a pole or zero to the basic transfer function

$$G(s) = \frac{K}{s^2 + 2\zeta\omega_n s + \omega_n^2} \qquad \zeta < 1 \qquad (7\text{-}9)$$

In analyzing Fig. 7-5 it is seen that the addition of a pole or a zero has the same general effect as in the previous example of Fig. 7-4. Thus the following general conclusions can be drawn.

1. The addition of a zero to a system has the effect of pulling its root locus to the left, tending to make it a more stable and a faster-responding (a shorter T_s) system.
2. The addition of a pole to a system has the effect of pulling the root locus to the right, tending to make it a less stable and a slower-responding system.

Figure 7-6 illustrates the root-locus configurations for negative feedback control systems having the following transfer functions:

$$G(s)H(s) = \frac{K(s + 1/T_2)(s + 1/T_4)}{s(s + 1/T_1)(s + 1/T_3)} \qquad (7\text{-}10)$$

$$G(s)H(s) = \frac{K(s + 1/T_2)(s + 1/T_4)}{(s + 1/T_1)(s + 1/T_3)(s + 1/T_5)} \qquad (7\text{-}11)$$

$$G(s)H(s) = \frac{K(s + 1/T_2)}{s(s - 1/T)(s^2 + 2\zeta\omega_n s + \omega_n^2)} \qquad (7\text{-}12)$$

Note that the third system contains a pole in the right half of the s plane. It represents the performance of an airplane with an autopilot in the logitudinal mode.

The root-locus method is a graphical technique for readily determining the location of all possible roots of a characteristic equation as the gain is varied from zero to infinity. Also, the manner in which the locus should be altered in order to improve the system's performance can be readily determined, based upon the knowledge of the effect of the addition of poles or zeros.

7-4 Procedure Outline

So that the reader can better visualize the order of the root-locus approach, the procedure to be followed in applying this method is first outlined.

1. Derive the open-loop transfer function $G(s)H(s)$ of the system.
2. Factor the numerator and denominator of the transfer function into linear factors of the form $s + a$.
3. Plot the zeros and poles of the open-loop transfer function in the $s = \sigma + j\omega$ plane.
4. The plotted zeros and poles of the open-loop function determine the roots of the characteristic equation of the closed-loop system $[1 + G(s)H(s) = 0]$. By use of the geometrical shortcuts and the Spirule,* or any other convenient method or device, determine the locus that describes the roots of the closed-loop characteristic equation.
5. Calibrate the locus in terms of the static loop sensitivity K. If the gain of the open-loop system is predetermined, the location of the exact roots of $1 + G(s)H(s)$ is immediately known. If the location of the roots is specified, then the required value of K can be determined.
6. Once the roots have been found, in step 5, the equation of the system's response can be calculated by taking the inverse Laplace transform, either by direct or by graphical calculations.
7. If the response does not meet the desired specifications, determine the shape that the root locus must have to meet these specifications.
8. Synthesize the network that must be inserted into the system, if other than gain adjustment is required, to make the required modification on the original locus. This process is called *compensation*.

7-5 Open-loop Transfer Function

In securing the open-loop transfer function, keep the terms in the factored form of $s + a$ or $s^2 + 2\zeta\omega_n s + \omega_n^2$. For unity feedback the open-loop function is equal to the forward transfer function $G(s)$. For nonunity feedback it also includes the transfer function of the feedback network. This open-loop transfer function is of the form

$$G(s)H(s) = \frac{K(s + a_1) \cdots (s + a_m) \cdots (s + a_w)}{s^n(s + b_1)(s + b_2) \cdots (s + b_c) \cdots (s + b_u)} \qquad (7\text{-}13)$$

where a_m and b_c may be real or complex numbers and may lie in either the left-half or right-half s plane. The value of K may be either positive or negative. For example, consider

$$G(s)H(s) = \frac{K(s + a_1)}{s(s + b_1)(s + b_2)} \qquad (7\text{-}14)$$

---

* Appendix D describes the construction and use of the Spirule for obtaining points on the root locus.

When the transfer function is in this form (with the coefficients of s all equal to unity), the K is defined as the *static loop sensitivity*. By inspection it can be seen that for this example a zero of the open-loop transfer function exists at $s = -a_1$ and the poles are at $s = 0$, $s = -b_1$, and $s = -b_2$. Now let the zeros and poles be denoted by the letters z and p, respectively, i.e.,

$$z_1 = -a_1 \qquad z_2 = -a_2 \qquad \cdots \qquad z_w = -a_w$$
$$p_1 = -b_1 \qquad p_2 = -b_2 \qquad \cdots \qquad p_u = -b_u$$

Then Eq. (7-13) can be rewritten as

$$G(s)H(s) = \frac{K(s - z_1) \cdots (s - z_w)}{s^n(s - p_1) \cdots (s - p_u)} = \frac{K \prod_{m=1}^{w} (s - z_m)}{s^n \prod_{c=1}^{u} (s - p_c)} \qquad (7\text{-}15)$$

7-6   *Poles of the Control Ratio* $C(s)/R(s)$

It is stated earlier in the chapter that the underlying principle of the root-locus method is based upon the fact that the poles of the control ratio $C(s)/R(s)$ are related to the zeros and poles of the $G(s)H(s)$ function and to the static loop sensitivity. This can be shown as follows. Let

$$G(s) = \frac{N_1(s)}{D_1(s)} \qquad (7\text{-}16)$$

and

$$H(s) = \frac{N_2(s)}{D_2(s)} \qquad (7\text{-}17)$$

Then

$$G(s)H(s) = \frac{N_1 N_2}{D_1 D_2} \qquad (7\text{-}18)$$

Thus

$$\frac{C(s)}{R(s)} = M(s) = \frac{A(s)}{B(s)} = \frac{G(s)}{1 + G(s)H(s)} = \frac{N_1/D_1}{1 + N_1 N_2/D_1 D_2} \qquad (7\text{-}19)$$

where

$$B(s) \equiv 1 + G(s)H(s) = 1 + \frac{N_1 N_2}{D_1 D_2} = \frac{D_1 D_2 + N_1 N_2}{D_1 D_2} \qquad (7\text{-}20)$$

Rationalizing Eq. (7-19) gives

$$\frac{C(s)}{R(s)} = M(s) = \frac{N_1 D_2}{D_1 D_2 + N_1 N_2} = \frac{P(s)}{Q(s)} \qquad (7\text{-}21)$$

From Eqs. (7-20) and (7-21) it is seen that the zeros of $B(s)$ are the poles of $M(s)$ and determine the form of the system's transient response. In terms of $G(s)H(s)$ given by Eq. (7-15), the degree of $B(s)$ is equal to $n + u$; therefore $B(s)$ has $n + u$ finite zeros. As shown in Chap. 4, Laplace Transforms, the factors of $Q(s)$ fall into the following three categories:

Pole of $C(s)$	Corresponding inverse is of the form
s	$u(t)$—unit step
$s + 1/T$	$e^{-t/T}$
$s^2 + 2\zeta\omega_n s + \omega_n^2$	$e^{-\zeta\omega_n t} \sin(\omega_n \sqrt{1 - \zeta^2}\, t + \phi)$, where $\zeta < 1$

The numerator $P(s)$ of Eq. (7-21) merely modifies the constant multiplier of these transient components. The roots of $B(s) = 0$, which is the characteristic equation of the system, can be determined as follows.

From Eq. (7-20), the characteristic equation is

$$B(s) = 1 + G(s)H(s) = 0 \qquad (7\text{-}22)$$

or

$$G(s)H(s) = \frac{K(s - z_1) \cdots (s - z_w)}{s^n(s - p_1) \cdots (s - p_u)} = -1 \qquad (7\text{-}23)$$

Thus, as the static loop sensitivity K assumes values from zero to infinity, the transfer function $G(s)H(s)$ must always be equal to -1. The corresponding values of s, for any value of K, which satisfy Eq. (7-23) are the poles of $M(s)$. The plots of these values of s are defined as the root locus of $M(s)$.

Conditions that determine the root locus for *positive* values of static loop sensitivity are now determined. The general form of $G(s)H(s)$ is

$$G(s)H(s) = Fe^{-j\beta}$$

and -1 can be written as

$$-1 = e^{j(1+2m)\pi} \qquad m = 0, \pm 1, \pm 2, \ldots$$

Equation (7-23) is rewritten as

$$Fe^{-j\beta} = e^{j(1+2m)\pi} \qquad (7\text{-}24)$$

where

$$F = |G(s)H(s)| = 1$$

and

$$-\beta = (1 + 2m)\pi \qquad (7\text{-}25)$$

From the above it can be concluded that the magnitude of $G(s)H(s)$, a function of the complex variable s, must always be unity and its phase angle must be an odd multiple of π if the particular value of s is to be a zero of $B(s) = 1 + G(s)H(s)$. Consequently, the following two conditions are formalized for all positive values of K from zero to infinity:

$$\boldsymbol{K > 0}$$

$$|G(s)H(s)| = 1 \qquad \text{\textit{magnitude condition}} \qquad (7\text{-}26)$$

$$\underline{/G(s)H(s)} = (1 + 2m)180° \qquad \text{for } m = 0, \pm 1, \pm 2, \ldots$$

$$\text{\textit{angle condition}} \qquad (7\text{-}27)$$

Conditions that determine the root locus for *negative* values of loop sensitivity ($-\infty < K < 0$) are now determined. The substitution of

$K = -|K|$ is made into Eq. (7-23) as follows:

$$G(s)H(s) = \frac{-|K|(s - z_1) \cdots (s - z_w)}{s^n(s - p_1) \cdots (s - p_u)} = -1 \qquad (7\text{-}28)$$

This equation can be rewritten as follows:

$$G(s)H(s) = \frac{|K|(s - z_1) \cdots (s - z_w)}{s^n(s - p_1) \cdots (s - p_u)} = 1 = \underline{1/2\pi m} \qquad (7\text{-}29)$$

where $m = 0, \pm 1, \pm 2, \ldots$. Equation (7-29) may also be expressed as follows:

$$G(s)H(s) = Fe^{-j\beta} = e^{j2\pi m} \qquad (7\text{-}30)$$

Consequently, from the above equation, the following two conditions are formalized for all negative values of K from zero to minus infinity:

$$K < 0$$

$$|G(s)H(s)| = 1 \qquad \textit{magnitude condition} \qquad (7\text{-}31)$$
$$\underline{/G(s)H(s)} = (m)360° \qquad \text{for } m = 0, \pm 1, \pm 2, \ldots$$
$$\textit{angle condition} \qquad (7\text{-}32)$$

Thus the root-locus method provides a plot of the variation of each of the poles of $C(s)/R(s)$ in the complex s plane as the gain is varied from $K = 0$ to $K = \pm \infty$.

7-7   Application of the Magnitude and Angle Conditions

Once the open-loop transfer function $G(s)H(s)$ has been determined and put into the proper form, the poles and zeros of this function are plotted in the $s = \sigma + j\omega$ plane.   As an example, consider

$$G(s)H(s) = \frac{K(s + 1/T_1)}{s(s + 1/T_2)(s^2 + 2\zeta\omega_n s + \omega_n^2)} \qquad (7\text{-}33)$$

For the quadratic factor $s^2 + 2\zeta\omega_n s + \omega_n^2$ with the damping ratio $\zeta < 1$ the zeros [poles of Eq. (7-33)] are

$$s_{1,2} = -\zeta\omega_n \pm j\omega_n \sqrt{1 - \zeta^2} = \sigma \pm j\omega_d$$

The plot of the poles and zeros of Eq. (7-33) is shown in Fig. 7-7.   Remember that complex poles or zeros always occur in conjugate pairs, that σ is the damping constant, and ω_d is the damped natural frequency of oscillation.

With the open-loop poles and zeros plotted, they are now used in the graphical construction of the locus of the poles (the roots of the characteristic equation) of the closed-loop control ratio.

Consider the following general open-loop transfer function:

$$G(s)H(s) = \frac{K(s - z_1) \cdots (s - z_w)}{s^n(s - p_1)(s - p_2) \cdots (s - p_u)} \qquad (7\text{-}34)$$

For any particular value (real or complex) of s the terms s, $s - p_1$, $s - p_2$, $s - z_1$, etc., are complex numbers designating directed line segments. For example, if $s = -4 + j4$ and $p_1 = -1$, then $s - p_1 = -3 + j4$ or

$$|s - p_1| = 5$$

and $\qquad \phi_1 = \underline{/s - p_1} = 126.8°$ (see Fig. 7-8)

A multiple pole or zero is indicated by $]_q$, where $q = 1, 2, 3$, etc., is the order of the pole or zero.

In the preceding section it is shown that the zeros of $1 + G(s)H(s)$ are the roots of the characteristic equation

$$1 + G(s)H(s) = 0 \qquad (7\text{-}35)$$

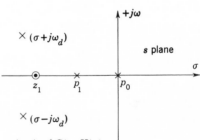

Fig. 7-7 *A plot of the poles and zeros of* $G(s)H(s)$ *given by Eq. (7-33).*

$\times$ A pole of $G(s)H(s)$
$\odot$ A zero of $G(s)H(s)$

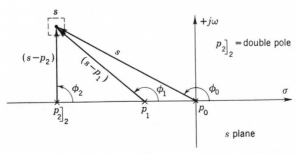

Fig. 7-8 *Graphical notation in the application of the magnitude and angle conditions.*

and that for all values of s which satisfy this equation

$$|G(s)H(s)| = 1 \tag{7-36}$$

$$\text{and} \quad /\underline{G(s)H(s)} = \begin{cases} (1 + 2m)180° & \text{for } K > 0 \\ (m)360° & \text{for } K < 0 \end{cases}$$

$$\text{for } m = 0, \pm 1, \pm 2, \ldots \tag{7-37}$$

These are labeled as the magnitude and angle conditions, respectively. Therefore, applying these two conditions to the general equation (7-34) results in

$$\frac{|K| \cdot |s - z_1| \cdot \cdot \cdot |s - z_w|}{|s^n| \cdot |s - p_1| \cdot |s - p_2| \cdot \cdot \cdot |s - p_u|} = 1 \tag{7-38}$$

and

$$-\beta = /\underline{s - z_1} + \cdot \cdot \cdot + /\underline{s - z_w} - n/\underline{s} - /\underline{s - p_1} -$$

$$\cdot \cdot \cdot - /\underline{s - p_u} = \begin{cases} (1 + 2m)180° & \text{for } K > 0 \\ (m)360° & \text{for } K < 0 \end{cases} \tag{7-39}$$

Rewriting these equations,

$$|K| = \frac{|s^n| \cdot |s - p_1| \cdot |s - p_2| \cdot \cdot \cdot |s - p_u|}{|s - z_1| \cdot \cdot \cdot |s - z_w|} = \text{static loop sensitivity} \tag{7-40}$$

and

$$-\beta = \Sigma(\text{angles of numerator terms}) - \Sigma(\text{angles of denominator terms})$$

$$= \begin{cases} (1 + 2m)180° & \text{for } K > 0 \\ (m)360° & \text{for } K < 0 \end{cases} \tag{7-41}$$

All angles are considered positive, measured in the counterclockwise sense. Multiplying Eq. (7-41) by -1 and rearranging terms,

$$\beta = /\underline{\text{denominator}} - /\underline{\text{numerator}}$$

$$= \begin{cases} -(1 + 2m)180° & \text{for } K > 0 \\ -(m)360° & \text{for } K < 0 \end{cases} \tag{7-42}$$

Equations (7-40) and (7-42) are in the form in which they are generally used in the graphical construction of the root locus. In other words, there are particular values of s for which $G(s)H(s)$ satisfies the angle condition. For a given static loop sensitivity only certain of these values of s simultaneously satisfy the magnitude condition. These values of s that satisfy both the angle and the magnitude conditions are the roots of the characteristic equation and are $n + u = v$ in number. Thus, corresponding to step 4 in Sec. 7-4, the locus of all possible roots is obtained by applying the angle condition, and for a known or desired value of static loop sensitivity the particular roots are determined from the magnitude condition.

Example

With

$$G(s)H(s) = \frac{K_0(1 + 0.25s)}{(1 + s)(1 + 0.5s)(1 + 0.2s)} \tag{7-43}$$

determine the locus of all possible closed-loop poles.

$$G(s)H(s) = \frac{K_0(0.25)}{(0.5)(0.2)} \frac{s + 4}{(s + 1)(s + 2)(s + 5)}$$

$$= \frac{K(s + 4)}{(s + 1)(s + 2)(s + 5)} \qquad K = 2.5K_0 \tag{7-44}$$

1. Plot the poles and zeros as shown in Fig. 7-9.
2. In Fig. 7-9 the ϕ's are denominator angles and ψ's are numerator angles. Also, the l's are the lengths of the directed segments stemming from the denominator factors, and (l)'s are the lengths of the directed segments stemming from the numerator factors. After plotting the poles and zeros of the open-loop transfer function, arbitrarily choose a search point. To this point, draw directed line segments from all the open-loop poles and zeros and label as indicated. For this search point to be a point on the locus, the following angle condition must be true:

$$\beta = \phi_1 + \phi_2 + \phi_3 - \psi_1 = \begin{cases} -(1 + 2m)180° & \text{for } K > 0 \\ -(m)360° & \text{for } K < 0 \end{cases} \tag{7-45}$$

If this equation is not satisfied, select another search point until it is satisfied. Locate a sufficient number of points in the s plane that satisfy the angle condition. In the next section additional information is given that lessens the work involved in this trial-and-error approach.

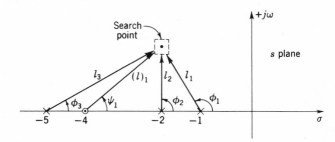

Fig. 7-9 *Construction of the root locus.*

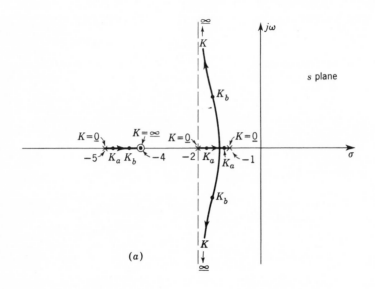

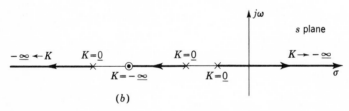

Fig. 7-10 *The complete root locus of Eq. (7-43), (a) for $K > 0$. (b) for $K < 0$.*

3. Once the complete locus has been determined by this trial-and-error method, the locus can be calibrated in terms of the static loop sensitivity for any root s_1 as follows:

$$|K| = \frac{l_1 l_2 l_3}{(l)_1} \tag{7-46}$$

where
$$l_1 = |s_1 + 1|$$
$$l_2 = |s_1 + 2|$$
$$l_3 = |s_1 + 5|$$
$$(l)_1 = |s_1 + 4|$$

In other words, the values of l_1, l_2, l_3, and $(l)_1$ are known for a given point s_1 that satisfies the angle condition; thus the value of $|K|$ for this point can be calculated. The appropriate sign must be given to the magnitude of K compatible with the particular angle condition being utilized to obtain the root locus. Note that since complex roots must occur in conjugate pairs the locus is symmetrical

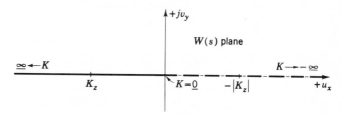

Fig. 7-11 *Plot of K in the W(s) plane.*

about the real axis. Thus the bottom half of the locus can be drawn, once the locus above the real axis has been determined. The root locus for this system is shown in Fig. 7-10.

$W(s)$ *Plane*

From Eq. (7-44) the following is obtained:

$$W(s) = u_x + jv_y = \frac{(s + 1)(s + 2)(s + 5)}{s + 4} = -K \qquad (7\text{-}47)$$

To better show the relationship of the variation of K to the root locus, a plot of Eq. (7-47) in the $W(s)$ plane is given in Fig. 7-11. The line $u_x = -K$ in the $W(s)$ plane, shown in Fig. 7-11, maps into the curves indicated in Fig. 7-10. That is, for each value of u_x in the $W(s)$ plane there is a particular value or a set of values of s in the s plane.

7-8   Geometrical Properties (Construction Rules)

To facilitate the application of the root-locus method, the following rules are established for $K > 0$. These rules are based upon the interpretation of the angle condition and the analysis of the characteristic equation. The reader should be able to extend these rules for the case where $K < 0$. The rules for both $K > 0$ and $K < 0$ are listed in Sec. 7-14 for easy reference.

The rules presented aid in obtaining the root locus by expediting the manual plotting of the locus. The root locus may also be obtained by various automatic methods using the analog or digital computer.[10,11] For automatic plotting these rules provide *check points* to ensure that the computer solution is correct. Another plotting device, the ESIAC algebraic computer,[12] can be used to obtain root-locus plots and, in addition, frequency-response plots, transient response, etc.

Rule 1.   Number of branches of the locus

The characteristic equation $B(s) = 0$ is of the degree $n + u = v$; therefore there are v roots, which are continuous functions of the static loop sensitivity K.   As K is varied from zero to infinity, each root traces a continuous curve. Since there are v roots, there are the same number of curves or branches in the complete root locus.   Since the degree of the polynomial $B(s)$ is determined by the poles of the open-loop transfer function, it can be stated that *the number of branches of the root locus is equal to the number of poles of the open-loop transfer function.*

Rule 2.   Real-axis locus

In Fig. 7-12 are shown a number of open-loop poles and zeros.   If the angle condition is applied to any search point such as s_1 on the real axis, it is seen that the angular contribution of all the poles and zeros on the real axis to the left of this point is zero.   The angular contribution of the complex conjugate poles to this point is 360°.   (This is also true for complex conjugate zeros.) Finally, the poles and zeros on the real axis to the right of this point each contribute 180° (with the appropriate sign included).   From Eq. (7-42) the angle of $G(s)H(s)$ to the point s_1 is given by

$$\phi_0 + \phi_1 + \phi_2 + \phi_3 + (\phi_4)_{+j} + (\phi_4)_{-j} - (\psi_1 + \psi_2) = (1 + 2m)180° \quad (7\text{-}48)$$
$$\text{or} \qquad\qquad 180° + 360° = (1 + 2m)180° \qquad\qquad (7\text{-}49)$$

Therefore s_1 is a point on a locus.   Similarly, it can be shown that the point s_2 is not a point on the locus.   It can be seen that the poles and zeros to the left of the s point and the 360° contributed by the complex conjugate poles or zeros do not affect the odd-multiple-of-180° requirement.   Thus the

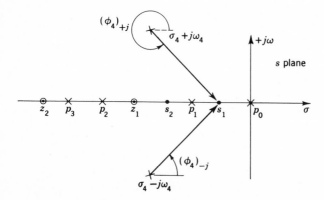

Fig. 7-12   *Determination of the real-axis locus.*

necessary angle condition for determining the real-axis locus is:

$$(R_p - R_z)180° = (1 + 2m)180° \qquad (7\text{-}50)$$

where R_p is the number of poles, on the real axis, to the right of the search point and R_z is the number of zeros, on the real axis, to the right of the search point. When Eq. (7-50) is analyzed, it is seen that $R_p - R_z$ must be an odd number if an odd multiple of 180° is to be obtained. By applying Eq. (7-50) to Fig. 7-12, the real-axis locus is determined to exist between the following points: 0 to p_1, z_1 to p_2, and p_3 to z_2. In other words, *if the total number of poles and zeros to the right of the s point on the real axis is odd, then this point lies on the locus.*

All points on the real axis between z_1 and p_2 in Fig. 7-12 satisfy the angle condition and are therefore points on the root locus. There is no guarantee that this section of the real axis is part of one branch. Figure 7-6b and Prob. 7-4 illustrate the situation where the part of the real axis between a pole and a zero is divided into three sections which are parts of three different branches.

Rule 3. Locus end points

Equations (7-40) and (7-47) are of the following general form:

$$W(s) = \frac{\displaystyle\prod_{c=1}^{v} (s - p_c)}{\displaystyle\prod_{m=1}^{w} (s - z_m)} = -K \qquad (7\text{-}51)$$

where Π is a product symbol used like the symbol Σ, which represents a summation. In terms of absolute values, this becomes

$$|W(s)| = K = \frac{\displaystyle\prod_{c=1}^{v} |s - p_c|}{\displaystyle\prod_{m=1}^{w} |s - z_m|} \qquad (7\text{-}52)$$

When it is remembered that the numerator and denominator factors of Eq. (7-52) are the poles and zeros, respectively, of the open-loop transfer function, the following conclusions can be drawn:

1. When $s = p_c$, the sensitivity factor K is zero.
2. When $s = z_m$, the sensitivity factor K is infinite. When the numerator of Eq. (7-52) is of higher order than the denominator, $s = \infty$ also makes K infinite, thus being equivalent in effect to a zero.

Thus it can be said that the locus starting points $(K = 0)$ are at the open-loop poles and that the locus ending points $(K = \infty)$ are at the open-loop zeros (the point at infinity being considered as an equivalent zero of multiplicity equal to $v - w$).

Rule 4.   Asymptotes of locus as *s* approaches infinity

Plotting of the locus is greatly facilitated if one can determine the asymptotes approached by the various branches as s takes on large values.   The characteristic equation, when rearranged, is

$$G(s)H(s) = K \frac{\prod\limits_{m=1}^{w} (s - z_m)}{\prod\limits_{c=1}^{v} (s - p_c)} = -1 \tag{7-53}$$

Taking the limit of $G(s)H(s)$ as s approaches infinity yields

$$\lim_{s \to \infty} G(s)H(s) = \lim_{s \to \infty} \left[K \frac{\prod\limits_{m=1}^{w} (s - z_m)}{\prod\limits_{c=1}^{v} (s - p_c)} \right] = \frac{K}{s^{v-w}} = -1 \tag{7-54}$$

It must be remembered that K in Eq. (7-54) is still a variable in the manner prescribed previously, thus allowing the magnitude condition to be met. Therefore,

$$-K = s^{v-w} \tag{7-55}$$
$$|-K| = |s^{v-w}| \qquad \text{magnitude condition} \tag{7-56}$$
$$\underline{/-K} = \underline{/s^{v-w}} = (1 + 2m)180° \qquad \text{angle condition} \tag{7-57}$$

Rewriting Eq. (7-57),

$$(v - w) \underline{/s} = (1 + 2m)180°$$

or
$$\gamma = \frac{(1 + 2m)180°}{v - w} \qquad \text{as } s \to \infty \tag{7-58}$$

The asymptotes of the root locus are given by

$$\gamma = \frac{(1 + 2m)180°}{[\text{number of poles of } G(s)H(s)] - [\text{number of zeros of } G(s)H(s)]} \tag{7-59}$$

Equation (7-59) reveals that, no matter what magnitude s may have, after a sufficiently large value has been reached, the argument (angle) of s

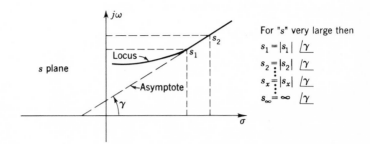

Fig. 7-13 *Asymptotic condition for large values of s.*

remains constant. For a search point that has a sufficiently large magnitude the open-loop poles and zeros appear to it as if they have collapsed into a single point. Since these values of s constitute possible values of the zeros of $1 + G(s)H(s)$, they are points on the locus of roots. Therefore the branches are asymptotic to straight lines whose slopes and directions are given by Eq. (7-59). (See Fig. 7-13.)

One may be led to believe from the limit condition of Eq. (7-54) that the asymptotes go through the origin. This is not true, and the correct position can be found as shown next by Rule 5. However, when it is desired to obtain rapidly the general shapes of the root locus, the asymptotes can be drawn through the origin. When the true locus is to be drawn precisely, the correct location of the asymptotes must be determined.

Rule 5.   Real-axis intercept of the asymptotes

When both the numerator and the denominator of Eq. (7-51) are expanded,

$$W(s) = \frac{s^v - s^{v-1} \sum_{c=1}^{v} (p_c) + \cdots}{s^w - s^{w-1} \sum_{m=1}^{w} (z_m) + \cdots} = -K \qquad (7\text{-}60)$$

Dividing the numerator by the denominator yields the following equation:

$$s^{v-w} + s^{v-w-1} \left[\sum_{m=1}^{w} (z_m) - \sum_{c=1}^{v} (p_c) \right] + \cdots = -K \qquad (7\text{-}61)$$

The angle condition $\gamma = (1 + 2m)180°/(v - w)$ reveals only the slope of the asymptotes. It does not indicate which one of the infinite number of

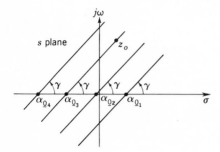

Fig. 7-14 *Possible locations of the asymptotes.*

possible lines, as shown in Fig. 7-14, is the true asymptote.   By use of a technique in the theory of linear equations, the true asymptotes are found by determining the real-axis intercept, as follows:

1.   Taking the $\lim\limits_{s \to \infty} G(s)H(s)$ reveals that s approaches an angle

$$\gamma = \frac{180°(1 + 2m)}{v - w}$$

2.   The straight lines given by the equation

$$s^{v-w} = -K$$

go through the origin of the s plane and have the same slope as the asymptotes.

3.   The true asymptotes must satisfy Eq. (7-61), which is of the form

$$s^{v-w} + as^{v-w-1} + bs^{v-w-2} + \cdots = -K$$

as well as the angle condition in step 1.

4.   Assume that a particular true asymptote goes through a point z_o in the s plane.   If the origin is moved to z_o by a change of variable $s = s_1 + z_o$, then the asymptote in terms of s satisfies the equation in step 3.   The asymptotes in terms of s_1 are determined by $s_1^{v-w} = -K$.

5.   The change of variable is made by substituting $s = s_1 + z_o$ into the equation of step 3, and the new equation is

$$s_1^{v-w} + [(v - w)z_o + a]s_1^{v-w-1} + [\quad]s_1^{v-w-2} + \cdots = -K$$

6.   For the equation in step 5 to be equal to the equation for the asymptotes in step 4, all the coefficients of powers of s_1 less than $v - w$ must simultaneously be equal to zero.

7. Setting the coefficient of s_1^{v-w-1} equal to zero and solving for z_o,

$$z_o = \frac{\sum_{c=1}^{v} (p_c) - \sum_{m=1}^{w} (z_m)}{v - w}$$

This value of z_o is always a real number since all complex poles and zeros occur in conjugate pairs.

8. Since z_o is independent of $-K$, all the asymptotes cross the real axis at $z_o = \sigma_o$. The value σ_o is the *centroid* of the pole-zero configuration.

The results can be summarized by saying that, since the portion of the locus away from the asymptotes and near the axes is generally of prime concern, it is necessary to find the real-axis crossing of the asymptotes, which is

$$\sigma_o = \frac{\sum_{c=1}^{v} \text{Re}\,(p_c) - \sum_{m=1}^{w} \text{Re}\,(z_m)}{v - w} \tag{7-62}$$

It should be pointed out that a locus may cross its asymptote. Where high-precision plotting is desired, it may be valuable to know from which side the root locus approaches its asymptote. Lorens and Titsworth present a method for obtaining this information[7] which shows that the locus lies exactly along the asymptote if the pole-zero pattern is symmetric about the asymptote line extended through the point σ_o.

Rule 6. Breakaway point on the real axis

It has been shown that the locus starts at the poles where $K = 0$ and ends at the zeros (which are finite or at $s = \infty$). Thus, consider the case where the locus has branches on the real axis between two poles (see Fig. 7-15a). There must be a point at which the two branches break away from the real axis and enter the complex region of the s plane in order to approach zeros or the point at infinity. For the case of two zeros (see Fig. 7-15c) the branches are coming from poles in the complex region and enter the real axis. In Fig. 7-15a for the case of two poles, there is a point s_1 for which the sensitivity K_z is greater than for any other point on either side of s_1 on the real axis. In other words, since K starts with a value of zero at the poles and increases in value as the locus moves away from the poles, there is a point somewhere in between where the K's for the two branches simultaneously reach a maximum value. This point is called the *breakaway point*. The plot of $|K|$ versus σ is shown in Fig. 7-15b. The variation in the value

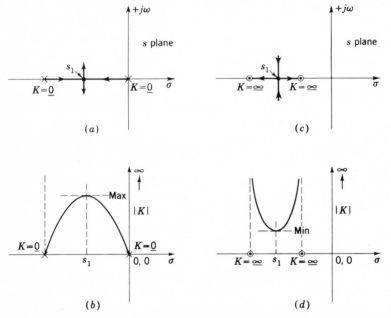

Fig. 7-15 *Real-axis loci for two consecutive poles or zeros on the real axis and the corresponding plot of $|K|$ versus σ.*

of $|K|$ along a real-axis locus between two zeros is shown in Fig. 7-15d.   The value of σ for which the value of $|K|$ is a minimum between these two zeros is called the *break-in point*.   The breakaway and break-in points can easily be calculated for an open-loop pole-zero combination for which the derivative of $W(s) = -K$ is of the second order.   As an example, if

$$G(s)H(s) = \frac{K}{s(s+1)(s+2)} \tag{7-63}$$

then

$$W(s) = s(s+1)(s+2) = -K \tag{7-64}$$

Multiplying the factors together,

$$W(s) = s^3 + 3s^2 + 2s = -K \tag{7-65}$$

When $s^3 + 3s^2 + 2s$ is a minimum, $-K$ is a minimum; and K is a maximum. Thus by taking the derivative of this function and setting it equal to zero, the points s_1 can be determined:

$$\frac{dW(s)}{ds} = 3s^2 + 6s + 2 = 0 \tag{7-66}$$

or

$$s_{1,2} = -1 \pm 0.5743 = -0.4257, -1.5743$$

Since the breakaway point for $K > 0$ must lie between $s = 0$ and $s = -1$ in order to satisfy the angle condition, the value of s_1 is

$$s_1 = -0.4257 \tag{7-67}$$

(The other point, $s_2 = -1.5743$, is the break-in point on the root locus for $K < 0$.) Substituting $s_1 = -0.4257$ into Eq. (7-65) gives the value of K at the breakaway point for $K > 0$ as

$$-K = (-0.426)^3 + (3)(-0.426)^2 + (2)(-0.426) \tag{7-68}$$
$$K = 0.385 \tag{7-69}$$

When the derivative of $W(s)$ is of higher order than 2, the method indicated above to calculate the breakaway point is laborious. This is so because of the relative difficulty in factoring a polynomial higher than second order. This objection can be overcome by the use of the "Hills-and-dales" method, i.e., by plotting the value of $|K|$ versus real values of s. Figure 7-15b and d illustrates the plots for the simple cases of a locus between two poles or two zeros. This plot permits the determination of the maxima and minima without differentiation. The value of the static loop sensitivity K is determined as shown by Eq. (7-40) for real values of s. The curve of $|K|$ has a maximum at a breakaway point and a minimum at a break-in point. There is always a breakaway point when a locus exists on the real axis between two poles; a break-in point must be present when the locus exists between two zeros on the real axis. It is also possible to have breakaway and break-in points between a pole and a zero (finite or infinite) on the real axis, as shown in Fig. 7-6. Applying the Hills-and-dales method to the real-axis locus of Fig. 7-6a and b yields, respectively, the plots of Fig. 7-16a and b. In these figures the plots of $|K|$ between two poles and two zeros (including the zero at ∞) clearly indicate maxima and minima, respectively. The root locus has corresponding breakaway and break-in points. The Hills-and-dales plot for a locus between a pole and zero falls into one of the following categories:

1. The plot clearly indicates a peak and a dip, as illustrated between p_1 and z_1 in Fig. 7-16b. The peak represents a "maximum" value of $|K|$ that satisfies the condition for a breakaway point. The dip represents a "minimum" value of $|K|$ that satisfies the condition for a break-in point.

2. The plot does not indicate a dip-and-peak combination but contains an inflection point. An inflection point occurs when the breakaway and break-in points coincide, as is the case between p_2 and z_1 in Fig. 7-16a. By a visual inspection of the plot one cannot always distinguish an inflection point from a maximum and a minimum which are close together. Plotting points on the root locus by means of the Spirule *may* indicate the presence of a break-in and breakaway point on the real-axis locus between a pole and a zero.

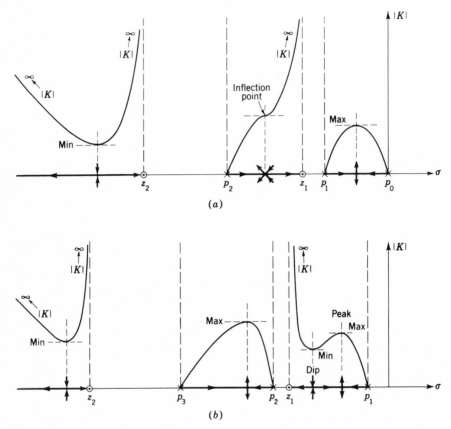

Fig. 7-16 *Hills-and-dales plots of* $|K|$ *and the corresponding real-axis locus for* (a) *Fig. 7-6a and* (b) *Fig. 7-6b.*

3.   The plot does not indicate a dip-and-peak combination and clearly indicates the absence of any possibility of the existence of an inflection point.   For this situation there are no break-in and breakaway points.

The partition method for determining breakaway and break-in points is presented in Sec. 8-6.   The next rule shows a graphical method for determining these points.

Rule 7.   Angular condition applied in the vicinity of the breakaway point

The use of the Spirule in the vicinity of the breakaway point is not very accurate in applying the angular condition.   Because of this fact and the

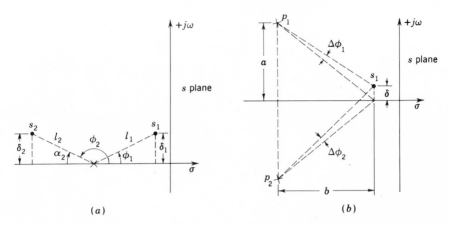

Fig. 7-17 *Typical angular conditions of poles (or zeros) in the vicinity of the break-away point.*

complexity in calculating the breakaway point, an analytical-graphical method used in this region is described below.

Figure 7-17 indicates the typical angular conditions encountered in plotting the root locus for real and complex poles in the vicinity of the breakaway point. Although poles are considered in the present discussion, the results apply equally well for zeros.

In Fig. 7-17a, if the search point s_1 is to the right of the pole,

$$\sin \phi_1 = \frac{\delta_1}{l_1} \tag{7-70}$$

If δ_1 is very small compared with l_1,

$$\phi_1 \approx \frac{\delta_1}{l_1} \tag{7-71}$$

For the case where s_2 is to the left of the pole,

$$\sin \alpha_2 = \frac{\delta_2}{l_2} \tag{7-72}$$

or

$$\alpha_2 \approx \frac{\delta_2}{l_2} \tag{7-73}$$

and

$$\phi_2 = \pi - \alpha_2 \tag{7-74}$$

or

$$\phi_2 = \pi - \frac{\delta_2}{l_2} \tag{7-75}$$

Therefore it can be concluded that the angle contribution for a pole (or zero) on the real axis must be in the form of Eq. (7-71) if the search point is to its right, or in the form of Eq. (7-75) if the search point is to its left.

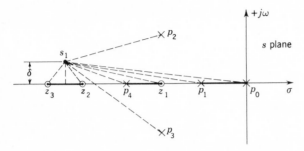

Fig. 7-18 *Example of the calculation of the angular condition in the vicinity of the breakaway point.*

When zeros are also present, the distance from a zero to the breakaway point is written in the form $(l)_1$ to distinguish it from the corresponding distance from a pole.

As shown in Fig. 7-17b, the angular contribution of the complex poles (or zeros) when the search point is on the real axis is always 360° for a conjugate pair. When δ is small, it is possible to calculate the angular contribution of the complex poles when the search point is in the vicinity of the breakaway point. From the consideration of the areas of the triangles in Fig. 7-17b, the change in the angle contribution of p_1 and p_2 is

$$\Delta\phi = \Delta\phi_1 + \Delta\phi_2 = \frac{2\delta b}{a^2 + b^2} \tag{7-76}$$

(See Appendix E for derivation.)

Thus the total angular contribution of p_1 and p_2 is $360° \pm \Delta\phi$. The plus sign is used when the complex poles or zeros lie to the left of s_1, since the angles increase upon departure from the real axis. In like manner, the minus sign is chosen when p_1 and p_2 are to the right of s_1.

An illustration is now given to demonstrate the use of the principles just stated. Figure 7-18 has been enlarged for visual purposes, and the standard notation is used although the figure is not completely labeled. The root locus exists on the real axis between the zeros z_2 and z_3. The problem is to find the location of the break-in point between them. From the angle condition,

$$\phi_0 + \phi_1 + \phi_{2,3} + \phi_4 - \psi_1 - \psi_2 - \psi_3 = (1 + 2m)\pi \tag{7-77}$$

or

$$\left(\pi - \frac{\delta}{l_0}\right) + \left(\pi - \frac{\delta}{l_1}\right) + \left(2\pi - \frac{2\delta b}{a^2 + b^2}\right) + \left(\pi - \frac{\delta}{l_4}\right)$$
$$- \left(\pi - \frac{\delta}{(l)_1}\right) - \left(\pi - \frac{\delta}{(l)_2}\right) - \frac{\delta}{(l)_3} = (1 + 2m)\pi \tag{7-78}$$

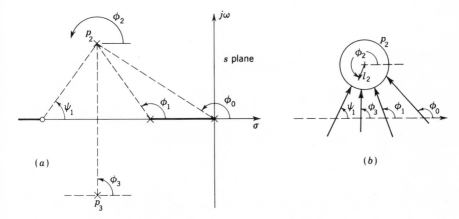

Fig. 7-19 *Angular condition in the vicinity of a complex pole.*

In the limit as $\delta \to 0$, the left-hand side of Eq. (7-78) must equal the right-hand side.   Thus

$$-\frac{\delta}{l_0} - \frac{\delta}{l_1} - \frac{2\delta b}{a^2 + b^2} - \frac{\delta}{l_4} + \frac{\delta}{(l)_1} + \frac{\delta}{(l)_2} - \frac{\delta}{(l)_3} = 0 \qquad (7\text{-}79)$$

or $\quad\quad \dfrac{1}{(l)_1} + \dfrac{1}{(l)_2} - \dfrac{1}{(l)_3} - \dfrac{1}{l_0} - \dfrac{1}{l_1} - \dfrac{1}{l_4} - \dfrac{2b}{a^2 + b^2} = 0 \qquad (7\text{-}80)$

By trial and error the point between z_2 and z_3 that satisfies Eq. (7-80) can be found.   Note that if the complex poles were on the left side of s_1, the sign of the angle due to these complex poles would change.   Therefore, for any plot of the poles and zeros the angular condition in the vicinity of a breakaway point has the form of Eq. (7-80) when δ is small.   It is advisable to determine the appropriate signs for each term from the plot.

Rule 8.   Complex pole (or zero): angle of departure

The next geometrical shortcut is the rapid determination of the direction in which the locus leaves a complex pole or enters a complex zero.   Although in Fig. 7-19a a complex pole is considered, the results also hold for a complex zero.

In Fig. 7-19a, an area about p_2 is chosen so that l_2 is very much smaller than l_0, l_1, l_3, and $(l)_1$.   For illustrative purposes, this area has been enlarged many times in Fig. 7-19b.   Under these conditions the angular contributions from all the other poles and zeros, except p_2, to a search point anywhere in this area are approximately constant.   They can be considered to have values determined as if the search point is right at p_2.   When the angle condition

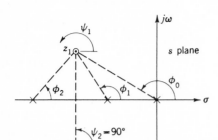

Fig. 7-20 *Angular condition in the vicinity of a complex zero.*

is applied to this small area, the angle equation is

$$\phi_0 + \phi_1 + \phi_2 + \phi_3 - \psi_1 = (1 + 2m)180° \qquad (7\text{-}81)$$
$$\phi_2 = (1 + 2m)180° - (\phi_0 + \phi_1 + 90° - \psi_1) \qquad (7\text{-}82)$$

In a similar manner the angle of approach to a complex zero may be determined. For an open-loop transfer function having the pole-zero arrangement shown in Fig. 7-20, the angle of approach ψ_1 to z_1 is given by

$$\psi_1 = (\phi_0 + \phi_1 + \phi_2 - 90°) - (1 + 2m)180° \qquad (7\text{-}83)$$

In other words, the direction of the locus as it leaves a pole or zero can be determined by adding up, according to the angle condition, all the angles of all vectors from all the other poles and zeros to the pole or zero in question. Subtracting this sum from 180° gives the required direction.

Rule 9. Imaginary-axis crossing point

In cases where the locus crosses the imaginary axis into the right-half s plane, the crossover point can usually be determined by Routh's method or by similar means. For example, if the closed-loop characteristic equation $D_1D_2 + N_1N_2$ is of the form

$$s^3 + bs^2 + cs + Kd = 0$$

the Routhian array is

The coefficient table is

$$
\begin{array}{c|ll}
s^3 & 1 & c \\
s^2 & b & Kd \\
s^1 & (bc - Kd)/b & \\
s^0 & Kd &
\end{array}
$$

An undamped oscillation exists if the s^1 row in the table equals zero.   For this condition the auxiliary equation obtained from the s^2 row is

$$bs^2 + Kd = 0 \tag{7-84}$$

and its roots are

$$s_{1,2} = \pm j\sqrt{\frac{Kd}{b}} = \pm j\omega_n \tag{7-85}$$

The sensitivity term K is determined by setting the s^1 row to zero:

$$K = \frac{bc}{d} \tag{7-86}$$

Equation (7-85) gives the natural frequency of the undamped oscillation. This corresponds to the point on the imaginary axis where the locus crosses over into the right-half s plane, denoting instability.   Also, the value of K from Eq. (7-86) determines the value of the static sensitivity at the cross-over point.   For values of $K < 0$ the term in the s^0 row is negative, thus characterizing an unstable response.   The limiting values for a stable response are, therefore,

$$0 < K < \frac{bc}{d} \tag{7-87}$$

In like manner, the crossover point can be determined for higher-order characteristic equations.   For these higher-order systems care must be exercised in *analyzing all terms in the first column that contain the term K in* order to obtain the correct limiting values of gain.

Rule 10.   Nonintersection or intersection of root locus branches[8]

By utilizing the theory of complex variables, the following properties are evolved:

1.   A value of s which satisfies the angle condition of Eq. (7-25) is a point on the root locus.   If $dW(s)/ds \neq 0$ at this point, there is one and only one branch of the root locus through the point.   Thus it can be said that there are no root-locus intersections at this point.
2.   If the first $y - 1$ derivatives of $W(s)$ vanish at a given point on the root locus, there will be y branches approaching and y branches leaving this point.   Thus, it can be said that there are root-locus intersections at this point.   The angle between two adjacent *approaching* branches is given by

$$\lambda_y = \pm \frac{360°}{y} \tag{7-88}$$

Also, the angle between a branch *leaving* and an adjacent branch

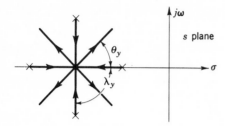

Fig. 7-21 *Root locus for* $G(s)H(s) = K/[(s+2)(s+4)(s^2+6s+10)]$.

that is *approaching* the same point is given by

$$\theta_y = \pm \frac{180°}{y} \tag{7-89}$$

Figure 7-21 illustrates these angles.

Rule 11.   Conservation of the sum of the system roots[9]

The technique described by this rule aids the determination of the general shape of the root locus.   Consider the general open-loop transfer function in the form

$$G(s)H(s) = \frac{K \displaystyle\prod_{m=1}^{w} (s - z_m)}{s^n \displaystyle\prod_{c=1}^{u} (s - p_c)} \tag{7-90}$$

By recalling that for physical systems $w \le v = u + n$, the denominator of $C(s)/R(s)$ can be written

$$B(s) = 1 + G(s)H(s) = \frac{\displaystyle\prod_{j=1}^{v} (s - r_j)}{s^n \displaystyle\prod_{c=1}^{u} (s - p_c)} \tag{7-91}$$

where r_j are the roots described by the root locus.

Substituting from Eq. (7-90) into Eq. (7-91) and equating numerators on each side of the resulting equation yields

$$s^n \prod_{c=1}^{u} (s - p_c) + K \prod_{m=1}^{w} (s - z_m) = \prod_{j=1}^{v} (s - r_j) \tag{7-92}$$

Expanding both sides of this equation,

$$\left(s^v - \sum_{c=1}^{u} p_c s^{v-1} + \cdots\right) + K\left(s^w - \sum_{m=1}^{w} z_m s^{w-1} + \cdots\right)$$
$$= s^v - \sum_{j=1}^{v} r_j s^{v-1} + \cdots \quad (7\text{-}93)$$

For those open-loop transfer functions that satisfy the condition $w \leq v - 2$ the following is obtained by equating the coefficients of s^{v-1} of Eq. (7-93):

$$\sum_{c=1}^{u} p_c = \sum_{j=1}^{v} r_j$$

This equation can also be written as

$$\sum_{j=1}^{v} p_j = \sum_{j=1}^{v} r_j \quad (7\text{-}94)$$

where p_j now represents all the open-loop poles, including those at the origin, and r_j are the roots of the characteristic equation. This equation reveals that as the system gain is varied from zero to infinity, the sum of the system roots is constant. In other words, the sum of the system roots is conserved and is independent of K. When a system has several root-locus branches which go to infinity (as $K \to \infty$), the directions of the branches are such that the sum of the roots is constant. A branch going to the right therefore requires that there will be a branch going to the left. The root locus of Fig. 7-2 satisfies the conservancy law for the root locus. The sum of the roots is a constant for all values of K.

Rule 12. Determination of roots on the root locus

After the root locus has been plotted, the specifications for system performance are used to determine the dominant roots (the roots closest to the imaginary axis). For the dominant roots the required static loop sensitivity can be determined by applying the magnitude condition, as shown in Eq. (7-40). The remaining roots on each of the other branches may be determined by any of the following methods:

1. Trial-and-error search for a point on each locus that satisfies the static loop sensitivity for the dominant roots. (See Fig. 7-10.)
2. If all except one real or a complex pair of roots are known, either of the following procedures can be used:
   a. Divide the characteristic equation by the factors representing the known roots. The remainder gives the remaining roots.

One should not expect an exact division because of the inaccuracies introduced in graphically determining the poles and the static loop sensitivity.

  b.   Equation (7-94), known as Grant's rule, can be used to find some of the roots. A necessary condition is that the denominator of $G(s)H(s)$ be at least of degree 2 higher than the numerator. If all the roots except one real root are known, application of Eq. (7-94) yields directly the value of the real root. However, for complex roots of the form $r = \sigma \pm j\omega_d$ it yields only the value of its real component σ. The magnitude of these roots is determined from the relationship

$$|r|^2 = K_n(-1)^n \frac{\displaystyle\prod_{c=1}^{u} (p_c)}{\displaystyle\prod_{j=1}^{v-2} (r_j)} \qquad n \geq 1 \tag{7-95}$$

where K_n = static error coefficient (gain)

  p_c = poles of $G(s)H(s)$, excluding poles at origin

  r_j = known roots of characteristic equation

The imaginary part of the roots can now be determined from

$$\omega_d = (|r|^2 - \sigma^2)^{1/2} \tag{7-96}$$

7-9   *Example 1*

Given:

$$G(s) = \frac{K(s+1)}{s^2 + 3s + 3.25} = \frac{K(s+1)}{[s + (1.5 - j1)][s + (1.5 + j1)]}$$

$$H(s) = 1$$

Draw the complete root locus for this system.

  1.   Plot p's and z's of $G(s)$ (see Fig. 7-22). From the plot of the p's and z's of $G(s)$ it is evident that the system has a stable response for $\pm$ values of K. Applying Rule 9 shows that the system is stable for $-3 < K < \infty$. This example is divided into two parts: the first for positive K and the second for negative K.

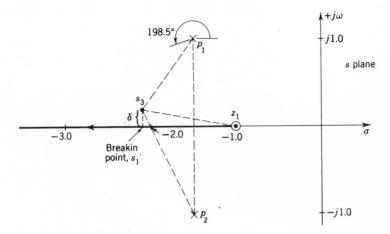

Fig. 7-22 *Sample problem.*

For $K > 0$

2. Real-axis locus exists to the left of z_1.
3. Number of branches of locus: two.
4. Asymptotes:

$$\text{Angle of asymptotes} = \frac{(1 + 2m)180°}{v - w} = (1 + 2m)180°$$

Thus the asymptote is on the real axis.

5. Break-in point s_1 on the real axis:

$$W(s) = -K = \frac{s^2 + 3s + 3.25}{s + 1}$$

$$\frac{dW(s)}{ds} = \frac{(s + 1)(2s + 3) - (s^2 + 3s + 3.25)}{(s + 1)^2} = 0$$

$$s^2 + 2s - 0.25 = 0$$

$$s_{1,2} = -1 \pm \tfrac{1}{2}\sqrt{4 + 1} = -1 \pm 1.12$$

The break-in point is

$$s_1 = -2.12$$

The value of the static loop sensitivity at $s = s_1$ is

$$-K = \frac{(-2.12)^2 + (3)(-2.12) + 3.25}{-2.12 + 1}$$

$$K = 1.23$$

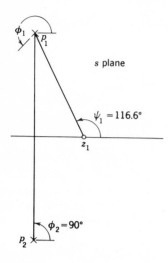

Fig. 7-23   Determination of departure angle.

6. Locus departure from the complex poles (see Fig. 7-23): for a point near p_1,

$$\phi_1 + \phi_2 - \psi_1 = 180°$$
$$\phi_1 + 90° - 116.6° = 180°$$
$$\phi_1 = 206.6°$$

7. Search point s_2 just off the real axis near s_1: the angle equation is

$$\phi_1 + \phi_2 - \psi_1 = (1 + 2m)\pi$$

or

$$\left(2\pi - \frac{2\delta b}{a^2 + b^2}\right) - \left[\pi - \frac{\delta}{(l)_1}\right] = (1 + 2m)\pi$$
$$\frac{-2b}{a^2 + b^2} + \frac{1}{(l)_1} = 0$$
$$\frac{1}{(l)_1} = \frac{2b}{a^2 + b^2}$$

By trial and error the point $s_2 = -2.12 + j0.125$ is selected, where $a = 1.0$, $b = 0.62$, $\delta = 0.125$, and $(l)_1 \approx 1.12$. These values give

$$0.893 \approx 0.895$$

Within the accuracy of the plot, the point $s_2 = -2.12 + j0.125$ is a point on the locus.

8. A few more search points can be taken in this area until a distance from the real axis is reached where the Spirule can be used with accuracy. Step 6 indicates the general direction in which the search points must be chosen since they must link up with the locus coming from the poles. Thus, all the geometrical shortcuts mini-

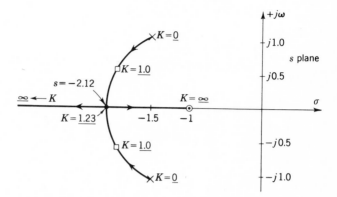

Fig. 7-24 *Root locus of sample problem for $K > 0$.*

mize the trial and error in plotting the locus. In Appendix F is illustrated a more refined method, using the phase angle loci, for determining the root locus. Once a student understands the fundamentals of the original method, it is to his advantage to learn and utilize the phase angle loci since this method yields the locus more rapidly for the more complex transfer functions. The root locus for this problem is drawn in Fig. 7-24.

9. By using the magnitude condition, the locus can be calibrated, and in Fig. 7-24 a few points have been so determined.

10. If $K = 1$, the form of the time solution can be immediately determined for a unit step input by reading the zeros of $1 + G(s)$ from the locus. Thus, for unity feedback,

$$c(t) = \mathcal{L}^{-1}\left[\frac{G(s)}{1 + G(s)} R(s) \right] = \mathcal{L}^{-1}\left[\frac{s + 1}{s(s + 2 + j0.5)(s + 2 - j0.5)} \right]$$

$$c(t) = 0.235 + 1.09e^{-2t} \sin{(0.5t - 0.218)} \tag{7-97}$$

The coefficients of the terms in $c(t)$ can be determined from the pole-zero diagram of $C(s)$ and the method of Sec. 4-10.

From Fig. 7-24 it can be seen that this system is stable for any positive value of gain since all possible roots lie in the left-half s plane. Notice in the solution $c(t)$ that the real part of the complex roots $-\zeta\omega_n = -2$ corresponds to the rate of decay and that the imaginary part $\omega_d = 0.5$ is the damped natural frequency of oscillation of the transient.

The root locus shows the response of this system as the static loop sensitivity K is varied. A smaller sensitivity reduces the damping ratio, with the result that the peak transient overshoot is greater. A larger sensitivity increases the damping ratio, with the result that the overshoot is smaller and

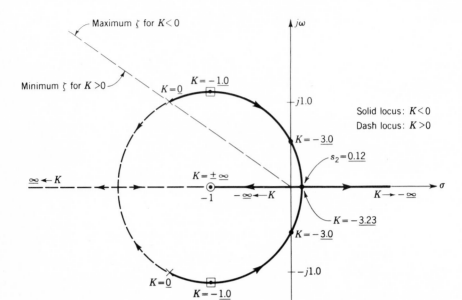

Fig. 7-25   *Root locus of sample problem for both positive and negative values of K.*

a shorter time is required to reach the steady-state value. A sensitivity larger than 1.23 makes the system overdamped. Depending on the use to be made of this system, the sensitivity can be adjusted to give the optimum response. If the desired response cannot be obtained with any setting of static loop sensitivity, compensation techniques must be used. Compensators and their use to improve the performance of a system are discussed in Chaps. 12 through 14.

For $K < 0$

By repeating the detailed solution for $K > 0$ but with the appropriate modifications, the following corresponding data are obtained:

1. Real-axis locus exists to the right of z_1.
2. Number of branches of locus: two.
3. Asymptote: $\gamma = 0°$.
4. Break-in point on the real axis: $s_2 = 0.12$ with a value of sensitivity $K = -3.23$.
5. Departure angle from p_1: $\phi_1 = 26.6°$.
6. Utilizing the auxiliary equation in Routh's stability criterion yields the imaginary-axis crossing points as $\pm j0.5$ for $K = -3.0$.
7. A sufficient number of additional points are determined by the use of the angle condition $\beta = m360°$ to obtain the root locus (solid locus)

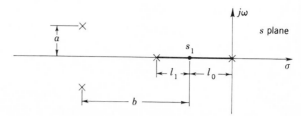

Fig. 7-26 *Location of the breakaway point.*

of Fig. 7-25. For comparison, the locus for $K > 0$ is shown in the figure as the dashed locus. The root locus is a circle with its center at z_1 and passing through the poles.

For $K = -1$ the control ratio for this unity feedback system is given by

$$\frac{C(s)}{R(s)} = \frac{-(s+1)}{(s+1-j1.12)(s+1+j1.12)} \tag{7-98}$$

Note that the minus sign in the numerator of the above equation is a result of the fact that $K < 0$. For nonunity feedback systems the minus sign associated with the value of K, for $K < 0$, is the result of *either* a minus value for the constant term in N_1 or N_2. Therefore the control ratio takes on the $+$ or $-$ sign of the constant associated with N_1.

7-10   Example 2

Given:

$$G(s) = \frac{K_1}{s(s^2/2{,}600 + s/26 + 1)} \quad \text{and} \quad H(s) = \frac{1}{0.04s + 1}$$

Rearranging gives

$$G(s) = \frac{2{,}600K_1}{s(s^2 + 100s + 2{,}600)} = \frac{N_1}{D_1} \quad \text{and} \quad H(s) = \frac{25}{s + 25} = \frac{N_2}{D_2}$$

Thus

$$G(s)H(s) = \frac{65{,}000K_1}{s(s + 25)(s^2 + 100s + 2{,}600)}$$

Find $C(s)/R(s)$ with $\zeta = 0.5$ for the dominant roots (roots closest to the imaginary axis).

1. The poles of $G(s)H(s)$ are plotted on the s plane in Fig. 7-26. The values of these poles are $s = 0,\ -25,\ -50 + j10,\ -50 - j10$.

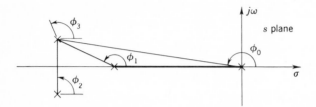

Fig. 7-27 *Determination of the departure angle.*

The system is completely unstable for $K < 0$. Therefore this example is solved only for the condition $K > 0$.

2. There are four branches of the root locus.

3. The locus exists on the real axis between 0 and -25.

4. The angles of the asymptotes are

$$\gamma = \frac{(1 + 2m)180°}{4} = \pm45°,\ \pm135°$$

5. The real-axis intercept of the asymptotes is

$$\sigma_o = \frac{0 - 25 - 50 - 50}{4} = -31.25$$

6. The breakaway point s_1 (see Fig. 7-26) on the real axis between 0 and -25 is found from

$$-\frac{1}{l_0} + \frac{1}{l_1} + \frac{2b}{a^2 + b^2} = 0$$

By trial and error,

$$l_0 = 9$$

7. The angle of departure ϕ_3 (see Fig. 7-27) from the pole $(-50 + j10)$ is obtained from

$$\phi_0 + \phi_1 + \phi_2 + \phi_3 = (1 + 2m)180°$$
$$168.7° + 158.2° + 90° + \phi_3 = (1 + 2m)180°$$
$$\phi_3 = 123.1°$$

Similarly, the angle of departure from the pole $(-50 - j10)$ is $-123.1°$.

8. The imaginary-axis intercepts are obtained from:

$$\frac{C(s)}{R(s)} = \frac{2{,}600K_1(s + 25)}{s^4 + 125s^3 + 5{,}100s^2 + 65{,}000s + 65{,}000K_1}$$

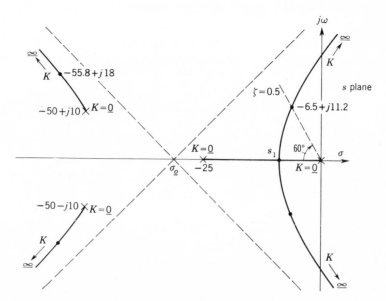

Fig. 7-28   *Root locus for* $G(s)H(s) = 65{,}000K_1/s(s + 25)(s^2 + 100s + 2{,}600)$.

The Routhian array for the denominator of $C(s)/R(s)$, which is the characteristic equation, is

s^4	1	5,100	65,000K_1
s^3	1	520 (after division by 125)	
s^2	1	14.2K_1 (after division by 4,580)	
s^1	$520 - 14.2K_1$		
s^0	14.2K_1		

Pure imaginary roots exist when the s^1 row is zero. This occurs when $K_1 = 520/14.2 = 36.7$. The auxiliary equation is

$$s^2 + 14.2K_1 = 0$$

and the imaginary roots are

$$s = \pm j\sqrt{14.2K_1} = \pm j\sqrt{520} = \pm j22.8$$

9. Additional points on the root locus are found by use of the Spirule, locating points that satisfy the angle condition

$$\underline{/s} + \underline{/s + 25} + \underline{/s + 50 - j10} + \underline{/s + 50 + j10} = (1 + 2m)180°$$

The locus is shown in Fig. 7-28.

10.   The radial line for $\zeta = 0.5$ is drawn on the graph of Fig. 7-28 at the angle (see Fig. 4-3 for definition of η)

$$\eta = \cos^{-1} 0.5 = 60°$$

The dominant roots obtained from the graph are

$$s = -6.5 \pm j11.2$$

11.   The gain is obtained graphically from the expression

$$K = 65{,}000K_1 = |s| \cdot |s + 25| \cdot |s + 50 - j10| \cdot |s + 50 + j10|$$

For $s = -6.5 + j11.2$,

$$K = 65{,}000K_1 = (13)(21.6)(43.5)(48.4)$$
$$K_1 = 9.1$$

12.   The other roots are evaluated to satisfy the magnitude condition ($K_1 = 9.1$). By trial and error the value $s = -55.8 + j18.0$ gives

$$(65{,}000)(9.1) = (58.7)(35.9)(10.0)(28.8)$$
$$592{,}000 = 606{,}000$$

This is sufficiently close. Therefore the other roots of the characteristic equation are

$$s = -55.8 \pm j18.0$$

12.   (Alternative methods.) The other roots can also be obtained by dividing the characteristic equation by the quadratic factor representing the two dominant roots. With $K_1 = 9.1$, the characteristic equation is

$$s^4 + 125s^3 + 5{,}100s^2 + 65{,}000s + 591{,}500 = 0$$

The quadratic factor representing the dominant roots is

$$(s + 6.5 - j11.2)(s + 6.5 + j11.2) = s^2 + 13s + 168$$

Dividing the characteristic equation by this quadratic factor leaves the quadratic representing the other roots as

$$s^2 + 112s + 3{,}476$$

There is a remainder when this division is performed which is due to the graphical inaccuracy. The other roots obtained from this factor are

$$s = -56 \pm j18.4$$

The values of the other roots obtained by both methods are almost equal.

The additional roots can also be determined by using Grant's method.   From Eq. (7-94),

$$0 - 25 + (-50 + j10) + (-50 - j10)$$
$$= (-6.5 + j11.2) + (-6.5 - j11.2) + (\sigma + j\omega_d) + (\sigma - j\omega_d)$$

This gives $\sigma = -56$

By using this value, the roots can be determined from the root locus as $(-56 \pm j18.2)$.   If these branches of the locus have not been drawn, proceed as shown below.

From Eq. (7-95),

$$|r|^2 = (9.1)(-1)^1 \frac{(-25)(-50 + j10)(-50 - j10)}{(-6.5 + j11.2)(-6.5 - j11.2)} = 3,525$$

From Eq. (7-96),

$$\omega_d = (3,525 - 3,140)^{1/2} = 19.6$$

This gives the roots as $(-56 \pm j19.6)$, which are within the expected graphical accuracy.

13.   The control ratio is

$$\frac{C(s)}{R(s)} = \frac{N_1 D_2}{\text{factors as determined from root locus}}$$

$$= \frac{23,680(s + 25)}{(s + 6.5 + j11.2)(s + 6.5 - j11.2)(s + 55.8 + j18)(s + 55.8 - j18)}$$

$$= \frac{23,680(s + 25)}{(s^2 + 13s + 168)(s^2 + 111.6s + 3,450)}$$

14.   The response $c(t)$ for a step input is found from

$$C(s) = \frac{23,680(s + 25)}{s(s^2 + 13s + 168)(s^2 + 111.6s + 3,450)}$$

$$= \frac{A_0}{s} + \frac{A_1}{s + 6.5 - j11.2} + \frac{A_2}{s + 6.5 + j11.2}$$

$$\qquad\qquad + \frac{A_3}{s + 55.8 - j18} + \frac{A_4}{s + 55.8 + j18}$$

The constants are

$$A_0 = 1.07$$
$$A_1 = 0.63\underline{/-202.7°}$$
$$A_3 = 0.14\underline{/-64.2°}$$

Note that A_0 should be exactly 1 but has an error due to the accumulated graphical errors.   This can be seen from the fact that $G(s)$ is Type 1 and the gain of $H(s)$ is unity.   It can also be

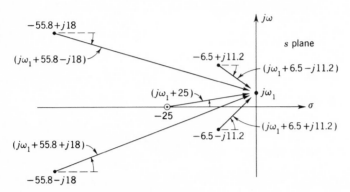

Fig. 7-29 *Frequency response from the plot, in the s plane, of the closed-loop poles and zeros.*

obtained from the equation of $C(s)/R(s)$ in step 8.   Inserting $R(s) = 1/s$ and finding the final value gives $c(t)_{ss} = 1.0$.   The value of $c(t)$ is

$$c(t) = 1 + 1.26e^{-6.5t} \sin{(11.2t - 112.7°)}$$
$$+ 0.28e^{-55.8t} \sin{(18t + 25.8°)}   (7\text{-}99)$$

7-11   *Frequency Response*[6]

Once the root locus has been determined and the system gain has been set for the desired performance, it is very easy to determine the steady-state frequency response.   The frequency response gives the ratio of phasor output to phasor input for sinusoidal inputs over a band of frequencies.   The plots of the magnitude M and the angle α of $C(j\omega)/R(j\omega)$ versus the frequency ω define the frequency response of a control system.   These curves are very useful in control system design:

1.   They enable a designer to minimize the effect of any undesirable noise within the system.
2.   They present a qualitative picture of the system's transient response. The correlation between the transient and frequency responses is discussed in detail in Chaps. 8 and 11.

For the problem of Sec. 7-10 the closed-loop control ratio is

$$\frac{C(s)}{R(s)} = \frac{23{,}680(s + 25)}{(s + 6.5 - j11.2)(s + 6.5 + j11.2)(s + 55.8 - j18)(s + 55.8 + j18)}$$
$$(7\text{-}100)$$

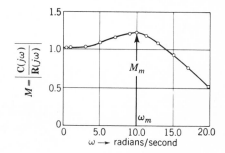

Fig. 7-30 *Closed-loop frequency response for the system of Sec. 7-10.*

To obtain the steady-state frequency response, let s assume values equal to $j\omega$. The control ratio as a function of frequency is

$$\frac{\mathbf{C}(j\omega)}{\mathbf{R}(j\omega)} = \frac{23{,}680(j\omega + 25)}{(j\omega + 6.5 - j11.2)(j\omega + 6.5 + j11.2)(j\omega + 55.8 - j18)(j\omega + 55.8 + j18)} \quad (7\text{-}101)$$

For any frequency ω_1 each of the factors of Eq. (7-101) is a directed line segment and can be measured on the pole-zero plot, as shown in Fig. 7-29. When the magnitudes and angles for each term of Eq. (7-101) have been obtained, the value of $\mathbf{C}(j\omega)/\mathbf{R}(j\omega)$ can be determined. Note that the angles are measured in the manner shown in Fig. 7-29, where clockwise angles are negative and counterclockwise angles are positive. This procedure is repeated for a sufficient number of values of frequency to draw a plot of $\mathbf{C}(j\omega)/\mathbf{R}(j\omega)$ versus ω. Figure 7-30 shows a plot of the magnitude $|\mathbf{C}(j\omega)/\mathbf{R}(j\omega)|$ versus ω obtained from the pole-zero plot of Fig. 7-29. The angle curve of $\mathbf{C}(j\omega)/\mathbf{R}(j\omega)$ is also evaluated graphically but is not shown here. The maximum value of the magnitude curve is labeled M_m, and the frequency at which this occurs is labeled ω_m. In Chap. 11 these quantities are related to the time response.

7-12   *Performance Characteristics*

As pointed out early in this chapter, the root-locus method incorporates the more desirable features of both the classical method and the steady-state sinusoidal phasor analysis. In the example of Sec. 7-9 a direct relationship is noted between the root locus and the time solution. This section is devoted to strengthening this relationship to enable the designer to synthesize and/or compensate a system.

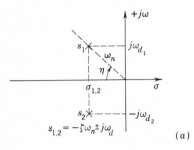

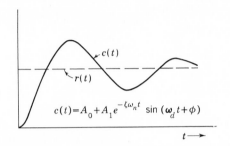

(a)

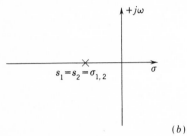

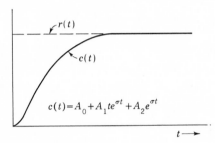

(b)

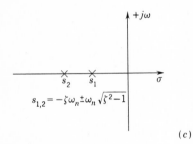

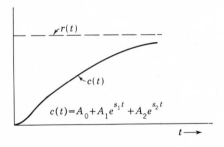

(c)

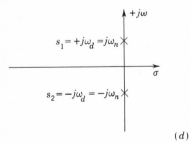

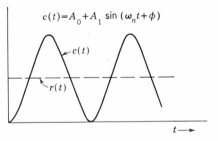

(d)

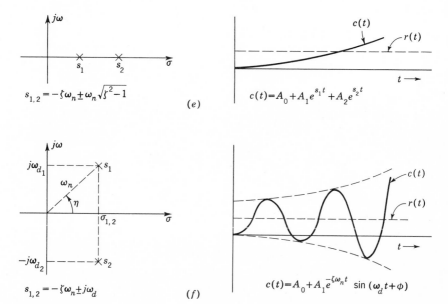

Fig. 7-31 *Plot of roots of a second-order characteristic equation in the s plane and their correlation to the transient solution in the time domain for a step input.* (a) *Underdamped–stable,* $0 < \zeta < 1.$ (b) *Critically damped–stable,* $\zeta = 1.$ (c) *Overdamped–stable,* $\zeta > 1.$ (d) *Undamped–sustained oscillations,* $\zeta = 0.$ (e) *Real roots–unstable.* (f) *Underdamped–unstable,* $0 > \zeta > -1.$

General Introduction

In review, consider a simple second-order system whose control ratio is

$$\frac{C(s)}{R(s)} = \frac{K}{s^2 + 2\zeta\omega_n s + \omega_n^2} \tag{7-102}$$

and whose transient component of the response to a step input is

$$c(t)_t = C_1 e^{s_1 t} + C_2 e^{s_2 t} \tag{7-103}$$

where
$$s_1 = -\zeta\omega_n + j\omega_n\sqrt{1 - \zeta^2} = \sigma + j\omega_d \qquad \zeta < 1 \tag{7-104}$$
$$s_2 = -\zeta\omega_n - j\omega_n\sqrt{1 - \zeta^2} = \sigma - j\omega_d$$

Thus
$$c(t)_t = A e^{\sigma t} \sin(\omega_d t + \phi) \tag{7-105}$$

Consider now a plot of the roots in the s plane and their correlation with the transient solution in the time domain for a step input. In Fig. 7-31 are illustrated six cases for different values of damping, showing both the locations of the roots and the corresponding transient plots.

In the case of $\zeta = 0$ the roots lie on the $\pm j\omega$ axis. For this case there

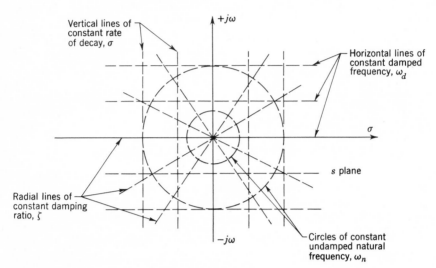

Fig. 7-32   *Constant-parameter curves on the s plane.*

are sustained oscillations. Those portions of the locus that yield roots in the right-half s plane result in unstable operation. Thus, the desirable roots are on that portion of the locus in the left-half s plane. As shown in Chap. 10, the $\pm j\omega$ axis of the s plane corresponds to the $(-1 + j0)$ point of the Nyquist plot.

Table 4-1 summarizes the information available from Fig. 7-31, i.e., the correlation between the location of the closed-loop poles and the corresponding transient component of the response. Thus the value of the root-locus method, i.e., it is possible to determine all the forms of the transient component of the response that a control system may have.

Plot of Characteristic Roots for $0 < \zeta < 1$

The important desired roots of the time solution lie in the region in which $0 < \zeta < 1$ (generally between 0.4 and 0.8). In Fig. 7-31a, the radius r from the origin to the root s_1 is

$$r = \sqrt{\omega_{d_1}^2 + \sigma_{1,2}^2} = \sqrt{\omega_n^2(1 - \zeta^2) + \omega_n^2\zeta^2} = \omega_n \qquad (7\text{-}106)$$

and

$$\cos \eta = \left| \frac{-\sigma_{1,2}}{r} \right| = \frac{\zeta\omega_n}{\omega_n} = \zeta \qquad (7\text{-}107)$$

or

$$\eta = \cos^{-1} \zeta \qquad (7\text{-}108)$$

From the above equations the constant-parameter loci are drawn in Fig. 7-32.

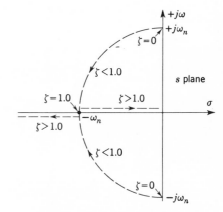

Fig. 7-33 *Variation of roots of a simple second-order system for a constant ω_n.*

From Figs. 7-31 and 7-32 and Eqs. (7-106) to (7-108) the following conclusions are drawn pertaining to the s plane:

1. Horizontal lines represent lines of constant damped natural frequency ω_d. The closer these lines are to the real axis, the lower is the value of ω_d. For roots lying on the real axis ($\omega_d = 0$) there is no oscillation.

2. Vertical lines represent lines of constant damping or constant speed of response of the transient. The closer these lines are (or the characteristic roots are) to the imaginary axis, the longer it takes for the transient response to die out.

3. Circles about the origin are circles of constant undamped natural angular frequency ω_n. Since $\sigma^2 + \omega_d^2 = \omega_n^2$, it is seen that the locus of the roots of constant ω_n is a circle in the s plane, that is, $|s_1| = |s_2| = \omega_n$. Circles in the s plane with their center at the origin are loci of constant ω_n. The smaller the circles are, the lower is ω_n.

4. Radial lines passing through the origin at angle η are lines of constant damping ratio ζ. The angle η is measured clockwise from the negative real axis for positive ζ, as shown in Fig. 7-31a.

Variation of Roots with ζ

Note in Fig. 7-33 the following:

1. For $\zeta > 1$, the roots $s_{1,2} = -\zeta\omega_n \pm \omega_n \sqrt{\zeta^2 - 1}$ are real.
2. For $\zeta = 1$, the roots $s_{1,2} = -\zeta\omega_n$ are real and equal.
3. For $\zeta < 1$, the roots $s_{1,2} = -\zeta\omega_n \pm j\omega_n \sqrt{1 - \zeta^2}$ are complex conjugates.
4. For $\zeta = 0$, the roots $s_{1,2} = \pm j\omega_n$ are imaginary.

Higher-order Systems

The control ratio of a system of order v is given by

$$\frac{C(s)}{R(s)} = \frac{P(s)}{s^v + a_{v-1}s^{v-1} + \cdots + a_0} \tag{7-109}$$

In a system having one or more sets of complex conjugate roots in the characteristic equation, each quadratic factor is of the form

$$s^2 + 2\zeta\omega_n s + \omega_n^2 \tag{7-110}$$

and the roots are

$$s_{1,2} = \sigma \pm j\omega_d \tag{7-111}$$

The relationships developed for ω_n and ζ earlier in this section apply equally well for each complex conjugate pair of roots of a vth-order system. The distinction is that the dominant ζ and ω_n apply for that pair of complex conjugate roots which lie closest to the imaginary axis. These values of ζ and ω_n are dominant because the corresponding transient term has the longest settling time. Thus the dominant values of ζ and ω_n are set or adjusted for the desired response. It must be remembered that, in setting the values of the dominant ζ and ω_n, the other ζ's and ω_n's are automatically set. Depending on the location of the other roots, they modify the solution obtained from the dominant roots.

In the example of Sec. 7-10 the response $c(t)$ with a step input given by Eq. (7-99) is

$$c(t) = 1 + 1.26e^{-6.5t}\sin(11.2t - 112.7°)$$
$$+ 0.28e^{-55.8t}\sin(18t + 25.8°) \tag{7-112}$$

Note that the transient term due to the roots of $s = -55.8 \pm j18$ dies out in approximately one-tenth the time of the transient term due to the dominant roots $s = -6.5 \pm j11.2$. Therefore, the solution can be approximated by the simplified equation

$$c(t) \approx 1 + 1.26e^{-6.5t}\sin(11.2t - 112.7°) \tag{7-113}$$

This expression is valid except for a short initial period of time while the other transient term dies out. Equation (7-113) suffices if an approximate solution is desired, i.e., for determining M_p, t_p, and T_s. This is so since the neglected term does not appreciably affect these three quantities. In general, the designer can judge from the location of the roots which ones may be neglected. If an exact solution is desired, then all roots must be considered.

The transient response of any complex system, with the effect of all the roots taken into account, often can be considered to be the result of an equivalent second-order system. On this basis, it is possible to define effective (or equivalent) values of ζ and ω_n. Realize that, in order to alter either effective quantity, the location of one or more of the roots must be altered.

7-13 Synthesis

The root-locus method lends itself very readily to the problem of synthesis because of the direct relationship between the frequency and the time domains to the s domain. The desired response can be achieved by keeping in mind the following eight points.

1. First plot the locus; then, based on the known information, the location of the dominant closed-loop poles must be specified.
2. If the gain has been specified, then, by applying the magnitude condition to each branch of the locus, all the roots can be determined.
3. In the event that the gain is not known but must be determined, enough information must be known about the desired response to proceed with the synthesis.
4. Since the desired time response, in general, has a peak value M_p between 1.0 and 1.4, the locus has at least one set of complex conjugate poles. Often there is a pair of complex poles near the imaginary axis such as to dominate the time response of the system; if so, these poles are referred to as the dominant-pole pair.
5. To be able to determine or set the system gain, any of the following items must be specified for the dominant-pole pair: the damping ratio ζ, the response time (for 2 per cent: $T_s = 4/\zeta\omega_n$), the undamped natural frequency ω_n, or the damped natural frequency ω_d. As previously determined, each of these factors corresponds to either a line or a circle in the s domain.
6. When the line or circle corresponding to the given information has been drawn, as stated in point 5, its intersection with the locus corresponding to the dominant-pole pair fixes the value of the system gain. By applying the magnitude condition to this point the value of the gain required can be determined. Setting the value of the gain to this value results in the transient-response terms, corresponding to the dominant-pole pair, having the desired characterizing parameter.
7. Having fixed the value of the gain on the dominant branch, thus fixing the dominant poles, automatically fixes the location of all remaining roots on the other branches.
8. If the root locus does not yield the desired response, the locus must be altered by means of compensation to achieve the desired results.

The subject of compensation in the s domain is discussed in later chapters.

7-14   Summary of Root-locus Construction Rules for Negative Feedback

Rule 1.   The number of branches of the root locus is equal to the number of poles of the open-loop transfer function.

Rule 2.   For *positive* values of K, the root locus exists on those portions of the real axis for which the sum of the poles and zeros to the right is an odd number.   For *negative* values of K, the root locus exists on those portions of the real axis for which the sum of the poles and zeros to the right is an even number (including zero).

Rule 3.   The root locus starts $(K = 0)$ at the open-loop poles and ends $(K = \pm \infty)$ at the open-loop zeros or at infinity.

Rule 4.   The angles of the asymptotes of the root locus that end at infinity are determined by

$$\gamma = \frac{(1 + 2m)180° \text{ for } K > 0 \quad \text{ or } \quad m360° \text{ for } K < 0}{[\text{number of poles of } G(s)H(s)] - [\text{number of zeros of } G(s)H(s)]}$$

Rule 5.   The real-axis intercept of the asymptotes is

$$\sigma_o = \frac{\sum_{c=1}^{v} \text{Re}(p_c) - \sum_{m=1}^{w} \text{Re}(z_m)}{v - w}$$

Rule 6.   The breakaway point for the locus between two poles on the real axis (or the break-in point for the locus between two zeros on the real axis) can be determined by taking the derivative of the static loop sensitivity K with respect to s.   Equate this derivative to zero and find the roots of the resulting equation.   The root that occurs between the poles (or the zeros) is the breakaway (or break-in) point.   If the resulting equation is of the third order or higher, it may be easier to apply the Hills-and-dales method or Rule 7 to determine the breakaway or break-in point.

Rule 7.   The breakaway point can be determined by choosing a trial point a small distance σ from the real axis between two poles.   The incremental angle to this trial point due to a pole or zero on the real axis is approximately equal to δ/l, where l is the distance from the real pole or zero to the trial point.   The incremental angle from conjugate poles or zeros is

$$\Delta\phi = \frac{2\delta b}{a^2 + b^2}$$

where a is the imaginary value of the pole or zero and b is the real component of the distance between the pole or zero and the trial point. The sum of the incremental angles due to all the poles and zeros, with the correct sign included, must equal zero. Trial points are inserted into the equation until it satisfies this condition. The value of the trial point that satisfies this condition is the breakaway point.

Rule 8. For $K > 0$ the angle of departure from a complex pole is equal to 180° minus the sum of the angles from the other poles plus the sum of the angles from the zeros. Any of these angles may be positive or negative. For $K < 0$ the departure angle is 180° from that obtained for $K > 0$.

For $K > 0$ the angle of approach to a complex zero is equal to the sum of the angles from the poles minus the sum of the angles from the other zeros minus 180°. For $K < 0$ the approach angle is 180° from that obtained for $K > 0$.

Rule 9. The imaginary-axis crossing of the root locus can be determined by setting up the Routhian array from the closed-loop characteristic equation. Equate the s^1 row to zero and form the auxiliary equation from the s^2 row. The roots of the auxiliary equation are the imaginary-axis crossover points.

Rule 10. The selection of the dominant roots of the characteristic equation is based on the specifications that give the required system performance. The static loop sensitivity for these roots is determined by means of the magnitude condition. The remaining roots are then determined to satisfy the same magnitude condition.

Rule 11. For those open-loop transfer functions for which $w \leq v - 2$, Grant's rule states that the sum of the closed-loop roots is equal to the sum of the open-loop poles.

Rule 12. Once the dominant roots have been located, Grant's rule

$$\sum_{j=1}^{v} p_j = \sum_{j=1}^{v} r_j$$

may be used to find one real or two complex roots. Factoring known roots from the characteristic equation can simplify the work of finding the remaining roots. A trial-and-error procedure can also be used to locate all the roots for a specified value of K.

7-15 Conclusions

In this chapter the root-locus method is developed to solve graphically for the roots of the characteristic equation. This method can be extended to solving for the roots of any polynomial. The next chapter illustrates how any polynomial may be rearranged and put into the mathematical form of a ratio of factored polynomials which is equal to plus or minus unity, i.e.,

$$\frac{N(s)}{D(s)} = \pm 1$$

Once this form has been obtained, the procedures given in this chapter can be utilized to locate the roots of the polynomial.

The root locus permits the analysis of the performance of feedback systems and provides a basis for selecting the gain in order to best meet the performance specifications. Since the closed-loop poles are obtained explicitly, the form of the time response is directly available. If the performance specifications cannot be met, the root locus can be analyzed to determine the appropriate compensation to yield the desired results. This is covered in Chap. 12.

Bibliography

1. Saunders, R. M.: The Dynamo Electric Amplifier—Class A Operation, *Trans. AIEE*, vol. 68, pp. 1368–1373, 1949.
2. Truxal, J. G.: "Automatic Feedback Control System Synthesis," pp. 221–277, McGraw-Hill Book Company, New York, 1955.
3. Smith, O. J. M.: "Feedback Control Systems," chap. 9, McGraw-Hill Book Company, New York, 1958.
4. Savant, C. J., Jr.: "Basic Feedback Control System Design," chap. 4, McGraw-Hill Book Company, New York, 1958.
5. Evans, W. R.: "Control-system Dynamics," McGraw-Hill Book Company, New York, 1954.
6. Jackson, A. S.: An Extension of the Root-locus Method to Obtain Closed-loop Frequency Response of Feedback Control Systems, *Trans. AIEE*, vol. 73, pt. II, pp. 176–179, September, 1954.
7. Lorens, C. S., and R. C. Titsworth: Properties of Root Locus Asymptotes, letter in *IRE Trans. Auto. Control*, vol. AC-5, pp. 71–72, January, 1960.
8. Wilts, C. H.: "Principles of Feedback Control," Addison-Wesley Publishing Company, Inc., Reading, Mass., 1960.
9. Grant, A. J., North American Aviation, Inc.: "The Conservation of the Sum of the System Roots as Applied to the Root Locus Method," unpublished paper, Apr. 10, 1953.
10. Liethen, F. E., C. H. Houpis, and J. J. D'Azzo: An Automatic Root Locus Plotter Using an Analog Computer, *Trans. AIEE*, vol. 79, pt. II, pp. 523–527, January, 1961.
11. Paskin, H. M.: "Automatic Computation of Root Loci Using a Digital Computer," M.Sc. Thesis, Air Force Institute of Technology, Dayton, Ohio, March, 1962.
12. Morgan, M. L., and J. C. Looney: Design of the ESIAC Algebraic Computer, *IRE Trans. Electron. Computers*, vol. EC-10, no. 3, September, 1961.

8

Specialized Pole-Zero Topics

8-1 Introduction

This chapter presents a number of topics which enhance the analysis of a control system. Graphical means, utilizing a pole-zero diagram, are presented for calculating the figures of merit of the system and the effect of additional significant nondominant poles. Typical pole-zero patterns are employed to demonstrate the correlation of the pole-zero diagram with the frequency and time responses. The partition method of Sec. 8-5 utilizes the root-locus principle for factoring a polynomial, which is useful in finding break-in and breakaway points. The last section illustrates the analysis of a system with a simple nonlinearity by means of the root-locus method, which is a linear technique.

8-2   *Transient Response—Dominant Complex Poles*[1]

The root-locus plot permits selection of the best poles for the control ratio. The criteria for determining which poles are best must come from the specifications of system performance and from practical considerations. For example, the presence of backlash, dead zone, and Coulomb friction can produce a steady-state error with a step input, even though the linear portion of the system is Type 1 or higher. The response obtained from the pole-zero locations of the control ratio does not show the presence of a steady-state error because the nonlinearities have been neglected. The effect of nonlinear characteristics on system performance is taken up in Chap. 18. For the present, it is stated that the effect of some nonlinearities can be minimized by making the system underdamped. Therefore, the system gain is adjusted so that there is a dominant pair of complex poles and the response to a step input has the form shown in Fig. 8-1. In this case the transient contributions from the other poles must be small. By reference to Eq. (4-74), the necessary conditions for the time response to be dominated by one pair of complex poles require the pole-zero pattern of Fig. 8-2 and have the following characteristics:

1. The other poles must be far to the left of the dominant poles, so that the transients due to these other poles are small in amplitude and die out rapidly.
2. Any other pole which is not far to the left of the dominant complex poles must be near a zero so that the magnitude of the transient term due to that pole is small.

With the system designed so that the response to a unit step input has the underdamped form shown in Fig. 8-1, the following figures of merit are used to judge its performance. They are described in Secs. 3-11 and 3-12.

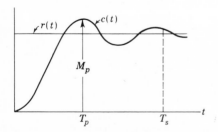

Fig. 8-1 *Transient response to a step input with dominant complex poles.*

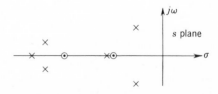

Fig. 8-2 *Pole-zero pattern of C/R for the desired response.*

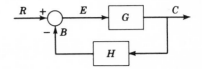

Fig. 8-3 *Feedback system.*

1. M_p, peak overshoot, is the amplitude of the first overshoot.
2. T_p, peak time, is the time to reach peak overshoot.
3. T_s, settling time, is the time for response to first reach and thereafter remain within 2 per cent of the final value.
4. N is the number of oscillations in the response up to the settling time.

Consider the nonunity feedback system shown in Fig. 8-3, using

$$G(s) = \frac{N_1}{D_1} = \frac{K_G \prod (s - z_g)}{\prod (s - p_g)} \tag{8-1}$$

$$H(s) = \frac{N_2}{D_2} = \frac{K_H \prod (s - z_h)}{\prod (s - p_h)} \tag{8-2}$$

$$G(s)H(s) = \frac{N_1 N_2}{D_1 D_2} = \frac{K_G K_H \prod_{m=1}^{w} (s - z_m)}{\prod_{c=1}^{v} (s - p_c)} \tag{8-3}$$

The product $K_G K_H = K$ is defined as the static loop sensitivity. For $G(s)H(s)$ the degree of the numerator is w and the degree of the denominator is v. These symbols are used throughout Chap. 7. The control ratio is

$$\frac{C(s)}{R(s)} = \frac{P(s)}{Q(s)} = \frac{N_1 D_2}{D_1 D_2 + N_1 N_2} = \frac{K_G \prod_{m=1}^{w'} (s - z_m)}{\prod_{c=1}^{v} (s - p_c)} \tag{8-4}$$

Note that the constant K_G in Eq. (8-4) is not the same as the static loop sensitivity unless the system has unity feedback. The degree v of the denominator of $C(s)/R(s)$ is the same as for $G(s)H(s)$, regardless of whether the system has unity or nonunity feedback. The degree w' of the numerator of $C(s)/R(s)$ is equal to the sum of the degrees of N_1 and D_2. For a unity feedback system the degree of the numerator is $w' = w$.

For a unit step input the output of Fig. 8-3 is

$$C(s) = \frac{P(s)}{sQ(s)} = \frac{K_G \displaystyle\prod_{m=1}^{w'} (s - z_m)}{s \displaystyle\prod_{c=1}^{v} (s - p_c)}$$

$$= \frac{A_0}{s} + \frac{A_1}{s - p_1} + \cdots + \frac{A_k}{s - p_k} + \cdots + \frac{A_v}{s - p_v} \qquad (8\text{-}5)$$

The pole $s = 0$ which comes from $R(s)$ must be included as a pole of $C(s)$, as shown in Eq. (8-5). The values of the coefficients can be obtained graphically from the pole-zero diagram by the method described in Sec. 4-10 and Eq. (4-74). Assume that the system represented by Eq. (8-5) has a dominant complex pole $p_1 = \sigma + j\omega_d$. The complete time solution is given by

$$c(t) = \frac{P(0)}{Q(0)} + 2 \left| \frac{K_G \displaystyle\prod_{m=1}^{w'} (p_1 - z_m)}{p_1 \displaystyle\prod_{c=2}^{v} (p_1 - p_c)} \right| e^{\sigma t} \cos \left[\omega_d t + \underline{/P(p_1)} - \underline{/p_1} - \underline{/Q'(p_1)}\right]$$

$$+ \sum_{k=3}^{v} \left[\frac{P(p_k)}{p_k Q'(p_k)} \right] e^{p_k t} \qquad (8\text{-}6)$$

where

$$Q'(p_k) = \frac{dQ(s)}{ds}\bigg]_{s=p_k} = \frac{Q(s)}{s - p_k}\bigg]_{s=p_k}$$

By assuming the pole-zero pattern of Fig. 8-2, the last term of Eq. (8-6) may be neglected. The time response is therefore approximated by

$$c(t) \approx \frac{P(0)}{Q(0)} + 2 \left| \frac{K_G \displaystyle\prod_{m=1}^{w'} (p_1 - z_m)}{p_1 \displaystyle\prod_{c=2}^{v} (p_1 - p_c)} \right| e^{\sigma t} \cos \left[\omega_d t + \underline{/P(p_1)} - \underline{/p_1} - \underline{/Q'(p_1)}\right]$$

$$(8\text{-}7)$$

Note that although the transient terms due to the other poles have been neglected, the effect of those poles on the amplitude and phase angle of the dominant transient has not been neglected.

The peak time T_p is obtained by setting the derivative with respect to time of Eq. (8-7) equal to zero. This gives

$$T_p = \frac{1}{\omega_d} \left[\frac{\pi}{2} - \underline{/P(p_1)} + \underline{/Q'(p_1)} \right] \qquad (8\text{-}8)$$

which may be stated as follows:

$$T_p = \frac{1}{\omega_d} \left\{ \frac{\pi}{2} - \begin{bmatrix} \text{sum of angles from} \\ \text{zeros of } C(s)/R(s) \\ \text{to dominant pole } p_1 \end{bmatrix} + \begin{bmatrix} \text{sum of angles from all} \\ \text{other poles of } C(s)/R(s) \\ \text{to dominant pole } p_1, \\ \text{including conjugate pole} \end{bmatrix} \right\}$$

(8-9)

From a physical consideration of Eq. (8-7) it can be seen that the phase angle $\phi = \underline{/P(p_1)} - \underline{/p_1} - \underline{/Q'(p_1)}$ cannot have a value greater than 2π. This same limitation must also be applied to the angles of Eq. (8-9), i.e., the value T_p must occur within one "cycle" of the transient. Inserting this value of T_p into Eq. (8-7) gives the peak overshoot M_p. By using the value $\cos(\pi/2 - \underline{/p_1}) = \omega_d/\omega_n$, the value M_p can be expressed as

$$M_p = \frac{P(0)}{Q(0)} + \frac{2\omega_d}{\omega_n^2} \left| \frac{K_G \prod\limits_{m=1}^{w'} (p_1 - z_m)}{\prod\limits_{c=2}^{v} (p_1 - p_c)} \right| e^{\sigma T_p}$$

(8-10)

The first term in Eq. (8-10) represents the final value and the second term represents the overshoot M_o. For a unity feedback system which is Type 1 or higher, the equation for M_o can be put in an alternative form. For such a system there is zero steady-state error when the input is a step function. Therefore the first term on the right side of Eq. (8-7) is equal to unity. The same expression can also be obtained by applying the final-value theorem to Eq. (8-5). This equality is used to solve for K_G. Note that under these conditions $K_G = K$ and $w' = w$:

$$K_G = \frac{\prod\limits_{c=1}^{v} (-p_c)}{\prod\limits_{m=1}^{w'} (-z_m)}$$

(8-11)

The value M_o can therefore be expressed as

$$M_o = \frac{2\omega_d}{\omega_n^2} \left| \frac{\prod\limits_{c=1}^{v} (-p_c) \prod\limits_{m=1}^{w'} (p_1 - z_m)}{\prod\limits_{m=1}^{w'} (-z_m) \prod\limits_{c=2}^{v} (p_1 - p_c)} \right| e^{\sigma T_p}$$

(8-12)

Some terms inside the brackets cancel the terms in front so that M_o can be expressed in words as

$$M_o = \frac{\begin{bmatrix} \text{product of distances} \\ \text{from all poles of} \\ C(s)/R(s) \text{ to origin,} \\ \text{excluding distances} \\ \text{of two dominant poles} \\ \text{from origin} \end{bmatrix} \begin{bmatrix} \text{product of distances} \\ \text{from all zeros of} \\ C(s)/R(s) \text{ to dominant} \\ \text{pole } p_1 \end{bmatrix}}{\begin{bmatrix} \text{product of distances} \\ \text{from all other poles} \\ \text{of } C(s)/R(s) \text{ to} \\ \text{dominant pole } p_1, \\ \text{excluding distance} \\ \text{between dominant} \\ \text{poles} \end{bmatrix} \begin{bmatrix} \text{product of distances} \\ \text{from all zeros of} \\ C(s)/R(s) \text{ to origin} \end{bmatrix}} e^{\sigma T_p} \qquad (8\text{-}13)$$

If there are no finite zeros of $C(s)/R(s)$, the factors in M_o involving zeros become unity. Equation (8-13) is valid only for a unity feedback system that is Type 1 or higher. It may be sufficiently accurate for a Type 0 system that has a large value for K_0. The value of M_o can be calculated either from the right-hand term of Eq. (8-10) or from Eq. (8-13), whichever is more convenient. The effect on M_o of other poles, which cannot be neglected, is discussed in the next section.

The values of T_s and N can be obtained approximately from the dominant roots. Section 3-11 shows that T_s is four time constants for 2 per cent error:

$$T_s = \frac{4}{|\sigma|} = \frac{4}{\zeta \omega_n} \qquad (8\text{-}14)$$

$$N = \frac{\text{settling time}}{\text{period}} = \frac{T_s}{2\pi/\omega_d} = \frac{2\omega_d}{\pi|\sigma|} = \frac{2}{\pi} \frac{\sqrt{1 - \zeta^2}}{\zeta} \qquad (8\text{-}15)$$

An examination of Eq. (8-8) or (8-9) reveals that zeros of the control ratio cause a decrease in the peak time T_p, whereas poles increase the peak time. Peak time can also be decreased by shifting zeros to the right or poles (other than the dominant poles) to the left. Equation (8-13) shows that the larger the value of T_p, the smaller the value of M_o because $e^{\sigma T_p}$ decreases. M_o can also be decreased by reducing the ratios $p_c/(p_c - p_1)$ and $(z_m - p_1)/z_m$. But this can have an adverse effect on T_p. The conditions on pole-zero locations that lead to a small M_o may therefore produce a large T_p. Conversely, the conditions that lead to a small T_p may be obtained at the expense of a large M_o. Thus a compromise is required in the peak time and peak overshoot that are attainable. Some improvements can be obtained by introducing additional poles and zeros into the system and locating them appropriately. This is covered in Chap. 12 on compensation.

The approximate equations, given in this section, for the response of a system to a step-function input are based on the fundamental premise that there are dominant complex poles and that the effect of other poles is small. Gustafson[9] gives another method for approximating the time response. A more complete analysis which is not subject to these limitations and which gives additional performance characteristics is contained in the technical literature.[3,4] The following section covers some particular cases.

8-3 *Additional Significant Poles*[2-5]

When there are two dominant complex poles, the approximations developed in Sec. 8-2 give accurate results. However, there are cases where an additional pole of $C(s)/R(s)$ is significant. Figure 8-4 shows a pole-zero diagram which contains dominant complex poles and an additional real pole p_3. The control ratio is given by

$$\frac{C(s)}{R(s)} = \frac{K}{(s^2 + 2\zeta\omega_n s + \omega_n^2)(s - p_3)} \tag{8-16}$$

With a unit step input the time response is given by

$$c(t) = 1 + 2|A_1|e^{-\zeta\omega_n t} \sin\left(\omega_n \sqrt{1 - \zeta^2}\, t + \phi\right) + A_3 e^{p_3 t} \tag{8-17}$$

The transient term due to the real pole p_3 has the form $A_3 e^{p_3 t}$, *where A_3 is always negative.* Therefore the peak overshoot M_p is reduced, but the settling time T_s is increased. This is the typical effect of an additional real pole. The magnitude A_3 depends on the location of p_3 relative to the complex poles. The further to the left the pole p_3 is located, the smaller is the magnitude of A_3; therefore the smaller is its effect on the total response. A pole which is six times as far to the left as the complex poles has negligible effect on the time response. The typical time response is shown in Fig. 8-5a. As the pole p_3 moves to the right, the magnitude of A_3 increases and the overshoot becomes smaller. As p_3 approaches but is still to the left of the complex poles, the first maximum in the time response is less than the final value. Overshoot of the final value can occur at the second or a later maximum. Figure 8-5b shows such a time response.

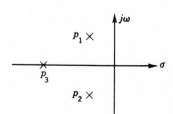

Fig. 8-4 *Pole-zero diagram of $C(s)/R(s)$.*

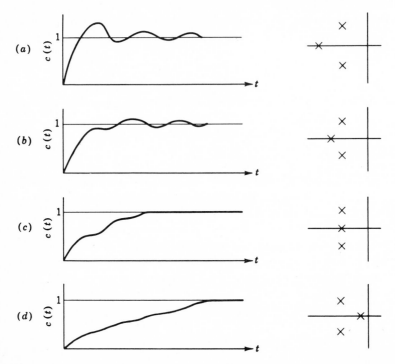

Fig. 8-5   *Typical time responses as a function of the real-pole location.*

When p_3 is located at the real-axis projection of the complex poles, the response is monotonic, i.e., there is no overshoot.   This represents the critically damped situation, as shown in Fig. 8-5c.

When p_3 is located to the right of the complex poles, it is the dominant pole, and the response is overdamped.   The complex poles contribute a "ripple" to the time response, as shown in Fig. 8-5d.

When the real pole p_3 is to the left of the complex poles, the peak time T_p is approximately given by Eq. (8-9).   Although the effect of the *real* pole is to increase the peak time, this change is small if the real pole is fairly far to the left.   A first-order correction can be made to the peak overshoot M_o given by Eq. (8-13) by adding the value of $A_3 e^{p_3 T_p}$ to the peak overshoot due to the complex poles.   The effect of the real pole on the actual settling time t_s can be estimated by calculating $A_3 e^{p_3 T_s}$ and comparing it with the size and sign of the underdamped transient at time T_s obtained from Eq. (8-14). If both are negative at T_s, then the true settling time t_s is increased.   If they have opposite signs, then the value $T_s = 4T$ based on the complex roots is a good approximation for t_s.

The presence of a real zero in addition to the real pole further modifies the transient response.   Figure 8-6 shows a possible pole-zero diagram for

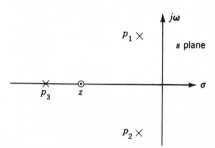

Fig. 8-6 *Pole-zero diagram of* $C(s)/R(s)$.

which the control ratio is

$$\frac{C(s)}{R(s)} = \frac{K(s - z)}{(s^2 + 2\zeta\omega_n s + \omega_n^2)(s - p_3)} \tag{8-18}$$

The complete time response to a unit-step-function input still has the form given in Eq. (8-17). However, the sign of A_3 depends on the relative locations of the real pole and the real zero. A_3 is negative if the zero is to the left of p_3, and it is positive if the zero is to the right of p_3. Also, the magnitude of A_3 is proportional to the distance from p_3 to z [see Eq. (4-74)]. Therefore, if the zero is close to the pole, A_3 is small and the contribution of this transient term is correspondingly small. Compared with the response for Eq. (8-16), as shown in Figure 8-5:

1. If the zero z is to the left of the real pole p_3, the response is qualitatively the same as that for a system with only complex poles, but the peak overshoot is smaller.
2. If the zero z is to the right of the real pole p_3, the peak overshoot is greater than that for a system with only complex poles.

When the real pole p_3 and the real zero z are close together, the change in the time response is small compared with that of a system which has only complex poles. The value of peak time T_p obtained from Eq. (8-9) can therefore be considered essentially correct. Actually T_p decreases if the zero is to the right of the real pole, and vice versa. A first-order correction to M_o from Eq. (8-13) can be obtained by adding the contribution of $A_3 e^{p_3 t}$ at the time T_p. The analysis of the effect on t_s is similar to that described above for the case of just an additional real pole.

Many control systems can be approximated by one having the following characteristics: (1) two complex poles; (2) two complex poles and one real pole; and (3) two complex poles, one real pole, and one real zero. For case 1 the relations T_p, M_o, T_s, and N developed in Sec. 8-2 give an accurate representation of the time response. For case 2 these approximate values can be corrected if the real pole is far enough to the left. Then the contribution of the additional transient term is small, and the total response remains essentially the sum of a constant and an underdamped sinusoid. For case 3 the approximate values can be corrected provided the zero is near the real pole so that the amplitude of the additional transient term is small. More exact calculation of the figures of merit can be obtained by plotting the exact response as a function of time.   This is conveniently accomplished by simulating the system on an analog computer or using the method of Koenig.[8]   Estimates of the figures of merit can also be obtained from graphs and families of curves that are available.[3-5]

8-4   Correlation of Pole-Zero Diagram with Frequency and Time Responses

The equations given in Sec. 8-2 readily permit the evaluation of the figures of merit for a control ratio of the form

$$\frac{C(s)}{R(s)} = \frac{\omega_n^2}{s^2 + 2\zeta\omega_n s + \omega_n^2} \qquad \text{for } \zeta < 1 \qquad (8\text{-}19)$$

For control ratios that have poles (in addition to the dominant complex pair) that are far to the left of the dominant complex pair and/or very close to zeros, the equations in Sec. 8-2 are still applicable.   When these conditions are not satisfied, additional significant poles may be analyzed with respect to the figures of merit, as indicated in Sec. 8-3.   A guide that may be used in conjunction with these two sections to determine time-response characteristics is the frequency-response plot of $|C(j\omega)/R(j\omega)|$ versus ω.

To illustrate the correlation of the pole-zero diagram with the frequency and time responses, the following three control ratios are utilized:

$$\frac{C(s)}{R(s)} = \frac{1}{s^2 + s + 1} \qquad (8\text{-}20)$$

$$\frac{C(s)}{R(s)} = \frac{3.13(s + 0.8)}{(s + 0.25)(s^2 + 0.3s + 1)} \qquad (8\text{-}21)$$

$$\frac{C(s)}{R(s)} = \frac{4}{(s^2 + s + 1)(s^2 + 0.4s + 4)} \qquad (8\text{-}22)$$

The pole-zero diagram, the frequency response, and the time response to a step input for each of these equations are shown in Fig. 8-7.

From Fig. 8-7a, which represents Eq. (8-20), the following characteristics are noted:

1.   The control ratio has only two complex poles, which are dominant, and no zeros.

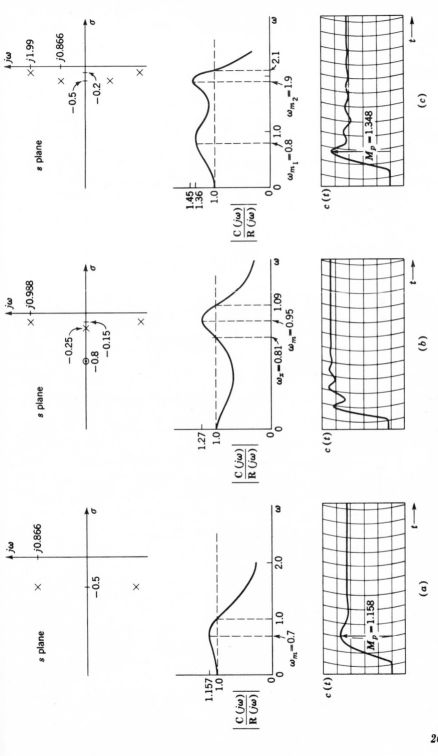

Fig. 8-7 *Comparison of frequency and time responses for three pole-zero patterns.*

261

2. The frequency-response curve has the following characteristics:
   a. A single peak, $M_m = 1.157$.
   b. $1.0 < M < M_m$ in the frequency range $0 < \omega < 1.0$.
3. The time response has the typical waveform described in Chap. 3. That is, the first maximum of $c(t)$ due to the oscillatory term is greater than $c(t)_{ss}$ and the $c(t)$ response after this maximum oscillates around the value of $c(t)_{ss}$.

From Fig. 8-7b, for Eq. (8-21), the following characteristics are noted:

1. The control ratio has two complex poles and one real pole, all dominant, and one real zero.
2. The frequency-response curve has the following characteristics:
   a. A single peak, $M_m = 1.27$.
   b. $M < 1.0$ in the frequency range $0 < \omega < \omega_x$.
   c. The peak M_m occurs at $\omega_m = 0.95 > \omega_x$.
3. The time response does not have the conventional waveform. That is, the first maximum of $c(t)$ due to the oscillatory term is less than $c(t)_{ss}$.

From Fig. 8-7c, for Eq. (8-22), the following characteristics are noted:

1. The control ratio has four complex poles, all dominant, and no zeros.
2. The frequency-response curve has the following characteristics:
   a. There are two peaks, $M_{m1} = 1.36$ and $M_{m2} = 1.45$.
   b. $1.0 < M < 1.45$ in the frequency range of $0 < \omega < 2.1$.
   c. The time response does not have the simple second-order waveform. That is, the first maximum of $c(t)$ in the oscillation is greater than $c(t)_{ss}$, and the oscillatory portion of $c(t)$ does not oscillate about a value of $c(t)_{ss}$.

Another control-ratio pole-zero pattern is shown in Fig. 8-4. The time responses to a unit step input, for various locations of the real pole, are shown in Fig. 8-5. For case (a), where the real pole is sufficiently far to the left that the complex poles are truly the dominant poles, the frequency-response plot has the form shown in Fig. 8-7a. For case (b), where all poles are dominant and the real pole is to the left of the complex poles, the frequency-response plot has the form shown in Fig. 8-7b. For case (c), where all poles are dominant and have the same value of σ, the frequency-response plot has the form shown in Fig. 8-8a. The *maximum* value is less than unity. Lastly, the frequency response of case (d), where the real pole is dominant and the complex poles are to its left, has the form shown in Fig. 8-8b.

The examples discussed in this section show that the time-response waveform is closely related to the system's frequency response. In other

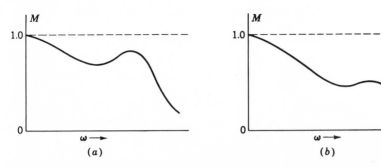

Fig. 8-8   *Form of the frequency-response waveform for cases (c) and (d) of Fig. 8-5, respectively.*

words, the system's time-response waveform may be predicted from the shape of the frequency-response plot, as summarized in the following chart.

If a system's M versus ω plot has the form shown in	*The system's anticipated time-response waveform has the form shown in*
Fig. 8-7a	Fig. 8-7a
Fig. 8-7b	Fig. 8-7b
Fig. 8-7c	Fig. 8-7c
Fig. 8-8a	Fig. 8-5c
Fig. 8-8b	Fig. 8-5d

Thus, as illustrated in this section, the frequency-response plot may be utilized as a guide in determining (or predicting) time-response characteristics. This may be of value for those systems in which two complex poles are not clearly the dominant poles.

8-5   *Factoring of Polynomials by Use of Root Locus (Partition Method)*

Chapter 7 deals with the development of the root locus as a graphical method for determining the roots of a characteristic equation (zeros of a polynomial) as a function of the static loop sensitivity. The development was associated with single-loop systems in which the characteristic equation has the form $1 + I(s) = 0$, where $I(s) = G(s)H(s)$ is available in factored form. This section illustrates how the root-locus method can also be applied for factoring polynomials which are not initially in that form. Some situations in which this problem arises in this book are:

1. Determination of breakaway and break-in points on a root locus by use of $dW(s)/ds = 0$ (Sec. 7-8).
2. Control systems which have minor loops. Such minor loops may exist inherently in the system or may be introduced for compensation purposes. (See Secs. 14-8 to 14-10.) In Fig. 5-26 the loop containing the a-c rate generator is called the *minor loop*, while the over-all loop containing unity feedback is called the *major loop*.
3. Control systems which have multiple inputs and/or multiple outputs (see Sec. 15-13).

In each of these situations, the root-locus method can be utilized to determine the zeros of a polynomial. The only distinction between them is the particular purpose for which the roots are desired. The application of the root-locus method for factoring polynomials described here is called the *partition method*.

When the degree of a polynomial is higher than two, the solution for its roots can become tedious. The partition method reduces the tediousness in many cases. The roots of a third-degree polynomial can usually be found directly by synthetic division. However, to illustrate the partition method, a third-degree polynomial is used which has the form

$$s^3 + as^2 + bs + c = 0 \tag{8-23}$$

This equation can be rearranged or partitioned at three places, as indicated above. They are

Partition ①:
$$s^3 = -(as^2 + bs + c) \tag{8-24}$$

$$-1 = \frac{a[s^2 + (b/a)s + c/a]}{s^3}$$

$$= \frac{a(s + \alpha)(s + \beta)}{s^3} = I_1(s) \tag{8-25}$$

where α and β may be either real or complex.

Partition ②:
$$s^3 + as^2 = -(bs + c) \tag{8-26}$$

$$-1 = \frac{b(s + c/b)}{s^2(s + a)} = I_2(s) \tag{8-27}$$

Partition ③:
$$s^3 + as^2 + bs = -c \tag{8-28}$$

$$-1 = \frac{c}{s(s^2 + as + b)}$$

$$= \frac{c}{s(s + \gamma)(s + \delta)} = I_3(s) \tag{8-29}$$

where γ and δ may be either real or complex.

Note: The resulting equation in each case has the mathematical form given by $I(s) = -1$, where $I(s)$ is in factored form. Therefore, it must satisfy the magnitude and angle conditions expressed by Eqs. (7-26) and (7-27), respectively. *Therefore, since the resulting equation after partitioning has the identical mathematical form as $G(s)H(s) = -1$, the root-locus method of Chap. 7 can be utilized to determine the roots of a polynomial.*

For the third-degree polynomial of Eq. (8-23) partition ② may be preferable because it does not require the factoring of a quadratic. The following two examples illustrate how the roots of a third-degree polynomial can be determined.

Example 1

$$s^3 + as^2 + bs + c = s^3 + 2s^2 + 2s + 2 = 0 \qquad (8\text{-}30)$$

Using partition ② yields

$$I_2(s) = \frac{b(s + c/b)}{s^2(s + a)} = \frac{2(s + 1)}{s^2(s + 2)} = -1 \qquad (8\text{-}31)$$

First replace the constant coefficient in the numerator of Eq. (8-31) by K, that is, replace 2 by K, where K is permitted to vary from 0 to ∞. The angle condition is given by

$$\bigg/\!\frac{K(s + 1)}{s^2(s + 2)} = (1 + 2m)180° \qquad (8\text{-}32)$$

A root-locus plot can be drawn using the standard techniques of Sec. 7-8. After the root locus is obtained, the point corresponding to $K = 2$ locates the roots of the original polynomial, Eq. (8-30). The pertinent data for this example are: $\sigma_o = -\frac{1}{2}$ and $\gamma = \pm 90°$. Application of the Routh criterion to the equation $s^2(s + 2) + K(s + 1) = s^3 + 2s^2 + Ks + K = 0$, which is obtained from Eq. (8-31), yields the result that all roots are in the left-half s plane for $K > 0$. The root locus is drawn in Fig. 8-9.

The magnitude condition that must be satisfied is

$$K = \frac{|s|^2|s + 2|}{|s + 1|} \qquad (8\text{-}33)$$

Utilizing this condition locates the points on the root locus at which $K = 2$; they represent the roots of the original polynomial, Eq. (8-30), as indicated in Fig. 8-9.

The work involved in locating the roots of a polynomial on the root locus may be minimized by an additional technique. Consider that the root locus has been obtained accurately for each of two different partitionings of a

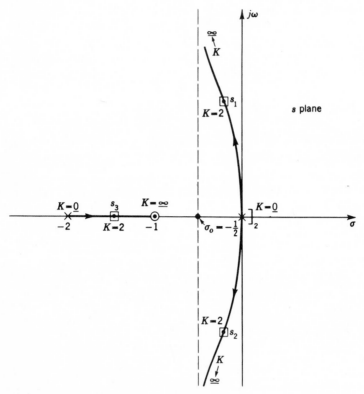

Fig. 8-9 *Determination of the roots of $s^3 + 2s^2 + 2s + 2 = 0$ by the root-locus method.*

polynomial.   These two loci are then drawn on the same graph.   Since these loci represent the same polynomial, the points of intersection must be the roots of the polynomial.   In other words, they are the only points that simultaneously satisfy both partitioned equations.   In practice, it is generally sufficient to first draw sketches of the root locus for each partition based upon accurately using the information obtained from the geometrical characteristics.   One then obtains an idea of the general location of the roots and can apply the angle condition to either of the two partitioned equations to locate the roots more accurately.   Note that the values of K in each partitioned equation are generally not identical.   This is to be expected since the value of K in each case is a function of the manner in which the polynomial is partitioned.   To illustrate this technique, partition ③ is applied to Eq. (8-30).   Letting $K = c$ yields

$$I_3(s) = \frac{c}{s(s^2 + as + b)} = \frac{K}{s(s^2 + 2s + 2)} = -1 \qquad (8\text{-}34)$$

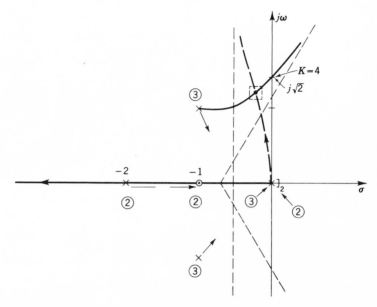

Fig. 8-10 *Sketch of the root locus for partitions two and three of* $s^3 + 2s^2 + 2s + 2 = 0$.

The geometrical shortcuts yield the following information: $\sigma_o = -\frac{2}{3}$; $\gamma = \pm 60°$, $180°$; departure angle from the complex poles is $\pm 45°$, roots are in the left-half s plane for $0 < K < 4$, and the imaginary-axis crossing occurs at $s = \pm j \sqrt{2}$. The sketch, using the above data, is drawn for the root locus of partition ③ in Fig. 8-10, as represented by the solid lines. The long dashed lines represent the sketch of the root locus for partition ②.

From the sketch in Fig. 8-10 the point of intersection is estimated. In a small number of trials the root is determined to be $-0.23 + j1.115$. The remaining real root can be readily determined by dividing

$$(s + 0.23 + j1.115)(s + 0.23 - j1.115) = (s^2 + 0.46s + 1.293)$$

into $s^3 + 2s^2 + 2s + 2$. Thus $s_3 \approx -1.54$.

Example 2

$$s^3 + as^2 + bs + c = s^3 + 2s^2 - 2s + 2 = 0 \qquad (8\text{-}35)$$

Using partition ② yields

$$I_2(s) = \frac{b(s + c/b)}{s^2(s + a)} = \frac{-2(s - 1)}{s^2(s + 2)} = -1 \qquad (8\text{-}36)$$

or

$$\frac{2(s - 1)}{s^2(s + 2)} = \frac{K(s - 1)}{s^2(s + 2)} = 1 \qquad (8\text{-}37)$$

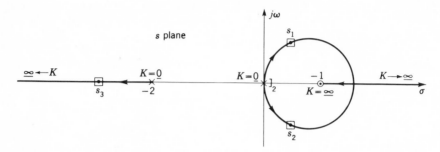

Fig. 8-11 *Determination of the roots of $s^3 + 2s^2 - 2s + 2 = 0$ by the root-locus method.*

The angle condition which must be satisfied for this equation is

$$\left/\frac{K(s-1)}{s^2(s+2)}\right. = \underline{/1} = m360° \tag{8-38}$$

The generalized magnitude condition for this example is

$$K = \frac{|s|^2|s+2|}{|s-1|} \tag{8-39}$$

Applying the conditions of Eqs. (8-38) and (8-39) yields the root locus of Fig. 8-11 and the following roots of Eq. (8-35):

$$s_{1,2} \approx 0.46 \pm j0.689 \qquad s_3 \approx -2.92$$

Solving for the roots of fourth- and fifth-degree polynomials by the partitioning method is just as straightforward as the third-degree examples used for illustration. Table 8-1 shows the possible partitions of these two polynomials.

The roots of polynomials of higher than the fifth degree can also be determined by repeated application of the partition method. Consider the polynomial

$$s^r + a_{r-1}s^{r-1} + a_{r-2}s^{r-2} + \cdots + a_0 = 0 \tag{8-40}$$

The following chart indicates the number of times the partitioning must be applied to solve for the roots of a polynomial.

Range of the degree, r   *Partitions required*

$3 \le r \le 5$	single
$6 \le r \le 8$	double
$9 \le r \le 11$	triple
$\cdots$	$\cdots$

Table 8-1.   Standard partitions of third-, fourth-, and fifth-degree polynomials*

Polynomial	Partitions		
	①	②	③
$s^3 + as^2 + bs + c$ ① ② ③	$\dfrac{a[s^2 + (b/a)s + c/a]}{s^3} = -1$	$\dfrac{b(s + c/b)}{s^2(s + a)} = -1$	$\dfrac{c}{s(s^2 + as + b)} = -1$
$s^4 + as^3 + bs^2 + cs + d$ ① ②	$\dfrac{b[s^2 + (c/b)s + d/b]}{s^3(s + a)} = -1$	$\dfrac{c(s + d/c)}{s^2(s^2 + as + b)} = -1$	
$s^5 + as^4 + bs^3 + cs^2 + ds + e$ ①	$\dfrac{c[s^2 + (d/c)s + e/c]}{s^3(s^2 + as + b)} = -1$		

* The coefficients can have either positive or negative values.

Examples 1 and 2 fall into the single-partition category, that is, only one partition of the polynomial is necessary.   The following example falls into the double-partition category.

Example 3

$$Q(s) = s^6 + as^5 + bs^4 + cs^3 + ds^2 + es + f = 0 \qquad (8\text{-}41)$$
$$\underset{①}{\big|}$$

Partitioning the equation in the manner shown yields

$$\frac{d[s^2 + (e/d)s + f/d]}{s^3(s^3 + as^2 + bs + c)} = \frac{d[s^2 + (e/d)s + f/d]}{s^3 Q_1(s)} = -1 \qquad (8\text{-}42)$$

To obtain the root locus of $Q(s)$ requires first the determination of the zeros and poles of Eq. (8-42).   The zeros of this equation are readily determined. To determine the poles, the partition method can be applied to $Q_1(s)$.   Since $Q_1(s)$ falls into one of the categories in Table 8-1, it is seen that the partition method must be used twice to determine the roots of $Q(s)$, which accounts for the notation *double partition*.   Once the roots of $Q_1(s)$ are determined in the manner illustrated by Examples 1 and 2, all the zeros and poles of Eq. (8-42) are known.   This permits the utilization of Eq. (8-42) to determine the root locus of $Q(s)$.   The reader can readily prove that $9 \leq r \leq 11$ requires a triple partition, and so forth for $r > 11$.

For polynomials up to the fifth degree, the partition is made at a point which permits direct evaluation of the poles and zeros of $I(s)$. For polynomials of degree higher than 5, the partition is made to produce a quadratic, which minimizes the total number of partitions required.

8-6   *Existence and Location of Breakway and Break-in Points (Partition Method)*

Rules 6 and 7 in Sec. 7-8 present methods for locating real-axis breakaway and break-in points on the root locus. Except for the simplest case, where the numerator polynomial of $dW(s)/ds$ is of degree 2 or less, these methods have one or more of the following drawbacks:

1. The detection of the maximum and the minimum points in terms of the magnitude K may be difficult because the Hills-and-dales may not be pronounced.
2. One does not know in advance whether breakaway and break-in points exist on a real-axis locus between a pole and a zero (finite or at infinity). If they do exist, the number of such points is not known in advance. (See Fig. 7-6a and b.)
3. One does not know in advance whether there exist more than one breakaway (break-in) point between two poles (zeros).

Item 2 concerns the case where a root locus exists on the real axis between a pole and a zero. If there are any break-in points in this region, there must be an equal number of breakaway points. For items 2 and 3, a real-axis locus may be divided into a number of sections so that each section is part of a different branch. The maximum number of such sections is equal to the total number of branches. The sum of the break-in and breakaway points in any real-axis locus section is less than the total number of branches. The partition method may simplify detection of the breakaway and break-in points. The effectiveness of this method is brought out by the following example.

Consider a unity feedback system that has the following transfer function:

$$G(s) = \frac{K(s + 1.1)(s + 5)}{s(s + 1)(s + 4)} \tag{8-43}$$

Figure 8-12 shows the location of the real-axis locus for $K > 0$.

By analyzing the figure, the following conclusions can be made with

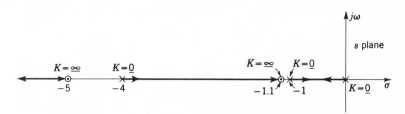

Fig. 8-12 *Real-axis locus of Eq. (8-43) for K > 0.*

respect to breakaway and break-in points:

1. The real-axis locus may be divided into a maximum of 3 parts.
2. At least one breakaway point must exist between 0 and -1.
3. At least one break-in point must exist between -5 and $-\infty$.
4. *a.* There may be no breakaway and break-in points between -1.1 and -4.
   b. If a breakaway point exists between -1.1 and -4, then there must also be a break-in point in this region.

Taking the derivative of

$$W(s) = \frac{s(s+1)(s+4)}{(s+1.1)(s+5)} = \frac{s^3 + 5s^2 + 4s}{s^2 + 6.1s + 5.5} = -K \qquad (8\text{-}44)$$

and setting it equal to zero yields

$$\frac{dW(s)}{ds} = \frac{s^4 + 12.2s^3 + 43s^2 + 55s + 22}{(s^2 + 6.1s + 5.5)^2} = 0 \qquad (8\text{-}45)$$

The real zeros of the above equation yield the breakaway and break-in points if they exist on the real-axis locus shown in Fig. 8-12.

The partition method is now applied to the numerator of Equation (8-45) to determine its zeros. Utilizing partition ② from Table 8-1 yields

$$I(s) = \frac{55(s + {}^{22}\!/_{55})}{s^2(s^2 + 12.2s + 43)} = \frac{K'(s + 0.4)}{s^2(s + 6.1 - j2.4)(s + 6.1 + j2.4)} = -1 \qquad (8\text{-}46)$$

The root locus must be drawn for $0 < K' < \infty$. The points on the locus for which $K' = 55$ yield the roots of this partitioned equation. The root locus for Eq. (8-46) is shown in Fig. 8-13a. The magnitude condition is expressed by

$$K' = \frac{|s|^2|s + 6.1 - j2.14|^2}{|s + 0.4|} \qquad (8\text{-}47)$$

The locus of Fig. 8-13a has four real roots for which $K' = 55$; they are

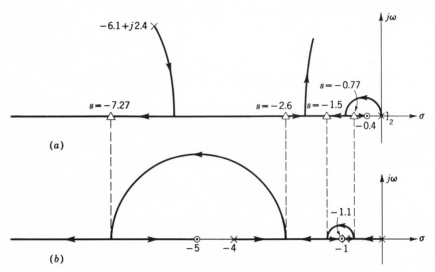

Fig. 8-13   (a) *Root locus for Eq. (8-46).*   (b) *Root locus for Eq. (8-43).*

denoted by the symbol △.   From Fig. 8-13b it is seen that these zeros lie on the real-axis locus; thus the zeros $s = -0.77$ and $s = -2.6$ are breakaway points and the zeros $s = -1.5$ and $s = -7.27$ are break-in points.

The partition method has the following distinctive features in finding break-in and breakaway points:

1. It conclusively shows those portions of the real axis where break-in and breakaway points cannot exist.
2. One is searching for points having a known value of K'.   This is a distinct advantage over searching for unknown maximum and minimum values of K in the Hills-and-dales method.   Utilizing the information from the geometrical short-cuts, one can determine the regions of σ that should be searched for the specific value of K'.
3. Some or all the zeros of $dW/ds = 0$ may be complex.   If the location of these zeros in the root-locus plot of $1 + G(s)H(s)$ satisfies the angle condition, they represent breakaway or break-in points in the complex region of the s plane.   The following example has this property.

Consider a unity feedback system that has the transfer function

$$G(s) = \frac{K}{s(s + 2)(s^2 + 2s + 10)} \tag{8-48}$$

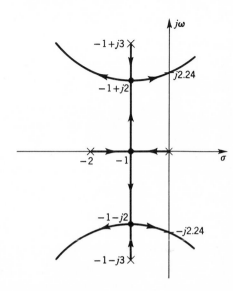

Fig. 8-14 *A root locus with complex breakaway points.*

whose root locus is shown in Fig. 8-14.   The zeros of

$$\frac{dW}{ds} = s^3 + 3s^2 + 7s + 5 = (s + 1)(s + 1 - j2)(s + 1 + j2) = 0 \quad (8\text{-}49)$$

are points on the root locus that satisfy the angle conditions and are thus breakaway points.

8-7   Root-locus Plots for Parameter Variation Other than Gain

A parameter other than gain may also be a variable within a control system; that is, poles and zeros of the open-loop transfer function can vary.   This may be due to wear in a mechanical system, drifting of electric-circuit parameters with age or temperature changes, etc.   The effect of these variations on the closed-loop control ratio can be investigated by use of the root locus; in other words, the root locus of the poles of the control ratio is determined as a function of a system parameter other than gain.

Consider the control ratio

$$\frac{C(s)}{R(s)} = \frac{G(s)}{1 + G(s)H(s)} \quad (8\text{-}50)$$

Letting $G(s) = N_1/D_1$ and $H(s) = N_2/D_2$ yields

$$\frac{C(s)}{R(s)} = \frac{N_1 D_2}{D_1 D_2 + N_1 N_2} \tag{8-51}$$

The characteristic equation of the control system is

$$D_1 D_2 + N_1 N_2 = 0 \tag{8-52}$$

The variable parameter δ may exist in a nonrepeated factor of N_1, N_2, D_1, or D_2. Consider that δ represents the time constant of a first-order term, that is, $\delta s + 1$, so that

$$\delta = \delta_0 + \delta' \tag{8-53}$$

where δ_0 is the fixed component of δ, and δ' is the variable component of δ. Replacing δ by $\delta_0 + \delta'$ permits any one of the terms N_1, N_2, D_1, or D_2 to be written in the following form:

$$\begin{aligned} A(s) &= A_1(s)(\delta s + 1) = A_1(s)[(\delta_0 s + 1) + \delta' s] \\ &= [A_1(s)(\delta_0 s + 1)] + [A_1(s)\,\delta' s] = U(s) + V(s) \end{aligned} \tag{8-54}$$

$A_1(s)$ is the original term excluding the factor $\delta s + 1$, and $V(s)$ contains the variable parameter δ'.

As an example, assume that the variable parameter is contained in the denominator term D_1. Thus

$$D_1(s) = A(s) = U(s) + V(s) \tag{8-55}$$

Substituting this into the system's characteristic equation (8-52) yields

$$(U + V)D_2 + N_1 N_2 = 0 \tag{8-56}$$

This equation is rearranged to isolate the term $V(s)$ that contains the variable δ':

$$V D_2 + (U D_2 + N_1 N_2) = 0 \tag{8-57}$$

To determine the roots of this characteristic equation, it is partitioned as shown. The partitioning process yields

$$\frac{V D_2}{U D_2 + N_1 N_2} = \frac{\delta' s A_1 D_2}{U D_2 + N_1 N_2} = -1 \tag{8-58}$$

which is in the mathematical form permitting the plotting of a root locus as a function of δ' for the range of values $-\infty < \delta' < +\infty$. In order to obtain this root locus the poles of Eq. (8-58) must first be determined. By analyzing the denominator of this equation, it is seen that

$$U D_2 + N_1 N_2 = [D_1 D_2 + N_1 N_2]_{\delta'=0} = 0 \tag{8-59}$$

which in turn yields the equation

$$\left.\frac{N_1 N_2}{D_1 D_2}\right]_{\delta'=0} = -1 \tag{8-60}$$

whose zeros and poles are known. This equation yields the normal root locus of the control system as a function of the system's static loop sensitivity K when the variable parameter is assumed to have a constant value ($\delta' = 0$ and $\delta = \delta_0$). From this root locus the roots of the characteristic equation are selected to yield the desired dominant roots, which in turn determine the value of the system gain. For this value of gain the roots of Eq. (8-59) become the poles of Eq. (8-58). The root locus of Eq. (8-58) can now be drawn as a function of δ' (with constant gain).

For the above example, it should be noted that the variable parameter δ affects only the poles of Eq. (8-51). This is also true if $A(s) = N_2$. In those situations where either N_1 or D_2 contains the variable parameter, it is noted that $C(s)/R(s)$ also contains a variable zero. Whether the variable parameter is the time constant of a first-order term or the damping ratio ζ of a second-order term and occurs in either N_1, N_2, D_1, or D_2, the procedure for analyzing this system is identical to the one illustrated.

Example

To illustrate this new application of the root-locus technique, consider the unity feedback position control system whose forward transfer function is given by

$$G(s) = \frac{C(s)}{E(s)} = \frac{K_1}{s(T_m s + 1)} = \frac{N_1}{D_1} \tag{8-61}$$

where T_m is the motor time constant. Consider that K_1 is held constant and that T_m varies over a given range. Thus

$$\delta = T_m = \delta_0 + \delta' \tag{8-62}$$

where δ_0 is the reference value, and δ' represents the change in T_m from the reference value. Since

$$\begin{aligned} D_1(s) = A(s) = s(\delta s + 1) &= s[(\delta_0 s + 1) + \delta' s] \\ &= [s(\delta_0 s + 1)] + (\delta' s^2) = U(s) + V(s) \end{aligned} \tag{8-63}$$

then Eq. (8-58) yields

$$\frac{\delta' s^2}{s(\delta_0 s + 1) + K_1} = -1 \tag{8-64}$$

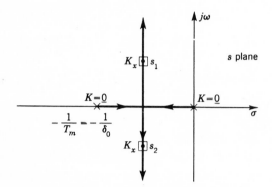

Fig. 8-15 *Root locus for Eq. (8-67), where*
$\delta' = 0$, $K \geq 0$.

Equation (8-64) is rearranged so the coefficient of the highest power of s of the resulting denominator polynomial is unity.   Thus

$$\frac{(\delta'/\delta_0)s^2}{s(s + 1/\delta_0) + K_1/\delta_0} = -1 \tag{8-65}$$

At this point it is noted that the denominator of Eq. (8-65) is equal to the denominator of $C(s)/R(s)$ with $\delta' = 0$:

$$\frac{C(s)}{R(s)} = \frac{G(s)}{1 + G(s)}\bigg]_{\delta'=0} = \frac{K_1/T_m}{s(s + 1/T_m) + K_1/T_m} = \frac{K_1/\delta_0}{s(s + 1/\delta_0) + K_1/\delta_0} \tag{8-66}$$

The poles of $C(s)/R(s)$ with $\delta' = 0$ can be obtained from the root locus by using

$$G(s) = \frac{K}{s(s + 1/T_m)} = \frac{K}{s(s + 1/\delta_0)} = -1 \tag{8-67}$$

where $K = K_1/T_m = K_1/\delta_0$.   The root locus is shown in Fig. 8-15.   The roots s_1 and s_2 in Fig. 8-15 are selected on the basis of performance requirements and determine the value of K_1, which is then fixed.   The values s_1 and s_2 are the poles of Eq. (8-65); thus the root locus for Eq. (8-65) can be drawn as a function of δ'.   This locus has the form shown in Fig. 8-16; the solid lines represent the locus for $\delta' > 0$ and the dashed lines for $\delta' < 0$.

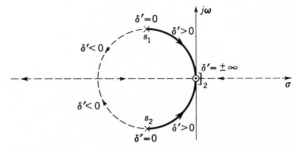

Fig. 8-16 *Root locus of Eq. (8-65) for* $-\infty < \delta' < \infty$, $K_1 = constant.$

The procedure outlined above permits the determination of the effect of a variable parameter T or ζ which appears in a nonrepeated factor of $G(s)H(s)$. This is a simple nonlinearity which can be analyzed by linear techniques.

Bibliography

1. Chu, Y.: Synthesis of Feedback Control System by Phase-angle Loci, *Trans. AIEE*, vol. 71, pt. II, pp. 330–339, November, 1952.
2. Truxal, J. G.: "Automatic Feedback Control System Synthesis," pp. 37–43, McGraw-Hill Book Company, New York, 1955.
3. Narendra, K. S.: "Pole-Zero Configurations and Transient Response of Linear Systems," Technical Report 299, Cruft Laboratory, Harvard University, Cambridge, Mass., May 5, 1959.
4. Elgerd, O. I., and W. C. Stephens: Effect of Closed-loop Transfer Function Pole and Zero Locations on the Transient Response of Linear Control Systems, *Trans. AIEE*, vol. 78, pt. II, pp. 121–127, May, 1959.
5. Clark, R. N.: "Introduction to Automatic Control Systems," pp. 135–146, John Wiley & Sons, Inc., New York, 1962.
6. Murphy, G. J.: "Basic Automatic Control Theory," pp. 433–434, D. Van Nostrand Company, Inc., Princeton, N.J., 1957.
7. Houpis, C. H.: Parameter Sensitivity by Root Locus, *Control Engineering*, vol. 12, pp. 111–112, April, 1965.
8. Koenig, J. F.: A Method for Obtaining the Time Response of Any Linear System, *IEEE Trans. on Automatic Control*, vol. AC-9, pp. 556–565, October, 1964.
9. Gustafson, R. D.: A Paper and Pencil Control System Design Technique, Conference paper, 1965. Joint Automatic Control Conference, Rensselaer Polytechnic Institute, June, 1965.

9

Frequency
Response

9-1   Introduction

In Chap. 7 it is pointed out that there are two basic methods for
predicting and adjusting a system performance without resorting
to the actual solving of the system's differential equation.   One
of these methods, the root-locus, is discussed in detail in that
chapter.   The other method, based upon the frequency response,
to which the Nyquist stability criterion is applied, is developed in
this and the next chapter.[1,2]   For the comprehensive study of a
system it is necessary to use both methods of analysis.   The
principal advantage of the root-locus method is that it indicates
the actual time response.   However, it is sometimes necessary to
have performance requirements in terms of the response at specific
frequencies.   Also, the noise present in a system can result in
poor over-all performance.   Having the frequency-response
curves of a system permits analysis with respect to these two items.

278

The curves permit the design of a passband for the system response that excludes the noise and therefore improves system performance. It is also important in situations in which the actual transfer functions of some or all of the blocks in a block diagram are not known analytically. The frequency response of the open-loop transfer function can be determined experimentally for these situations. An approximate expression for the transfer function can be obtained from the plot of the experimentally determined frequency response. Thus no particular representation can be judged supreme above the rest. Each has its particular use and advantage in a particular situation. As the reader becomes fully acquainted with all methods, he becomes aware of the potentialities of each method and knows when each should be used.

In this chapter two graphical representations of transfer functions are presented: the logarithmic and the polar plots. In the next chapter these plots are used to develop Nyquist's stability criterion.

9-2   *Correlation of the Sinusoidal and Time Responses*[3]

Earlier in the text it was pointed out that solving for $c(t)$ by the classical method is rather laborious and impractical for synthesis purposes, especially when the input is not a simple analytical function. The use of Laplace-transform theory lessens the work involved and permits the engineer to synthesize and improve a system. The last two chapters on the root locus have just illustrated this fact. The advantages of the graphical representations in the frequency domain of the transfer functions are developed in the following pages.

Once the frequency response of a system has been determined, the time response can be determined by the use of the Fourier integral. The behavior in the frequency domain for a given driving function $r(t)$ can be determined by the Fourier integral as

$$\mathbf{R}(j\omega) = \int_{-\infty}^{\infty} r(t)e^{-j\omega t}\, dt \tag{9-1}$$

Thus for a given control system the frequency response of the controlled variable is

$$\mathbf{C}(j\omega) = \frac{\mathbf{G}(j\omega)}{1 + \mathbf{G}(j\omega)\mathbf{H}(j\omega)}\, \mathbf{R}(j\omega) \tag{9-2}$$

By use of the inverse Fourier integral the controlled variable as a function of time is

$$c(t) = \frac{1}{2\pi}\int_{-\infty}^{\infty} \mathbf{C}(j\omega)e^{j\omega t}\, d\omega \tag{9-3}$$

This approach is much used in practice. If the design engineer cannot evaluate Eq. (9-3) by reference to a table of definite integrals, this equation can be evaluated by using numerical or graphical integration. This is necessary if $C(j\omega)$ is available only as a curve and cannot be simply expressed in analytical form, as is often the case. The procedure for evaluating the time response of Eq. (9-3) when $C(j\omega)$ is available only as a graph is described in several books.[5]

Various methods have been developed, based on the Fourier integral and a step input signal, relating $C(j\omega)$ qualitatively to the time solution without actually taking the inverse Fourier integral. These methods permit the engineer to determine approximately the response of his system through the interpretation of graphical plots in the frequency domain. This makes possible the design and improvement of feedback systems with a minimum effort.

Sections 4-12 and 7-11 have shown that the frequency response is a function of the pole-zero pattern in the s plane. It therefore provides significant information concerning the time response of the system. Two features of the frequency response are the maximum value M_m and the resonant frequency ω_m. The time response is qualitatively related to the values M_m and ω_m, which can be determined from the frequency-response plots.

9-3   Types of Plots

From experience one finds that, if analytical solutions are tedious, then, in general, graphical solutions are easier to use. The plots in the frequency domain that have found great use in graphical analysis in the field of feedback control systems are of two categories. The first category is the plot of the magnitude of the output-input ratio versus frequency in rectangular coordinates, as illustrated in Secs. 4-12 and 7-11, or in logarithmic coordinates. Associated with this plot is a second plot of the corresponding phase angle versus frequency. In the second category the output-input ratio may be plotted in polar coordinates, with frequency as a parameter. For this latter category there are two types of polar plots, direct and inverse. Polar plots are generally used only for the open-loop response. They are commonly referred to as Nyquist plots, because of H. Nyquist's work, about 1932, relating the stability of a system to the form of these plots.[4,5]

9-4   *Frequency-response Curves*

The frequency response, magnitude versus frequency, for the control ratio

$$\frac{C(j\omega)}{R(j\omega)} = \frac{G(j\omega)}{1 + G(j\omega)H(j\omega)} \qquad (9\text{-}4)$$

can be determined for any given value of frequency.   For each value of frequency, Eq. (9-4) yields a phasor quantity whose magnitude is $|C(j\omega)/R(j\omega)|$ and whose phase angle α is the angle between $C(j\omega)$ and $R(j\omega)$.

For a given sinusoidal input signal, the input and steady-state output are of the following forms:

$$r(t) = R \sin \omega t \qquad (9\text{-}5)$$
$$c(t) = C \sin (\omega t + \alpha) \qquad (9\text{-}6)$$

An ideal system may be one in which

$$\begin{aligned} \alpha &= 0° \\ R &= C \end{aligned} \qquad (9\text{-}7)$$

for $0 < \omega < \infty$.   Curves 1 in Fig. 9-1 represent the ideal system.   If the above equations are analyzed, it is found that an instant transfer of energy must occur from the input to the output in zero time.   This is a prerequisite for faithful reproduction of a step input signal.   In reality, this generally cannot be achieved, since in any physical system there is energy dissipation

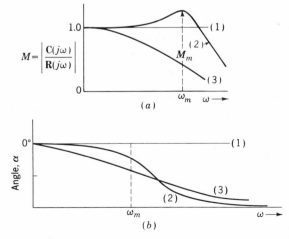

Fig. 9-1 *Frequency-response characteristics of* $C(j\omega)/$ $R(j\omega)$ *in rectangular coordinates.*

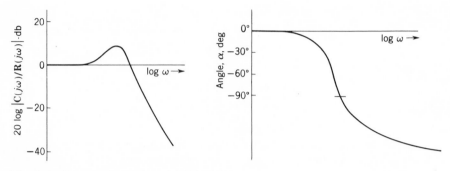

Fig. 9-2 *Log magnitude plots of* $\mathbf{C}(j\omega)/\mathbf{R}(j\omega)$.

and there are energy-storage elements.   Curves 2 and 3 in Fig. 9-1 represent actual systems.   Even if the ideal system were possible, it would not be desirable from the noise-reproduction standpoint.   In any system the input signal may contain spurious signals in addition to the true signal input, or there may be sources of noise within the closed-loop system.   This noise is generally in a band of frequencies above the dominant frequency band of the true signal.   Thus, in order to reproduce the true signal and attenuate the noise, feedback control systems are designed to have a definite cutoff frequency.   In certain cases the noise frequency may exist in the same frequency band as the true signal.   When this occurs, the problem of eliminating it becomes complex.

Physically realizable systems represented by curves 2 and 3 of Fig. 9-1 have the following characteristics:

1.   The passband or bandwidth is defined as the range of frequencies from 0 up to the frequency ω_b, where $M = 0.707$ of the value at $\omega = 0$.   The frequency ω_m is more easily obtained than the passband frequency ω_b and is often used as a figure of merit.
2.   As $\omega \to 0$, the ratio $|\mathbf{C}(j\omega)/\mathbf{R}(j\omega)| \to 1$ and $\alpha \to 0°$.
3.   As $\omega \to \infty$, the ratio $|\mathbf{C}(j\omega)/\mathbf{R}(j\omega)| \to 0$ and $\alpha \to -k\pi/2$ radians, where $-k$ is equal to the degree of $\mathbf{C}(j\omega)/\mathbf{R}(j\omega)$ (see Sec. 4-12).

As discussed more fully later, a practical system generally has a frequency-response characteristic as illustrated by curve 2 in Fig. 9-1.   Note that for a given frequency band the ratio $|\mathbf{C}(j\omega)/\mathbf{R}(j\omega)|$ is greater than 1.

Curve 2 in Fig. 9-1 may also be represented by log magnitude plots, as shown in Fig. 9-2, which are introduced in this chapter.   The control ratio $\mathbf{C}(j\omega)/\mathbf{R}(j\omega)$ is generally represented graphically in the form of either Fig. 9-1 or 9-2.

Another method of representing the open-loop steady-state sinusoidal response is the polar plot, which is often used for the open-loop transfer

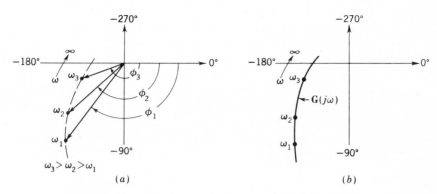

Fig. 9-3 *Phasor-locus development in the complex plane.*

function. From this polar plot the stability and the frequency response of the closed-loop system may be obtained. Also, the transient response of the system may be predicted from it.

Consider the forward transfer function

$$G(j\omega) = \frac{C(j\omega)}{E(j\omega)} \tag{9-8}$$

For a given frequency, $G(j\omega)$ is a phasor quantity whose magnitude is $|C(j\omega)/E(j\omega)|$ and whose phase angle is $\phi(\omega)$. Representing this quantity in polar coordinates is a phasor whose length corresponds to the magnitude $|C(j\omega)/E(j\omega)|$ and whose angle with respect to the positive real axis corresponds to $\phi(\omega)$, as shown in Fig. 9-3a. Thus a set of values of $G(j\omega)$ for values of ω between 0 and ∞ yields a set of phasors whose tips are connected by a smooth curve, as shown in Fig. 9-3b. The polar plot in Fig. 9-3b contains all the necessary information and is the form used.

The reader has now been introduced to several forms of graphical representations employed in synthesis work. The remainder of this chapter makes use of both the logarithmic plot and the polar plot.

9-5   Introduction to Logarithmic Plots[5]

Logarithmic plots of the magnitude and angle of the transfer function systematize and simplify the designing of a closed-loop system. The advantage of logarithmic plots is that the functions of multiplication and division become addition and subtraction and the work of obtaining the transfer function is largely graphical instead of analytical. The basic factors of the transfer function fall into three categories, and these may be plotted easily by means of straight-line asymptotic approximations. The use of semilog paper eliminates the need actually to take logarithms of very many numbers and also expands the low-frequency range, which is of primary importance.

In preliminary design studies the straight-line approximations are used to obtain rough performance characteristics very quickly. As the design becomes more firmly specified, the straight-line curves can be corrected for greater accuracy. The effect of compensators in cascade for improvement of performance is easily determined. Also, from these logarithmic plots enough data in the frequency range of concern can be readily obtained to determine the corresponding polar plots.

9-6   *Definitions of Logarithmic Terms*

1. The logarithm to the base e of a complex number is itself a complex number. The symbol ln is used to indicate the logarithm to the base e:

$$\ln |\mathbf{G}(j\omega)|e^{j\phi(\omega)} = \ln |\mathbf{G}(j\omega)| + \ln e^{j\phi(\omega)}$$
$$= \ln |\mathbf{G}(j\omega)| + j\phi(\omega) \tag{9-9}$$

The logarithm of the complex number $|\mathbf{G}(j\omega)|e^{j\phi(\omega)}$ has a real part equal to the logarithm of the magnitude $\ln |\mathbf{G}(j\omega)|$ and an imaginary part equal to the angle $\phi(\omega)$, expressed in radians. Both of these components are a function of the frequency ω.

2. In a similar fashion, the logarithm to the base 10 of a complex number is itself a complex number. The symbol log is used to indicate the logarithm to the base 10:

$$\log |\mathbf{G}(j\omega)|e^{j\phi(\omega)} = \log |\mathbf{G}(j\omega)| + \log e^{j\phi(\omega)}$$
$$= \log |\mathbf{G}(j\omega)| + j0.434\phi(\omega) \tag{9-10}$$

As expected, the real part is equal to the logarithm of the magnitude, $\log |\mathbf{G}(j\omega)|$, and the imaginary part is proportional to the angle, $0.434\phi(\omega)$. In the rest of this chapter the factor 0.434 is omitted, and only the angle $\phi(\omega)$ is used.

3. The unit commonly used in feedback-system work for the logarithm of the magnitude is the decibel. The term *decibel* as used originally in communications engineering referred to the ratio of two values of power, but in recent years it has been used for other relations such as characterizing the ratio of two voltages. This latter ratio is therefore a dimensionless quantity; but as used in feedback-system work the ratio of two quantities is not necessarily dimensionless. Logarithms of transfer functions are used, where the transfer func-

tion is the ratio of output to input of a block. They are not necessarily in the same units. For example, the output may be speed and the input may be voltage. It is necessary only to be consistent in the units used.

The logarithm of the magnitude of a transfer function $G(j\omega)$ expressed in decibels is

$$20 \log |G(j\omega)| \quad \text{db}$$

This quantity is called the *log magnitude* and is abbreviated Lm. Thus

$$\text{Lm } G(j\omega) = 20 \log |G(j\omega)| \quad \text{db} \qquad (9\text{-}11)$$

Since the transfer function is a function of frequency, the log magnitude is also a function of frequency.

4. The manner in which multiplication and division become addition and subtraction in the logarithmic process is shown in the following example:

$$G(j\omega) = \frac{A_1(j\omega)A_2(j\omega)}{B_1(j\omega)B_2(j\omega)}$$

$$\begin{aligned}
\text{Lm } G(j\omega) &= 20 \log |G(j\omega)| \\
&= \text{Lm } A_1(j\omega) + \text{Lm } A_2(j\omega) \\
&\quad - \text{Lm } B_1(j\omega) - \text{Lm } B_2(j\omega) \\
&= 20 \log |A_1(j\omega)| + 20 \log |A_2(j\omega)| \\
&\quad - 20 \log |B_1(j\omega)| - 20 \log |B_2(j\omega)|
\end{aligned}$$

5. Two units used to express frequency bands or frequency ratios are the octave and the decade. An octave is a frequency band from f_1 to f_2, where $f_2/f_1 = 2$. Thus, the frequency band from 1 to 2 cps is 1 octave in width and the frequency band from 2 to 4 cps is 1 octave in width. Also, the band from 17.4 to 34.8 cps is 1 octave in width. Note that 1 octave is not a fixed frequency bandwidth but depends on the frequency range being considered. The number of octaves in the frequency range from f_1 to f_2 is given by

$$\frac{\log (f_2/f_1)}{\log 2} = 3.32 \log \frac{f_2}{f_1} \quad \text{octaves} \qquad (9\text{-}12)$$

There is an increase of 1 decade from f_1 to f_2, where $f_2/f_1 = 10$. The frequency band from 1 to 10 cps or from 2.5 to 25 cps is 1 decade in width. The number of decades from f_1 to f_2 is given by

$$\log \frac{f_2}{f_1} \quad \text{decades} \qquad (9\text{-}13)$$

6. The decibel values of some common numbers are given in Table 9-1. Note that the reciprocals of numbers differ only in sign. Thus, the decibel value of 2 is $+6$ db and the decibel value of $\frac{1}{2}$ is -6 db.

Table 9-1. Decibel values of some common numbers

Number	Decibels
0.01	-40
0.1	-20
0.5	-6
1.0	0
2.0	6
10.0	20
100.0	40
200.0	46

Based on the characteristics of logarithms and Table 9-1, the following conclusions are drawn:

1. As a number doubles, the decibel value increases by 6 db. The number 2 is twice as big as 1, and its decibel value is 6 db more. The number 200 is twice as big as 100, and its decibel value is 6 db greater.
2. As a number increases by a factor of 10, the decibel value increases by 20. The number 100 is ten times as large as the number 10, and its decibel value is 20 db more. The number 200 is one hundred times larger than the number 2, and its decibel value is 40 db greater.

9-7   General Frequency-transfer-function Relationships

The frequency transfer function may be written in generalized form as the ratio of polynomials:

$$G(j\omega) = \frac{K_n(1 + j\omega T_1)(1 + j\omega T_2)^m \, \cdots}{(j\omega)^n(1 + j\omega T_a)[1 + (2\zeta/\omega_n)j\omega + (1/\omega_n^2)(j\omega)^2] \, \cdots} = K_n G'(j\omega)$$

(9-14)

where K_n is the gain. The logarithm of the transfer function is a complex quantity; the real portion is proportional to the log of the magnitude, and the complex portion is proportional to the angle. Two separate equations are written, one for the log magnitude and one for the angle. Appendix G shows a simple graph for converting magnitudes to decibels, and vice versa.

$$\text{Lm } \mathbf{G}(j\omega) \;=\; \text{Lm } K_n + \text{Lm } (1 + j\omega T_1)$$
$$+\, m \text{ Lm } (1 + j\omega T_2) + \cdots - n \text{ Lm } j\omega$$
$$-\, \text{Lm } (1 + j\omega T_a) - \text{Lm } \left[1 + \frac{2\zeta}{\omega_n} j\omega + \frac{1}{\omega_n^2} (j\omega)^2 \right] - \cdots \quad (9\text{-}15)$$

$$\underline{/\mathbf{G}(j\omega)} \;=\; \underline{/K_n} + \underline{/1 + j\omega T_1} + m\underline{/1 + j\omega T_2} + \cdots - n\underline{/j\omega}$$
$$-\, \underline{/1 + j\omega T_a} - \underline{\bigg/1 + \frac{2\zeta}{\omega_n} j\omega + \frac{1}{\omega_n^2} (j\omega)^2} - \cdots \quad (9\text{-}16a)$$

The angle equation may be rewritten as

$$\underline{/\mathbf{G}(j\omega)} \;=\; \underline{/K_n} + \tan^{-1} \omega T_1 + m \tan^{-1} \omega T_2 + \cdots - n90°$$
$$-\, \tan^{-1} \omega T_a - \tan^{-1} \frac{2\zeta\omega/\omega_n}{1 - \omega^2/\omega_n^2} - \cdots \quad (9\text{-}16b)$$

The gain K_n is a real number but may be positive or negative; therefore its angle is correspondingly 0° or 180°. Unless otherwise indicated, a positive value of gain is assumed in this book. Both the log magnitude and the angle given by these equations are functions of frequency. When the log magnitude and the angle are plotted as functions of the log of frequency, the resulting curves are referred to as the *log magnitude diagram and the phase diagram*. Equations (9-15) and (9-16) show that the resultant curves are obtained by the addition and subtraction of the corresponding individual terms in the transfer-function equation.

The log magnitude and phase diagrams are used to analyze the closed-loop response of a feedback control system. From these curves certain qualitative information about relative stability is readily obtained. The two curves can be combined into a single curve of log magnitude versus angle with frequency as a parameter. This curve, called the *log magnitude–angle diagram*, corresponds to the Nyquist polar plot and is used for the quantitative design of the feedback system to meet specifications of required performance.

9-8   *Drawing the Log Magnitude and Phase Diagram*

The generalized form of the transfer function as given by Eq. (9-14) shows that the numerator and denominator have four basic types of factors. These are

$$K_n \quad (9\text{-}17)$$
$$(j\omega)^{\pm n} \quad (9\text{-}18)$$
$$(1 + j\omega T)^{\pm m} \quad (9\text{-}19)$$
$$\left[1 + \frac{2\zeta}{\omega_n} j\omega + \frac{1}{\omega_n^2} (j\omega)^2 \right]^{\pm p} \quad (9\text{-}20)$$

Each of these terms may appear raised to an integral power other than 1. A negative exponent means that the factor appears in the denominator, and a positive exponent means that it appears in the numerator. The log magnitude of each factor is obtained by multiplying the log magnitude of the basic factor by the power to which it is raised. Similarly, the angle is obtained by multiplying the angle of the basic factor by the power to which it is raised.

The curves of log magnitude and angle versus the log frequency can be drawn very easily for each factor. Then these curves for each factor can be added together graphically to get the curves for the complete transfer function. The procedure can be further simplified by using asymptotic approximations to these curves, as shown in the following pages.

Constants

The constant K_n is frequency-invariant, so the plot of

$$\mathrm{Lm}\ K_n = 20\ \log\ K_n \qquad \mathrm{db}$$

is a horizontal straight line. Basically, the constant raises or lowers the log magnitude curve of the complete transfer function by a fixed amount. The angle, of course, is zero as long as K_n is positive.

$j\omega$ Factors

The factor $j\omega$ appearing in the denominator has a log magnitude

$$\mathrm{Lm}\ \frac{1}{j\omega} = 20\ \log\ \left|\frac{1}{j\omega}\right| = -20\ \log\ \omega \qquad (9\text{-}21)$$

When plotted against $\log\ \omega$, this curve is a straight line with a negative slope of 6 db/octave or 20 db/decade. Values of this function can be obtained from Table 9-1 for several values of ω. The angle is constant and equal to $-90°$.

When the factor $j\omega$ appears in the numerator, the log magnitude is

$$\mathrm{Lm}\ (j\omega) = 20\ \log\ |j\omega| = 20\ \log\ \omega \qquad (9\text{-}22)$$

This curve is a straight line with a positive slope of 6 db/octave or 20 db/decade. The angle is constant and equal to $+90°$. Notice that the only difference between the curves for $j\omega$ and for $1/j\omega$ is a change in the sign of the slope of the log magnitude and a change in the sign of the angle. Both curves go through the point 0 db at $\omega = 1$.

For the factor $(j\omega)^{\pm n}$ the log magnitude curve has a slope of $\pm 6n$ decibels per octave or $\pm 20n$ decibels per decade, and the angle is constant and equal to $\pm n90°$.

$1 + j\omega T$ *Factors*

The factor $1 + j\omega T$ appearing in the denominator has a log magnitude

$$\text{Lm } \frac{1}{1 + j\omega T} = 20 \log \left| \frac{1}{1 + j\omega T} \right| = -20 \log \sqrt{1 + \omega^2 T^2} \quad (9\text{-}23)$$

For very small values of ω, that is, $\omega T \ll 1$,

$$\text{Lm } \frac{1}{1 + j\omega T} \approx \log 1 = 0 \quad (9\text{-}24)$$

The plot of the log magnitude at small frequencies is the 0-db line. For very large values of ω, that is, $\omega T \gg 1$,

$$\text{Lm } \frac{1}{1 + j\omega T} \approx 20 \log \left| \frac{1}{j\omega T} \right| = -20 \log \omega T \quad (9\text{-}25)$$

At $\omega = 1/T$ the value of Eq. (9-25) is 0, and for values of $\omega > 1/T$ this function is a straight line with a negative slope of 6 db/octave. The asymptotes of the plot of Lm $[1/(1 + j\omega T)]$ are two straight lines—one of zero slope below $\omega = 1/T$ and one of -6 db/octave slope above $\omega = 1/T$. These asymptotes are drawn in Fig. 9-4.

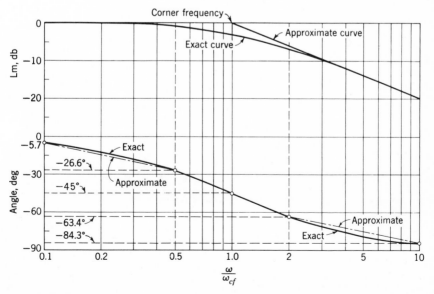

Fig. 9-4 *Log magnitude and phase diagram for $(1 + j\omega T)^{-1} = [1 + j(\omega/\omega_{cf})]^{-1}$.*

Table 9-2.   Values of Lm $(1 + j\omega T)^{-1}$ for several frequencies

$\dfrac{\omega}{\omega_{cf}}$	Exact value, db	Value of the asymptote, db	Error db
1/10	0.04	0	−0.04
1/4	−0.26	0	−0.26
1/2	−0.97	0	−0.97
0.76	−2.00	0	−2.00
1	−3.01	0	−3.01
1.31	−4.35	−2.35	−2.00
2	−6.99	−6.02	−0.97
4	−12.30	−12.04	−0.26
10	−20.04	−20.0	−0.04

The frequency at which the asymptotes to the log magnitude curve intersect is defined as the corner frequency ω_{cf}. The value $\omega_{cf} = 1/T$ is the corner frequency for the function $1/(1 + j\omega T) = 1/(1 + j\omega/\omega_{cf})$.

The exact values of Lm $[1/(1 + j\omega T)]$ are given in Table 9-2 for several frequencies in the range a decade above and below the corner frequency.

The exact curve is also drawn in Fig. 9-4. The error, in decibels, between the exact curve and the asymptotes is approximately as follows:

1. At the corner frequency: 3
2. One octave above and below the corner frequency: 1
3. Two octaves from the corner frequency: 0.26

Frequently the preliminary design studies are made by using the asymptotes only. The correction to the straight-line approximation to yield the true log magnitude curve is shown in Fig. 9-5. For more exact studies the

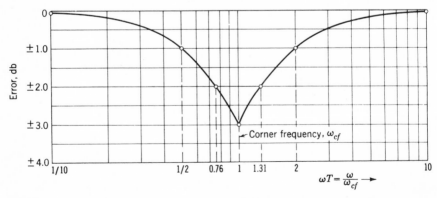

Fig. 9-5 *Log magnitude correction for $(1 + j\omega T)^{\pm 1}$.*

corrections are put in at the corner frequency and at 1 octave above and below the corner frequency, and the new curve is drawn freehand or with a french curve. If a more accurate curve is desired, corrections at other frequencies may be determined readily. As an example, consider the factor

$$\frac{1}{1 + j2.5\omega} \qquad \text{where} \qquad \omega_{cf} = \frac{1}{T} = \frac{1}{2.5}$$

If the correction at $\omega = 1$ is desired, then for $\omega/\omega_{cf} = 2.5$ Fig. 9-5 yields an approximate correction of -0.62 db.

The phase curve for this function is plotted in Fig. 9-4. At zero frequency the angle is $0°$; at the corner frequency $\omega = \omega_{cf}$ the angle is $-45°$; and at infinite frequency the angle is $-90°$. The angle curve is symmetrical about the corner-frequency value when plotted against log (ω/ω_{cf}) or log ω. Since the abscissa of the curves in Fig. 9-4 is ω/ω_{cf}, the shapes of the angle and log magnitude curves are independent of the time constant T. Thus when the curves are plotted with the abscissa in terms of ω, changing T just slides the log magnitude and the angle curves left or right so that the -3 db and the $45°$ points occur at the frequency $\omega = \omega_{cf}$.

Templates of the log magnitude and of the angle can be drawn to the same scale as the semilog paper used and can then be used for all functions of the form $1 + j\omega T$. When templates are not available, fairly accurate log magnitude and phase curves are obtained by drawing smooth curves through the key points given in Tables 9-2 and 9-3.

Table 9-3. *Angles of* $(1 + j\omega/\omega_{cf})^{-1}$ *for key frequency points*

ω/ω_{cf}	Angle, deg
0.1	-5.7
0.5	-26.6
1.0	-45.0
2.0	-63.4
10.0	-84.3

For preliminary design studies, straight-line approximations of the phase curve may be utilized. The approximations illustrated in Fig. 9-4 consist of straight lines drawn between the points:

1. $-5.7°$ at $\omega/\omega_{cf} = 0.1$ and $-26.6°$ at $\omega/\omega_{cf} = 0.5$ and the extension of this line to the $0°$ line
2. $-26.6°$ at $\omega/\omega_{cf} = 0.5$ and $-63.4°$ at $\omega/\omega_{cf} = 2.0$
3. $-63.4°$ at $\omega/\omega_{cf} = 2.0$ and $-84.3°$ at $\omega/\omega_{cf} = 10$ and the extension of this line to the $-90°$ line

The maximum error resulting from the above approximation is about $\pm 4°$. Note that in the region between $0.5 \leq \omega/\omega_{cf} \leq 2.0$ the exact and approximate curves are identical. A simplified approximation is to draw only one straight line through the following three points:

ω/ω_{cf}	0.1	1.0	10
Angle	$0°$	$-45°$	$-90°$

The maximum error resulting from this approximation is about $\pm 6°$.

The factor $1 + j\omega T$ appearing in the numerator has the log magnitude

$$\text{Lm } (1 + j\omega T) = 20 \log \sqrt{1 + \omega^2 T^2}$$

This is the same function as its inverse $\text{Lm } [1/(1 + j\omega T)]$ except that it is positive. The corner frequency is the same, and the angle varies from 0 to 90° as the frequency increases from zero to infinity. The log magnitude and angle curves for the function $1 + j\omega T$ are symmetrical about the abscissa to the curves for $1/(1 + j\omega T)$. Therefore the same templates can be used for both.

Quadratic Factors

Quadratic factors of the form

$$\left[1 + \frac{2\zeta}{\omega_n} j\omega + \frac{1}{\omega_n^2} (j\omega)^2 \right]^{-1} \tag{9-26}$$

often occur in feedback-system transfer functions. For $\zeta > 1$ the quadratic can be factored into two first-order factors with real zeros which can be plotted in the manner shown previously. But for $\zeta < 1$ the factors are conjugate complex factors, and the entire quadratic is plotted without factoring:

$$\text{Lm } \left[1 + \frac{2\zeta}{\omega_n} j\omega + \frac{1}{\omega_n^2} (j\omega)^2 \right]^{-1} = -20 \log \left[\left(1 - \frac{\omega^2}{\omega_n^2} \right)^2 + \left(\frac{2\zeta\omega}{\omega_n} \right)^2 \right]^{1/2} \tag{9-27}$$

$$\text{Angle } \left[1 + \frac{2\zeta}{\omega_n} j\omega + \frac{1}{\omega_n^2} (j\omega)^2 \right]^{-1} = -\tan^{-1} \frac{2\zeta\omega/\omega_n}{1 - \omega^2/\omega_n^2} \tag{9-28}$$

By analyzing Eq. (9-27), it is seen that for very small values of ω the low-frequency asymptote is represented by log magnitude $= 0$. For very high values of frequency, the log magnitude is approximately

$$-20 \log \frac{\omega^2}{\omega_n^2} = -40 \log \frac{\omega}{\omega_n}$$

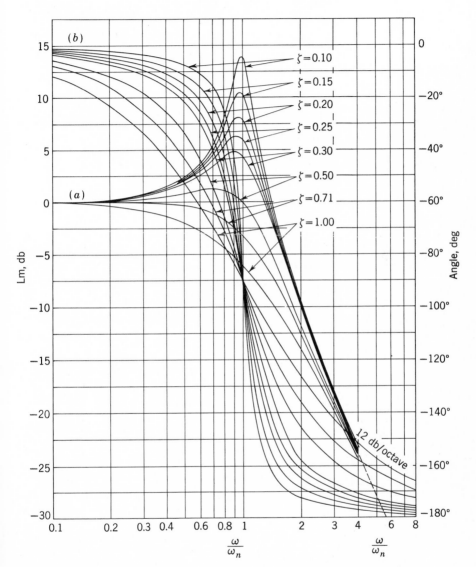

Fig. 9-6 *Log magnitude and phase diagram for* $[1 + j2\zeta\omega/\omega_n + (j\omega/\omega_n)^2]^{-1}$.

and the high-frequency asymptote has a slope of -40 db/decade. The corner frequency for this quadratic is $\omega = \omega_n$, and the asymptotes cross at this frequency.

From Eq. (9-27) it can be seen that a resonant condition is exhibited in the vicinity of $\omega = \omega_n$. Therefore there may be a substantial deviation of the log magnitude curve from the straight-line asymptotes, depending on

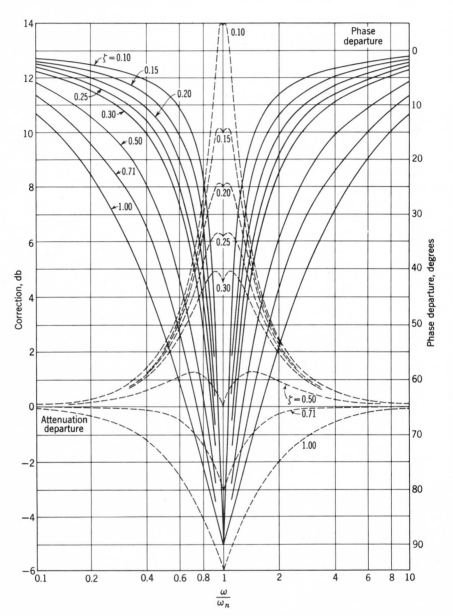

Fig. 9-7 *Corrections to the log magnitude and phase diagram for* $[1 + j2\zeta\omega/\omega_n + (j\omega/\omega_n)^2]^{-1}$. *(Reproduced from H. M. James, N. B. Nichols, and R. S. Phillips, "Theory of Servomechanisms," McGraw-Hill Book Company, New York, 1947.)*

the value of ζ.   A family of curves for several values of $\zeta < 1$ is plotted in Fig. 9-6.   The corrections of each actual curve from the straight-line asymptotes are shown in Fig. 9-7.

If these curves are used often enough, it is worthwhile to make templates for several values of ζ.   Then the appropriate correct curve can be traced directly on the graph being drawn.   If templates are not used, sufficient points can be picked from Fig. 9-6 to draw the curve.

The phase angle curve for this function also varies with ζ.   At zero frequency the angle is $0°$, at the corner frequency the angle is $-90°$, and at infinite frequency the angle is $-180°$.   A family of curves for various values of $\zeta < 1$ is plotted in Fig. 9-6.   Templates can be made of these curves, or enough values to draw the appropriate curve can be taken from Fig. 9-6.

When the factor

$$1 + \frac{2\zeta}{\omega_n} j\omega + \frac{1}{\omega_n^2} (j\omega)^2$$

appears in the numerator, the magnitudes of the log magnitude and phase angle are the same as those in Fig. 9-6 except that they are changed in sign.

The Lm $[1 + j2\zeta\omega/\omega_n + (j\omega)^2/\omega_n^2]^{-1}$ with $\zeta < 1$ has a peak value.   The magnitude of this peak value and the frequency at which it occurs are important terms.   These values, derived in Sec. 11-3, are

$$M_m = \frac{1}{2\zeta \sqrt{1 - \zeta^2}} \tag{9-29}$$

$$\omega_m = \omega_n \sqrt{1 - 2\zeta^2} \tag{9-30}$$

Note that the peak value depends only on the damping ratio ζ.   Equation (9-30) is meaningful only for real values of ω_m.   Therefore the curve of M versus ω has a peak value greater than unity only for $\zeta < 0.707$.   The frequency at which the peak value occurs depends on both the damping ratio ζ and the undamped natural frequency ω_n.   This information is used when adjusting a control system for good response characteristics.   These characteristics are discussed in Chap. 11.

The discussion thus far deals with minimum-phase factors (all poles and zeros are in the left-half s plane).   The log magnitude curves for non-minimum-phase factors (poles and zeros in the right-half s plane) are the same as those for the corresponding minimum-phase factors.   However, the angle curves are different.   For the factor $1 - j\omega T$ the angle varies from 0 to $-90°$ as ω varies from zero to infinity.   If ζ is negative, the quadratic factor of expression (9-26) becomes a non-minimum-phase term.   Its angle varies from $-360°$ at $\omega = 0$ to $-180°$ at $\omega = \infty$.   This information can be obtained from the pole-zero diagram discussed in Sec. 4-12.

9-9   Example of Drawing the Log Magnitude and Phase Diagram

Figure 9-8 shows the block diagram of a feedback control system with unity feedback.   The log magnitude and phase diagram is drawn for the open-loop transfer function of this system.   The log magnitude curve is drawn both for the straight-line approximation and for the exact curve.

Table 9-4 lists the pertinent characteristics for each factor.   The log magnitude asymptotes and angle curves for each factor are shown in Figs. 9-9 and 9-10, respectively.   They are added algebraically to obtain the composite curve.   In general, it is not necessary to draw the asymptotes for each factor.   The composite log magnitude curve using straight-line approximations is drawn directly, as outlined below.

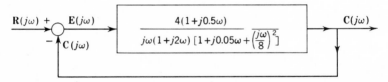

Fig. 9-8   *Block diagram of a control system with unity feedback.*

Table 9-4.   *Characteristics of log magnitude and angle diagram for various factors*

Factor	Corner frequency ω_{cf}	Log magnitude	Angle characteristics
4	None	Constant magnitude of $+12$ db	Constant $0°$
$(j\omega)^{-1}$	None	Constant slope of -20 db/decade	Constant $-90°$
$(1 + j2\omega)^{-1}$	$\omega_1 = 0.5$	0 slope below the corner frequency -20 db/decade slope above the corner frequency	Varies from $0°$ to $-90°$
$1 + j0.5\omega$	$\omega_2 = 2.0$	0 slope below the corner frequency $+20$ db/decade slope above the corner frequency	Varies from $0°$ to $+90°$
$\{1 + j0.05\omega + [j\omega/8]^2\}^{-1}$ $\zeta = 0.2$ $\omega_n = 8$	$\omega_3 = 8.0$	0 slope below the corner frequency -40 db/decade slope above the corner frequency	Varies from $0°$ to $-180°$

1. At frequencies less than ω_1, the first corner frequency, only the factors Lm 4 and Lm $(j\omega)^{-1}$ are effective. All the other factors have zero value. At ω_1, Lm 4 = 12 db and Lm $(j\omega_1)^{-1}$ = 6 db; thus at this frequency the composite curve has the value of 18 db. Below ω_1 the composite curve has a slope of -20 db/decade because of the Lm $(j\omega)^{-1}$ term.

2. Above ω_1, the factor Lm $(1 + j2\omega)^{-1}$ has a slope of -20 db/decade and must be added to the terms in step 1. When the slopes are added, the composite curve has a total slope of -40 db/decade in the frequency band from ω_1 to ω_2. Since this bandwidth represents 2 octaves, the value of the composite curve at ω_2 is -6 db.

3. Above ω_2, the factor Lm $(1 + j0.5\omega)$ is effective. This factor has a slope of $+20$ db/decade above ω_2 and must be added to obtain the composite curve. The composite curve now has a total slope of -20 db/decade in the frequency band from ω_2 to ω_3. The bandwidth from ω_2 to ω_3 is 2 octaves; therefore the value of the composite curve at ω_3 is -18 db.

4. Above ω_3 the last term Lm $[1 + j0.05\omega + (j\omega/8)^2]^{-1}$ must be added. This factor has a slope of -40 db/decade; therefore the total slope of the composite curve above ω_3 is -60 db/decade.

5. Once the asymptotic plot of the log magnitude of $\mathbf{G}(j\omega)$ has been drawn, the corrections can be added if desired. The corrections at each corner frequency and at an octave above and below the corner frequency are usually sufficient. For first-order terms the corrections are ± 3 db at the corner frequencies and ± 1 db at an octave above and below the corner frequency. The corrections for quadratic terms at these points can be obtained from Fig. 9-7 since they are a function of the damping ratio. If the correction curve for the quadratic factor is not available, the correction at the frequencies $\omega = \omega_n$ and $\omega = 0.707\omega_n$ can be calculated easily from the expression

$$1 - \left(\frac{\omega}{\omega_n}\right)^2 + j2\zeta\,\frac{\omega}{\omega_n} \tag{9-31}$$

The corrected log magnitude curves for each factor and for the composite curve are also shown in Fig. 9-9.

Determination of the phase angle curve of $\mathbf{G}(j\omega)$ can be simplified by using the following procedure:

1. For the $(j\omega)^{-n}$ term, draw a line at the angle of $(-n)90°$.
2. For each $(1 + j\omega T)^{\pm 1}$ term, locate the following points:
   a. $\pm 45°$ at the corner frequency
   b. $\pm 26.6°$ an octave below the corner frequency
   c. $\pm 5.7°$ a decade below the corner frequency
   d. $\pm 63.4°$ an octave above the corner frequency
   e. $\pm 84.3°$ a decade above the corner frequency

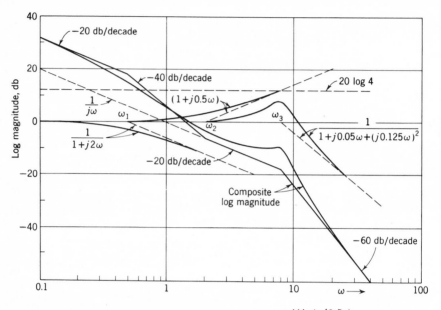

Fig. 9-9   *Log magnitude curve for* $\mathbf{G}(j\omega) = \dfrac{4(1 + j0.5\omega)}{j\omega(1 + j2\omega)[1 + j0.05\omega + (j0.125\omega)^2]}$

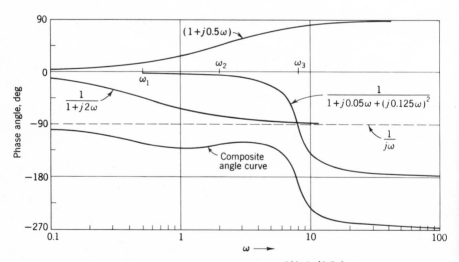

Fig. 9-10   *Phase angle curve for* $\mathbf{G}(j\omega) = \dfrac{4(1 + j0.5\omega)}{j\omega(1 + j2\omega)[1 + j0.05\omega + (j0.125\omega)^2]}$

Through these points, with the aid of a french curve or by a free-hand sketch, draw a curve representing the phase angle plot for each $(1 + j\omega T)^{\pm 1}$ term.

3. For each $[1 + j2\zeta\omega/\omega_n + (j\omega/\omega_n)^2]$ term,
   a. Locate the $\pm 90°$ point at the corner frequency.
   b. From Fig. 9-6, for the respective ζ, obtain a few points to draw the phase plot for each term with the aid of a french curve. The angle at $\omega = 0.707\omega_n$ may be sufficient and can be evaluated easily from

$$\phi = \tan^{-1} \frac{2\zeta(\omega/\omega_n)}{1 - \omega^2/\omega_n^2} \tag{9-32}$$

4. Once the phase plot of each term of $\mathbf{G}(j\omega)$ has been drawn, the phase plot of $\mathbf{G}(j\omega)$ is easily determined with the aid of a pair of dividers.
   a. Use the line representing the angle equal to

$$\underline{/\lim_{\omega \to 0} \mathbf{G}(j\omega)} = (-n)90°$$

   as the base line, and add or subtract the angles of each factor from this reference line.
   b. At a particular frequency on the graph, measure the angle for each single-order and quadratic factor with the dividers. Add and/or subtract them by use of the dividers from this base line until all terms have been accounted for.
   c. The number of frequency points for which the above should be done is determined by the desired accuracy of the phase plots.

9-10   System Type and Gain as Related to Log Magnitude Curves

The steady-state error of a closed-loop system depends on the system type and the gain. The static error coefficients are determined by these two characteristics, as was noted in Chap. 6. For any given log magnitude curve the system type and gain can be determined. Also, with the transfer function given so that the system type and gain are known, they can expedite drawing the log magnitude curve. This is described for Type 0, 1, and 2 systems.

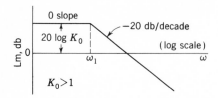

Fig. 9-11 *Log magnitude plot for* $G(j\omega) = K_0/(1 + j\omega T_a)$.

Type 0 System

A Type 0 system has a transfer function of the form

$$G(j\omega) = \frac{K_0}{1 + j\omega T_a}$$

At low frequencies, $\omega < 1/T_a$, Lm $G(j\omega) = 20 \log K_0$, which is a constant. The corner frequency is $\omega_1 = 1/T_a$. The slope of the log magnitude curve is zero below the corner frequency and -20 db/decade above the corner frequency. The log magnitude curve is shown in Fig. 9-11.

For a Type 0 system the characteristics are as follows:

1. The slope at low frequencies is zero.
2. The magnitude at low frequencies is $20 \log K_0$.
3. The gain K_0 is the static step error coefficient.

Type 1 System

A Type 1 system has a transfer function of the form

$$G(j\omega) = \frac{K_1}{j\omega(1 + j\omega T_a)}$$

At low frequencies ($\omega < 1/T_a$) LmG $(j\omega) = $ Lm $K_1 -$ Lm $j\omega$, which has a slope of -20 db/decade. At $\omega = K_1$, Lm $G(j\omega)_{\omega=K_1} = 0$. If the corner frequency $\omega_1 = 1/T_a$ is greater than K_1, the low-frequency portion of the curve of slope -20 db/decade crosses the 0-db axis at a value of $\omega_x = K_1$. If the corner frequency is less than K_1, the low-frequency portion of the curve of slope -20 db/decade does not cross the 0-db axis. However, by extending the low-frequency initial slope of -20 db/decade until it does cross the 0-db axis, the value of the frequency at which the extension crosses the 0-db axis is $\omega_x = K_1$. In other words, the plot of $K_1/j\omega$ crosses the 0-db value at $\omega = K_1$. The log magnitude curve for these two cases is shown in Fig. 9-12.

At $\omega = 1$, Lm $j\omega = 0$; therefore Lm $G(j\omega)_{\omega=1} = 20 \log K_1$. For $T_a < 1$ this value is a point on the slope of -20 db/decade. For $T_a > 1$ this value is a point on the extension of the initial slope, as shown in Fig. 9-12. The

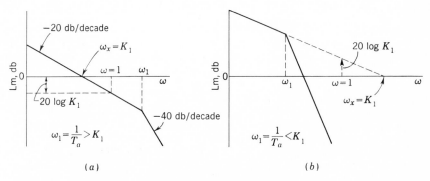

Fig. 9-12 *Log magnitude plot for* $\mathbf{G}(j\omega) = K_1/j\omega(1 + j\omega T_a)$.

frequency ω_x is smaller or larger than unity, depending on whether K_1 is smaller or larger than unity.

For a Type 1 system the characteristics are as follows:

1. The slope at low frequencies is -20 db/decade.
2. The intercept of the low-frequency slope of -20 db/decade (or its extension) with the 0-db axis occurs at a frequency ω_x, where $\omega_x = K_1$.
3. The value on the low-frequency slope of -20 db/decade (or its extension) at the frequency $\omega = 1$ is equal to 20 log K_1.
4. The gain K_1 is the static ramp error coefficient.

Type 2 System

A Type 2 system has a transfer function of the form

$$\mathbf{G}(j\omega) = \frac{K_2}{(j\omega)^2(1 + j\omega T_a)}$$

At low frequencies, $\omega < 1/T_a$, the Lm $\mathbf{G}(j\omega) = \text{Lm } K_2 - \text{Lm } (j\omega)^2$, for which the slope characteristic is -40 db/decade. At $\omega^2 = K_2$, Lm $\mathbf{G}(j\omega) = 0$; therefore the intercept of the initial slope of -40 db/decade (or its extension, if necessary) with the 0-db axis occurs at a frequency ω_y so that $\omega_y^2 = K_2$.

At $\omega = 1$, Lm $(j\omega)^2 = 0$; therefore, Lm $\mathbf{G}(j\omega)_{\omega=1} = 20 \log K_2$. This point occurs on the initial slope or on its extension, depending on whether $\omega_1 = 1/T_a$ is larger or smaller then $\sqrt{K_2}$. If $K_2 > 1$ the quantity 20 log K_2 is positive, and if $K_2 < 1$ the quantity 20 log K_2 is negative.

The log magnitude curve for a Type 2 transfer function is shown in Fig. 9-13. The determination of gain K_2 from the graph is shown.

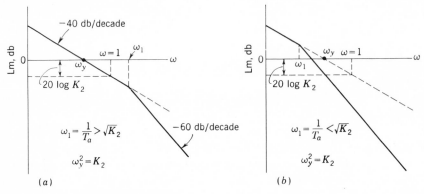

Fig. 9-13 *Log magnitude plot for* $G(j\omega) = K_2/(j\omega)^2(1 + j\omega T_a)$.

For a Type 2 system the characteristics are as follows:

1. The slope at low frequencies is -40 db/decade.
2. The intercept of the low-frequency slope of -40 db/decade (or its extension, if necessary) with the 0-db axis occurs at a frequency ω_y, where $\omega_y^2 = K_2$.
3. The value on the low-frequency slope of -40 db/decade (or its extension) at the frequency $\omega = 1$ is equal to 20 log K_2.
4. The gain K_2 is the static parabolic error coefficient.

9-11   Experimental Determination of Transfer Functions[6]

The log magnitude and phase angle diagram is of great value in the situation where the transfer function of a given system is not known.   The magnitude and angle of the ratio of the output to the input can be obtained experimentally for a steady-state sinusoidal input signal for a number of frequency points.   These data are used to obtain the exact log magnitude and angle diagram.   Asymptotes are drawn on the exact log magnitude curve by utilizing the fact that they have to be multiples of ± 20 db/decade.   From these asymptotes the system type and the approximate time constants are determined.   Thus, in this manner, the transfer function of the system can be synthesized.

Care must be exercised in determining whether any poles or zeros of the transfer function are in the right-half s plane.   A system that has no open-loop poles or zeros in the right-half s plane is defined as a minimum-phase system.[7]   For this type all factors are of the form $1 + Ts$ and/or $1 + As + Bs^2$.   A system that has open-loop poles or zeros in the right-half *s* plane is defined as a non-minimum-phase system. Thus, for this situation, one or more terms in the transfer function have the form $1 - Ts$ and/or $1 \pm As \pm Bs^2$. As an example, consider the functions $1 + j\omega T$ and $1 - j\omega T$. The log magnitude plots of these functions are identical, but the angle diagram for the former goes from 0 to 90° whereas for the latter it goes from 0 to −90°. Therefore care must be exercised in interpreting the angle plot to determine whether any factors of the transfer function lie in the right-half *s* plane. Many practical systems are in the minimum-phase category.

9-12   *Introduction to Polar Plots*[5]

In the earlier sections of this chapter a simple graphical method is presented for plotting the characteristic curves of transfer functions. These curves were the log magnitude and the angle of $G(j\omega)$ versus ω, plotted on semilog graph paper. The reason for presenting this method first is that these curves can be constructed easily and rapidly. Frequently, the polar plot of the transfer function is desired. The magnitude and angle of $G(j\omega)$, for sufficient frequency points, are readily obtainable from the Lm $G(j\omega)$ and $\underline{/G(j\omega)}$ versus log ω curves. The conversion of decibels to magnitude is facilitated by use of the graph in Appendix G. This approach often takes less time than calculating $|G(j\omega)|$ and $\phi(\omega)$ analytically for each frequency point desired, unless a digital computer is available.[10]

The data for drawing the polar plot of the frequency response can also be obtained from the pole-zero diagram, as described in Sec. 4-12. It is possible to visualize the complete shape of the frequency-response curve from the pole-zero diagram because the angular contribution of each pole and zero is readily apparent. The Spirule can be used to evaluate the precise magnitude and angle of the frequency response for any value of frequency.

The rest of this chapter is devoted to the *direct* and *inverse* polar plots of $G(j\omega)$ and the association of the shape of these curves with system types. The polar plot of $G(j\omega)$ is called the *direct polar plot*, and the polar plot of $[G(j\omega)]^{-1}$ is called the *inverse polar plot*.

9-13   *Direct Polar Plots*

In this section a number of examples of direct polar plots are given.

Simple RL Circuit (*Lag Compensator*)

Figure 9-14 shows a simple series *RL* circuit with an applied sinusoidal voltage $e_{in}(t)$ of constant amplitude and an output voltage $e_o(t)$ across resistor

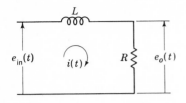

Fig. 9-14 *A simple RL circuit with a sinusoidal voltage* $e_{in}(t)$ *applied.*

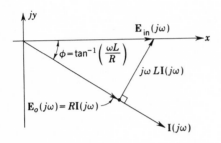

Fig. 9-15 *Phasor representation of the effective voltages in the circuit of Fig. 9-14.*

R.   The transfer function of this circuit, considering effective quantities, is

$$\mathbf{G}(j\omega) = \frac{\mathbf{E}_o(j\omega)}{\mathbf{E}_{in}(j\omega)} = \frac{R}{R + j\omega L} \tag{9-33}$$

or

$$\mathbf{G}(j\omega) = \frac{R}{[R^2 + (\omega L)^2]^{1/2}} \bigg/ -\tan^{-1}\frac{\omega L}{R} \tag{9-34}$$

Note that in Eq. (9-34) both the magnitude and the angle are functions of the excitation frequency ω.

If the phasor $\mathbf{E}_{in}$ shown in Fig. 9-15 is of constant magnitude and is taken as the reference for all values of frequency, the tip of the phasor $\mathbf{E}_o(j\omega)$ lies on a semicircle whose diameter is equal to $\mathbf{E}_{in}$.   Dividing each phasor in Fig. 9-15 by $\mathbf{E}_{in}$ results in a phasor representing the transfer function of Eq. (9-34) for all values of frequency, as shown in Fig. 9-16.

For the polar plot of the circuit in Fig. 9-14 to be applicable for all values of R and L, Eq. (9-33) is rewritten in the following nondimensional (standard) form:

$$\mathbf{G}(j\omega T) = \frac{\mathbf{E}_o(j\omega T)}{\mathbf{E}_{in}(j\omega T)} = \frac{1}{1 + j\omega T} \tag{9-35}$$

where ωT is a numeric and $T = L/R$ is the time constant of the circuit.

A plot of Eq. (9-35) as the frequency is varied from 0 to ∞ is shown in Fig. 9-17.   When $\omega = 0$, the ratio $\mathbf{E}_o(j\omega)/\mathbf{E}_{in}(j\omega) = 1/0°$; and when $\omega \to \infty$, $\mathbf{E}_o(j\omega)/\mathbf{E}_{in}(j\omega) \to 0/-90°$.   Thus the polar plot is a semicircle with its center at 0.5 on the real axis and with a radius equal to $\frac{1}{2}$.   Since the values used to obtain this polar plot are in terms of ωT, the plot can be interpolated for the particular values of R and L in a given circuit.

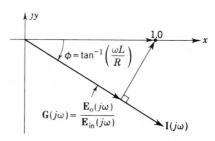

Fig. 9-16 *Phasor representation of Eq.* (9-34).

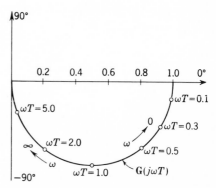

Fig. 9-17 *Polar plot of* $G(j\omega T) = 1/(1 + j\omega T)$.

The transfer function of Eq. (9-33) can also be put in the form given in Eq. (4-98). The pole-zero diagram and the frequency response are shown in Fig. 4-12. This can be converted to the polar form shown in Fig. 9-17.

Simple *RC* Circuit (*Lead Compensator*)

A sinusoidal voltage e_{in} of constant amplitude is applied to the simple series *RC* circuit shown in Fig. 9-18. This circuit is the one considered in Sec. 5-4. The ratio $E_o(j\omega)/E_{in}(j\omega)$ results in the following transfer function for this circuit:

$$G(j\omega) = \frac{E_o(j\omega)}{E_{in}(j\omega)} = \frac{R}{R + 1/j\omega C} \tag{9-36}$$

or

$$G(j\omega) = \frac{E_o(j\omega)}{E_{in}(j\omega)} = \frac{R}{\sqrt{R^2 + (1/\omega C)^2}} \bigg/ \tan^{-1}\frac{1}{\omega RC} \tag{9-37}$$

Both the magnitude and the angle of Eq. (9-37) for the *RC* circuit are functions of the frequency ω.

Equation (9-36) is now put into the standard form so that its polar plot (Fig. 9-19) may represent any *RC* circuit:

$$G(j\omega T) = \frac{E_o(j\omega T)}{E_{in}(j\omega T)} = \frac{j\omega T}{1 + j\omega T} \tag{9-38}$$

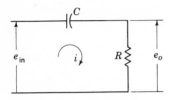

Fig. 9-18 *A simple RC circuit with a sinusoidal voltage e_{in} applied.*

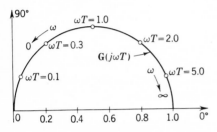

Fig. 9-19 *Polar plot of $G(j\omega T) = j\omega T/(1 + j\omega T)$.*

where $T = RC$, the time constant of the circuit. Figure 9-19 shows how the nondimensionalized equation varies as ω is increased from 0 to ∞. When $\omega = 0$, the ratio $E_o(j\omega)/E_{in}(j\omega) = 0/90°$ and when $\omega \to \infty$, $E_o(j\omega)/E_{in}(j\omega) \to 1/0°$. This is exactly opposite to what happens for the RL circuit as the frequency is varied from 0 to ∞. Note that the polar plot is a semicircle with its center on the real axis and with a radius equal to $\frac{1}{2}$. Since this plot is in the standard nondimensionalized form, it can be interpolated for the particular values of R and C in a given circuit.

Complex RC Network (Lag-Lead Compensator)

The properties of the networks of the last two examples may be combined and produced by the use of a single RC circuit, as shown in Fig. 9-20. The transfer function of this circuit is derived in Sec. 5-13 as

$$G(s) = \frac{E_o(s)}{E_{in}(s)} = \frac{1 + (T_1 + T_2)s + T_1T_2s^2}{1 + (T_1 + T_2 + T_{12})s + T_1T_2s^2} \tag{9-39}$$

where the time constants are

$$T_1 = R_1C_1 \qquad T_2 = R_2C_2 \qquad T_{12} = R_1C_2$$

As a function of frequency, the transfer function is

$$G(j\omega) = \frac{E_o(j\omega)}{E_{in}(j\omega)} = \frac{(1 - \omega^2T_1T_2) + j\omega(T_1 + T_2)}{(1 - \omega^2T_1T_2) + j\omega(T_1 + T_2 + T_{12})} \tag{9-40}$$

By the proper choice of the time constants, the circuit acts as a lag network in the lower-frequency range of 0 to ω_x and as a lead network in the higher-frequency range of ω_x to ∞. It is left to the reader to show that the polar plot of this circuit is a circle with its center on the real axis and lying in the

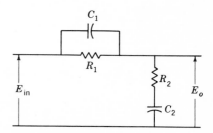

Fig. 9-20 *A complex RC circuit with an effective voltage E_{in} applied.*

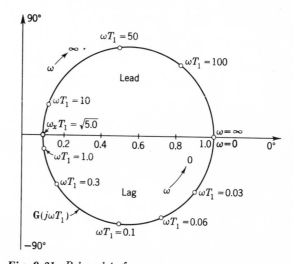

Fig. 9-21 *Polar plot of*

$$G(j\omega T_1) = \frac{(1 + j\omega T_1)(1 + j0.2\omega T_1)}{(1 + j11.1\omega T_1)(1 + j0.0179\omega T_1)}$$

for $T_2 = 0.2T_1$, $T_{12} = 10.0T_1$.

first and fourth quadrants. Figure 9-21 illustrates such a polar plot in non-dimensionalized form for a typical circuit.

When T_2 and T_{12} in Eq. (9-40) are expressed in terms of T_1, the following features are revealed:

1. $\lim\limits_{\omega \to 0} G(j\omega T_1) \to 1\underline{/0°}$

2. $\lim\limits_{\omega \to \infty} G(j\omega T_1) \to 1\underline{/0°}$

3. For the frequency $\omega = \omega_x$,

$$\omega_x^2 T_1 T_2 = 1 \tag{9-41}$$

and Eq. (9-40) becomes

$$G(j\omega_x T_1) = \frac{T_1 + T_2}{T_1 + T_2 + T_{12}} = |G(j\omega_x T_1)|\underline{/0°} \tag{9-42}$$

Note that Eq. (9-42) represents the minimum value of the transfer function in the whole frequency spectrum. From Fig. 9-21 it is seen that for frequencies below ω_x the circuit acts like the RL circuit of Fig. 9-17. For frequencies above ω_x it acts like the RC circuit of Fig. 9-19. The applications and advantages of this circuit are discussed in later chapters on compensation.

Type 0 Feedback Control System

Figure 6-8 illustrates a typical Type 0 system whose block-diagram representation is shown in Fig. 6-9 and whose forward transfer function is

$$\mathbf{G}(j\omega) = \frac{\mathbf{C}(j\omega)}{\mathbf{E}(j\omega)} = \frac{K_0}{(1 + j\omega T_f)(1 + j\omega T_m)} \qquad (9\text{-}43)$$

It is noted from Eq. (9-43) that when

$$\omega = 0 \qquad \mathbf{G}(j\omega) = K_0\underline{/0^\circ} \qquad (9\text{-}44)$$
$$\omega \to \infty \qquad \mathbf{G}(j\omega) \to 0\underline{/-180^\circ} \qquad (9\text{-}45)$$

Also, for each term in the denominator the angular contribution to $\mathbf{G}(j\omega)$, as ω goes from 0 to ∞, goes from 0 to -90°. Thus the polar plot of this transfer function must start at $\mathbf{G}(j\omega) = K_0\underline{/0^\circ}$ for $\omega = 0$ and proceed first through the fourth and then through the third quadrants to

$$\lim_{\omega \to \infty} \mathbf{G}(j\omega) = 0\underline{/-180^\circ}$$

as the frequency approaches infinity. In other words, the angular variation of $\mathbf{G}(j\omega)$ is continuously decreasing in one direction: 0 to -180°. The polar plot of Eq. (9-43) is illustrated in Fig. 9-22. The exact shape of this plot is determined by the particular values of the time constants. A further analysis of Eq. (9-43) reveals that the closer the two time constants come to being equal, the more the curve is "pulled into" the third quadrant.

If Eq. (9-43) had another term of the form $1 + j\omega T$ in the denominator, the $\mathbf{G}(j\omega)\big]_{\omega = \infty}$ point would rotate clockwise by an additional 90°. In other words, when $\omega \to \infty$, $\mathbf{G}(j\omega) \to 0\underline{/-270^\circ}$. In this case the curve crosses the

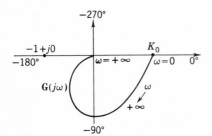

Fig. 9-22 *Polar plot for a typical Type 0 transfer function [Eq. (9-43)].*

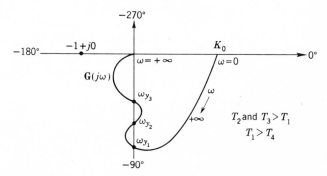

Fig. 9-23 *Polar plot for a more complex Type 0 transfer function* [*Eq. (9-46)*].

real axis at a frequency ω_x for which the imaginary part of the transfer function is zero. When terms of this form appear in the numerator, they each result in an angular variation of 0 to 90° (a counterclockwise rotation) as the frequency is varied from 0 to ∞. Thus the angular variation $G(j\omega)$ may not continuously change in one direction. Also, the resultant polar plot may not be as smooth as the one shown in Fig. 9-22. As an example, consider the transfer function

$$G(j\omega) = \frac{K_0(1 + j\omega T_1)^2}{(1 + j\omega T_2)(1 + j\omega T_3)(1 + j\omega T_4)^2} \qquad (9\text{-}46)$$

whose polar plot is indented as shown in Fig. 9-23. The time constants T_2 and T_3 are greater than T_1, and T_1 is greater than T_4. From the angular contribution of each factor and from the analysis above, it can be surmised that the polar plot has the general shape shown in the figure. In the event that T_1 is smaller than all the others, its polar plot is similar in shape to the one shown in Fig. 9-22.

In the same manner, a quadratic in either the numerator or the denominator of a transfer function results in an angular contribution of 0 to $\pm 180°$, respectively. Consequently, the polar plot of $G(j\omega)$ is affected accordingly. It can be seen from both examples that the polar plot of a Type 0 system always starts at a value K_0 (step error coefficient) on the positive real axis for $\omega = 0$ and ends at zero magnitude (for $v > w$) and tangent to one of the major axes at $\omega = \infty$.

Type 1 Feedback Control System

In Sec. 6-5 a typical Type 1 system is illustrated whose forward transfer function is

$$G(j\omega) = \frac{C(j\omega)}{E(j\omega)} = \frac{K_1}{j\omega(1 + j\omega T_m)(1 + j\omega T_c)(1 + j\omega T_q)} \qquad (9\text{-}47)$$

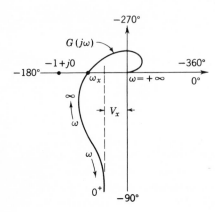

Fig 9-24 *Polar plot for a typical Type 1 transfer function [Eq. (9-47)].*

For Eq. (9-47), as

$$\omega \to 0^+ \qquad G(j\omega) \to \infty \underline{/-90°} \qquad (9\text{-}48)$$

$$\omega \to \infty \qquad G(j\omega) \to 0\underline{/-360°} \qquad (9\text{-}49)$$

Note that the $j\omega$ term in the denominator always contributes a constant $-90°$ to the total angle of $G(j\omega)$ for all frequencies. Thus the basic difference between Eqs. (9-43) and (9-47) is the presence of the term $j\omega$ in the denominator of the latter equation. Since all the $1 + j\omega T$ terms of Eq. (9-47) appear in the denominator, its polar plot, as shown in Fig. 9-24, has no dents. From the remarks of this and previous sections, it can be seen that the angular variation of $G(j\omega)$ decreases continuously in the same direction from -90 to $-360°$ as ω increases from 0 to ∞. The presence of any frequency-dependent factor in the numerator has the same general effect on the polar plot as that described previously.

It is seen from Eq. (9-48) that the magnitude of the function $G(j\omega)$ approaches infinity as the value of ω approaches zero. This equation does not indicate whether the function approaches infinity asymptotically to the $-90°$ axis or to some line parallel to it. The true asymptote is determined by finding the value of the real part of $G(j\omega)$ as ω approaches zero. Thus

$$V_x = \lim_{\omega \to 0} \text{Re}\,[G(j\omega)] \qquad (9\text{-}50)$$

or, for this particular transfer function,

$$V_x = -K_1(T_q + T_c + T_m) \qquad (9\text{-}51)$$

Equation (9-51) shows that the magnitude of $G(j\omega)$ approaches infinity asymptotically to a vertical line whose real-axis intercept equals V_x, as illustrated in Fig. 9-24. Note that the value of V_x is always a direct function of the ramp error coefficient. The significance of this fact is discussed more fully in a later section.

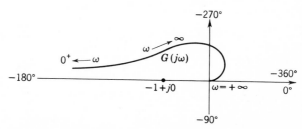

Fig. 9-25 *Polar plot for a typical Type 2 transfer function, resulting in an unstable feedback control system [Eq. (9-54)].*

The frequency of the crossing point on the negative real axis of the $G(j\omega)$ function is that value of frequency ω_x for which the imaginary part of $G(j\omega)$ is equal to zero. Thus

$$\text{Im } [G(j\omega_x)] = 0 \tag{9-52}$$

or, for this particular transfer function,

$$\omega_x = (T_c T_q + T_q T_m + T_m T_c)^{-\frac{1}{2}} \tag{9-53}$$

The significance of this value of frequency is pointed out in later chapters on compensation.

Type 2 Feedback Control System

The transfer function of the Type 2 system illustrated in Sec. 6-5, prior to its modification, is

$$G(j\omega) = \frac{C(j\omega)}{E(j\omega)} = \frac{K_2}{(j\omega)^2(1 + j\omega T_f)(1 + j\omega T_m)} \tag{9-54}$$

For Eq. (9-54), as

$$\omega \to 0^+ \qquad G(j\omega) \to \infty \underline{/-180°} \tag{9-55}$$

$$\omega \to +\infty \qquad G(j\omega) \to 0 \underline{/-360°} \tag{9-56}$$

The presence of the $(j\omega)^2$ term in the denominator contributes a constant $-180°$ to the total angle of $G(j\omega)$ for all frequencies. The form of the transfer function of Eq. (9-54) is such that its polar plot (Fig. 9-25) is a smooth curve whose angle $\phi(\omega)$ decreases continuously from -180 to $-360°$.

The introduction of an additional pole and a zero can alter the shape of the polar plot. Consider the transfer function

$$G_o(j\omega) = \frac{K_2'(1 + j\omega T_1)}{(j\omega)^2(1 + j\omega T_f)(1 + j\omega T_m)(1 + j\omega T_2)} \tag{9-57}$$

where $T_1 > T_2$. The polar plot can be obtained from the pole-zero diagram shown in Fig. 9-26. At $s = j\omega = j0^+$ the angle of each factor is zero except

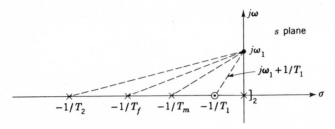

Fig. 9-26 *Pole-zero diagram for*

$$G_o(s) = \frac{K_2'T_1}{T_fT_mT_2}\frac{s+1/T_1}{s^2(s + 1/T_f)(s + 1/T_m)(s + 1/T_2)}$$

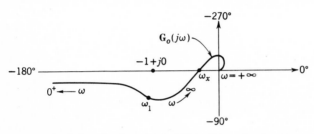

Fig. 9-27 *Polar plot for a typical Type 2 transfer function, resulting in a stable system [Eq. (9-57)].*

for the double pole at the origin. The angle at $\omega = 0^+$ is therefore $-180°$, as given by Eq. (9-55), which is still applicable. As ω increases from zero, the angle of $j\omega + 1/T_1$ increases faster than the angles of the other poles. In fact, at low frequencies the angle due to the zero is larger than the sum of the angles due to the poles located to the left of the zero. This is shown qualitatively at the frequency ω_1 in Fig. 9-26. Therefore, the angle of $G(j\omega)$ at low frequencies is greater than $-180°$. As the frequency increases to a value ω_x, the sum of the component angles of $G(j\omega)$ is $-180°$ and the polar plot crosses the real axis, as shown in Fig. 9-27. As ω increases further, the angle of $j\omega + 1/T_1$ shows only a small increase, but the angles from the poles increase rapidly. In the limit as $\omega \to \infty$, the angles of $j\omega + 1/T_1$ and $j\omega + 1/T_2$ are equal and opposite in sign, so the angle of $G(j\omega)$ approaches $-360°$, as given by Eq. (9-56).

Figure 9-27 shows the complete polar plot of $G(j\omega)$. A comparison of Figs. 9-25 and 9-27 shows that both curves approach $-180°$ at $\omega = 0^+$, which is typical of Type 2 systems. As $\omega \to \infty$, the angle approaches $-360°$ because both $G(j\omega)$ and $G_o(j\omega)$ are of the same degree. As shown in the next chapter, the feedback system containing $G(j\omega)$ is unstable, while the system containing $G_o(j\omega)$ is stable.

It can be shown that as $\omega \to 0+$ the polar plot for a Type 2 system is below the real axis if

$$\Sigma(T_{\text{numerator}}) - \Sigma(T_{\text{denominator}})$$

is a positive value and is above the real axis if it is a negative value. Thus for this example the necessary condition is $T_1 > (T_f + T_m + T_2)$.

9-14   Summary of Direct Polar Plots

To obtain the direct polar plot of a system's forward transfer function, the following criteria are used to determine the key parts of the curve:

1. The forward transfer function has the general form

$$\mathbf{G}(j\omega) = \frac{K_n(1 + j\omega T_a)(1 + j\omega T_b) \cdots (1 + j\omega T_w)}{(j\omega)^n(1 + j\omega T_1)(1 + j\omega T_2) \cdots (1 + j\omega T_u)} \qquad (9\text{-}58)$$

From the transfer function the system type is determined. Then the portion of the polar plot representing the $\lim_{\omega \to 0} \mathbf{G}(j\omega)$ is approximately located. The low-frequency polar-plot characteristics (as $\omega \to 0$) of the different system types are summarized in Fig. 9-28. The angle at $\omega = 0$ is $n(-90°)$.

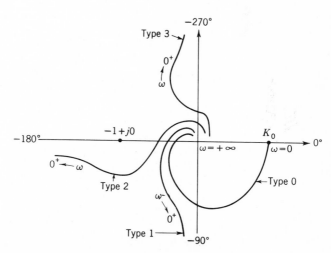

Fig. 9-28   *A summary of direct polar plots of different types of systems.*

2. The high-frequency end of the polar plot can be determined as follows:

$$\lim_{\omega \to +\infty} \mathbf{G}(j\omega) = 0 / \underline{(w - n - u)90°} \qquad (9\text{-}59)$$

Note that, since the degree of the denominator of Eq. (9-58) is always greater than the degree of the numerator, the high-frequency point ($\omega = \infty$) is approached (i.e., the angular condition) in the clockwise sense. The plot ends at the origin tangent to the axis determined by Eq. (9-59). Tangency may occur on either side of the axis.

3. The asymptote that the low-frequency end approaches, for a Type 1 system, is determined by taking the limit as $\omega \to 0$ of the real part of the transfer function.

4. The frequencies at the points of intersection of the polar plot with the negative real axis and the imaginary axis are determined, respectively, by setting

$$\text{Im } [\mathbf{G}(j\omega)] = 0 \qquad (9\text{-}60)$$
$$\text{Re } [\mathbf{G}(j\omega)] = 0 \qquad (9\text{-}61)$$

5. If there are no time constants in the numerator of the transfer function, the curve is a smooth one in which the angle of $\mathbf{G}(j\omega)$ continuously decreases as ω goes from 0 to ∞. With time constants in the numerator, and depending upon their values, the angle may not continuously vary in the same direction, thus creating "dents" in the polar plot.

6. As is seen later in this chapter, it is important to know the exact shape of the polar plot in the vicinity of the $(-1 + j0)$ point. Enough points of $\mathbf{G}(j\omega)$ should be accurately determined in this area.

7. By reviewing Sec. 6-4, it is seen that there exists a correlation of the polar plot with the steady-state solution of a stable feedback control system. This correlation may be described as follows:

    a. For a constant actuating signal in a Type 0 system

    $$c(t)_{ss} = K_0 E_0 \qquad (9\text{-}62)$$

    Thus, for a given value of $c(t)_{ss}$, the larger K_0 is made, the smaller the actuating signal E_0 is. In other words, for a system with unity feedback the error existing between the output and the input is smaller. Therefore, in terms of the polar plot, the farther away the point $\lim_{\omega \to 0^+} \mathbf{G}(j\omega) = K_0$ is from the origin, the greater is the steady-state accuracy for a step input signal.

    b. In a Type 1 system, a constant actuating signal results in a constant derivative of the input:

    $$Dc_{ss} = K_1 E_0 \qquad (9\text{-}63)$$

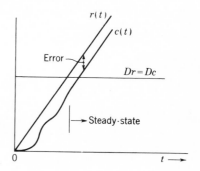

Fig. 9-29 *Performance of a feedback control system for a constant rate of change of the input* $Dr = R_1$.

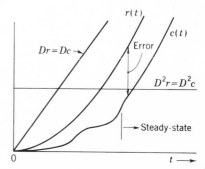

Fig. 9-30 *Performance of a feedback control system for a constant second derivative of the input signal* $D^2r = R_2$.

For a given rate of change of the output, the higher K_1 can be made, the smaller the actuating signal is. In Sec. 6-2.it is pointed out that it may be necessary to analyze the system for a ramp input function to determine its response to an irregular input signal of the form illustrated in Fig. 6-2. Thus for a unity feedback system, the higher K_1 can be made, the smaller is the error between the input and output for a ramp input signal, as shown in Fig. 9-29. It is seen in Sec. 9-13 that V_x is directly proportional to K_1. Therefore it can be concluded, in terms of the polar plot, that the larger V_x is made, or the farther away the portion $\lim_{\omega \to 0} \text{Re} \,[\mathbf{G}(j\omega)]$ is from the imaginary axis, the greater is the steady-state accuracy of the system to a ramp input signal. The fact that the system is identified as Type 1 from the polar plot reveals that a zero error exists for a step input signal.

c. By following similar reasoning for the Type 2 system, it is seen that the larger K_2 is made, the smaller is the error between the input and output for a parabolic input signal. Figure 9-30 illustrates the performance of a control system for a constant second derivative of the input signal. The fact that a plot shows a system to be Type 2 indicates that, for a step or a ramp input signal, zero error exists between the output and input signals.

Thus there has been established a correlation between the direct polar plot and the steady-state solution of a stable control system.

9-15   Inverse Polar Plots

The direct polar plots have certain drawbacks when they are used for systems that utilize elements in the feedback path.   In these cases it is found that the graphical analysis of the inverse polar plots is much simpler and therefore advantageous.   The inverse polar plot is obtained by plotting the phasor quantity

$$G^{-1}(j\omega) = \frac{1}{G(j\omega)} = \frac{E(j\omega)}{C(j\omega)} \tag{9-64}$$

as a function of frequency.

Compensators

The inverse plots of the lag, lead, and lag-lead compensators discussed in Sec. 9-13 are illustrated in Fig. 9-31.   They are used in the chapters on compensation.

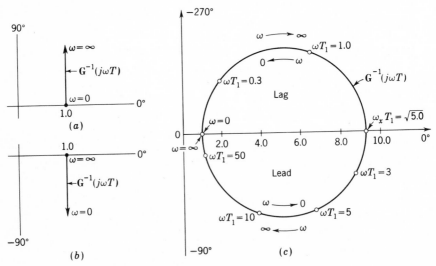

Fig. 9-31 *The inverse polar plots of lag, lead, and lag-lead networks.   (a) Plot of* $G^{-1}(j\omega T) = 1 + j\omega T$.   *(b) Plot of* $G^{-1}(j\omega T) = 1 + 1/j\omega T$.   *(c) Plot of*

$$G^{-1}(j\omega T_1) = \frac{(1 + j11.1\omega T_1)(1 + j0.0179\omega T_1)}{(1 + j\omega T_1)(1 + j0.2\omega T_1)}$$

for $T_2 = 0.2T_1$, $T_{12} = 10.0T_1$.

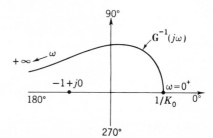

Fig. 9-32 *Inverse polar plot of a typical Type 0 feedback control system* [*Eq. (9-65)*].

Type 0 Feedback Control System

The inverse of the transfer function used in Sec. 9-13 for the Type 0 system of Fig. 6-8 is

$$\mathbf{G}^{-1}(j\omega) = \frac{(1 + j\omega T_f)(1 + j\omega T_m)}{K_0} \tag{9-65}$$

For the two limiting points,

$$\lim_{\omega \to 0^+} \mathbf{G}^{-1}(j\omega) = \frac{1}{K_0} \tag{9-66}$$

$$\lim_{\omega \to +\infty} \mathbf{G}^{-1}(j\omega) = \lim_{\omega \to +\infty} (j\omega)^2 T_f T_{ma} = \infty \,\underline{/180°} \tag{9-67}$$

As ω goes from 0 to ∞, the angle of each term in the numerator of $\mathbf{G}^{-1}(j\omega)$ goes from 0 to 90°. Since there are no frequency-dependent terms in the denominator, the angular condition of $\mathbf{G}^{-1}(j\omega)$ is continuously increasing from 0 to 180° as ω goes from 0 to ∞. The inverse plot of this Type 0 system is shown in Fig. 9-32. Each additional $1 + j\omega T$ factor in the numerator of $\mathbf{G}^{-1}(j\omega)$ rotates the high-frequency portion of the plot counterclockwise by 90°.

Type 1 Feedback Control System

The inverse of the transfer function used in Sec. 9-13 for the Type 1 system of Fig. 6-12 is

$$\mathbf{G}^{-1}(j\omega) = \frac{j\omega(1 + j\omega T_m)(1 + j\omega T_c)(1 + j\omega T_q)}{K_1} \tag{9-68}$$

Thus for Eq. (9-68)

$$\lim_{\omega \to 0^+} \mathbf{G}^{-1}(j\omega) = 0 \,\underline{/90°} \tag{9-69}$$

$$\lim_{\omega \to +\infty} \mathbf{G}^{-1}(j\omega) = \infty \,\underline{/360°} \tag{9-70}$$

With the above end points and based upon previous analyses, the inverse polar plot of Eq. (9-68) is a smooth curve, as shown in Fig. 9-33. Additional

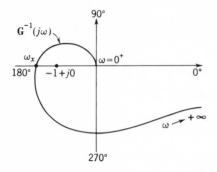

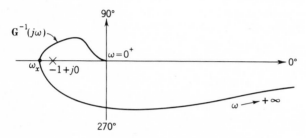

Fig. 9-33 *Inverse polar plot of a typical Type 1 feedback control system [Eq. (9-68)].*

Fig. 9-34 *Inverse polar plot of a typical Type 2 feedback control system [Eq. (9-71)].*

terms in the numerator or the denominator alter the location of the end points and the shape of the plot accordingly.

Type 2 Feedback Control System

In the same manner, the inverse polar plot for the modified Type 2 system of Sec. 6-5 can be obtained. The inverse of the transfer function of Eq. (9-57) for this system is

$$G^{-1}(j\omega) = \frac{(j\omega)^2(1 + j\omega T_f)(1 + j\omega T_m)(1 + j\omega T_2)}{K_2'(1 + j\omega T_1)} \qquad (9\text{-}71)$$

and its polar plot is illustrated in Fig. 9-34.

9-16   Summary of Inverse Polar Plots

The inverse polar plots of a forward transfer function are obtained with the aid of the following criteria which determine the key parts of the curve.

1. Once the system type is determined from the forward transfer function, the portion of the inverse polar plot representing the $\lim_{\omega \to 0^+}$ $G^{-1}(j\omega)$ is located. Note that in taking the limit the zero-frequency point is approached in the clockwise sense.

2. By using the general form of the inverse forward transfer function as given by

$$G^{-1}(s) = \frac{s^n(1 + T_1s)(1 + T_2s) \cdots (1 + T_us)}{K_n(1 + T_as) \cdots (1 + T_ws)} \qquad (9\text{-}72)$$

the high-frequency end of the polar plot is determined as follows:

$$\lim_{\omega \to +\infty} G^{-1}(j\omega) = \infty \underline{/(n + u - w)90°} \qquad (9\text{-}73)$$

3. The frequencies at the points of intersection of the polar plot with the negative real axis and the imaginary axis are determined, respectively, by setting

$$\text{Im } [G^{-1}(j\omega)] = 0 \qquad (9\text{-}74)$$
$$\text{Re } [G^{-1}(j\omega)] = 0 \qquad (9\text{-}75)$$

4. With no time constants in the denominator of the inverse transfer function, the plot is a smooth curve in which the angle of $G^{-1}(j\omega)$ continuously increases as ω goes from 0 to ∞. If there are time constants in the denominator, and depending upon their values, the angle may not continuously vary in the same direction, thus creating "dents" in the polar plot.

5. The exact shape of the polar plot in the vicinity of the negative real axis crossing is important, and points on $G^{-1}(j\omega)$ should be accurately determined in this area with the aid of step 3.

A summary of the inverse polar plots for the different types of systems is made in Fig. 9-35. Just as for the direct polar plots, the low-frequency characteristic (as $\omega \to 0^+$) distinguishes the different system types.

By reviewing Sec. 9-14, the following statements can be made with respect to the correlation of the inverse polar plot with the steady-state solution of a stable feedback control system.

1. For a Type 0 system, the closer the point $\lim_{\omega \to 0^+} G^{-1}(j\omega) = 1/K_0$ is made to the origin, the greater is its steady-state accuracy for a step input signal.

2. The fact that a Type 1 system is identified from the inverse plot reveals that a zero error exists for a step input signal.

3. The fact that an inverse plot indicates a system to be Type 2 shows that, for a step or a ramp input signal, zero error exists between the output and input signals.

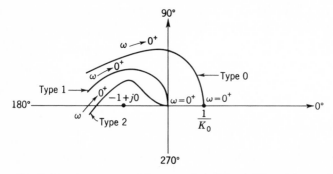

Fig. 9-35   *A summary of inverse polar plots of different types of systems.*

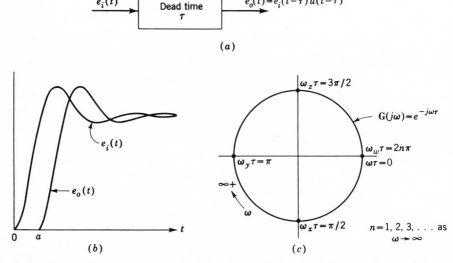

Fig. 9-36   (*a*) *Transport lag,* τ.   (*b*) *Time characteristic.*   (*c*) *Polar-plot characteristic.*

9-17   *Dead Time*

Some elements in control systems are characterized by dead time, or transport lag, indicated by a dead period when no output is present in response to an input.   Figure 9-36a shows the block diagram of a transport-lag element. Figure 9-36b shows the resultant output $e_o(t) = e_i(t - \tau)u(t - \tau)$, which has the same form as the input but is delayed by a time τ.   Dead time is a non-linear characteristic which, fortunately, can be represented precisely in terms of the Laplace transform (see Sec. 4-4). Thus

$$E_o(s) = e^{-\tau s} E_i(s) \tag{9-76}$$

The transfer function of the transport-lag element is

$$G(s) = \frac{E_o(s)}{E_i(s)} = e^{-\tau s} \tag{9-77}$$

The frequency transfer function of an element with dead time only is

$$\mathbf{G}(j\omega) = e^{-j\omega\tau} = 1\underline{/-\omega\tau} \tag{9-78}$$

It has a magnitude of unity and a negative angle which increases directly in proportion to frequency. The polar plot of Eq. (9-78) is a unit circle which is traced indefinitely, as shown in Fig. 9-36c. The log magnitude and phase angle diagram shows a constant value of 0 db and a phase angle which increases with frequency.

9-18 *Conclusions*

In this chapter different types of frequency-response plots are introduced. From the log magnitude and phase angle diagram the polar plot can be obtained with ease and rapidity. All these plots indicate the type of system under consideration and the necessary adjustments that must be made to improve its response. The manner in which these adjustments are made is discussed in the following chapters.

The methods presented for obtaining the frequency-response plots have stressed graphical techniques. For greater accuracy the curves can be determined analytically. A digital-computer program can be used to calculate these data. An analog computer can also be used;[8-10] in this case, plots of the frequency-response curves are produced directly.

The methods described in this chapter for obtaining frequency-response plots are based upon the condition that the transfer function of a given system is known. These plots can also be obtained from experimental data, which does not require that the analytical expression for the transfer function of a system be known. Chapter 20 discusses the methods of obtaining these experimental data, as well as methods of synthesizing the transfer functions from a given system's frequency-response plot.

Bibliography

1. Maccoll, L. A.: "Fundamental Theory of Servomechanisms," D. Van Nostrand Company, Inc., Princeton, N.J., 1945.

2. Harris, H.: The Frequency Response of Automatic Control Systems, *Trans. AIEE*, vol. 65, pp. 539–545, 1946.
3. James, H. M., N. B. Nichols, and R. S. Phillips: "Theory of Servomechanisms," McGraw-Hill Book Company, New York, 1947.
4. Nyquist, H.: Regeneration Theory, *Bell System Tech. J.*, vol. 11, pp. 126–147, 1932.
5. Brown, G. S., and D. P. Campbell: "Principles of Servomechanisms," John Wiley & Sons, Inc., New York, 1948.
6. Bruns, R. A., and R. M. Saunders: "Analysis of Feedback Control Systems," chap. 14, McGraw-Hill Book Company, New York, 1955.
7. Nixon, F. E.: "Principles of Automatic Controls," pp. 121–122, 172–173, 239–242, Prentice-Hall, Inc., Englewood Cliffs, N.J., 1953.
8. Ogar, G. W.: Obtaining the Frequency Response of Physical Systems by Analog Computer Techniques, *IRE Intern. Conv. Record*, vol. 9, pt. 2, pp. 196–210, 1961.
9. Vickers, D. B.: "Feedback System Analysis Using Frequency Response Curves from an Analog Computer," M.Sc. Thesis, Air Force Institute of Technology, Dayton, Ohio, 1960.
10. Dorrity, J. L.: "Frequency Response Program," Air Force Institute of Technology, Dayton, Ohio, 1965.

10

Nyquist's Stability Criterion

It has been stated often that a designer likes to have a procedure
that yields the maximum amount of information with a minimum
of effort. Also, there must be a method to indicate readily the
manner in which the system should be compensated to yield the
desired response. The development of the root-locus method
met the desires of the designer and has been a tremendous aid in
feedback control system synthesis. Another method that has
proved satisfactory to the designer is Nyquist's stability criterion,[1-3] which is applied to the polar plot of $G(j\omega)$ or $G^{-1}(j\omega)$
and the log magnitude–angle plots of $G(j\omega)$.

This method has many desirable features when applied to a
given plot of the $G(j\omega)$ function:

1. It is a simple graphical procedure for determining
   whether a system is stable or unstable.

2. The degree of stability can be readily ascertained by use of the M-α circles, which are developed in Chap. 11. A correlation can be made from the sinusoidal response to the step time response for a signal applied to the system.
3. It points the way for improving both the transient and the steady-state response of a system.
4. The performance of the closed-loop system to steady-state sinusoidally varying inputs can be determined from this same plot.

A knowledge of the effect of the static error coefficients for various types of inputs, coupled with the characteristics listed above, makes Nyquist's stability criterion a highly desirable method.

10-2   Development of Nyquist's Stability Criterion

It has been demonstrated previously that for a stable system none of the roots of the characteristic equation

$$B(s) = 1 + G(s)H(s) = 0 \tag{10-1}$$

can lie in the right-half s plane or on the $j\omega$ axis, as shown in Fig. 10-1. When Eq. (7-18) is substituted into Eq. (10-1), the denominator of the control ratio is

$$B(s) = 1 + \frac{N_1 N_2}{D_1 D_2} = \frac{D_1 D_2 + N_1 N_2}{D_1 D_2} \tag{10-2}$$

The condition for stability may be restated: For a stable system none of the zeros of $B(s)$ can lie in the right-half s plane or on the imaginary axis. It is seen that *the poles of the open-loop transfer function $G(s)H(s)$ are the poles of $B(s)$.*

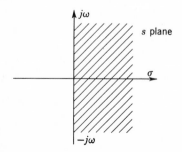

Fig. 10-1 *Prohibited region in the s plane.*

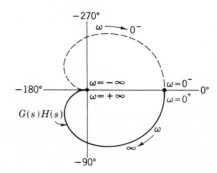

Fig. 10-2 A plot of a typical $G(s)H(s)$
transfer function for $-\infty < \omega < \infty$.

Briefly, it can be stated at this point that Nyquist's stability criterion relates the number of zeros and poles of $B(s)$ that lie in the right-half s plane to the polar plot of $G(s)H(s)$. The criterion is developed first for the direct plot and then for the inverse plot. In the preceding chapter the polar plots were drawn for frequencies in the range of 0^+ to $+\infty$. In utilizing the $G(s)H(s)$ plots for Nyquist's criterion, they must be drawn for a range of frequencies from $-\infty$ to $+\infty$ (this corresponds to the $j\omega$ axis of the s plane), as illustrated in Fig. 10-2. To complete the plot, the values of s traverse the infinite semicircle that encircles the entire right half of the s plane. The latter may map into a single point for $G(s)H(s)$ of modulus zero but variable phase as the semicircle is traversed.

The first thing to observe in Fig. 10-2 is that the plot is drawn for the open-loop transfer function $G(s)H(s)$ and not for the forward transfer function $G(s)$. Secondly, the plot is drawn for both positive and negative frequency values. *The polar plot drawn for negative frequencies is the conjugate of the plot drawn for positive frequencies.* This means that the curve for negative frequencies is symmetrical to the curve for positive frequencies, with the real axis as the axis of symmetry.

The Generalized Nyquist Stability Criterion Limitations

In this textbook it is assumed that all the control systems are inherently linear or that their limits of operation are confined to give a linear operation. This yields a set of linear differential equations with constant coefficients for the systems.

It has been stated previously that, because of the physical nature of feedback control systems, the order of the denominator D_1D_2 is equal to or greater than the order of the numerator N_1N_2 of the open-loop transfer function $G(s)H(s)$. Mathematically, this means that $\lim_{s \to \infty} G(s)H(s) \to 0$ or a constant.

These two factors satisfy the necessary limitations to the generalized Nyquist stability criterion.

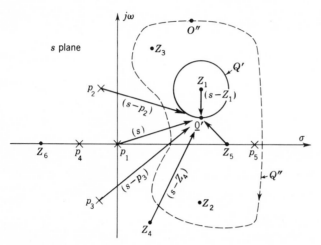

Fig. 10-3 *A plot of some poles and zeros of Eq. (10-3).*

Mathematical Basis for Nyquist's Stability Criterion

A rigorous mathematical derivation of Nyquist's stability criterion involves complex-variable theory.   Since many readers are not sufficiently versed in these areas, the rigorous derivation is given in Appendix I.   Fortunately, the result of the derivation is simple and is readily applied.   However, a complete knowledge of its derivation ensures its use with greater facility and sureness.   For those readers who do not refer to Appendix I a qualitative approach is now presented for the special case that $B(s)$ is a rational fraction. The function $B(s)$ given by Eq. (10-2) is rationalized, factored, and then written in the form

$$B(s) = \frac{(s - Z_1)(s - Z_2) \cdots (s - Z_v)}{(s - p_1)(s - p_2) \cdots (s - p_v)} \tag{10-3}$$

where $Z_1, Z_2, \ldots, Z_v$ are the zeros of the characteristic function and p_1, $p_2, \ldots, p_v$ are the poles of the characteristic function.   [These poles are the same as the poles of the open-loop transfer function $G(s)H(s)$ and include the s term for which $p = 0$, if it is present.]

In Fig. 10-3 some of the poles and zeros of a generalized function $B(s)$ are arbitrarily drawn on the s plane.   Also, an arbitrary *closed* curve Q' is drawn in the right-half plane about the zero Z_1.   To the point O' on Q', whose coordinates are $s = \sigma + j\omega$, are drawn directed line segments from all the poles and zeros.   The lengths of these directed line segments are given by $s - Z_1$, $s - Z_2$, $s - p_1$, $s - p_2$, etc.   Not all the directed segments from the poles and zeros are indicated in the figure, as they are not necessary to proceed with this development.   *As the point O' is rotated clockwise once*

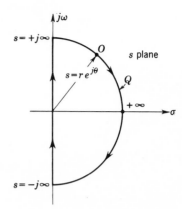

Fig. 10-4 *The contour that encloses the entire right-half s plane.*

around the closed curve Q', the length $s - Z_1$ rotates through a net angle of 360°. All the other directed segments have rotated through a net angle of 0°. Thus, by referring to Eq. (10-3), it is seen that the clockwise rotation of 360° for the length $s - Z_1$ must simultaneously be realized by the function $B(s)$ for the enclosure of the zero Z_1 by the path Q'.

Consider now a larger closed contour Q'' which includes the zeros Z_1, Z_2, Z_3, and Z_5 and the pole p_5. As a point O'' is rotated clockwise once around the closed curve Q'', each of the directed line segments from the enclosed pole and zeros rotates through a net angle of 360°. Since the angular rotation of the pole is experienced by the characteristic function in its denominator, the net angular rotation realized by Eq. (10-3) must be equal to the net angular rotations due to the pole p_5 minus the net angular rotations due to the zeros Z_1, Z_2, Z_3, and Z_5. In other words, the net angular rotation experienced by $1 + G(s)H(s)$ is $360° - (4)(360°) = -1,080°$. Therefore, for this case it can be stated that the net number of rotations N experienced by $1 + G(s)H(s)$ by the clockwise movement of point O'' once about the closed contour Q'' is equal to -3; i.e.,

(Number of poles enclosed) $-$ (number of zeros enclosed) $= 1 - 4 = -3$

where the minus sign denotes clockwise (cw) rotation. Note that if the contour Q'' includes only the pole p_5, $B(s)$ experiences one counterclockwise (ccw) rotation as the point O'' is moved around the contour. Also, for *any* closed path that may be chosen in the *right*-half s plane, all the poles and zeros that lie outside the closed path each contribute a net angular rotation of 0° to $B(s)$ as a point is moved once around this contour.

Generalizing Nyquist's Stability Criterion

Consider now a closed contour Q such that the whole right-half s plane is encircled (see Fig. 10-4), thus encircling all zeros and poles of $B(s)$ that may

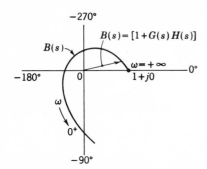

Fig. 10-5   *A polar plot of $B(s)$ for a feedback control system.*

have positive real parts.   As a consequence of the theory of complex variables used in the derivation, it is necessary that the contour Q not pass through any poles and zeros of $B(s)$.   When the results previously discussed are applied to the contour Q, the following conclusions are made:

1.   The total number of clockwise rotations of $B(s)$ due to its zeros is equal to its total number of zeros Z_R in the right-half s plane.
2.   The total number of counterclockwise rotations of $B(s)$ due to its poles is equal to its total number of poles P_R in the right-half s plane.
3.   The net number of rotations N of $B(s) = 1 + G(s)H(s)$ about the origin is equal to its total number of poles P_R minus its total number of zeros Z_R in the right-half s plane. N may be positive (ccw), negative (cw), or zero.

The essence of the above three conclusions can be represented by the equation

$$N = \frac{\text{change in phase of } 1 + G(s)H(s)}{2\pi} = P_R - Z_R \qquad (10\text{-}4)$$

where counterclockwise rotation is defined as being positive and clockwise rotation is negative.

Note that, for the characteristic function to realize a net rotation N, the directed line segment representing $B(s)$ (see Fig. 10-5) must rotate about the origin $360N$ degrees, or N complete revolutions.

A stable system $B(s)$ can have no zeros Z_R in the right-half s plane; therefore it can be concluded that, *for a stable system, the net number of rotations of $B(s)$ about the origin must be counterclockwise and equal to the number of poles P_R that lie in the right-half s plane; that is,*

$$N = P_R \qquad (10\text{-}5)$$

In other words, if $B(s)$ experiences a net clockwise rotation, this indicates that $Z_R > P_R$ where $P_R \geq 0$, and that the system is unstable.   If there

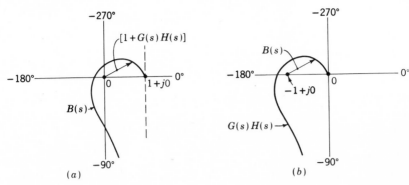

Fig. 10-6   *A change of reference for $B(s)$.*

are zero net rotations, then $Z_R = P_R$ and the system may or may not be stable, depending on whether $P_R = 0$ or $P_R > 0$, respectively.

The plot of $B(s) = 1 + G(s)H(s)$ for values s on the path Q surrounding the entire right-half s plane can be simplified.   Figure 10-6a and b show a plot of $B(s)$ and a plot of $G(s)H(s)$.   By moving the origin of Fig. 10-6b to the $(-1 + j0)$ point, the curve is now equal to $1 + G(s)H(s)$, which is $B(s)$. Since $G(s)H(s)$ is known, it is easier to plot this function and then move the origin to the -1 point to obtain $B(s)$.

In general, the open-loop transfer functions of many physical systems do not have any poles P_R in the right-half s plane.   In this case, $N = Z_R$. *Thus for a stable system the net number of rotations about the $(-1 + j0)$ point must be zero when there are no poles of $G(s)H(s)$ in the right-half s plane.*

In the event that the function $G(s)H(s)$ has some poles in the right-half s plane and the denominator is not in factored form, the number P_R can be determined by applying Routh's criterion to D_1D_2.   The Routhian array gives the number of roots in the right-half s plane by the number of sign changes in the first column.

Analysis of Path Q

In applying Nyquist's criterion, the whole right-half s plane must be encircled to ensure the inclusion of all poles or zeros in this portion of the plane.   In Fig. 10-4, the entire right-half s plane is included by considering the closed path Q to be composed of the following two segments:

1.   One segment is the imaginary axis from $-j\infty$ to $+j\infty$.
2.   The other segment, which completes the closed path, is a semicircle of infinite radius that encloses the entire right-half s plane.

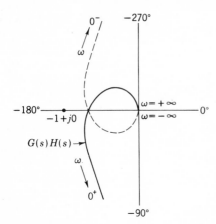

Fig. 10-7 *A plot of the transfer function of Eq. (10-7).*

The portion of the path along the imaginary axis is represented mathematically by $s = j\omega$. Thus, replacing s by $j\omega$ in Eq. (10-3) and letting ω take on all values from $-\infty$ to $+\infty$ gives that portion of the $B(s)$ plot corresponding to that portion of the closed contour Q on the imaginary axis.

One of the limitations of the Nyquist criterion is that $\lim_{s \to \infty} G(s)H(s) \to 0$ or a constant. Thus, for $B(s) = 1 + G(s)H(s)$, $\lim_{s \to \infty} [1 + G(s)H(s)] \to 1$ or 1 plus the constant. As a consequence, as the point O moves along the segment of the closed contour represented by the semicircle of infinite radius, the corresponding portion of the $B(s)$ plot must undergo no rotation. That is, $B(\infty) = $ constant for $s = re^{j\theta}$, where $r \to \infty$ and θ goes from $\pi/2$ to $-\pi/2$.

As a result of the above, considering the movement of point O only along the imaginary axis from $-\infty$ to $+\infty$ results in giving the same net rotation of $B(s)$ as if the whole contour Q is considered. *In other words, all the rotation of $B(s)$ occurs while the point O goes from $-j\infty$ to $+j\infty$ along the imaginary axis* if the denominator of $B(s)$ is at least one degree greater than the degree of the numerator. More generally, this statement applies only to those transfer functions $G(s)H(s)$ that conform to the limitations stated earlier in this section.*

Effect of Poles at the Origin on the Rotation of $B(s)$

For transfer functions $G(s)H(s)$ that have an s^n factor in the denominator, $s = 0$ is a zero of the denominator. Since no poles or zeros can lie on the contour, the contour shown in Fig. 10-4 must be modified. For these cases the manner in which the $\omega = 0^-$ and $\omega = 0^+$ portions of the plot are joined

---

* A transfer function that does not conform to these limitations and to which the above italicized statement does not apply is $G(s)H(s) = s$.

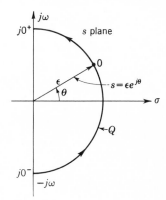

Fig. 10-8 *Portion of the closed contour Q in the vicinity of s = 0.*

is now investigated. This can best be done by taking an example. Consider the transfer function

$$G(s)H(s) = \frac{K_1}{s(1 + T_1 s)(1 + T_2 s)} \tag{10-6}$$

To obtain the direct polar plot of this function, substitute $s = j\omega$ into Eq. (10-6):

$$\mathbf{G}(j\omega)\mathbf{H}(j\omega) = \frac{K_1}{j\omega(1 + j\omega T_1)(1 + j\omega T_2)} \tag{10-7}$$

Figure 10-7 represents the plot of $G(s)H(s)$ for values of s along the imaginary axis; that is, $j0^+ < j\omega < j\infty$ and $-j\infty < -j\omega < j0^-$.

To determine the system stability, the nature of the $B(s)$ plot in the vicinity of $\omega = 0$ must be investigated. The closed contour Q in the vicinity of $s = 0$ is modified to avoid passing through the origin, as shown in Fig. 10-8. In other words, the point O is moved along the negative imaginary axis from $s = -j\infty$ to a point where $s = -j\omega = 0^-\underline{/-\pi/2}$ becomes very small, that is, $s = j\epsilon$. Then the point O moves along a semicircular path in the right-half s plane with a very small radius ϵ until it reaches the positive imaginary axis at $s = +j\omega = j0^+ = 0^+\underline{/\pi/2}$. From here the point O proceeds along the positive imaginary axis to $s = +j\infty$. With this modified closed contour Q of the right-half plane, only a very small portion of this plane is not encircled. The area of this portion approaches zero as the radius $\epsilon \to 0$, thus ensuring the inclusion of all poles and zeros in the right-half s plane. Therefore, to complete the plot of $B(s)$ the effect of moving point O on this semicircle must be investigated.

For the semicircular portion of the path Q,

$$s = \epsilon e^{j\theta} \tag{10-8}$$

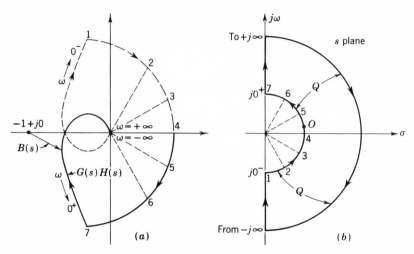

Fig. 10-9 *Complete contour of Eq. (10-6) and its correlation to the contour Q which encircles the right-half s plane.*

where $\epsilon \to 0$ and $-\pi/2 \leq \theta \leq \pi/2$.   Equation (10-6) becomes

$$G(s)H(s) = \frac{K_1}{s} = \frac{K_1}{\epsilon e^{j\theta}} = \frac{K_1}{\epsilon} e^{-j\theta} = \frac{K_1}{\epsilon} e^{j\psi} \qquad (10\text{-}9)$$

where $K_1/\epsilon \to \infty$ as $\epsilon \to 0$ and $\psi = -\theta$ goes from $\pi/2$ to $-\pi/2$ as the directed segment s goes from $\epsilon \underline{/-\pi/2}$ to $\epsilon \underline{/+\pi/2}$.   Thus, in Fig. 10-7, the end points from $\omega \to 0^-$ and $\omega \to 0^+$ are joined by a semicircle of infinite radius in the first and fourth quadrants.   Figure 10-9 illustrates the above procedure and shows the completed contour of $G(s)H(s)$ as the point O moves along the modified contour Q in the s plane in the clockwise direction.   By moving the origin to the $(-1 + j0)$ point, the curve becomes $B(s)$.   The plot in Fig. 10-9a does not encircle the $(-1 + j0)$ point and therefore the encirclement N is zero.

Since P_R is found to be zero from Eq. (10-6), $Z_R = 0$ and the system is stable:

$$N = P_R - Z_R \qquad (10\text{-}10)$$
$$Z_R = P_R - N = 0 - 0 = 0 \qquad (10\text{-}11)$$

Although it is evident from Fig. 10-9a that $N = 0$, it requires, in general, the determination of the net angular change of the directed line segment $B(s)$ as it is rotated in the direction of increasing frequency.   The movement of $B(s)$ in the direction of increasing frequency in Fig. 10-9a corresponds to the clockwise movement of point O in Fig. 10-9b.   Thus, each time the directed segment sweeps out an angle of 2π in the clockwise, or negative, direction, a negative encirclement of the $(-1 + j0)$ point is made. Likewise a 2π counterclockwise rotation results in a positive encirclement of the $(-1 + j0)$ point.

Transfer functions that have more than one s factor in the denominator have the general form, as $\epsilon \rightarrow 0$,

$$G(s)H(s) = \frac{K_n}{s^n} = \frac{K_n}{(\epsilon^n)e^{jn\theta}} = \frac{K_n}{\epsilon^n}\,e^{-jn\theta} = \frac{K_n}{\epsilon^n}\,e^{jn\psi} \qquad (10\text{-}12)$$

where $n = 1, 2, 3, 4, \ldots$. With the reasoning used in the preceding example, it is seen from Eq. (10-12) that, as ω passes from 0^- to 0^+, the plot of $G(s)H(s)$ makes n clockwise semicircles of infinite radius about the origin. If $n = 2$, then, as θ goes from $-\pi/2$ to $\pi/2$ in the s plane, $G(s)H(s)$ experiences a net rotation, in the vicinity of $\omega = 0$, of $(2)(180°)$, or $360°$.

Since the polar plots are symmetrical about the real axis, it is only necessary to determine the shape of the plot of $G(s)H(s)$ for a range of values of $0 < \omega < +\infty$. The net rotation of the plot for the range of $-\infty < \omega < +\infty$ is twice that of the plot for the range of $0 < \omega < +\infty$.

When $G(j\omega)H(j\omega)$ *Passes through the Point* $(-1 + j0)$

When the curve of $G(j\omega)H(j\omega)$ passes through the $(-1 + j0)$ point, the number of encirclements N is indeterminate. This corresponds to the condition where $B(s)$ has zeros on the imaginary axis. A necessary condition for the Nyquist criterion is that the path encircling the specified area must not pass through any poles or zeros of $B(s)$. When this condition is violated, the value for N becomes indeterminate and the Nyquist stability criterion cannot be applied. Simple imaginary zeros of $B(s)$ mean that the closed-loop system will have a continuous steady-state sinusoidal component in its output which is independent of the form of the input. Unless otherwise stated, this condition is considered unstable.

10-3   *Examples of Nyquist's Criterion Applied to Direct Polar Plot*

Several polar plots are illustrated in this section. These plots can be obtained with the aid of pole-zero diagrams. In other words, the angular variation of each term of a $G(s)H(s)$ function, as the contour Q is traversed, is readily determined from its pole-zero diagram. Both minimum-and nonminimum-phase systems are illustrated in the following examples.

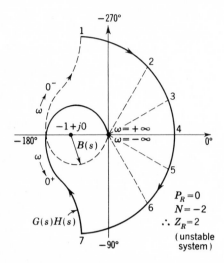

Fig. 10-10 *Complete polar plot of Fig. 10-9 with increased gain.*

Example 1 (Type 0)

The direct polar plot shown in Fig. 10-2 is for the following transfer function:

$$G(s)H(s) = \frac{K_0}{(1 + T_1 s)(1 + T_2 s)} \tag{10-13}$$

When Nyquist's criterion is applied to this plot, it is seen that $N = 0$, and from Eq. (10-13) the value $P_R = 0$. The value of Z_R is

$$Z_R = P_R - N = 0 - 0 = 0 \tag{10-14}$$

Therefore the system represented by Eq. (10-13) is stable. Thus the criterion indicates that, no matter how much the gain K_0 is increased, this system is always stable.

Example 2 (Type 1)

The example in Fig. 10-9 for the transfer function of the form

$$G(s)H(s) = \frac{K_1}{s(1 + T_1 s)(1 + T_2 s)} \tag{10-15}$$

is shown in the preceding section to be stable. If the gain of Eq. (10-15) is increased, the system is made unstable, as seen from Fig. 10-10. Note that the rotation of $B(s)$, in the direction of increasing frequency, produces a net angular rotation of $-720°$, or two complete clockwise rotations ($N = -2$).

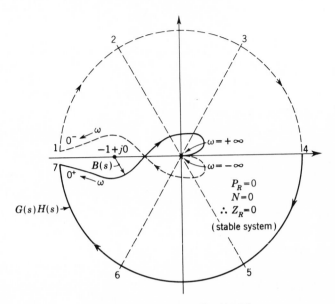

Fig. 10-11 *The complete polar plot of Eq. (10-16).*

Example 3 (Type 2)

$$G(s)H(s) = \frac{K_2(1 + T_4 s)}{s^2(1 + T_1 s)(1 + T_2 s)(1 + T_3 s)} \tag{10-16}$$

where $T_4 > T_1$, T_2, and T_3. Figure 10-11 shows the mapping of $G(s)H(s)$ for the contour Q of the s plane. The word *mapping*, as used here, means that for a given point in the s plane there corresponds a given point in the $G(s)H(s)$ or $B(s)$ plane.

As pointed out in the preceding section, the presence of the s^2 term in the denominator of Eq. (10-16) results in a net rotation of 360° in the vicinity of $\omega = 0$, as shown in Fig. 10-11. For the complete range of frequencies the net rotation is zero; thus with $P_R = 0$ the system is stable. Like the previous example, this system can be made unstable by increasing the gain sufficiently so that the $G(s)H(s)$ plot crosses the negative real axis to the left of the $(-1 + j0)$ point.

Example 4 (Conditionally Stable System)

$$G(s)H(s) = \frac{K_0(1 + T_1 s)^2}{(1 + T_2 s)(1 + T_3 s)(1 + T_4 s)(1 + T_5 s)^2} \tag{10-17}$$

where $T_5 < T_1 < T_2$, T_3, and T_4. The complete polar plot of Eq. (10-17) is illustrated in Fig. 10-12 for a particular value of gain.

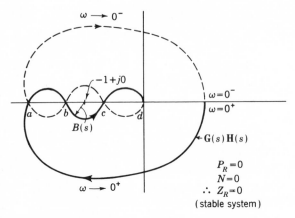

Fig. 10-12   *The complete polar plot of Eq. (10-17).*

In this example it is seen that the system can be made unstable not only by increasing the gain but by decreasing the gain.   If the gain is increased sufficiently that the $(-1 + j0)$ point lies between the points c and d of the polar plot, the net clockwise rotation is equal to 2.   Therefore $Z_R = 2$ and the system is unstable.   On the other hand, if the gain is decreased so that the $(-1 + j0)$ point lies between the points a and b of the polar plot, the net clockwise rotation is again 2 and the system is unstable.   The gain can be further decreased so that the $(-1 + j0)$ point lies to the left of point a of the polar plot, resulting in a stable system.   *This system is therefore conditionally stable.*   A conditionally stable system is stable for a given range of values of gain but becomes unstable if the gain is either reduced or increased sufficiently.   Such a system places a greater restriction on the stability and drift of amplifier-gain characteristics.   In addition, an effective gain reduction occurs in amplifiers that reach saturation with large input signals. This, in turn, may result in an unstable operation for the conditionally stable system.

Example 5 (non-minimum phase)

$$G(s)H(s) = \frac{K_1(T_2s + 1)}{s(T_1s - 1)} \tag{10-18}$$

The complete polar plot of the above equation is illustrated in Fig. 10-13 for a particular value of gain.   In this example it is seen that the system is unstable for low values of gain $(0 < K_1 < K_{1x})$ and that the system is stable for large values of gain $(K_{1x} < K_1 < \infty)$.   For the range $0 < K_1 < K_{1x}$ the $(-1 + j0)$ point is located, as shown in Fig. 10-13, between the points

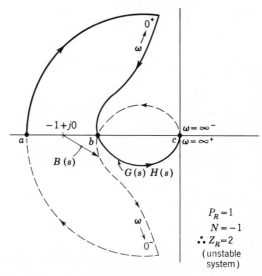

Fig. 10-13   *The complete polar plot of*

$$G(s)H(s) = \frac{K_1(T_2s + 1)}{s(T_1s - 1)}$$

a and b.   For the range $K_{1x} < K_1 < \infty$ the $(-1 + j0)$ point lies between the points b and c, which yields $N = +1$, thus resulting in $Z_R = 0$ and a stable system.

10-4   *Nyquist's Stability Criterion Applied to Systems Having Dead Time*

An example can best illustrate the application of the Nyquist criterion to a control system having dead time, or transport lag.   Figure 10-14a illustrates the complete polar plot of a stable system having a specified value of gain and the transfer function

$$G_x(s)H(s) = \frac{K_1}{s(1 + T_1s)(1 + T_2s)} \tag{10-19}$$

If dead time is added to this system, its transfer function becomes

$$G(s)H(s) = \frac{K_1e^{-\tau s}}{s(1 + T_1s)(1 + T_2s)} \tag{10-20}$$

and the resulting complete polar plot is shown in Fig. 10-14b.   In traversing the contour Q and including the polar-plot characteristic of dead time, shown

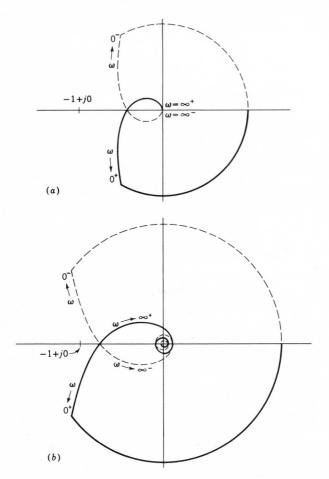

Fig. 10-14 *Complete polar plots of* (a) *Eq.* (10-19); (b) *Eq.*
(10-20).

in Fig. 9-36c, the effects on the complete polar plot are:

1. In traversing the imaginary axis of the contour Q between $0^+ <$
    $\omega < +\infty$, the polar plot of $\mathbf{G}(j\omega)\mathbf{H}(j\omega)$ in the third quadrant is
    shifted clockwise, closer to the $(-1 + j0)$ point. Thus, if the dead
    time is increased sufficiently, the $(-1 + j0)$ point will be enclosed
    by the polar plot and the system becomes unstable.
2. As $\omega \to +\infty$ along the imaginary axis, a spiraling effect results as
    $|\mathbf{G}(j\omega)\mathbf{H}(j\omega)| \to 0$.

A transport lag therefore tends to make a system less stable.

10-5 *Regional Location of Roots*[6]

It is possible to determine the location of roots of a characteristic equation by utilizing the Nyquist criterion. Consider a stable system whose open-loop transfer function is given by $G(s)H(s)$. For this system the contour Q of Fig. 10-15a results in the value $Z_R = 0$. In order to determine whether there are any roots in the cross-hatched region of the s plane, the transformation $s = s_1 - a$, where $a > 0$, is utilized. Substituting this value of s into the $G(s)H(s)$ function yields a new transfer function

$$G(s_1 - a)H(s_1 - a) = G(s_1)H(s_1)$$

For this new transfer function Nyquist's criterion is applied for the contour Q' of the translated half-plane shown in Fig. 10-15b. The resulting value of $Z_{R,1}$ indicates the number of roots of the characteristic equation, if any, that lie in the cross-hatched region. By a repeated process of translating the half-plane, the location of all roots can be determined. This method of locating roots is similar to the one discussed in Sec. 4-15, which use Routh's criterion.

Another method of determining the location of roots is to use the contour Q'', shown in Fig. 10-16, which is bounded by a ζ line. If the system is initially stable, that is, $Z_R = 0$ for the contour Q of Fig. 10-15a, the polar form of s,

$$s = r\underline{/\phi} \tag{10-21}$$

is used to determine the number of roots, if any, that lie in the cross-hatched region shown in Fig. 10-16. Equation (10-21) is substituted into the transfer function $G(s)H(s)$. By utilizing the new form of transfer function $G(r\underline{/\phi})H(r\underline{/\phi})$, the complete polar plot may be obtained. The contour Q''

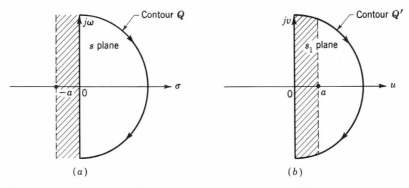

(a) (b)

Fig. 10-15 *Translated half-planes.*

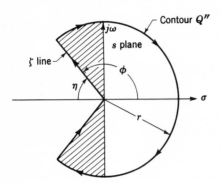

Fig. 10-16 *Contour Q'' containing the ζ line.*

must be modified to exclude any poles of $G(s)H(s)$ at the origin.   This method is theoretically sound but is very tedious.

10-6   Nyquist's Stability Criterion Applied to the Inverse Polar Plots

The criterion applied to the inverse plots is the same as that for the direct plots except for one minor modification.   As previously stipulated, none of the roots of the characteristic equation

$$B(s) = 1 + G(s)H(s) = 0 \qquad (10\text{-}22)$$

can lie in the right-half s plane or on the $j\omega$ axis for a system to be stable. Dividing this equation by $G(s)H(s)$,

$$B'(s) = \frac{1}{G(s)H(s)} + 1 = 0 \qquad (10\text{-}23)$$

Substituting Eq. (7-18) into Eq. (10-23),

$$B'(s) = \frac{D_1 D_2}{N_1 N_2} + 1 = \frac{D_1 D_2 + N_1 N_2}{N_1 N_2} = 0 \qquad (10\text{-}24)$$

Thus the zeros of $B'(s)$ are the roots of the characteristic equation

$$D_1 D_2 + N_1 N_2 = 0$$

which is seen to be the same as the zeros of $B(s)$.   Equation (10-24) is expressed as

$$B'(s) = \frac{(s - Z_1)(s - Z_2) \cdots (s - Z_v)}{(s - p_1')(s - p_2') \cdots (s - p_w')} \qquad (10\text{-}25)$$

where Z_1, Z_2, . . . , Z_v are the zeros of the functions $B(s)$ and $B'(s)$, and p'_1, p'_2, . . . , p'_w are the poles of $B'(s)$ which are the same as the poles of $1/G(s)H(s)$ or zeros of $G(s)H(s)$.

The mathematical basis for Nyquist's stability criterion as applied to the inverse plots is the same as that stipulated in Sec. 10-2 for the direct plots. The only difference is that, in Fig. 10-3, p_1, p_2, . . . , p_5 are replaced by p'_1, p'_2, . . . , p'_5. Therefore, when the results of Sec. 10-2 are applied for a path Q which encircles the entire right-half s plane and with this modification, the following conclusions can be drawn:

1. The total number of clockwise rotations of $B'(s)$ due to its zeros is equal to its total number of zeros Z_R in the right-half s plane.
2. The total number of counterclockwise rotations of $B'(s)$ due to its poles is equal to its total number of poles P'_R in the right-half s plane. Remember that P'_R equals the poles of $1/G(s)H(s)$ or the zeros of $G(s)H(s)$ in the right-half s plane.
3. The net number of rotations N' of $B'(s) = [1/G(s)H(s)] + 1$ about the origin [or about the $(-1 + j0)$ point of $1/G(s)H(s)$] is equal to its total number of poles P'_R minus its total number of zeros Z_R in the right-half s plane,

$$N' = P'_R - Z_R \qquad (10\text{-}26)$$

where counterclockwise rotation is positive and clockwise rotation is negative.

Therefore it can be concluded that, *for a stable system, the net number of rotations of $B'(s)$ about the origin must be counterclockwise and equal to the poles P'_R that lie in the right-half s plane; that is,*

$$N' = P'_R \qquad (10\text{-}27)$$

As pointed out in Sec. 10-2 for the direct plots, it is simpler to count the rotations about the $(-1 + j0)$ point, using the $G(s)H(s)$ plot. The same is true for the inverse plots. Since $1/G(s)H(s)$ is known, it is easier to plot this function and then move the origin to the $(-1 + j0)$ point to obtain $B'(s)$. In general, the open-loop transfer function of many physical systems does not have any zeros (and hence no poles P'_R) in the right-half s plane. In this case, $N' = Z_R$. *Thus for a stable system the net number of rotations about the $(-1 + j0)$ point must be zero where there are no poles P'_R in the right-half s plane.*

In the event that the function $G(s)H(s)$ has some zeros in the right-half s plane, the number of poles P'_R can be determined by factoring N_1N_2 or by applying Routh's criterion to N_1N_2.

Analysis of Path Q

Replacing s by $j\omega$ in Eq. (10-25) and letting ω take on all values from $-\infty$ to $+\infty$ gives the portion of the $B'(s)$ plot corresponding to that portion of the closed contour Q on the imaginary axis of Fig. 10-4. For the case of $\omega = 0$ the value $B'(s)$, resulting from the limitations of the Nyquist criterion, is a constant. Thus, as the point O on the closed contour passes through the point $0 + j0$, the corresponding portion of the $B'(s)$ plot crosses the real axis.

To complete the plot of $B'(s)$, the effect of moving point O on the semicircle of infinite radius on the contour, which ensures the inclusion of all poles and zeros in the right-half plane, must be investigated. For this semicircle,

$$s = re^{j\theta} \tag{10-28}$$

where $r \to \infty$ and $\pi/2 \geq \theta \geq -\pi/2$. Referring to the general open-loop transfer function

$$G(s)H(s) = \frac{K_n(s - z_1) \cdots (s - z_w)}{s^n(s - p_1)(s - p_2) \cdots (s - p_u)} \tag{10-29}$$

it can be seen that $B'(s)$ has a magnitude of infinity and undergoes a net rotation of $(n + u - w)\pi$ for this portion of the contour Q. Mathematically this is stated as follows:

$$\lim_{s \to \infty} \frac{1}{G(s)H(s)} = \frac{s^n s^u}{K_n s^w} = \frac{(r)^{(n+u-w)}}{K_n} e^{j(n+u-w)\theta} \tag{10-30}$$

From this equation it is easily seen that, as the point O, for $r = \infty$, goes from $\pi/2$ to $-\pi/2$ on the path Q in the s plane, the plot of $1/G(s)H(s)$ makes $n + u - w$ semicircles of infinite radius about the origin.

In other words, the rotation of $B'(s)$ occurs while the point O goes from 0^+ to $+\infty$ (on the $+j\omega$ axis), then from $+\infty$ to $-\infty$ (on the semicircle of infinite radius), and then from $-\infty$ to 0^- (on the $-j\omega$ axis).

10-7   Examples of Nyquist's Criterion Applied to the Inverse Polar Plots

Example 1 (Type 0)

The inverse polar plot of

$$[G(s)H(s)]^{-1} = \frac{(1 + T_1 s)(1 + T_2 s)}{K_0} \tag{10-31}$$

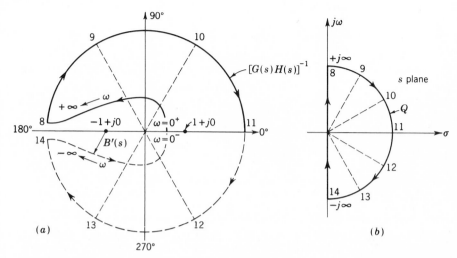

Fig. 10-17 *Complete contour of Eq. (10-31) and its correlation to the Q contour in the s plane.*

is illustrated in Fig. 10-17. For this example

$$\lim_{s \to \infty} [G(s)H(s)]^{-1} = \frac{T_1 T_2}{K_0} r^2 e^{j2\theta} \tag{10-32}$$

By referring to Fig. 10-17b and using Eq. (10-32), the complete polar plot of $[G(s)H(s)]^{-1}$ shown in Fig. 10-17a is obtained. By moving the origin to the $(-1 + j0)$ point, the curve becomes $B'(s)$. When Nyquist's criterion is applied to this plot, the net encirclement of the origin is zero. From Eq. (10-31) the value $P'_R = 0$; therefore,

$$Z_R = P'_R - N' = 0 - 0 = 0 \tag{10-33}$$

In other words, the system is stable. The criterion indicates that, no matter how much the gain K_0 is increased, this particular system is never unstable.

Example 2 (Type 1)

The inverse polar plot of

$$[G(s)H(s)]^{-1} = \frac{s(1 + T_1 s)(1 + T_2 s)}{K_1} \tag{10-34}$$

given in Fig. 10-18 shows that the system is stable. When the gain is increased to the value K'_1, the plot drawn in Fig. 10-19 shows that the system is unstable. The stability of this system therefore depends on the gain.

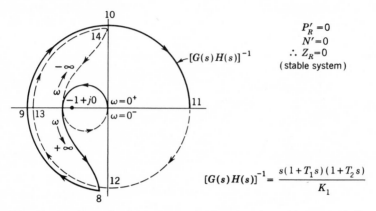

Fig. 10-18   *A completed inverse transfer function of a stable system, indicating its correlation to Fig. 10-17b.*

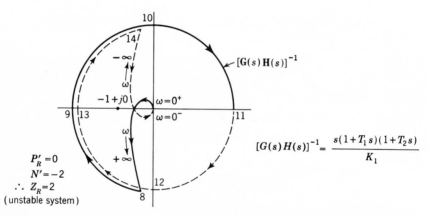

Fig. 10-19   *A completed inverse transfer function of an unstable system, indicating its correlation to Fig. 10-17b.*

10-8   *Definitions of Phase Margin and Gain Margin and Their Relation to Stability*[4,5]

The stability and approximate degree of stability can be determined from the log magnitude and phase diagram.   The stability characteristic is specified in terms of the following quantities:

1. *Gain Crossover.* This is the point on the plot of the transfer function at which the magnitude is unity [Lm $G(j\omega) = 0$ db]. The frequency at gain crossover is called the phase-margin frequency ω_ϕ.
2. *Phase Margin.* This is 180° plus the negative trigonometrically considered angle of the transfer function at the gain-crossover point. It is designated as the angle γ, which can be expressed as $\gamma = 180° + \phi$, where ϕ is negative.
3. *Phase Crossover.* This is the point on the plot of the transfer function at which the phase angle is $-180°$. The frequency at which phase crossover occurs is called the gain-margin frequency ω_c.
4. *Gain Margin.* This is the additional gain a that just makes the system unstable. Expressed in terms of the transfer function at the frequency ω_c, it is

$$|G(j\omega_c)|a = 1 \qquad (10\text{-}35)$$

On the polar plot of $G(j\omega)$ the value at ω_c is

$$|G(j\omega_c)| = \frac{1}{a} \qquad (10\text{-}36)$$

In terms of the log magnitude, in decibels, this is

$$\text{Lm } a = - \text{ Lm } G(j\omega_c) \qquad (10\text{-}37)$$

which identifies the gain margin on the log magnitude diagram.

These quantities are illustrated in Fig. 10-20 on both the log and the polar curves. Note the algebraic sign associated with these two quantities as marked on the curves. Figure 10-20*a*, *b*, and *c* represent a stable system, and Fig. 10-20*d*, *e*, and *f* represent an unstable system.

The phase margin is the amount of phase shift at the frequency ω_ϕ that would just produce instability. For minimum-phase networks, the phase margin must be positive for a stable system, whereas a negative phase margin means that the system is unstable.

It can be shown that the phase margin is related to the effective damping ratio ζ of the system. This is discussed qualitatively in the next chapter. Satisfactory response is usually obtained with a phase margin of 45 to 60°. As an individual gains experience and develops his own particular technique, the value of γ to be used for a particular system becomes more evident. This guide line for system performance applies only to those systems where behavior is equivalent to that of a second-order system.

The gain margin is the factor by which the gain must be changed in order to produce instability. The gain margin must be positive when expressed in decibels (greater than unity as a numeric) for a stable system A negative gain margin means that the system is unstable.

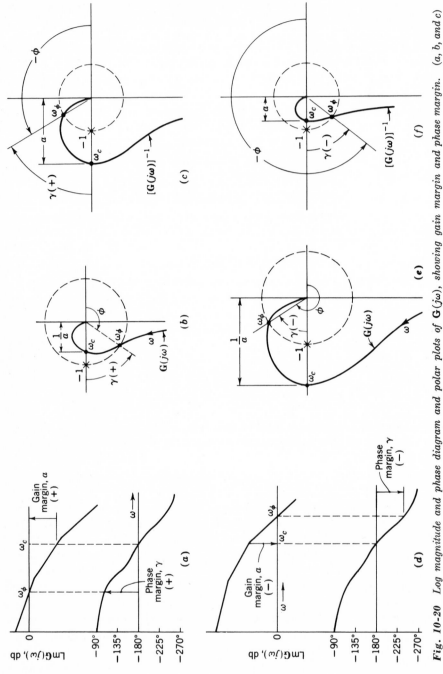

Fig. 10-20 *Log magnitude and phase diagram and polar plots of* $\mathbf{G}(j\omega)$, *showing gain margin and phase margin.* (*a, b, and c*) *Stable.* (*d, e, and f*) *Unstable.*

The damping ratio ζ of the system is also related to the gain margin. However, the phase margin gives a better estimate of damping ratio, and therefore of the transient overshoot of the system, than the gain margin.

Further information about the speed of response of the system can be obtained from the log magnitude–angle diagram which defines the maximum value of the control ratio and the frequency at which this maximum occurs.

10-9   Stability Characteristics of the Log Magnitude and Phase Diagram

The preceding chapter tells how the asymptotes of the log magnitude curve are related to each factor of the transfer function. For example, factors of the form $(1 + j\omega)^{-1}$ have a negative slope equal to -20 db/decade at high frequencies. Also, the angle for this factor varies from $0°$ at low frequencies to $-90°$ at high frequencies.

The total phase angle of a transfer function at any frequency is closely related to the slope of the log magnitude curve at that frequency. A slope of -20 db/decade is related to an angle of $-90°$; a slope of -40 db/decade is related to an angle of $-180°$; a slope of -60 db/decade is related to an angle of $-270°$; etc. Changes of slope at higher and lower frequencies, around the particular frequency being considered, contribute to the total angle at that frequency. The farther away the changes of slope are from the particular frequency, the less they contribute to the total angle at that frequency.

By observing the asymptotes of the log magnitude curve, it is possible to estimate the approximate value of the angle. With reference to the example of Sec. 9-9, the angle at $\omega = 4$ is now investigated. The slope of the curve at $\omega = 4$ is -20 db/decade; therefore the angle is near $-90°$. The slope changes at $\omega = 2$ and $\omega = 8$, so the slopes beyond these frequencies contribute to the total angle at $\omega = 4$. The actual angle, as read from the graph of Fig. 9-10, is $-122°$. The farther away the corner frequencies occur, the closer the angle is to $-90°$.

As has been seen, the stability of a system requires that the phase margin be positive for a minimum-phase system. For this to be true, the angle at the gain crossover [Lm $\mathbf{G}(j\omega) = 0$ db] must be greater than $-180°$. This places a limit on the slope of the log magnitude curve at the gain crossover. *The slope at the gain crossover should be more positive than -40 db/decade if the adjacent corner frequencies are not close.* A slope of -20 db/decade is preferable. This is derived from the consideration of a theorem by Bode. However, it can be seen qualitatively from the association of the slope of the log magnitude curve to the value of the phase. This guide should be used to assist in system design.

The log magnitude and phase diagram reveals some pertinent information, just as the polar plots do. For example, the gain can be adjusted (this raises or lowers the log magnitude curve) to produce a phase margin in the desirable range of 45 to 60°. The shape of the low-frequency portion of the curve determines system type and therefore the degree of steady-state accuracy. The system type and the gain determine the static error coefficients and therefore the steady-state error. The phase-margin frequency ω_ϕ gives a qualitative indication of the speed of response of a system. However, this is only a qualitative relationship. A more detailed analysis of this relationship is made in the next chapter.

10-10   *Log Magnitude–Angle Diagram*

The log magnitude–angle diagram is drawn by picking for each frequency the values of log magnitude and angle from the log magnitude and phase diagram. The resultant curve has frequency as a parameter. The curve for the example of Sec. 9-9 is sketched in Fig. 10-21.

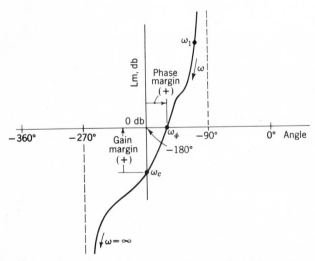

Fig. 10-21   *Log magnitude–angle diagram obtained from the curves of Figs. 9-9 and 9-10.*

10-11   Stability from the Log Magnitude–Angle Diagram

The curve of Fig. 10-21 shows a positive gain margin and phase margin; therefore this represents a stable system. Changing the gain raises or lowers the curve without changing the angle characteristics. Increasing the gain raises the curve, thereby decreasing the gain margin and phase margin, with the result that the stability is decreased. Increasing the gain so that the curve has a positive log magnitude at $-180°$ results in a negative gain and phase margin; therefore an unstable system results. Decreasing the gain lowers the curve and increases stability. However, a large gain is desired to reduce steady-state errors, as shown in Chap. 6.

The log magnitude–angle diagram for $G(s)H(s)$ may be drawn for all values of s on the contour Q of Fig. 10-9*b*. For minimum-phase systems the resultant curve is a closed contour. Nyquist's criterion can be applied to this contour by determining the number of points (having the values 0 db and odd multiples of 180°) enclosed by the curve of $G(s)H(s)$. This number is the value of N'' which is used in the equation $N'' = Z_R$ to determine the value of Z_R. As an example, consider a control system whose transfer function is given by

$$G(s) = \frac{K_1}{s(1 + Ts)} \qquad (10\text{-}38)$$

Its log magnitude–angle diagram, for the contour Q, is shown in Fig. 10-22. From this figure it is seen that the value of N is zero and the system is stable. For a non-minimum-phase system the log magnitude–angle contour does not close; thus it is difficult to determine the value of N. For these cases the polar plot must be utilized to determine stability.

For minimum-phase systems it is not necessary to obtain the complete log magnitude–angle contour to determine stability. Only that portion of the contour is drawn representing $G(j\omega)$ for the range of values $0^+ < \omega < \infty$. The stability is then determined from the position of the curve of $G(j\omega)$ relative to the $(0\text{-db}, -180°)$ point. In other words, the curve is traced in the direction of increasing frequency. The system is stable if the $(0\text{-db}, -180°)$ point is to the right of the curve. This is a simplified rule of thumb which is based on Nyquist's stability criterion.

A conditionally stable system (as defined in Sec. 10-3, Example 4) is one in which the curve crosses the $-180°$ axis at more than one point. Figure 10-23 shows the transfer-function plot for such a system with two stable and two unstable regions. The gain determines whether the system is stable or unstable. Additional stability information can be obtained from the log magnitude–angle diagram. This is shown in the next chapter.

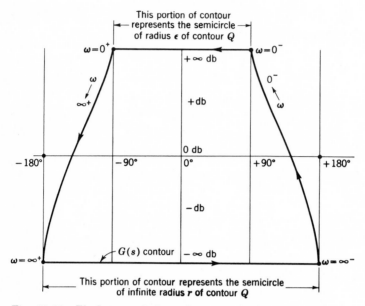

Fig. 10-22   *The log magnitude–angle contour for the minimum-phase system of Eq. (10-38).*

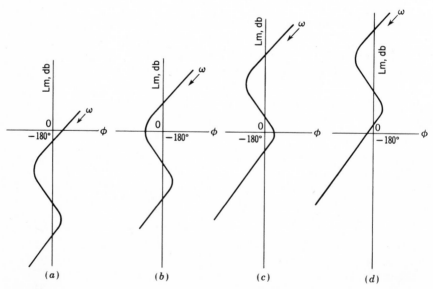

Fig. 10-23   *Log magnitude–angle diagram for a conditionally stable system.   (a and c) Stable.   (b and d) Unstable.*

10-12 *Effect on Stability of the Addition of a Pole or Zero*

The material of this chapter is made more meaningful by the consideration of Sec. 7-3, which presents the qualitative effects of adding a pole or zero to the root locus of a basic control system. The validity of these effects is demonstrated in a later chapter on root-locus compensation. With the above knowledge it is possible to correlate the transient response of a closed-loop system with the steady-state sinusoidal frequency response of the open-loop transfer function.

A polar plot of the frequency response of an open-loop transfer function

$$\mathbf{G}_x(j\omega) = \frac{K_x}{j\omega(1 + j\omega T_1)} \tag{10-39}$$

for a unity feedback control system is shown in Fig. 10-24. Adding a zero to the above system, so that its new open-loop transfer function is

$$\mathbf{G}_z(j\omega) = \frac{K_x(1 + j\omega T_2)}{j\omega(1 + j\omega T_1)} \tag{10-40}$$

results in the rotation of the frequency plot of the original system in a counter-clockwise direction. If a pole is added, instead of zero, so that the new open-loop transfer function is

$$\mathbf{G}_p(j\omega) = \frac{K_x}{j\omega(1 + j\omega T_1)(1 + j\omega T_2)} \tag{10-41}$$

the frequency plot of the original system is rotated in the clockwise direction.

Thus the addition of a zero or pole causes a corresponding shift of the frequency plot just as it produces a shift of the root-locus plot. Since the root locus yields quantitative data on the effects on the time response,

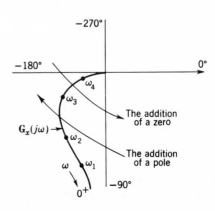

Fig. 10-24 *A plot of* $\mathbf{G}_x(j\omega) = K_x/j\omega(1 + j\omega T_1)$ *for several values of frequency.*

because of the addition of a zero or pole, this information is utilized as the basis for the following correlation between the sinusoidal and time responses.

1. The addition of a zero to a system has the effect of rotating the original polar plot in a counterclockwise direction, tending to make it a more stable and a faster-responding system. (As is seen in Chap. 13 on frequency-response compensation, a higher ω_m results, which is indicative of a faster-responding system, i.e., a shorter T_s.)

2. The addition of a pole to a system has the effect of rotating the polar plot in the clockwise direction, tending to make it a less stable and a slower-responding system. (In this case a lower value of ω_m results, which is indicative of a slower-responding system.) These statements correspond to the ones given in Sec. 7-3 in relation to the root-locus plots.

Sections 7-11 and 8-4 illustrated the correlation between the steady-state frequency response $\mathbf{C}(j\omega)/\mathbf{R}(j\omega)$ and the location in the s plane of the poles and zeros of $C(s)/R(s)$. The location of the poles was determined from the root locus. The root locus and frequency response are therefore directly related. The material in the following chapters demonstrates the effect of adding a zero or a pole to the open-loop transfer function, and the corresponding values of M_m and ω_m. In turn, the values of M_m and ω_m are related to the transient response.

10-13   Conclusions

In this chapter it is shown that the polar plot of the transfer function $G(s)H(s)$ or its inverse, in conjunction with Nyquist's stability criterion, gives a rapid means of determining whether a system is stable or unstable. The same information is obtained from the log magnitude and phase angle diagrams and the log magnitude–angle diagram; the phase margin and gain margin are also used as a means of indicating stability. In the next chapter it is shown that other key information related to the time domain is readily obtained from any of these plots. Thus the designer can determine whether the given system is satisfactory or not.

Other useful applications of the Nyquist criterion are the following: (1) analyzing systems having the characteristic of dead time and (2) locating roots in the s plane by means of the translated half-planes and the ζ-line bounded contours.

In certain situations, the exact expression of $G(s)H(s)$ may not be known and the only available information is an experimental curve representing this function.　For these cases, if the system is minimum phase, the starting point is the log magnitude and phase angle diagrams, from which system stability may be determined.

The Nyquist criterion has one disadvantage in that poles or zeros on the $j\omega$ axis require special treatment.　This treatment is not expounded upon, since, in general, most problems are free of poles or zeros on the $j\omega$ axis.

Bibliography

1. Nyquist, H.: Regeneration Theory, *Bell System Tech. J.*, vol. II, pp. 126–147, 1932.
2. James, H. M., N. B. Nichols, and R. S. Phillips: "Theory of Servomechanisms," chap. 2, McGraw-Hill Book Company, New York, 1947.
3. Bode, H. W.: "Network Analysis and Feedback Amplifier Design," chap. 8, D. Van Nostrand Company, Inc., Princeton, N.J., 1945.
4. Brown, G. S., and D. P. Campbell: "Principles of Servomechanisms," pp. 260–262, John Wiley & Sons, Inc., New York, 1948.
5. Chestnut, H., and R. W. Mayer: "Servomechanisms and Regulating System Design," 2d ed., vol. 1, pp. 317–322, John Wiley & Sons, Inc., New York, 1959.
6. Koenig, J.: On the Zeros of Polynomials and the Degree of Stability of Linear Systems, *J. Appl. Phys.*, vol. 24, pp. 476–482, April, 1953.

11

Feedback Control System Performance Based on the Frequency Response

11-1 Introduction

In Secs. 7-12 and 8-4 there is summarized the correlation between the location of the roots of the characteristic equation in the s plane and the time response. Also, Chaps. 7 to 9 discuss qualitatively the correlation between the frequency and time responses. This chapter is devoted to expounding on the latter and to the method of gain setting for a specified closed-loop frequency response.[1]

11-2 Direct Polar Plot

In the preceding chapter the steady-state performances of many systems with sinusoidal inputs are described. It also is shown

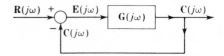

Fig. 11-1 *A block diagram of a simple control system.*

earlier that all the dependent variables, after the transient has subsided, are operating at the same frequency as the input signal but with different magnitudes and phase angles. This response is plotted as a log or polar plot, to which Nyquist's stability criterion can be applied. The polar plot is now further analyzed to obtain additional information relating the frequency and time responses.

Consider a simple control system with unity feedback, as shown in Fig. 11-1. The following equations describe the performance of this system with a sinusoidal input $R(j\omega)$:

$$R(j\omega) - C(j\omega) = E(j\omega) \tag{11-1}$$

$$\frac{C(j\omega)}{E(j\omega)} = G(j\omega) = |G(j\omega)|e^{j\phi} \tag{11-2}$$

$$\frac{C(j\omega)}{R(j\omega)} = \frac{G(j\omega)}{1 + G(j\omega)} = M(j\omega) = M(\omega)e^{j\alpha(\omega)} \tag{11-3}$$

$$\frac{E(j\omega)}{R(j\omega)} = \frac{1}{1 + G(j\omega)} \tag{11-4}$$

In Fig. 11-2 is shown the forward transfer function of this control system. As pointed out in Chap. 10, the directed segment drawn from the $(-1 + j0)$ point to any point on the $G(j\omega)H(j\omega)$ curve represents the quantity $B(j\omega) = 1 + G(j\omega)H(j\omega)$. For the system under consideration $H(j\omega)$ is unity; therefore

$$B(j\omega) = |B(j\omega)|e^{j\lambda(\omega)} = 1 + G(j\omega) \tag{11-5}$$

The frequency control ratio $C(j\omega)/R(j\omega)$ is therefore equal to the ratio of $A(j\omega)$ to $B(j\omega)$. In other words,

$$\frac{C(j\omega)}{R(j\omega)} = \frac{A(j\omega)}{B(j\omega)} = \frac{|A(j\omega)|e^{j\phi(\omega)}}{|B(j\omega)|e^{j\lambda(\omega)}} = \frac{G(j\omega)}{1 + G(j\omega)} \tag{11-6}$$

$$\frac{C(j\omega)}{R(j\omega)} = \frac{|A(j\omega)|}{|B(j\omega)|} e^{j(\phi-\lambda)} = M(\omega)e^{j\alpha} \tag{11-7}$$

Note that the angle $\alpha(\omega)$ can be determined directly from the construction shown in Fig. 11-2. Since the magnitude of the angle $\phi(\omega)$ is greater than the magnitude of the angle $\lambda(\omega)$, the value of the angle $\alpha(\omega)$ is negative. Remember that counterclockwise rotation is taken as positive.

Combining Eqs. (11-4) and (11-5) gives

$$\frac{E(j\omega)}{R(j\omega)} = \frac{1}{1 + G(j\omega)} = \frac{1}{|B(j\omega)|e^{j\lambda}} \tag{11-8}$$

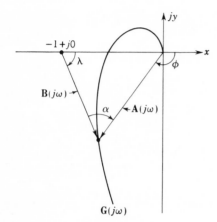

Fig. 11-2 *Polar plot of the forward transfer function of the control system of Fig. 11-1.*

From Eq. (11-8) it can be seen that the greater the distance from the $(-1 + j0)$ point to a point on the $\mathbf{G}(j\omega)$ locus, for a given frequency, the smaller is the steady-state sinusoidal error for a stated sinusoidal input. Thus the usefulness and importance of the polar plot of $\mathbf{G}(j\omega)$ have been enhanced.

11-3   Determination of M_m and ω_m for a Simple Second-order System

The frequency at which the maximum value of $|\mathbf{C}(j\omega)/\mathbf{R}(j\omega)|$ occurs is referred to as the resonant frequency ω_m. This maximum value is labeled M_m. These two quantities are mentioned in Chap 7 as being important figures of merit of a system. The methods of compensation applied to the polar and log plots, to be shown in later chapters, are based upon a knowledge of these two factors.

Only for a *simple second-order system* can a direct relationship be obtained for M_m and ω_m in terms of the system parameters. Consider the position control system of Fig. 11-3 whose forward and closed-loop transfer functions are, respectively,

$$\frac{C(s)}{E(s)} = \frac{K}{s(Js + B)} \qquad (11\text{-}9)$$

$$\frac{C(s)}{R(s)} = \frac{1}{(J/K)s^2 + (B/K)s + 1} = \frac{1}{s^2/\omega_n^2 + (2\zeta/\omega_n)s + 1} \qquad (11\text{-}10)$$

where
$$K = AK_m$$

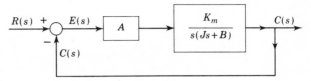

Fig. 11-3   *A position control system.*

The damping ratio and the undamped natural frequency for this system, as determined from Eq. (11-10), are

$$\zeta = \frac{B}{2\sqrt{KJ}} \tag{11-11}$$

$$\omega_n = \sqrt{\frac{K}{J}} \tag{11-12}$$

The control ratio as a function of frequency is

$$\frac{\mathbf{C}(j\omega)}{\mathbf{R}(j\omega)} = \frac{1}{(1 - \omega^2/\omega_n^2) + j2\zeta(\omega/\omega_n)} = M(\omega)e^{j\alpha(\omega)} \tag{11-13}$$

For a particular value of ω_n, plots of $M(\omega)$ and $\alpha(\omega)$ versus ω may be obtained for different values of ζ. In order that these plots may be applicable to all simple second-order systems with different values of ω_n, they are plotted versus ω/ω_n, as shown in Fig. 11-4a and b.

For a unit step input the inverse Laplace transform of $C(s)$ gives the time response as

$$c(t) = 1 - \frac{1}{\sqrt{1 - \zeta^2}} e^{-\zeta\omega_n t} \sin(\omega_n \sqrt{1 - \zeta^2}\, t + \cos^{-1}\zeta) \tag{11-14}$$

This time solution is derived in Sec. 3-9. The plot of $c(t)$ for several values of damping ratio ζ is drawn in Fig. 11-4c.

Next, consider the magnitude M^2, as derived from Eq. (11-13):

$$M^2 = \frac{1}{(1 - \omega^2/\omega_n^2)^2 + 4\zeta^2(\omega^2/\omega_n^2)} \tag{11-15}$$

To find the maximum value of M and the frequency at which it occurs, Eq. (11-15) is differentiated with respect to frequency and set equal to zero:

$$\frac{dM^2}{d\omega} = -\frac{-4(1 - \omega^2/\omega_n^2)(\omega/\omega_n^2) + 8\zeta^2(\omega/\omega_n^2)}{[(1 - \omega^2/\omega_n^2)^2 + 4\zeta^2(\omega^2/\omega_n^2)]^2} = 0 \tag{11-16}$$

The frequency ω_m at which the value M exhibits a peak (see Fig. 11-5), as found from Eq. (11-16), is

$$\omega_m = \omega_n \sqrt{1 - 2\zeta^2} \tag{11-17}$$

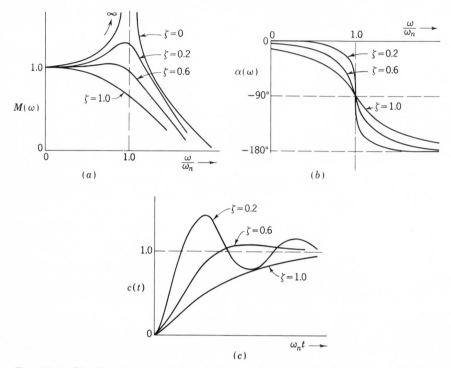

Fig. 11-4 *The M and α versus ω/ωₙ plots for a simple second-order system, with the corresponding time plots.*

When this value of frequency is substituted into Eq. (11-15),

$$M_m = \frac{1}{2\zeta \sqrt{1 - \zeta^2}} \qquad (11\text{-}18)$$

From these equations it is seen that M versus ω has a peak value, other than at $\omega = 0$, only for $\zeta < 0.707$.

Figure 11-6 shows a plot of M_m versus ζ for a simple second-order system. For values of $\zeta < 0.4$ it is seen that M_m increases very rapidly in magnitude; the transient oscillatory response is, therefore excessively large and might damage the physical equipment.

In Sec. 7-12 it is determined that the damped natural frequency ω_d for the transient of the simple second-order system is

$$\omega_d = \omega_n \sqrt{1 - \zeta^2} \qquad (11\text{-}19)$$

Also, in Sec. 3-11 there is determined the peak value c_p for a unit step input to this simple system. The ratio of the peak overshoot to the magnitude of

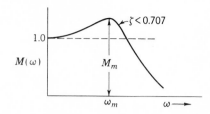

Fig. 11-5 *A closed-loop frequency-response curve indicating M_m and ω_m.*

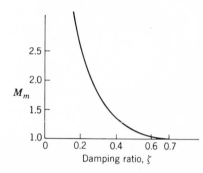

Fig. 11-6 *A plot of M_m versus ζ for a simple second-order system.*

the step input is called M_p:

$$M_p = \frac{c_p}{r} = 1 + e^{-\zeta\pi/\sqrt{1-\zeta^2}} \tag{11-20}$$

Therefore for this simple second-order system the following conclusions are made with respect to the correlation between the frequency and time responses.

1. Inspection of Eq. (11-17) reveals that ω_m is a function of both ω_n and ζ. For a given ζ, the larger the value of ω_m, the larger is ω_n and the faster is the transient time of response for this system.

2. Inspection of Eqs. (11-18) and (11-20) shows that both M_m and M_p are functions of ζ. The smaller ζ becomes, the larger in value M_m and M_p become. Thus it is concluded that the larger the value of M_m, the larger is the value of M_p.

   For values of $\zeta < 0.4$ the correspondence between M_m and M_p is only qualitative for this simple case. In other words, for $\zeta = 0$ in the time domain, $M_p = 2$; whereas in the frequency domain, $M_m = \infty$. In practice, systems with $\zeta < 0.4$ are not utilized. When $\zeta > 0.4$, there is a close correspondence between M_m and M_p. As an example, for ζ equal to 0.6, $M_m = 1.04$ and $M_p = 1.09$.

3. In Fig. 11-7 are shown polar plots for different damping ratios for the simple second-order system. Note that the shorter the distance between the $(-1 + j0)$ point and a particular $\mathbf{G}(j\omega)$ plot, as indicated by the dashed lines, the smaller is the damping ratio. Thus M_m is larger and consequently M_p is also larger.

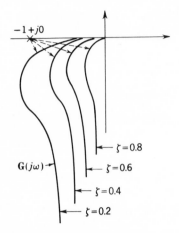

$-1+j0$

$\zeta = 0.8$

$G(j\omega)$

$\zeta = 0.6$

$\zeta = 0.4$

$\zeta = 0.2$

Fig. 11-7 Polar plots of $G(j\omega)$ for different damping ratios for the system shown in Fig. 11-3.

As a result of these characteristics, a designer can obtain a good approximation of the time response of a simple second-order system by knowing only the M_m and ω_m of its frequency response.

The procedure used above of setting the derivative of $C(j\omega)/R(j\omega)$ with respect to ω equal to zero works very well with a simple system. But as $C(j\omega)/R(j\omega)$ becomes more complicated, the differentiation and solution become tedious. This analytical procedure can be simplified, but a graphic procedure is generally used, as shown in the following sections.[2]

11-4   Correlation of Sinusoidal and Time Responses[3]

Although the correlation in the preceding section is for a simple second-order system, it has been found by experience that M_m is also a function of the *effective* ζ and ω_n for higher-order systems. The effective ζ and ω_n of a higher-order system (see Sec. 7-12) is dependent upon the ζ and ω_n of each second-order term and the values of the real roots in the characteristic equation of $C(s)/R(s)$. Thus, in order to alter the M_m, the location of some of the roots must be changed. The ones that should be altered depend on which are dominant in the time domain.

Based on the analysis for a simple second-order system, whenever the frequency response has the shape shown in Fig. 11-5, the following correlation exists between the frequency and time responses for any-order system.

1. The larger ω_m is made, the faster is the time of response for the system.

2. The value of M_m gives a good approximation of M_p within the acceptable range of the *effective* damping ratio $0.4 < \zeta < 0.707$. In terms of M_m, the acceptable range is $1.0 < M_m < 1.4$.
3. The closer the $G(j\omega)$ curve comes to the $(-1 + j0)$ point, the larger is the value of M_m.

To the three items above can be added one more factor that was brought out in Chap. 6, i.e., the larger K_0, K_1, or K_2 is made, the greater is the steady-state accuracy for a step, a ramp, and a parabolic input, respectively. In terms of the polar plot, the farther the point $G(j\omega)\big]_{\omega=0} = K_0$ for a Type 0 system is from the origin, the more accurate is the steady-state time response for a step input. For a Type 1 system, the farther the low-frequency asymptote (as $\omega \to 0$) is from the imaginary axis, the more accurate is the steady-state time response for a ramp input.

It must be remembered that all the factors mentioned above are merely *guideposts* in the *frequency domain* to assist the designer in obtaining an *approximate* idea of the time response of a particular system. This serves as a "stop-and-go signal" with respect to whether one is headed in the right direction in achieving the desired time response. If the desired performance specifications are not satisfactorily met, compensation techniques (see later chapters) must be used. After compensation the exact time response can be obtained by taking the inverse Laplace transform of $C(s)$, if desired. It should be obvious by now that the approximate approach saves much valuable time. Exceptions to the above analysis occur for higher-order systems.

11-5 Development of Constant $M(\omega)$ and $\alpha(\omega)$ Contours of $C(j\omega)/R(j\omega)$ on the Complex Plane (Direct Plot)

The open-loop transfer function and its polar plot for a given feedback control system have provided the following information, so far:

1. The stability or instability of the system
2. If the system is stable, the degree of its stability
3. The system type
4. The degree of steady-state accuracy
5. A graphical method of determining $C(j\omega)/R(j\omega)$

All these items permit a qualitative idea about the system's time response. The value of the polar plot would be greatly enhanced if the values of M_m and ω_m could be readily obtained from the plot.

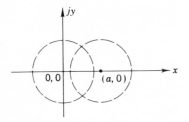

Fig. 11-8   Circles in the complex plane.

The importance of knowing these two values for a given system is stressed in the previous sections.   The contours of constant values of M drawn in the complex plane yield a rapid means for determining the values of M_m and ω_m.   In conjunction with the contours of constant values of $\alpha(\omega)$, also drawn in the complex plane, the plot of $\mathbf{C}(j\omega)/\mathbf{R}(j\omega)$ can be obtained more rapidly than by the graphical method indicated in Sec. 11-2.   The M and α contours are developed only for unity feedback systems.   At the end of this section it is shown how these contours can be applied to a nonunity feedback system.

Equation of a Circle

The equation of a circle with its center at the origin of the complex plane (see Fig. 11-8) is described by

$$x^2 + y^2 = r^2 \tag{11-21}$$

A circle with its center on the x axis but displaced from the origin has an equation of the form

$$(x - a)^2 + y^2 = r^2 \tag{11-22}$$

When the center of the circle has its center at the point a,b, the equation of the circle is

$$(x - a)^2 + (y - b)^2 = r^2 \tag{11-23}$$

It should be realized that the x and y axes have identical scales.   For a minus value of a the circle is displaced to the left of the origin.   These equations of a circle in the xy plane are utilized later in this section to express contours of constant M and α.

$M(\omega)$ Contours

In Fig. 11-9 is drawn the polar plot of a forward transfer function of a *unity feedback system*.   Applying the results of Sec. 11-2 to this figure, that is,

$$\frac{\mathbf{C}(j\omega)}{\mathbf{R}(j\omega)} = \frac{\mathbf{A}(j\omega)}{\mathbf{B}(j\omega)} = \frac{\mathbf{G}(j\omega)}{1 + \mathbf{G}(j\omega)} \tag{11-24}$$

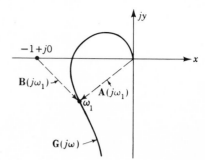

Fig. 11-9   *A plot of the transfer function* $G(j\omega)$.

aids in the development of the constant M contours.   First, consider the magnitude portions of Eq. (11-24):

$$M(\omega) = \frac{|A(j\omega)|}{|B(j\omega)|} = \frac{|G(j\omega)|}{|1 + G(j\omega)|} \tag{11-25}$$

The question at hand is: How many points are there in the complex plane for which the ratios of the magnitudes of the phasors $A(j\omega)$ and $B(j\omega)$ have the same value of $M(\omega)$?   For example, referring to Fig. 11-9, for the frequency $\omega = \omega_1$, M has a value of M_a.   It is desirable to determine all the other points in the complex plane for which

$$\frac{|A(j\omega)|}{|B(j\omega)|} = M_a(\omega)$$

To derive the constant M locus, express the transfer function in the rectangular coordinates.   That is,

$$G(j\omega) = x + jy \tag{11-26}$$

Substituting this equation into Eq. (11-25),

$$M = \frac{|x + jy|}{|1 + x + jy|} = \left[\frac{x^2 + y^2}{(1 + x)^2 + y^2} \right]^{\frac{1}{2}}$$

or

$$M^2 = \frac{x^2 + y^2}{(1 + x)^2 + y^2}$$

Rearranging the terms of this equation yields

$$\left(x + \frac{M^2}{M^2 - 1} \right)^2 + y^2 = \frac{M^2}{(M^2 - 1)^2} \tag{11-27}$$

By comparison with Eq. (11-22) it is seen that Eq. (11-27) is the equation of a circle with its center on the real axis with M as a parameter.   The center is located at

$$x_0 = - \frac{M^2}{M^2 - 1} \tag{11-28}$$

$$y_0 = 0 \tag{11-29}$$

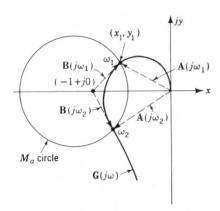

Fig. 11-10 *A plot of the transfer function* $G(j\omega)$ *and the* M_a *circle.*

and the radius is

$$r_0 = \left| \frac{M}{M^2 - 1} \right| \tag{11-30}$$

Inserting a given value of $M = M_a$ into Eq. (11-27) results in a circle in the complex plane having a radius r_0 and its center at x_0,y_0. This circle is called a constant M contour of $M = M_a$.

For all points on the M_a circle, the ratio of magnitudes of the phasors $A(j\omega)$ and $B(j\omega)$ drawn to any one of these points has the same value. That is,

$$\frac{|A(j\omega)|}{|B(j\omega)|} = M_a(\omega)$$

As an example, refer to Fig. 11-10, in which the M_a circle and the $G(j\omega)$ function are plotted. It is seen that the circle intersects the $G(j\omega)$ plot at the two frequencies ω_1 and ω_2. Thus in this example there are two frequencies for which

$$\frac{|A(j\omega_1)|}{|B(j\omega_1)|} = \frac{|A(j\omega_2)|}{|B(j\omega_2)|} = M_a(\omega)$$

In other words, a given point x_1,y_1 is simultaneously a point on a particular transfer function $G(j\omega)$ and a point on the M circle passing through it.

The plot of Fig. 11-10 is redrawn in Fig. 11-11 with two more M circles added. In this figure the circle $M = M_b$ is just tangent to the $G(j\omega)$ plot. Thus it is seen that there is only one point (x_3,y_3) for $G(j\omega_3)$ in the complex plane for which the ratio $|A(j\omega)/B(j\omega)|$ is equal to M_b. Also, the M_c circle does not intersect and is not tangent to the $G(j\omega)$ plot. This indicates that there are no points in the plane that can simultaneously satisfy Eqs. (11-25) and (11-27) for an $M = M_c$.

In Fig. 11-12 is drawn a family of circles in the complex plane for different values of M. In this figure it is noticed that the larger the value M,

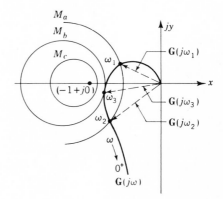

Fig. 11-11 *M contours and a* $\mathbf{G}(j\omega)$ *plot.*

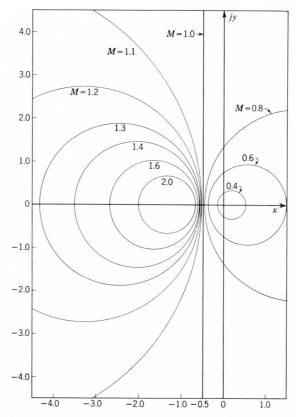

Fig. 11-12 *Loci of constant M contours.*

the smaller is its corresponding M circle. Thus, for the example shown in Fig. 11-11 the ratio $\mathbf{C}(j\omega)/\mathbf{R}(j\omega)$, for a unity feedback control system, has a maximum value of M equal to

$$M_m = M_b$$

A further inspection of Fig. 11-12 and Eq. (11-30) reveals the following:

1. For $M \to \infty$, which represents a condition of oscillation ($\zeta \to 0$), the center of the M circle $x_0 \to (-1 + j0)$ and the radius $r_0 \to 0$. This agrees with the statement made previously that as the $\mathbf{G}(j\omega)$ plot comes closer to the $(-1 + j0)$ point the system's effective ζ becomes smaller and the degree of its stability becomes less.
2. For $M(\omega) = 1$, which represents the condition where $\mathbf{C}(j\omega) = \mathbf{R}(j\omega)$, $r_0 \to \infty$ and the M contour becomes a straight line perpendicular to the real axis at $x = -\frac{1}{2}$.
3. For $M \to 0$, the center of the M circle $x_0 \to 0$ and the radius $r_0 \to 0$.
4. For $M > 1$ the centers of the circles lie to the left of $x = -1 + j0$, and for $M < 1$ the centers of the circles lie to the right of $x = 0$. All centers are on the real axis.

$\alpha(\omega)$ *Contours*

The $\alpha(\omega)$ contours, representing constant values of phase angle $\alpha(\omega)$, can be determined in the same manner as the M contours. For a system with unity feedback,

$$\frac{\mathbf{C}(j\omega)}{\mathbf{R}(j\omega)} = M(\omega)e^{j\alpha(\omega)} = \frac{\mathbf{G}(j\omega)}{1 + \mathbf{G}(j\omega)} = \frac{\mathbf{A}(j\omega)}{\mathbf{B}(j\omega)} \qquad (11\text{-}31)$$

By substituting Eq. (11-26) into this equation it becomes

$$M(\omega)e^{j\alpha(\omega)} = \frac{x + jy}{(1 + x) + jy} = \frac{\mathbf{A}(j\omega)}{\mathbf{B}(j\omega)} \qquad (11\text{-}32)$$

The question at hand now is: How many points are there in the complex plane for which the ratio of the phasors $\mathbf{A}(j\omega)$ and $\mathbf{B}(j\omega)$ yields the same angle α? To answer this question, express the angle α obtained from Eq. (11-32) as follows:

$$\alpha = \tan^{-1}\frac{y}{x} - \tan^{-1}\frac{y}{1 + x} \qquad (11\text{-}33)$$

$$\alpha = \tan^{-1}\frac{y/x - y/(1 + x)}{1 + (y/x)[y/(1 + x)]} = \tan^{-1}\frac{y}{x^2 + x + y^2} \qquad (11\text{-}34)$$

$$\tan \alpha = \frac{y}{x^2 + x + y^2} = N \qquad (11\text{-}35)$$

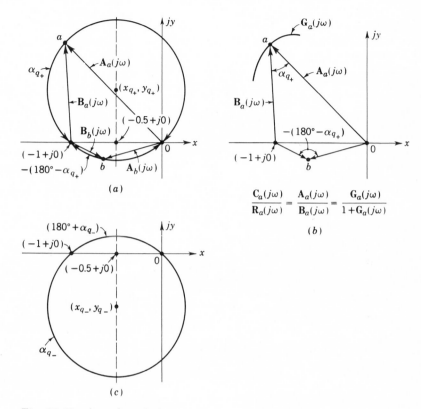

Fig. 11-13   *Arcs of constant α.*

For a constant value of the angle α, $\tan \alpha$, as expressed by the letter N, is also constant.   Rearranging Eq. (11-35) results in

$$\left(x + \frac{1}{2}\right)^2 + \left(y - \frac{1}{2N}\right)^2 = \frac{1}{4}\frac{N^2 + 1}{N^2} \tag{11-36}$$

By comparing with Eq. (11-23) it is seen that Eq. (11-36) is an equation of a circle with N as a parameter.   It has its center at

$$x_q = -\tfrac{1}{2} \tag{11-37}$$

$$y_q = \frac{1}{2N} \tag{11-38}$$

and has a radius of

$$r_q = \frac{1}{2}\left(\frac{N^2 + 1}{N^2}\right)^{\frac{1}{2}} \tag{11-39}$$

Inserting a given value of $N_q = \tan \alpha_q$ into Eq. (11-36) results in a circle (see Fig. 11-13) of radius r_q with its center at x_q, y_q.

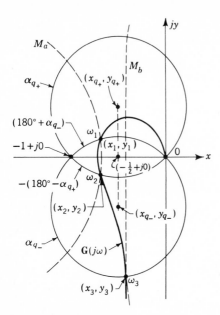

Fig. 11-14 *α contours and a* $\mathbf{G}(j\omega)$ *plot.*

The tangent N of angles in the first and third quadrant is positive. Therefore the y_q coordinate given by Eq. (11-38) is the same for an angle in the first quadrant and for the negative of its supplement, which is in the third quadrant. As a result, the constant α contour is only an arc of the circle. In other words, the $\alpha = 50$ and $-130°$ arcs are parts of the same circle, as shown in Fig. 11-13a. Similarly, angles α in the second and fourth quadrants have the same value y_q if they are negative supplements of each other. The constant α contours for these angles are shown in Fig. 11-13c.

For all points on the α_q arc the ratio of the complex quantities $\mathbf{A}(j\omega)$ and $\mathbf{B}(j\omega)$ drawn to any one of these points yields the same phase angle α_q, that is,

$$\underline{/\mathbf{A}(j\omega)} - \underline{/\mathbf{B}(j\omega)} = \alpha_q$$

Figure 11-13b shows the interrelationship between the M and α contours, $\mathbf{G}(j\omega)$ and $\mathbf{C}(j\omega)/\mathbf{R}(j\omega)$, for a unity feedback system. In Fig. 11-13a and b, if the $\mathbf{G}(j\omega)$ plot passes through the point a, the angle is α_{q+}. The designer must then interpret this value from the graphical construction of the plot as to whether the angle of $\mathbf{C}(j\omega)/\mathbf{R}(j\omega)$ is α_{q+} or $\alpha_{q+} \pm k(360°)$, where $k = 0$, $1, 2, \ldots$. A good procedure in the determination of the angle α is to start at zero frequencies where $\alpha = 0$ and to proceed to higher frequencies, knowing that the angle is a continuous function.

The point x_2, y_2 in Fig. 11-14 yields a value of α that is equal to the negative value of the supplementary angle of α_{q+}. As an example, in Fig. 11-14 it is seen that the α_{q-} circle intersects the $\mathbf{G}(j\omega)$ plot at the two points

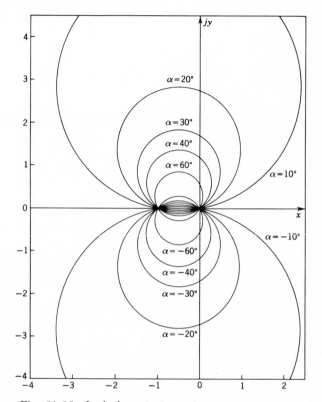

Fig. 11-15 *Loci of constant α contours.*

ω_1 and ω_3. The value of $\alpha(\omega)$ at ω_1 is

$$\alpha_1 = \underline{/\mathbf{A}(j\omega_1)} - \underline{/\mathbf{B}(j\omega_1)} = 180° + \alpha_{q-}$$

The value of $\alpha(\omega)$ at ω_3 is

$$\alpha_3 = \underline{/\mathbf{A}(j\omega_3)} - \underline{/\mathbf{B}(j\omega_3)} = \alpha_{q-}$$

For different values of α there results a family of circles in the complex plane with centers on the line represented by $(-\frac{1}{2},y)$, as illustrated in Fig. 11-15.

By referring to Fig. 11-14, it is noted that the $\mathbf{G}(j\omega)$ curve and the M_a and α_{q-} contours intersect at the point x_1,y_1 corresponding to the frequency ω_1. By reviewing the development of the M and α contours it is seen that an interrelationship exists between these curves at this common point. In other words, if the coordinates (x_1,y_1) are inserted into Eqs. (11-26), (11-27), and (11-36), the quantities $\mathbf{G}(j\omega_1)$, M_a, and α_{q-} are obtained. By taking

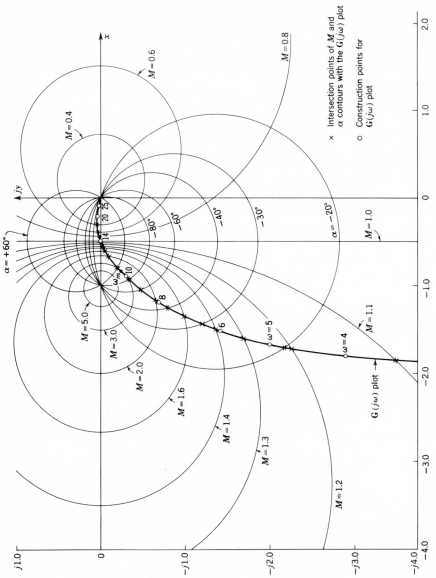

Fig. 11-16 *Determination of $|C(j\omega)/R(j\omega)|$ and α from the $G(j\omega)$ plot with the aid of M-α contours.*

× Intersection points of M and α contours with the $G(j\omega)$ plot

○ Construction points for $G(j\omega)$ plot

$\alpha = +60°$

$M = 0.6$

$M = 0.4$

$M = 0.8$

$M = 1.0$

$M = 1.1$

$M = 1.2$

$M = 1.3$

$M = 1.4$

$M = 1.6$

$M = 2.0$

$M = 3.0$

$M = 5.0$

$-80°$

$-60°$

$-40°$

$-30°$

$\alpha = -20°$

ω_m

$\omega = 5$

$\omega = 4$

$G(j\omega)$ plot

14 20 25

Table 11-1.   Values obtained from Fig. 11-16 to plot the curves in Fig. 11-17

Values for $|\mathbf{C}(j\omega)/\mathbf{R}(j\omega)|$ plot		Values for $|\mathbf{C}(j\omega)/\mathbf{R}(j\omega)|$ plot		Values for α plot	
ω, radians/sec	M	ω, radians/sec	M	ω, radians/sec	α, deg
3.8	1.1	13.4	1.4	3	-10
4.7	1.2	13.6	1.3	4.8	-20
5.5	1.3	13.8	1.2	6.5	-30
6.2	1.4	13.9	1.1	7.7	-40
7.0	1.6	14.0	1.0	8.8	-60
8.1	2.0	15.0	0.8	9.7	-80
11.0	3.0	16.5	0.6	11.3	-120
12.6	2.0	18.0	0.4	14.0	-180
13.2	1.6				

additional points the plots of M and α versus ω can readily be determined. This can be done more rapidly by superimposing the contours on the $\mathbf{G}(j\omega)$ curve and finding the points of intersection.   Thus for

$$\omega = \omega_1: \qquad \frac{\mathbf{C}(j\omega_1)}{\mathbf{R}(j\omega_1)} = M_a e^{j(180°+\alpha_{q-})}$$

$$\omega = \omega_2: \qquad \frac{\mathbf{C}(j\omega_2)}{\mathbf{R}(j\omega_2)} = M_a e^{-j(180°-\alpha_{q+})}$$

$$\omega = \omega_3: \qquad \frac{\mathbf{C}(j\omega_3)}{\mathbf{R}(j\omega_3)} = M_b e^{j\alpha_{q-}}$$

Figure 11-16 shows a plot of $\mathbf{G}(j\omega)$ with constant M and α contours superimposed.   From this figure, Table 11-1 gives the magnitude M and the angle α of the control ratio $\mathbf{C}(j\omega)/\mathbf{R}(j\omega)$ for a number of values of frequency. These control-ratio characteristics are plotted in Fig. 11-17.

Tangents to the M Circles

The line drawn through the origin of the complex plane tangent to a given M circle plays an important part in the gain setting of the $\mathbf{G}(j\omega)$ function. By referring to Fig. 11-18 and remembering that $bc = r_0$ is the radius and $ob = x_0$ is the distance to the center of the particular M circle, then

$$\sin \psi = \frac{bc}{ob} = \frac{M/(M^2 - 1)}{M^2/(M^2 - 1)} = \frac{1}{M} \qquad (11\text{-}40)$$

This relationship is utilized in the section on gain adjustment.

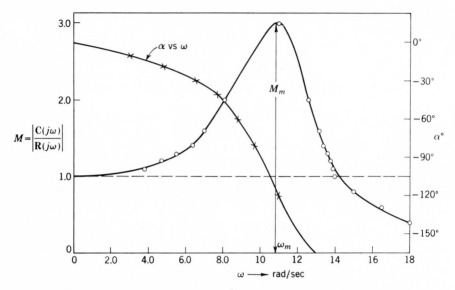

Fig. 11-17   *Resultant* $|C(j\omega)/R(j\omega)|$ *and* α *versus* ω, *obtained from Fig. 11-16.*

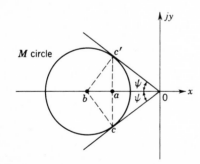

Fig. 11-18   *Determination of* $\sin \psi$.

Utilized along with the tangent is the fact that the point a in Fig. 11-18 is the $(-1 + j0)$ point.   This is proved as follows:

$$(oc)^2 = (ob)^2 - (bc)^2 \tag{11-41}$$
$$ac = oc \sin \psi \tag{11-42}$$
and
$$(oa)^2 = (oc)^2 - (ac)^2 \tag{11-43}$$

Combining Eqs. (11-41) to (11-43) yields

$$(oa)^2 = \frac{M^2 - 1}{M^2} [(ob)^2 - (bc)^2] \tag{11-44}$$

Table 11-2. *Values for constructing M circles*

M	Center $x_0 = -[M^2/(M^2 - 1)]$	Radius $r_0 = \|M/(M^2 - 1)\|$	Angle, deg, $\psi = \sin^{-1}(1/M)$
0.5	0.333	0.67	—
0.7	0.960	1.37	—
0.9	4.26	4.74	—
1.0	∞	∞	90
1.05	−10.74	10.24	72.3
1.1	−5.76	5.24	65.4
1.15	−4.1	3.57	60.3
1.2	−3.27	2.73	56.4
1.25	−2.78	2.22	53.2
1.3	−2.45	1.88	50.3
1.35	−2.215	1.64	47.7
1.4	−2.04	1.46	45.6
1.5	−1.8	1.20	41.8
1.6	−1.64	1.03	38.7
1.7	−1.53	0.90	36.0
1.8	−1.45	0.80	33.7
1.9	−1.38	0.729	31.7
2.0	−1.33	0.67	30.0
2.5	−1.19	0.48	23.6
3.0	−1.13	0.38	19.5
3.5	−1.10	0.34	16.6
4.0	−1.07	0.27	14.5
5.0	−1.04	0.21	11.5

The values of the distances to the center ob and of the radius bc of the M circle are then substituted into Eq. (11-44), which results in

$$oa = 1 \qquad (11\text{-}45)$$

The values necessary for constructing M and α contours, for typical values of M and α, are given in Tables 11-2 and 11-3, respectively.

Nonunity Feedback Control System

To apply the M and α contours to a nonunity feedback system, the system is first represented by the equivalent block diagram of Appendix J, manipulation 12. Its control ratio can therefore be put in the form:

$$\frac{C(j\omega)}{R(j\omega)} = \frac{G(j\omega)}{1 + G(j\omega)H(j\omega)} = \frac{1}{H(j\omega)} \frac{G(j\omega)H(j\omega)}{1 + G(j\omega)H(j\omega)} \qquad (11\text{-}46)$$

or

$$\frac{C(j\omega)}{R(j\omega)} = \frac{1}{H(j\omega)} \frac{G_0(j\omega)}{1 + G_0(j\omega)} \qquad (11\text{-}47)$$

Table 11-3.   *Values for constructing* α *contours*

$\alpha \pm 180°m$, deg	N	Radius $r_q = (1/2N) \sqrt{N^2 + 1}$	Center $y_q = 1/2N$
−90	−∞	0.500	0
−80	−5.67	0.528	−0.0882
−75	−3.732	0.518	−0.134
−70	−2.75	0.531	−0.182
−60	−1.73	0.577	−0.289
−50	−1.19	0.656	−0.420
−45	−1.00	0.707	−0.500
−40	−0.838	0.775	−0.596
−30	−0.577	1.000	−0.866
−20	−0.364	1.460	−1.370
−15	−0.268	1.931	−1.866
−10	−0.176	2.88	−2.84
0	0.0	∞	∞
10	0.176	2.88	2.84
30	0.577	1.000	0.866
50	1.19	0.656	0.42
70	2.75	0.531	0.182
90	∞	0.5	0

where

$$\mathbf{G}_0(j\omega) = \mathbf{G}(j\omega)\mathbf{H}(j\omega)$$

Thus

$$\frac{\mathbf{C}_0(j\omega)}{\mathbf{R}_0(j\omega)} = \frac{\mathbf{G}_0(j\omega)}{1 + \mathbf{G}_0(j\omega)} \qquad (11\text{-}48)$$

Equation (11-48) is of the standard form for a unity feedback system. Thus the M and α contours can be applied to $\mathbf{G}_0(j\omega)$ to obtain $\mathbf{C}_0(j\omega)/\mathbf{R}_0(j\omega)$. Multiplying $\mathbf{C}_0(j\omega)/\mathbf{R}_0(j\omega)$ by $1/\mathbf{H}(j\omega)$ gives $\mathbf{C}(j\omega)/\mathbf{R}(j\omega)$.

11-6   *Contours in the Inverse Polar Plane*

The use of direct polar plots for feedback control systems that utilize non-unity feedback is rather tedious when compared with the use of inverse plots. Also, the effect of the $\mathbf{H}(j\omega)$ term is more evident in the inverse plot. Even for unity feedback systems the value of required gain can be determined just as rapidly as on the direct plot, as seen in Sec. 11-7.

Inverse Polar Plot

Figure 11-19 illustrates a control system with a feedback network $\mathbf{H}(j\omega)$, whose performance is described by the following equations:

$$R(j\omega) - B(j\omega) = E(j\omega) \tag{11-49}$$

$$B(j\omega) = H(j\omega)C(j\omega) \tag{11-50}$$

$$\frac{C(j\omega)}{E(j\omega)} = G(j\omega) = |G(j\omega)|e^{j\phi(\omega)} \tag{11-51}$$

$$\frac{C(j\omega)}{R(j\omega)} = \frac{G(j\omega)}{1 + G(j\omega)H(j\omega)} = Me^{j\alpha} \tag{11-52}$$

$$\frac{R(j\omega)}{C(j\omega)} = \frac{1}{G(j\omega)} + H(j\omega) = \frac{1}{M}e^{-j\alpha} \tag{11-53}$$

Note that the last equation is composed of two complex quantities that can be readily plotted in the complex plane, as shown in Fig. 11-20. Thus, by plotting the complex quantities $1/G(j\omega)$ and $H(j\omega)$, the $R(j\omega)/C(j\omega)$ term can be calculated graphically. It is now seen that the effect on $R(j\omega)/C(j\omega)$ of changing the $H(j\omega)$ term is more evident than it would be with direct plots.

Constant $1/M$ and α Contours

Constant $1/M$ and α contours for the inverse plots are developed for both unity and nonunity feedback systems. The $R(j\omega)/C(j\omega)$ equation for a unity feedback system is

$$\frac{R(j\omega)}{C(j\omega)} = \frac{1}{G(j\omega)} + 1 = \frac{1}{M}e^{-j\alpha} \tag{11-54}$$

Figure 11-21 illustrates the quantities in Eq. (11-54). Note that, because of the geometry of construction, the directed line segment drawn

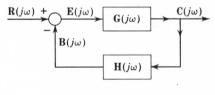

Fig. 11-19 *Block diagram of a control system with a feedback network.*

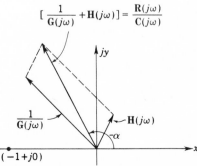

Fig. 11-20 *Complex representation of the quantities in Eq. (11-53).*

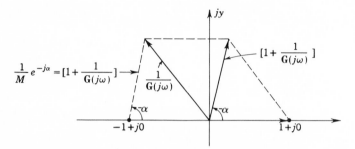

Fig. 11-21   *Representation of the quantities in Eq. (11-54).*

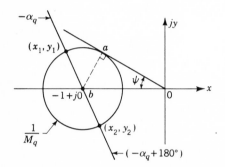

Fig. 11-22   *Contours for a particular magnitude $\mathbf{R}(j\omega)/\mathbf{C}(j\omega)$.*

from the $(-1 + j0)$ point is also the quantity $1 + 1/\mathbf{G}(j\omega)$ or $(1/M)e^{-j\alpha}$. Thus it is seen that the inverse plot yields very readily the form and location of the contours of constant $1/M$ and α.   In other words, in the inverse plane:

1.   Contours of constant values of M are circles whose centers are at the $(-1 + j0)$ point, and the radii are equal to $1/M$.
2.   Contours of constant values of $-\alpha$ are radial lines passing through the $(-1 + j0)$ point.

Figure 11-22 indicates the contours for a particular magnitude of $\mathbf{R}(j\omega)/\mathbf{C}(j\omega)$ and a tangent drawn to the M circle from the origin.   In Fig. 11-22, the following is seen:

(1) $$\sin\psi = \left|\frac{ab}{ob}\right| = \frac{1/M}{1} = \frac{1}{M}$$

which is the same as for the contours in the direct plane.

(2)   The directed line segment from $(-1 + j0)$ to the point x_1, y_1 has a value of $1/M_q$ and an angle of $-\alpha_q$.   If the polar plot of $1/\mathbf{G}(j\omega)$ passes through the x_1, y_1 point, then $\mathbf{R}(j\omega_1)/\mathbf{C}(j\omega_1)$ has these values.

(3) The directed line segment from $(-1 + j0)$ to the point x_2, y_2 has a magnitude of $1/M_q$ and an angle of $(-\alpha_q + 180°)$. If the polar plot of $1/\mathbf{G}(j\omega)$ passes through the x_2, y_2 point, then $\mathbf{R}(j\omega_2)/\mathbf{C}(j\omega_2)$ has these values.

Note in Fig. 11-23 that the $1/\mathbf{G}_v(j\omega)$ plot intersects the $1/M_q$ circle at two places. Thus, at the respective frequencies for the points x_2, y_2 and x_3, y_3 the magnitude of $\mathbf{R}(j\omega)/\mathbf{C}(j\omega)$ is the same but the value of α is different. Since the $1/\mathbf{G}_u(j\omega)$ plot is tangent to the $1/M_q$ circle, the maximum value of $\mathbf{C}(j\omega)/\mathbf{R}(j\omega)$ is M_q, or the minimum value of $\mathbf{R}(j\omega)/\mathbf{C}(j\omega)$ is $1/M_q$. In other words, for larger values of M_q the circles are smaller and do not intersect the $1/\mathbf{G}_u(j\omega)$ plot. Figure 11-24 illustrates typical M and α contours.

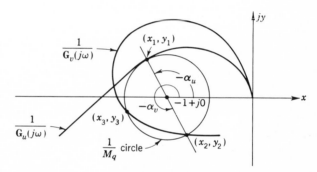

Fig. 11-23 *Correlation between $1/\mathbf{G}(j\omega)$ plots and M and α contours.*

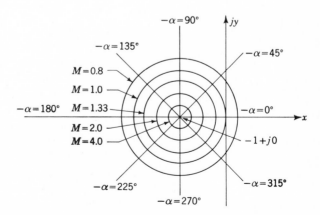

Fig. 11-24 *Typical M and α contours on the inverse polar plane for unity feedback.*

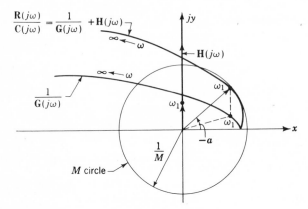

Fig. 11-25 *An example of the correlation of M and α contours for nonunity feedback.*

The values of $C(j\omega)/R(j\omega)$ for different values of frequency can be determined in the manner given in the example in Sec. 11-5. By superimposing the M and α contours on the $1/G(j\omega)$ plot, the points of intersection provide data for plotting M and α versus ω.

Nonunity Feedback Control System

The case of nonunity feedback is handled more simply on the inverse plot than on the direct plot since

$$\frac{R(j\omega)}{C(j\omega)} = \frac{1 + G(j\omega)H(j\omega)}{G(j\omega)} = \frac{1}{G(j\omega)} + H(j\omega) \qquad (11\text{-}55)$$

The curves $1/G(j\omega)$ and $H(j\omega)$ can be drawn separately and then added, as shown in Fig. 11-20. From the origin the quantity $R(j\omega)/C(j\omega)$ is measured directly in both magnitude and angle. The $1/M$ circles and α lines are the same as those for the unity feedback case but are now drawn from the origin.

Example

$$G(s) = \frac{K_0}{(1 + T_1 s)(1 + T_2 s)} \qquad (11\text{-}56)$$

$$H(s) = K_t s \qquad (11\text{-}57)$$

The curves $1/G(j\omega)$ and $H(j\omega)$ are plotted in Fig. 11-25. Their sum, $1/G(j\omega) + H(j\omega) = R(j\omega)/C(j\omega)$, is also shown in this figure.

11-7 Gain Setting by Use of the Contours for Direct and Inverse Polar Plots

The root-locus method provides a means by which locus of all possible roots is determined, with gain as the independent variable. Once the locus is determined, a particular value of gain is selected that meets, as closely as possible, the given set of desired response specifications. Note that setting the gain is the first and easiest step in adjusting the system for the optimum performance.

This is also true with respect to polar plots. If satisfactory response cannot be achieved by gain adjustment alone, then compensation techniques must be utilized, as described in later chapters.

11-8 Gain Adjustment for a Desired M_m of a Unity Feedback System: Direct Polar Plot

In Fig. 11-26 is drawn $\mathbf{G}_x(j\omega)$ with its respective M_m circle in the complex plane. Since

$$\mathbf{G}_x(j\omega) = x + jy = K_x\mathbf{G}'_x(j\omega) = K_x(x' + jy') \qquad (11\text{-}58)$$

then

$$x' + jy' = \frac{x}{K_x} + j\frac{y}{K_x}$$

and

$$\mathbf{G}'_x(j\omega) = \frac{\mathbf{G}_x(j\omega)}{K_x}$$

$\mathbf{G}'_x(j\omega)$ is defined as the frequency-sensitive portion of $\mathbf{G}_x(j\omega)$ with unity gain. Note that changing the gain merely changes the amplitude and not the angle of the locus of points of $\mathbf{G}_x(j\omega)$. Thus, if in Fig. 11-26 a change of scale is made by dividing the (x,y) coordinates by K_x so that the new coordinates are (x',y'), the following are true:

1. The $\mathbf{G}_x(j\omega)$ plot becomes the $\mathbf{G}'_x(j\omega)$ plot.
2. The M_m circle becomes *a circle* which is simultaneously tangent to $\mathbf{G}'_x(j\omega)$ and the line representing $\sin \psi = 1/M_m$.
3. The $(-1 + j0)$ point becomes the $(-1/K_x + j0)$ point.
4. The radius r_0 becomes $r'_0 = r_0/K_x$.

In other words, if $\mathbf{G}'_x(j\omega)$ is drawn on a separate graph sheet from that of $\mathbf{G}_x(j\omega)$ (see Fig. 11-27), by superimposing the two graph papers so that the axes coincide, the circles and the $\mathbf{G}_x(j\omega)$ and $\mathbf{G}'_x(j\omega)$ plots also coincide. Referring to Figs. 11-26 and 11-27, note that $oa = -1$ and $oa' = -1/K_x$.

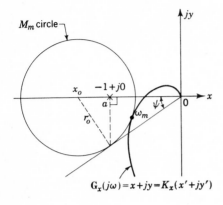

$$G_x(j\omega) = x+jy = K_x(x'+jy')$$

Fig. 11-26 *A* $G_x(j\omega)$ *plot with the respective* M_m *circle.*

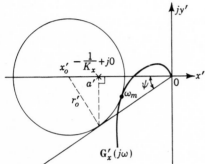

Fig. 11-27 *A* $G'_x(j\omega)$ *plot with circle drawn tangent to it and to the line representing the angle* $\sin \psi = 1/M_m$.

As a consequence, it is possible to determine the required gain to achieve a desired M_m for a given system by the following graphical procedure:

1. If the original system has a transfer function

$$G_x(j\omega) = \frac{K_x(1 + j\omega T_1)(1 + j\omega T_2) \cdots}{(j\omega)^n(1 + j\omega T_a)(1 + j\omega T_b)(1 + j\omega T_c) \cdots} \qquad (11\text{-}59)$$

plot only the frequency-sensitive portion

$$G'_x(j\omega) = \frac{(1 + j\omega T_1)(1 + j\omega T_2) \cdots}{(j\omega)^n(1 + j\omega T_a)(1 + j\omega T_b)(1 + j\omega T_c) \cdots} \qquad (11\text{-}60)$$

where K_x is the original gain.

2. Draw the line representing the angle $\psi = \sin^{-1}(1/M_m)$.
3. By trial and error, find a circle whose center lies on the negative real axis and is simultaneously tangent to the $G'_x(j\omega)$ plot and the line representing the angle ψ.
4. Having found this circle, draw a line from the point of tangency on the ψ-angle line perpendicular to the real axis. Label the point where this line intersects the axis as a'.

5. For this circle to be an M circle representing M_m, the point a' must be the $(-1 + j0)$ point. Thus, the (x',y') coordinates must be multiplied by a gain factor K_n in order to convert this plot to a plot of $G(j\omega)$. From the graphical construction the value of K_n is

$$K_n = \frac{1}{oa'}$$

6. Thus the original gain is changed by a factor

$$A = \frac{K_n}{K_x}$$

Note that, if $G_x(j\omega)$ is already plotted, this plot can become $G'_x(j\omega)$ by merely changing the scale. That is, divide the (x,y) coordinates by K_x. However, it is possible to work directly with the plot of the function $G_x(j\omega)$ which includes a gain K_x. Following the procedure outlined above results in the determination of the *additional* gain required to produce the specified M_m; that is, the additional gain is

$$A = \frac{K_n}{K_x} = \frac{1}{oa''}$$

Example

It is desired that the system which has the transfer function

$$G_x(j\omega) = \frac{1.47}{j\omega(1 + j0.25\omega)(1 + j0.1\omega)}$$

have an $M_m = 1.3$. The problem is to determine the actual gain K_1 needed and the amount by which the original gain K_x must be changed to obtain this M_m. The procedure previously outlined is applied to this problem. $G'_x(j\omega)$ is plotted in Fig. 11-28 and results in

$$K_1 = \frac{1}{oa'} \approx \frac{1}{0.34} = 2.94 \text{ sec}^{-1}$$

The additional gain required is

$$A = \frac{K_1}{K_x} = \frac{2.94}{1.47} = 2.0$$

In other words, the original gain must be doubled to obtain an M_m of 1.3 for $C(j\omega)/R(j\omega)$.

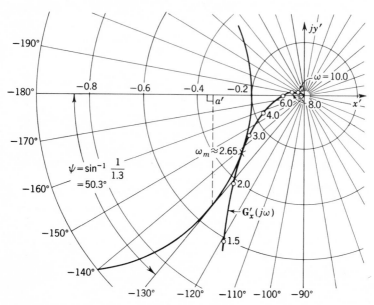

Fig. 11-28 *An example utilizing the procedure outlined in Sec. 11-8 for* $G'_x(j\omega) = 1/j\omega(1 + j0.25\omega)(1 + j0.1\omega)$.

11-9   Gain Adjustment for a Desired M_m of a Unity Feedback System: Inverse Polar Plot

The approach is similar to that for the direct plots in establishing the procedure for gain setting for the inverse plot. Referring to Fig. 11-29,

$$\frac{1}{G_x(j\omega)} = \frac{1}{K_x G'_x(j\omega)} = G_x^{-1}(j\omega) = x + jy \tag{11-61}$$

Then
$$\frac{K_x}{G_x(j\omega)} = \frac{1}{G'_x(j\omega)} = K_x x + j K_x y = x' + jy' \tag{11-62}$$

Thus, if in Fig. 11-29 a change of scale is made, by multiplying the (x,y) coordinates by K_x so that the new coordinates are (x',y'), the following are true:

1. The $G_x^{-1}(j\omega)$ plot becomes the $[G'_x(j\omega)]^{-1}$ plot.
2. The $1/M_m$ circle becomes *a circle* which is simultaneously tangent to $[G'_x(j\omega)]^{-1}$ and the line representing $\sin \psi = 1/M_m$.
3. The $(-1 + j0)$ point becomes the $(-K_x + j0)$ point.
4. The radius r_0 becomes $r'_0 = K_x r_0$.

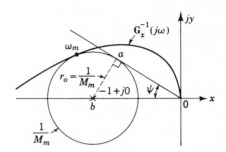

Fig. 11-29 *A* $G_x^{-1}(j\omega)$ *plot with its respective* $1/M_m$ *circle.*

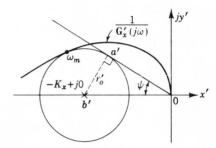

Fig. 11-30 *A* $[G_x'(j\omega)]^{-1}$ *plot with circle drawn tangent to it and to the line representing the angle* $\sin \psi = 1/M_m$.

In the same manner as with the direct plot, if $[G_x'(j\omega)]^{-1}$ is drawn on a separate sheet from that of $G_x^{-1}(j\omega)$ (see Fig. 11-30), by superimposing the two graph papers so that the axes coincide, the circles and the $G_x^{-1}(j\omega)$ and $[G_x'(j\omega)]^{-1}$ plots also coincide. Referring to Figs. 11-29 and 11-30, note that $ob = -1$ and that $ob' = -K_x$.

To determine the gain necessary for a given M_m from the inverse plots, the following graphical procedure can be used:

1. If the original system has an inverse transfer function of

$$G_x^{-1}(j\omega) = \frac{(j\omega)^n(1 + j\omega T_a)(1 + j\omega T_b)(1 + j\omega T_c) \cdots}{K_x(1 + j\omega T_1)(1 + j\omega T_2) \cdots} \qquad (11\text{-}63)$$

   plot only the frequency-sensitive portion

$$[G_x'(j\omega)]^{-1} = \frac{(j\omega)^n(1 + j\omega T_a)(1 + j\omega T_b)(1 + j\omega T_c) \cdots}{(1 + j\omega T_1)(1 + j\omega T_2) \cdots} \qquad (11\text{-}64)$$

2. Draw the line representing the angle $\psi = \sin^{-1}(1/M_m)$.
3. By trial and error, find a circle whose center lies on the negative real axis and is simultaneously tangent to both the $[G_x'(j\omega)]^{-1}$ plot and the line representing the angle ψ.
4. For this circle to be a $1/M_m$ circle representing M_m, the point b' must be the $(-1 + j0)$ point. Thus, the (x',y') coordinates must

be divided by a gain factor of K_n in order to convert this plot to a plot of $\mathbf{G}^{-1}(j\omega)$. From the graphical construction the value of K_n is

$$K_n = ob'$$

5. Thus the original gain is changed by a factor of

$$A = \frac{K_n}{K_x}$$

Just as for the direct plot, if $\mathbf{G}_x^{-1}(j\omega)$ is already plotted, this plot can become $[\mathbf{G}_x'(j\omega)]^{-1}$ by merely changing the scale, that is, multiplying the (x,y) coordinates by K_x. However, in the same manner as for the direct plot, one can use the $\mathbf{G}_x^{-1}(j\omega)$ plot for the determination of the *additional* gain required for the desired M_m. The value of the additional gain is equal to ob''.

Example

Assume that a system having the inverse transfer function

$$\mathbf{G}_x^{-1}(j\omega) = \frac{j\omega(1 + j0.8\omega)(1 + j0.25\omega)}{0.5}$$

must have an $M_m = 1.5$. In Fig. 11-31 the above procedure is utilized in order to determine the required gain K_1. Thus

$$K_1 = 1.25 \text{ sec}^{-1}$$

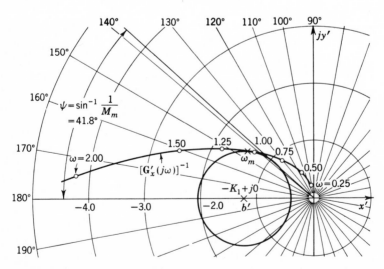

Fig. 11-31 *An example utilizing the procedure outlined in Sec. 11-9 for* $[\mathbf{G}_x'(j\omega)]^{-1} = j\omega(1 + j0.8\omega)(1 + j0.25\omega).$

Then, the amount by which the original gain K_x must be changed is

$$A = \frac{K_1}{K_x} = \frac{1.25}{0.5} = 2.5$$

Therefore, increasing the original gain by a factor of 2.5 gives the desired $M_m = 1.5$ for $\mathbf{C}(j\omega)/\mathbf{R}(j\omega)$.

11-10   *Constant M and α Curves on the Log Magnitude–Angle Diagram (Nichols Chart)*[5]

As derived earlier in this chapter, the constant M curves on the direct and inverse polar plots are found to be circles. The next necessary step is to transform these curves to the log magnitude–angle diagram. This is done more easily by starting from the inverse polar plot since all the M circles have the same center. This requires a change of sign of the log magnitude and angle obtained, since the transformation is from the inverse transfer function on the polar plot to the direct transfer function on the log magnitude plot.

A constant M circle is shown in Fig. 11-32 on the inverse polar plot. The magnitude ρ and angle ψ drawn to any point on this circle are shown.

The equation for this M circle is

$$y^2 + (1 + x)^2 = \frac{1}{M^2} \tag{11-65}$$

where $$x = \rho \cos \lambda \qquad y = \rho \sin \lambda \tag{11-66}$$

Combining these equations produces

$$\rho^2 M^2 + 2\rho M^2 \cos \lambda + M^2 - 1 = 0 \tag{11-67}$$

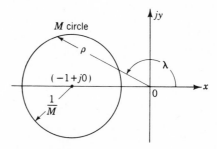

Fig. 11-32 *Constant M circle on the inverse polar plot.*

Solving for ρ and λ,

$$\rho = -\cos \lambda \pm \left(\cos^2 \lambda - \frac{M^2 - 1}{M^2}\right)^{1/2} \qquad (11\text{-}68)$$

$$\lambda = \cos^{-1} \frac{1 - M^2 - \rho^2 M^2}{2\rho M^2} \qquad (11\text{-}69)$$

These equations are derived from the inverse polar plot $1/G(j\omega)$. Since the log magnitude–angle diagram is drawn for the transfer function $G(j\omega)$ and not for its reciprocal, a change in the equations must be made by substituting

$$r = \frac{1}{\rho} \qquad \phi = -\lambda \qquad (11\text{-}70)$$

Since Lm $r = -$Lm ρ, it is necessary only to change the sign of Lm ρ. For any value of M a series of values of angle λ can be inserted in Eq. (11-68) to solve for ρ. This magnitude must be changed to decibels. Alternatively, for any value of M a series of values of ρ can be inserted in Eq. (11-69) to solve for the corresponding angle λ. Therefore the constant M curve on the log magnitude–angle diagram can be plotted by using either of these two equations.

In a similar fashion the constant α curves can be drawn on the log magnitude–angle diagram. The α curves on the inverse polar plot are semi-infinite straight lines terminating on the -1 point and are given by the following equation:

$$\tan \alpha + x \tan \alpha + y = 0 \qquad (11\text{-}71)$$

Combining with Eq. (11-66) produces

$$\tan \alpha + \rho \cos \lambda \tan \alpha + \rho \sin \lambda = 0 \qquad (11\text{-}72)$$

$$\tan \alpha = \frac{-y}{1 + x} = \frac{-\rho \sin \lambda}{1 + \rho \cos \lambda} \qquad (11\text{-}73)$$

For constant values of α a series of values of λ can be inserted in this equa-

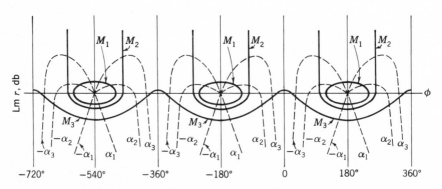

Fig. 11-33 *Constant M and α curves.*

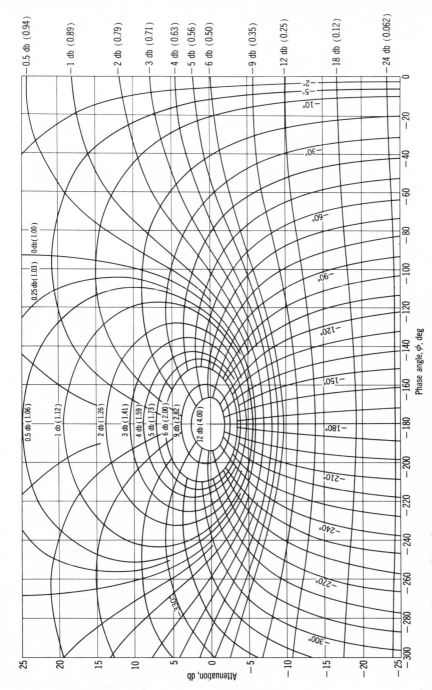

Fig. 11-34 Constant M and α curves.

387

tion to solve for ρ.   The constant α semi-infinite curves can then be plotted on the log magnitude–angle diagram.

Constant M and α curves, as shown in Fig. 11-33, repeat for every 360°. Also, there is symmetry at every 180° interval.   An expanded 300° section of the constant M and α graph is shown in Fig. 11-34.   Note that the $M = 1$ (0 db) curve is asymptotic to $\phi = -90°$ and $-270°$ and that the curves for $M < \frac{1}{2}$ (−6 db) are always negative. $M = \infty$ is the point at 0 db, $-180°$, and the curves for $M > 1$ are closed curves inside the limits $\phi = -90°$ and $\phi = -270°$.

11-11   *Adjustment of Gain by Use of the Log Magnitude–Angle Diagram*

The log magnitude–angle diagram for

$$G(j\omega) = \frac{2.04(1 + j2\omega/3)}{j\omega(1 + j\omega)(1 + j0.2\omega)(1 + j0.2\omega/3)} \tag{11-74}$$

is drawn as the solid curve in Fig. 11-35 on graph paper which has the constant M and α contours.   (The log magnitude and phase angle diagram must be drawn first to get the data for this curve.)   It is convenient to use the same log magnitude and angle scales on both the log magnitude and phase diagram and the log magnitude–angle diagram.   A pair of dividers can then be used to move the transfer-function locus to the log magnitude–angle diagram.   The $M = 1.12$ (1-db) curve is tangent to the curve at $\omega_{m1} = 1.1$. These values are the maximum value of the control ratio, M_m, and the resonant frequency ω_m.

It is specified that the system be adjusted to produce an $M_m = 1.26$ (2 db) by changing the gain.   The dashed curve is obtained by raising the transfer-function curve until it is tangent to the $M_m = 1.26$ (2 db) curve. The resonant frequency is now equal to $\omega_{m2} = 2.2$.   The curve has been raised by the amount Lm $A = 4.5$ db, meaning that an additional gain of $A = 1.68$ must be put into the system.

In practice it is easier to make a template of the desired M_m curve. Then the template can be moved up or down until it is tangent to the transfer-function curve.   The amount that the template is moved represents the gain change required.   Graph paper with the constant M and α curves superimposed is available commercially,* which simplifies system analysis and design.

* "Nichols Chart," Boonshaft and Fuchs, Inc., Hatboro, Pa.

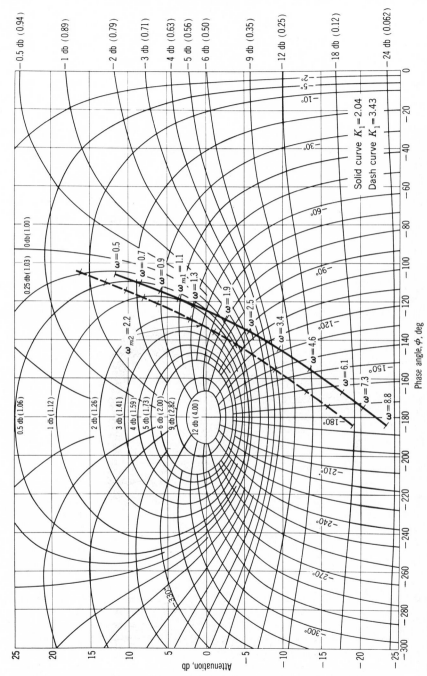

Fig. 11-35 Log magnitude–angle diagram for Eq. (11-74).

Phase angle, ϕ, deg

Attenuation, db

Solid curve $K_1 = 2.04$
Dash curve $K_1 = 3.43$

$\omega = 0.5$
$\omega = 0.7$
$\omega = 0.9$
$\omega_{m1} = 1.1$
$\omega = 1.3$
$\omega = 1.9$
$\omega = 2.5$
$\omega = 3.4$
$\omega = 4.6$
$\omega = 6.1$
$\omega = 7.3$
$\omega = 8.8$
$\omega_{m2} = 2.2$

389

After the gain has been adjusted for the desired M_m, there are several methods described in the literature for obtaining the approximate values of the closed-loop poles.[4]

11-12   *Closed-loop Frequency Response from the Log Magnitude–Angle Diagram*

When the gain has been adjusted for the desired M_m, the closed-loop frequency response can be found.   For any frequency point on the transfer-function curve the values of M and α can be read from the graph.   For example, with M_m made equal to 1.26, the value of M at $\omega = 3.4$ is -0.5 db and α is $-110°$.   The c'osed-loop frequency response, both log magnitude and angle, obtained from Fig. 11-35 is plotted in Fig. 11-36 for both values of gain.

The feedback system which has now been adjusted may be a portion of a larger system.   Analysis and design of the larger system can proceed in a similar fashion.

The closed-loop response with the resultant gain adjustment may have too low a resonant frequency for the desired performance.   The next design step would be to compensate the system to improve the frequency response. Compensation is studied in detail in a later chapter.

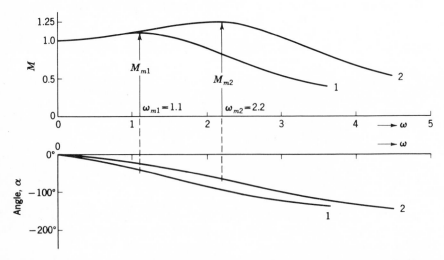

Fig. 11-36   *Control ratio versus frequency obtained from the log magnitude–angle diagram of Fig. 11-35.*

11-13 Gain Adjustment for a Desired M_m of a Nonunity Feedback System

The frequency control ratio of the feedback control system with nonunity feedback, as shown in Fig. 11-37, is

$$\frac{\mathbf{C}(j\omega)}{\mathbf{R}(j\omega)} = \frac{\mathbf{G}(j\omega)}{1 + \mathbf{G}(j\omega)\mathbf{H}(j\omega)} = \frac{1}{\mathbf{H}(j\omega)}\frac{\mathbf{G}(j\omega)\mathbf{H}(j\omega)}{1 + \mathbf{G}(j\omega)\mathbf{H}(j\omega)} \quad (11\text{-}75)$$

or

$$\frac{\mathbf{C}(j\omega)}{\mathbf{R}(j\omega)} = \frac{1}{\mathbf{H}(j\omega)}\frac{\mathbf{G}_0(j\omega)}{1 + \mathbf{G}_0(j\omega)} = \frac{1}{\mathbf{H}(j\omega)}\frac{\mathbf{C}_0(j\omega)}{\mathbf{R}(j\omega)} \quad (11\text{-}76)$$

Letting

$$\mathbf{G}_1(j\omega) = \frac{\mathbf{C}_0(j\omega)}{\mathbf{R}(j\omega)} \quad \text{and} \quad \mathbf{G}_2(j\omega) = \frac{1}{\mathbf{H}(j\omega)}$$

then

$$\frac{\mathbf{C}(j\omega)}{\mathbf{R}(j\omega)} = \mathbf{G}_1(j\omega)\mathbf{G}_2(j\omega) \quad (11\text{-}77)$$

By use of Eq. (11-76) the system of Fig. 11-37 can be represented by its mathematical equivalent, as shown in Fig. 11-38a. A further simplification can be made by considering $\mathbf{C}_0(j\omega)/\mathbf{R}(j\omega)$ and $1/\mathbf{H}(j\omega)$ as equivalent to two cascaded transfer functions represented by $\mathbf{G}_1(j\omega)$ and $\mathbf{G}_2(j\omega)$, respectively, as shown in Fig. 11-38b.

If $\mathbf{H}(j\omega)$ is non-frequency-sensitive, then either the direct or inverse plot or the log plot can be used with equal ease for gain adjustments for a specified M_m. The methods outlined in Secs. 11-8, 11-9, and 11-11 are applicable only to the $\mathbf{C}_0(j\omega)/\mathbf{R}(j\omega)$ portion of Fig. 11-38a. The $1/\mathbf{H}(j\omega)$ term serves only to modify the magnitude of the ratio $\mathbf{C}_0(j\omega)/\mathbf{R}(j\omega)$. The transient response and M_m of the control system are determined only by the $\mathbf{C}_0(j\omega)/\mathbf{R}(j\omega)$ portion. Thus the system gain is adjusted to give the desired M_m for $\mathbf{C}_0(j\omega)/\mathbf{R}(j\omega)$.

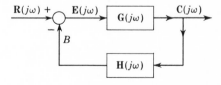

Fig. 11-37 *A nonunity feedback control system.*

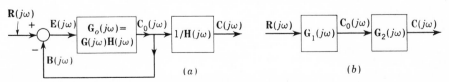

(a)   *(b)*

Fig. 11-38 *Equivalent diagrams of the nonunity feedback control system of Fig. 11-37.*

For the case where $H(j\omega)$ is a frequency-sensitive element, the inverse polar plot is most easily used, since $R(j\omega)/C(j\omega) = 1/G(j\omega) + H(j\omega)$. The gain required for a given M_m is determined by the following procedure:

1. Plot $1/G(j\omega)$, $H(j\omega)$, and the M_m circle.
2. By trial and error, adjust the gain so that

$$\frac{R(j\omega)}{C(j\omega)} = \frac{1}{G(j\omega)} + H(j\omega)$$

is tangent to the desired M_m circle.

This method is used in the chapter on feedback compensation.

11-14 Conclusions

In summary, this chapter is devoted to indicating the correlation between the frequency and time responses. The figures of merit M_m and ω_m are established as guideposts for evaluating a system's performance. The addition of a pole to an open-loop transfer function produces a clockwise shift of the direct polar plot, which results in a larger value of M_m. Chapter 13 on frequency plot compensation shows that the response also suffers because ω_m becomes smaller. The reverse is true if a zero is added to the open-loop transfer function. This agrees with the analysis of the root locus, which shows that the addition of a pole or zero results in a less stable or more stable system, respectively. Thus the qualitative correlation between the root locus and the frequency response is enhanced. The M and α contours are developed as a graphical aid in obtaining the closed-loop frequency response and in adjusting the gain to obtain a desired M_m. The methods described for setting the gain for a desired M_m are based on the fact that generally the values of concern of M_m are greater than 1. The procedure for gain adjustment may not yield a satisfactory value of ω_m. In this case the system must be compensated in order to increase ω_m without changing the value of M_m. Compensation procedures are covered in the following chapters.

The procedure used in the frequency-response method is summarized as follows:

1. Derive the open-loop transfer function $G(s)H(s)$ of the system.
2. Put the transfer function into the form $G(j\omega)H(j\omega)$.
3. Arrange the various factors of the transfer function so that they are in the complex form $1 + jb$ and jb.
4. Plot the log magnitude and phase angle diagram for $G(j\omega)H(j\omega)$, using the graphical methods of Chap. 10.

5. Transfer the data from the plots in step 4 to any of the following: (*a*) log magnitude–angle diagram, (*b*) inverse polar plot, or (*c*) direct polar plot.
6. Apply the Nyquist stability criterion and adjust the gain for the desired degree of stability M_m of the system. Then check the correlation to the time response for a step input signal. This correlation reveals some qualitative information about the time response.
7. If the qualitative response does not meet the desired specifications, determine the shape that the plot must have to meet these specifications.
8. Synthesize the compensator that must be inserted into the system, if other than just gain adjustment is required, to make the necessary modification on the original plot.

Bibliography

1. Brown, G. S., and D. P. Campbell: "Principles of Servomechanisms," chaps. 6 and 8, John Wiley & Sons, Inc., New York, 1948.
2. Higgins, T. J., and C. M. Siegel: Determination of the Maximum Modulus, or the Specified Gain, of a Servomechanism by Complex Variable Differentiation, *Trans. AIEE*, vol. 72, pt. II, p. 467, January, 1954.
3. Chu, Y.: Correlation between Frequency and Transient Responses of Feedback Control Systems, *Trans. AIEE*, vol. 72, pt. II, pp. 81–92, May, 1953.
4. Chen, K.: A Quick Method for Estimating Closed-loop Poles of Control Systems, *Trans. AIEE*, vol. 76, pt. II, pp. 80–87, May, 1957.
5. James, H. M., N. B. Nichols, and R. S. Phillips: "Theory of Servomechanisms," chap. 4, McGraw-Hill Book Company, New York, 1947.

12

Cascade Compensation: Root Locus

12-1   Introduction

The preceding chapters have dealt with basic feedback control systems comprising the minimum amount of hardware in adjunct to the controlled element.   Refinements to a system are not made until the designer has made an analysis of the performance of the basic system.   As a result of this analysis, equipment may be added to achieve the desired time response.   The next few chapters are devoted to the manner of achieving the necessary refinements.

The process of introducing additional equipment into a system to reshape its root locus in order to improve system performance is called *compensation*, or stabilization.   When the system is compensated, it is stable, has a satisfactory transient, and has a large enough gain to ensure that the steady-state error does not exceed the specified maximum.   The compensation devices may

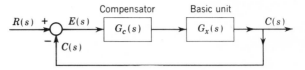

Fig. 12-1   *Block diagram showing the location of a cascade compensator.*

consist of electric networks or mechanical equipment containing levers, springs, dashpots, etc.   The compensator (also called a filter) may be placed in the following locations:

1. In cascade with the forward transfer function (cascade or series compensation)
2. In the feedback path (feedback or parallel compensation)

The selection of the location for inserting the compensator is dependent largely on the control system, the physical modifications that are necessary, and the results desired.   This chapter covers the design of compensators inserted in cascade, as shown in Fig. 12-1.   The cascade compensator is inserted at the low-energy point in the forward path so that power dissipation is very small.   This also requires that the input impedance be high. The use of isolation amplifiers may be necessary to avoid loading of or by the compensating network.

The cascade networks used for compensation are generally called lag, lead, and lag-lead compensators.   The examples used for each of these compensators are not the only ones available.   They are intended primarily to show the methods of applying compensation.

12-2   *Reshaping the Root Locus*[1]

The root-locus plots described in Chap. 7 show the relationship between the gain of the system and the time response.   Depending on the specifications established for the system, the gain that best achieves the desired performance is selected.   The performance specifications may be based on the desired damping ratio, undamped natural frequency, time constant, or steady-state error.   The root locus may show that the desired performance cannot be achieved just by adjustment of the gain.   In fact, the system may be unstable for all values of gain.   The control systems engineer must then investigate the methods for reshaping the root locus to meet the performance specifications.

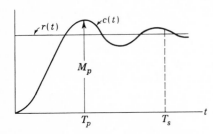

Fig. 12-2 *Desired form of the response.*

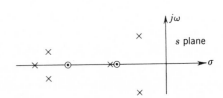

Fig. 12-3 *Pole-zero pattern of C/R for the desired response.*

The purpose of reshaping the root locus generally falls into one of the following categories:

1. A given system is stable and its transient response is satisfactory, but its steady-state error is too large. Thus, the gain must be increased to reduce the steady-state error (see Chap. 6). This must be accomplished without appreciably reducing the system stability.
2. A given system is stable but its transient response is unsatisfactory. Thus the root locus must be reshaped so that it is moved farther to the left, away from the imaginary axis.
3. A given system is stable but both its transient response and its steady-state response are unsatisfactory. Thus the locus must be moved to the left, and the gain must be increased.
4. A given system is unstable for all values of gain. Thus, the root locus must be reshaped so that part of each branch falls in the left-half s plane, thereby making the system stable.

Compensation of a system by the introduction of poles and zeros is used to improve the operating performance. However, each additional compensator pole increases the number of roots of the closed-loop characteristic equation. To make a comparison of the time responses, the contribution of each root must be taken into account. The larger the number of roots, the more laborious is the work of obtaining an exact solution.

A unit step function is widely used as a standard input. If an underdamped response of the form shown in Fig. 12-2 is desired, the system gain must be adjusted so that there is a pair of dominant complex poles. This requires that any other pole is far to the left or near a zero so that its transient has a small amplitude and therefore has a small effect on the total time response. The required pole-zero diagram is shown in Fig. 12-3. The approximate values of peak overshoot (M_o), peak time (T_p), settling time (T_s), and the number of oscillations up to settling time (N) can be obtained from the pole-zero pattern described in Sec. 8-2. The effect of compensator poles and zeros on these quantities can therefore be evaluated fairly rapidly.

12-3 Ideal Integral Compensation

When the transient response of a feedback control system is considered satisfactory but the steady-state error is too large, it is possible to eliminate the error by increasing the system type. This must be accomplished *without appreciably changing the dominant roots of the characteristic equation.* The system type can be increased by operating on the actuating signal to produce one that is proportional to both the magnitude and the integral of this signal. This is shown in Fig. 12-4, where

$$E_1(s) = \left(1 + \frac{K_i}{s}\right) E(s) \tag{12-1}$$

$$G_c(s) = \frac{E_1(s)}{E(s)} = \frac{s + K_i}{s} \tag{12-2}$$

In this system the quantity $e_1(t)$ continues to increase as long as an error $e(t)$ is present. Eventually $e_1(t)$ becomes large enough to produce an output $c(t)$ equal to the input $r(t)$. The error $e(t)$ is then equal to zero. The constant K_i (generally very small) and the over-all gain of the system must be selected to produce satisfactory roots of the characteristic equation. The resulting system performance has been improved. Since the system type has been increased, the corresponding static error coefficient is equal to infinity. Provided that the new roots of the characteristic equation can be satisfactorily located, the transient response is still acceptable.

The locations of the pole and zero of this ideal integral plus proportional compensator are shown in Fig. 12-5. The pole alone would move the root locus to the right, thereby slowing down the time response. The zero must be near the origin in order to minimize the increase in response time of the complete system.

The main limitation on the use of ideal integral plus proportional control is the equipment required to obtain the integral signal. This generally requires an amplifier in a positive feedback system.[5] Mechanically the integral signal can be obtained by an integrating gyroscope, which is used

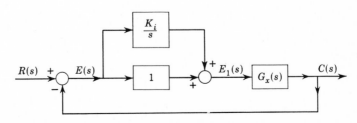

Fig. 12-4 *Ideal integral plus proportional control.*

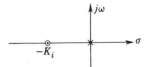

Fig. 12-5   Location of the pole and zero of an ideal integral plus proportional compensator.

in aircraft where the improved performance justifies the cost. Frequently, however, a static electric network consisting of resistors and capacitors sufficiently approximates the proportional plus integral action. This type of lag compensation, which is much cheaper, is covered in the next section.

12-4   *Lag Compensation Using Passive Elements*

In Sec. 5-13 there is shown a network which approximates a proportional plus integral output and is used as a lag compensator. Putting an amplifier of gain A in series with this network, as shown in Fig. 12-6, yields the transfer function

$$G_c(s) = A\, \frac{1 + Ts}{1 + \alpha Ts} = \frac{A}{\alpha}\, \frac{s + 1/T}{s + 1/\alpha T} \tag{12-3}$$

where $\alpha > 1$. The pole $s = -1/\alpha T$ is therefore to the right of the zero $s = -1/T$. The locations of the pole and zero of $G_c(s)$ on the s plane can be made close to those of the ideal compensator. Besides furnishing the necessary gain, the amplifier also acts as an isolating unit to prevent any loading effects between the compensator and the original system.

Assume that the original forward transfer function is

$$G_x(s) = \frac{K \prod\limits_{m=1}^{w} (s - z_m)}{\prod\limits_{c=1}^{v} (s - p_c)} \tag{12-4}$$

For the original system the static loop sensitivity for the selected root s is

$$K = \frac{\prod\limits_{c=1}^{v} |s - p_c|}{\prod\limits_{m=1}^{w} |s - z_m|} \tag{12-5}$$

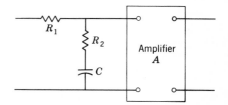

Fig. 12-6 Integral or lag compensator.

With the addition of the lag compensator in cascade, the new forward transfer function is

$$G(s) = G_c(s)G_x(s) = \frac{AK}{\alpha} \frac{s + 1/T}{s + 1/\alpha T} \frac{\prod\limits_{m=1}^{w} (s - z_m)}{\prod\limits_{c=1}^{v} (s - p_c)} \tag{12-6}$$

When the desired roots of the characteristic equation are located on the root locus, the magnitude of the static loop sensitivity for the new root s' becomes

$$K' = \frac{AK}{\alpha} = \frac{|s' + 1/\alpha T|}{|s' + 1/T|} \frac{\prod\limits_{c=1}^{v} |s' - p_c|}{\prod\limits_{m=1}^{w} |s' - z_m|} \tag{12-7}$$

As an example, a lag compensator is applied to a Type 0 system. To improve the steady-state accuracy, the static error coefficient must be increased. The value of K_0 before and after the addition of the compensator is calculated by using the definition $K_0 = \lim\limits_{s \to 0} G(s)$: from Eqs. (12-4) and (12-6), respectively,

$$K_0 = \frac{\prod\limits_{m=1}^{w} (-z_m)}{\prod\limits_{c=1}^{v} (-p_c)} K \tag{12-8}$$

$$K_0' = \frac{\prod\limits_{m=1}^{w} (-z_m)}{\prod\limits_{c=1}^{v} (-p_c)} \alpha K' \tag{12-9}$$

The following procedure is used to design the passive lag compensator. First, the pole $s = -1/\alpha T$ and the zero $s = -1/T$ of the compensator are placed very close together. This means that most of the original root locus remains practically unchanged. If the angle contributed by the compensator *at the original closed-loop dominant root* is less than 5°, the new locus is displaced only slightly. This 5° figure is only a guide and should not be applied arbitrarily. The new closed-loop pole s' is therefore essentially unchanged from the uncompensated value. This satisfies the restriction that the transient response must not change appreciably. As a result, the values $s' + 1/\alpha T$ and $s' + 1/T$ are almost equal, and the values K and K' in Eqs. (12-5) and (12-7) are approximately equal. The values K_0 and K'_0 in Eqs. (12-8) and (12-9) now differ only by the factor α so that $K'_0 \approx \alpha K_0$. The gain required to produce the new root s' therefore increases approximately by the factor α, which is the ratio of the compensator zero and pole. Summarizing, the necessary conditions on the compensator are that (1) the pole and zero must be close together and (2) the ratio α of the zero and pole must approximately equal the desired increase in gain. These requirements can be achieved by placing the compensator pole and zero very close to the origin. The size of α is limited by the physical parameters required in the network. A value $\alpha = 10$ is often used.

Although the statements above are based on a Type 0 system, the same conditions apply equally well for a Type 1 system.

Example of Lag Compensation Applied to a Type 1 System

A control system with unity feedback has a forward transfer function

$$G_x(s) = \frac{K_1}{s(1+s)(1+0.2s)} = \frac{K}{s(s+1)(s+5)} \qquad (12\text{-}10)$$

which yields the root locus shown in Fig. 12-7. For the basic control system a damping ratio $\zeta = 0.45$ yields the following pertinent data:

Dominant Roots:

$$s_{1,2} = -0.40 \pm j0.81$$

Static Loop Sensitivity:

$$\begin{aligned}
K &= |s| \cdot |s+1| \cdot |s+5| \\
&= |-0.40 + j0.81| \cdot |0.60 + j0.81| \cdot |4.60 + j0.81| \\
&= 4.17
\end{aligned}$$

Ramp Error Coefficient:

$$\begin{aligned}
K_1 &= \lim_{s \to 0} sG(s) \\
&= \frac{K}{5} = \frac{4.17}{5} = 0.83 \text{ sec}^{-1}
\end{aligned}$$

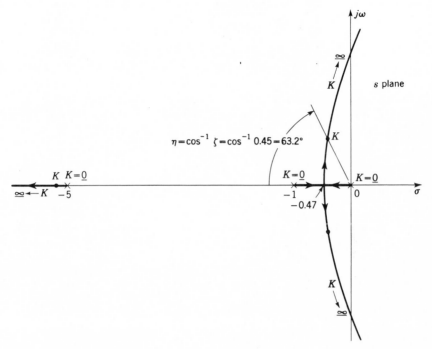

Fig. 12-7 *Root locus of* $G_x(s) = K/s(s + 1)(s + 5)$.

Undamped Natural Frequency:

$$\omega_n = 0.90 \text{ radian/sec}$$

Third Root:

$$s = -5.19$$

The values of peak time T_p, peak overshoot M_o, and settling time T_s are obtained from Eqs. (8-9), (8-13), and (8-14) as $T_p = 4.25$ sec, $M_o = 0.20$, and $T_s = 10.0$ sec. This system is simulated on the analog computer in Sec. 19-6, Example 1. The time response for several values of gain is shown in Fig. 19-27; increasing the gain results in an increase in the peak overshoot.

To increase the gain, a lag compensator is put in cascade with the forward transfer function. With the criteria discussed in the preceding section and $\alpha = 10$, the compensator pole is located at $s = -0.01$ and the zero at $s = -0.10$. This selection is based on the requirements of a practical network. For the network of Fig. 12-6, the values $R_1 = 9$ megohms, $R_2 = 1$ megohm, and $C = 10$ μf produce the pole and zero required. The input

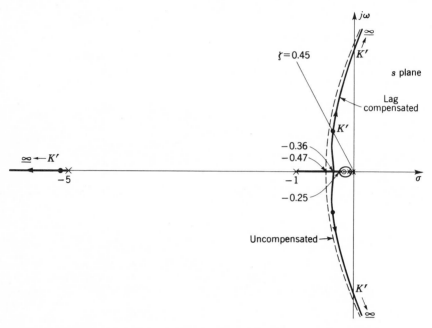

Fig. 12-8   *Root locus of $G(s) = K'(s + 0.10)/s(s + 1)(s + 5)(s + 0.01)$.*

impedance is greater than 10 megohms.   The angle of the compensator at the original dominant roots is about 5.4° and is acceptable.   Since the compensator pole and zero are very close together, they make only a small change in the new root locus in the vicinity of the original roots.   The compensator transfer function is

$$G_c(s) = \frac{A}{\alpha} \frac{s + 1/T}{s + 1/\alpha T} = \frac{A}{10} \frac{s + 0.10}{s + 0.01} \qquad (12\text{-}11)$$

Figure 12-8 (not to scale) shows the new root locus for $G(s) = G_x(s)G_c(s)$ as solid lines and the original locus as dashed lines.   Note that for the damping ratio $\zeta = 0.45$ the new locus and the original locus are close together.   The new dominant roots are $s_{1,2} = -0.36 \pm j0.72$; thus the roots are essentially unchanged.

For the compensated system adjusted to a damping ratio 0.45 the following results are obtained:

Dominant Roots:

$$s_{1,2} = -0.36 \pm j0.72$$

Static Loop Sensitivity:

$$K' = \frac{|s| \cdot |s + 1| \cdot |s + 5| \cdot |s + 0.01|}{|s + 0.10|}$$

$$= \frac{|-0.36 + j0.72| \cdot |0.64 + j0.72| \cdot |4.64 + j0.72| \cdot |-0.35 + j0.72|}{|-0.25 + j0.72|}$$

$$= 3.75$$

Ramp Error Coefficient:

$$K_1' = \frac{K'\alpha}{5} = \frac{(3.75)(10)}{5} = 7.5 \text{ sec}^{-1}$$

Increase in Gain:

$$A = \frac{K_1'}{K_1} = \frac{7.5}{0.83} = 9.03$$

Undamped Natural Frequency:

$$\omega_n = 0.80 \text{ radian/sec}$$

Other Roots:

$$s_3 = -5.17 \quad \text{and} \quad s_4 = -0.1135$$

The value of T_p for the compensated system is determined by use of Eq. (8-9), with the values shown in Fig. 12-9:

$$T_p = \frac{1}{\omega_d}\left[\frac{\pi}{2} - \phi_1 + (\phi_2 + \phi_3 + \phi_4)\right]$$

$$= \frac{1}{0.72}[90° - 110° + (109.4° + 4.6° + 90°)]\frac{\pi \text{ radians}}{180°}$$

$$= 4.46 \text{ sec}$$

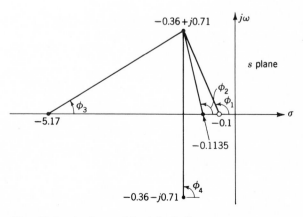

Fig. 12-9 *Angles used to evaluate T_p.*

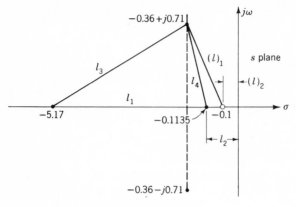

Fig. 12-10   *Lengths used to evaluate M_o.*

The value of M_o for the compensated system is determined by use of Eq. (8-13), with the values shown in Fig. 12-10:

$$
\begin{aligned}
M_o &= \frac{l_1 l_2}{l_3 l_4} \frac{(l)_1}{(l)_2} e^{\sigma T_p} \\
&= \frac{(5.17)(0.1135)}{(4.46)(0.752)} \frac{0.757}{0.1} e^{(-0.36)(4.46)} \\
&= 0.24
\end{aligned}
$$

A comparison of the uncompensated system and the system with integral compensation shows that the ramp error coefficient K_1' has increased by a factor of 9 but the undamped natural frequency has been decreased from 0.90 to 0.80 radian/sec.   This means that the steady-state error with a ramp input has decreased but the settling time has been increased.   Provided that the increased settling time is acceptable, the system has been improved. Although the complete root locus is drawn in Fig. 12-8, in practice this is not necessary.   Only the general shape is sketched and the particular needed points are accurately determined.

12-5   Ideal Derivative Compensation

When the transient response of a feedback system must be improved, it is necessary to reshape the root locus so that it is moved farther to the left of the imaginary axis.   Sec. 7-3 shows that introducing an additional zero in the forward transfer function produces this effect.   A zero can be produced by operating on the actuating signal to produce one that is proportional to both the magnitude and the derivative (rate of change) of this signal.   This is shown in Fig. 12-11, where

$$E_1(s) = (1 + K_d s)E(s) \qquad (12\text{-}12)$$

Physically the effect can be described as introducing anticipation into the system.   The system reacts not only to the magnitude of the error but also to its probable value in the future.   If the error is changing rapidly, then $e_1(t)$ is large and the system responds faster.   The net result is to speed up the response of the system.

On the root locus the introduction of a zero has the effect of shifting the curves to the left.   This is shown in the following examples.

Example 1

The root locus for a Type 1 system which is stable for a small static loop sensitivity $K < K_a$ is shown in Fig. 12-12.   Adding a zero by the compensator $G_c(s)$ (an ideal derivative plus proportional compensator) has the effect of moving the locus to the left, as shown in Fig. 12-13.   The addition

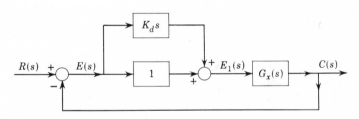

Fig. 12-11   *Ideal derivative plus proportional control.*

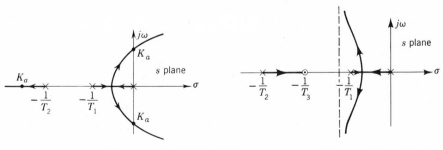

Fig. 12-12   *Root locus of*

$$G_x(s) = \frac{K}{s(s + 1/T_1)(s + 1/T_2)}$$

Fig. 12-13   *Root locus of*

$$G(s) = \frac{K(s + 1/T_3)}{s(s + 1/T_1)(s + 1/T_2)}$$

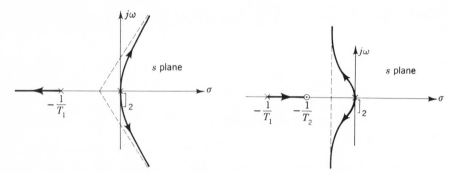

Fig. 12-14   Root locus of

$$G_x(s) = \frac{K}{s^2(s + 1/T_1)}$$

Fig. 12-15   *Root locus of*

$$G(s) = \frac{K(s + 1/T_2)}{s^2(s + 1/T_1)}$$

of the zero, in the location shown, makes the system stable for all values of gain and decreases the value of the time constant $T = 1/\zeta\omega_n$ of the closed-loop transient response.   An analysis of the effect on the performance of various positions for the zero is suggested as an exercise for the reader.

Example 2

The root locus for a Type 2 system having a forward transfer function

$$G_x(s) = \frac{K}{s^2(s + 1/T_1)}$$

is shown in Fig. 12-14.   This locus shows that the system is unstable for all values of static loop sensitivity K.   The introduction of a zero $s = -1/T_2$ between the origin and the pole at $-1/T_1$ stabilizes the system, as shown in Fig. 12-15.

Both of these examples have shown how the introduction of a zero in the forward transfer function by an ideal derivative plus proportional compensator has modified and improved the feedback control system.   However, an ideal differentiator is difficult to construct and requires much equipment that must be critically adjusted.   In addition, the derivative action amplifies any spurious signal or noise that may be present in the actuating signal. This noise amplification may actually saturate electronic amplifiers so that the system does not operate properly.   Therefore ideal derivative compensators are not used.   Instead, passive networks are used to approximate the proportional plus derivative action needed.

12-6 *Lead Compensation Using Passive Elements*

A derivative or lead compensator made up of electrical elements was shown in Sec. 5-13. To this network an amplifier of gain A is added in series, as shown in Fig. 12-16. The transfer function is

$$G_c(s) = A\alpha \frac{1 + Ts}{1 + \alpha Ts} = A \frac{s + 1/T}{s + 1/\alpha T} = A \frac{s - z_c}{s - p_c} \qquad (12\text{-}13)$$

where $\alpha < 1$. This lead compensator introduces a zero at $s = -1/T$ and a pole at $s = -1/\alpha T$. By making α sufficiently small, the location of the pole is far to the left and has small effect on the important part of the root locus. Near the zero the net angle of the compensator is due predominantly to the zero. Figure 12-17 shows the location of the lead-compensator pole and zero and the angles contributed by each at a point s_1. The best location of the zero must be determined by trial and error. It is found that the gain of the compensated system is often increased. The maximum increases in gain and in the real part ($\zeta\omega_n$) of the dominant root of the characteristic equation do not coincide. The compensator zero location must then be determined, by making several trials, for the desired optimum performance.

It can be shown that the static loop sensitivity is proportional to the ratio of $|s + 1/\alpha T|$ to $|s + 1/T|$. Therefore, as α decreases, the sensitivity increases. The minimum value of α is limited by the size of the parameters needed in the network to obtain the minimum input impedance required. Note also from Eq. (12-13) that a small α requires a large value of additional gain A from the amplifier. The value $\alpha = 0.1$ is a common choice.

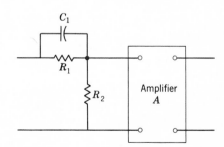

Fig. 12-16 *Derivative or lead compensator.*

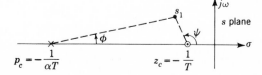

Fig. 12-17 *Location of pole and zero of a lead compensator.*

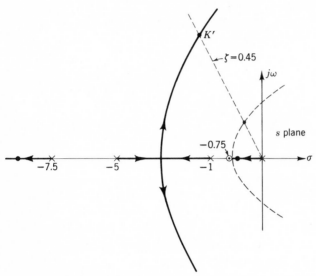

Fig. 12-18 *Root locus of* $G(s) = AK(s + 0.75)/s(s + 1)$
$(s + 5)(s + 7.5)$.

Example of lead compensation applied to a Type 1 system

The same type of system used in Sec. 12-4 is now used with lead compensation. The locations of the pole and zero of the compensator are first selected by trial. At the conclusion of this section some rules are given to show the best location.

The forward transfer function of the original system is

$$G_x(s) = \frac{K_1}{s(1 + s)(1 + 0.2s)} = \frac{K}{s(s + 1)(s + 5)} \qquad (12\text{-}14)$$

The transfer function of the compensator, using a value $\alpha = 0.1$, is

$$G_c(s) = 0.1A \frac{1 + Ts}{1 + 0.1Ts} = A \frac{s + 1/T}{s + 1/0.1T} \qquad (12\text{-}15)$$

The new forward transfer function is

$$G(s) = G_x(s)G_c(s) = \frac{AK(s + 1/T)}{s(s + 1)(s + 5)(s + 1/0.1T)} \qquad (12\text{-}16)$$

Three selections for the position of the zero $s = -1/T$ are made. A comparison of the results shows the relative merits of each. The three locations of the zero are $s = -0.75$, $s = -1.00$, and $s = -1.50$. Figures 12-18 to 12-20 show the resultant loci for these three cases. In each figure the dashed lines are the locus of the original uncompensated system.

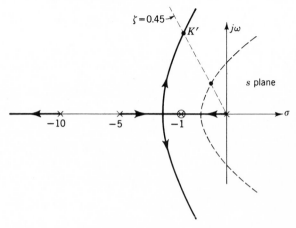

Fig. 12-19 *Root locus of $G(s) = AK(s + 1)/s(s + 1)$*
$(s + 5)(s + 10)$.

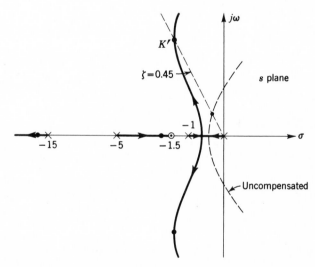

Fig. 12-20 *Root locus of $G(s) = AK(s + 1.5)/s(s + 1)$*
$(s + 5)(s + 15)$.

The static loop sensitivity is evaluated in each case from the expression

$$K' = AK = \frac{|s| \cdot |s + 1| \cdot |s + 5| \cdot |s + 1/\alpha T|}{|s + 1/T|} \qquad (12\text{-}17)$$

and the ramp error coefficient is evaluated from the equation

$$K_1' = \frac{K'\alpha}{5} \qquad (12\text{-}18)$$

Table 12-1. Results of lead compensation

Compensator	Dominant complex roots	ω_n	K_1	Additional roots
Uncompensated	$-0.40 + j0.81$	0.90	0.83	-5.19
$(s + 0.75)/(s + 7.5)$	$-1.52 + j3.03$	3.39	2.06	$-0.69;\; -9.77$
$(s + 1.0)/(s + 10)$	$-1.58 + j3.15$	3.52	2.96	-11.84
$(s + 1.50)/(s + 15.0)$	$-1.53 + j3.04$	3.40	4.38	$-1.76;\; -16.18$

The results, given in Table 12-1, show that a large increase in ω_n has been obtained by adding the lead compensators. Since the value of ζ is held constant, the system has a much faster response time. A comparison of the individual lead compensators shows the following:

1. A zero of the compensator to the right of the pole at $s = -1$ results in a root on the negative real axis close to the origin (see Fig. 12-18). From Eq. (4-74) it can be seen that the transient term due to this root is negative. Therefore it has the very desirable effect of decreasing the peak overshoot, but it has the undesirable characteristic of having a long time constant. But if the size of this transient term is small (the magnitude is proportional to the distance from the real root to the zero), the effect on the settling time is small.

   The decrease in M_o can be determined by calculating the contribution of the transient term due to the real root, as discussed in Sec. 8-3. At the time $t = T_s$ the sign of the underdamped transient can be determined. If it is negative, the actual settling time has increased. If it is positive, the settling time may be decreased. In this example both transients are negative at T_s; therefore the actual settling time is increased. The exact settling time can be obtained from a complete solution of the time response, which is shown in Fig. 12-21. The correct settling time is 2.8 sec.

2. The best transient response due to the complex roots occurs when the zero of the compensator cancels the pole $s = -1$ of the original transfer function (see Fig. 12-19). There is no real root near the imaginary axis in this case. The actual setting time is $t_s = 2.46$ sec.

3. The largest gain occurs when the zero of the compensator is located at the left of the pole $s = -1$ of the original transfer function (see Fig. 12-20). The transient due to the root $s = -1.76$ is positive. Therefore it increases the peak overshoot of the response. The value of the transient due to the real root can be added to the peak value of the underdamped sinusoid to obtain the correct value of the peak overshoot. The real root may affect the settling time. The change of T_s can be estimated as outlined in case 1. In this case the real root

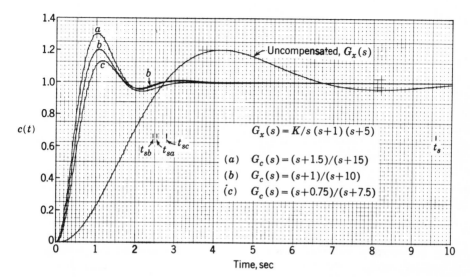

Fig. 12-21 *Time responses with a step input for the uncompensated and the compensated cases of Table 12-1.*

decreased the peak and settling times. From Fig. 12-21 the actual values are $t_p = 1.03$ sec and $t_s = 2.55$ sec.

The time response of the original system for several values of gain is shown in Fig. 19-27. It is evident that increasing the gain increases the peak overshoot and the settling time. The response with the three lead compensators is shown in Fig. 12-21. The effect of the additional real root is to change the peak overshoot and the settling time. The designer must make a choice between the largest gain attainable, the smallest peak overshoot, and the shortest settling time.

Based on this example, the following rules are used to apply cascade lead compensators to a Type 1 or higher system:

1. If the zero $s = -1/T$ of the lead compensator is superimposed and cancels the largest real pole (excluding the pole at zero) of the original transfer function, a good improvement in the transient response is obtained.

2. If a larger gain is desired than that obtained by rule 1, several trials should be made with the zero of the compensator moved to the left or right. The location of the zero that results in the desired gain is selected.

For a Type 0 system it will often be found that a better time response and a larger gain can be obtained by placing the compensator zero so that it cancels or is close to the second largest real pole of the original transfer function.

12-7   General Lead-compensator Design

Additional methods are now demonstrated for the design of an appropriate lead compensator G_c which is placed in cascade with a basic transfer function G_x, as shown in Fig. 12-1. Figure 12-22 shows the original root locus of a control system. For a specified damping ratio ζ the dominant root of the uncompensated system is s_1. Also shown is s_2, which is the desired root of the system's characteristic equation. Selection of s_2 as a desired root is based on the performance required for the system. The design problem is to select a lead compensator that results in s_2 being a root. The first step is to find the sum of the angles at the point s_2 due to the poles and zeros of the original system. This angle is $180° + \phi$. For s_2 to be on the new root locus, it is necessary for the lead compensator to contribute an angle $\phi_c = -\phi$ at this point. The total angle at s_2 is then $180°$, and it is a point on the new root locus. A simple lead compensator represented by Eq. (12-13), with its zero to the right of its pole, can be used to provide the angle ϕ_c at the point s_2.

Method 1

Reference to Fig. F-3 in Appendix F shows that the constant-angle curve ϕ_c for a lead compensator is an arc of a circle going through the pole and zero. The center of this circle is given by Eq. (F-4). One procedure is to arbitrar-

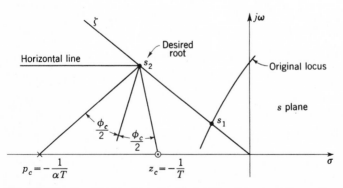

Fig. 12-22 *Graphical construction for locating the pole and zero of a simple lead compensator.*

ily select the spacing of the compensator pole and zero and to draw the constant-angle locus by using Eq. (F-4). Then slide this locus along the real axis until it goes through s_2. This location for the compensator pole and zero makes s_2 a point on the complete root locus of the compensated system. The other roots must also be determined. It is then possible to obtain the time response and to make sure that it meets the performance specifications.

Method 2

Actually, there are many possible locations of the compensator pole and zero that will produce the necessary angle ϕ_c at the point s_2. Cancellation of poles of the original open-loop transfer function by means of compensator zeros may simplify the root locus and thereby reduce the complexity of the problem. The compensator zero is simply placed over a real pole. Then the compensator pole is placed further to the left at a location which makes s_2 a point on the new root locus. The pole to be canceled depends on the system type. For a Type 1 system the largest real pole (excluding the pole at zero) should be canceled. For a Type 0 system the second largest pole should be canceled.

Method 3

The following construction is based on obtaining the maximum value of α, which is the ratio of the compensator zero and pole. The steps in the location of the lead compensator pole and zero (see Fig. 12-22) are:

1. Locate the desired root s_2. Draw a line from this root to the origin and a horizontal line to the left of s_2.
2. Bisect the angle between the two lines drawn in step 1.
3. Measure the angle $\phi_c/2$ on either side of the bisector drawn in step 2.
4. The intersections of these lines with the real axis locate the compensator pole p_c and zero z_c.

The construction outlined above is shown in Fig. 12-22. It is left for the reader to show that this construction results in the largest possible value of α. Since α is less than unity and appears as the gain of the compensator [see Eq. (12-13)], the largest value of α will require the smallest additional gain A.

Method 4

The procedures outlined above give the desired location of a root. However, no conditions have been placed on the desired system gain. A minimum value of gain may be specified to restrict the maximum steady-state error.

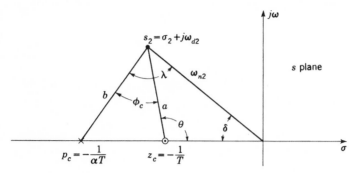

Fig. 12-23 *Pole-zero diagram for a lead compensator.*

Changing the locations of the compensator pole and zero can produce a range of values for the gain while maintaining the desired root s_2. A specific procedure for achieving both the desired root s_2 and the system gain K_n is given by reference 12.

The basic open-loop transfer function $G_x(s)$ has the form of Eq. (12-4), and the compensator transfer function is given by Eq. (12-13). The compensator pole and zero are shown in Fig. 12-23. The following definitions and conditions are used:

s_2 = desired root

$K_n = K_x A \alpha$ = desired gain of $G(s)$, where $G(s) = G_c(s) G_x(s)$*

K_u = static loop sensitivity evaluated at s_2, from the pole-zero diagram by use of the Spirule or analytically, for basic system $G_x(s)$

K = desired static loop sensitivity, determined from basic system $G_x(s)$ utilizing desired value of K_n

A = gain of amplifier used with compensator

$\alpha = z_c/p_c$ = attenuation coefficient of compensator

$\phi_c = (1 + 2m)180° - /G_x(s_2)$ = angle required from compensator to make s_2 a point on root locus of compensated system

δ = angle between directed line segment s_2 and negative real axis

The condition

$$A = \frac{1}{\alpha} = \frac{p_c}{z_c} \tag{12-19}$$

is used, with the result that $K_n = K_x$. The value of K can be expressed as

$$K = \frac{K_u|s_2 - p_c|}{A|s_2 - z_c|} = \frac{K_u\alpha|s_2 - p_c|}{|s_2 - z_c|} \tag{12-20}$$

* $K_x = \lim_{s \to 0} s^n G_x(s)$

From Fig. 12-23 the following relationships are obtained by the law of sines:

$$\frac{|z_c|}{|s_2 - z_c|} = \frac{\sin (\lambda - \phi_c)}{\sin \delta} \tag{12-21}$$

$$\frac{|p_c|}{|s_2 - p_c|} = \frac{\sin \lambda}{\sin \delta} \tag{12-22}$$

These relationships can be solved for the compensator pole and zero locations

$$|p_c| = \frac{1}{\alpha T} = \frac{|s_2| \sin \lambda}{\sin (\theta - \phi_c)} \tag{12-23}$$

$$|z_c| = \frac{1}{T} = \frac{|s_2| K \sin \lambda}{K_u \sin \theta} \tag{12-24}$$

where

$$\lambda = \cot^{-1} \left(\cot \phi_c - \frac{K}{K_u \sin \phi_c} \right) \tag{12-25}$$

$$\theta = 180° + \phi_c - \delta - \lambda \tag{12-26}$$

The angle λ is greater than ϕ_c (see Fig. 12-23) and has a positive value. *A single lead section can be used when* $\theta > \phi_c$. *If this inequality is not satisfied, more than one compensator section is required.*

As an example, consider the Type 1 system given by

$$G_x(s) = \frac{K_1}{s(1 + 0.2s)} = \frac{K}{s(s + 5)} \tag{12-27}$$

A gain $K_1 = 20$ is specified. The roots for this gain are $s_1 = -2.5 \pm j9.2$. However, it is desired to relocate the poles at $s_2 = -14.14 \pm j14.14$ to reduce the settling time. The computations necessary to locate the compensator pole and zero are:

$$\underline{/G_x(s_2)} = -258°$$

$$\phi_c = 78°$$

$$K = \frac{K_1}{0.2} = \frac{20}{0.2} = 100$$

$$|s_2| = 20$$

$$K_u = (20)(16.9) = 338$$

$$\delta = \tan^{-1}\frac{14.14}{14.14} = 45°$$

$$\lambda = \cot^{-1}\left(\cot 78° - \frac{100}{338 \sin 78°} \right) = 94.9°$$

$$\theta = 180° + 78° - 45° - 94.9° = 118.1°$$

Applying Eqs. (12-23) and (12-24)

$$|p_c| = \frac{1}{\alpha T} = \frac{20 \sin 94.9°}{\sin 40.1°} = 31.0$$

$$|z_c| = \frac{1}{T} = \frac{(20)(100) \sin 94.9°}{338 \sin 118.1°} = 6.67$$

Therefore

$$\alpha = \frac{z}{p} = 0.215$$

$$A = \frac{1}{\alpha} = 4.65$$

The compensated transfer function has the required gain $K_1 = 20$:

$$G(s) = \frac{465(s + 6.67)}{s(s + 5)(s + 31)} = \frac{20(1 + 0.15s)}{s(1 + 0.2s)(1 + 0.0323s)}$$

The closed-loop system has an additional pole at $s = -7.7$ so that

$$\frac{C(s)}{R(s)} = \frac{465(s + 6.67)}{(s + 14.14 - j14.14)(s + 14.14 + j14.14)(s + 7.7)}$$

$$= \frac{465(s + 6.67)}{(s^2 + 28.28s + 400)(s + 7.7)}$$

The root at $s = -7.7$ has the effect of increasing the peak overshoot, and it may increase the settling time. The time response must be compared with the complete specifications in order to determine the suitability of this compensated system.

12-8   *Lag-Lead Compensation*

The preceding sections have shown the following:

1. The insertion of an integral or lag compensator in cascade results in a large increase in gain and a small reduction in the undamped natural frequency.
2. The insertion of a derivative or lead compensator in cascade results in a small increase in gain and a large increase in the undamped natural frequency.

By inserting both the lag and the lead compensators in cascade with the original transfer function, the advantages of both can be realized simultaneously; that is, a large increase in gain and a large increase in the undamped

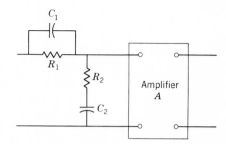

Fig. 12-24 *Lag-lead compensator.*

natural frequency can be obtained. Instead of using two separate networks, it is possible to use one network that acts as both a lead and a lag compensator. Such a network, shown in Fig. 5-30, is called a lag-lead compensator. With an amplifier of gain A added in cascade with this network, the compensator appears in Fig. 12-24. The transfer function of this compensator is

$$G_c(s) = A \frac{(1 + T_1 s)(1 + T_2 s)}{(1 + \alpha T_1 s)[1 + (T_2/\alpha)s]} = A \frac{(s + 1/T_1)(s + 1/T_2)}{(s + 1/\alpha T_1)(s + \alpha/T_2)} \quad (12\text{-}28)$$

where $\alpha > 1$ and $T_1 > T_2$.

The fraction $(1 + T_1 s)/(1 + \alpha T_1 s)$ represents the lag compensator, and the fraction $(1 + T_2 s)/[1 + (T_2/\alpha)s]$ represents the lead compensator. The values T_1, T_2, and α are selected to achieve the desired improvement in system performance. The specific values of the compensator components are obtained from the relationships shown in Eqs. (5-53) and (12-28). The requirements are that $T_1 = R_1 C_1$, $T_2 = R_2 C_2$, $\alpha T_1 + T_2/\alpha = R_1 C_1 + R_2 C_2 + R_1 C_2$. It may also be desirable to specify a minimum value of the input impedance over the passband frequency range.

The procedure for applying the lag-lead compensator is a combination of the procedures for the individual units.

1. For the integral or lag component:
   a. The zero $s = -1/T_1$ and the pole $s = -1/\alpha T_1$ are selected close together, with α set to a large value such as $\alpha = 10$.
   b. The pole and zero are located to the left of and close to the origin. This results in an increased gain.
2. For the derivative or lead component:
   a. The zero $s = -1/T_2$ is superimposed on the pole of the original system. This results in moving the root locus to the left and therefore increases the undamped natural frequency.

The relative positions of the poles and zeros of the lag-lead compensators are shown in Fig. 12-25.

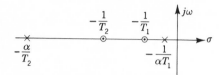

Fig. 12-25 *Location of the poles and zeros of the lag-lead compensator.*

Example of Lag-Lead Compensation Applied to a Type 1 System

The system described in Secs. 12-4 and 12-6 is now used with a lag-lead compensator. The forward transfer function is

$$G_x(s) = \frac{K_1}{s(1+s)(1+0.2s)} = \frac{K}{s(s+1)(s+5)} \tag{12-29}$$

The lag-lead compensator used is shown in Fig. 12-24, and the transfer function $G_c(s)$ is given by Eq. (12-28). The new complete forward transfer function is

$$G(s) = G_x(s)G_c(s) = \frac{AK(s+1/T_1)(s+1/T_2)}{s(s+1)(s+5)(s+1/\alpha T_1)(s+\alpha/T_2)} \tag{12-30}$$

The poles and zeros of the compensator are selected in accordance with the principles outlined previously in this section. They coincide with the values used for the integral and derivative compensators when applied individually in Secs. 12-4 and 12-6. With an $\alpha = 10$, the poles and zeros are

$$\frac{1}{T_1} = 0.10 \qquad \frac{1}{T_2} = 1$$

$$\frac{1}{\alpha T_1} = 0.01 \qquad \frac{\alpha}{T_2} = 10$$

The forward transfer function becomes

$$G(s) = G_x(s)G_c(s) = \frac{K'(s+0.10)}{s(s+0.01)(s+5)(s+10)} \tag{12-31}$$

The root locus (not to scale) for this system is shown in Fig. 12-26.

For a damping ratio $\zeta = 0.45$ the following pertinent data are obtained for the compensated system:

Dominant Roots:

$$s_{1,2} = -1.55 \pm j3.08$$

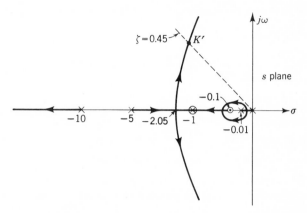

Fig. 12-26 *Root locus of $G(s) = K'(s + 1)(s + 0.10)/$ $s(s + 1)(s + 0.01)(s + 5)(s + 10)$.*

Static Loop Sensitivity (evaluated at $s = s_1$):

$$K' = \frac{|s| \cdot |s + 0.01| \cdot |s + 5| \cdot |s + 10|}{|s + 0.10|} = 145$$

Ramp Error Coefficient:

$$K'_1 = \frac{K'(0.1)}{(0.01)(5)(10)} = 29.0 \text{ sec}^{-1}$$

Undamped Natural Frequency:

$$\omega_n = 3.45 \text{ radians/sec}$$

The additional roots are $s_3 = -11.81$ and $s_4 = -0.103$. These results, when compared with those of the uncompensated system, show a large increase in both the gain and the undamped natural frequency. They indicate that the advantages of the lag-lead compensator are equivalent to the combined improvement obtained by the lag and the lead compensators. Note the practicality in the choice of $\alpha = 10$, which is based on experience. The time responses of the original and compensated systems are shown in Figs. 19-27 and 19-31, respectively.

12-9   Comparison of Cascade Compensators

Table 12-2 shows a comparison of the system response obtained when a lag, a lead, and a lag-lead compensator are added in cascade with the same

Table 12-2.   Comparison of performance of several cascade compensators for the system of Eq. (12-10) with $\zeta = 0.45$

Compensator	Dominant complex roots	Undamped natural frequency ω_n	Other roots	Ramp error coefficient K_1	T_p, sec	M_o	T_s, sec	Additional gain required A
Lag: Uncompensated	$-0.40 + j0.81$	0.90	-5.19	0.83	4.25	0.20	10.0	—
$\dfrac{s + 0.1}{s + 0.01}$	$-0.36 + j0.72$	0.80	-0.114 -5.17	7.5	4.46	0.24	11.1	9.03
Lead: $\dfrac{s + 1.5}{s + 15}$	$-1.53 + j3.04$	3.40	-1.76 -16.18	4.38	1.12	0.24	2.62	52.7
Lead: $\dfrac{s + 1}{s + 10}$	$-1.58 + j3.15$	3.52	-11.84	2.96	1.09	0.19	2.53	35.7
Lag-lead: $\dfrac{(s + 0.1)(s + 1)}{(s + 0.01)(s + 10)}$	$-1.55 + j3.08$	3.45	-0.103 -11.81	29.0	1.08	0.203	2.58	34.9

basic feedback system.   These results are for the examples of Secs. 12-4, 12-6, and 12-8.   The locations of the poles of the closed-loop system roughly satisfy the conditions of Sec. 8-2.   Therefore the approximate equations given in Sec. 8-2 are used to calculate M_o, T_p, and T_s.   The time responses of the original lead-compensated and lag-lead-compensated systems are shown in Figs. 19-27, 19-29, 19-31, and 12-21.

The results shown are typical of the changes introduced by the use of cascade compensators.   These changes can be summarized as follows:

1.   Lag compensator:
    a.   Results in a large increase in gain K_n (by a factor almost equal to α), which means a much smaller steady-state error.
    b.   Decreases ω_n and therefore has the disadvantage of producing a small increase in the settling time.
2.   Lead compensator:
    a.   Results in a moderate increase in gain K_n, thereby improving steady-state accuracy.
    b.   Results in a large increase in ω_n and therefore reduces the settling time considerably.
    c.   The transfer function of the lead compensator, using the passive network of Fig. 12-16, contains the gain α, which is less than unity.   Therefore the additional gain A, which must be added, is larger than the increase in K_n for the system.

3. Lag-lead compensator (essentially combines the desirable characteristics of the lag and the lead compensators):
   a. Results in a large increase in gain K_n, which improves the
      steady-state response.
   b. Results in a large increase in ω_n, which improves the transient-
      response settling time.

Figure 12-27 shows the frequency response for the original and compensated systems corresponding to Table 12-2. There is a close qualitative
correlation between the time-response characteristics listed in Table 12-2
and the corresponding frequency-response curves shown in Fig. 12-27.
The peak value M_p is directly proportional to the maximum value M_m, and
the settling time T_s is a direct function of the passband frequency. Curves
3, 4, and 5 have a much wider passband than curves 1 and 2; therefore these
systems have a faster settling time.

For systems other than the one used as an example, these simple compensators may not produce the desired changes. However, the basic principles
described here are applicable to any system for which the open-loop poles
closest to the imaginary axis are on the negative real axis.

The next section shows two systems which have two real and two complex
poles. When the real open-loop poles are dominant, the conventional compensators may be used and good results are obtained. When the complex

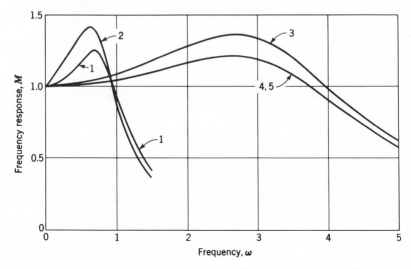

Fig. 12-27 *Frequency response for systems of Table 12-2:* (1) *original basic system;* (2) *lag-compensated;* (3) *lead-compensated—zero at* −1.5; (4) *lead-compensated—zero at* −1.0; (5) *lag-lead-compensated*

open-loop poles are dominant, the conventional compensators produce small improvements. Different compensators must be designed in this case.

12-10   *Special Problems*

The feedback system which has the forward transfer function

$$G(s) = \frac{K_0}{(1 + T_1 s)(1 + T_2 s)(1 + Bs + As^2)} \qquad T_1 < T_2$$

is considered for two sets of values of the system constants.

For the first case the open-loop real poles are larger than the real part of the complex poles, and the root locus is drawn in Fig. 12-28a. In the second case the sizes of the poles are reversed and the root locus is drawn in Fig. 12-28b. For both of these cases the gain can be adjusted, based on the system specification. For example, specifying the damping ratio ζ of the dominant roots fixes the gain required. The damping-ratio line is drawn in Fig. 12-28.

Consider now that an improvement in transient response is desired. This indicates that with the damping ratio held constant the undamped natural frequency ω_n must be increased. A conventional cascade lead compensator is used since previous experience has shown that it results in an increase in ω_n.

The conventional lead compensator has the transfer function

$$G_c(s) = \alpha \frac{1 + Ts}{1 + \alpha Ts} \tag{12-32}$$

where $\alpha < 1$. When this compensator is applied to the first case, the time constant T is made equal to the larger real time constant T_2 of the original

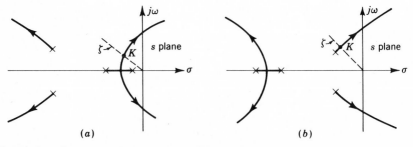

Fig. 12-28 *Two different root-locus plots for*
$G_x(s) = K_0/(1 + T_1 s)(1 + T_2 s)(1 + Bs + As^2)$

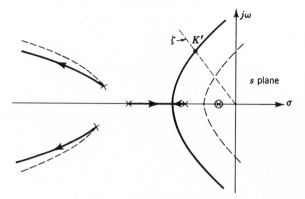

Fig. 12-29 *Root locus of* $G_x(s)G_c(s) = K_0'(1 + Ts)/$ $(1 + T_1s)(1 + Bs + As^2)(1 + \alpha Ts)$.

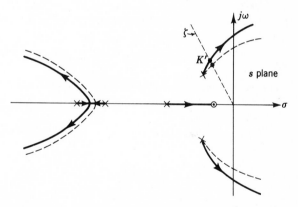

Fig. 12-30 *Root locus of* $G_x(s)G_c(s) = K_0'(1 + T/s)$ $(1 + T_1s)(1 + T_2s)(1 + Bs + As^2)(1 + \alpha Ts)$.

system. The root locus of the compensated system is shown in Fig. 12-29, where the original locus is shown in dashed lines and the new locus is drawn in solid lines. For the same damping ratio the graph shows that ω_n has been increased as desired. Thus, the objective has been achieved.

Consider now the second case, shown in Fig. 12-28b. The criterion previously applied for a lead compensator no longer can be used, because the complex open-loop poles are closer to the imaginary axis. The pole and zero of the conventional compensator can be located by trial and error, and one example is shown in Fig. 12-30. The new locus is drawn in solid lines and the original locus is drawn in dashed lines.

The modified locus of Fig. 12-30 shows that very little improvement has been achieved by inserting the conventional lead compensator. The undamped natural frequency of the dominant roots has increased by only a small amount. Also, a branch has been added on the real axis between the pole and zero of the compensator. With a step input the transient due to this additional root of the characteristic equation can be shown to be positive. This increases the total transient peak overshoot. The change of settling time depends on the period of the transient due to the complex roots. While a small improvement in settling time may be achieved, the addition of more equipment for the compensator is questionable. The design of compensators that achieve desirable results is discussed in the next section.

12-11   Other Compensators

The preceding section has shown that the conventional lead compensator has little effect on an original system when the dominant open-loop poles are complex, as represented by the root locus of Fig. 12-28b. A pair of complex poles near the imaginary axis is often encountered in aircraft and missile transfer functions. Effective compensation would remove the open-loop dominant complex poles and replace them with poles located farther to the left. This means that the compensator should have a transfer function of the form

$$G_c(s) = \frac{1 + Bs + As^2}{1 + Ds + Cs^2} \tag{12-33}$$

With exact cancellations, the zeros of $G_c(s)$ would coincide with the complex poles of the original transfer function. The composite forward transfer function would then be

$$G(s) = G_c(s)G_x(s) = \frac{K_0'}{(1 + T_1s)(1 + T_2s)(1 + Ds + Cs^2)} \tag{12-34}$$

and the root locus is shown in Fig. 12-31. The solid lines are the new locus and the dashed lines are the original locus.

The new poles introduced by the compensator can be real instead of complex. In this case the root locus is shown in Fig. 12-32. In both cases the transient response has been improved.

The problem remaining to be solved is the synthesis of a network with the desired poles and zeros. An example of a simple RLC network which has the desired transfer function is shown in Fig. 12-33. The objection to this network is that it requires an inductor, which is heavier and tends to be less stable than resistors and capacitors.

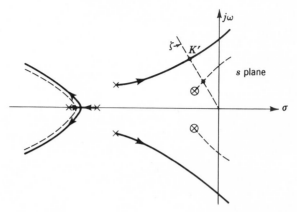

Fig. 12-31 Root locus of $G_x(s)G_c(s) = K'_0/(1 + T_1s)$
$(1 + T_2s)(1 + Ds + Cs^2)$.

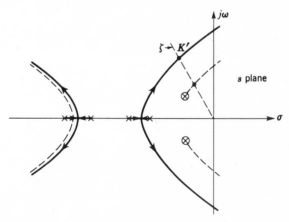

Fig. 12-32 Root locus of $G_c(s)G_x(s) = K'_0/(1 + T_1s)$
$(1 + T_2s)(1 + T_3s)(1 + T_4s)$.

A bridged-T network of the form shown in Fig. 12-34 also has a transfer function of the desired form and uses only resistors and capacitors. For other types of compensators that can be used the reader is referred to books that cover network synthesis.

It is not necessary that the zeros of the compensator cancel exactly the poles of the original system. If the zeros are near the poles, the effect is almost the same as exact cancellation. Figure 12-35 shows the application of a compensator in which the zeros do not exactly cancel the poles. There

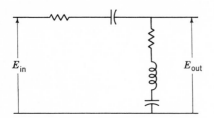

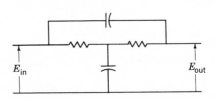

Fig. 12-33 *Simple RLC network which has the transfer function* $G_c(s) = (1 + Bs + As^2)/(1 + Ds + Cs^2)$.

Fig. 12-34 *Bridged-T network which has the transfer function* $G_c(s) = (1 + Bs + As^2)/(1 + Ds + Cs^2)$.

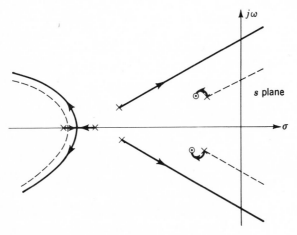

Fig. 12-35 *Root locus when the compensator zeros do not exactly cancel the original complex poles:*

$$G_x(s)G_c(s)$$
$$= \frac{K_0'}{(1 + T_1s)(1 + T_2s)(1 + Bs + As^2)} \frac{1 + Ds + Cs^2}{1 + Fs + Es^2}$$

are now two additional branches, but these roots have a very small effect on the response of the system.

12-12   *Realization of Transfer Functions*[14]

As mentioned in the last section, the synthesis of a network with desired poles and zeros may be very difficult. In this section a design procedure employing active networks is presented. It simplifies the synthesis of a wide range of transfer functions which can be used as compensators. Consider the transfer functions of the form

$$G_c(s) = \frac{\prod_{m=1}^{w} (s - z_m)}{\prod_{k=1}^{v} (s - p_k)} \qquad (12\text{-}35)$$

where $v \geq w$, Re $(p_k) < 0$, and there are no multiple poles. The foregoing equation may be expanded into a finite series by employing partial fractions. Thus

$$G_c(s) = \lim_{s \to \infty} G_{\dot{c}}(s) + \sum_{k=1}^{v} \frac{A_k}{s - p_k} \qquad (12\text{-}36)$$

where

$$A_k = [(s - p_k)G_c(s)]_{s=-p_k} \qquad (12\text{-}37)$$

A_k and p_k are both real or are both complex. In the latter case, the conjugates of A_k and of p_k both appear so that a second-order term arises in the form

$$\frac{A_k' + A_{k+1}'s}{(s - p_k)(s - p_k^*)} \qquad (12\text{-}38)$$

where A_k' and A_{k+1}' are real.

Although the realization procedure is easily applied to high-order transfer functions as well as to nonelectric signals, the discussion is limited by considering an electric network of the mathematical form

$$G_c(s) = \frac{(s - z_1)(s - z_2)}{(s - p_1)(s - p_2)} = 1 + \frac{A_1}{s - p_1} + \frac{A_2}{s - p_2} \qquad (12\text{-}39)$$

Four cases exist for this equation. They are:

Case 1. p_1, p_2, z_1, and z_2 are all real.
Case 2. p_1 and p_2 are real, and z_1 and z_2 are complex conjugates.
Case 3. p_1 and p_2 are complex conjugates, and z_1 and z_2 are real.
Case 4. p_1 and p_2 are complex conjugates, and z_1 and z_2 are complex conjugates.

The design procedure is applicable to all four cases. Case 2 is utilized as an example. Consider that a particular system requires the compensator

$$G_c(s) = \frac{s^2 + s + 3}{(s + 1)(s + 3)} \qquad (12\text{-}40)$$

From Eqs. (12-36) through (12-39)

$$G_c(s) = 1 + \frac{1.5}{s + 1} - \frac{4.5}{s + 3} = 1.5 \left(\frac{2}{3} + \frac{1}{1 + s} - \frac{1}{1 + s/3} \right) \qquad (12\text{-}41)$$

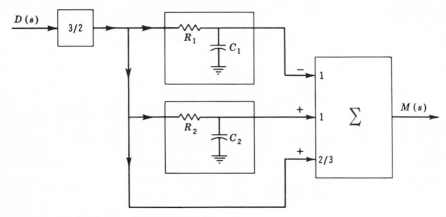

Fig. 12-36   *Mechanization of Eq. (12-41).*

Note that the two terms involving s represent transfer functions of simple RC lag networks.   The mechanization of Eq. (12-41) is shown in Fig. 12-36. This is a simpler mechanization than the ones required by other network synthesis procedures.

   The loading effect of the input impedance of the summing amplifier may alter the parameters of the lag circuit.   These parameters may be computed by using the modified form

$$\frac{A}{Ts + 1} = \frac{R}{R + R_i} \frac{1}{1 + [RR_i/(R + R_i)]Cs} \tag{12-42}$$

where R_i is the input resistance of the summer and R is the resistance in the RC lag network.   The gain must also be increased by a factor of $(R + R_i)/R$. Depending upon the particular system in which $G_c(s)$ is to be utilized, other difficulties may need to be overcome.

   The design procedure illustrated in this section has two salient features:

1.   The realization of zeros anywhere in the s plane.   When the poles are real (Cases 1 and 2), $G_c(s)$ can be accomplished with RC networks. When the poles are complex, it is necessary to use RLC networks.
2.   The realization of high-order functions with relative ease.

12-13   Conclusions

This chapter has shown the applications of lag, lead, and lag-lead compensators in cascade.   The procedures for applying the compensators are given so that they can be used with any system.   These procedures involve a certain amount of trial and error and the use of judgment based on past experience.   The improvements produced by each compensator are shown by application to a specific system.   In cases where the conventional compensators do not produce the desired results, the method of applying other compensators is shown.

It is important to emphasize that the conventional compensators cannot be applied indiscriminately.   For example, the necessary improvements may not fit the four categories listed in Sec. 12-2, or the use of the conventional compensator may cause a deterioration of other performance characteristics. In such cases the designer must use his ingenuity in locating the compensator poles and zeros to achieve the desired results.   The use of several compensating networks in cascade may be required to achieve the desired root locations.

Realization of compensator pole-zero locations by means of passive networks requires an investigation of network synthesis techniques.   Detail design procedures are available for specialized networks which produce two poles and two zeros.[1,9–11]   The design method for the bridged-T network, which produces two complex zeros and two real poles, is contained in Appendix H.   It is also possible to simulate such compensators by use of an analog computer, as described in Sec. 12-12.

It may occur to the reader that the methods of compensation in this and the following chapters can become laborious because there is some trial and error involved.   The use of computers[8] to solve design problems of this type is becoming more widespread.   By the use of a computer the optimum design of the compensator can be determined in a shorter time.   However, a computer is not always available.   If one is available, it is still necessary to obtain by an analytical or graphical method a few values of the solution to ensure that the problem has been properly set up and that the computer is operating correctly.

Bibliography

1. Truxal, J. G.: "Automatic Feedback Control System Synthesis," McGraw-Hill Book Company, New York, 1955.
2. Chu, Y.: Synthesis of Feedback Control System by Phase-angle Loci, *Trans. AIEE*, vol. 71, pt. II, pp. 330–339, November, 1952.
3. Mulligan, J. H., Jr.: The Effect of Pole and Zero Locations on the Transient Response of Linear Dynamic Systems, *Proc. IRE*, vol. 37, pp. 516–529, 1949.
4. Zemanian, A. H.: Further Effects of the Pole and Zero Locations on the Step Response of Fixed Linear Systems, *Trans. AIEE*, vol. 74, pt. II, pp. 52–55, March, 1955.

5. Brown, G. S., and D. P. Campbell: "Principles of Servomechanisms," p. 206, John Wiley & Sons, Inc., New York, 1948.
6. Evans, W. R.: "Control-system Dynamics," p. 117, McGraw-Hill Book Company, New York, 1954.
7. Smith, O. J. M.: "Feedback Control Systems," pp. 272–274, McGraw-Hill Book Company, New York, 1958.
8. Johnson, C. L.: "Analog Computer Techniques," 2d ed., McGraw-Hill Book Company, New York, 1963.
9. Chandaket, P., and A. B. Rosenstein: Notes on Bridged-T Complex Conjugate Compensation and 4-Terminal Network Loading, *Trans. AIEE*, vol. 78, pt. II, pp. 148–163, July, 1959.
10. Slaughter, J. B., and A. B. Rosenstein: Twin-T Compensation Using Root Locus Methods, *Trans. IEEE*, vol. 81, pt. II, pp. 339–349, January, 1963.
11. Lazear, T. J., and A. B. Rosenstein: On Pole-Zero Synthesis and the General Twin-T, *Trans. IEEE*, vol. 83, pt. II, pp. 389–393, November, 1964.
12. Ross, E. R., T. C. Warren, and G. J. Thaler: Design of Servo Compensation Based on the Root Locus Approach, *Trans. AIEE*, vol. 79, pt. II, pp. 272–277, September, 1960.
13. Pena, L. Q.: Designing Servocompensators Graphically, *Control Eng.*, vol. 82, pp. 79–81, January, 1964.
14. Smith, M.: Realization of Transfer Functions, *Trans. AIEE*, vol. 80, pt. II, pp. 332–336, January, 1962.

13

Cascade Compensation: Frequency-response Plots

13-1   Introduction

General methods have been developed for improving the frequency response of a feedback control system when it is analyzed by use of either the polar plot or the log plot.[1,2]   In the preceding chapter the determining factors for the selection of a lag, lead, or lag-lead network and their corresponding effects on the time response are presented.   The choice of the poles and zeros of a compensator by the use of the root-locus method is often readily determinable for any system.   The designer can place the poles and zeros of the proposed compensator on the s plane and determine the poles of the closed-loop system, which in turn permits evaluation of the closed-loop time response.   Similarly, he can determine the new values of M_m, ω_m, and K_n by the frequency-response method, using either the polar plot or the log plot.   However, the closed-loop poles are not explicitly determined. Furthermore, the correlation between the frequency-response parameters M_m and ω_m and the time response is only qualitative, as discussed in Secs. 8-4, 11-3, and 11-4. The presence of real roots near the complex dominant roots further changes the correlation between the frequency-response parameters and the time response.[3] This is a disadvantage of the frequency-response approach.

This chapter is concerned with the changes that can be made in the frequency-response characteristics by use of the three types of cascade compensators (lag, lead, and lag-lead) presented in the preceding chapter. As previously mentioned, they are only representative of the compensators that can be used. A study is made of the effect of these compensators on the overall system. Design procedures to obtain improvement of the system performance are described. With practical experience behind him, a designer can extend or modify these design procedures, which are based on the presence of one pair of complex dominant roots, to those systems where there are dominant roots besides the main complex pair.[1,4]

Both the log plots and the polar plots are utilized in this chapter in applying the frequency-response compensation criteria to show that either type of plot can be used; the choice depends on individual preference.

13-2   Reshaping the Frequency-response Plot

The performance of a closed-loop system can be described in terms of M_m, ω_m, and the static error coefficient. The value of M_m essentially describes the damping ratio and therefore the amount of overshoot in the transient response. If M_m is fixed, ω_m determines the undamped natural frequency ω_n, which in turn determines the response time of the system. The error coefficient is important because it determines the steady-state error with the appropriate standard input. The design procedure is usually based on selecting a value for M_m and using the methods described in Chap. 11 to find the corresponding values of ω_m and gain K_n required. Once this is accomplished, if the desired performance specifications are not met, compensating devices must be used. These devices alter the shape of the frequency-response plot to try to meet the performance specifications. Also, for those systems that are unstable for all values of gain it is mandatory that a stabilizing or compensating network be inserted in the system.

The reasons for reshaping the frequency-response plot generally fall into the following categories:

1. A given system is stable and its M_m and ω_m (and therefore the transient response) are satisfactory, but its steady-state error is too large.

The gain must therefore be increased to reduce the steady-state error (see Chap. 6) without appreciably altering the values of M_m and ω_m. It is shown later that in this case the high-frequency portion of the frequency-response plot is satisfactory but the low-frequency end is not.

2. A given system is stable but its transient response is unsatisfactory; that is, the M_m is set to a satisfactory value but the ω_m is too low. The gain may be satisfactory or a small increase may be desirable. The high-frequency portion of the frequency-response plot must be altered in order to increase the value of ω_m.

3. A given system is stable and has a desired M_m, but both its transient response and its steady-state response are unsatisfactory. Therefore the values of both ω_m and K_n must be increased. The portion of the frequency plot in the vicinity of the $(-1 + j0)$ or $(-180°,0\text{-db})$ point must be altered to yield the desired ω_m, and the low-frequency end must be changed to obtain the increase in gain desired.

4. A given system is unstable for all values of gain. The frequency-response plot must be altered in the vicinity of the $(-1 + j0)$ or $(-180°,0\text{-db})$ point to produce a stable system with a desired M_m and ω_m.

Thus the objective of compensation is to reshape, by means of a compensator, the frequency-response plot of the basic system to try to obtain the performance specifications. Examples which demonstrate this objective are presented in the following sections.

13-3   Selection of a Compensator

Consider the following:

$G_x(j\omega)$ = basic feedback control system forward transfer function
$G(j\omega)$ = desired over-all forward transfer function having the required stability and steady-state accuracy.

Dividing $G(j\omega)$ by $G_x(j\omega)$ gives the necessary performance requirements of the compensator:

$$G_c(j\omega) = \frac{G(j\omega)}{G_x(j\omega)} \tag{13-1}$$

The physical network for the compensator described by Eq. (13-1) can be synthesized by the techniques of Foster, Cauer, Brune, and Guillemin. In the preceding and present chapters the compensators are limited to rela-

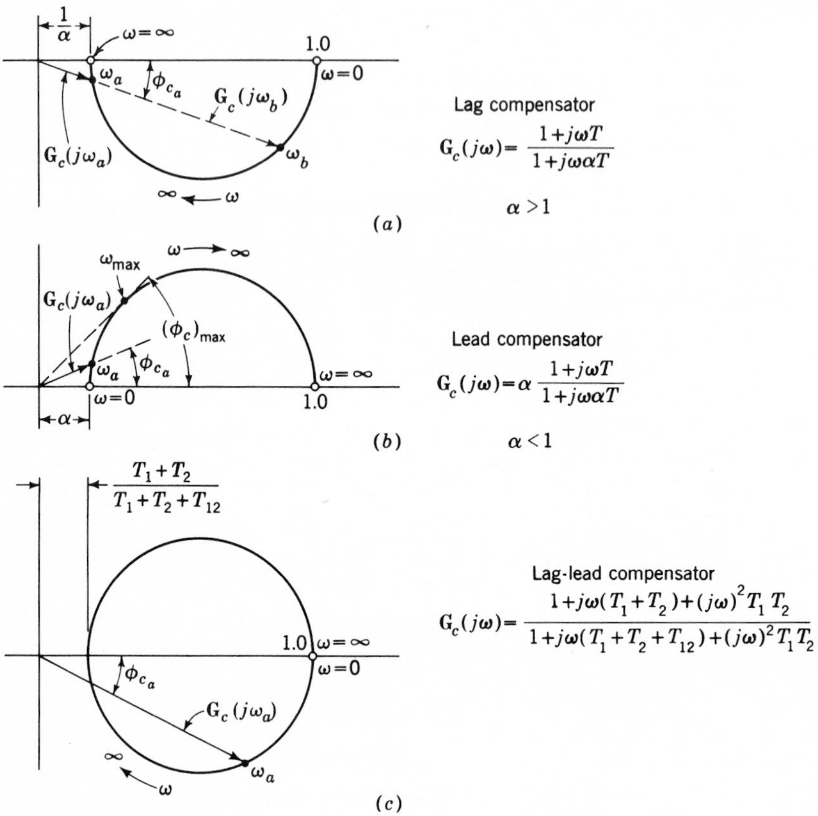

Lag compensator

$$G_c(j\omega) = \frac{1+j\omega T}{1+j\omega\alpha T}$$

$$\alpha > 1$$

(a)

Lead compensator

$$G_c(j\omega) = \alpha\frac{1+j\omega T}{1+j\omega\alpha T}$$

$$\alpha < 1$$

(b)

Lag-lead compensator

$$G_c(j\omega) = \frac{1+j\omega(T_1+T_2)+(j\omega)^2 T_1 T_2}{1+j\omega(T_1+T_2+T_{12})+(j\omega)^2 T_1 T_2}$$

(c)

Fig. 13-1 *Polar plots.* (a) *Lag compensator.* (b) *Lead compensator.* (c) *Lag-lead compensator.*

tively simple RC networks.   Based upon the design criteria presented, the parameters of these simple networks can be evaluated in a rather direct manner.   The simple networks quite often provide the desired improvement in the control system's performance.   When they do not, more complex networks must be synthesized.[5]   Physically realizable networks are limited in their performance over the whole frequency spectrum; this makes it difficult to obtain the transfer-function characteristic given by Eq. (13-1).   This is actually not a serious limitation, since the desired performance specifications are interpreted in terms of a comparatively small bandwidth of the frequency spectrum.   A compensator can be constructed to have the desired magnitude and angle characteristics over this bandwidth.

The characteristics of the lag and lead compensators and their effects upon a given polar plot are summarized in Fig. 13-1a and b and in Table 13-1.   Where the characteristics of both the lag network and the lead net-

Table 13-1. Characteristics of lag and lead compensators and their corresponding effects upon a given system

Basic compensator $G_c(s)$	Magnitude of $G_c(j\omega)$			Angle of $G_c(j\omega)$ for increasing ω	Effect on phasor $G_x(j\omega_1)$	Magnitude of new ω_m	Resultant over-all gain characteristic with compensator amplifier inserted
	$\omega = 0^+$	$\omega = \infty^+$	For increasing ω				
Lag: $\mathbf{G}_c(j\omega) = \dfrac{1 + j\omega T}{1 + j\omega\alpha T}$ $\quad \alpha > 1$	1	$\dfrac{1}{\alpha}$	Decreasing	0 to $-90°$	Rotates it clockwise	Lower	Large increase
Lead: $\mathbf{G}_c(j\omega) = \alpha\,\dfrac{1 + j\omega T}{1 + j\omega\alpha T}$ $\quad \alpha < 1$	α	1	Increasing	0 to $+90°$	Rotates it counter-clockwise	Higher	Usually a small increase

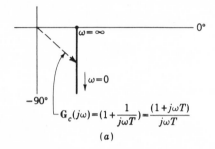

$$G_c(j\omega) = \left(1 + \frac{1}{j\omega T}\right) = \frac{(1 + j\omega T)}{j\omega T}$$

(a)

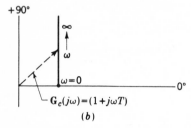

$$G_c(j\omega) = (1 + j\omega T)$$

(b)

Fig. 13-2 (a) *Ideal integral plus proportional compensator.* (b) *Ideal derivative plus proportional compensator.*

work are necessary to achieve the desired specifications, a lag-lead compensator must be used. The polar plot of this lag-lead compensator is shown in Fig. 13-1c. The corresponding log plots for each of these compensators are included with the design procedures later in the chapter.

The choice of a compensator depends on the characteristics of the given (or basic) feedback control system and the desired specifications. Since these networks are to be inserted in cascade, they must be of the (1) proportional plus integral and (2) proportional plus derivative types. In the vicinity of $\omega = \infty$ the plot of the lag compensator shown in Fig. 13-1a approximates the ideal integral plus proportional compensator shown in Fig. 13-2a. Also, in the vicinity of $\omega = 0$ the plot of the lead compensator shown in Fig. 13-1b approximates the ideal derivative plus proportional compensator shown in Fig. 13-2b. These are the frequency ranges where the proportional plus integral and proportional plus derivative controls are needed to provide the necessary compensation.

The equations of $G_c(j\omega)$ in Table 13-1, which represent the lag and the lead compensators used in this chapter, can be expressed as

$$\mathbf{G}_c(j\omega) = |\mathbf{G}_c(j\omega)|\underline{/\phi_c} \tag{13-2}$$

Solving for the angle ϕ_c yields

$$\phi_c = \tan^{-1}\omega T - \tan^{-1}\omega\alpha T = \tan^{-1}\frac{\omega T - \omega\alpha T}{1 + \omega^2\alpha T^2} \tag{13-3}$$

or

$$T^2 + \frac{\alpha - 1}{\omega\alpha\tan\phi_c}T + \frac{1}{\omega^2\alpha} = 0 \tag{13-4}$$

where $\alpha > 1$ for the lag compensator and $\alpha < 1$ for the lead compensator. It is shown in later sections that a frequency ω is specified at which a particular value of ϕ is desired. For selected values of α, ϕ_c, and ω it is possible to determine, by use of Eq. (13-4), the value T required for the compensator. For a lag compensator a small value of ϕ_c is required, whereas for a lead compensator a large value of ϕ_c is required. For a lead compensator the value ω_{max} at which the maximum phase shift occurs, for a given T and α, can be determined by setting the derivative of ϕ_c with respect to the frequency ω equal to zero. Therefore, from Eq. (13-3) the maximum angle occurs at the frequency

$$\omega_{max} = \frac{1}{T\sqrt{\alpha}} \tag{13-5}$$

Inserting this value of frequency into Eq. (13-3) gives the maximum phase shift

$$(\phi_c)_{max} = \sin^{-1}\frac{1-\alpha}{1+\alpha} \tag{13-6}$$

Typical values for a lead compensator are

α	0	0.1	0.2	0.3	0.4	0.5
$(\phi_c)_{max}$	90°	54.9°	41.8°	32.6°	25.4°	19.5°

A knowledge of the maximum phase shift is useful in the application of lead compensators.

The design procedures developed in the following sections utilize the characteristics of compensators discussed in this section.

13-4   Lag Compensator

When the M_m and ω_m (that is, the transient response) of a feedback control system are considered satisfactory but the steady-state error is too large, it is necessary to increase the gain without appreciably altering that portion of the given Lm–angle plot in the vicinity of the ($-180°$,0-db) point (category 1). This can be done by introducing an integral or lag compensator in cascade with the original forward transfer function. Shown in Fig. 12-6 is a representative lag network having the transfer function

$$G_c(s) = A\frac{1+Ts}{1+\alpha Ts} \tag{13-7}$$

where $\alpha > 1$. As an example of its application, consider a basic Type 0

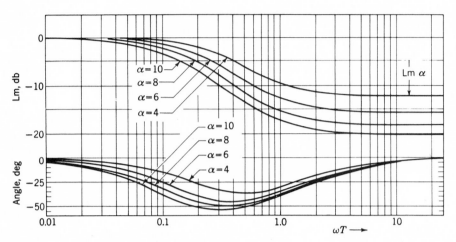

Fig. 13-3 *Log magnitude and phase angle diagram for lag compensator* $\mathbf{G}'_c(j\omega) = (1 + j\omega T)/(1 + j\omega\alpha T)$.

system. For an 'improvement in the steady-state accuracy the static step error coefficient K_0 must be increased. Its values before and after the addition of the compensator are

$$K_0 = \lim_{s \to 0} G_x(s) \tag{13-8}$$

$$K'_0 = \lim_{s \to 0} G_c(s)G_x(s) = AK_0 \tag{13-9}$$

The compensator amplifier gain A must have a value which gives the desired increase in the value of the static step error coefficient. Thus

$$A = \frac{K'_0}{K_0} \tag{13-10}$$

This increase in gain must be achieved while maintaining the same M_m and without appreciably changing the value of ω_m. However, the lag compensator has a negative angle which moves the original Lm-angle plot to the left, or closer to the $(-180°,0\text{-db})$ point. This has a destabilizing effect and reduces ω_m. To limit this decrease in ω_m, the lag compensator is designed so that it introduces a small angle, generally no more than $-5°$, at the original resonant frequency. The value of $-5°$ at $\omega = \omega_{m1}$ is an empirical value, determined from practical experience in applying this type of network. A slight variation in method, which gives equally good results, is to select the lag compensator so that the magnitude of its angle is 5° or less at the original phase-margin frequency.

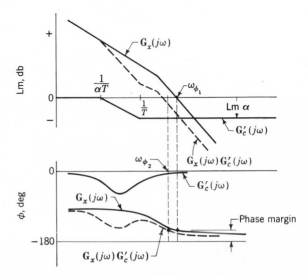

Fig. 13-4 *Original and lag-compensated curves.*

The over-all characteristics of lag compensators can best be visualized from the log plot. The transfer function in terms of ω is

$$G_c(j\omega) = A\,\frac{1 + j\omega T}{1 + j\omega \alpha T} \qquad (13\text{-}11)$$

The log magnitude and phase angle equations for the compensator are

$$\text{Lm } G_c'(j\omega) = \text{Lm } (1 + j\omega T) - \text{Lm } (1 + j\omega \alpha T) \qquad (13\text{-}12)$$
$$\underline{/G_c(j\omega)} = \underline{/1 + j\omega T} - \underline{/1 + j\omega \alpha T} \qquad (13\text{-}13)$$

For various values of α there is a family of curves for the log magnitude and phase angle diagram, as shown in Fig. 13-3. Engineers who use this compensator repeatedly may find it convenient to make templates of these curves. From the shape of these curves it is seen that an attenuation equal to Lm α has been introduced above $\omega T = 1$. Assume that a system has the original forward transfer function $G_x(j\omega)$ and that the gain has been adjusted for the desired phase margin. The log magnitude and phase angle diagram of $G_x(j\omega)$ is sketched in Fig. 13-4. The addition of the lag compensator reduces the log magnitude by Lm α for the frequencies above $\omega = 1/T$. The value of T of the compensator is selected so that the attenuation Lm α occurs at the original phase-margin frequency $\omega_{\phi 1}$. Notice that the phase margin at $\omega_{\phi 1}$ has been reduced. To maintain the specified phase margin, the phase-margin frequency has been reduced to $\omega_{\phi 2}$. It is now possible to increase the gain of $G_x(j\omega)G_c'(j\omega)$ by the value A in order that the Lm curve will have the value 0 db at the frequency $\omega_{\phi 2}$.

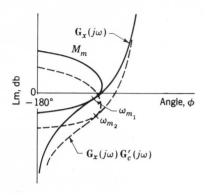

Fig. *13-5* *Original and lag-compensated curves.*

The effects of the lag compensator can now be analyzed. The reduction in Lm at the higher frequencies due to the compensator is desirable. This permits an increase in the gain to maintain the desired phase margin. Unfortunately, the negative angle of the compensator lowers the angle curve. To maintain the desired phase margin, the phase-margin frequency is reduced. To keep this reduction small, the value of $\omega_{cf} = 1/T$ should be made as small as possible.

The log magnitude–angle diagram permits gain adjustment for the desired M_m. The curves for $G_x(j\omega)$ and $G_x(j\omega)G'_c(j\omega)$ are shown in Fig. 13-5. The increase in gain to obtain the same M_m is the amount that the curve of $G_x(j\omega)G'_c(j\omega)$ must be raised in order to be tangent to the M_m curve. Because of the negative angle introduced by the lag compensator, the new resonant frequency ω_{m2} is smaller than the original resonant frequency ω_{m1}. Provided that the decrease in ω_m is small and the increase in gain is sufficient to meet the specifications, the system can be considered satisfactorily compensated.

In summary, the lag compensator is basically a low-pass filter; the low frequencies are passed and the higher frequencies are attenuated. The attenuation characteristic of the lag compensator is useful and permits an increase in the gain. The negative-phase-shift characteristic is detrimental to system performance but must be tolerated. Because the predominant and useful feature of the compensator is attenuation, a more appropriate name for it is *high-frequency attenuation compensator.*

13-5   Example of Lag Compensation

To give a basis for comparison of all methods of compensation, the examples of Chap. 12 are used in this chapter. For the unity feedback Type 1 system of Sec. 12-4 the forward transfer function is

$$G_x(s) = \frac{K_1}{s(1 + s)(1 + 0.2s)} \tag{13-14}$$

By the methods of Secs. 11-8 and 11-11, the gain required for this system to have an M_m of 1.25 (1.94 db) is found to be $K_1 = 0.88$ sec^{-1}, and the resulting resonant frequency is $\omega_{m1} = 0.72$ radian/sec. With these values the transient response is considered to be satisfactory, but the steady-state error is too large. It has been pointed out previously that putting a lag compensator in cascade with a basic system (falling into category 1) permits an increase in gain and therefore reduces the steady-state error.

The cascade compensator to be used has the transfer function

$$G_c'(j\omega) = \frac{1 + j\omega T}{1 + j\omega\alpha T} \tag{13-15}$$

Selection of an $\alpha = 10$ means that an increase in gain of almost 10 is desired. Based on the criterion discussed in the last section (a phase shift of $-5°$ at the original ω_m), the compensator time constant can be determined. Note that in Fig. 13-3, $\phi_c = -5°$ occurs for values of ωT approximately equal to 0.01 and 10. These two points correspond, respectively, to points ω_b and ω_a in Fig. 13-1a. The choice $\omega T = 10 = \omega_{m1}T$ (or $\omega_a = \omega_{m1}$) is made since this provides the maximum attenuation at ω_{m1}. This ensures a maximum possible gain increase while maintaining the desired M_m. Thus for $\omega_{m1} = 0.72$ the value $T \approx 13.9$ is obtained. Note that the value of T can also be obtained from Eq. (13-4). For ease of calculation, the value of the time constant is rounded off to 14.0, which does not appreciably affect the results.

In Figs. 13-6 and 13-7 the log magnitude and phase angle diagrams for the compensator are shown, together with the curves for $G_x'(j\omega)$. The sum of the curves representing the function $G_x'(j\omega)G_c'(j\omega)$ is also shown.

The new curves of $G_x'(j\omega)G_c'(j\omega)$ from Figs. 13-6 and 13-7 are used to draw the log magnitude–angle diagram of Fig. 13-8. By using this curve, the gain is adjusted to obtain an $M_m = 1.25$. The results are $K_1' = 6.68$ and $\omega_m = 0.54$. Thus an improvement in K_1 has been obtained, but a penalty of a smaller ω_m results. An increase in gain of approximately $A = 7.6$ has resulted from the use of a lag compensator with $\alpha = 10$, which is typical of the gain increase permitted by a given α. In actual practice this additional gain may be obtained from the amplifier in the basic system if it is not already operating at maximum gain. Note that the value of gain is somewhat less than the chosen value of α. Provided that the decrease in ω_m and the increase in gain are acceptable, the system has been suitably compensated.

To further show the effects of compensation, the polar plots representing the basic and compensated systems are shown in Fig. 13-9. Note that there

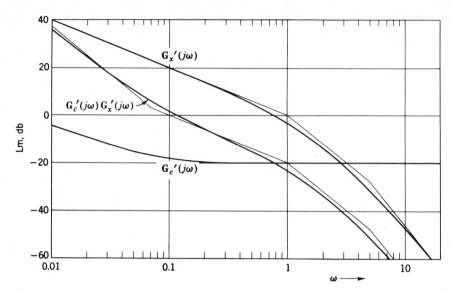

Fig. 13-6 *Log magnitude diagram for*

$$\mathbf{G}'_x(j\omega) = \frac{1}{j\omega(1 + j\omega)(1 + j0.2\omega)}$$

$$\mathbf{G}'_c(j\omega)\mathbf{G}'_x(j\omega) = \frac{1}{j\omega(1 + j\omega)(1 + j0.2\omega)} \frac{1 + j14\omega}{1 + j140\omega}$$

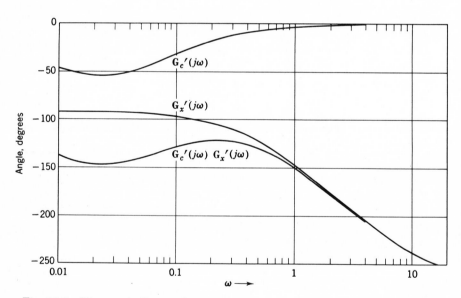

Fig. 13-7 *Phase angle diagram for*

$$\mathbf{G}'_x(j\omega) = \frac{1}{j\omega(1 + j\omega)(1 + j0.2\omega)}$$

$$\mathbf{G}'_c(j\omega)\mathbf{G}'_x(j\omega) = \frac{1}{j\omega(1 + j\omega)(1 + j0.2\omega)} \frac{1 + j14\omega}{1 + j140\omega}$$

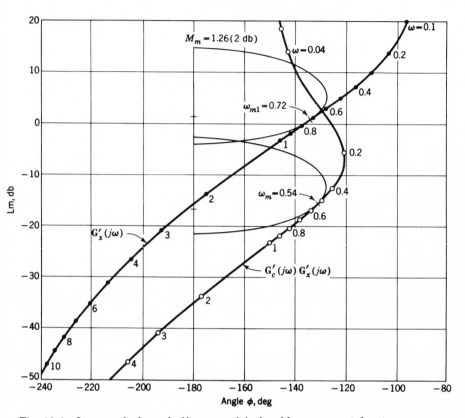

Fig. 13-8 *Log magnitude–angle diagram: original and lag-compensated system.*

Table 13-2. , *Comparison between root-locus and frequency methods of applying a lag compensator to a basic control system*

Method	$G'_c(s)$	ω_{m1}	ω_m	ω_{n1}	ω_n	K_1	K'_1	Damping ratio (held constant)	Maximum peak overshoot
Root locus	$\dfrac{1 + 10s}{1 + 100s}$	—	—	0.90	0.80	0.83	7.5	ζ of dominant roots = 0.45	$M_p = 1.24$ $M_m = 1.41$
Frequency plot	$\dfrac{1 + 14s}{1 + 140s}$	0.72	0.54	—	—	0.88	6.68	Effective $\zeta = 0.45$	$M_m = 1.26$
Frequency plot	$\dfrac{1 + 10s}{1 + 100s}$	0.72	0.50	—	—	0.88	6.17	Effective $\zeta = 0.45$	$M_m = 1.26$

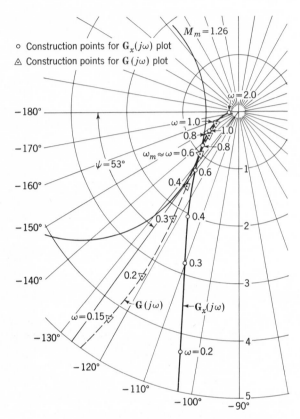

Fig. 13-9 *Polar plots of a Type 1 control system, with and without lag compensation:*

$$G_x(j\omega) = \frac{0.88}{j\omega(1 + j\omega)(1 + j0.2\omega)}$$

$$G(j\omega) = G_x(j\omega)G_c(j\omega)$$

$$= \frac{6.45(1 + j14\omega)}{j\omega(1 + j\omega)(1 + j0.2\omega)(1 + j140\omega)}$$

is essentially no rotation of the high-frequency portion of $G(j\omega)$ compared with $G_x(j\omega)$. However, in the low-frequency region there is a clockwise rotation of $G(j\omega)$ due to the lag compensator. The gain K_1 and the resonant frequency ω_m shown in Fig. 13-9 differ from those obtained from the log plots. This difference is due to the graphical accuracy of working with the polar or log plots.

Table 13-2 presents a comprehensive comparison between the root-locus method and the frequency-plot method of applying a lag compensator to a basic control system. Note the similarity of results obtained by these methods. The selection of the method depends upon individual preference. The compensators developed in the root-locus and frequency-plot methods are different, but the difference in the results is small.   To show this explicitly, Table 13-2 includes the case of the compensator developed by the root-locus method and applied to the frequency-response method.   In the frequency-response method, either the log or the polar plot may be used.

13-6   Lead Compensation

Figure 12-16 shows a lead compensator made up of passive elements which has the transfer function

$$\mathbf{G}'_c(j\omega) = \alpha \frac{1 + j\omega T}{1 + j\omega\alpha T} \qquad \alpha < 1 \tag{13-16}$$

This equation is marked with a prime since it does not contain the gain A. The log magnitude and phase angle equations for this compensator are

$$\text{Lm } \mathbf{G}'_c(j\omega) = \text{Lm } \alpha + \text{Lm } (1 + j\omega T) - \text{Lm } (1 + j\omega\alpha T) \tag{13-17}$$
$$\underline{/\mathbf{G}'_c(j\omega)} = \underline{/1 + j\omega T} - \underline{/1 + j\omega\alpha T} \tag{13-18}$$

A family of curves for various values of α is shown in Fig. 13-10. It may be convenient to make templates of these log magnitude and phase angle curves

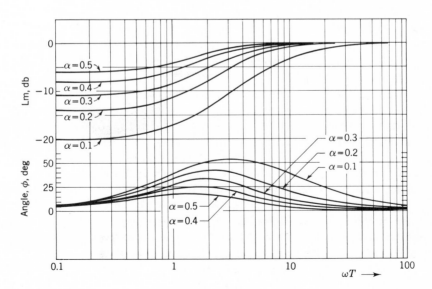

Fig. 13-10   *Log magnitude and phase angle diagram for lead compensator*

$$\mathbf{G}'_c(j\omega) = \alpha \frac{1 + j\omega T}{1 + j\omega\alpha T}$$

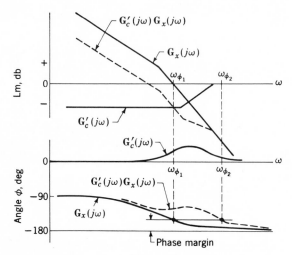

Fig. 13-11 *Original and lead-compensated curves.*

for the lead compensator. It is seen from the shape of the log magnitude curves that an attenuation equal to Lm α has been introduced at frequencies below $\omega T = 1$. Thus the lead network is basically a high-pass filter: the high frequencies are passed and the low frequencies are attenuated. Also, an appreciable lead angle is introduced in the frequency range from $\omega = 1/T$ to $\omega = 1/\alpha T$ ($\omega T = 1$ to $\omega T = 10$). Because of its angle characteristic a lead network can be used to improve the performance of a system that falls in category 2.

Application of the lead compensator can be based on adjusting the phase margin and the phase-margin frequency. Assume that a system has the original forward transfer function $\mathbf{G}_x(j\omega)$ and that the gain has been adjusted for the desired phase margin or for the desired M_m. The log magnitude and phase angle diagram of $\mathbf{G}_x(j\omega)$ is sketched in Fig. 13-11. The purpose of the lead compensator is to increase the phase-margin frequency and therefore to increase ω_m. The lead compensator introduces a positive angle over a relatively narrow bandwidth. By properly selecting the value of T, the phase-margin frequency can be increased from $\omega_{\phi 1}$ to $\omega_{\phi 2}$. Selection of T can be accomplished by physically placing the angle curve for the compensator on the same graph with the angle curve of the original system. The location of the compensator angle curve must be such as to produce the specified phase margin at the highest possible frequency; this location determines the value of the time constant T. The gain of $\mathbf{G}_x(j\omega)\mathbf{G}'_c(j\omega)$ must be increased for the Lm curve to have the value 0 db at the frequency $\omega_{\phi 2}$. An analysis of the Lm curve of Fig. 13-10 shows that, for a given α, the farther to the right

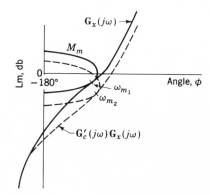

Fig. 13-12 *Original and lead-compensated log magnitude–angle curves.*

the compensator curves are placed (i.e., the smaller T is made), the larger is the new gain of the system.

It is also possible to use the criterion derived in the previous chapter for the selection of T. This criterion is to select T, for Type 1 or higher systems, equal to or slightly smaller than the largest time constant of the original forward transfer function. This is the procedure used in the example of the next section. For a Type 0 system the compensator time constant T is made equal to or slightly smaller than the second largest time constant of the original system. Several locations of the compensator curves should be tested and the best results selected.

More accurate application of the lead compensator is based on adjusting the gain to obtain a desired M_m by use of the log magnitude–angle diagram. Figure 13-12 shows the original curve $\mathbf{G}_x(j\omega)$ and the new curve $\mathbf{G}_x(j\omega)\mathbf{G}'_c(j\omega)$. The increase in gain to obtain the same M_m is the amount that the curve $\mathbf{G}_x(j\omega)\mathbf{G}'_c(j\omega)$ must be raised to be tangent to the M_m curve. Because of the positive angle introduced by the lead compensator, the new resonant frequency ω_{m2} is larger than the original resonant frequency ω_{m1}. The increase in gain is not as large as that obtained by use of the lag compensator.

13-7   Example of Lead Compensation Applied to a Type 1 System

For the unity feedback Type 1 system of Sec. 12-4 the forward transfer function is

$$G_x(s) = \frac{K_1}{s(1 + s)(1 + 0.2s)} \tag{13-19}$$

From Sec. 13-5 it is found that, for $M_m = 1.25$, $K_1 = 0.88$ sec^{-1} and $\omega_{m1} = 0.72$ radian/sec. In this example the value of ω_m is considered to be

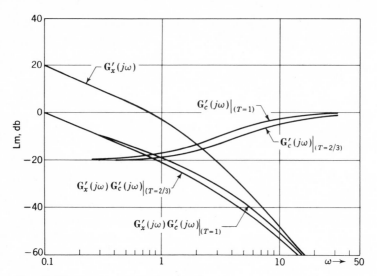

Fig. 13-13 *Log magnitude diagram of*

$$\mathbf{G}'_x(j\omega) = \frac{1}{j\omega(1 + j\omega)(1 + j0.2\omega)}$$

$$\mathbf{G}'_x(j\omega)\mathbf{G}'_c(j\omega) = \frac{1}{j\omega(1 + j\omega)(1 + j0.2\omega)} \frac{0.1(1 + j\omega T)}{1 + j0.1\omega T}$$

$$T = 1 \qquad T = \tfrac{2}{3}$$

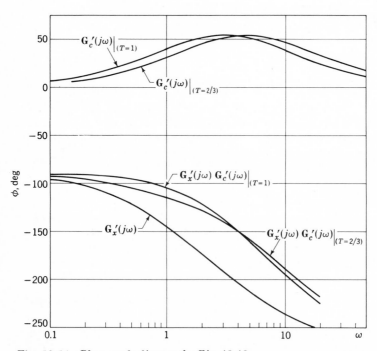

Fig. 13-14 *Phase angle diagram for Fig. 13-13.*

too low. From previous considerations, putting a lead compensator in cascade with the basic system increases the value of ω_m and therefore improves the time response of the system. This situation falls in category 2.

The choice of values of α and T of the compensator must be such that its resulting angle ϕ_c, at frequency ω_{m1}, adds a sizable positive phase shift to the over-all forward transfer function. The nominal value of $\alpha = 0.1$ is often utilized; thus only the determination of the value of T remains. The values of T selected in Sec. 12-6 give satisfactory results. Therefore these values of T are used in this example in order that a comparison can be made between the root-locus and frequency-response techniques. This example is worked with only two of the three values of T used in Sec. 12-6. The results are used to establish the design criteria for lead compensators.

In Figs. 13-13 and 13-14 are drawn the log magnitude and phase angle diagrams for the basic system $G'_x(j\omega)$, the two compensators $G'_c(j\omega)$, and the two composite transfer functions $G'_x(j\omega)G'_c(j\omega)$. Using the curves of $G'_x(j\omega)G'_c(j\omega)$ from Figs. 13-13 and 13-14, the log magnitude–angle diagrams are drawn in Fig. 13-15. The gain is again adjusted to obtain an $M_m = 1.25$

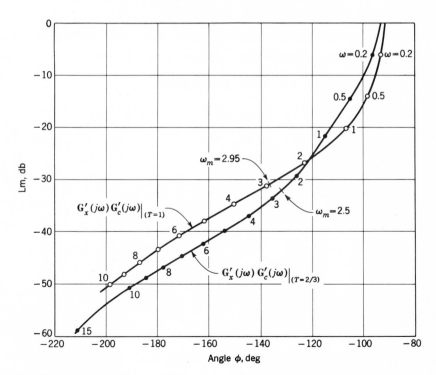

Fig. 13-15 *Log magnitude–angle diagrams with lead compensators from Figs. 13-13 and 13-14.*

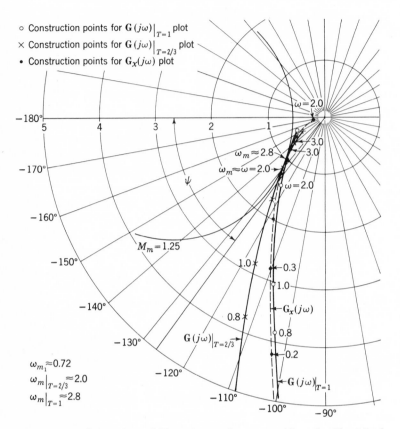

○ Construction points for $\mathbf{G}(j\omega)\big|_{T=1}$ plot
× Construction points for $\mathbf{G}(j\omega)\big|_{T=2/3}$ plot
• Construction points for $\mathbf{G}_x(j\omega)$ plot

$\omega_{m_1} \approx 0.72$
$\omega_m\big|_{T=2/3} \approx 2.0$
$\omega_m\big|_{T=1} \approx 2.8$

Fig. 13-16 *Polar plots of Type 1 control system with and without lead compensation:*

$$\mathbf{G}_x(j\omega) = \frac{0.87}{j\omega(1 + j\omega)(1 + j0.2\omega)}$$

$$\mathbf{G}(j\omega)\Big]_{T=\frac{2}{3}} = \frac{3.42(1 + j2\omega/3)}{j\omega(1 + j\omega)(1 + j0.2\omega)(1 + j0.2\omega/3)}$$

$$\mathbf{G}(j\omega)\Big]_{T=1} = \frac{3.15}{j\omega(1 + j0.2\omega)(1 + j0.1\omega)}$$

(1.94 db). The 2-db M curve is used and gives sufficiently accurate results. The corresponding polar plots of the original system $\mathbf{G}_x(j\omega)$ and the compensated systems $\mathbf{G}(j\omega) = \mathbf{G}_x(j\omega)\mathbf{G}_c(j\omega)$ are shown in Fig. 13-16. The different gain and ω_m shown in this figure are due to the graphical accuracies achieved with the polar-plot technique.

Table 13-3 presents a comprehensive comparison between the root-locus method and the frequency-response method of applying a lead compensator to a basic control system. The log-plot results show that a lead compensator

Table 13-3. **Comparison between root-locus and frequency-response methods of applying a lead compensator to a basic control system**

Method	$G'_c(s)$	ω_{m1}	ω_m	ω_{n1}	ω_n	K_1	K_1'	Damping ratio (held constant)	Maximum peak overshoot
Root locus	$0.1\dfrac{1+s}{1+0.1s}$	—	—	0.90	3.52	0.83	2.98	ζ of dominant roots = 0.45	$M_p = 1.19$ $M_m = 1.22$
	$0.1\dfrac{1+0.667s}{1+0.0667s}$	—	—	0.90	3.40	0.83	4.38		$M_p = 1.24$ $M_m = 1.37$
Frequency plot	$0.1\dfrac{1+s}{1+0.1s}$	0.72	2.95	—	—	0.88	3.10	Effective ζ = 0.45	$M_m = 1.25$
	$0.1\dfrac{1+0.667s}{1+0.0667s}$	0.72	2.5	—	—	0.88	4.30		$M_m = 1.25$

increases the resonant frequency ω_m. However, a range of values of ω_m is possible, depending on the compensator time constant used. The larger the value of ω_m, the smaller the value of K_1. The characteristics selected must be based on the system specifications. In each case the selection of the compensator must be based on the performance desired. The increase in gain is not as large as that obtained with a lag compensator. Note in the table the similarity of results obtained by the root-locus and frequency-response methods. Because of this similarity, the design rules of Sec. 12-6 can be interpreted in terms of the frequency-response method as follows.

Rule 1

The time constant T in the numerator of $\mathbf{G}_c(j\omega)$ given by Eq. (13-16) should be set equal to the value of the largest time constant in the denominator of the original transfer function for a Type 1 or higher system. This usually results in the largest increase in ω_m (best time response), and there is an increase in system gain K_n.

Rule 2

The time constant T in the numerator of $\mathbf{G}_c(j\omega)$ should be set to a value slightly smaller than the largest time constant in the denominator of the original transfer function for a Type 1 or higher system. This results in achieving the largest gain increase for the system with an appreciable increase in ω_m.

If the maximum improvement in the time response is desired with whatever gain increase is obtainable, Rule 1 is applied to the design of the lead compensator. Where maximum gain increase and a good improvement in the time of response are desired, Rule 2 is applicable.

For a Type 0 system, Rules 1 and 2 are modified so that the lead-compensator time constant is selected either equal to or slightly smaller than the second largest time constant of the original system.

Remember that the lag and lead networks are designed for the same basic system; thus from Tables 13-2 and 13-3 it is seen that both result in a gain increase. The distinction between the two types of compensation is that the lag network gives the largest increase in gain (with the best improvement in steady-state accuracy) at the expense of increasing the response time, whereas the lead network gives an appreciable improvement in the time of response and a small improvement in steady-state accuracy. The particular problem at hand dictates the type of compensation to be used.

13-8   Lag-Lead Compensator

The previous sections have demonstrated that the introduction of a lag compensator results in an increase in the gain and a lead compensator results in an increase in the resonant frequency ω_m. These changes in the time response reduce the steady-state error and improve the response time, respectively. If both changes are required, then both a lag and a lead compensator must be utilized. This improvement may be accomplished by inserting the individual lag and lead networks in cascade. However, it is more economical in equipment to use a new network that has both the lag and the lead characteristics. Such a network, shown in Fig. 12-24, is called a lag-lead compensator. Its transfer function is

$$\mathbf{G}_c(j\omega) = \frac{A(1 + j\omega T_1)(1 + j\omega T_2)}{(1 + j\omega\alpha T_1)(1 + j\omega T_2/\alpha)} \tag{13-20}$$

where $\alpha > 1$ and $T_1 > T_2$. The first half of this transfer function produces the lag effect and the second half produces the lead effect.

Based upon an individual's experience, various design procedures may be developed for a lag-lead compensator. In this section two "typical" design procedures are given for the use of beginners. These procedures can be varied to suit the needs of the designer.

One method is first to draw the lag-lead compensator curves for a given value of α and a ratio of T_1 to T_2 that has been found empirically to be satisfactory. One such set of curves is drawn in Fig. 13-17 for a value $\alpha = 10$, $T = T_2$, and $T_1 = 5T_2$. The curves for the composite lag-lead

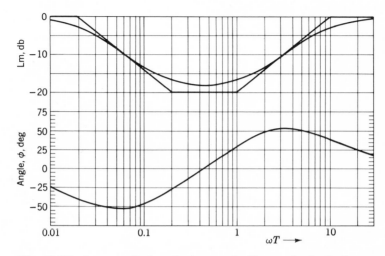

Fig. 13-17 *Log magnitude and phase angle diagram for lag-lead compensator*

$$\mathbf{G}'_c(j\omega) = \frac{(1 + j5\omega T)(1 + j\omega T)}{(1 + j50\omega T)(1 + j0.1\omega T)}$$

compensator are placed on and added to the log magnitude and phase angle diagram of the original transfer function.   The selection of the position for the compensator curves can be based on making T_2 equal to an appropriate time constant in the denominator of the original system (based upon the system type, as discussed in the previous section).   An alternative procedure is to locate the compensator curves on the log plots of the original system to produce the highest possible phase-margin frequency.   Figure 13-18 shows a sketch of the log magnitude and phase angle diagram for the original forward transfer function $\mathbf{G}_x(j\omega)$, the compensator $\mathbf{G}'_c(j\omega)$, and the combination $\mathbf{G}_x(j\omega)\mathbf{G}'_c(j\omega)$.   The phase-margin frequency has been increased from $\omega_{\phi 1}$ to $\omega_{\phi 2}$, and the additional gain is labeled Lm A.   The new value of M_m can be obtained from the log magnitude–angle diagram in the usual manner.

The second method is simply the combination of the design procedures given for the use of the lag and lead networks, respectively.   Basically, the log magnitude and phase angle curves of the lag compensator can be located as described in Sec. 13-4.   The lag-compensator curves are given by Fig. 13-3 and are located so that the negative angle introduced at either the original phase-margin frequency $\omega_{\phi 1}$ or the original resonant frequency ω_{m1} is small—of the order of $-5°$.   This permits determination of the time constant T_1.   The log magnitude and phase angle curves of the lead compensator can be located as described in Sec. 13-6.   The lead-compensator curves are given by Fig. 13-10 except that the log magnitude curve should be

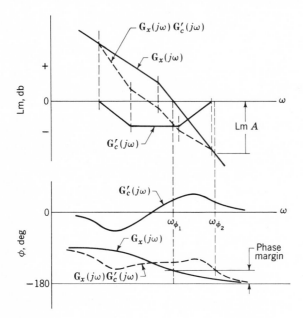

Fig. 13-18 *Original and lag-lead compensated curves.*

raised to have a value of 0 db at the low frequencies.   This is necessary because the lag-lead compensator transfer function does not contain α as a factor.   Either the time constant T_2 of the compensator can be made equal to the largest time constant of the original system (for a Type 1 or higher system), or the angle curve is located to produce the specified phase margin at the highest possible frequency.   In the latter case, the location chosen for the angle curve permits determination of the time constant T_2.   To utilize the lag-lead network of Fig. 12-24, the value of α used for the lag compensation must be the reciprocal of the α used for the lead compensation.   The example given in the next section utilizes this second method since it is more flexible.

13-9   *Example of Lag-Lead Compensation*

For the basic system treated in Secs. 13-5 and 13-7, the forward transfer function is

$$\mathbf{G}_x(j\omega) = \frac{K_1}{j\omega(1 + j\omega)(1 + j0.2\omega)} \tag{13-21}$$

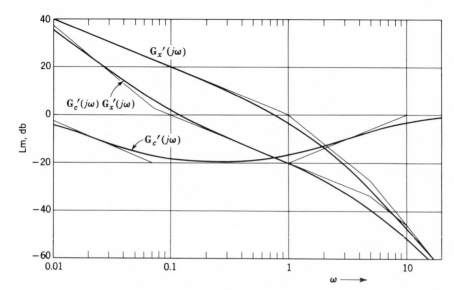

Fig. 13-19   *Log magnitude diagrams of Eqs. (13-21) to (13-23).*

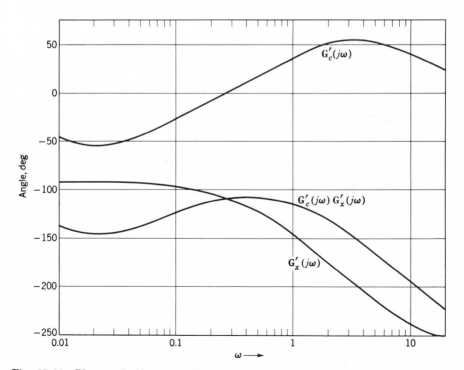

Fig. 13-20   *Phase angle diagrams of Eqs. (13-21) to (13-23).*

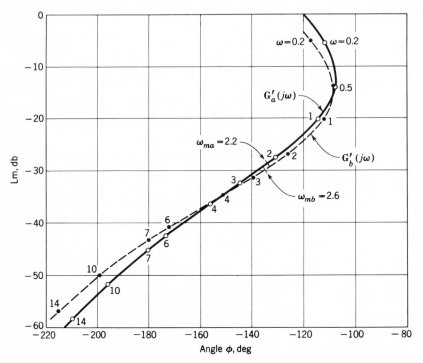

Fig. 13-21   *Log magnitude–angle diagram of $G'(j\omega)$ from Eq. (13-23).*

For an $M_m = 1.25$ the result is $K_1 = 0.88$ and $\omega_{m1} = 0.72$ radian/sec. Based upon the second method described above, the lag-lead compensator is the combination of the individual lag and lead compensators designed in Secs. 13-5 and 13-7.   Thus the compensator transfer function is

$$G_c(j\omega) = A\,\frac{(1 + j14\omega)(1 + j\omega)}{(1 + j140\omega)(1 + j0.1\omega)} \qquad \text{for } \alpha = 10 \qquad (13\text{-}22)$$

The new forward transfer function is

$$G(j\omega) = G_x(j\omega)G_c(j\omega) = \frac{AK_1(1 + j14\omega)}{j\omega(1 + j0.2\omega)(1 + j140\omega)(1 + j0.1\omega)} \qquad (13\text{-}23)$$

In Figs. 13-19 and 13-20 are drawn the log magnitude and phase angle diagrams of the basic system $G'_x(j\omega)$, the compensator $G'_c(j\omega)$, and the composite transfer function $G'_x(j\omega)G'_c(j\omega)$.   Using these curves, the log magnitude–angle diagram is drawn in Fig. 13-21.   By adjusting for an $M_m = 1.25$, the results are $K_1' = 26.3$ and $\omega_m = 2.2$. When these values are compared with those of Tables 13-2 and 13-3 it is seen that lag-lead compensation results in a value of K_1 approximately equal to the product of the K_1's of the lag- and lead-compensated systems. Also, the ω_m is only slightly less than that obtained with the lead-compensated system. Thus a larger increase in gain is achieved than with lag compensation, and the value of ω_m is almost as large as the value obtained by using the lead-compensated system. One may therefore be inclined to use a lag-lead compensator exclusively.

The resulting error coefficient (gain) due to the insertion of a lag-lead compensator is approximately equal to the product of the following:

1. The original error coefficient K_n of the uncompensated system
2. The increase in K_n due to the insertion of a lag compensator
3. The increase in K_n due to the insertion of a lead compensator

Table 13-4 presents a comparison of the root-locus and log-plot methods of applying a lag-lead compensator to a basic control system. The compensator designed from the log plots is different from the one designed by the root-locus method. To show that the difference in performance is small, Table 13-4 includes both compensators applied to the frequency plots. Both plots are shown in Fig. 13-21. By referring to Table 12-2, it is seen that the ω_n for the lag-lead compensator obtained from the root locus is a little smaller than that obtained with a lead compensator. Similarly, in the frequency-response method it should be expected that the same would be true for the values of ω_m. This is confirmed in Table 13-5, which gives a composite summary of the results achieved with all three types of compensators.

Table 13-4. Comparison between root-locus and frequency-response methods of applying a lag-lead compensator to a basic control system

Method	$G_c'(s)$	ω_{m1}	ω_m	ω_{n1}	ω_n	K_1	K_1'	Damping ratio (held constant)	Maximum peak overshoot
Root locus	$\dfrac{(1+s)(1+10s)}{(1+0.1s)(1+100s)}$	—	—	0.90	3.45	0.83	29.0	ζ of dominant roots = 0.45	$M_p = 1.23$ $M_m = 1.22$
Frequency plot	$\dfrac{(1+s)(1+10s)}{(1+0.1s)(1+100s)}$	0.72	2.6	—	—	0.88	28.8	Effective ζ = 0.45	$M_m = 1.25$
Frequency plot	$\dfrac{(1+s)(1+14s)}{(1+0.1s)(1+140s)}$	0.72	2.2	—	—	0.88	26.3	Effective ζ = 0.45	$M_m = 1.25$

Table 13-5. Summary of results of lag, lead, and lag-lead compensation of a basic control system using the frequency-response method

Compensator	ω_m	K_1	Effective ζ	M_m	Additional gain required
Uncompensated:	0.72	0.88	0.45	1.25	—
Lag:					
$\dfrac{1 + 14s}{1 + 140s}$	0.54	6.68	0.45	1.25	7.6
$\dfrac{1 + 10s}{1 + 100s}$	0.50	6.17	0.45	1.25	7.0
Lead:					
$0.1\,\dfrac{1 + s}{1 + 0.1s}$	2.95	3.10	0.45	1.25	35.2
$0.1\,\dfrac{1 + 2s/3}{1 + 0.2s/3}$	2.5	4.30	0.45	1.25	48.9
Lag-lead:					
$\dfrac{(1 + s)(1 + 14s)}{(1 + 0.1s)(1 + 140s)}$	2.2	26.3	0.45	1.25	29.9
$\dfrac{(1 + s)(1 + 10s)}{(1 + 0.1s)(1 + 100s)}$	2.6	28.8	0.45	1.25	32.7

13-10   *Example of Lead Compensation Applied to a Type 2 System*

The polar plot $\mathbf{G}_x(j\omega)$ of a Type 2 system which is completely unstable is shown in Fig. 13-22.   This case falls in category 4, and only a lead compensator can be used to achieve the desired stability.   The maximum speed of response that can be obtained after compensation of this system is determined by the minimum value that α can have in the transfer function

$$G_c(s) = A\alpha\,\frac{1 + Ts}{1 + \alpha Ts} \qquad (13\text{-}24)$$

An α much less than 0.1 is not practical.   This factor is determined by the maximum phase shift that a lead network can produce, as observed in Figs. 13-1b and 13-10.   This maximum phase shift is given by Eq. (13-6), and typical values are given in Sec. 13-3.

Compensation methods based on the frequency response can be adapted for use with either the polar plot or log plots.   The polar plot is used here

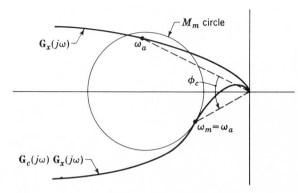

Fig. 13-22   *Polar plot of an unstable Type 2 system,* $\mathbf{G}_x(j\omega)$, *and the stabilized transfer function* $\mathbf{G}_c(j\omega)\mathbf{G}_x(j\omega)$.

as an illustration; the log plots are equally applicable, but the procedure would require some modification.

The first step in the design procedure requires drawing $\mathbf{G}_x(j\omega)$ and the desired M_m circle.   The new approximate shape of the compensated transfer function $\mathbf{G}(j\omega) = \mathbf{G}_c(j\omega)\mathbf{G}_x(j\omega)$ is then sketched, as shown in Fig. 13-22, with an estimated point of tangency to the M_m circle.   For the system to be stable, the point of tangency usually occurs in the third quadrant.   The desired resonant frequency $\omega_a = \omega_m$ must fall at this point of tangency.   The necessary phase shift ϕ_c to be introduced by the lead compensator at the frequency ω_a can be measured from the graph, as shown in Fig. 13-22. Appropriate values of α, T, and A for the compensator are then selected. This design procedure is based on practical experience and gives satisfactory results.   As one becomes more versed in compensation techniques, satisfactory variations in this procedure may become evident.   The detailed procedure is shown in the following example.   Consider the system having the forward transfer function

$$\mathbf{G}_x(j\omega) = \frac{1.5}{(j\omega)^2(1 + j0.2\omega)} \tag{13-25}$$

The specifications for the compensated system are $M_m = 1.3$ and $\omega_m = 0.8$ radian/sec.

Case 1.   (First trial)

After the plot of $\mathbf{G}_x(j\omega)$ is drawn in Fig. 13-23, the $M_m = 1.3$ circle is drawn. From the graph, for $\omega_a = 0.8$,

$$\mathbf{G}_x(j0.8) = 2.3\underline{/-190°} \tag{13-26}$$

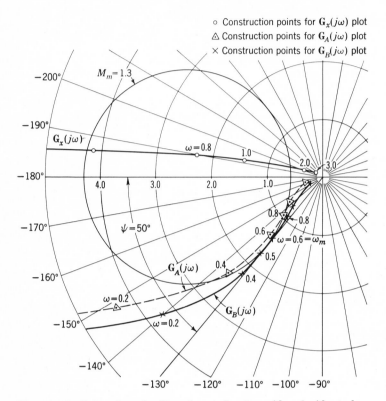

Fig. 13-23 *Polar plots of a Type 2 control system with and without phase lead compensation:*

$$G_x(j\omega) = \frac{1.5}{(j\omega)^2(1 + j0.2\omega)}$$

$$G_A(j\omega) = \frac{0.205(1 + j4.0\omega)}{(j\omega)^2(1 + j0.2\omega)(1 + j0.4\omega)} \qquad \alpha = 0.1, \; \omega_m = 0.6, \; M_m = 1.35$$

$$G_B(j\omega) = \frac{0.153(1 + j5.55\omega)}{(j\omega)^2(1 + j0.2\omega)(1 + j0.5\omega)} \qquad \alpha = 0.09, \; \omega_m = 0.6, \; M_m \approx 1.3$$

After inspection of the plot of $G_x(j\omega)$ with respect to the approximate location of the compensated plot, the decision is made that the maximum phase shift obtainable from a lead network will be utilized at $\omega = 0.8$.   By selecting an $\alpha = 0.1$ and assuming $\omega_m = \omega_a = 0.8$, the angle at the tangency point is determined as follows:

$$\underline{/G_A(j\omega_a)} = \underline{/G_c(j\omega_a)} + \underline{/G_x(j\omega_a)} = 55° - 190° = -135°$$

This angle crosses the $M = 1.3$ circle at a magnitude of unity: therefore $G_A(j0.8) = 1.0\underline{/-135°}$.   Enough data are now known to determine the values of A and T of the compensator. Thus, from Eq. (13-5),

$$T = \frac{1}{\omega_a \sqrt{\alpha}} = \frac{1}{0.8 \sqrt{0.1}} \approx 4.0 \tag{13-27}$$

For $\omega = \omega_m = \omega_x = 0.8$,

$$\begin{aligned}
\mathbf{G}_c(j\omega) &= \frac{\mathbf{G}(j\omega_m)}{\mathbf{G}_x(j\omega)} = A\alpha \, \frac{1 + j\omega T}{1 + j\omega\alpha T} \\
&= \frac{1.0/-135°}{2.3/-190°} = 0.1A \, \frac{1 + j(0.8)(4.0)}{1 + j(0.8)(0.1)(4.0)} \\
A &= 1.36
\end{aligned} \tag{13-28}$$

Therefore the transfer function of the compensated system is

$$G_A(s) = \frac{0.205(1 + 4s)}{s^2(1 + 0.2s)(1 + 0.4s)} \tag{13-29}$$

The polar plot of $\mathbf{G}_A(j\omega)$ of this compensated system is shown as dashed lines in Fig. 13-23. Note that this curve is not quite tangent to the M_m circle but crosses it. Although not shown in the figure, $\mathbf{G}_A(j\omega)$ has an approximate M_m of 1.35 and $\omega_m = 0.6$. If a larger gain K_2 is required, either a smaller value of α must be selected or two lead networks can be placed in cascade.

Case 2.   (Second trial)

In order that the compensated curve should not intersect the M_m circle, a second approximation is made for the tangency point. The point

$$\mathbf{G}_B(j\omega_m) = 1.4/-130.3°$$

is chosen; note that it is closer to the tangency point between the M_m circle and the line representing the angle ψ than for Case 1. In order not to go much below a value of $\alpha = 0.1$ and to obtain a sufficient angular shift, a lower value of $\omega_a = \omega_m$ is chosen. Thus for $\omega = \omega_a = \omega_m = 0.6$,

$$(\phi_c)_{\max} = /\mathbf{G}_B(j\omega_m) - /\mathbf{G}_x(j\omega) = -130.3° + 186.9° = 56.6°$$

An $\alpha = 0.09$ gives this value of $(\phi_c)_{\max}$ and a T equal to 5.55. This results in the over-all transfer function

$$G_B(s) = \frac{0.102(1 + 5.55s)}{s^2(1 + 0.2s)(1 + 0.5s)} \tag{13-30}$$

whose polar plot is shown in Fig. 13-23. This curve is tangent to the M_m circle and has an $\omega_m \approx 0.6$.

The results of Case 2 are not very different from those of Case 1.   The second case is inserted here to demonstrate that, although one initially attempts to make the compensated curve tangent to the M_m circle, a slight intersection of this circle is not too critical.   The results of Case 1 could have been accepted without a second trial.   Whether the intersection is slight or not can be determined by the relative spacing of the compensated curve $G_A(j\omega)$ between the circles for $M = 1.3$ and 1.4.

Furthermore, the results indicate that for this simple lead network a value of $\omega_m = 0.8$ cannot be obtained.   By the use of a more complex lead network, a sufficient counterclockwise phase shift may be realized to produce the desired value of ω_m and at the same time obtain a larger value of gain.

In this section it has been demonstrated that a stable system may be obtained when a lead network is applied to an unstable Type 2 system. Some compromise in the desired M_m and gain may be necessary.   The design procedures illustrated above for this unstable system can also be applied to any type of system whose inherent gain cannot be reduced in order to make its polar plot tangent to the desired M_m circle.

13-11   *Conclusions*

Various methods of compensation have been presented in this chapter for the improvement of the performance of a basic feedback control system. As one becomes more adept in compensation techniques, he is able to make refinements to the methods illustrated and to reduce the amount of trial-and-error effort.   The main advantages of using the log plots are:

1.   Templates representing the compensators can be used.   This simplifies the work of applying the compensator.
2.   More flexibility is possible by using these templates, since many trials can be made in a shorter period of time than with the polar plots.   Other factors that a designer must keep in mind in the selection of a compensator are performance limitations, space, weight, environmental conditions, and cost, etc., of the equipment and components.

A summary of the design procedures for each type of compensation follows.

Lag Compensation Procedures with

$$G_c(s) = A \frac{1 + Ts}{1 + \alpha Ts} \qquad \alpha > 1$$

1. Adjust the gain of the basic system $G_x(s)$ for the desired M_m which results in an ω_{m1}.
2. At $\omega = \omega_{m1}$, the phase shift ϕ_c of the lag compensator is limited to approximately $-5°$. With a value of α chosen larger than the desired increase in gain, the compensator time constant T can be determined from

$$T^2 + \frac{\alpha - 1}{\alpha \omega_{m1} \tan \phi_c} T + \frac{1}{\omega_{m1}^2 \alpha} = 0 \qquad (13\text{-}31)$$

where the larger value of T is used. For $\alpha = 10$ this becomes $T \approx 10/\omega_{m1}$.
3. Plot the new over-all transfer function $\mathbf{G}'(j\omega) = \mathbf{G}'_c(j\omega)\mathbf{G}'_x(j\omega)$, and again apply the gain-adjustment technique for the desired M_m. The additional amplifier gain A required is the ratio of the gain after compensation to the gain before compensation.

Lead Compensation Procedures with

$$G_c(s) = A\alpha \frac{1 + Ts}{1 + \alpha Ts} \qquad \alpha < 1$$

1. Adjust the gain of the basic system $G_x(s)$ for the desired M_m.
2. Select a value for α, and set the value of the time constant T equal to or slightly less than the value of the largest time constant in the denominator of $G_x(s)$ for a Type 1 or higher system. For a Type 0 system, apply this procedure to the second largest time constant.
3. Plot the new over-all transfer function $\mathbf{G}'(j\omega) = \mathbf{G}'_c(j\omega)\mathbf{G}'_x(j\omega)$, and again apply the gain-adjustment technique for the desired M_m. The additional amplifier gain A is the value required to produce the new gain. Do not neglect the term α which is introduced by the lead network.

Lag-Lead Compensation Procedures for

$$G_c(s) = A \frac{(1 + T_1 s)(1 + T_2 s)}{(1 + \alpha T_1 s)[1 + (T_2/\alpha)s]}$$

1. Adjust the gain of the basic system $G_x(s)$ for the desired M_m which results in an ω_{m1}.
2. Combine the individual lag and lead compensators which are designed individually (see above). To achieve the combined results with the network of Fig. 12-24, the value of α for the lag compensator must equal $1/\alpha$ for the lead compensator.

3. Plot the new over-all transfer function $\mathbf{G}'(j\omega) = \mathbf{G}'_c(j\omega)\mathbf{G}'_x(j\omega)$, and again apply the gain-adjustment technique for the desired M_m. Thus the additional amplifier gain A is determined.

Bibliography

1. Brown, G. S., and D. P. Campbell: "Principles of Servomechanisms," chaps. 7 and 8, John Wiley & Sons, Inc., New York, 1948.
2. Chestnut, H., and R. W. Mayer: "Servomechanisms and Regulating System Design," 2d ed., vol. 1, chaps. 10 and 12, John Wiley & Sons, Inc., New York, 1959.
3. Finnigan, R. E.: "Transient Analysis of Nonlinear Servomechanisms Using Describing Functions with Root-locus Techniques," Ph.D. Thesis, University of Illinois, 1957.
4. Nixon, F. E.: "Principles of Automatic Controls," chaps. 9 and 11, Prentice-Hall, Inc., Englewood Cliffs, N.J., 1953.
5. Stewart, J. L.: "Circuit Theory and Design," John Wiley & Sons, Inc., New York, 1956.

14

Feedback
Compensation

14-1   Introduction[1]

The preceding chapters have dealt with the application of cascade (series) compensation for improving the performance of a given feedback control system.   It is often possible to achieve the same improvement by feedback (parallel) compensation, i.e., by inserting a compensator in the feedback path, as shown in Fig. 14-1.

Note that the input signal is being compared with a modified form of and not directly with the output signal.   The compensator inserted in the feedback path may be of the same form as those used for cascade compensation: lag, lead, and lag-lead networks.

The following factors must be considered when making a choice between cascade and feedback compensation:

1.   The design procedures for a cascade compensator are more direct than those for a feedback compensator.   The application of feedback compensators is sometimes more laborious.

465

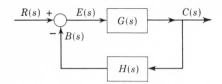

Fig. 14-1 *Block diagram of a feedback-compensated control system.*

2. Because of the physical form of the control system—i.e., whether it is electric, hydraulic, mechanical, etc.,—a cascade or feedback compensator may not exist or be practical for one or all of the systems.

3. The type of signal being fed into the compensator must be considered. For example, if the feedback signal utilizes a 400-cycle carrier, the design of a parallel compensator may be more difficult than that of a series compensator.

4. The economics in the use of either technique for a given control system involves items such as the size, weight, and cost of components and amplifiers. In the forward loop the signal goes from a low to a high energy level, whereas the reverse is true in the feedback loop. Thus, generally an amplifier may not be necessary in the feedback path. The cascade path generally requires an amplifier for gain and/or isolation purposes. Also, the size and weight of capacitors of a cascade-compensating network may be very large compared with those for the feedback compensator. These items are of great importance in aircraft, both commercial and military, where minimum size and weight of equipment are desired.

5. The environmental conditions in which the feedback control system is to be utilized affect the degree of accuracy and stability of the controlled quantity. This is a serious problem in an airplane, which is subjected to rapid changes in altitude and temperature. It is shown in the next section that the degree of accuracy and stability of a control system can be improved by the use of a parallel compensator.

6. The noise problem within a control system may determine the choice of compensator. In situations where a greater amplifier gain is required with a forward compensator than by the use of feedback circuits, the noise problem is accentuated. Also, the frequency characteristics of the feedback compensator to give the desired system improvement may be such that it attenuates the high-frequency portion of the noise content.

7. The time of response desired for a control system is a determining factor. Often a faster time of response can be achieved by the use of parallel compensation.

8. Some systems require "tight-loop" stabilization to isolate the dynamics of one portion of a control system from other portions of the complete system. This can be accomplished by introducing an inner feedback loop around the portion to be isolated.
9. Besides all these factors, the available components and the designer's experience and preferences influence the choice between a series and a parallel compensator.

14-2   *Environmental Conditions*

As stated in the preceding section, the environmental conditions to which a control system is subjected affect the degree of accuracy and stability of the system. The performance characteristics of most components are affected by their environment. Thus any change in the component characteristics causes a change in the transfer function and, therefore, in the controlled quantity.

To illustrate the effect of changes in the parameters of the transfer function, three cases are considered for which the input signal R and the frequency are constants.

Case 1.   Open-loop system of Fig. 14-2a

The effect of a change in G due to environmental conditions, for a fixed R, can be determined by differentiating

$$C_0 = RG \tag{14-1}$$

giving

$$dC_0 = R \, dG \tag{14-2}$$

Combining Eqs. (14-1) and (14-2),

$$dC_0 = \frac{dG}{G} \, C_0 \tag{14-3}$$

Therefore a change in G causes a corresponding change in the output C_0. This requires that the performance specifications of the components of G be such that any variation due to environmental conditions still results in the degree of accuracy within the prescribed limits.

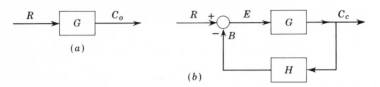

Fig. 14-2   *Control systems.   (a) Open-loop.   (b) Closed-loop.*

Case 2.   Closed-loop unity feedback system of Fig. 14-2b ($H = 1$)

Proceeding in the same manner as for Case 1,

$$C_c = R\,\frac{G}{1 + G} \tag{14-4}$$

$$dC_c = R\,\frac{dG}{(1 + G)^2} \tag{14-5}$$

$$dC_c = \frac{dG}{G(1 + G)}\,C_c = \left(\frac{1}{1 + G}\,\frac{dG}{G}\right)C_c \tag{14-6}$$

Comparing Eq. (14-6) with Eq. (14-3) readily reveals that the effect of parameter changes upon the output, when going from open-loop to closed-loop control, is reduced by the factor $1/(1 + G)$.

Case 3.   Closed-loop nonunity feedback system of Fig. 14-2b[4]

a.   First consider that the feedback function H is constant.   Then, proceeding as before,

$$C_c = R\,\frac{G}{1 + GH} \tag{14-7}$$

$$dC_c = R\,\frac{dG}{(1 + GH)^2} \tag{14-8}$$

$$dC_c = \frac{dG}{G(1 + GH)}\,C_c = \left(\frac{1}{1 + GH}\,\frac{dG}{G}\right)C_c \tag{14-9}$$

Within the range for which the degree of accuracy of C_c must be maintained, the term GH may be large enough that a greater reduction of the effect of parameter changes results over that for Case 2. If the term $1 + GH$ is greater than the term $1 + G$, this illustrates a possible advantage of parallel over series compensation.

b.   Secondly, consider that the forward function is constant and only the components of H are affected.   From Eq. (14-7),

$$dC_c = R\,\frac{-G^2\,dH}{(1 + GH)^2} \tag{14-10}$$

Multiplying and dividing Eq. (14-10) by H and also dividing by Eq. (14-7) results in

$$dC_c \approx - \frac{dH}{H} C_c \qquad (14\text{-}11)$$

for those values of frequency where it can be considered that $|\mathbf{G}(j\omega)\mathbf{H}(j\omega)| \gg 1$. When Eq. (14-11) is compared with Eq. (14-3), it is seen that a variation in the feedback function has approximately a direct effect upon the output, in the same manner as for the open-loop case. Thus the components of $\mathbf{H}(j\omega)$ must be selected upon the basis of maintaining the desired degree of accuracy and stability in the output signal $c(t)$.

The two situations in Case 3 serve to point out the advantage of parallel compensation when considered from the standpoint of environmental changes. Since feedback compensation minimizes the effect of variations in the components of $\mathbf{G}(j\omega)$, prime consideration can be given to obtaining the necessary power requirement in the forward loop rather than to accuracy and stability. $\mathbf{H}(j\omega)$ can then be designed so that the output $c(t)$ has the desired accuracy and stability. In other words, *by use of feedback compensation the performance of the system can be made more dependent on the feedback term than on the forward term.*

The sensitivity of the system's response to a system parameter variation can best be expressed by the *sensitivity function*[5] S_δ^M, which is defined as

$$S_\delta^M = \left[\frac{\text{change in system response } M}{\text{change in open-loop parameter } \delta} \right]_{\text{for specified parameter variations}} \qquad (14\text{-}12)$$

Change is defined as the ratio of the differential variation of the function to the function itself. Applying this equation to each of the three cases, where $M = C/R$, $\delta = G$ (for Cases 1, 2, and 3a), and $\delta = H$ (for Case 3b), yields the results shown in Table 14-1. The table readily reveals that S_δ^M *never exceeds a value of 1*, and the smaller this value, the less sensitive is the system to a parameter variation.

Equation (14-12) may also be applied to the situation in Sec. 8-7. The sensitivity function for the example in this section is given by

$$S_\delta^M = \frac{dC/C}{dV/V} = - \frac{1}{(U + N_1)/V + 1} \qquad (14\text{-}13)$$

Thus if $|S_\delta^M| \ll 1$ the variation of the open-loop pole has a negligible effect upon the system's output response.

Table 14-1.   Sensitivity functions

Case	System variable parameter	S_δ^M
1	G	$(dC_0/C_0)/(dG/G) = 1$
2	G	$(dC_c/C_c)/(dG/G) = 1/(1 + G)$
3a	G	$(dC_c/C_c)/(dG/G) = 1/(1 + GH)$
3b	H	$(dC_c/C_c)/(dH/H) = -1$

14-3   Time of Response

The effect of feedback upon the time of response of a given unit can best be illustrated by a numerical example.   Figure 14-3 shows various methods of improving the time of response of the output.   It is assumed that the system must be overdamped, that the input is $r(t) = 10u(t)$, and that the steady-state value of $c(t)$ is to be equal to 1.

Case 1

The basic unit is shown in Fig. 14-3a.   In order for $c_{ss}(t)$ to equal 1, the value of A must be equal to unity.   Thus

$$C(s) = \frac{0.1}{1 + 0.1s} R(s) = \frac{1}{s(1 + 0.1s)} \tag{14-14}$$

The system time constant is equal to 0.1 sec.

Case 2

The basic unit with unity feedback is shown in Fig. 14-3b.   To improve the time of response of the system, the output $C(s)$ is fed back and compared directly with the input.   The system's transfer function is

$$\frac{C(s)}{R(s)} = \frac{0.1A}{0.1s + (1 + 0.1A)} \tag{14-15}$$

Then

$$C(s) = \frac{A}{s[0.1s + (1 + 0.1A)]} \tag{14-16}$$

By applying the final-value theorem to Eq. (14-16), it is determined that $A = 10/9$ is required in order to maintain the desired value of the output:

$$C(s) = \frac{1}{s(1 + 0.09s)} \tag{14-17}$$

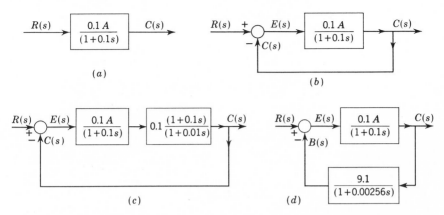

Fig. 14-3 *Various methods of improving the response* $C(s)$ *for a given input* $R(s)$.

By use of unity feedback the system time constant T has been reduced from 0.1 to 0.09 sec, thus improving the time of response of the system. Note that the constant is reduced by a factor of $1/(1 + 0.1A)$. Also, the increase in gain is used to produce a faster time of response and not to increase the magnitude of the output.

Case 3

The basic unit with series compensation and unity feedback is shown in Fig. 14-3c.

Further improvement of the system time constant can be achieved by inserting a lead compensator in series with the basic unit. The parameters of the compensator are chosen by using the design criteria of the preceding chapter; i.e., the open-loop time constant is replaced by a smaller time constant. This results in

$$\frac{C(s)}{R(s)} = \frac{0.01A}{0.01s + (1 + 0.01A)} \tag{14-18}$$

$$C(s) = \frac{0.1A}{s[0.01s + (1 + 0.01A)]} \tag{14-19}$$

The required value of A is determined from Eq. (14-19) to be approximately 11. Therefore,

$$C(s) = \frac{1}{s(1 + 0.009s)} \tag{14-20}$$

giving a system time constant T of 0.009 sec, which is smaller than the previous two cases. When this is compared with Case 1, it is seen that the

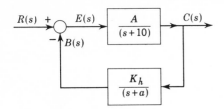

Fig. 14-4   *Feedback compensation using a lag network.*

increased gain and lead compensation are used to reduce T by a factor of $1/(10 + 0.1A)$.

Case 4

The basic unit with feedback compensation is shown in Fig. 14-3d.

If further improvement in T is desired over that obtained in Case 3, feedback compensation must be utilized. As this case demonstrates, to obtain the same effect as the series lead compensation of Case 3, a parallel lag compensator must be utilized. Similarly, to obtain the same effect as a series lag compensator, a parallel lead compensator must be used.

Figure 14-4 shows the basic system with a parallel lag compensator. The closed-loop response is given by

$$\frac{C(s)}{R(s)} = \frac{A(s + a)}{s^2 + (10 + a)s + (10a + AK_h)} \tag{14-21}$$

The values a, A, and K_h are determined from the requirement that $c_{ss}(t) = 1$ and that a closed-loop time constant $T = 0.005$ sec is desired. For a critically damped system the results are $a = 390$, $A = 11.2$, and $K_h = 3,550$, which give

$$H = \frac{9.1}{1 + 0.00256s} \tag{14-22}$$

The transform of the output is

$$C(s) = \frac{1 + 0.0026s}{s(1 + 0.005s)^2} \tag{14-23}$$

This case has demonstrated that it is possible to obtain a faster time of response by use of a parallel compensator.

14-4   Feedback-compensator Location

The location of a feedback compensator depends upon the complexity of the basic system, the accessibility of insertion points, the form of the signal

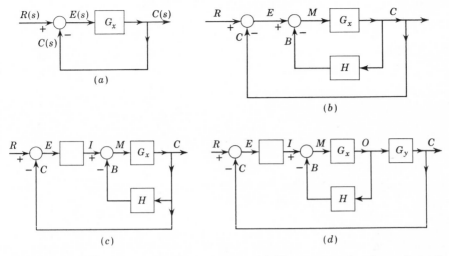

Fig. 14-5 *Various forms of feedback systems.* (a) *Uncompensated system.* (b, c, d) *Feedback-compensated systems.*

being fed back, the signal with which it is being compared, and the desired improvement.

Figure 14-5 shows various forms of feedback-compensated systems. For each case there is a unity feedback loop (major loop) besides the compensated loop (minor loop) in order to maintain a direct correspondence between the output and input.

14-5   Restrictions on Feedback Components

In applying feedback compensation, it is often necessary to maintain the type of the basic system. Care must therefore be exercised in determining the number of differentiating terms in the numerator of $H(s)$. Table 14-2 presents the basic forms that $H(s)$ may have so that the system type remains

Table 14-2. Basic forms of $H(s)$ to be used to prevent change of system type

Basic form of $H(s)$	Can be used with
K	Type 0
Ks	Type 0 and 1
Ks^2	Type 0, 1, and 2
Ks^3	Type 0, 1, 2, and 3

the same after the minor feedback-compensation loop is added. $H(s)$ may have additional factors in the numerator and denominator.

It can be shown that the order of the factor s in the numerator of $H(s)$ must be equal to or higher than the type of the forward transfer function $G_x(s)$ shown in Fig. 14-5.

14-6   *Transforming Non-Type 0 Control Elements to Type 0*

By the use of a direct or unity feedback loop any non-Type 0 control element can be converted to a Type 0 element.   This effect is advantageous when going from a low-energy-level input to a high-energy-level output for a given control element and, also, when the forms of the input and output energy are different.

In Fig. 14-6 the $G_x(s)$ function may represent either a single control element or a group of control elements that form part of a control system and whose transfer function is of the general form

$$G_x(s) = \frac{K_n(1 + T_1s) \cdot \cdot \cdot}{s^n(1 + T_as)(1 + T_bs) \cdot \cdot \cdot} \tag{14-24}$$

where $n \neq 0$.   The over-all ratio of output to input is given by

$$\frac{O(s)}{I(s)} = \frac{G_x(s)}{1 + G_x(s)} = G(s) \tag{14-25}$$

or $\qquad G(s) = \dfrac{K_n(1 + T_1s) \cdot \cdot \cdot}{s^n(1 + T_as)(1 + T_bs) \cdot \cdot \cdot + K_n(1 + T_1s) \cdot \cdot \cdot} \tag{14-26}$

Equation (14-26) represents an equivalent forward transfer function between $O(s)$ and $I(s)$ of Fig. 14-6 and has the form of a Type 0 control element. Thus the transformation of type has been effected.

As an example, consider a unit that has the transfer function

$$\mathbf{G}_x(j\omega) = \frac{K_1}{(j\omega)(1 + j\omega T_1)(1 + j\omega T_2)} \tag{14-27}$$

Fig. 14-6   *Transforming non-Type 0 control element to Type 0.*

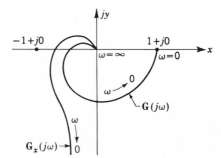

Fig. 14-7 *Plot of a Type 1 control element*
$G_x(j\omega)$ *and the Type 0 control element* $\mathbf{G}(j\omega)$
produced by adding unity feedback

By the use of unity feedback the transfer function becomes

$$G(j\omega) = \frac{\mathbf{O}(j\omega)}{\mathbf{I}(j\omega)} = \frac{K_1}{j\omega(1 + j\omega T_1)(1 + j\omega T_2) + K_1}$$

$$= \frac{1}{(j\omega)^3(T_1 T_2/K_1) + (j\omega)^2[(T_1 + T_2)/K_1] + j\omega/K_1 + 1}$$
(14-28)

Therefore, for a step input,

$$O_{ss}(t) = \lim_{s \to 0} [I(s)G(s)] = 1 \qquad (14\text{-}29)$$

In other words, for this equivalent single element an input of constant value produces an output of constant value, which satisfies the definition of a Type 0 element.   As a consequence, a given input signal tends to produce an output signal of equal magnitude but at a higher energy level and/or with a different form of energy.

The polar plots of Eqs. (14-27) and (14-28) are shown in Fig. 14-7. These plots show that for any given frequency the phase shift is less with feedback, thus indicating a more suitable system (a smaller value of M_m).

14-7   Feedback Compensation Using the Root Locus:[2] Design Procedures

Techniques are now described for applying feedback compensation to the root-locus method of system design.   The system to be investigated is shown in the block diagram of Fig. 14-8.   A minor feedback loop has been formed around the original transfer function $G_x(s)$.   The minor-loop feedback transfer function $H(s)$ consists of elements and networks described in the following pages.   Either the gain portion of $G_x(s)$ or the cascade amplifier A is adjusted to fix the system characteristics.   For more complex systems one can develop a design approach based on the knowledge obtained for this simple system.

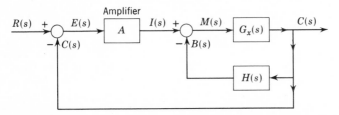

Fig. 14-8 *Block diagram for feedback compensation.*

The method of attack is similar to that used for series compensation. The characteristic equation of the complete system may be obtained and the root locus plotted, by use of the partition method, as a function of a gain that appears in the system. A more general approach is first to adjust the roots of the characteristic equation of the inner loop. These roots are then the poles of the forward transfer function and are used to draw the root locus for the over-all system. Both methods are used in the succeeding sections.

The techniques developed with series compensation should be kept in mind. Basically, the addition of poles and zeros changes the root locus. To improve the system time response the locus must be moved to the left, away from the imaginary axis. The following sections take up various types of feedback compensators.

14-8   Simplified Tachometer Feedback Compensation (Root Locus)

The feedback system shown in Fig. 14-8 provides the opportunity for varying three functions—$G_x(s)$, $H(s)$, and A. The larger the number of variables inserted into a system, the better is the opportunity for achieving a specified performance. However, the design problem becomes much more complex. In order to develop a *feel* for the changes that can be introduced by feedback compensation, a simplification is made by letting $A = 1$. The system of Fig. 14-8 then reduces to that shown in Fig. 14-9a. It can be further simplified as shown in Fig. 14-9b.

The first and simplest case of feedback compensation to be considered is the use of a tachometer which has the transfer function

$$H(s) = K_t s \qquad (14\text{-}30)$$

The original forward transfer function is

$$G_x(s) = \frac{K}{s(s+1)(s+5)} \qquad (14\text{-}31)$$

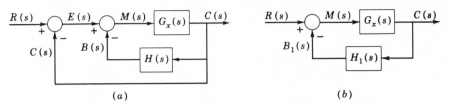

Fig. 14-9   (a) *Minor loop compensation.*   (b) *Simplified block diagram, $H_1(s) = 1 + H(s)$.*

The control ratio is

$$\frac{C(s)}{R(s)} = \frac{G_x(s)}{1 + G_x(s)H_1(s)} = \frac{G_x(s)}{1 + G_x(s)[1 + H(s)]} \tag{14-32}$$

The characteristic equation is

$$1 + G_x(s)[1 + H(s)] = 1 + \frac{K(1 + K_t s)}{s(s+1)(s+5)} = 0 \tag{14-33}$$

Equation (14-33) shows that the tachometer has introduced the ideal proportional plus derivative control that can also be achieved by a cascade compensator.   The results should therefore be qualitatively the same.

The root locus can now be drawn for

$$\frac{KK_t(s + 1/K_t)}{s(s+1)(s+5)} = -1 \tag{14-34}$$

It is seen from Eq. (14-34) that introduction of tachometer feedback results in the introduction of an open-loop zero.   It therefore has the effect of moving the locus to the left and improving the time response.   *This is not the same as introducing a cascade zero, because no zero appears in the rationalized equation of the control ratio:*

$$\frac{C(s)}{R(s)} = \frac{K}{s(s+1)(s+5) + KK_t(s + 1/K_t)} \tag{14-35}$$

The tachometer constant must be chosen to improve the location of the closed-loop roots.   *One must be careful in selecting the value of K_t to avoid being deceived.*   For example, if $K_t = 1$ is used, Eq. (14-34) reduces to

$$\frac{K}{s(s+5)} = -1 \tag{14-36}$$

and one can get the impression that $C(s)/R(s)$ has only two poles.   This is not the case, as can be seen from Eq. (14-35).   Letting $K_t = 1$ provides the common factor $s + 1$ in the characteristic equation so that it becomes

$$(s+1)[s(s+5) + K] = 0 \tag{14-37}$$

Table 14-3.   Comparison of performances

System	Dominant complex poles	Other roots	Ramp error coefficient K_1	T_p, sec	M_o	T_s, sec
Original	$-0.40 \pm j0.81$	-5.19	0.83	4.25	0.20	10.0
Tachometer feedback	$-1.0 \pm j2.0$	-4.0	1.54	2.83	0.17	4.0

One closed-loop pole has been made equal to $s = -1$, and the other two poles are obtained from the root locus of Eq. (14-36).   If complex roots are selected from the root locus of Eq. (14-36), it is easily shown that the pole $s = -1$ is dominant and the response of the closed-loop system is overdamped (see Fig. 8-5).   The proper procedure is to select a number of trial locations for the zero $s = -1/K_t$, to tabulate or plot the system characteristics that result, and then to select the best combination.   As an example, using $\zeta = 0.45$ as the criterion for the dominant closed-loop poles, a comparison of the original system and the tachometer-compensated system with $K_t = 0.4$ is given in Table 14-3.   The gain K_1 must be obtained from the forward transfer function:

$$K_1 = \lim_{s \to 0} sG(s) = \lim_{s \to 0} \frac{sG_x(s)}{1 + G_x(s)H(s)} = \lim_{s \to 0} \frac{sK}{s[(s+1)(s+5) + KK_t]}$$

$$= \frac{K}{5 + KK_t} \tag{14-38}$$

An analysis of Table 14-3 shows that the effects of tachometer feedback are similar to those of a cascade lead compensator but the gain K_1 is not so large (see Sec. 12-9).   The control ratio of the system with tachometer feedback is

$$\frac{C(s)}{R(s)} = \frac{20}{(s + 1 - j2)(s + 1 + j2)(s + 4)} \tag{14-39}$$

14-9   Tachometer Feedback (Root Locus)

The more general case of feedback compensation shown in Fig. 14-8, with the amplifier gain A not equal to unity, is considered in this section.   Two methods of attacking this problem are shown.   With the addition of the inner feedback loop, the forward transfer function of the complete system is

$$G(s) = \frac{AG_x(s)}{1 + G_x(s)H(s)} = \frac{AK}{s(s + 1)(s + 5) + KK_t s} \tag{14-40}$$

Method 1

The first method involves the characteristic equation of the complete system which is

$$1 + G(s) = 1 + \frac{AK}{s(s + 1)(s + 5) + KK_ts} = 0 \qquad (14\text{-}41)$$

By clearing fractions and rewriting, the characteristic equation is

$$s(s + 1)(s + 5) + KK_t\left(s + \frac{A}{K_t}\right) = 0 \qquad (14\text{-}42)$$

This equation can be factored by using the partition method described in Sec. 8-5. This consists in dividing the equation by $s(s + 1)(s + 5)$ and putting it in the form

$$\frac{KK_t(s + A/K_t)}{s(s + 1)(s + 5)} = -1 \qquad (14\text{-}43)$$

Since the numerator and denominator are factored, the roots can be obtained by plotting a root locus. The angle condition of $(1 + 2m)180°$ must be satisfied for the root locus of Eq. (14-43). It should be kept in mind, however, that Eq. (14-43) is *not* the forward transfer function of this system; it is *just an equation whose roots are the poles of the control ratio.* The three roots of this characteristic equation vary as K, K_t, and A are varied. Obviously, it is necessary to fix two of these quantities and to determine the roots as a function of the third. A sketch of the root locus as a function of K is shown in Fig. 14-10 for arbitrary values of A and K_t. It should be noted that the tachometer feedback has resulted in an equivalent open-loop zero, so that the system is stable for all values of gain for the value of A/K_t chosen. This is typical of the stabilizing effect which can be achieved by derivative feedback and corresponds to the effect achieved with an ideal derivative plus proportional compensator in cascade.

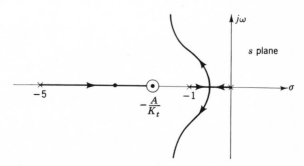

Fig. 14-10 *Root locus for Eq. (14-43).*

The asymptotes of two of the branches of Eq. (14-43) are $\pm 90°$. For the selection A/K_t shown in Fig. 14-10, these asymptotes are in the left half of the s plane and the system is stable for any value of K. However, if $(A/K_t) > 5$, the asymptotes are in the right half of the s plane and the system can become unstable if K is large enough. For the best transient response the root locus should be moved as far to the left as possible. This is determined by the location of the zero $s = -A/K_t$. To avoid having a small real root, the value of A/K_t is made greater than unity. The selection of these two parameters also determines the ramp error coefficient, since Eq. (14-40) gives the value $K_1' = AK/(5 + KK_t)$. The designer must select K_t to meet the system specifications. However, tachometer sensitivities normally do not exceed $K_t = 0.2$ volt/radian/sec unless an amplifier is used.

The values $A = 2.0$ and $K_t = 0.8$ result in a zero at $s = -2.5$, the same as the example in Sec. 14-8. For $\zeta = 0.45$, the control ratio is given by Eq. (14-39) and the performance is the same as that listed in Table 14-3. The use of the additional amplifier A does not change the performance; rather, it permits greater flexibility in the allocation of gains. For example, K can be decreased and both A and K_t increased by a corresponding amount with no change in performance.

Method 2

A more general method is to start with the inner loop, adjusting the gain to select the poles. Then successive loops are adjusted, setting the poles for each loop to desired values. For the system being studied, the transfer function of the inner loop is

$$\frac{C(s)}{I(s)} = \frac{K}{s(s+1)(s+5) + KK_ts} = \frac{K}{s[(s+1)(s+5) + KK_t]}$$
$$= \frac{K}{s(s-p_1)(s-p_2)} \tag{14-44}$$

The equation for which the root locus is to be plotted to find the roots p_1 and p_2 is

$$(s+1)(s+5) + KK_t = 0$$

which can be rewritten as

$$\frac{KK_t}{(s+1)(s+5)} = -1 \tag{14-45}$$

The locus is plotted in Fig. 14-11, where several values of damping ratio ζ_i for the roots are shown. Selection of a ζ_i permits evaluation of the product KK_t by use of Eq. (14-45). Upon assignment of a value to one of these quantities, the other can then be determined. The poles of the inner loop are the points on the locus that are determined by the ζ_i selected. They also

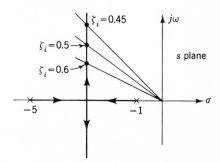

Fig. 14-11 *Poles of the inner loop for several values of ζ_i.*

are the poles of the open-loop transfer function for the over-all system. It is therefore necessary that the damping ratio of the inner loop, ζ_i, have a higher value than the damping ratio of the over-all system. The forward transfer function of the complete system becomes

$$G(s) = \frac{C(s)}{E(s)} = \frac{AK}{s(s^2 + 2\zeta_i\omega_{n,i}s + \omega_{n,i}^2)}$$

$$= \frac{AK}{s(s + \zeta_i\omega_{n,i} - j\omega_{n,i}\sqrt{1 - \zeta_i^2})(s + \zeta_i\omega_{n,i} + j\omega_{n,i}\sqrt{1 - \zeta_i^2})} \quad (14\text{-}46)$$

The root locus of the complete system is shown in Fig. 14-12 for several values of the inner-loop damping ratio ζ_i. These loci are plotted as a function of AK.

The control ratio has the form

$$\frac{C(s)}{R(s)} = \frac{AK}{s(s^2 + 2\zeta_i\omega_{n,i}s + \omega_{n,i}^2) + AK}$$

$$= \frac{AK}{(s + \alpha)(s^2 + 2\zeta\omega_n s + \omega_n^2)}$$

$$= \frac{AK}{(s + \alpha)(s + \zeta\omega_n - j\omega_n\sqrt{1 - \zeta^2})(s + \zeta\omega_n + j\omega_n\sqrt{1 - \zeta^2})} \quad (14\text{-}47)$$

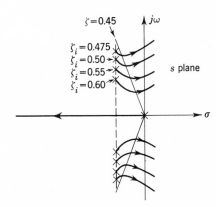

Fig. 14-12 *Root locus of complete system for several values of ζ_i.*

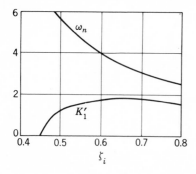

Fig. 14-13   *Variation of K_1' and ω_n with ζ_i.*

In this problem the damping ratio ζ of the complex poles of the control ratio is used as the basis for design. Both the undamped natural frequency ω_n and the ramp error coefficient $K_1' = AK/(5 + KK_t)$ depend on the selection of the inner-loop damping ratio ζ_i. Several trials are necessary to show how these quantities vary with the selection of ζ_i. This information is shown graphically in Fig. 14-13. The system performance specifications must be used to determine the values of K_1' and ω_n used. Once the values of K_1' and ω_n are selected, the values of A, K, and K_t can be determined.

14-10   *Tachometer plus RC Feedback (Root Locus)*

To further improve the system performance, the output of the tachometer may be fed to the RC circuit shown in Fig. 14-14. The use of this high-pass filter will result in an increased error coefficient.

The feedback transfer function for the inner loop is

$$H(s) = \frac{B(s)}{C(s)} = \frac{K_t s^2}{s + 1/RC} \tag{14-48}$$

Two methods are used to show the application of this compensator.

Method 1

The characteristic equation of the complete system is

$$1 + G(s) = 1 + \frac{AK(s + 1/RC)}{s(s + 1)(s + 5)(s + 1/RC) + KK_t s^2} \tag{14-49}$$

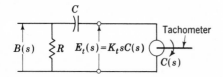

Fig. 14-14   *Tachometer and RC feedback network.*

By clearing fractions, this equation becomes

$$s(s + 1)(s + 5)\left(s + \frac{1}{RC}\right) + KK_t\left(s^2 + \frac{A}{K_t}s + \frac{A}{K_tRC}\right) = 0 \quad (14\text{-}50)$$

The root locus can be drawn as the gain K varies from zero to infinity from the following equation:

$$\frac{KK_t(s^2 + As/K_t + A/K_tRC)}{s(s + 1)(s + 5)(s + 1/RC)} = -1 \quad (14\text{-}51)$$

Since A/K_t and $1/RC$ must be selected before the locus can be drawn, it is well to look at the expression for K_1' to determine the limits to be placed on these values:

$$K_1' = \frac{AK}{5} \quad (14\text{-}52)$$

It appears that K_1' is independent of RC; therefore this quantity can be chosen to produce the most desirable root locus. There are two zeros which are the factors of $s^2 + (A/K_t)s + A/K_tRC$, and they may be either real or complex. The zeros are located at

$$s_{a,b} = -\frac{A}{2K_t} \pm \sqrt{\left(\frac{A}{2K_t}\right)^2 - \frac{A}{K_tRC}} \quad (14\text{-}53)$$

Two possible sketches of the root locus are shown in Fig. 14-15 for an arbitrary selection of A, K_t, and RC.

The values $A = 1$, $K_t = 0.344$, $1/RC = 0.262$ give zeros at

$$s_a = -0.29 \quad \text{and} \quad s_b = -2.62$$

By specifying a damping ratio $\zeta = 0.35$ for the dominant roots, the result is $K = 53.3$, $K_1' = 10.6$, and $\omega_n = 3.79$, and the over-all control ratio is

$$\frac{C(s)}{R(s)} = \frac{53.3(s + 0.262)}{(s + 1.32 + j3.55)(s + 1.32 - j3.55)(s + 3.34)(s + 0.291)} \quad (14\text{-}54)$$

Note that a passive lead network in the feedback loop acts like a passive lag network in cascade and has resulted in a large increase in K_1'. Several trials may be necessary to determine the best selection of the closed-loop poles and ramp error coefficient. A $\zeta = 0.35$ is chosen instead of $\zeta = 0.45$ to obtain a larger value for K_1'. The pole at $s = -3.34$ reduces the peak overshoot. The polar plot method Case 3, Section 14-15 yields similar results. Fig. 19-29 shows the time response.

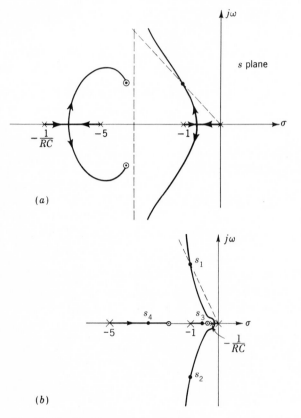

Fig. 14-15 *Root locus of Eq. (14-51).*

Method 2

Starting with the inner loop, the transfer function is

$$\frac{C(s)}{I(s)} = \frac{K(s + 1/RC)}{s(s + 1)(s + 5)(s + 1/RC) + KK_t s^2}$$

$$= \frac{K(s + 1/RC)}{s[(s + 1)(s + 5)(s + 1/RC) + KK_t s]} \tag{14-55}$$

There are three roots to be selected as a function of KK_t. The root locus is shown in Fig. 14-16. A large value of $1/RC$ is selected in order to move the locus far to the left. Also, K_t must be specified before K can be evaluated.

The damping ratio ζ_i of the inner loop is chosen consistent with the damping ratio of the over-all system and the location of the inner loop roots. These roots become the poles of the forward transfer function of the over-all system and therefore determine the root locus for the outer loop. The

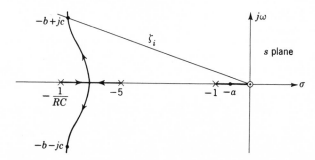

Fig. 14-16 *Root locus of the poles of Eq. (14-55).*

inner-loop roots are shown in Fig. 14-16 as $s = -a$, $s = -b \pm jc$. The characteristic equation of the over-all system is now given by

$$1 + G(s) = 1 + \frac{AK(s + 1/RC)}{s(s + a)(s + b - jc)(s + b + jc)} \qquad (14\text{-}56)$$

The root locus is shown in Fig. 14-17 and is a function of the amplifier gain A.

Several trials may be necessary to obtain the specified performance and the best selection for the constants such as RC, K_t, and A. As more experience is obtained, the design engineer establishes limits on these constants that reduce the number of trials necessary.

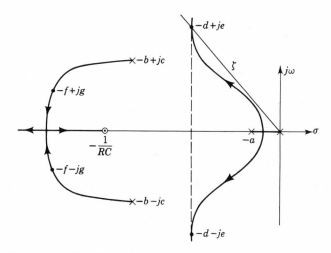

Fig. 14-17 *Root locus of Eq. (14-56).*

*Table 14-4.   Comparison of cascade and feedback
compensation using the root locus*

System	K_1	ω_n	ζ
Uncompensated	0.83	0.90	0.45
Tachometer feedback	1.8	3.92	0.45
Tachometer plus RC feedback	10.6	3.79	0.35
Cascade lag compensator	7.5	0.80	0.45
Cascade lead compensator	3.0	3.52	0.45
Cascade lag-lead compensator	29.0	3.45	0.45

14-11   Results of Feedback Compensation Using the Root Locus

The preceding sections have shown the application of feedback compensation by the root-locus method.   Basically, it has been shown that tachometer feedback is similar to ideal derivative control in cascade.   The addition of a high-pass RC filter in the feedback permits a large increase in the static error coefficient.   For good results the feedback compensator $H(s)$ should have a zero, $s = 0$, of higher order than the type of the original transfer function $G_x(s)$.

The method of adjusting the poles of the inner loop and then adjusting successively larger loops is applicable to all multiloop systems.

The results of feedback compensation applied in the preceding sections are summarized in Table 14-4.   Results of cascade compensation from Chap. 12 are also listed for comparison.   This tabulation shows that comparable results are obtained by cascade and feedback compensation.

14-12   Feedback Compensation Using Polar Plots: Design Procedures[1,3]

For cascade compensation direct polar plots and inverse polar plots are equally applicable.   When feedback compensation is to be applied to a control system it is more convenient to use the inverse polar plots, as demonstrated below.

By referring to Fig. 14-18, the following equation can be derived:

$$\frac{C(s)}{E(s)} = \frac{G_x(s)}{1 + G_x(s)H(s)} = G(s) \tag{14-57}$$

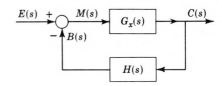

Fig. 14-18 *Feedback compensation of a control element $G_x(s)$.*

The reciprocal of this equation is

$$\frac{E(s)}{C(s)} = \frac{1 + G_x(s)H(s)}{G_x(s)} = \frac{1}{G_x(s)} + H(s) = \frac{1}{G(s)} \qquad (14\text{-}58)$$

When this equation is analyzed, the effect of the feedback compensator upon the original system can be seen directly in the inverse polar plots. Figure 14-19 shows the graphical phasor addition

$$\frac{\mathbf{E}(j\omega)}{\mathbf{C}(j\omega)} = \frac{1}{\mathbf{G}_x(j\omega)} + \mathbf{H}(j\omega) \qquad (14\text{-}59)$$

In Fig. 14-20 the original inverse transfer function $1/\mathbf{G}_x(j\omega)$ is plotted. On this same plot one can sketch the approximate shape of the desired inverse plot, $1/\mathbf{G}(j\omega)$, that the compensated system must have. The difference between these two plots must represent the value of the compensator $\mathbf{H}(j\omega)$.

By comparing the above procedure for determining $1/\mathbf{G}(j\omega)$ with that of determining $\mathbf{G}(j\omega)$ by direct plots, the simplicity of inverse plots can be appreciated. The examples worked out in the next three sections are solved only by inverse polar plots.

A typical control system shown in Fig. 14-21 has been adjusted for an $M_m = 1.25$ and has an $\omega_{m1} = 0.72$ radian/sec and $K_1 = 0.88$. Design procedures for four different cases of feedback compensation of this system are now presented. These cases are intended to indicate possible design

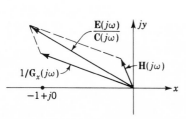

Fig. 14-19 *Graphical representation of*

$$\frac{\mathbf{E}(j\omega)}{\mathbf{C}(j\omega)} = \frac{1}{\mathbf{G}_x(j\omega)} + \mathbf{H}(j\omega)$$

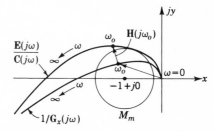

Fig. 14-20 *A plot of the original and desired inverse transfer functions.*

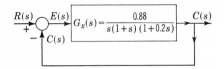

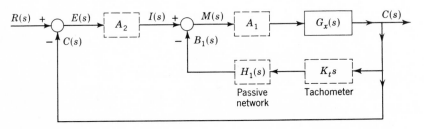

Fig. 14-21   A typical control system having an $M_m = 1.25$, $\omega_{m,1} = 0.72$, and $K_1 = 0.88$.

Fig. 14-22. *Feedback compensation with gain adjustments in both forward and feedback paths.*

approaches and the effect of feedback compensation in achieving a desired output response. Each case makes use of some or all of the additional elements shown in Fig. 14-22. An $M_m = 1.25$ is maintained for all cases, while the values of ω_m and the static error coefficient are improved.

14-13   Tachometer Feedback (Inverse Polar Plot)

A tachometer having the transfer function $K_t s$ is used in a minor feedback loop as shown in Fig. 14-22. In the first case the values $A_2 = H_1 = 1/\underline{0°}$ are used and the values of K_t and A_1 are adjusted to obtain $M_m = 1.25$. In the second case the values $A_1 = H_1 = 1/\underline{0°}$ are used and the values K_t and A_2 are adjusted. In both cases the procedure includes the selection of the desired value of ω_m. The various gains are then adjusted to obtain this performance.

Case 1

$$A_2 = 1 \qquad \text{and} \qquad \mathbf{H}_1(j\omega) = 1/\underline{0°}$$

Compensation of the system shown in Fig. 14-21 is achieved by inserting an amplifier A_1 in cascade with $G_x(s)$ and a tachometer in the minor feedback loop, as shown in Fig. 14-22.

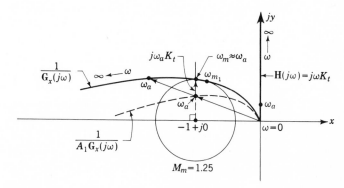

Fig. 14-23 *Polar plots of various portions of the control system of Fig. 14-22 with $A_2 = 1$ and $H_1(s) = 1$.*

The transfer function of the minor loop is

$$G(s) = \frac{C(s)}{I(s)} = \frac{A_1 G_x(s)}{1 + A_1 G_x(s) K_t s} \qquad (14\text{-}60)$$

For this system the new ramp error coefficient is

$$K_1' = \lim_{s \to 0} sG(s) = \frac{0.88 A_1}{1 + 0.88 A_1 K_t} \qquad (14\text{-}61)$$

This coefficient can be improved by satisfying the condition

$$\frac{0.88 A_1}{1 + 0.88 A_1 K_t} > 0.88$$

or
$$\frac{1}{A_1} + 0.88 K_t < 1.0 \qquad (14\text{-}62)$$

In other words, the larger A_1 and the smaller K_t can be made, the larger K_1' becomes. By experience a designer knows the value of $\omega_m = \omega_a$ that is satisfactory for the compensated system. Once this value of ω_m is selected, the values of A_1 and K_t can be determined.

The design procedure for determining A_1 and K_t is as follows:

1. Draw the phasor of $1/G_x(j\omega_a)$, as shown in Fig. 14-23.
2. Draw the M_m circle with the center at -1 and of radius $1/M_m$.
3. Draw a line through the $(-1 + j0)$ point parallel to the j axis until it intersects the phasor $1/G_x(j\omega_a)$ and the top half of the M_m circle. It is assumed that this point of intersection with the M_m circle is approximately the ω_m point of the desired $\mathbf{I}(j\omega)/\mathbf{C}(j\omega)$ plot.
4. The intersection of 1 and 3 is the point $1/A_1 G_x(j\omega_a)$. Thus, dividing the phasor $1/G_x(j\omega_a)$ by $1/A_1 G_x(j\omega_a)$ gives the value of A_1 required.

5. The length of the phasor between the approximate point of $\mathbf{I}(j\omega_m)/$ $\mathbf{C}(j\omega_m)$ and the point $1/A_1\mathbf{G}_x(j\omega_a)$ is equal to $K_t\omega_a$. Thus the value of K_t is determined.

6. With the calculated values of K_t and A_1, draw the curve $1/\mathbf{G}(j\omega)$. This curve should be tangent to and should not intersect the M_m circle. If the curve of $1/\mathbf{G}(j\omega)$ crosses the M_m circle, it is necessary to repeat the procedure above with a new point of tangency.

7. If the values of K_t and A_1 satisfy Eq. (14-62), an improvement has been made in the error coefficient. In the event that the values of K_1' and/or ω_m are not satisfactory, choose a different value of ω_a and/or point of tangency on the M_m circle.

Case 2

$$A_1 = 1 \quad \text{and} \quad \mathbf{H}_1(j\omega) = 1\underline{/0^\circ}$$

The compensation of the system is achieved for this case by inserting an amplifier A_2 in cascade with the minor loop and a tachometer in the minor feedback loop, as shown in Fig. 14-22. For this figure, the new over-all forward transfer function is

$$G(s) = \frac{C(s)}{E(s)} = \frac{A_2G_x(s)}{1 + G_x(s)K_ts} \tag{14-63}$$

and for this particular system the new ramp error coefficient is

$$K_1' = \lim_{s \to 0} sG(s) = \frac{0.88A_2}{1 + 0.88K_t} \tag{14-64}$$

This coefficient can be improved by satisfying the condition

$$\frac{0.88A_2}{1 + 0.88K_t} > 0.88 \quad \text{or} \quad A_2 > 1 + 0.88K_t \tag{14-65}$$

The first step is to choose an acceptable value of $\omega_m = \omega_a$.

The design procedure for determining K_t and A_2 is as follows:

1. Draw a line through the point $1/\mathbf{G}_x(j\omega_a)$ parallel to the j axis until it intersects the line representing the angle $\psi = \sin^{-1}(1/M_m)$ (see Fig. 14-24).

2. Select a trial point on this vertical line. A point between one-half to three-quarters of the distance up from $1/\mathbf{G}_x(j\omega_a)$ is a good first trial. This point is made $\mathbf{I}(j\omega_a)/\mathbf{C}(j\omega_a)$.

3. The length of the phasor between the point $1/\mathbf{G}_x(j\omega_a)$ and $\mathbf{I}(j\omega_a)/$ $\mathbf{C}(j\omega_a)$ is equal to $K_t\omega_a$. Thus the value of K_t is now determined.

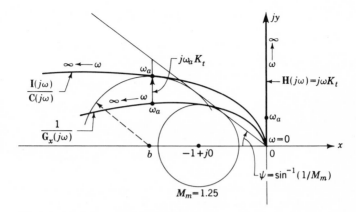

Fig. 14-24 *Polar plots of various portions of the control system of Fig. 14-22 with $A_1 = 1$ and $H_1(s) = 1$.*

4. Using this value of K_t, plot $\mathbf{I}(j\omega)/\mathbf{C}(j\omega) = 1/\mathbf{G}_x(j\omega) + K_t j\omega$. By the gain-adjustment techniques of Chap. 11, determine the required value of A_2, that is, $A_2 = ob$.

5. If these values of K_t and A_2 satisfy Eq. (14-65), an improvement has been made in the error coefficient. In the event that the values of K_1' and/or ω_m are not satisfactory, another trial point must be selected.

14-14 Tachometer plus *RC* Feedback (Inverse Polar Plot)

Further improvement in system performance can be obtained by adding a high-pass filter in the minor feedback loop together with the tachometer. The following examples show that a larger value of the static error coefficient results from the use of this lead network. The procedure is based on selecting the desired value of ω_m and then adjusting the gains and time constants to give this performance.

Case 3

$$A_2 = 1 \quad \text{and} \quad H_1(s) = \frac{Ts}{1 + Ts}$$

The feedback transfer function of the minor loop is

$$\mathbf{H}(j\omega) = j\omega K_t \mathbf{H}_1(j\omega) = \frac{TK_t(j\omega)^2}{1 + j\omega T} \tag{14-66}$$

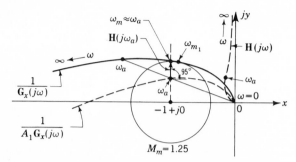

Fig. 14-25 *Polar plots of various portions of the control system of Fig. 14-22 with $A_2 = 1$ and $H_1(s) = Ts/(1 + Ts)$.*

The approach in this case is similar to that of Case 1.   With a lead network inserted into the feedback loop, the new ramp error coefficient is

$$K_1' = 0.88A_1 \tag{14-67}$$

When this result is compared with Eq. (14-61), it is seen that a larger improvement in the error coefficient is obtained than in Case 1.

The design procedure for determining A_1, K_t, and T follows:

1.   Select the desired value of $\omega_m = \omega_a$.
2.   For $\omega_m = \omega_a$, assume that the angle of $1 + j\omega_a T$ is a large value such as 85°.   The value of the time constant T can now be determined since

$$\tan 85° = \omega_a T$$

and $$T = \frac{\tan 85°}{\omega_a} = \frac{11.43}{\omega_a} \tag{14-68}$$

3.   Draw the phasor $G_x(j\omega_a)^{-1}$ and the M_m circle, as shown in Fig. 14-25.
4.   Through the $(-1 + j0)$ point draw a line parallel to the j axis until it intersects the top half of the M_m circle.   It is assumed that this point of intersection is approximately the point of tangency of the desired $I(j\omega)/C(j\omega)$ plot.   Draw the phasor $H(j\omega_a)$ at a 95° angle from this point until it intersects the phasor $1/G_x(j\omega_a)$, as shown in Fig. 14-25.
5.   The intersection of step 4 locates the point $1/A_1 G_x(j\omega_a)$.   Dividing the phasor $1/G_x(j\omega_a)$ by $1/A_1 G_x(j\omega_a)$ yields the value A_1.
6.   The phasor between the point of tangency on the M_m circle and the end of the phasor $1/A_1 G_x(j\omega_a)$ is equal to $H(j\omega_a)$.   This phasor is equal to

$$H(j\omega_a) = \frac{TK_t(j\omega_a)^2}{1 + j\omega_a T}$$

Because the angle of $1 + j\omega_a T$ is set at 85°, the magnitude $\omega_a T \gg 1$, and $\mathbf{H}(j\omega_a)$ can be approximated by

$$\mathbf{H}(j\omega_a) \approx K_t \omega_a \underline{/95°} \tag{14-69}$$

From the measured length of this phasor the value of K_t can be determined.

7. With these values of A_1, K_t, and T draw the plot

$$\frac{\mathbf{I}(j\omega)}{\mathbf{C}(j\omega)} = \frac{1}{A_1\mathbf{G}_x(j\omega)} + \mathbf{H}(j\omega)$$

If this plot is not approximately tangent to the desired M_m circle, choose a different value of ω_a and/or point of tangency on the M_m circle.

Case 4

$$A_1 = 1 \qquad \text{and} \qquad H_1(s) = \frac{Ts}{1 + Ts}$$

The feedback transfer function of the minor loop is given by

$$\mathbf{H}(j\omega) = j\omega K_t \mathbf{H}_1(j\omega) = \frac{TK_t(j\omega)^2}{1 + j\omega T} \tag{14-70}$$

The approach in this case is similar to that of Case 2. With a lead network inserted into the feedback loop the new ramp error coefficient is

$$K_1' = 0.88A_2 \tag{14-71}$$

When this is compared with Eq. (14-64), it is seen that a larger improvement in the error coefficient is obtained over Case 2.

The design procedure for determining A_2, K_t, and T is as follows:

1. Select the desired value of $\omega_m = \omega_a$.
2. For $\omega = \omega_a$, assume that the angle of the denominator term $1 + j\omega_a T$ is 85°. The value of the time constant T can be determined, using

$$T = \frac{\tan 85°}{\omega_a} = \frac{11.43}{\omega_a} \tag{14-72}$$

3. Draw the $\mathbf{G}_x(j\omega)^{-1}$ curve and mark the phasor $\mathbf{G}_x(j\omega_a)^{-1}$.
4. The phasor $\mathbf{H}(j\omega_a)$ can be represented by

$$\mathbf{H}(j\omega_a) \approx K_t \omega_a \underline{/95°}$$

Draw a line through the point $1/\mathbf{G}_x(j\omega_a)$ at an angle of 95°, representing $\mathbf{H}(j\omega_a)$, until it intersects the line representing the angle

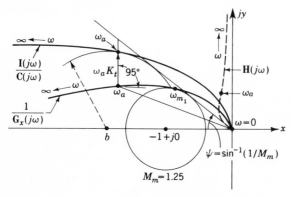

Fig. 14-26 *Polar plots of various portions of the control system of Fig. 14-22 with $A_1 = 1$ and $H_1(s) = Ts/(1 + Ts)$.*

$\psi = \sin^{-1}(1/M_m)$ (see Fig. 14-26).   Then select a trial point on this line.   A point one-half to three-quarters of the distance up from the $1/\mathbf{G}_x(j\omega_a)$ point is a good first trial.   This point is made

$$\frac{\mathbf{I}(j\omega_a)}{\mathbf{C}(j\omega_a)} = \frac{1}{\mathbf{G}_x(j\omega_a)} + \mathbf{H}(j\omega_a)$$

The length of the phasor between points $1/\mathbf{G}_x(j\omega_a)$ and $\mathbf{I}(j\omega_a)/\mathbf{C}(j\omega_a)$ is equal to $K_t\omega_a$.   Thus the value of K_t can now be determined.

5.   Using this value of K_t, plot $\mathbf{I}(j\omega)/\mathbf{C}(j\omega)$.   By the gain-adjustment techniques of Chap. 11, determine the required value of A_2, that is, $A_2 = ob$.

14-15   *Results of Feedback Compensation Using the Inverse Polar Plot*

The original forward transfer function is

$$\mathbf{G}_x(j\omega) = \frac{0.88}{j\omega(1 + j\omega)(1 + j0.2\omega)}$$

For all four cases a value of $\omega_m = \omega_a = 3$ is selected as the desired resonant frequency.   Applying the design procedure of Case 1, as shown in Fig. 14-27, gives $K_t\omega_a \approx 1.033$, $K_t = 0.344$, and $A_1 \approx 12.3$.   With these values of K_t and A_1 determined, the complete plot of $\mathbf{I}(j\omega)/\mathbf{C}(j\omega)$ can be graphically drawn, as indicated in Fig. 14-27.

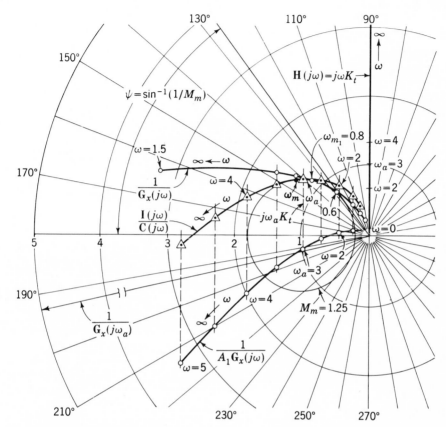

Fig. 14-27 *The graphical design procedures of Case 1 applied to the control system of Fig. 14-22, where $A_2 = 1$ and $\mathbf{H}_1(j\omega) = 1\underline{/0°}$.*

For Case 2, Fig. 14-28 shows the construction and evaluation of the constants as $K_t\omega_a = 12$, $K_t = 4$, and $A_2 \approx 11.4$. The resulting plot of $\mathbf{I}(j\omega)/\mathbf{C}(j\omega)$ is shown.

For Cases 3 and 4, having chosen the value of $\omega_m = \omega_a = 3$ yields

$$T = \frac{11.43}{3} = 3.81 \text{ sec}$$

From Fig. 14-29 for Case 3 the following data are obtained: $K_t\omega_a \approx 1.033$, $K_t \approx 0.344$, and $A_1 \approx 13.8$. The resulting feedback transfer function for Case 3 is

$$\mathbf{H}(j\omega) = \mathbf{H}_1(j\omega)K_tj\omega = \frac{1.31(j\omega)^2}{1 + j3.81\omega} \qquad (14\text{-}73)$$

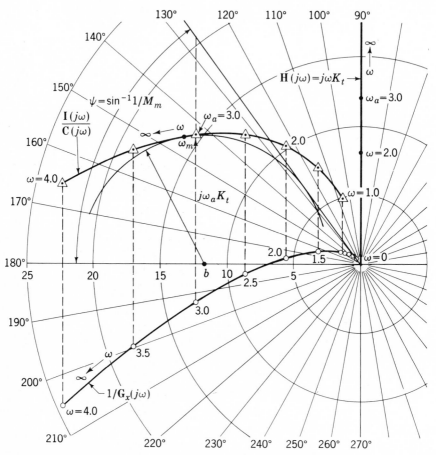

Fig. 14-28 *The graphical design procedures of Case 2 applied to the control system of Fig. 14-22, where $A_1 = 1$ and $\mathbf{H}_1(j\omega) = 1\underline{/0°}$.*

For Case 4, Fig. 14-30 yields $K_t\omega_a = 13.5$, $K_t = 4.5$, and $A_2 = 13.4$. Thus

$$\mathbf{H}(j\omega) = \mathbf{H}_1(j\omega)K_t j\omega = \frac{17.2(j\omega)^2}{1 + j3.81\omega} \tag{14-74}$$

Table 14-5 gives a summary of the results for all four cases. Comparing the results of Cases 1 and 2 shows that, for approximately the same value of gain in A_2 and A_1, approximately the same value of ramp error coefficient is obtained. A value of K_t greater than 1 generally requires an additional amplifier in conjunction with the tachometer. Thus, when comparing the values of K_t a designer would select Case 1 over Case 2 since it requires minimum equipment and results in minimum noise effects.

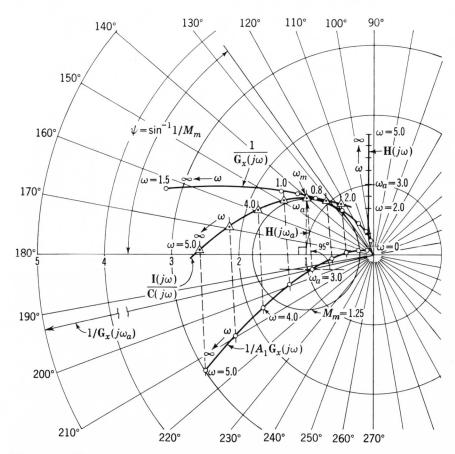

Fig. 14-29 *The graphical design procedures of Case 3 applied to the control system of Fig. 14-22, where $A_2 = 1$ and $\mathbf{H}_1(j\omega) = j\omega T/(1 + j\omega T)$.*

Table 14-5. **Results of applying feedback compensation design procedures by the inverse polar plot to the control system of Fig. 14-22**

Design	A_1	A_2	K_t	K_1	T	ω_m	$H(s)$
Uncompensated	—	—	—	0.88	—	0.72	—
Case 1	12.3	1.0	0.344	2.29	—	3.0	1.0
Case 2	1.0	11.4	4.0	2.21	—	3.0	1.0
Case 3	13.8	1.0	0.344	12.0	3.81	3.0	$\dfrac{1.31s^2}{1 + 3.81s}$
Case 4	1.0	13.4	4.5	11.65	3.81	3.0	$\dfrac{17.2s^2}{1 + 3.81s}$

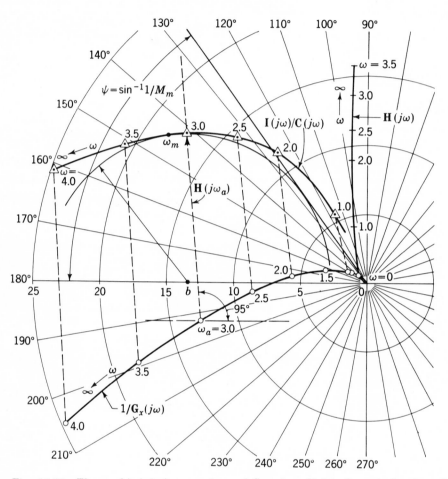

Fig. 14-30 *The graphical design procedures of Case 4 applied to the control system of Fig. 14-22, where $A_1 = 1$ and $\mathbf{H}_1(j\omega) = j\omega T/(1 + j\omega T)$.*

Table 14-6. Results of lag and lead cascade compensation and feedback compensation of the same system by the frequency response method

Type of compensation	K_1	ω_m	A_1
Cascade lag compensator	6.68	0.54	7.6
Cascade lead compensator	3.10	2.95	35.2
Cascade lag-lead compensator	26.3	2.2	29.9
Feedback tachometer compensation, Case 1	2.3	3.0	12.3
Feedback tachometer plus lead compensation, Case 3	12.0	3.0	13.8

The same effects as above can be noted when comparing Cases 3 and 4. For the same reasons, Case 3 would be chosen as the preferred solution. Note that the addition of a lead compensator in the feedback loop gives a large improvement in the ramp error coefficient (the ω_m being held the same for all cases). This effect is similar to that obtained when inserting a lag compensator in cascade.

Table 14-6 presents a summary of the results of the cascade lag, lead, and lag-lead compensation of Chap. 13. Also included are the results of Cases 1 and 3 using a minor feedback loop for compensation of the same basic system. These results indicate that improvements in system performance can be obtained by either cascade or feedback compensation. In all cases a compromise must be made between improvement of the error coefficient and the resonant frequency. These results also show, as was shown by the root-locus method, that the feedback compensator should have a zero at $s = 0$ which is of higher order than the original system type.

The examples given are intended to be typical solutions illustrating feedback compensation by means of the inverse polar plot. Other networks and techniques of feedback compensation exist. Once an individual has obtained sufficient experience, he can devise his own design procedures for a particular situation to meet specified performance requirements, utilizing techniques similar to the ones used here.

14-16   *Feedback Compensation Using Log Plots*[1]

The previous sections have shown that the procedures for applying feedback compensation by the root-locus and polar-plot methods are more involved than those for applying cascade compensation. The same is true when using log plots. However, it should be restated that once the feedback compensation has been designed it may have distinct advantages in the control system. For example, it may be easier to build, install, and adjust the feedback compensator. The feedback compensator may be physically smaller, may require the use of smaller gains, or may better achieve the desired performance.

It is difficult to make generalizations for applying feedback compensation that are simple enough to apply universally. Therefore, the methods applied in this chapter by the log plots are based on specific examples. These methods can then be extended and modified to new cases. The case to be studied is a system with a minor loop, as shown in Fig. 14-31. The transfer function $\mathbf{G}_x(j\omega)$ represents the basic system, $\mathbf{H}(j\omega)$ is the feedback compensator forming a minor loop around $\mathbf{G}_x(j\omega)$, and A is an amplifier

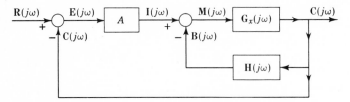

Fig. 14-31 *Block diagram for feedback compensation.*

which is used to adjust the over-all system performance. The forward transfer function of this system is

$$G_1(j\omega) = \frac{C(j\omega)}{I(j\omega)} = \frac{G_x(j\omega)}{1 + G_x(j\omega)H(j\omega)} \tag{14-75}$$

For the case of cascade compensation the effect of the compensator on the forward transfer function is directly determined. However, the effect of the feedback $H(j\omega)$ on the forward transfer function of the system is not easily determined. Therefore new techniques must be developed. This is done by first using some approximations and the straight-line log magnitude curves and then developing an exact procedure. Consider the case when $|G_x(j\omega)H(j\omega)| \ll 1$. The forward transfer function can be approximated by

$$G_1(j\omega) \approx G_x(j\omega) \qquad \text{for } |G_x(j\omega)H(j\omega)| \ll 1 \tag{14-76}$$

The next condition is $|G_x(j\omega)H(j\omega)| \gg 1$. Then the forward transfer function can be approximated by

$$G_1(j\omega) \approx \frac{1}{H(j\omega)} \qquad \text{for } |G_x(j\omega)H(j\omega)| \gg 1 \tag{14-77}$$

There is still undefined the condition when $|G_x(j\omega)H(j\omega)| \approx 1$, in which case neither Eq. (14-76) nor Eq. (14-77) is applicable. In the approximate procedure this condition is neglected and Eqs. (14-76) and (14-77) are used when $|G_x(j\omega)H(j\omega)| < 1$ and $|G_x(j\omega)H(j\omega)| > 1$, respectively. This approximation allows investigation of the qualitative results to be obtained. After these results are found to be satisfactory, the refinements for an exact solution are introduced.

An example illustrates the use of these approximations. Assume that $G_x(j\omega)$ represents a motor having inertia and damping. The transfer function can be represented by

$$G_x(j\omega) = \frac{K_1}{j\omega(1 + j\omega T)} \tag{14-78}$$

Let the feedback $H(j\omega) = 1/\underline{0°}$. This is a sufficiently simple problem that it can be solved exactly algebraically. However, use is made of the log

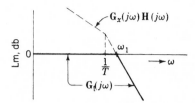

Fig. 14-32   *Log magnitude curve for*

$$\mathbf{G}_x(j\omega)\mathbf{H}(j\omega) = K_1/j\omega(1 + j\omega T)$$

magnitude curve and the approximate conditions.   In Fig. 14-32 is sketched the log magnitude curve for $\mathbf{G}_x(j\omega)\mathbf{H}(j\omega)$.

From Fig. 14-32 it is seen that $|\mathbf{G}_x(j\omega)\mathbf{H}(j\omega)| > 1$ for all frequencies below ω_1. With the approximation, the value of $\mathbf{G}_1(j\omega)$ can be represented by $1/\mathbf{H}(j\omega)$ for frequencies below ω_1. Also, $|\mathbf{G}_x(j\omega)\mathbf{H}(j\omega)| < 1$ for all frequencies above ω_1. Therefore $\mathbf{G}_1(j\omega)$ can be represented as shown by the line of zero slope and 0 db for frequencies up to ω_1 and the line of -12 db slope above ω_1. The equation of $\mathbf{G}_1(j\omega)$ therefore has a quadratic in the denominator with $\omega_n = \omega_1$:

$$\mathbf{G}_1(j\omega) = \frac{1}{1 + 2\zeta j(\omega/\omega_1) + [j(\omega/\omega_1)]^2} \tag{14-79}$$

Of course this could be obtained algebraically for this simple case from Eq. (14-75), with the result given as

$$\mathbf{G}_1(j\omega) = \frac{1}{1 + (1/K_1)j\omega + (j\omega)^2 T/K_1} \tag{14-80}$$

where $\omega_1 = \omega_n = (K_1/T)^{1/2}$ and $\zeta = 1/[2(K_1 T)^{1/2}]$. Note that the approximate result is basically correct but some detail information, in this case the value of ζ, is missing. The approximate angle curve can be drawn to correspond to the log magnitude curve or to Eq. (14-79).

For any case the exact curve of $\mathbf{G}_1(j\omega)$ can always be obtained in the following way.   First the forward transfer function is put in the form

$$\mathbf{G}_1(j\omega) = \frac{1}{\mathbf{H}(j\omega)}\left[\frac{\mathbf{G}_x(j\omega)\mathbf{H}(j\omega)}{1 + \mathbf{G}_x(j\omega)\mathbf{H}(j\omega)}\right] = \frac{1}{\mathbf{H}(j\omega)}\left[\frac{\mathbf{G}_0(j\omega)}{1 + \mathbf{G}_0(j\omega)}\right] \tag{14-81}$$

The quantity in brackets in Eq. (14-81) can be obtained by use of the Nichols chart.   In other words, the log magnitude and angle of

$$\mathbf{G}_0(j\omega) = \mathbf{G}_x(j\omega)\mathbf{H}(j\omega)$$

are plotted on the Nichols chart.   From the intersections of the $\mathbf{G}_0(j\omega)$ curve with the M and α curves the data are taken to plot $\mathbf{G}_0(j\omega)/[1 + \mathbf{G}_0(j\omega)]$ in both log magnitude and phase angle.   The curve of $1/\mathbf{H}(j\omega)$ is then added to the curve of $\mathbf{G}_0(j\omega)/[1 + \mathbf{G}_0(j\omega)]$ to produce the curve of $\mathbf{G}(j\omega)$.

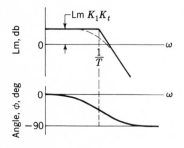

Fig. 14-33 *Log magnitude and phase angle diagram of* $\mathbf{G}_x(j\omega)\mathbf{H}(j\omega) = K_1K_t/(1 + j\omega T)$.

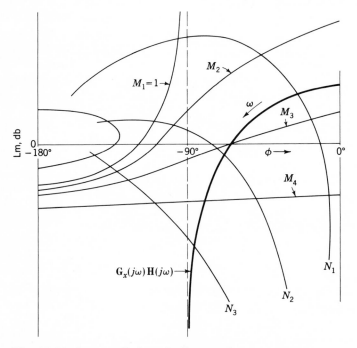

Fig. 14-34 *Log magnitude–angle diagram of*
$$\mathbf{G}_x(j\omega)\mathbf{H}(j\omega) = K_1K_t/(1 + j\omega T)$$

By applying the exact method to the following example,

$$\mathbf{G}_x(j\omega) = \frac{K_1}{j\omega(1 + j\omega T)} \tag{14-82}$$

$$\mathbf{H}(j\omega) = K_t j\omega \tag{14-83}$$

the log magnitude and phase angle diagram of $\mathbf{G}_0(j\omega) = \mathbf{G}_x(j\omega)\mathbf{H}(j\omega)$ is drawn in Fig. 14-33. The data from these curves are transferred to the Nichols chart in Fig. 14-34. From this figure, data are obtained from the intersections with M and α curves for the log magnitude and phase angle

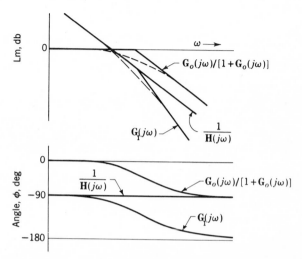

Fig. 14-35   *Log magnitude and phase angle diagrams of*

$$\frac{G_0(j\omega)}{1 + G_0(j\omega)} \qquad \frac{1}{H(j\omega)} \qquad G_1(j\omega) = \frac{1}{H(j\omega)} \frac{G_0(j\omega)}{1 + G_0(j\omega)}$$

of $G_0(j\omega)/[1 + G_0(j\omega)]$, which are plotted in Fig. 14-35, along with the log magnitude and phase angle curves of $1/H(j\omega)$. The curves are then combined to obtain the exact curves for $G_1(j\omega)$. This gives the exact curves but is more work than the approximate procedure.

14-17   *Application of Feedback Compensation (Log Plots)*

The system of Fig. 14-31 is investigated with the value of $G_x(j\omega)$ given by

$$G_x(j\omega) = \frac{K_x}{j\omega(1 + j\omega)(1 + j0.2\omega)} \qquad (14\text{-}84)$$

The system having this transfer function has been used throughout this text for the various methods of compensation. The log magnitude and phase angle diagram using the straight-line Lm curve is drawn in Fig. 14-36. A phase margin of 45° occurs at the frequency $\omega_{\phi 1} = 0.8$ with a gain of -2 db. The object in applying compensation is to increase both the phase-margin frequency and the gain. In order to select a feedback compensator $H(j\omega)$, recall the following facts:

1.   The system type should be maintained. In accordance with the conditions outlined in Sec. 14-5 and the results given in Sec. 14-11

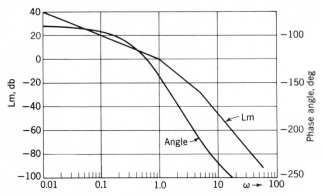

Fig. 14-36 *Log magnitude–phase angle diagram of* $\mathbf{G}'_x(j\omega) = 1/j\omega(1 + j\omega)(1 + j0.2\omega)$.

for the root-locus method and Sec. 14-15 for the polar-plot method, it is known that good improvement is achieved by using a feedback compensator $\mathbf{H}(j\omega)$ which has a zero, $s = 0$, of order equal to or, preferably, higher than the type of the original transfer function $\mathbf{G}_x(j\omega)$.

2. It is shown in Sec. 14-16 that the new forward transfer function can be approximated by

$$\mathbf{G}_1(j\omega) \approx \begin{cases} \mathbf{G}_x(j\omega) & \text{for } |\mathbf{G}_x(j\omega)\mathbf{H}(j\omega)| < 1 \\ \dfrac{1}{\mathbf{H}(j\omega)} & \text{for } |\mathbf{G}_x(j\omega)\mathbf{H}(j\omega)| > 1 \end{cases}$$

Now apply this information to a consideration of the proper $\mathbf{H}(j\omega)$ to use. Consider the possibility of replacing a portion of the curves of $\mathbf{G}_x(j\omega)$ for a range of frequencies by the curves of $1/\mathbf{H}(j\omega)$ with the intent of increasing the values of ω_ϕ and the gain. This requires that the value of $\mathbf{H}(j\omega)$ be such that $|\mathbf{G}_x(j\omega)\mathbf{H}(j\omega)| > 1$ for that range of frequencies. A feedback unit using a tachometer and an RC derivative network is considered. The circuit is shown in Fig. 14-14, and the transfer function is

$$\mathbf{H}(j\omega) = \frac{K_t T(j\omega)^2}{1 + j\omega T} \tag{14-85}$$

The log magnitude and phase angle diagram for $1/\mathbf{H}'(j\omega)$ is shown in Fig. 14-37 with a value of $T = 1$. These curves shift to the left or to the right as T is increased or decreased, respectively. The angle curve $1/\mathbf{H}(j\omega)$ shows that the phase margin can be obtained as a function of the compensator time constant T. This shows promise for increasing the phase margin, provided that the magnitude $|\mathbf{G}_x(j\omega)\mathbf{H}(j\omega)|$ can be made greater than unity

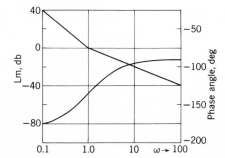

Fig. 14-37 *Log magnitude and phase angle diagram of* $1/\mathbf{H}'(j\omega) = (1 + j\omega)/(j\omega)^2$.

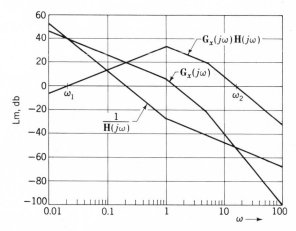

Fig. 14-38 *Log magnitude plots of* $1/H(j\omega)$, $G_x(j\omega)$, *and* $G_x(j\omega)H(j\omega)$.

over the correct range of frequencies. Also, the larger K_tT is made, the smaller becomes the magnitude of $1/\mathbf{H}(j\omega)$ at the phase-margin frequency. This permits a large value of A and therefore a large error coefficient.

The selection of T and K_t must be based on a trial-and-error procedure. The object is to produce a section of the $\mathbf{G}_1(j\omega)$ log magnitude curve with a slope of -20 db/decade and with a smaller magnitude than (i.e., it is below) the $\mathbf{G}_x(j\omega)$ log magnitude curve. This new section of the $\mathbf{G}_1(j\omega)$ curve must occur at a higher frequency than the original phase-margin frequency. Section 9-9 describes the desirability of a slope of -20 db/decade to produce a large phase margin. To achieve compensation, the log magnitude curve $1/\mathbf{H}(j\omega)$ is placed over the $\mathbf{G}_x(j\omega)$ curve so that there are one or two points of intersection. There is no reason to restrict the value of K_x. One such arrangement is shown in Fig. 14-38, with $K_x = 2$, $T = 1$, $K_tT = 25.1$, and intersections between the two curves occur at $\omega_1 = 0.021$ and $\omega_2 = 16$.

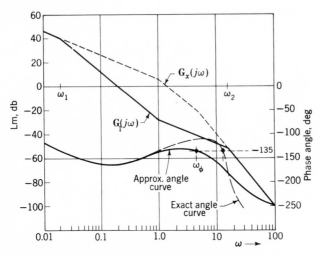

Fig. 14-39 *Log magnitude and phase angle diagram of*

$$G_1(j\omega) = \frac{2(1 + j\omega)}{j\omega(1 + j47.5\omega)(1 + j0.0625\omega)^2}$$

The log magnitude plot of $G_x(j\omega)H(j\omega)$ is also drawn in Fig. 14-38 and shows that

$$|G_x(j\omega)H(j\omega)| < 1 \qquad \text{for } \omega_2 < \omega < \omega_1$$
$$\text{and} \qquad |G_x(j\omega)H(j\omega)| > 1 \qquad \text{for } \omega_2 > \omega > \omega_1$$

Therefore $G_1(j\omega)$ can be represented approximately by

$$G_1(j\omega) \approx \begin{cases} G_x(j\omega) & \text{for } \omega_2 < \omega < \omega_1 \\ \dfrac{1}{H(j\omega)} & \text{for } \omega_2 > \omega > \omega_1 \end{cases}$$

The composite curve can therefore be represented by

$$G_1(j\omega) = \frac{2(1 + j\omega)}{j\omega(1 + j47.5\omega)(1 + j0.0625\omega)^2} \qquad (14\text{-}86)$$

The log magnitude and phase angle diagram for Eq. (14-86) is shown in Fig. 14-39. The new value of phase-margin frequency is $\omega_\phi = 4.5$, and the gain is 41 db. This shows a considerable improvement of the system performance.

An exact curve of $G_1(j\omega)$ can be obtained in either of two ways. The first method is graphical and consists in plotting $G_x(j\omega)H(j\omega)$ on the Nichols chart. From this chart the function $G_x(j\omega)H(j\omega)/[1 + G_x(j\omega)H(j\omega)]$ can be determined from the intersections of the $G_x(j\omega)H(j\omega)$ curve and the M

and α curves. The $1/\mathbf{H}(j\omega)$ curves can be added to the

$$\frac{\mathbf{G}_x(j\omega)\mathbf{H}(j\omega)}{1 + \mathbf{G}_x(j\omega)\mathbf{H}(j\omega)}$$

curves to obtain the composite curves of $\mathbf{G}_1(j\omega)$. This is necessarily a tedious process and is not done here. A second method is analytical. The functions $\mathbf{G}_x(j\omega)$ and $\mathbf{H}(j\omega)$ were found to be

$$\mathbf{G}_x(s) = \frac{2}{s(1 + s)(1 + 0.2s)} \qquad H(s) = \frac{25s^2}{1 + s}$$

The function $\mathbf{G}_1(j\omega)$ is obtained as follows:

$$
\begin{aligned}
\mathbf{G}_1(j\omega) &= \frac{\mathbf{G}_x(j\omega)}{1 + \mathbf{G}_x(j\omega)\mathbf{H}(j\omega)} \\
&= \frac{2(1 + j\omega)}{j\omega(1 + j\omega)^2(1 + j0.2\omega) + 50(j\omega)^2} \\
&= \frac{2(1 + j\omega)}{j\omega[0.2(j\omega)^3 + 1.4(j\omega)^2 + 52.2(j\omega) + 1]} \\
&= \frac{2(1 + j\omega)}{j\omega(1 + j52\omega)[0.00385(j\omega)^2 + 0.027(j\omega) + 1]} \qquad (14\text{-}87)
\end{aligned}
$$

A comparison of Eqs. (14-86) and (14-87) shows that there is little difference between the approximate and the exact equations. The main difference is that the approximate equation has the term $(1 + j0.0625\omega)^2$, which assumes a damping ratio $\zeta = 1$ for this quadratic factor. The actual damping ratio is $\zeta = 0.217$ in the correct quadratic factor $[0.00385(j\omega)^2 + 0.027(j\omega) + 1]$. The corner frequency is the same for both cases, $\omega = 16$. As a result, the exact angle curve, given in Fig. 14-39, shows a higher phase-margin frequency $(\omega = 13)$ than that obtained for the approximate curve $(\omega = 4.5)$. If the basis of design is the value of M_m, the exact curve should be used.

The adjustment of the amplifier gain A is based on either the phase margin or M_m. An approximate value for A is the gain necessary to raise the curve of Lm $\mathbf{G}_1(j\omega)$ so that it crosses the 0-db line at ω_ϕ. A more precise value for A is obtained by plotting $\mathbf{G}_1(j\omega)$ on the log magnitude–angle diagram and adjusting the system for a desired value of M_m. The exact Lm $\mathbf{G}_1(j\omega)$ curve should be used for best results.

14-18   Results of Feedback Compensation Using the Logarithmic Plots

The example of Sec. 14-17 has demonstrated qualitatively that feedback compensation can be used to produce a section of the log magnitude curve of the forward transfer function $G(j\omega)$ with a slope of -20 db/decade. This section of the curve can be placed in the desired frequency range, with the result that the new phase-margin frequency is larger than that of the original system. However, to obtain the correct quantitative results the exact curves should be used instead of the straight-line approximations. Also, the use of the log magnitude–angle diagram as the last step in the design process, to adjust the gain for a specified M_m, gives a better indication of performance than does the phase margin.

The procedure in applying the feedback compensator essentially is based on a number of trials to select the best results. This is necessarily the case since there are three constants that must be adjusted. Figures 14-38 and 14-39 show that the phase-margin frequency is dependent on the frequency ω_2 at which the $1/H(j\omega)$ and $G_x(j\omega)$ curves cross. The larger ω_2 is made, the larger ω_ϕ becomes. The frequency ω_2 can be changed by adjusting any one of the three coefficients K_x, K_t, and T. The designer must therefore use his judgment in order to limit the number of trials that are necessary. Use of the logarithmic plots is probably more flexible than use of the polar plots and gives a better indication of the adjustments in the magnitudes of K_x, K_t, and T that improve the performance.

The system designer now has the tools for extending and modifying the procedures of applying feedback compensation by the use of log plots.

Bibliography

1. Chestnut, H., and R. W. Mayer: "Servomechanisms and Regulating System Design," 2d. ed., vol. 1, chaps. 8, 10, and 12, John Wiley & Sons, Inc., New York, 1959.
2. Truxal, J. G.: "Automatic Feedback Control System Synthesis," McGraw-Hill Book Company, New York, 1955.
3. Brown, G. S., and D. P. Campbell: "Principles of Servomechanisms," chaps. 7 and 8, John Wiley & Sons, Inc., New York, 1948.
4. Carter, W. C.: Accuracy and Control System Gain, *Control Eng.*, vol. 10, pp. 102–104, March, 1963.
5. "Sensitivity and Modal Response for Single-loop and Multiloop Systems," Technical Documentary Report ASD-TDR-62-812, Flight Control Laboratory, ASD, AFSC, Wright-Patterson AFB, Ohio, January, 1963.

15

Complex
Control
Systems

15-1   Introduction

Up to this point only simple control systems with a single desired input signal have been considered. Although systems are generally more complex, the simple ones have been used for the basic purpose of establishing definitions, a method of analysis, and the techniques of compensation. With this basic purpose accomplished, attention can now be focused on more complex systems and systems with more than one input signal (desired and/or unwanted signals).[1-3]

A study of complex control systems can be divided into the two general categories of simple and intercoupled multiple-loop systems. A simple system has only one loop, whereas a multiple-loop system comprises more than one loop. Multiple loops may arise from the inherent nature of the system being controlled. This means that the system is not a simple one but is made up of several interconnected loops. These additional loops may be due to the physical aspects or the nature of the control problem. Also, to obtain satisfactory performance, it may be necessary to use additional loops intentionally.

An airplane has motion around three axes, as described in Sec. 5-11. Because of the inherent interrelationship of the rolling and yawing motions, the control of these two motions cannot be considered separately. Achieving stability of the airplane and suitable performance must be based on considering the airplane as a whole.

15-2 *Multiple-loop Control Systems with a Single Desired Input*

A multiloop control system having a single input and a single output is shown by the block diagram of Fig. 15-1. The simple Type 1 control system of Chap. 14, for which two different cases of feedback compensation are considered, is of the form represented by this block diagram. In these cases compensation is achieved through the use of a frequency-sensitive feedback unit in conjunction with appropriate amplifiers. If further improvement is desired, consideration can be given to the insertion of a frequency-sensitive network plus amplifier in cascade with the minor loop, as shown in Fig. 15-1. This results in a complex control system. This complex system can also arise if G_1 and G_2 are part of the basic system and the feedback compensation network can be inserted only in the manner shown in Fig. 15-1.

Table 14-6 gives the performance obtained for the Type 1 control system when one of several cascade and feedback compensators is applied. The increase in the resonant frequency ω_m when feedback compensation is used is as good as or better than that obtained by either a lead or a lag-lead cascade network. However, the improvement in the error coefficient K_1 is not as good as that provided by the use of the lag-lead cascade network. Thus, if

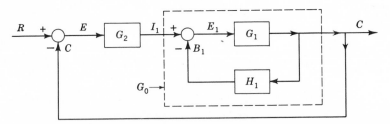

Fig. 15-1 *A complex control system.*

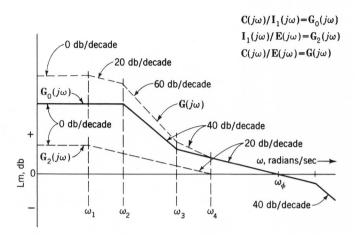

Fig. 15-2 *The over-all forward transfer function of the control system of Fig. 15-1, utilizing a lag compensator for* $G_2(j\omega)$.

G_2, in Fig. 15-1, is composed of a lag network and an amplifier, a further improvement in K_1 can be made without appreciably affecting the stability of the system. The method suggested above for further improvement in K_1 does not bar other possible methods. For example, it may be possible to obtain the desired improvement by the use of a double RC network for the feedback function H_1, instead of the lag network for G_2. The choice of one method over the other is determined by the practical reasons given at the beginning of Chap. 14. In addition to these, one must consider loading and interaction between circuits. Also, the magnitude of the signals in the system may be of such a large value as to cause saturation of some of the system components, thus resulting in nonlinear operation.

As another example, consider a Type 0 system that can also be represented by the block diagram of Fig. 15-1. Figure 15-2 illustrates the use of the cascaded lag network $G_2(s)$, having the form

$$G_2(s) = A \frac{1 + Ts}{1 + \alpha Ts} \qquad \alpha > 1 \tag{15-1}$$

The use of a log magnitude diagram clearly indicates the decided improvement in the step error coefficient for this Type 0 control system.

From Fig. 15-1 the following equivalent transfer function is obtained for the minor loop bracketed by the dashed lines:

$$\frac{C(s)}{I_1(s)} = \frac{G_1(s)}{1 + G_1(s)H_1(s)} = G_0(s) \tag{15-2}$$

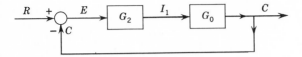

Fig. 15-3   *Equivalent block diagram of Fig. 15-1.*

This results in the equivalent block diagram of Fig. 15-3, whose forward and closed-loop transfer functions are

$$\frac{C(s)}{E(s)} = G(s) = G_2(s)G_0(s) \tag{15-3}$$

and

$$\frac{C(s)}{R(s)} = \frac{G_2(s)G_0(s)}{1 + G_2(s)G_0(s)} = \frac{G(s)}{1 + G(s)} \tag{15-4}$$

15-3   A Complex Control System

Each simple control system discussed in the previous chapters is considered to be a complete or over-all control system.   Quite often these simple control systems form a part of a large and more comprehensive control system.   As an example, Sec. 14-6 illustrates how a high-gain Type 1 power element can be converted to a Type 0 element by use of a unity feedback loop.   This permits an input signal of low energy to be reproduced in the output but at a much higher power level.   This simple system could form a part of a more complete control system, as shown in Fig. 15-4.

The transfer function for the minor loop involving only the power element is

$$G_a = \frac{C}{I_1} = \frac{G_1}{1 + G_1} \tag{15-5}$$

The over-all forward transfer function is given by

$$G = \frac{C}{E} = \frac{G_3 G_2[G_1/(1 + G_1)]}{1 + G_2 H_2[G_1/(1 + G_1)]} \tag{15-6}$$

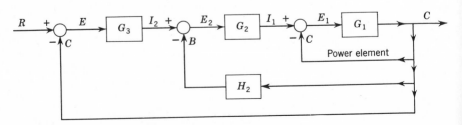

Fig. 15-4   *A multiloop control system.*

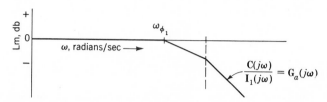

Fig. 15-5 *Log magnitude plot for* $C(j\omega)/I_1(j\omega)$ *of Fig. 15-4.*

As one can realize, this equation becomes quite complex when the actual expressions for the G functions are inserted. To determine the exact solution of $c(t)$ would be tedious and difficult. From the various methods of analysis presented, it has been shown that an exact solution is not necessary to obtain a satisfactory idea of how the control system will perform. An approximate idea of the response of the system in Fig. 15-4 can be obtained, as in most engineering problems, by making intelligent engineering assumptions.

The function $G_1(j\omega) = C(j\omega)/E_1(j\omega)$ has a phase-margin frequency $\omega_{\phi 1}$ at which the magnitude of $G_1(j\omega)$ is unity. In the frequency range below $\omega_{\phi 1}$ the value of $|G_1(j\omega)| > 1$. For frequencies above $\omega_{\phi 1}$ the value of $|G_1(j\omega)| < 1$. Therefore the value of $|C(j\omega)/I_1(j\omega)|$ can be approximated by

$$\left| \frac{C(j\omega)}{I_1(j\omega)} \right| \approx \begin{cases} 1 & \omega < \omega_{\phi 1} \\ |G_1(j\omega)| & \omega > \omega_{\phi 1} \end{cases} \tag{15-7}$$

This approximation makes the solution of the system comparatively simple when using the straight-line approximations for the log magnitude plots. The Lm plot for $C(j\omega)/I_1(j\omega)$, utilizing straight-line approximations, is shown in Fig. 15-5, which indicates the Type 0 characteristic of the minor loop, $C(j\omega)/I_1(j\omega)$. Also, it shows that good accuracy with no phase shift is maintained up to the frequency $\omega_{\phi 1}$. Above this frequency both attenuation and phase shift take place.

The response shown by Fig. 15-5 for the minor-loop response $C(j\omega)/I_1(j\omega)$ is a function of the power element selected to perform the control function. The range in which the frequency response is constant can be extended above $\omega_{\phi 1}$ by the use of $G_2(j\omega)$ in cascade and $H_2(j\omega)$ in a minor feedback loop. For example, $H_2(j\omega)$ may represent tachometric feedback and $G_2(j\omega)$ may represent an amplifier. The resultant frequency response can again be approximated by the straight-line log magnitude curves. The response of the minor loop from $C(j\omega)$ to $I_2(j\omega)$ is given by

$$G_b = \frac{C}{I_2} = \frac{G_1 G_2/(1 + G_1)}{1 + G_1 G_2 H_2/(1 + G_1)} = \frac{G_2 G_a}{1 + G_2 G_a H_2} \tag{15-8}$$

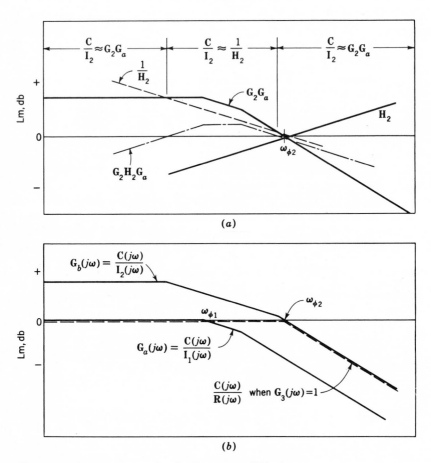

Fig. 15-6 *Log magnitude plots for the system of Fig. 15-4.*

When $|G_1G_2H_2/(1 + G_1)| < 1$, the response $\mathbf{C}(j\omega)/\mathbf{I}_2(j\omega)$ can be approximated by

$$\frac{C}{I_2} \approx \frac{G_1G_2}{1 + G_1} = G_2G_a \tag{15-9}$$

When $|G_1G_2H_2/(1 + G_1)| > 1$, the response $\mathbf{C}(j\omega)/\mathbf{I}_2(j\omega)$ can be approximated by

$$\frac{C}{I_2} \approx \frac{1}{H_2} \tag{15-10}$$

Figure 15-6a shows the plots of G_2G_a, H_2, $1/H_2$, and $G_2H_2G_a$ for the case where $\mathbf{G}_2(j\omega)$ is an amplifier and $\mathbf{H}_2(j\omega)$ is a tachometer. By utilizing the approximations of Eqs. (15-9) and (15-10), the approximate frequency response of $G_b = C/I_2$ is obtained as shown in Fig. 15-6a. The effect of G_2

and H_2 is to produce a log magnitude curve of $\mathbf{C}(j\omega)/\mathbf{I}_2(j\omega)$ which has a higher gain than the curve of $\mathbf{C}(j\omega)/\mathbf{I}_1(j\omega)$ and therefore has a higher phase-margin frequency $\omega_{\phi 2}$. This effect is shown in Fig. 15-6*b*.

With $\mathbf{G}_3(j\omega) = 1$, the over-all system response can be approximated by the methods previously used. In other words, the value of $|\mathbf{C}(j\omega)/\mathbf{R}(j\omega)|$ is equal to unity up to frequency $\omega_{\phi 2}$ and is equal to $|\mathbf{C}(j\omega)/\mathbf{I}_2(j\omega)|$ for frequencies above $\omega_{\phi 2}$. The minor loop that contains $\mathbf{G}_2(j\omega)$ and $\mathbf{H}_2(j\omega)$ has been used to improve the frequency response; this has been accomplished. The element $\mathbf{G}_3(j\omega)$ can now be used to determine the over-all open-loop frequency response for the range $0 < \omega < \omega_{\phi 2}$. This can result in a larger static error coefficient without appreciably affecting the closed-loop frequency response.

The straight-line technique of the Lm plots provides a relatively simple manner of obtaining and improving the performance of complex control systems. Once the approximate desired improvement has been made, any of the methods discussed in the previous chapters can be utilized to obtain an exact solution. This requires use of the exact curves and the Nichols chart.

15-4 *Intercoupled Multiple-loop Control Systems*[1]

Interaction, desired or undesired, between parts of the control system can result in a multiple-loop system. Figure 15-7 illustrates what is meant by the phrase *an intercoupled multiple-loop control system*. As the circuit now stands, it is very difficult to analyze and to determine the necessary improvement. Fortunately, the laws of association, distribution, and commutation permit the simplification of this system by use of the proper transformation.

For the actual system in Fig. 15-7*a*,

$$B_2 = H_2 I_2 \qquad (15\text{-}11)$$

To simplify this control system, move the feedback take-off point *a* (the input to H_2) to point *b*. The signal at point *b* is

$$C = I_2 G_1 \qquad (15\text{-}12)$$

In order not to alter the value of B_2 given by Eq. (15-11), an equivalent block $Q_2(s)$ must be inserted in cascade with H_2, as shown in Fig. 15-7*b*. The value of Q_2 can be determined as follows:

$$B_2 = H_2 Q_2 C = H_2 Q_2 G_1 I_2 \qquad (15\text{-}13)$$

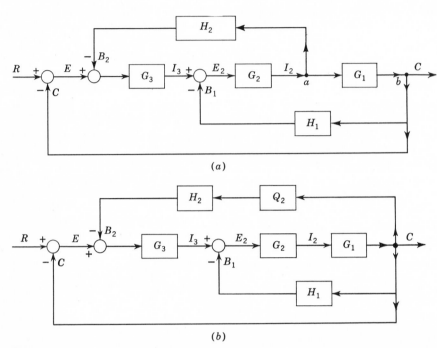

(a)

(b)

Fig. 15-7 *(a) An intercoupled multiple-loop control system.   (b) An equivalent block diagram after simplification.*

But B_2 must have the value given by Eq. (15-11); thus

$$B_2 = H_2 Q_2 G_1 I_2 = H_2 I_2$$

Therefore, $$Q_2 = \frac{1}{G_1} \qquad (15\text{-}14)$$

With this value of Q_2 inserted in cascade with H_2, as shown in Fig. 15-7b, this figure becomes the equivalent of Fig. 15-7a.

As long as the system remains within its linear range of operation, the equations that represent its performance have not been altered by this simplification.   In other words, the performance equations obtained from Fig. 15-7a and b are identical.   With this simplification the system can now be analyzed and improved by the techniques previously discussed.

By a similar approach, other transformations can be devised; they are summarized in Appendix J.[5]

The use of a signal flow graph and Mason's rule eliminates the need to rearrange the block diagram.   The signal flow graph for the system of Fig. 15-7a is shown in Fig. 15-8.   The system determinant is

$$\Delta = 1 - (-G_2 G_3 H_2 - G_1 G_2 H_1 - G_1 G_2 G_3)$$

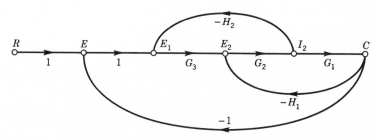

Fig. 15-8 *Signal flow graph for Fig. 15-7a.*

The over-all transmittance is

$$\frac{C}{R} = \frac{G_1 G_2 G_3}{1 + G_2 G_3 H_2 + G_1 G_2 H_1 + G_1 G_2 G_3} \tag{15-15}$$

15-5   *Multiple Inputs*

The problem of analysis and compensation for optimum performance of a control system has been relatively simplified, up to this point, by the consideration of only a single input. This signal is the desired reference input signal to which the system must respond. In many control problems this simplification does not exist. A system may have a single desired input signal but one or more unwanted input signals due to undesired disturbances. For example, a gust of wind on an airplane is an unwanted disturbance. A system may have more than one desired input signal but no or negligible unwanted input signals, or it may have more than one desired and one or more unwanted input signals.

The hydraulic position control system discussed in Sec. 5-12 is one example. Figures 5-26 and 5-27 indicate the undesired input signal resulting from a varying load-torque disturbance T_L. With the desired input signal constant, a variation of the torque load, on the output, causes the output position to change. This torque variation is an undesired disturbance and can be treated mathematically as an equivalent input signal.

The derivation of the differential equation of an amplidyne in Sec. 2-9 is based on a constant speed of rotation. However, if the amplidyne is motor-driven, its speed may drop as the load is increased. The output varies as a function of the amplidyne speed. The variation of the speed from the reference value can be treated as an additional input to the system.

A one-stage vacuum-tube amplifier can be treated in a similar manner. For example, if the d-c supply voltage varies, the operating range may be altered and the output changes. Such a change in the energizing voltage source can be treated as an additional input.

Since the systems considered are assumed to be linear, the method of analysis and compensation is relatively simplified by use of the linear superposition theorem. A more recent approach is characterizing a multiple-loop system by a set of mesh or nodal operational equations. In certain cases the technique of using *conditional feedback* systems,[6] as defined by G. Lang and J. M. Ham, may be useful in minimizing disturbances. In the sections to follow, each of the above methods is discussed.

15-6   *Treatment of Unwanted Disturbances*

When an unwanted disturbance is of sufficient magnitude, it reacts as an equivalent input signal to the control system so that it adversely affects the controlled variable. Examples of unwanted disturbances are a gust of wind upon a controlled directional antenna or on an airplane, recoil action of a gun, and noise resulting from the control-system components or from nearby magnetic fields inducing a signal within the system.

Consider the case of a gust of wind acting on the rotating antenna of a radar set. Figure 5-18 shows the forward loop of the speed control system for the antenna, excluding an amplifier gain A. The loading effect of the antenna is represented by the inertia and damping. It is highly desirable to attempt to modify the speed control system to minimize the effect of the wind load on the speed. Consider that this gust of wind is represented by an unwanted torque load T_L acting on the shaft. The direction in which it is acting determines the sign of its value. In this example it is assumed that it acts in the same direction as the motor rotation. A tachometer with a sensitivity factor K_t is utilized in the feedback to compare the output with the input. Equations (5-20) to (5-24) are now modified to take into account the unwanted torque load—neglecting L_g and L_m, which are small—and the addition of a cascade amplifier. They are written as transformed equations; thus,

$$E_f(s) = (L_f s + R_f)I_f(s) \tag{15-16}$$
$$E_g(s) = K_g I_f(s) \tag{15-17}$$
$$E_g(s) = (R_g + R_m)I(s) + K_b\Omega(s) \tag{15-18}$$
$$T(s) = K_T I(s) = (Js + B)\Omega(s) - T_L(s) \tag{15-19}$$
$$E_f(s) = AE(s) \tag{15-20}$$

At this point it is desirable to manipulate the above equations to treat T_L as another input signal along with the desired input signal. Equations (15-16) and (15-17) yield the portions of the forward path shown in the block diagram of Fig. 15-9.

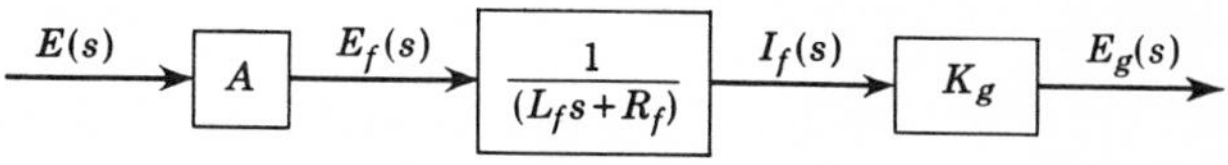

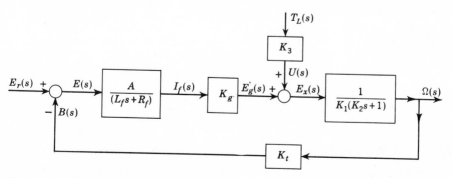

Fig. 15-9 *A portion of the forward transfer function of a speed control system.*

Solving Eq. (15-19) for $I(s)$ gives

$$I(s) = \left(\frac{J}{K_T} s + \frac{B}{K_T}\right) \Omega(s) - \frac{1}{K_T} T_L(s) \tag{15-21}$$

Substituting this equation into Eq. (15-18) gives

$$E_g(s) = (R_g + R_m)\left(\frac{J}{K_T} s + \frac{B}{K_T}\right) \Omega(s) - \frac{R_g + R_m}{K_T} T_L(s) + K_b\Omega(s) \tag{15-22}$$

Rearranging and simplifying Eq. (15-22) results in

$$E_g(s) = [K_1(K_2 s + 1)]\Omega(s) - K_3 T_L(s) \tag{15-23}$$

where

$$K_1 = \frac{(R_g + R_m)B + K_T K_b}{K_T}$$

$$K_2 = \frac{J(R_g + R_m)}{B(R_g + R_m) + K_T K_b}$$

$$K_3 = \frac{R_g + R_m}{K_T}$$

Moving the torque-load component of Eq. (15-23) to the left-hand side results in

$$E_x(s) = E_g(s) + K_3 T_L(s) = K_1(K_2 s + 1)\Omega(s) \tag{15-24}$$

Equation (15-24) shows that the controlled variable is now a function of both the signal $E_g(s)$ and the torque disturbance signal $T_L(s)$. The units

Fig. 15-10 *Block diagram of a speed control system with an undesired load disturbance.*

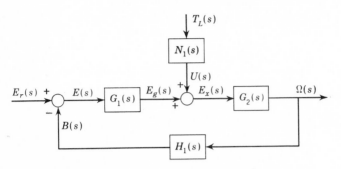

Fig. 15-11   *Simplified block diagram of Fig. 15-10.*

of K_3 are such that the quantity $K_3 T_L(s)$ has the unit of volts.   The block diagram in Fig. 15-9 can be completed by adding a summation point, to represent the left-hand side of Eq. (15-24), whose output $E_x(s)$ feeds into a block representing the right-hand side of this equation.   The complete diagram of this speed control system is shown in Fig. 15-10. $E_r(s)$ in this diagram represents the desired signal to which the system should respond.

For the purpose of analyzing this control system, the diagram of Fig. 15-10 is simplified in Fig. 15-11.   The equation relating the controlled variable to each of the two inputs is determined by the use of the linear superposition theorem.

First consider that the undesired input $T_L(s)$ is equal to zero.   This results in

$$\Omega(s)\Big]_{T_L=0} = \frac{G_1(s)G_2(s)}{1 + G_1(s)G_2(s)H_1(s)} \, E_r(s) \tag{15-25}$$

Next, consider that the desired input is equal to zero.   This yields

$$\Omega(s)\Big]_{E_r=0} = \frac{N_1(s)G_2(s)}{1 + G_1(s)G_2(s)H_1(s)} \, T_L(s) \tag{15-26}$$

Combining Eqs. (15-25) and (15-26) gives

$$\Omega(s) = \Omega(s)\Big]_{T_L=0} + \Omega(s)\Big]_{E_r=0} = \frac{G_1(s)G_2(s)}{1 + G_1(s)G_2(s)H_1(s)} \, E_r(s)$$
$$+ \frac{N_1(s)G_2(s)}{1 + G_1(s)G_2(s)H_1(s)} \, T_L(s) \tag{15-27}$$

Equation (15-27) must now be analyzed to determine what modifications must be made to the system in order to minimize the effect of the undesired input $T_L(s)$ upon the controlled variable $\Omega(s)$.

The analysis, as effected by frequency-response calculation, reveals the following:

1. Equations (15-25) and (15-26) have the identical factor

$$\frac{G_2(s)}{1 + G_1(s)G_2(s)H_1(s)}$$

2. To minimize the effect of $\mathbf{T}_L(j\omega)$, the ratio

$$\frac{\mathbf{\Omega}(j\omega)\big]_{E_r=0}}{\mathbf{\Omega}(j\omega)\big]_{T_L=0}} = \frac{\mathbf{N}_1(j\omega)}{\mathbf{G}_1(j\omega)}\frac{\mathbf{T}_L(j\omega)}{\mathbf{E}_r(j\omega)} \tag{15-28}$$

should be made as small as possible. For a given controlled element the function $\mathbf{N}_1(j\omega)$ represents a physical characteristic and thus cannot be altered. Therefore $\mathbf{G}_1(j\omega)$ should be made as large as possible.

A method of minimizing the effect of $\mathbf{T}_L(j\omega)$ (an undesirable input) upon the system performance can be determined as shown in the following example.

15-7 Example of Minimizing an Unwanted Disturbance

For the antenna control system discussed above, the following transfer functions are typical values for the physical system represented by the block diagram in Fig. 15-11:

$$G_1(s) = \frac{280}{1 + 0.08s} \tag{15-29}$$

$$G_2(s) = \frac{0.2}{1 + 0.6s} \tag{15-30}$$

$$H_1(s) = 0.2 \tag{15-31}$$

$$N_1(s) = 2 \times 10^{-3} \tag{15-32}$$

For $T_L(t) = 0$, the block diagram can be simplified to that of Fig. 15-12, for which Eq. (15-25) can be written as follows:

$$\begin{aligned}
\frac{\Omega(s)}{E_r(s)} &= \frac{G_1(s)G_2(s)}{1 + G_1(s)G_2(s)H_1(s)} = \frac{1}{H_1(s)}\frac{G_1(s)G_2(s)H_1(s)}{1 + G_1(s)G_2(s)H_1(s)} \\
&= \frac{1}{H_1(s)}\frac{G_0(s)}{1 + G_0(s)} = \frac{1}{H_1(s)}\frac{\Omega(s)}{E_{r0}(s)}
\end{aligned} \tag{15-33}$$

where $G_0(s) = G_1(s)G_2(s)H_1(s)$

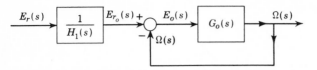

Fig. 15-12 *Simplification of Fig. 15-11 for $T_L = 0$.*

For $e_r(t) = 0$, the block diagram can be simplified to that of Fig. 15-13, for which Eq. (15-26) can be written as follows:

$$\frac{\Omega(s)}{U(s)} = \frac{G_2(s)}{1 + G_1(s)G_2(s)H_1(s)} = \frac{1}{G_1(s)H_1(s)} \frac{G_1(s)G_2(s)H_1(s)}{1 + G_1(s)G_2(s)H_1(s)}$$

$$= \frac{1}{G_1(s)H_1(s)} \frac{G_0(s)}{1 + G_0(s)} = \frac{1}{G_1(s)H_1(s)} \frac{\Omega(s)}{U_0(s)} \qquad (15\text{-}34)$$

where
$$\frac{\Omega(s)}{U_0(s)} = \frac{\Omega(s)}{E_{r_0}(s)}$$

The block diagrams of Figs. 15-12 and 15-13 can now be combined to show the simultaneous effect of both inputs, as shown in Fig. 15-14.

Figure 15-14 illustrates the results of the analysis made previously, namely, that only the $G_1(s)$ portion of the control system need be altered for an improvement in system performance. Utilizing the mathematical models of the control system, as represented by these figures, simplifies the work involved.   The logarithmic plots on the following pages demonstrate the ease with which a solution to the improvement of the system's performance by the use of these models is achieved.   A compensator $G_c(s)$ is inserted in cascade with $G_1(s)$ to improve the response to the desired input $[\Omega(s)/E_r(s)]$ and to attenuate the response to the undesired input $[\Omega(s)/U(s)]$. In the first trial a lag compensator of the form

$$[G_c(j\omega)]_a = 6.3 \frac{1 + j0.5\omega}{1 + j5.0\omega} \qquad (15\text{-}35)$$

is inserted.   On the second trial a lag-lead compensator is inserted whose transfer function is

$$[G_c(j\omega)]_b = 20 \frac{(1 + j\omega)(1 + j0.2\omega)}{(1 + j10\omega)(1 + j0.02\omega)} \qquad (15\text{-}36)$$

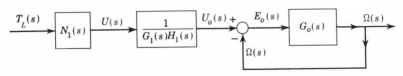

Fig. 15-13 *Simplification of Fig. 15-11 for $E_r = 0$.*

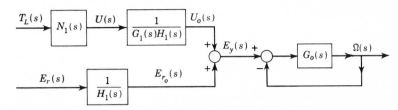

Fig. 15-14 *Modification of Fig. 15-11.*

For each case, respectively,

$$[\mathbf{G}_1(j\omega)]_a = A_a\mathbf{G}_1(j\omega)[\mathbf{G}_c'(j\omega)]_a \qquad (15\text{-}37)$$

and

$$[\mathbf{G}_1(j\omega)]_b = A_b\mathbf{G}_1(j\omega)[\mathbf{G}_c'(j\omega)]_b \qquad (15\text{-}38)$$

The plots involved in the solution appear in the following order:

Plot of	Figure number
$\mathbf{G}_0(j\omega) = \mathbf{G}_1(j\omega)\mathbf{G}_2(j\omega)\mathbf{H}_1(j\omega) = \dfrac{11.2}{(1 + j0.6\omega)(1 + j0.08\omega)}$	15-15
$\mathbf{M}_0(j\omega) = \dfrac{\mathbf{\Omega}(j\omega)}{\mathbf{E}_y(j\omega)} = \dfrac{\mathbf{G}_0(j\omega)}{1 + \mathbf{G}_0(j\omega)}$	15-15
$\dfrac{1}{\mathbf{G}_1(j\omega)\mathbf{H}_1(j\omega)} = \dfrac{1 + j0.08\omega}{56}$	15-16
$\mathbf{M}_u(j\omega) = \dfrac{\mathbf{\Omega}(j\omega)}{\mathbf{U}(j\omega)} = \dfrac{1}{\mathbf{G}_1(j\omega)\mathbf{H}_1(j\omega)}\dfrac{\mathbf{\Omega}(j\omega)}{\mathbf{E}_y(j\omega)}$	15-16
$[\mathbf{G}_0(j\omega)]_a = [\mathbf{G}_1(j\omega)]_a\mathbf{G}_2(j\omega)\mathbf{H}_1(j\omega) = \dfrac{70.6(1 + j0.5\omega)}{(1 + j0.6\omega)(1 + j0.08\omega)(1 + j5\omega)}$	15-17
$[\mathbf{M}_0(j\omega)]_a = \dfrac{\mathbf{\Omega}(j\omega)}{\mathbf{E}_y(j\omega)}\bigg]_a = \dfrac{[\mathbf{G}_0(j\omega)]_a}{1 + [\mathbf{G}_0(j\omega)]_a}$	15-17
$\dfrac{1}{[\mathbf{G}_1(j\omega)]_a\mathbf{H}_1(j\omega)} = \dfrac{(1 + j0.08\omega)(1 + j5\omega)}{353(1 + j0.5\omega)}$	15-18
$[\mathbf{M}_u(j\omega)]_a = \dfrac{\mathbf{\Omega}(j\omega)}{\mathbf{U}(j\omega)}\bigg]_a = \dfrac{1}{[\mathbf{G}_1(j\omega)]_a\mathbf{H}_1(j\omega)}\left[\dfrac{\mathbf{\Omega}(j\omega)}{\mathbf{E}_y(j\omega)}\right]_a$	15-18
$[\mathbf{G}_0(j\omega)]_b = [\mathbf{G}_1(j\omega)]_b\mathbf{G}_2(j\omega)\mathbf{H}_1(j\omega)$ $= \dfrac{224(1 + j\omega)(1 + j0.2\omega)}{(1 + j0.08\omega)(1 + j10\omega)(1 + j0.02\omega)(1 + j0.6\omega)}$	15-19
$[\mathbf{M}_0(j\omega)]_b = \dfrac{\mathbf{\Omega}(j\omega)}{\mathbf{E}_y(j\omega)}\bigg]_b = \dfrac{[\mathbf{G}_0(j\omega)]_b}{1 + [\mathbf{G}_0(j\omega)]_b}$	15-19
$\dfrac{1}{[\mathbf{G}_1(j\omega)]_b\mathbf{H}_1(j\omega)} = \dfrac{(1 + j0.08\omega)(1 + j10\omega)(1 + j0.02\omega)}{1{,}120(1 + j\omega)(1 + j0.2\omega)}$	15-20
$[\mathbf{M}_u(j\omega)]_b = \dfrac{\mathbf{\Omega}(j\omega)}{\mathbf{U}(j\omega)}\bigg]_b = \dfrac{1}{[\mathbf{G}_1(j\omega)]_b\mathbf{H}_1(j\omega)}\left[\dfrac{\mathbf{\Omega}(j\omega)}{\mathbf{E}_y(j\omega)}\right]_b$	15-20
$\dfrac{\mathbf{N}_1(j\omega)}{\mathbf{G}_1(j\omega)} = \dfrac{1 + j0.08\omega}{140 \times 10^3}$	15-21
$\dfrac{\mathbf{N}_1(j\omega)}{[\mathbf{G}_1(j\omega)]_b} = \dfrac{(1 + j0.08\omega)(1 + j10\omega)(1 + j0.02\omega)}{(1 + j\omega)(1 + j0.2\omega)(28 \times 10^5)}$	15-21

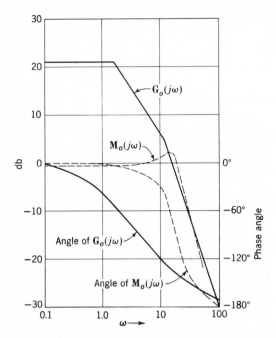

Fig. 15-15 *Log magnitude and phase diagrams of*

$$G_0(j\omega) = \frac{11.2}{(1 + j0.6\omega)(1 + j0.08\omega)}$$

$$M_0(j\omega) = \frac{G_0(j\omega)}{1 + G_0(j\omega)}$$

Inserting the lag compensator $[G_c(j\omega)]_a$ in cascade with $G_1(j\omega)$ has produced the usual effects upon the output with respect to the desired input. It has reduced the steady-state effect of a step input of load torque upon the output and has reduced the steady-state error for a step input of the desired signal. However, for the frequency range of 1 to 20 radians/sec the frequency response is not as good as that of the original system. The transient response is therefore not as good for both input signals. These effects can be observed by referring to Figs. 15-15 to 15-21.

An improvement in both the step input and the frequency response of the undesired signal can be achieved by inserting the lag-lead compensator $[G_c(j\omega)]_b$ in cascade with $G_1(j\omega)$. This improvement can be noted by referring to Figs. 15-15, 15-16, 15-19, and 15-20. It has been stated earlier that the ratio $N_1(j\omega)/G_1(j\omega)$ must be made smaller to minimize the effect of T_L. By referring to Fig. 15-21 it is seen that the ratio $N_1(j\omega)/[G_1(j\omega)]_b$ is smaller than the ratio $N_1(j\omega)/G_1(j\omega)$. Thus the effect of the load torque T_L has been reduced.

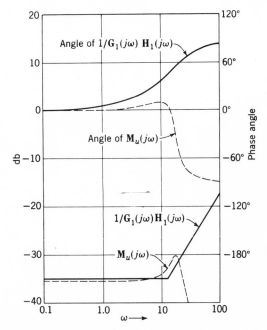

Fig. 15-16 *Log magnitude and phase diagrams of*

$$\frac{1}{G_1(j\omega)H_1(j\omega)} = \frac{1 + j0.08\omega}{56}$$

$$M_u(j\omega) = \frac{1}{G_1(j\omega)H_1(j\omega)} \frac{G_0(j\omega)}{1 + G_0(j\omega)}$$

Although the system used in the example is fairly simple, by the techniques discussed earlier in the chapter one can often reduce a more complex system to the mathematical model of Fig. 15-11. Once this has been done, one can proceed in the same manner as in the above example to minimize the effect of the undesired signal. If more than one undesired signal exists, the problem becomes more involved. By use of the superposition theorem, each unwanted signal can be analyzed separately. The difficulty arises in trying to locate the optimum place for insertion of any additional network and for a change of gain to produce a desired improvement for all unwanted signals without adversely affecting the desired response.

For the case of no undesired and two desired input signals, the approach is similar to the case of one undesired and one desired input signal. In the latter case it is necessary to minimize the response to $U(s)$ and optimize the response to $R(s)$. In the former case one must alter the basic system to optimize the response to both desired input signals. The greater the number

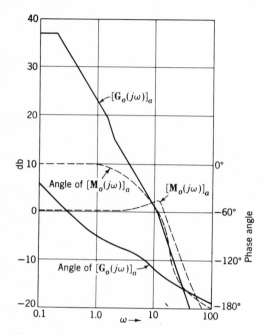

Fig. 15-17 *Log magnitude and phase diagrams with compensator $[G_c(j\omega)]_a$ inserted.*

$$[G_0(j\omega)]_a = \frac{70.6(1 + j0.5\omega)}{(1 + j0.6\omega)(1 + j0.08\omega)(1 + j5\omega)}$$

$$[M_0(j\omega)]_a = \frac{[G_0(j\omega)]_a}{1 + [G_0(j\omega)]_a}$$

of desired inputs, the greater the difficulty of optimizing the performance. Naturally, the complexity of optimizing for a combination of desired and undesired input signals becomes greater as the number of each becomes greater. In like manner one can consider the feedback control system that has more than one desired output signal.

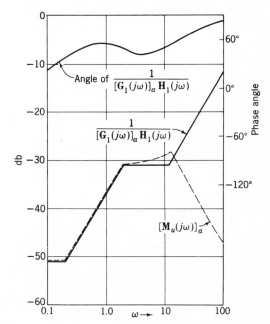

Fig. 15-18 *Log magnitude and phase diagram with compensator* $[\mathbf{G}_c(j\omega)]_a$ *inserted.*

$$\frac{1}{[\mathbf{G}_1(j\omega)]_a\mathbf{H}_1(j\omega)}$$
$$= \frac{(1 + j0.08\omega)(1 + j5\omega)}{353(1 + j0.5\omega)}$$

$$[\mathbf{M}_u(j\omega)]_a$$
$$= \frac{1}{[\mathbf{G}_1(j\omega)]_a\mathbf{H}_1(j\omega)}\frac{[\mathbf{G}_0(j\omega)]_a}{1 + [\mathbf{G}_0(j\omega)]_a}$$

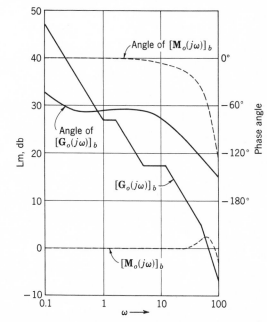

Fig. 15-19 *Log magnitude and phase diagrams with compensator* $[\mathbf{G}_c(j\omega)]_b$ *inserted.*

$$[\mathbf{G}_0(j\omega)]_b$$
$$= \frac{224(1 + j\omega)(1 + j0.2\omega)}{(1 + j0.6\omega)(1 + j0.08\omega)}$$
$$(1 + j10\omega)(1 + j0.02\omega)$$

$$[\mathbf{M}_0(j\omega)]_b = \frac{[\mathbf{G}_0(j\omega)]_b}{1 + [\mathbf{G}_0(j\omega)]_b}$$

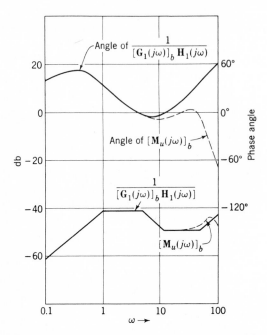

Fig. 15-20 *Log magnitude and phase diagram with compensator* $[\mathbf{G}_c(j\omega)]_b$ *inserted.*

$$\frac{1}{[\mathbf{G}_1(j\omega)]_b\mathbf{H}_1(j\omega)}$$
$$= \frac{(1 + j0.08\omega)(1 + j10\omega)(1 + j0.02\omega)}{1{,}120(1 + j\omega)(1 + j0.2\omega)}$$

$$[\mathbf{M}_u(j\omega)]_b = \frac{1}{[\mathbf{G}_1(j\omega)]_b\mathbf{H}_1(j\omega)} \frac{[\mathbf{G}_0(j\omega)]_b}{1 + [\mathbf{G}_0(j\omega)]_b}$$

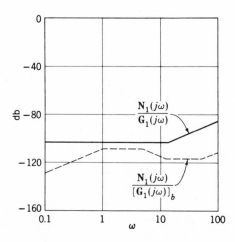

Fig. 15-21 *Log magnitude and phase diagram of*

$$\frac{\mathbf{N}_1(j\omega)}{\mathbf{G}_1(j\omega)} = \frac{1 + j0.08\omega}{140 \times 10^3}$$

$$\frac{\mathbf{N}_1(j\omega)}{[\mathbf{G}_1(j\omega)]_b}$$
$$= \frac{(1 + j0.08\omega)(1 + j10\omega)(1 + j0.02\omega)}{(1 + j\omega)(1 + j0.2\omega)(28 \times 10^5)}$$

15-8 Feed Forward Compensation

In the chapters on cascade compensation it has been stated that it is possible to minimize the steady-state error of a control system by the utilization of a cascade lag compensator. Figure 15-22 illustrates another means of improving the steady-state performance. The solid lines represent the block diagram of the basic system, and the dashed lines represent the location of a compensator. Note that in Fig. 12-1 the input to G_c is a function of both the input and the feedback quantities, that is, $E(s) = R(s) - C(s)$. In Fig. 15-22 the input to G_c is just the input quantity $R(s)$. Thus in this figure $G_c(s)$ can be said to be a forward-acting compensator which is external to the closed loop of the basic system and accomplishes error reduction by nonfeedback means. The utilization of a compensator in the manner shown in Fig. 15-22 is called *feed forward compensation*.[8,9] The control ratio for this system is

$$\frac{C(s)}{R(s)} = \frac{G_1(s)G_2(s) + G_2(s)G_c(s)}{1 + G_1(s)G_2(s)H(s)} \tag{15-39}$$

If $G_2(s)$ represents only the controlled element, then $G_1(s)$ and $H(s)$ can be designed to achieve the degree of stability and the speed of response desired; whereas $G_c(s)$ can be designed to achieve the desired steady-state response without affecting the response adversely for a Type O system. Problems illustrating this concept are given at the end of the text.

15-9 Conditional Feedback Systems

The example of Sec. 15-7 illustrates the manner of improving the output response to a desired input and minimizing the effect on the output by an unwanted input signal or disturbance. This requires consideration of the effect of the modification of the system upon both signals. In other words, the desired improvements for each signal cannot be considered independently. In this example the controlled element is represented by the transfer function $G_2(s)$ (see Fig. 15-11), with its associated disturbance $T_L(s)$. By altering the manner in which the controlling elements are connected to the controlled element and adding the necessary components, as shown in Fig. 15-23, it is possible to simplify the problem of improving the system performance in the manner suggested by Lang and Ham.[6] Note that the

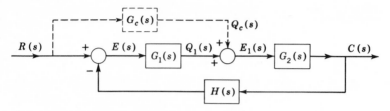

Fig. 15-22   *Feed forward compensation.*

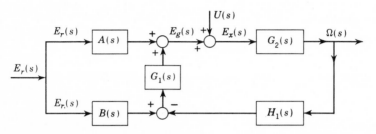

Fig. 15-23   *The feedback control system of Fig. 15-11 converted into a conditional feedback control system.*

system shown in this figure utilizes the concept of feed forward compensation discussed in the last section.

For the modified control system of Fig. 15-23 the output response is given by

$$\Omega = \frac{G_2}{1 + G_1G_2H_1} U + AG_2 \frac{1 + (B/A)G_1}{1 + G_1G_2H_1} E_r \qquad (15\text{-}40)$$

Setting

$$1 + \frac{B}{A} G_1 = 1 + G_1G_2H_1$$

yields the necessary requirement that

$$B = AG_2H_1 \qquad (15\text{-}41)$$

Equation (15-40) is now simplified to the form

$$\Omega = \frac{G_2}{1 + G_1G_2H_1} U + AG_2E_r \qquad (15\text{-}42)$$

The above equation indicates that in this modified control system the feedback term H_1 is effective only upon the undesired signal. This is called a *conditional feedback control system*. With this type of system the specifications of desired output to desired input and output to disturbance input can be met independently, which can simplify the design problem.   For a more detailed analysis of conditional feedback control systems and their applications the reader is referred to the original article and the references given therein.

15-10   *Principle of Invariance*

The method of minimizing an unwanted disturbance illustrated in Sec. 15-7 required the insertion of a compensator in cascade with $G_1(s)$ in Fig. 15-11 to achieve the desired results. Another possible method of minimizing an unwanted disturbance is to insert into the system a feed forward branch whose input is the unwanted disturbance. This method is applicable only for those systems in which the disturbance can be physically detected in such a manner that a feed forward branch can be utilized. A feed forward branch is inserted into the basic system of Fig. 15-11 as shown in Fig. 15-24. From this figure, for sinusoidal inputs, the following equations are obtained:

$$\frac{\Omega(j\omega)}{T_L(j\omega)}\bigg]_{E_r=0} = \frac{[\mathbf{N}_1(j\omega) + \mathbf{G}_1(j\omega)\mathbf{N}_2(j\omega)]\mathbf{G}_2(j\omega)}{1 + \mathbf{G}_1(j\omega)\mathbf{G}_2(j\omega)\mathbf{H}_1(j\omega)} \qquad (15\text{-}43)$$

$$\frac{\Omega(j\omega)}{E_r(j\omega)}\bigg]_{T_L=0} = \frac{\mathbf{G}_1(j\omega)\mathbf{G}_2(j\omega)}{1 + \mathbf{G}_1(j\omega)\mathbf{G}_2(j\omega)\mathbf{H}_1(j\omega)} \qquad (15\text{-}44)$$

The ideal case for the minimization of the disturbance is the condition

$$\frac{\Omega(j\omega)}{T_L(j\omega)}\bigg]_{E_r=0} = 0 \qquad (15\text{-}45)$$

for $0 \leq \omega < \infty$. It can be stated that the ideal case yields an invariant condition in the control system, that is, $\Omega(j\omega) = 0$ for $0 \leq \omega < \infty$ for all $T_L(j\omega)$. To produce the invariance specified by Eq. (15-45), the bracketed

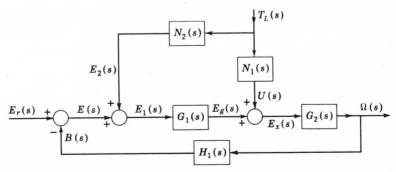

Fig. 15-24 *Minimizing an unwanted disturbance by use of feed forward compensation.*

term in the numerator of Eq. (15-43) is set equal to zero as

$$\mathbf{N}_1(j\omega) + \mathbf{G}_1(j\omega)\mathbf{N}_2(j\omega) = 0$$

or
$$\mathbf{N}_2(j\omega) = -\frac{\mathbf{N}_1(j\omega)}{\mathbf{G}_1(j\omega)} \tag{15-46}$$

Thus the $\mathbf{N}_2(j\omega)$ unit is designed to have the transfer function specified by the above equation in order to yield the specified invariant condition. In practice it may be difficult to maintain this condition. The principle of invariance is dealt with in detail in an article written by Dr. B. N. Petrov[10] and a Ph.D. thesis by I. G. Sarmo.[14]

15-11   Preliminary Synthesis of a Complex Multiloop Control System

A method of handling complex multiloop control systems is described in an article[7] by D. J. Povejsil and A. M. Fuchs. Because of its merit, the essence of this article is presented in the following pages. For a more complete analysis of the method the reader is referred to the original article.

The block diagram of a complex multiloop system, with more than one input and output signal, can be reduced (as in the preceding sections) to a familiar multiloop form when considering the ratio between any one output and one input signal. In many cases, because of the complexity of the system, there cannot be developed a logical synthesis procedure of this reduced form. This is often the case because an improvement in the response due to one input may simultaneously impair the response due to another input.

The initial synthesis on a complex system such as that shown in Fig. 15-25 requires the following steps:

1. *Differential Equations.* Write the differential equations that represent the open-loop uncontrolled system, i.e., the system to be controlled.
2. *Specifications.* Determine the set of over-all performance specifications desired for the closed-loop system.
3. *Feedback Quantities.* Postulate the necessary feedback quantities that will bring about the desired over-all performance. This requires the establishment of gain settings for the feedback quantities.
4. *Compensating Networks.* Postulate the necessary compensating networks and cross-coupling networks and set their time constants.

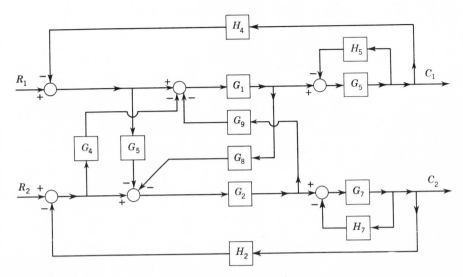

Fig. 15-25 *Complex multiloop control system.*

Implied in the above procedure is a continuous monitoring of the stability of the system.

Differential Equations

For a complex multiloop open-loop control system, n integrodifferential equations relating each of the controlled variables (output signals) to the input forcing functions, external disturbances, and initial conditions can be written in the following general form:

$$A_{11}C_1 + A_{12}C_2 + \cdots + A_{1n}C_n + F_{11}R_1 + F_{12}R_2 + \cdots$$
$$+ F_{1n}R_n + \Sigma(\text{IC})_1 + \Sigma(U_1) = 0$$
$$A_{21}C_1 + A_{22}C_2 + \cdots + A_{2n}C_n + F_{21}R_1 + F_{22}R_2 + \cdots$$
$$+ F_{2n}R_n + \Sigma(\text{IC})_2 + \Sigma(U_2) = 0 \qquad (15\text{-}47)$$
$$A_{n1}C_1 + A_{n2}C_2 + \cdots + A_{nn}C_n + F_{n1}R_1 + F_{n2}R_2 + \cdots$$
$$+ F_{nn}R_n + \Sigma(\text{IC})_n + \Sigma(U_n) = 0$$

where A_{ij}, F_{ij} = analytic functions of complex variables s.

C_j = Laplace transform of controlled variable (output), neglecting initial conditions.

R_j = Laplace transform of input forcing function, neglecting initial conditions. R_j is limited to inputs which can be modified by feedback, as shown in Fig. 15-26, to form an actuating signal.

$\Sigma(U_j)$ = summation of external disturbances in jth equation. These quantities cannot be modified directly by feedback to form an error signal.

$\Sigma(\text{IC})_j$ = summation of initial conditions in jth equation.

Equations (15-47) indicate that the number of independent controlled variables $(C_1 \cdots C_n)$ must be equal to or greater than the number of independent command signals $(R_1 \cdots R_n)$. From this set of equations the determinant Δ_u of the uncontrolled system is expressed as

$$\Delta_u = \begin{vmatrix} A_{11} & A_{12} & \cdots & A_{1n} \\ A_{21} & A_{22} & \cdots & A_{2n} \\ \vdots & & & \vdots \\ A_{n1} & A_{n2} & \cdots & A_{nn} \end{vmatrix} \tag{15-48}$$

Thus the relationship between a particular controlled variable and one of the input forcing functions can be expressed as

$$\frac{C_j}{R_k} = \frac{G_{jk}}{\Delta_u} \tag{15-49}$$

Expanding the development of Eq. (15-48), equating it to zero, and collecting coefficients of corresponding powers of s yields the characteristic equation of the uncontrolled system as

$$s^m + B_{m-1}s^{m-1} + B_{m-2}s^{m-2} + \cdots + B_1 s + B_0 = 0 \tag{15-50}$$

Specifications

A set of performance specifications gives the desired relationship between the controlled variables and the command signals of the closed-loop system.

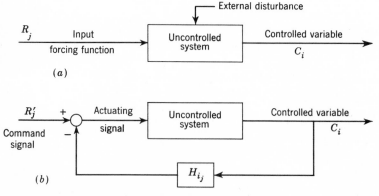

Fig. 15-26 *Explanation of terminology for synthesis of a complex multiloop system (a) before feedback loop is closed; (b) after feedback loop is closed.*

There may also be performance specifications giving the desired controlled-variable response to the disturbing function. The relationship between a particular controlled variable and one of the command signals may be completely specified as

$$\frac{C_j}{R'_k} = \frac{G'_{jk}}{\Delta} \tag{15-51}$$

In most instances, a key part of this specification is that the characteristic equation (i.e., the expansion of the system determinant Δ) of the controlled system have the specified form

$$s^m + B'_{m-1}s^{m-1} + B'_{m-2}s^{m-2} + \cdots + B'_1 s + B'_0 = 0 \tag{15-52}$$

Note that the formulation of the characteristic equation includes the stability (damping) and bandwidth of the final system.

Feedback Quantities

By referring to Eqs. (15-47), it is seen that closing the control loops around the uncontrolled system, as shown in Fig. 15-26, has the effect of changing the form of the input forcing functions $(R_1 \cdots R_n)$ to command inputs $(R'_1 \cdots R'_n)$. The input forcing functions R become the actuating signals when the feedbacks are inserted. The command input is related to the input forcing functions by

$$
\begin{aligned}
R_1 &= R'_1 + H_{11}C_1 + H_{12}C_2 + \cdots + H_{1n}C_n \\
R_2 &= R'_2 + H_{21}C_1 + H_{22}C_2 + \cdots + H_{2n}C_n \\
&\cdots\cdots\cdots\cdots\cdots\cdots\cdots\cdots\cdots\cdots\cdots\cdots \\
R_n &= R'_n + H_{n1}C_1 + H_{n2}C_2 + \cdots + H_{nn}C_n
\end{aligned} \tag{15-53}
$$

where R'_n is the input to the controlled system and H_{ij} is the transfer function between the measured controlled variable and the summing point.

Substituting Eqs. (15-53) into Eqs. (15-47) yields a set of equations relating the controlled variables to the command signals of the control system. The coefficients of the variables C_j assume the following form:

$$
\begin{aligned}
A'_{11} &= A_{11} + F_{11}H_{11} + F_{12}H_{21} + \cdots + F_{1n}H_{n1} \\
A'_{12} &= A_{12} + F_{11}H_{12} + F_{12}H_{22} + \cdots + F_{1n}H_{n2} \\
&\cdots\cdots\cdots\cdots\cdots\cdots\cdots\cdots\cdots\cdots\cdots\cdots \\
A'_{nn} &= A_{nn} + F_{n1}H_{1n} + F_{n2}H_{2n} + \cdots + F_{nn}H_{nn}
\end{aligned} \tag{15-54}
$$

The controlled-system determinant can be expressed as follows:

$$
\Delta = \begin{vmatrix}
A'_{11} & A'_{12} & \cdots & A'_{1n} \\
A'_{21} & A'_{22} & \cdots & A'_{2n} \\
\multicolumn{4}{c}{\cdots\cdots\cdots\cdots\cdots} \\
A''_{n1} & A'_{n2} & \cdots & A'_{nn}
\end{vmatrix} \tag{15-55}
$$

Expanding Eq. (15-55) yields the characteristic equation of the controlled system having the general form of Eq. (15-52). Comparing the determinant of Eq. (15-55) with Eq. (15-48) reveals that each term of the latter has been altered by the addition of feedback terms. These feedback quantities must be postulated for the expansion of the characteristic determinant Δ to fit the desired form of the specified characteristic equation given by Eq. (15-52). This requires the establishment of gain settings for the feedback quantities. Also, it may be necessary to postulate the required compensating and cross-coupling networks and to select their time constants. The step-by-step synthesis procedure can be outlined as follows:

1. Compare the coefficients B'_{m-1}, B'_{m-2}, . . . , B'_0 of the specified characteristic equation with the coefficients B_{m-1}, B_{m-2}, . . . , B_0 of the uncontrolled system.
2. Where the coefficients of the uncontrolled system are inadequate or nonexistent when compared with the desired control, examine the original-system equations for the form $F_{ij}H_{ji}$ that will establish the desired values of the coefficients. Careful tabulating procedures (of the factors B'_{m-1}, B'_{m-2}, . . . , B'_0) permit a rapid selection of the A_{ij} term whose modifications will produce the desired change in the determinant.
3. As each feedback term and its associated feedback transfer function are selected, it is necessary to reevaluate the characteristic equation of the partially controlled system. This reveals when the feedback quantity selected to improve one coefficient of the characteristic equation seriously interacts upon other coefficients. If the feedback quantity improves more than one coefficient of the partially controlled characteristic equation, it obviously represents a fortunate choice of feedback parameter, whereas if the feedback term's beneficial effect is negated by its action upon another coefficient, it is rejected and other feedback terms are examined.
4. This process is continued until the coefficients of the controlled system's characteristic equation are equal to the coefficients of the specified characteristic equation within the allowable tolerances.
5. Where G'_{jk} is also specified, as in Eq. (15-51), the process is repeated for the G'_{jk} determinant in the same manner as for the characteristic determinant.

Formulation of a Minimum Complexity System

A set of analytical rules that define the general nature of a control system of minimum complexity in relation to its performance specifications is given as follows:

1. Since the desired characteristic equation has m variables B'_{m-1}, B'_{m-2}, . . . , B'_1, B'_0, there must be at least m independent gains and time constants in the control systems.
2. The gains and time constants must be chosen so that each of the coefficients B'_j can be expressed as a function of at least one of the independent gains or time constants of the controlled system.
3. Any combination of kB''_j's must contain at least k number of the independent control-system gains and time constants in such a form that k independent equations can be written.

Similar rules apply in obtaining the specified G_{jk} equation.

15-12   *Multiloop Synthesis Example: Lateral Aircraft Control System*

To demonstrate this synthesis procedure, the design of an autopilot to control the lateral response of an aircraft is presented. This synthesis problem is selected as an example because it has large cross-coupling terms and is frequently specified by the equivalent of a desirable characteristic equation.

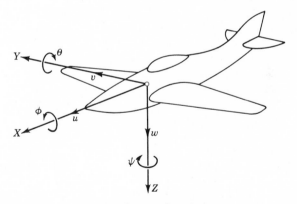

Positive directions are shown by arrows

Axis		Moment about axis		Angle		Velocities	
Designation	Force along axis	Designation	Positive direction	Designation	Definition	Linear along axis	Angular about axis
X (longitudinal)	X	L (rolling)	$Y \rightarrow Z$	ϕ (roll)	$\int p \, dt$	u	p
Y (lateral)	Y	M (pitching)	$Z \rightarrow X$	θ (pitch)	$\int q \, dt$	v	q
Z (normal)	Z	N (yawing)	$X \rightarrow Y$	ψ (yaw)	$\int r \, dt$	w	r

Fig. 15-27 *System of aircraft axes and symbols.*

The lateral motion of the aircraft may be described by three simultaneous linear equations of the form of Eqs. (15-47).   See Fig. 15-27 for the physical interpretation of the variables in these equations.

Roll Moments:

$$(s^2 - sl_p)\phi + (-s^2 i_{xz} - sl_r)\psi + (-l_\beta)\beta + (-l\delta_a)\ \delta_a$$
$$+ (-l\delta_r)\ \delta_r = 0 \quad (15\text{-}56)$$

Yaw Moments:

$$(-s^2 i_{zx} - sn_p)\phi + (s^2 - sn_r)\psi + (-n_\beta)\beta + (-n\delta_a)\ \delta_a$$
$$+ (-n\delta_r)\ \delta_r = 0 \quad (15\text{-}57)$$

Side Forces:

$$\left(-\frac{g}{V}\right)\phi + (1 - y_r)s\psi + (s - y_\beta)\beta + (-y\delta_a)\ \delta_a + (-y\delta_r)\ \delta_r = 0 \quad (15\text{-}58)$$

where ϕ = roll angle, radians
    ψ = yaw angle, radians
    β = sideslip angle, radians
    δ_a = aileron displacement, radians
    δ_r = rudder displacement, radians
    V = aircraft forward velocity, fps
    l, n, and y quantities = aerodynamic coefficients
    i_{xz} and i_{zx} terms = product-of-inertia quantities

By expanding the determinant Δ_u, the lateral motion gives a fifth-order characteristic equation of the form

$$B_5 s^5 + B_4 s^4 + B_3 s^3 + B_2 s^2 + B_1 s = 0 \quad (15\text{-}59)$$

where
$$B_5 = 1 - i_{xz} i_{zx} \quad (15\text{-}60)$$

$$B_4 = -l_p - n_r - y_\beta - n_p i_{xz} - l_r i_{xz} + i_{zx} i_{xz} y_\beta \quad (15\text{-}61)$$

$$B_3 = (1 - y_r)n_\beta + l_p n_r + l_\beta y_\beta + n_r y_\beta - n_p l_r + (1 - y_r)l_\beta i_{zx}$$
$$+ n_p i_{xz} y_\beta + l_r i_{zx} y_\beta \quad (15\text{-}62)$$

$$B_2 = -(1 - y_r)n_\beta l_p + (1 - y_r)l_\beta n_p - l_p n_r y_\beta + n_p l_r y_\beta$$
$$- \frac{g}{V} l_\beta - \frac{g}{V} n_\beta i_{xz} \quad (15\text{-}63)$$

$$B_1 = \frac{g}{V} l_\beta n_r - \frac{g}{V} n_\beta l_r \quad (15\text{-}64)$$

Lateral motion is a result of both yaw and roll motion.

The parameters used in this design are not equivalent to any particular airplane, nor do they necessarily represent a realizable aircraft design.   They have been selected to illustrate the synthesis procedure in a realistic fashion without introducing security restrictions:

$$l_p = -1.00 \qquad n_\beta = 6.41$$
$$l_r = 0.50 \qquad n\delta_a = -0.4$$
$$l\delta_a = -10.0 \qquad n\delta_r = -4.0$$
$$l\delta_r = 0.10 \qquad g/V = 0.04$$
$$l_\beta = -10.0 \qquad y\delta_a = 0$$
$$i_{zz} = 0.01 \qquad y\delta_r = 0.03$$
$$i_{xz} = 0.05 \qquad y_\beta = -0.10$$
$$n_p = 0.10 \qquad y_r = 0.003$$
$$n_r = -0.25$$

This gives an uncontrolled characteristic equation of

$$s^5 + 1.35s^4 + 6.61s^3 + 7.8s^2 - 0.028s = 0 \qquad (15\text{-}65)$$

The desired characteristic equation of the lateral mode of the aircraft is specified as

$$s(s^2 + 2\zeta_1\omega_{n1}s + \omega_{n1}^2)(s^2 + 2\zeta_2\omega_{n2}s + \omega_{n2}^2) = 0 \qquad (15\text{-}66)$$

where
$$\zeta_1 = 0.5 \qquad \omega_{n1} = 2.5 \text{ radians/sec}$$
$$\zeta_2 = 1.0 \qquad \omega_{n2} = 1.0 \text{ radian/sec}$$

It is not suggested that this is a definitive statement of a desirable lateral autopilot response. It is offered as representative of a desirable response and as convenient to illustrate the design procedure. Substituting into Eq. (15-66) the values of the parameters just given results in a desired characteristic equation

$$s^5 + 4.5s^4 + 12.25s^3 + 14.75s^2 + 6.25s = 0 \qquad (15\text{-}67)$$

Now, it is noticed that the desired characteristic equation has the same order as the characteristic equation given by Eq. (15-65) of the uncontrolled system. This fact suggests that all the required control may be obtained by modification of the existing characteristics of the uncontrolled system. Furthermore, since the desired characteristic equation possesses four independent coefficients, four independent parameters are required in the control. An inspection of Eqs. (15-60) to (15-64) shows that one combination of four parameters that yields the required control is n_r, l_p, n_β, and l_a. Thus, the proposed control equations are

$$\delta_a = H_{11}s\phi + H_{12}s\psi \qquad (15\text{-}68)$$
$$\delta_r = H_{22}s\psi + H_{23}\beta \qquad (15\text{-}69)$$

When these are substituted into the original airplane equations, the modified aircraft parameters become

$$l_p = -(1 + 10H_{11}) \tag{15-70}$$
$$l_r = -(-0.5 + 10H_{12} - 0.10H_{22}) \tag{15-71}$$
$$l_\beta = -(10 - 0.10H_{23}) \tag{15-72}$$
$$n_p = 0.10 - 0.40H_{11} \tag{15-73}$$
$$n_r = -(0.25 + 4H_{22} + 0.4H_{12}) \tag{15-74}$$
$$n_\beta = 6.41 - 4H_{23} \tag{15-75}$$
$$y_r = 0.003 - 0.03H_{22} \tag{15-76}$$
$$y_\beta = -(0.10 - 0.03H_{23}) \tag{15-77}$$

An exact solution for the values of H_{11}, H_{12}, H_{22}, and H_{23} can be obtained by substituting Eqs. (15-70) to (15-77) into Eqs. (15-60) to (15-64), with B_1 to B_5 defined by Eq. (15-67). When these substitutions have been made, these equations define the (B')'s of Eq. (15-52). For very complex problems a digital computer is usually applied. However, in many practical cases, a systematic manual calculation procedure is possible by closing the loops in sequence and employing an iteration process suited to the particular system under consideration. For example, the following process can be used in this case:

1. For the first iteration assume that only l_p, l_r n_r, and n_β are modified by the feedback terms. By using Eqs. (15-60) to (15-64), the proper values of l_p, l_r, n_r, and n_β needed to fit B_1 to B_5 to the coefficients of Eq. (15-67) can be calculated by closing the H_{12}, H_{22}, H_{23}, and H_{11} loops in sequence.
2. Then, with these values of l_p, l_r, n_r, and n_β, the required control parameters H_{ij} can be calculated from Eqs. (15-70), (15-71), (15-74), and (15-75).
3. New values of the parameters assumed to be constant in the first iteration, l_β, n_p, y_r, and y_β, can be calculated from Eqs. (15-72), (15-73), (15-76), and (15-77).
4. By using the parameters obtained in step 3 as constants in Eqs. (15-60) to (15-64), new values of l_p, l_r, n_r, and n_β can be found as the first step in the second iteration process.
5. Repeat steps 2 to 4 for as many iterations as are required to achieve the desired accuracy.

When this process was carried out for the example, the results of two iterations were

$$H_{11} = 0.125 \tag{15-78}$$
$$H_{12} = 2.21 \tag{15-79}$$
$$H_{22} = 0.2015 \tag{15-80}$$
$$H_{23} = 0 \tag{15-81}$$

giving a controlled-system determinant of

$$s^5 + 4.50s^4 + 12.15s^3 + 14.73s^2 + 6.31s = 0 \tag{15-82}$$

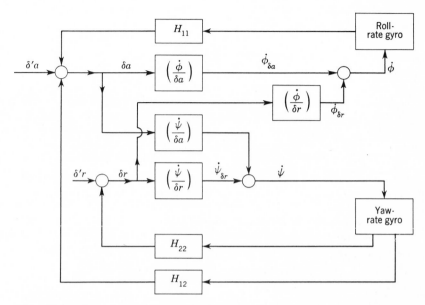

Fig. 15-28 *Lateral-control-system block diagram. Roll rate feedback to rudder; yaw rate cross-feed to aileron.*

Note that, for the particular case chosen, one of the control parameters is zero. This implies that one of the existing parameters of the uncontrolled system needs no modification to achieve the desired control. Also observe that one of the key control terms is the cross-feed term H_{12}.

The block diagram of the resulting system, shown in Fig. 15-28, reveals the startling complexity of this control system as seen by the frequency-response method.

The above solution neglects the dynamics of the autopilot in order to simplify the initial analysis. When an idea of the additional networks needed has been obtained, the autopilot dynamics can be included. The problem then would be put on a computer. With the aid of the results of the initial analysis, which has put the designer in the "right ball park," the final analysis can be made.

15-13 Analysis of Complex Multiloop Control System by the Root Locus Method [11]

The problem of controlling the interacting effects in linear multivariable control systems has proved formidable from both the theoretical and the design viewpoint.   Interaction within a multivariable control in response to an indeterminate disturbance may, in some applications, be desirable, the interaction being controlled rather than removed.   Recognizing that the physical construction of a completely noninteracting control is improbable, a designer is interested in knowing the economic and performance costs of a noninteracting design compared with a controlled interacting design.   The root-locus technique of analyzing a single-variable system can be extended to the analysis of multivariable systems which permits the determination of the effect of interaction.

Multivariable System Characteristics

Each response function of an interacting multivariable control is dependent in general on all system disturbances or inputs.   However, each response can be related to each input by a transfer function, a ratio of two polynomials of the variable s.   A common denominator, the system characteristic equation, exists for all these transfer functions while the numerator polynomials are different for each input-output pair.   Thus an investigation of the s-plane root locations of the system characteristic equation, while providing system stability information, will not suffice to predict transient-response characteristics, such as overshoot, settling time, and rise time, for the interacting effects.   This information follows from a relatively complete determination of the zero and pole locations for each possible transfer function.   A variation of any one system parameter may change the pole and zero locations of each transfer function. For example, a single gain change may modify all intercoupling effects throughout the entire system as well as change the transient characteristic of the input-output pair most directly associated with the particular gain change.

Consequently, the designer needs a root-locus plot not only of the system characteristic equation but also of the numerator polynomial of each possible transfer function.   While this implies a complicated design procedure, common factors and movements of the poles and zeros of the various transfer functions can be used to simplify plotting the loci.

To illustrate the application of root locus to a transfer function with a numerator and denominator dependent on a system parameter, consider

$$G(s) = \frac{(s + 4) + K(s + 2)}{(s + 1)(s + 5) + K(s + 1)}$$

For the locus of zeros of the numerator,

$$\frac{K(s + 2)}{s + 4} = -1$$

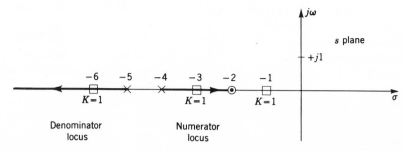

Fig. 15-29 *Pole and zero locations of example locus.*

and for the locus of zeros of the denominator,

$$\frac{K}{s+5} = -1$$

These loci, plotted in Fig. 15-29, indicate that for $K = 1$ the transfer function is

$$G(s) = \frac{2(s+3)}{(s+1)(s+6)}$$

Multivariable-system Root Loci

Consider the system illustrated in Fig. 15-30. The term G represents a matrix with components G_{ij} given by the equation

$$C = GR \qquad \text{or} \qquad C_i = \sum_j G_{ij}R_j \qquad (15\text{-}83)$$

where C and R are the output and input vectors with components C_i and R_i. Transfer functions can be expressed as

$$\frac{C_i}{R_j} = K \frac{s^w + a_{w-1}s^{w-1} + \cdots}{s^v + b_{v-1}s^{v-1} + \cdots} = G_{ij} \qquad v \geq w \qquad (15\text{-}84)$$

System compensation and feedback are included in G_{ij}. A root locus can be plotted for the numerator and denominator of each G_{ij} for a varying system parameter. The effect of the varying parameter on the characteristics of each response is then evident through the use of familiar root-locus techniques.

Fig. 15-30 *System notation.*

A disadvantage of the root-locus method as applied to single-variable systems is the limitation of representing the effects of one varying parameter per locus. It is quickly recognized that this disadvantage becomes proportionately greater for applications to multivariable systems. Rarely would a single parameter be subject to design modification in a multivariable system; rather, a set of parameters, such as the gain associated with each input-output pair C_i/R_i, would be involved. This complication is fundamental not to the method of design but to the complex nature of the design problem. Formidable as it may seem, it is probable that the design parameters can be selected in a sequence, from those most constrained to those least constrained. The statement of the system specifications can also influence the sequence of parameter selection.

Interaction

It has been suggested that the design specifications of an interacting control system would include a statement of permissible values for rise time and overshoot, for example, for each input-output pair C_i/R_i. If the inherent effects of interaction are not desirable, additional specifications might take the form of

$$\frac{C_i}{R_j}\bigg]_{R_i=0} \leq \epsilon_{ij} \frac{C_i}{R_i}\bigg]_{R_i=0} \qquad \text{for all } j \neq i \qquad (15\text{-}85)$$

at some frequency, for example, at $\omega = 0$, or at the frequency for which $[C_i/R_i]_{R_j=0}$ is a maximum, or at those frequencies for which each side of the equation is maximum. Alternatively or additionally,

$$C_i\bigg]_{\substack{R_j \neq 0 \\ R_i = 0}} \leq \epsilon'_{ij} \qquad \text{for all } j \neq i \qquad (15\text{-}86)$$

at some time, for example, at the time for which the response of C_i is a maximum. Requiring $\epsilon'_{ij} = 0$ in Eqs. (15-85) and (15-86) for all time is to specify no interaction between the ith response and the jth disturbance.

Illustration

The application of root locus to multivariable control system design procedures is directed by the specific form of the system and the nature of the specifications. For this reason the remainder of the discussion is devoted to an example of a possible problem formulation and development utilizing root loci.

The system in Fig. 15-31 represents an interacting plant with provision for compensation to be inserted as E_{11} and E_{22}. From the equations describ-

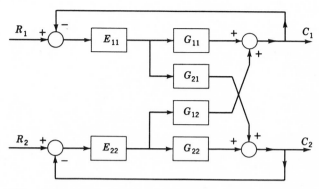

Fig. 15-31 *Multivariable control system.*

ing this block diagram,

$$(R_1 - C_1)E_{11}G_{11} + (R_2 - C_2)E_{22}G_{12} = C_1$$
$$(R_1 - C_1)E_{11}G_{21} + (R_2 - C_2)E_{22}G_{22} = C_2 \tag{15-87}$$

it follows that

$$C_1 = \frac{E_{11}G_{11}(1 + E_{22}G_{22}) - E_{11}E_{22}G_{12}G_{21}}{\Delta} R_1 + \frac{E_{22}G_{12}}{\Delta} R_2 \tag{15-88}$$

$$C_2 = \frac{E_{11}G_{21}}{\Delta} R_1 + \frac{E_{22}G_{22}(1 + E_{11}G_{11}) - E_{11}E_{22}G_{12}G_{21}}{\Delta} R_2 \tag{15-89}$$

$$\Delta = (1 + E_{11}G_{11})(1 + E_{22}G_{22}) - E_{11}E_{22}G_{12}G_{21} \tag{15-90}$$

C_1/R_1 and C_1/R_2 are defined from Eq. (15-88) with $R_2 = 0$ and $R_1 = 0$, respectively, etc. Assume that

$$G_{11}(s) = \frac{1}{s + 1} \qquad G_{22}(s) = \frac{1}{s + 2}$$

$$G_{12}(s) = G_{21}(s) = \frac{1}{s + 20}$$

and that the compensating transfer functions have the form

$$E_{11}(s) = \frac{K_1(s + a)}{s + ma}$$
$$E_{22}(s) = \frac{K_2(s + b)}{s + mb} \qquad m < 1$$

K_1, K_2, a, b, and m constitute the group of design variables, all positive real numbers.

The design specifications are that $c_1(t)$ in response to a step disturbance r_1 (with all initial conditions and r_2 equal to zero) reaches a maximum overshoot of 15 per cent in 0.6 sec or less, and $c_2(t)$ in response to a step disturbance

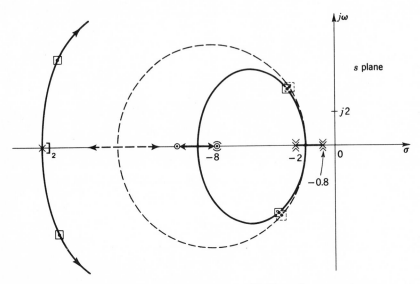

Fig. 15-32   *Locus of Eqs. (15-93) and (15-94) with K_2 as the parameter.*
Zeros of Eqs. (15-93) and (15-94) ○ ◌
Poles of Eqs. (15-93) and (15-94) ✕ ✕
Roots of Eqs. (15-93) and (15-94) ☐ ⊡

r_2 (with initial conditions and r_1 equal to zero) reaches a maximum overshoot of 20 per cent in 0.8 sec or less.   The interacting constraints are given by Eq. (15-85) for $\omega = 0$ and $\epsilon_{12} = \epsilon_{21} = 0.05$ and by Eq. (15-86) for $r(t)_j = u(t)$, c_1 evaluated at the time of maximum overshoot, and $\epsilon'_{12} = \epsilon'_{21} = 0.20$.

Initially the interacting constraint given by Eq. (15-85), $\omega = 0$ and $\epsilon_{12} = \epsilon_{21} = 0.05$, can be shown to reduce effectively to the inequalities

$$K_1 \geq \frac{K_2}{1 + K_2/2m}$$
$$K_2 \geq \frac{2K_1}{1 + K_1/m} \tag{15-91}$$

The root locus for the characteristic equation is obtained by setting Eq. (15-90) equal to zero.   Rearranging gives

$$\frac{E_{11}[G_{11} + E_{22}(G_{11}G_{22} - G_{12}G_{21})]}{1 + E_{22}G_{22}} = -1$$

By substituting the values for E and G, this becomes

$$\frac{K_1(s+a)\{(s+mb)(s+2)(s+20)^2 + K_2(s+b)[(s+20)^2 - (s+1)(s+2)]\}}{(s+ma)(s+1)(s+20)^2[(s+mb)(s+2) + K_2(s+b)]} = -1 \tag{15-92}$$

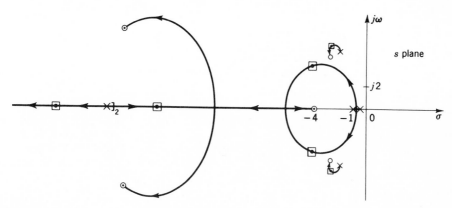

Fig. 15-33 *Locus of the characteristic equation with K_1 as the parameter.*

As the zeros and poles of Eq. (15-92) are functions of the design parameters, a root locus is plotted for the numerator and denominator. Setting the portion of the numerator within the braces equal to zero and rearranging,

$$\frac{37K_2(s + b)(s + 10.75)}{(s + mb)(s + 2)(s + 20)^2} = -1 \tag{15-93}$$

The locus is shown in Fig. 15-32 for $m = 0.1$ and $b = 8$. In practice a family of loci would be plotted for various b and m values; however, representative values are chosen for clarity and brevity. Assume at this point that a value of $K_2 = 2.5$ is desirable for further consideration. The corresponding roots, noted in Fig. 15-32, represent zero locations of Eq. (15-92).

Similarly, by setting the bracketed portion of the denominator of Eq. (15-92) equal to zero and rearranging,

$$\frac{K_2(s + b)}{(s + mb)(s + 2)} = -1 \tag{15-94}$$

a second family of loci may be plotted for various m and b values. Figure 15-32 also shows the locus corresponding to $m = 0.1$, $b = 8$, with root locations for $K_2 = 2.5$. These root locations represent poles of Eq. (15-92) and are functions of m, b, and K_2.

Figure 15-33 illustrates the root locus of the characteristic equation for the root locations selected in Fig. 15-32, for $a = 4$, and for the varying parameter K_1. Again for reasons of clarity the family of curves for alternative values of a is not shown. The solid-symbol root locations of Fig. 15-32 become pole locations in Fig. 15-33, along with poles at -20, -1, and $-ma$. The broken-symbol root locations of Fig. 15-32 become zero locations along with a zero at $-a$.

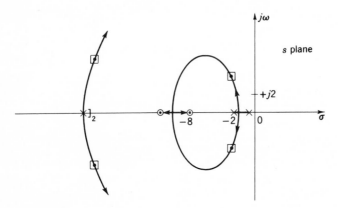

Fig. 15-34   *Locus of zeros of C_1/R_1 with K_2 as the parameter.*

The effects of varying K_1 are evident.   A value of $K_1 = 7$ is selected as a representative choice in view of the design specification and the loci which follow.   The corresponding roots, roots of the system characteristic equation, are indicated in Fig. 15-33.

In this example, the numerators of C_1/R_2 and C_2/R_1 do not contain both K_1 and K_2.   Consequently, it is only necessary to examine the root loci of the numerators of C_1/R_1 and C_2/R_2.   It can be shown that the roots of the numerator of C_1/R_1 correspond to a set of roots on the locus of

$$\frac{37K_2(s + b)(s + 10.75)}{(s + 20)^2(s + mb)(s + 2)} = -1$$

plus a root at $-a$.   This locus is sketched in Fig. 15-34, with numerator root locations corresponding to $K_2 = 2.5$, $m = 0.1$, and $b = 8$.   Notice the similarity between this locus and that given in Fig. 15-32 for Eq. (15-93). By comparing Figs. 15-33 and 15-34, the effective poles and zeros of the C_1/R_1 transfer function are evident, the dominant complex conjugate pole pair being located at $s = -4.4 \pm j3.4$.   The variations of K_1 on this portion of the design are also apparent.   Based upon an approximate method,[12] the transient response of C_1 for a step input at R_1 and with $R_2 = 0$ would be expected to satisfy the specification.   The corresponding root locus for the zeros of C_2/R_2 is shown in Fig. 15-35 for

$$\frac{37K_1(s + a)(s + 10.75)}{(s + 20)^2(s + ma)(s + 1)} = -1$$

By comparing Figs. 15-33 and 15-35, which contain the pole and zero locations of C_2/R_2, the dominant complex conjugate pole pair for C_2/R_2 is seen to be $s = -2.6 \pm j3.9$.   Again based on approximate methods, the transient

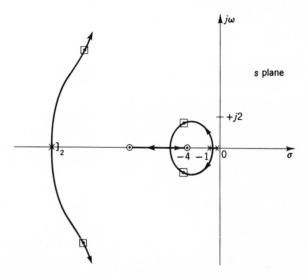

Fig. 15-35 *Locus of zeros of C_2/R_2 with K_1 as the parameter.*

response of C_2 for a step input at R_2 with $R_1 = 0$ would be expected to satisfy the specification.

The inequalities (15-91) are satisfied for $K_1 = 7$, $K_2 = 2.5$, and $m = 0.1$.

To consider the constraint on the interacting transient characteristic, the zero-pole locations of C_1/R_2 and C_2/R_1 are examined. These are readily derived from Eqs. (15-88) and (15-89) and are plotted in Fig. 15-36. On the basis of residue considerations the amplitude of the interacting responses would be expected to satisfy the specifications also.

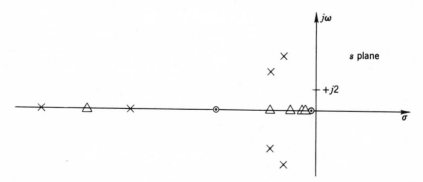

Fig. 15-36 *Poles and zeros of C_1/R_2 and C_2/R_1. Zeros of C_1/R_2 ◯ ; zeros of C_2/R_1 △ ; poles of C_1/R_2 and C_2/R_1 ✕*

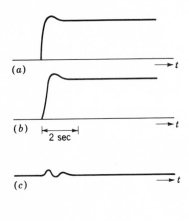

(a)

(b)

2 sec

(c)

(d)

Fig. 15-37 *Transient characteristics. (A) c_1 for step input at R_1. (B) c_2 for step input at R_2. (C) c_1 for step input at R_2. (D) c_2 for step input at R_1.*

Figure 15-37 shows a computer solution to the transient response characteristics of the design.

It must be recognized that this example design procedure is a simplification. In reality the procedure requires families of loci and possible modifications of initially assumed design values. The loci, however, clearly indicate the direction and general effect of each of these modifications.

While the number of input and output signals has been assumed to be equal in the preceding discussion, this is no limitation for the use of the loci. Generally, in addition to an examination of the roots of the characteristic equation, a locus is needed for the numerator of each input-output transfer function referred to in the design specifications.

Summary

A design procedure has been described which facilitates the control of all transient characteristics of an interacting system while considering the available compensation. The indication of transient characteristics and the effects of changing parameters on signal interactions are distinct advantages of the method. The procedure is extensive for a system with two input and output quantities, and the complexity increases with the number of input and output signals. It is necessary to recall, however, that this complexity is due to the detail of the specifications placed on the system and the number of design parameters; all possible input-output signal pairs are considered. This is in contrast to the simplifications possible through a noninteracting system design.

The example suggests that the writing of specifications for multivariable systems be preceded by a searching examination of the need of noninteracting controls, of the extent to which interaction must be limited, and of the freedom allowable for the transient characteristics of the primary input-output signal pairs.[13]

Practically, it is important to note that the advantage of the root-locus method applied to multivariable control system design can be realized only if the designer is able to make a rapid transition from the pole-zero locations to the time-domain characteristics. This is best implemented if an analog computer is available.

15-14   Conclusions

In this chapter different examples of complex control systems are given. A method of rearranging blocks to reduce the complex system to a simple mathematical model is presented, the use of the linear superposition theorem is applied to a complex system, and, finally, a more general approach to the synthesis of a complex multiloop control system is described. The problem of analyzing a complex system becomes involved and tedious. The use of a computer in conjunction with the above techniques can be of considerable aid in obtaining the optimum performance of a complex control system.

Bibliography

1. Chestnut, H., and R. W. Mayer: "Servomechanisms and Regulating System Design," 2d ed., vol. 1, chap. 14, John Wiley & Sons, Inc., New York, 1959.
2. Truxal, J. G.: "Automatic Feedback Control System Synthesis," chap. 6, McGraw-Hill Book Company, New York, 1955.
3. Smith, O. J. M.: "Feedback Control Systems," chap. 4, McGraw-Hill Book Company, New York, 1958.
4. "Fundamentals of Design of Piloted Aircraft Flight Control Systems," BuAer Report AE-61-4, vol. 1, chap. 2, U.S. Navy, Bureau of Aeronautics, 1952.
5. Graybeal, T. D.: Block Diagram Network Transformation, *Elec. Eng.*, vol. 70, no. 11, 1951.
6. Lang, G., and J. M. Ham: Conditional Feedback Systems—A New Approach to Feedback Control, *Trans. AIEE*, vol. 74, pt. II, pp. 152–161, July, 1955.
7. Povejsil, D. J., and A. M. Fuchs: A Method for the Preliminary Synthesis of a Complex Multiloop Control System, *Trans. AIEE*, vol. 74, pt. II, pp. 129–134, July, 1955.
8. Graham, R. E.: Linear Servo Theory, *Bell System Tech. J.*, vol. 25, pp. 616–651, 1946.

9. Moore, J. R.: Combination Open-cycle Closed-cycle System, *Proc. IRE*, vol. 39, pp. 1421–1432, 1951.
10. Petrov, B. N.: The Invariance Principle and the Conditions for Its Application during the Calculation of Linear and Nonlinear Systems, *Proc. Intern. Federation Autom. Control Congr., Moscow*, vol. 2, pp. 1123–1128, 1960. Published by Butterworth & Co. (Publishers), Ltd., London, 1961.
11. Kinnen, E., and D. Liu: Linear Multivariable Control System Design with Root Loci, *Trans. AIEE*, vol. 81, pt. II, pp. 41–44, 1962.
12. Chu, Y.: Synthesis of Feedback Control System by Phase-angle Loci, *Trans. AIEE*, vol. 71, pt. II, pp. 330–339, November, 1952.
13. Mesarovic, M. D.: "The Control of Multivariable Systems," pp. 65–66, John Wiley & Sons, Inc., New York, 1960.
14. Sarmo, I. G.: "Invariance Theory of Automatic Control Systems," Ph.D. Thesis, University of Wisconsin, 1964.

16

A-C Feedback Control Systems[10]

16-1   Introduction

The analysis made and the compensation techniques presented in previous chapters apply to d-c feedback control systems, so called because d-c signals exist in the system. For example, if the reference input and the controlled output are constant values, the actuating signal (for a Type 0 system) can be represented graphically by a straight line, as shown in Fig. 16-1. The signals in other parts of the system can be represented in a similar manner. For a d-c voltage control system the actuating signal $e(t)$ is a d-c voltage. Nonelectrical control systems whose actuating signals have the forms shown in Fig. 16-1 are also referred to as "d-c" feedback control systems. Description of the system as a d-c control system is based on the form of the signals present. These signals do not need to be electrical voltages; they may be position, torque, temperature, or any other quantity.

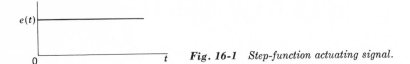

$e(t)$

0   t **Fig. 16-1** *Step-function actuating signal.*

In an "a-c" control system the signals are sinusoidal quantities. For example, even though the reference input R and the controlled output C are constants, the actuating signal is a sinusoidal function whose amplitude is proportional to the difference between R and C. The characteristics of a-c systems are described in the following section, along with those of d-c systems and of hybrid systems which contain both a-c and d-c elements.

16-2   General Types of Feedback Control

Figure 16-2 represents a typical d-c position control system in which the input is set by the position of the potentiometer, θ_i. The output shaft of the motor is mechanically connected to a potentiometer which indicates the output position θ_o. The potential difference between the slide arms of the two potentiometers is proportional to the difference between θ_i and θ_o. Note that in this system a constant input signal θ_i produces d-c signals throughout. This system can be compensated, if necessary, through the use of cascade, feedback, or load compensation techniques, as discussed in earlier chapters.

An a-c control system, which performs the same positioning function as in Fig. 16-2, is shown in Fig. 16-3. The sensing element in this case is composed of two synchros: a synchro generator (SG) and a control transformer (CT).[1,2] Synchros can be considered fundamentally as transformers in

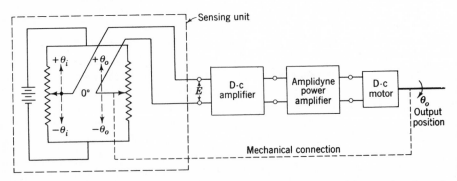

Fig. 16-2 *A d-c position control system with a d-c potentiometer position-sensing element.*

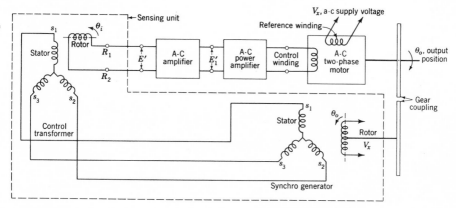

Fig. 16-3 *An a-c position control system with synchro sensing elements.*

which one winding can be rotated. The rotor of the synchro generator is energized by a voltage V_x of frequency ω_c. There are three stationary windings spaced 120 electrical degrees apart and connected as shown. The control transformer is similar to the synchro generator. The a-c voltages applied to the generator rotor set up an alternating magnetic field which induces a-c voltages in the generator stator windings. The magnitudes of these stator voltages are given by the equations

$$E_{12} = K \cos (\theta_o - 30°)$$
$$E_{23} = K \cos (\theta_o - 150°) \qquad (16\text{-}1)$$
$$E_{31} = K \cos (\theta_o - 270°)$$

The stator windings of the generator and control transformer are connected. Therefore an alternating flux is produced in the control transformer. In the relative position shown, the voltage induced in the rotor of the control transformer is zero. However, if the two rotors are displaced, the magnitude of the voltage induced in the rotor of the control transformer is given by

$$E_{\text{CT}} = E_m \sin (\theta_i - \theta_o) \qquad (16\text{-}2)$$

This is an a-c voltage of frequency ω_c.

In Fig. 16-3, if the gears are disengaged and one of the rotors is held stationary while the other rotor is rotated at a constant velocity ω_s, the voltage e_{CT} has the form in Fig. 16-4. This is a suppressed-carrier modulated

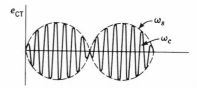

Fig. 16-4 *Form of CT rotor voltage for a constant rotation of the rotor.*

signal, where ω_c is the carrier frequency and ω_s is the signal frequency.   For this condition the magnitude of the CT output voltage is a sinusoidal function, given by the equation

$$e_{CT} = E_m \sin \omega_s t \tag{16-3}$$

The rotor voltage of the CT can be expressed as follows:

$$e' = e_{CT} \sin \omega_c t = (E_m \sin \omega_s t) \sin \omega_c t \tag{16-4}$$

or
$$e' = \frac{E_m}{2} \cos \omega_2 t - \frac{E_m}{2} \cos \omega_1 t \tag{16-5}$$

where
$$\omega_1 = \omega_c + \omega_s \qquad \text{upper side-band frequency}$$
$$\omega_2 = \omega_c - \omega_s \qquad \text{lower side-band frequency}$$

Thus the output voltage e' has two components: the upper and lower side-band frequency components.   Since the carrier frequency does not appear explicitly in Eq. (16-5), the term *suppressed carrier* is used.   For a more detailed explanation of synchros the reader is referred to the many texts on this subject.

For the control system of Fig. 16-3, whenever $\theta_i \neq \theta_o$, a voltage e' appears on the CT rotor winding.   This low-power voltage is amplified through voltage and power amplifiers and then applied to the control winding of the two-phase servomotor, which is the basic power unit in a-c feedback control systems.   The differential equation describing the performance of this type of motor is developed in Sec. 2-11.   To the reference winding of the motor is applied the same voltage V_x as that applied to the SG.   The two voltages applied to the windings of the servomotor must be approximately 90° out of phase.   This is frequently accomplished by inserting one capacitor in parallel with the reference winding of the motor and a second capacitor in series.   The a-c motor effectively acts as a demodulator unit followed by a d-c motor.   The signal component e' is thus effective in causing rotation of the motor.[3]   The system performance can be improved, if need be, through the use of cascade, feedback, or load compensation.   Load compensation can be applied in the same manner as for d-c systems.   However, the cascade-compensation procedure for d-c systems must be modified to take into account the suppressed-carrier modulated signal in the forward loop.   These modifications impose certain limitations on the networks used.   Both of these methods are illustrated in Fig. 16-5.   In feedback compensation (see Fig. 16-6), if only an a-c tachometer is needed to achieve the desired improvement, the techniques used for d-c feedback compensation apply.   This is so since the derivative action of the tachometer is dependent only upon $c(t)$ and is not affected by a change in carrier frequency.   Networks inserted into the feedback path in conjunction with the tachometer are limited in the same manner and require the same modifications in design as cascade net-

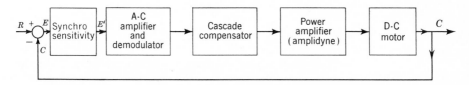

Fig. 16-5 *Control systems with load and cascade compensation.*

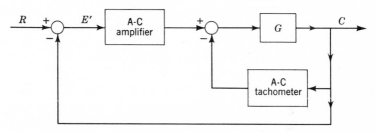

Fig. 16-6 *A-C control system with feedback compensation.*

works, because of the presence of the carrier frequency. Carrier frequencies often used are 60 and 400 cps. The former is used in ground systems because 60 cps is the commercial frequency available. Aircraft systems use 400 cps to reduce the size of components with magnetic cores. There are special applications where higher frequencies are used. A-c systems are used mostly for small instrument servos that do not require much power.

Another control system, illustrated by the block diagram of Fig. 16-7, can be labeled a hybrid system since it involves the use of both d-c and a-c signals.

It is to be noted that the synchro sensing unit of Fig. 16-3 can be represented symbolically, as shown in Fig. 16-7, in the same manner as d-c sensing elements. The use of a hybrid system permits the use of d-c compensation techniques. The location of a cascade compensator, as shown in Fig. 16-7, eliminates the need for taking into account the carrier frequency in its design.

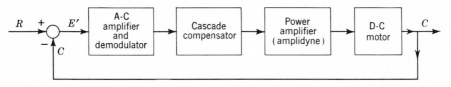

Fig. 16-7 *A hybrid control system.*

The choice of system depends on the following factors:

1. Use of d-c amplifiers with the required degree of stability necessitates a careful design. The ease of design and the inexpensive construction of a-c amplifiers, with no stability problem, make their use advantageous.
2. Maintenance of commutators and brushes, with the associated arcing, of d-c motors is a disadvantage when compared with the simplicity of construction of a-c motors.
3. Maintaining a constant carrier frequency for a-c systems involves difficulties.
4. The quadrature voltage in the a-c system may saturate the amplifiers.
5. The period of the carrier frequency must be sufficiently smaller than the response time of the system. Also, the response time of a-c systems may be longer than that of d-c systems.
6. The hybrid system combines the characteristics of the a-c and d-c systems, but more equipment may be necessary.
7. Miscellaneous factors:
   a. Type of power available
   b. Weight and space requirements
   c. Sensing elements available to compare the desired input and output quantities
   d. Compensating units available

16-3   Hybrid Control System

As an introduction to a-c control systems, a hybrid control system is discussed briefly. In Fig. 16-8 is shown a Type 1 position control system utilizing synchros as the sensing elements. The output of the control transformer is fed into an a-c preamplifier and then into a demodulator. Both of these functions may be performed in one unit. The output of the demodulator, E_1, is directly proportional to the quantity $\theta_i - \theta_o$.

Open-loop Analysis

It is assumed here that the shaft of the synchro generator is disconnected from the motor shaft and that the compensator unit is not included.

1. For $\theta_i(t) = u(t)$ radians, a constant actuating signal $e'(t)$ exists. Thus, for this Type 1 system, a constant rate of change of the output, $D\theta_o(t)$, results. The form of the signal appearing at various points in Fig. 16-8 is shown in Fig. 16-9.

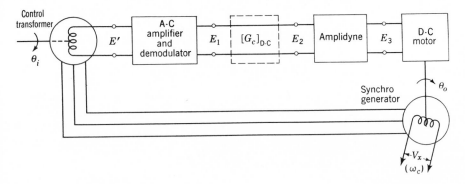

Fig. 16-8 *A hybrid position control system.*

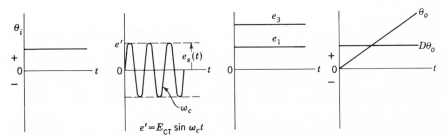

Fig. 16-9 *Form of the signal appearing at various points of Fig. 16-8 (open-loop) for a step input of θ_i.*

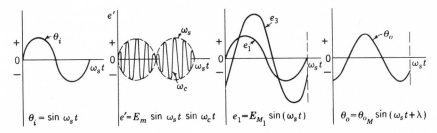

Fig. 16-10 *Form of the signal appearing at various points in Fig. 16-8 (open-loop) for a sinusoidal variation of θ_i.*

2. For $\theta_i(t) = \cos \omega_s t$ radians, the actuating signal e' has the form of a suppressed-carrier modulated signal. This results in θ_o varying sinusoidally at the signal frequency ω_s. The form of the signal appearing at various points in Fig. 16-8 is shown in Fig. 16-10. The phase lag of the output position compared with the input depends on the phase lag of the cascade components.

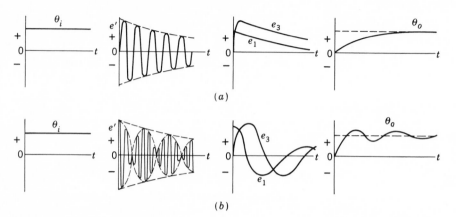

Fig. 16-11 *Form of the signal appearing at various points in Fig. 16-8 (closed-loop) for a step input of θ_i.   (a) Overdamped.   (b) Underdamped.*

Closed-loop Analysis

In this case it is assumed that the compensator unit is not included.   Figure 16-11 illustrates the form of the signals appearing at various points in the system for $\theta_i(t) = u(t)$ for both overdamped and underdamped situations.

Motor Transfer Function

It can be seen that the carrier frequency has no direct effect on the operation of the motor.   It is introduced by the nature of the synchro sensing unit for establishing the error signal and is removed by the demodulator.   The transfer function of the d-c motor is of the form

$$[G_M(s_s)]_{\text{d-c}} = \frac{K_M}{s_s(1 + Ts_s)} \tag{16-6}$$

where $s_s = j\omega_s$ for steady-state sinusoidal analysis, with a signal frequency ω_s.

If the system is unstable or not stable enough, a lead compensator can be inserted between the demodulator and the amplidyne of Fig. 16-8 to achieve the desired improvement.   A proportional plus derivative stabilizing network that can be used is shown in Fig. 16-12 and is described by

$$[G_c(s_s)]_{\text{d-c}} = \frac{E_2(s_s)}{E_1(s_s)} = \alpha \frac{1 + Ts_s}{1 + \alpha Ts_s} \tag{16-7}$$

where $\quad T = R_1 C \quad$ and $\quad \alpha = \dfrac{R_2}{R_1 + R_2}$

A plot of $|G_c(j\omega_s)|$ and ϕ_c versus ω_s is shown in Fig. 16-13.   This plot is shown for both positive and negative values of ω_s.   Although all frequencies

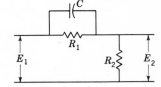

Fig. 16-12 *A d-c proportional plus derivative stabilizing network.*

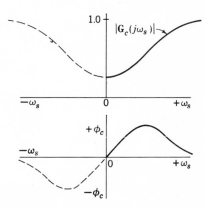

Fig. 16-13 *A plot of* $|G_c(\pm j\omega_s)|$ *and* ϕ_c *versus* ω_s *for a lead compensator.*

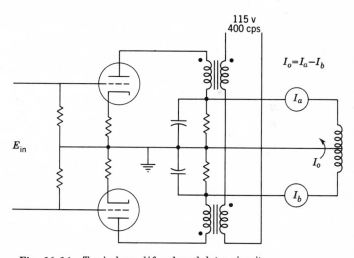

Fig. 16-14 *Typical amplifier-demodulator circuit.*

are present in the steady-state sinusoidal analysis, in practice the bandwidth of concern for compensation is approximately 3 to 10 cps.

Demodulator

The amplifier-demodulator unit has a form of push-pull arrangement whose output is connected to the control-field winding of the amplidyne, as shown

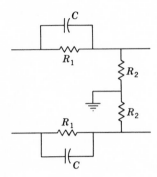

Fig. 16-15   *Balanced d-c lead compensator.*

in Fig. 16-14.   This arrangement requires a balanced (or push-pull) lead compensator, as shown in Fig. 16-15.   The necessary compensation is determined by considering only the d-c compensator network of Fig. 16-12.

16-4   A-C Control System[4,5]

Assume that the hybrid system of Fig. 16-8 is replaced by the a-c system of Fig. 16-16, which is also Type 1.   For the purpose of comparison, the physical quantity θ_o to be controlled is the same in both systems.   Note that in this case the sensing unit and the a-c preamplifier are common to both systems. The a-c power amplifier corresponds to the amplidyne, which functionally is also a power amplifier.

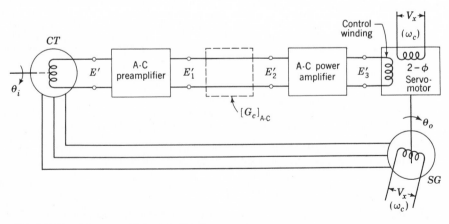

Fig. 16-16   *An a-c position control system.*

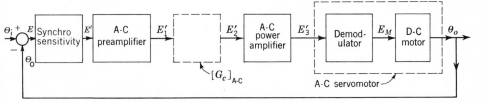

Fig. 16-17 *Mathematical model of Fig. 16-16.*

As shown in Fig. 16-16, on the reference winding of the servomotor is applied a constant voltage V_x from the same carrier-frequency source as that for the SG. On the other winding is applied the control signal $e_3(t)$. With these two voltages applied to the windings, it has been determined that the motor operation is a function only of the signal frequency ω_s[3]. The transfer function of the two-phase a-c servomotor has the form

$$[G_M(s_s)]_{\text{a-c}} = \frac{K_M}{s_s(1 + Ts_s)} \tag{16-8}$$

where
$$\omega_s \ll \omega_c$$

Thus the two-phase motor acts as a demodulator, and its transfer function has the same form as that for the d-c motor [Eq. (16-6)]. Practical values of the servomotor time constant lie in the range of 0.01 to 2.0 sec. For the purpose of analysis, the a-c motor could be considered to be replaced by an equivalent demodulator unit and a d-c motor. This equivalence is shown in Fig. 16-17.

Open- and Closed-loop Signal Analysis

As a result of the above, the functional difference between hybrid and a-c systems is the location of the demodulator action. When Figs. 16-8 and 16-16 are compared on a functional basis, it is seen that the demodulator action is shifted from occurring prior to the input of the power amplifier in the d-c system to occurring after the power amplifier in the a-c system. The mathematical model of the equivalent a-c system is illustrated in Fig. 16-17. Using this diagram, one can determine the forms of the signal appearing at various points in the system for step-function and sinusoidal inputs, respectively. The forms are similar to those indicated in Sec. 16-3. Particular attention should be paid to the form of the signal appearing in the input and output of the proposed compensator in each case. For the hybrid system it has the form of the signal shown for e_1 in Figs. 16-9 to 16-11.

For the a-c system it has the form of the signal shown for e' in the same figures.

A-C Compensation

The mathematical model of the a-c control system shown in Fig. 16-17 permits the use of cascade d-c compensation techniques. This is so since the amplifiers result in only a gain change at the signal frequency ω_s and the transfer function of the a-c motor is equivalent to that of the d-c motor. Thus, the basic system can be analyzed as an equivalent d-c control system to determine the necessary d-c compensation for the desired improvement. The required d-c compensator can be handled in either of two ways:

1. A demodulator can be inserted after the a-c preamplifier to produce a d-c signal. Then a d-c compensating unit can be inserted after the demodulator. A modulator must then be connected after the compensator to obtain a modulated signal, which is the required form of the input to the a-c power amplifier. The advantage of this method is that it permits the use of d-c compensating networks, which are easier to design and construct than a-c compensating networks. The disadvantages are the additional complexity, size, and noise added to the system by the use of the demodulator and modulator units.

2. The transformation can be made, through the low-pass-to-band-pass transformation technique, of the desired d-c compensating transfer function into an equivalent a-c compensator transfer function.[6] Another method is to approximate the a-c compensator by poles and zeros on the negative real axis when the d-c compensator represents a simple lead network. The resulting a-c compensator must produce the desired compensating effect on the signal envelope by operating directly on the suppressed-carrier modulated signal, as shown in Fig. 16-17. The advantages of this method are less weight, size, and noise as compared with the first method. The disadvantages are the possibility of excessive drift in the carrier frequency and the problem of design and construction of the desired compensating network. A certain small amount of drift in the carrier frequency can be tolerated without affecting the compensation. Depending upon the networks used, a drift of 10 to 15 per cent can be allowed.

The first approach has been thoroughly covered in earlier chapters on d-c compensation techniques and needs no further comment. The second approach is discussed in detail in the next section.

16-5   A-C Compensators[5]

From the mathematical model shown in Fig. 16-17 the desired d-c compensator (shown in Fig. 16-18a) operating on the signal of frequency ω_s can be synthesized by d-c techniques.

In the actual a-c system the necessary compensation must be done by an a-c compensator (shown in Fig. 16-18b) operating on the signal of frequency $\omega_c \pm \omega_s$. The problem now becomes one of transforming the desired characteristics of the d-c unit into the a-c unit. By referring to the figure, the transfer function for each compensator can be represented as follows:

$$G_c(s_s) = \frac{E_2(s_s)}{E_1(s_s)} \qquad \text{d-c} \tag{16-9}$$

$$G_c(s) = \frac{E_2'(s)}{E_1'(s)} \qquad \text{a-c} \tag{16-10}$$

Consider that the input signal to the d-c unit is $e_1(t)$ and its output is $e_2(t)$. Note that these two signals in the mathematical model represent the signal portion or the envelope of the suppressed-carrier modulated signal in the input and output of the a-c compensator. The ratio of the Laplace transforms of these two signals results in Eq. (16-9). The input to the a-c unit is the suppressed-carrier modulated signal

$$e_1'(t) = e_1(t) \cos \omega_c(t) \tag{16-11}$$

If this unit is to modify the signal portion of the suppressed carrier in exactly the same manner as the d-c unit, its output must have the form

$$e_2'(t) = e_2(t) \cos \omega_c(t) \tag{16-12}$$

Substituting the exponential form for the cosine into Eqs. (16-11) and (16-12) yields

$$e_1'(t) = e_1(t) \frac{e^{j\omega_c t} + e^{-j\omega_c t}}{2} \tag{16-13}$$

$$e_2'(t) = e_2(t) \frac{e^{j\omega_c t} + e^{-j\omega_c t}}{2} \tag{16-14}$$

Fig. 16-18 *D-c and a-c compensating units.* (a) *D-c compensator.* (b) *A-c compensator.*

Taking the Laplace transform of each of the above equations yields

$$E_1'(s) = \mathcal{L}[e_1'(t)] = \tfrac{1}{2} \int_0^\infty [e_1(t)e^{j\omega_c t}]e^{-st}\, dt + \tfrac{1}{2} \int_0^\infty [e_1(t)e^{-j\omega_c t}]e^{-st}\, dt$$

$$= \tfrac{1}{2} \int_0^\infty e_1(t)e^{-(s-j\omega_c)t}\, dt + \tfrac{1}{2} \int_0^\infty e_1(t)e^{-(s+j\omega_c)t}\, dt$$

$$= \tfrac{1}{2}E_1(s - j\omega_c) + \tfrac{1}{2}E_1(s + j\omega_c) \tag{16-15}$$

$$E_2'(s) = \tfrac{1}{2} \int_0^\infty e_2(t)e^{-(s-j\omega_c)t}\, dt + \tfrac{1}{2} \int_0^\infty e_2(t)e^{-(s+j\omega_c)t}\, dt$$

$$= \tfrac{1}{2}E_2(s - j\omega_c) + \tfrac{1}{2} E_2(s + j\omega_c) \tag{16-16}$$

Substituting Eqs. (16-15) and (16-16) into Eq. (16-10) results in

$$G_c(s) = \frac{E_2'(s)}{E_1'(s)} = \frac{E_2(s - j\omega_c) + E_2(s + j\omega_c)}{E_1(s - j\omega_c) + E_1(s + j\omega_c)} \tag{16-17}$$

or

$$\mathbf{G}_c(j\omega) = \frac{\mathbf{E}_2(j\omega - j\omega_c) + \mathbf{E}_2(j\omega + j\omega_c)}{\mathbf{E}_1(j\omega - j\omega_c) + \mathbf{E}_1(j\omega + j\omega_c)} \tag{16-18}$$

For the a-c system, $\omega = \omega_c \pm \omega_s$. By considering frequencies only on the $+j\omega$ axis, the $j\omega + j\omega_c$ terms in the above can be expressed as

$$\mathbf{E}_1(j\omega + j\omega_c) = \mathbf{E}_1(2j\omega_c \pm j\omega_s) \tag{16-19}$$

$$\mathbf{E}_2(j\omega + j\omega_c) = \mathbf{E}_2(2j\omega_c \pm j\omega_s) \tag{16-20}$$

On the basis that $\omega_s \ll \omega_c$ and that the system has a narrow passband around ω_c, the suppressed-carrier modulated input signal (and consequently the output signal) does not contain any component of frequency $(2\omega_c \pm \omega_s)$. Thus the terms expressed by Eqs. (16-19) and (16-20) are equal to zero. Equation (16-18) now reduces to

$$\mathbf{G}_c(j\omega) = \frac{\mathbf{E}_2(j\omega - j\omega_c)}{\mathbf{E}_1(j\omega - j\omega_c)} \qquad \text{for } \omega > 0 \tag{16-21}$$

Analyzing Eq. (16-21) shows that

$$\mathbf{E}_1(j\omega - j\omega_c) = \mathbf{E}_1(j\omega_c \pm j\omega_s - j\omega_c) = \mathbf{E}_1(\pm j\omega_s) \tag{16-22}$$

$$\mathbf{E}_2(j\omega - j\omega_c) = \mathbf{E}_2(j\omega_c \pm j\omega_s - j\omega_c) = \mathbf{E}_2(\pm j\omega_s) \tag{16-23}$$

Thus from Eq. (16-9),

$$\mathbf{E}_2(j\omega - j\omega_c) = \mathbf{E}_2(\pm j\omega_s) = \mathbf{E}_1(\pm j\omega_s)\mathbf{G}_c(\pm j\omega_s) \tag{16-24}$$

Substituting Eqs. (16-22) and (16-24) into Eq. (16-21) gives

$$\mathbf{G}_c(j\omega) = \frac{\mathbf{E}_1(\pm j\omega_s)\mathbf{G}_c(\pm j\omega_s)}{\mathbf{E}_1(\pm j\omega_s)} = \mathbf{G}_c(\pm j\omega_s) = \mathbf{G}_c(j\omega - j\omega_c) \text{ for } \omega > 0 \tag{16-25}$$

In the same manner, frequencies of ω along the $-j\omega$ axis yield the result

$$\mathbf{G}_c(j\omega) = \frac{\mathbf{E}_2(j\omega + j\omega_c)}{\mathbf{E}_1(j\omega + j\omega_c)} \qquad \text{for } \omega < 0 \tag{16-26}$$

or

$$\mathbf{G}_c(j\omega) = \mathbf{G}_c(\pm j\omega_s) = \mathbf{G}_c(j\omega + j\omega_c) \qquad \text{for } \omega < 0 \tag{16-27}$$

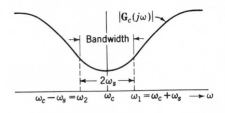

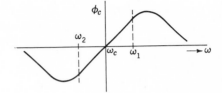

Fig. 16-19 *A plot of $|G_c(j\omega)|$ and ϕ_c versus ω for an a-c lead compensator.*

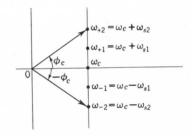

Fig. 16-20 *Required characteristics in the complex plane of an a-c lead compensator for the carrier frequency ω_c.*

By analyzing Eqs. (16-15) and (16-16) it is seen that, by the shifting theorem of Laplace theory (Theorem 3, Translation in Time, Sec. 4-4), the transform $G_c(\pm j\omega_s)$ has been shifted $\pm\omega_c$ units to the right and left, yielding the transform $G_c(j\omega)$. Because of the symmetrical conditions that exist for $\pm\omega$, it suffices to analyze the a-c compensator only along the $+j\omega$ axis. In other words, the curves in Fig. 16-13 for $G_c(\pm j\omega_s)$ have been shifted to the right ω_c units, as shown in Fig. 16-19.

If the required characteristics for the a-c compensator are drawn in polar form, the curve is symmetrical about the real axis, as shown in Fig. 16-20.

As a result of the preceding analysis, it can be concluded that, if an a-c compensator $G_c(j\omega)$ can be designed to have the frequency characteristic of Fig. 16-19 or 16-20, it will produce the desired lead compensation upon the signal of frequency ω_s. A physical network cannot be designed to have the above characteristic over the entire frequency spectrum. Where the band-width $2\omega_s$ is small in comparison with the carrier frequency ω_c (usually referred to as narrow-band operation), the problem of obtaining a physically realizable network for the a-c compensator is simplified. In other words, it is possible to design a physically realizable network so that it will have the desired characteristics over the narrow bandwidth of frequency $\omega_c \pm \omega_s$.

16-6   *Design of A-C Lead Compensators*[4,5]

It is stated in a previous section that two methods are available for synthesizing the a-c compensator. One is the low-pass-to-band-pass transformation technique applied to any form of $G(s_s)$. The other is the approximation of the a-c compensator by poles and zeros on the negative real axis when $G(s_s)$ represents a simple lead network. The latter method is discussed in detail in the following pages. The reader is referred to textbooks on network synthesis for discussions of the former method.

Case 1

Consider a d-c lead network of the form shown in Fig. 16-21a whose transfer function is given by

$$G(s_s) = \frac{Ts_s}{1 + Ts_s} \tag{16-28}$$

where $T = L/R$. Rewriting the above equation in terms of the frequency ω_s,

$$G(j\omega_s) = \frac{j\omega_s T^{\cdot}}{1 + j\omega_s T} \tag{16-29}$$

In deriving the equivalent a-c network, the size of the component R is the same for each network. The frequency-variant portion of the d-c lead network has the frequency-response characteristic (X versus ω_s) shown in Fig. 16-22a. For the a-c network to provide control identical to that of the d-c network, it must have the same frequency-response characteristic (X' versus ω) in the bandwidth of concern. To achieve this requirement the L portion of the d-c unit must be replaced with a series resonant $L'C'$ branch, as indicated in Fig. 16-21b, having identical magnitude and phase characteristics over the bandwidth $\omega_c \pm \omega_s$. The frequency-response characteristic of the $L'C'$ branch is shown in Fig. 16-22b. The resonant frequency of this branch must be equal to the carrier frequency. Thus,

$$\omega_c = \frac{1}{\sqrt{L'C'}} \tag{16-30}$$

Over the bandwidth of concern (see Fig. 16-22) the values of X and X' must be identical in the region of $\omega_s = 0$ and $\omega = \omega_c$, respectively. The

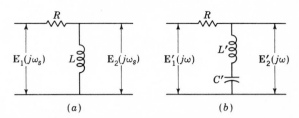

Fig. 16-21 *Lead compensators.* (*a*) *D-c compensator.* (*b*) *A-c compensator.*

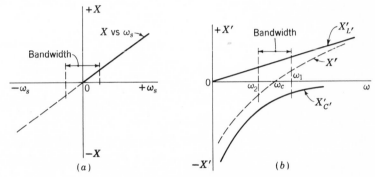

Fig. 16-22 *Frequency characteristics of the frequency-variant portions of the d-c and a-c networks of Fig. 16-21.*

respective reactances and the slopes of these curves are

$$X = \omega_s L \tag{16-31}$$

$$\frac{dX}{d\omega_s} = L \tag{16-32}$$

and

$$X' = \omega L' - \frac{1}{\omega C'} \tag{16-33}$$

$$\frac{dX'}{d\omega} = L' + \frac{1}{\omega^2 C'} \tag{16-34}$$

from which the slope in the vicinity of $\omega = \omega_c$ is

$$\frac{dX'}{d\omega}\bigg]_{\omega = \omega_c} = L' + \frac{1}{\omega_c^2 C'} \tag{16-35}$$

Combining Eqs. (16-30) and (16-35) yields

$$\frac{dX'}{d\omega}\bigg]_{\omega = \omega_c} = L' + L' = 2L' \tag{16-36}$$

Since the slopes in the vicinity of $\omega_s = 0$ and $\omega = \omega_c$ must be equal,

$$\frac{dX}{d\omega_s} = L = \frac{dX'}{d\omega}\bigg]_{\omega=\omega_c} = 2L'.$$

Therefore
$$L' = \frac{L}{2} \tag{16-37}$$

$$C' = \frac{1}{\omega_c^2 L'} = \frac{2}{\omega_c^2 L} \tag{16-38}$$

With the values of L' and C' determined by Eqs. (16-37) and (16-38), the a-c lead compensator of Fig. 16-21b produces the same derivative control on the signal as does the corresponding d-c lead compensator.

Case 2

Consider a d-c lead network of the form shown in Fig. 16-23a whose transfer function is given by

$$G(s_s) = \frac{Ts_s}{1 + Ts_s} \tag{16-39}$$

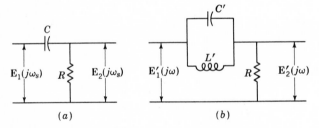

(a)   (b)

Fig. 16-23   *Lead compensators.   (a) D-c compensator.   (b) A-c compensator.*

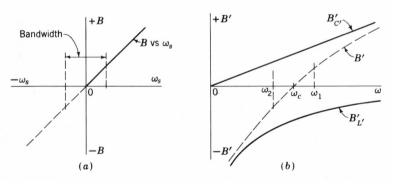

(a)   (b)

Fig. 16-24   *Frequency characteristics of the frequency-variant portions of the d-c and a-c networks of Fig. 16-23.*

In terms of the signal frequency ω_s this transfer function is

$$\mathbf{G}(j\omega_s) = \frac{j\omega_s T}{1 + j\omega_s T} \tag{16-40}$$

where

$$T = RC$$

In deriving the equivalent a-c network, the component R is the same for each network. The variant portion of the d-c lead network has the frequency-response characteristic (B versus ω_s) shown in Fig. 16-24a. B is the susceptance of the capacitor; that is, $B = \omega C$. For the a-c network to provide derivative control identical to that of the d-c network, it must have the same frequency-response characteristic (B' versus ω) in the bandwidth of concern. To achieve this requirement the capacitor C of the d-c unit must be replaced with a parallel resonant $L'C'$ branch, as indicated in Fig. 16-23b, having identical magnitude and phase characteristics over the bandwidth $\omega_c \pm \omega_s$. The frequency-response characteristic of the $L'C'$ branch is shown in Fig. 16-24b. The resonant frequency of this branch must be equal to the carrier frequency, as expressed by Eq. (16-30), when the resistances of C' and L' are very small. Over the bandwidth of concern (see Fig. 16-24a) the values of B and B' must be identical in the region of $\omega_s = 0$ and $\omega = \omega_c$, respectively. The respective values of susceptance and their slopes are

$$B = \omega_s C \tag{16-41}$$

$$\frac{dB}{d\omega_s} = C \tag{16-42}$$

$$B' = \omega C' - \frac{1}{\omega L'} \tag{16-43}$$

$$\frac{dB'}{d\omega} = C' + \frac{1}{\omega^2 L'} \tag{16-44}$$

The slope in the vicinity of $\omega = \omega_c$ is

$$\frac{dB'}{d\omega}\bigg]_{\omega = \omega_c} = C' + \frac{1}{\omega_c^2 L'} \tag{16-45}$$

Combining Eqs. (16-30) and (16-45) yields

$$\frac{dB'}{d\omega}\bigg]_{\omega = \omega_c} = C' + C' = 2C' \tag{16-46}$$

Since the slopes in the vicinity of $\omega_s = 0$ and $\omega = \omega_c$ must be equal,

$$\frac{dB}{d\omega_s} = C = \frac{dB'}{d\omega}\bigg]_{\omega = \omega_c} = 2C' \tag{16-47}$$

Therefore $C' = C/2$ and

$$L' = \frac{2}{\omega_c^2 C} \tag{16-48}$$

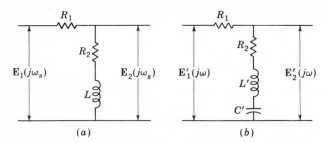

Fig. 16-25 *Lead compensators.* (a) *D-c compensator.* (b) *A-c compensator.*

With the values of L' and C' determined by Eqs. (16-47) and (16-48), the a-c lead compensator of Fig. 16-23b produces the same derivative control as does the corresponding d-c lead compensator.

Case 3

Consider a d-c lead network of the form shown in Fig. 16-25a whose transfer function is given by

$$G(s_s) = \frac{R_2 + s_s L}{R_1 + R_2 + s_s L} = \alpha \frac{1 + Ts_s}{1 + \alpha Ts_s} \tag{16-49}$$

where $T = L/R_2$ and $\alpha = R_2/(R_1 + R_2)$.

Rewriting the above equation in terms of $j\omega_s$,

$$\mathbf{G}(j\omega_s) = \frac{R_2 + j\omega_s L}{R_1 + R_2 + j\omega_s L} = \alpha \frac{1 + j\omega_s T}{1 + j\omega_s \alpha T} \tag{16-50}$$

In deriving the equivalent a-c network, the values of the resistors remain the same as for the d-c network and the inductor L is replaced by the series capacitor and inductor, C' and L'. The transfer function of the a-c compensator shown in Fig. 16-25b, as given by Eq. (16-51), is used in determining the values of L' and C':

$$
\begin{aligned}
\mathbf{G}(j\omega) &= \frac{\mathbf{E}_2'(j\omega)}{\mathbf{E}_1'(j\omega)} = \frac{R_2 + j\omega L' + 1/j\omega C'}{R_1 + R_2 + j\omega L' + 1/j\omega C'} \\
&= \frac{R_2}{R_1 + R_2} \frac{1 + j(1/R_2)(\omega L' - 1/\omega C')}{1 + j[R_2/(R_1 + R_2)](1/R_2)(\omega L' - 1/\omega C')}
\end{aligned} \tag{16-51}
$$

For the a-c compensator to perform the same compensation function as the d-c compensator, the resonant frequency must be

$$\omega_c = \frac{1}{\sqrt{L'C'}} \tag{16-52}$$

Equation (16-51), representing the transfer function of the a-c compensator, must be shown to be equivalent to Eq. (16-50) for the signal frequency ω_s. This is true if $\omega L' - 1/\omega C'$ is proportional to ω_s. Using the resonance condition of Eq. (16-52),

$$\omega L' - \frac{1}{\omega C'} = \sqrt{\frac{L'}{C'}} \left(\frac{\omega}{\omega_c} - \frac{\omega_c}{\omega} \right) = \sqrt{\frac{L'}{C'}} \frac{\omega^2 - \omega_c^2}{\omega \omega_c}$$

$$= \sqrt{\frac{L'}{C'}} \frac{(\omega - \omega_c)(\omega + \omega_c)}{\omega \omega_c} \tag{16-53}$$

Since $\omega = \omega_c \pm \omega_s$ and $\omega_s \ll \omega_c$,

$$\omega \approx \omega_c$$
$$\omega + \omega_c \approx 2\omega$$
$$\omega - \omega_c = \omega_s$$

With these approximations and the upper side band $\omega = \omega_c + \omega_s$, Eq. (16-53) can be represented by

$$\omega L' - \frac{1}{\omega C'} \approx \sqrt{\frac{L'}{C'}} \frac{2\omega_s}{\omega_c} = 2L' \omega_s \tag{16-54}$$

Equation (16-51) can therefore be written as

$$\mathbf{G}(j\omega_s) \approx \frac{R_2}{R_1 + R_2} \frac{1 + j(2L'/R_2)\omega_s}{1 + j[R_2/(R_1 + R_2)](2L'/R_2)\omega_s} \tag{16-55}$$

By using the definitions $\alpha = R_2/(R_1 + R_2)$ and $T = 2L'/R_2$ this equation can be written in the form

$$\mathbf{G}(j\omega_s) \approx \alpha \frac{1 + j\omega_s T}{1 + j\omega_s \alpha T} \tag{16-56}$$

Thus it is seen that $\mathbf{G}(j\omega)$ for the a-c network is a function only of ω_s and is independent of the carrier frequency within the bandwidth of concern. For the d-c and a-c networks to be equivalent, it can be seen by comparing Eqs. (16-49) and (16-55) that the necessary conditions are

$$L = 2L' \quad \text{or} \quad L' = \frac{L}{2}$$
$$C' = \frac{2}{\omega_c^2 L} \tag{16-57}$$

The results are the same as for Case 1 [see Eqs. (16-37) and (16-38)] and can be extended to any compensator containing an inductance in a shunt branch.

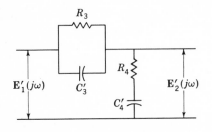

Fig. 16-26 *An RC derivative (lead) a-c network.*

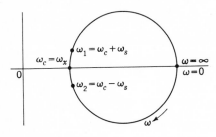

Fig. 16-27 *Polar plot of the transfer function of Eq. (16-58).*

Case 4

An a-c lead compensator using only resistors and capacitors is preferable to one requiring inductors, because inductors are much larger and heavier than capacitors and may not be linear. Therefore the circuit shown in Fig. 16-26 is suggested as a replacement for the circuit of Fig. 16-25b.

The ratio of the output to the input for this circuit is

$$\mathbf{G}(j\omega) = \frac{R_4 + 1/j\omega C_4'}{\dfrac{R_3/j\omega C_3'}{R_3 + 1/j\omega C_3'} + R_4 + 1/j\omega C_4'}$$

$$= \frac{(1 + j\omega R_4 C_4')(1 + j\omega R_3 C_3')}{j\omega R_3 C_4' + (1 + j\omega R_4 C_4')(1 + j\omega R_3 C_3')}$$

or $\qquad \mathbf{G}(j\omega) = \dfrac{(1 - \omega^2 R_3 R_4 C_3' C_4') + j\omega(R_3 C_3' + R_4 C_4')}{(1 - \omega^2 R_3 R_4 C_3' C_4') + j\omega(R_3 C_4' + R_3 C_3' + R_4 C_4')}$ $\qquad$ (16-58)

The polar plot of the transfer function given by Eq. (16-58) is drawn in Fig. 16-27. This plot is symmetrical about the real axis.

The compensator circuit elements can be selected so that the frequency ω_x equals ω_c, the carrier frequency. Then, since $\omega_s \ll \omega_c$, the portion of the curve for $\omega = \omega_c \pm \omega_s$ is a close approximation to the required characteristics shown in Fig. 16-20. Therefore, by the proper selection of the compensator circuit elements it is possible to use the circuit of Fig. 16-26 as an a-c lead compensator.

At the frequency $\omega_c = \omega_x$ the value of $\mathbf{G}(j\omega_c)$ obtained from Eq. (16-58) is real if the real terms in the numerator and denominator are equal to zero.

This condition yields

$$\omega_x^2 = \omega_c^2 = \frac{1}{R_3 R_4 C_3' C_4'} \tag{16-59}$$

For simplicity in writing the equations, let

$$T_1 = R_3 C_3' + R_4 C_4'$$
$$T_2 = R_3 C_4' + R_3 C_3' + R_4 C_4'$$

Equation (16-58) is now rewritten:

$$
\begin{aligned}
\mathbf{G}(j\omega) &= \frac{(1 - \omega^2/\omega_c^2) + j\omega T_1}{(1 - \omega^2/\omega_c^2) + j\omega T_2} \\
&= \frac{j\omega T_1}{j\omega T_2} \frac{1 + j\,\dfrac{(\omega^2/\omega_c^2) - 1}{\omega T_1}}{1 + j\,\dfrac{(\omega^2/\omega_c^2) - 1}{\omega T_2}}
\end{aligned}
\tag{16-60}
$$

It is necessary to show that this equation can be expressed approximately as a function of the signal frequency ω_s. This is done by using the condition $\omega_s \ll \omega_c$ and the upper side band $\omega = \omega_c + \omega_s$ so that $\omega - \omega_c = \omega_s$. The approximation can be made that $\omega \approx \omega_c$ and $\omega + \omega_c \approx 2\omega$. Hence,

$$\frac{\omega^2}{\omega_c^2} - 1 = \frac{\omega^2 - \omega_c^2}{\omega_c^2} = \frac{(\omega + \omega_c)(\omega - \omega_c)}{\omega_c^2} \approx \frac{2\omega\omega_s}{\omega_c^2} \tag{16-61}$$

The compensator transfer function is written as

$$\mathbf{G}(j\omega_s) \approx \frac{T_1}{T_2} \frac{1 + j\omega_s(2/T_1\omega_c^2)}{1 + j\omega_s(T_1/T_2)(2/T_1\omega_c^2)} \tag{16-62}$$

This transfer function can be put in the form of the lead compensator:

$$\mathbf{G}(j\omega_s) \approx \alpha \frac{1 + j\omega_s T}{1 + j\omega_s \alpha T} \tag{16-63}$$

where

$$\alpha = \frac{T_1}{T_2} = \frac{R_3 C_3' + R_4 C_4'}{R_3 C_4' + R_3 C_3' + R_4 C_4'} \tag{16-64}$$

$$T = \frac{2}{T_1\omega_c^2} = \frac{2 R_3 R_4 C_3' C_4'}{R_3 C_3' + R_4 C_4'} \tag{16-65}$$

There are four circuit parameters to be evaluated (R_3, R_4, C_3', C_4') and only three equations to be satisfied. The three equations are (16-59), (16-64), and (16-65). Therefore one of the four parameters must be selected arbitrarily. When this is done and the required values of ω_c, α, and T are known, the remaining terms can be determined.

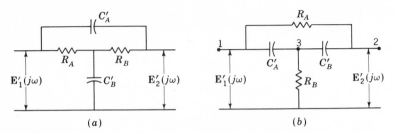

Fig. 16-28 *Bridged-T networks for 60-cps carrier systems (see Prob. 16-3f).*

Bridged-T Networks

The bridged-T networks that can be used as a-c lead networks are shown in Fig. 16-28. Solve for $\mathbf{E}_2'(j\omega)/\mathbf{E}_1'(j\omega)$ and then manipulate the equation to put it in the standard form. For the bridged-T network of Fig. 16-28b the equations for nodes 2 and 3 are

$$\left(\frac{1}{R_A}\right)\mathbf{E}_1' = \left(j\omega C_B' + \frac{1}{R_A}\right)\mathbf{E}_2' - (j\omega C_B')\mathbf{E}_3' \tag{16-66}$$

$$(j\omega C_A')\mathbf{E}_1' = -(j\omega C_B')\mathbf{E}_2' + \left(j\omega C_B' + j\omega C_A' + \frac{1}{R_B}\right)\mathbf{E}_3' \tag{16-67}$$

Solving these equations for $\mathbf{E}_2'/\mathbf{E}_1'$ gives

$$\frac{\mathbf{E}_2'(j\omega)}{\mathbf{E}_1'(j\omega)} = \frac{(j\omega)^2 R_A R_B C_A' C_B' + (j\omega) R_B(C_A' + C_B') + 1}{(j\omega)^2 R_A R_B C_A' C_B' + j\omega[R_B(C_A' + C_B') + R_A C_B'] + 1} \tag{16-68}$$

This equation has the same form as Eq. (16-58) for Case 4, therefore the same procedures are used. Resonance can be produced by making the real terms of both the numerator and denominator go to zero at $\omega = \omega_c$. The necessary relationship for the carrier frequency ω_c is

$$\omega_c^2 = \frac{1}{R_A R_B C_A' C_B'} \tag{16-69}$$

Using the condition that $\omega_s \ll \omega_c$ leads to the approximate form of Eq. (16-68) given by

$$\mathbf{G}_c(j\omega_s) \approx \alpha \frac{1 + j\omega_s T}{1 + j\omega_s \alpha T} \tag{16-70}$$

where

$$\alpha = \frac{R_B(C_A' + C_B')}{R_B C_A' + R_B C_B' + R_A C_B'} \tag{16-71}$$

$$T = \frac{2}{R_B(C_A' + C_B')\omega_c^2} \tag{16-72}$$

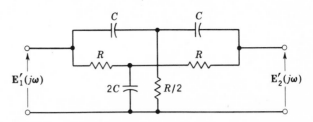

Fig. 16-29   *A symmetric parallel-T network.*

Symmetric Parallel-T Network

Another network that can be used as an a-c lead network is the symmetric parallel-T network shown in Fig. 16-29. Its transfer function can be derived in the same manner as that for the bridged-T network:

$$\mathbf{G}(j\omega) = \frac{\mathbf{E}_2'(j\omega)}{\mathbf{E}_1'(j\omega)} = \frac{1 - \omega^2 R^2 C^2}{(1 - \omega^2 R^2 C^2) + j4RC\omega} \qquad (16\text{-}73)$$

The polar plot of this transfer function is similar to that shown in Fig. 16-27, except that it goes through the origin at ω_x. This resonant frequency is made the carrier frequency, that is,

$$\omega_x = \omega_c = \frac{1}{RC} \qquad (16\text{-}74)$$

At $\omega = \omega_c$ the output of this network is zero.

To specify a minimum value other than zero for the output at $\omega = \omega_c$, a signal must be added to the output which is directly proportional to the input. This is obtained by using this network in the block diagram of Fig. 16-30. The output is given by

$$E_4' = E_2' + E_3' \qquad (16\text{-}75)$$

where $E_3' = \alpha E_1'$, $E_2' = \beta G E_1'$, $0 < \alpha < 1$, and $0 < \beta < 1$. Combining these equations and solving for the ratio E_4'/E_1' yields

$$\mathbf{G}_c(j\omega) = \frac{E_4'}{E_1'} = \frac{(1 - \omega^2 R^2 C^2) + j\omega 4RC\alpha}{(1 - \omega^2 R^2 C^2) + j\omega 4RC} \qquad (16\text{-}76)$$

As a check, note that when $\omega = \omega_c$, $\mathbf{G}_c(j\omega) = \alpha$. An RC network that produces the transfer function of Eq. (16-76) is shown in Table 16-2, where

$$\alpha = \frac{R_b}{R_a + R_b} \qquad \text{and} \qquad \beta = \frac{R_a}{R_a + R_b}$$

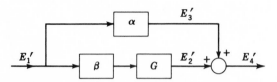

Fig. 16-30 *A block diagram using a symmetrical parallel-T, with a nonzero output at $\omega = \omega_c$.*

Note that a condition required by this derivation is that there be no loading on the β unit shown in Fig. 16-30.

The bridged-T and symmetric parallel-T networks may be designed to minimize the effect of carrier-frequency drift. A design procedure to accomplish this is presented in the literature.[8]

Summary

Table 16-1 summarizes the results of this section dealing with a-c compensators utilizing R, L, and C components. From this table it can be noted that the equivalent a-c network is obtained from the d-c network by doing either of the following:

1. Replacing each C by $C' = C/2$ and adding in parallel with the C' an L' to resonate at the carrier frequency ω_c
2. Replacing each L by $L' = L/2$ and adding in series with each L' a C' to resonate at the carrier frequency ω_c

The desired d-c compensation is determined from the required performance of the system. For simple d-c lead networks, as shown in Table 16-1, the equivalent a-c lead compensator can be found in the manner used in this chapter.

Table 16-2 summarizes the results of this section dealing with a-c compensators utilizing only R and C components. The a-c network is made equivalent to the d-c network by the proper choice of values for the R and C components so that the values of α and T are identical for both networks.

The characteristics of the a-c compensators described in this section assume a constant carrier frequency, but a design method which minimizes the effect of carrier drift is available.[8] A nonlinear lead compensator which is independent of drift in the carrier frequency is described in the literature.[7] Because of carrier drift, passive a-c lag compensators are normally not utilized. A method for designing nonpassive a-c lag compensators which permit satisfactory operation is also described in the literature.[9]

Table 16-1. *Comparison of d-c and a-c lead compensators utilizing R, L, and C components*

D-C compensators	A-C compensators $G(j\omega) = G(j\omega_s)$	
$$G(j\omega_s) = \frac{Ts_s}{1 + Ts_s}$$		$L' = \dfrac{L}{2}$ $\omega_c = \dfrac{1}{\sqrt{L'C'}}$ $C' = \dfrac{2}{\omega_c^2 L}$
$$G(j\omega_s) = \frac{Ts_s}{1 + Ts_s}$$		$C' = \dfrac{C}{2}$ $\omega_c = \dfrac{1}{\sqrt{L'C'}}$ $L' = \dfrac{2}{\omega_c^2 C}$
$\alpha = \dfrac{R_2}{R_1 + R_2}$ $T = \dfrac{L}{R_2}$ $$G(j\omega_s) = \alpha\frac{(1 + Ts_s)}{(1 + \alpha\, Ts_s)}$$	$L' = \dfrac{L}{2}$ $C' = \dfrac{2}{\omega_c^2 L}$ $\alpha = \dfrac{R_2}{R_1 + R_2}$	$\omega_c = \dfrac{1}{\sqrt{L'\,C'}}$ $T = \dfrac{2L'}{R_2}$
$\alpha = \dfrac{R_2}{R_1 + R_2}$ $T = R_1 C$ $$G(j\omega_s) = \alpha\frac{(1 + Ts_s)}{(1 + \alpha\, Ts_s)}$$		$C' = \dfrac{C}{2}$ $\omega_c = \dfrac{1}{\sqrt{L'C'}}$ $L' = \dfrac{2}{\omega_c^2 C}$

Table 16-2. *Comparison of d-c and a-c lead compensators utilizing R and C components*

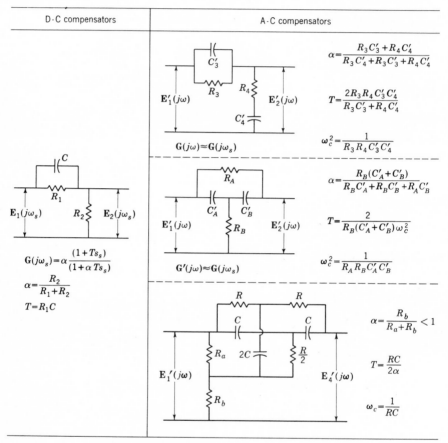

D-C compensators	A-C compensators

For the D-C compensator:

$$G(j\omega_s) = \alpha \frac{(1 + Ts_s)}{(1 + \alpha Ts_s)}$$

$$\alpha = \frac{R_2}{R_1 + R_2}$$

$$T = R_1 C$$

For the A-C compensators (top):

$$G(j\omega) \approx G(j\omega_s)$$

$$\alpha = \frac{R_3 C_3' + R_4 C_4'}{R_3 C_4' + R_3 C_3' + R_4 C_4'}$$

$$T = \frac{2 R_3 R_4 C_3' C_4'}{R_3 C_3' + R_4 C_4'}$$

$$\omega_c^2 = \frac{1}{R_3 R_4 C_3' C_4'}$$

(middle):

$$G'(j\omega) \approx G(j\omega_s)$$

$$\alpha = \frac{R_B(C_A' + C_B')}{R_B C_A' + R_B C_B' + R_A C_B'}$$

$$T = \frac{2}{R_B(C_A' + C_B')\omega_c^2}$$

$$\omega_c^2 = \frac{1}{R_A R_B C_A' C_B'}$$

(bottom):

$$\alpha = \frac{R_b}{R_a + R_b} < 1$$

$$T = \frac{RC}{2\alpha}$$

$$\omega_c = \frac{1}{RC}$$

16-7   Conclusions

In Sec. 16-2 a comparison of the three types of control systems is given, along with a list of the factors to be weighed in making a choice. A possible approach in the study of an a-c system is presented which is limited to the extent of the approximations made. The procedure for deriving a-c compensators is illustrated by several examples and is summarized in Tables 16-1 and 16-2. By the methods presented in textbooks on network synthesis, more complicated a-c compensating networks can be designed.

Bibliography

1. "Radar Circuit Analysis," AF Manual 52-8, chap. 13, U.S. Air Force, 1951.
2. Lauer, H., R. Lesnick, and L. E. Matson: "Servomechanism Fundamentals," chap. 2, McGraw-Hill Book Company, New York, 1947.
3. Sobczyk, A.: Stabilization of Carrier-frequency Servomechanisms, *J. Franklin Inst.*, vol. 246, pp. 21–44, July, 1948.
4. Chestnut, H., and R. W. Mayer: "Servomechanisms and Regulating System Design," vol. 2, chap. 6, John Wiley & Sons, Inc., New York, 1955.
5. Truxal, J. G.: "Automatic Feedback Control System Synthesis," pp. 390–409, McGraw-Hill Book Company, New York, 1955.
6. Bode, H. W.: "Network Analysis and Feedback Amplifier Design," D. Van Nostrand Company, Inc., Princeton, N.J., 1945.
7. Clegg, J. C.: A Time-dependent Nonlinear Compensating Network, *Trans. AIEE*, vol. 75, pt. II, pp. 306–308, November, 1956.
8. Nagrath, I. J., and V. K. Arya: Design of Lead Networks for A-C Servos with Carrier Frequency Drift, *Trans. AIEE*, vol. 81, pt. II, pp. 291–296, November, 1962.
9. Murphy, G. J., and A. Knox: A Method for the Design of a Phase-lag Demodulating Compensator for Use in Carrier Control Systems, *Trans. AIEE*, vol. 83, pt. II, pp. 252–257, July, 1964.
10. Ivey, K. A.: "A-C Carrier Control Systems," John Wiley & Sons, Inc., New York, 1964.

17

Optimum
Response*

17-1   Introduction

The synthesis and design of feedback control systems are treated
extensively in the literature, and the principal methods are
described in the earlier chapters of this book. The design
process includes the use of compensation or stabilization to
improve the transient and steady-state performance. Each of
the methods described has its particular advantages and disadvantages. The frequency-response method suffers from difficulty in determining the exact transient response from the frequency-response characteristics. While a correlation is made
between the frequency and time responses in Chaps. 7, 8, and 11,
this is based on representing the system by an approximately
equivalent second-order system. Effective damping ratio and
effective undamped natural frequency are related to the maximum value of the frequency response, M_m. This approximation

* The major portion of the material in this chapter is based on reference 1.

582

becomes less valid as the order of the system becomes higher. Precise methods for obtaining the time response from the frequency response are generally laborious. The root-locus method has the advantage that both the transient response and the frequency response may be obtained from the location of the closed-loop poles and zeros. Therefore the improvements obtained by additional poles and zeros can be evaluated.

The principal difficulty remaining in feedback system design is the establishment of an optimum criterion for performance. In other words, what shape of the frequency response or what location of the poles and zeros gives the best system performance?

Evaluation of performance, given in Sec. 3.12, is based on the time response to a step function input. The characteristics used to judge performance are as follows:

1. Maximum overshoot
2. Time for the error to reach its first zero (duplicating time)
3. Time to reach the maximum overshoot
4. Settling time (also called solution time), which is the time for the response to settle within a given percentage of the final value
5. Frequency of oscillation of the transient

These criteria or combinations of them are used to evaluate the optimum performance. It is desired to develop, using a step input signal as a reference, a single figure of merit or criterion to judge the "goodness" of the time response of the system. Such a criterion should have three basic properties: reliability, ease of application, and selectivity. It should be reliable for a given class of systems so that it can be applied with confidence, it should be easy to apply, and it should also be selective so that the "best" system is readily discernible. This chapter is devoted to presenting and comparing several proposed figures of merit for optimizing system response.

17-2   Development of the Solution-time Criterion[1]

To indicate the basis for establishing a criterion, the characteristics of a unity feedback system are studied. This criterion is then used to define optimum response. Consider first a simple linear second-order system described by the control ratio

$$\frac{C(s)}{R(s)} = \frac{\omega_n^2}{s^2 + 2\zeta\omega_n s + \omega_n^2} \tag{17-1}$$

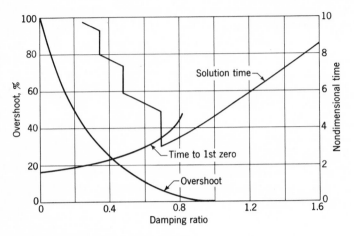

Fig. 17-1   *Transient performance of a simple second-order system with a step input.*

This system has a zero steady-state error with a step input. Only such systems are considered. This means that the open-loop transfer function must be Type 1 or higher.

Figure 17-1 shows the following three quantities as a function of the damping ratio ζ for the system represented by Eq. (17-1):

1. The time for the error to reach its first zero
2. The size of the first and largest overshoot expressed as a percentage of the final value
3. The solution time, which is the time to reach and remain within 5 per cent of the final value

A perfect response would require a step-function response identical to the input with no error at any time. Such a response is impossible, of course. Therefore the characteristics of Fig. 17-1 are studied to determine the best performance. It is seen that overshoot and time to first zero are conflicting characteristics; that is, their minimum values occur at different damping ratios. Therefore, if these two characteristics are to be used as criteria there must be an arbitrary compromise between the overshoot and rapid rise time. The solution time appears to combine both properties and can be used as a criterion of performance. The solution time for this second-order system specifies a damping ratio of 0.7 as optimum, which is a commonly accepted value. Therefore the solution time may be the figure of merit to use to optimize a system. Of course, it is necessary to extend the study of this characteristic to higher-order and to nonlinear systems to make sure that it is universally applicable. Studies along this line have been made and are presented after other criteria have been investigated.[2,3] There seems to be one possible weakness in the use of this figure of merit: its tendency to give an exaggerated picture of the optimum damping ratio compared with slightly larger or slightly smaller damping ratios.

17-3   Control-area Criterion[5]

Another measure of the transient response of a control system with a step input is the *control area*, as shown by the shaded area of Fig. 17-2. The "best" system is one that has the minimum control area. The value of the control area is given by the integral of the error:

$$I_1 = \int_0^\infty e \, dt \qquad (17\text{-}2)$$

The control ratio of the closed-loop system can be described by

$$\frac{C(s)}{R(s)} = \frac{(1 + A_1 s)(1 + A_2 s) \cdots (1 + A_w s)}{(1 + a_1 s)(1 + a_2 s) \cdots (1 + a_v s)} \qquad (17\text{-}3)$$

In this equation the values of a and A can be either real or complex. The value of the control area obtained from Eq. (17-2) is

$$I_1 = (a_1 + a_2 + \cdots + a_v) - (A_1 + A_2 + \cdots + A_w) \qquad (17\text{-}4)$$

If the values of a in the denominator of $C(s)/R(s)$ are complex, the minimum value of control area I_1 occurs when the damping ratio is zero. This is also shown in Fig. 17-3, which contains a plot of I_1 versus ζ for the system represented by Eq. (17-1). The minimum value of I_1 occurs at $\zeta = 0$, which is obviously not an optimum system. The effect of minimizing the control area without regard to the damping may therefore result in an oscillating response. Since this criterion is limited to overdamped systems, it does not serve the intended purpose and is rejected from further consideration.

A modification of control area as a criterion, using time as a weighting factor, has also been proposed.[6] This modified criterion is defined by the integral

$$I_2 = \int_0^\infty t e \, dt \qquad (17\text{-}5)$$

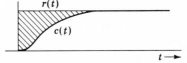

Fig. 17-2 *Control area for an overdamped system.*

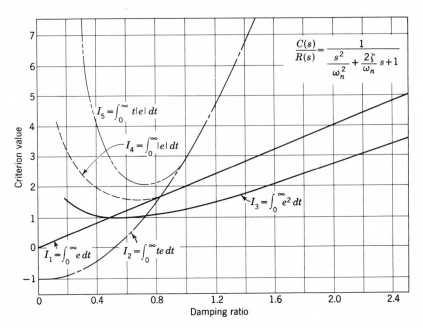

Fig. 17-3   *Criteria for second-order systems.*

The *weighted control area* adds a heavy penalty for errors that do not die out rapidly.   It is intended that a minimum value of I_2 should define an optimum system.   Figure 17-3 contains a plot of I_2 versus ζ for the system represented by Eq. (17-1).   The minimum value of I_2 also occurs at $\zeta = 0$.   This criterion fails in the same manner as the control-area criterion.

17-4   *Additional Criteria*

In this section a study is made of additional criteria that are investigated to demonstrate their suitability in defining optimum response.   These criteria are described below and are applied to the same second-order system used in Secs. 17-2 and 17-3:

$$\frac{C(s)}{R(s)} = \frac{\omega_n^2}{s^2 + 2\zeta\omega_n s + \omega_n^2} \tag{17-6}$$

To make a comparison between them, these additional criteria are also plotted versus the damping ratio in Fig. 17-3.

One proposed figure of merit is the *integral of squared error* (ISE)

$$I_3 = \int_0^\infty e^2 \, dt \tag{17-7}$$

Both positive and negative values of the error, which are present in under-damped response, increase the size of the integral. This eliminates the bad feature of the unmodified control area. Figure 17-3 contains a plot of I_3 versus ζ. The minimum value of I_3 occurs at $\zeta = 0.5$, which may be considered a reasonable value although it results in more overshoot and a longer solution time than $\zeta = 0.7$. Also, this criterion is not very selective since the plot of I_3 has a broad minimum region.

Another figure of merit is defined by the *integral of absolute value of error* (IAE)

$$I_4 = \int_0^\infty |e| \, dt \tag{17-8}$$

By using the magnitude of the error, the integral increases for either positive or negative error. This should result in a good underdamped system. Figure 17-3 contains the plot of this function versus ζ. The minimum value of I_4 occurs at about $\zeta = 0.7$, which is known to give a good response. It also gives a slightly better selectivity than I_3.

Time-weighting the magnitude of error gives the criterion defined by the integral

$$I_5 = \int_0^\infty t|e| \, dt \tag{17-9}$$

This function is known as the *integral of time multiplied by the absolute value of error* (ITAE) criterion and is plotted in Fig. 17-3. This curve has a minimum value at $\zeta = 0.7$ and is more selective than the others shown. This criterion shows promise; it is used later in this chapter for more complex systems to check its suitability for determining their optimum response. The analog-computer circuit [14] for obtaining I_5 is shown in Fig. 17-4; symbols utilized in this figure are discussed in Chap. 19.

Still other figures of merit can be formed with more complex combinations of error and time weighting. Three such criteria are:

Integral of Time Multiplied by Squared Error (ITSE):

$$I_6 = \int_0^\infty te^2 \, dt \tag{17-10}$$

Integral of Squared Time Multiplied by Squared Error (ISTSE):

$$I_7 = \int_0^\infty t^2 e^2 \, dt \tag{17-11}$$

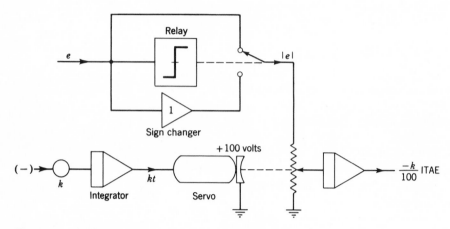

Fig. 17-4 *Analog-computer circuit for computing I_5.*

Table 17-1. **Performance comments for I_5 through I_8**

Criteria	Comments
I_5	Selects good Type 1 systems (systems with one integration in open-loop transfer function). Type 2 systems have excessive overshoot. Graham and Lathrop studied unity numerator systems through the eighth order.
I_6	Has not been studied extensively. However, all results available indicate that it is one of the more valuable performance indices.
I_7	Selects good Type 1 systems. Information is not available on optimum Type 2 systems.
I_8	The limited information available indicates that this criterion may be of general value.

Integral of Squared Time Multiplied by Absolute Value of Error (ISTAE):

$$I_8 = \int_0^\infty t^2 \, |e| \, dt \tag{17-12}$$

Table 17-1 presents comments[15] on the usefulness of criteria I_5 through I_8.

The generalization of optimizing criteria has been expanded by Schultz and Rideout[12] and by Gibson.[15]

17-5   *Definition of Zero Error Systems*[17]

In general, the control ratio for a feedback control system has the form

$$\frac{C(s)}{R(s)} = \frac{a_w s^w + a_{w-1} s^{w-1} + \cdots + a_2 s^2 + a_1 s + a_0}{s^v + b_{v-1} s^{v-1} + \cdots + b_2 s^2 + b_1 s + b_0} \tag{17-13}$$

The steady-state error for this system can be shown to be

$$e_{ss} = e_0 r + e_1 Dr + e_2 D^2 r + \cdots \qquad (17\text{-}14)$$

The form of the input $r(t)$ determines the size of the steady-state error.

Since e_0 is a function of $b_0 - a_0$, the requirement for zero steady-state error with a step-function input is that $a_0 = b_0$. This also means that in a unity feedback system the forward transfer function is Type 1 or higher. Since the order of the numerator of $C(s)/R(s)$ can be equal to or less than the order of the denominator, there are many possible forms of $C(s)/R(s)$ for which the steady-state error is zero with a step input. The system having the control ratio with only a constant in the numerator is referred to as the *zero steady-state step error system* and is given by

$$\frac{C(s)}{R(s)} = \frac{b_0}{s^v + b_{v-1}s^{v-1} + \cdots + b_2 s^2 + b_1 s + b_0} \qquad (17\text{-}15)$$

Since e_1 is a function of $b_0 - a_0$ and $b_1 - a_1$, zero steady-state error with a ramp-function input results when $a_0 = b_0$ and $a_1 = b_1$. This means that the system is Type 2 or higher. *Zero steady-state ramp error systems* are described by

$$\frac{C(s)}{R(s)} = \frac{b_1 s + b_0}{s^v + b_{v-1}s^{v-1} + \cdots + b_2 s^2 + b_1 s + b_0} \qquad (17\text{-}16)$$

Since e_2 is a function of $b_0 - a_0$, $b_1 - a_1$, and $b_2 - a_2$, the steady-state error is zero with a parabolic-function input if $a_0 = b_0, a_1 = b_1,$ and $a_2 = b_2$. This means that the system is Type 3 or higher. Systems that have a quadratic numerator are referred to as *zero steady-state parabolic error systems* and are described by

$$\frac{C(s)}{R(s)} = \frac{b_2 s^2 + b_1 s + b_0}{s^v + b_{v-1}s^{v-1} + \cdots + b_2 s^2 + b_1 s + b_0} \qquad (17\text{-}17)$$

17-6   Zero Steady-state Step Error Systems

The zero steady-state step error system is now studied in terms of the various optimizing methods. The control ratio of the closed-loop system is given by

$$\frac{C(s)}{R(s)} = \frac{b_0}{s^v + b_{v-1}s^{v-1} + \cdots + b_2 s^2 + b_1 s + b_0} \qquad (17\text{-}18)$$

The objective is to apply an optimizing figure of merit to produce a table of standard forms in which the best value is specified for each coefficient of the denominator. Such a table of standard forms would assist the designer since he would try to duplicate these coefficients in his system. He would then be sure of having the "best" system possible.

One criterion suggested as an optimum standard is to have the characteristic equation composed of equal critically damped modes.[4,8,9] This means that the coefficients are obtained from the binomial expansion for the control ratio given by

$$\frac{C(s)}{R(s)} = \frac{\omega_0^n}{(s + \omega_0)^n} \tag{17-19}$$

Table 17-2 gives the standard binomial forms for the denominator of $C(s)/R(s)$. This response is essentially slow and becomes slower as the order of the system is higher. It therefore does not represent an optimum but can be used for comparison purposes.

Another suggestion, made by Butterworth,[10] locates the poles uniformly on a circle with its center at the origin in the left-hand s plane. Table 17-3 gives the standard Butterworth forms of the denominator of $C(s)/R(s)$. The response to a step function using these standard forms is shown in Fig. 17-5 for systems of order 2 to 8.

The minimum ITAE criterion was applied to this system, resulting in the standard forms given in Table 17-4. The procedure used was to vary each coefficient separately until the ITAE value became a minimum. Then the successive coefficients were varied in sequence to minimize the ITAE value. The response to a step function using these standard forms is shown in Fig. 17-6 for systems of order 2 to 8.

Table 17-2.   Binomial standard forms

$$s + \omega_0$$
$$s^2 + 2\omega_0 s + \omega_0^2$$
$$s^3 + 3\omega_0 s^2 + 3\omega_0^2 s + \omega_0^3$$
$$s^4 + 4\omega_0 s^3 + 6\omega_0^2 s^2 + 4\omega_0^3 s + \omega_0^4$$
$$s^5 + 5\omega_0 s^4 + 10\omega_0^2 s^3 + 10\omega_0^3 s^2 + 5\omega_0^4 s + \omega_0^5$$
$$s^6 + 6\omega_0 s^5 + 15\omega_0^2 s^4 + 20\omega_0^3 s^3 + 15\omega_0^4 s^2 + 6\omega_0^5 s + \omega_0^6$$
$$s^7 + 7\omega_0 s^6 + 21\omega_0^2 s^5 + 35\omega_0^3 s^4 + 35\omega_0^4 s^3 + 21\omega_0^5 s^2 + 7\omega_0^6 s + \omega_0^7$$
$$s^8 + 8\omega_0 s^7 + 28\omega_0^2 s^6 + 56\omega_0^3 s^5 + 70\omega_0^4 s^4 + 56\omega_0^5 s^3 + 28\omega_0^6 s^2 + 8\omega_0^7 s + \omega_0^8$$

Table 17-3.   Butterworth standard forms

$$s + \omega_0$$
$$s^2 + 1.4\omega_0 s + \omega_0^2$$
$$s^3 + 2.0\omega_0 s^2 + 2.0\omega_0^2 s + \omega_0^3$$
$$s^4 + 2.6\omega_0 s^3 + 3.4\omega_0^2 s^2 + 2.6\omega_0^3 s + \omega_0^4$$
$$s^5 + 3.24\omega_0 s^4 + 5.24\omega_0^2 s^3 + 5.24\omega_0^3 s^2 + 3.24\omega_0^4 s + \omega_0^5$$
$$s^6 + 3.86\omega_0 s^5 + 7.46\omega_0^2 s^4 + 9.13\omega_0^3 s^3 + 7.46\omega_0^4 s^2 + 3.86\omega_0^5 s + \omega_0^6$$
$$s^7 + 4.5\omega_0 s^6 + 10.1\omega_0^2 s^5 + 14.6\omega_0^3 s^4 + 14.6\omega_0^4 s^3 + 10.1\omega_0^5 s^2 + 4.5\omega_0^6 s + \omega_0^7$$
$$s^8 + 5.12\omega_0 s^7 + 13.14\omega_0^2 s^6 + 21.84\omega_0^3 s^5 + 25.69\omega_0^4 s^4 + 21.84\omega_0^5 s^3 + 13.14\omega_0^6 s^2 + 5.12\omega_0^7 s + \omega_0^8$$

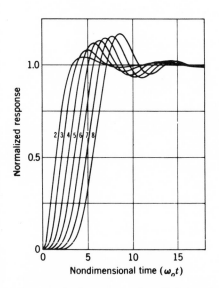

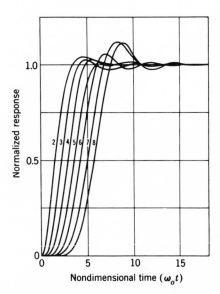

Fig. 17-5 *Response using Butterworth standard forms for zero step error systems.*

Fig. 17-6 *Response using ITAE standard forms for zero step error systems.*

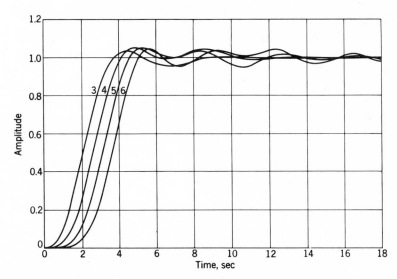

Fig. 17-7 *Response using the solution-time criterion for zero step error systems.*

Table 17-4.   ITAE standard forms for zero step error systems

$$s + \omega_0$$
$$s^2 + 1.4\omega_0 s + \omega_0^2$$
$$s^3 + 1.75\omega_0 s^2 + 2.15\omega_0^2 s + \omega_0^3$$
$$s^4 + 2.1\omega_0 s^3 + 3.4\omega_0^2 s^2 + 2.7\omega_0^3 s + \omega_0^4$$
$$s^5 + 2.8\omega_0 s^4 + 5.0\omega_0^2 s^3 + 5.5\omega_0^3 s^2 + 3.4\omega_0^4 s + \omega_0^5$$
$$s^6 + 3.25\omega_0 s^5 + 6.60\omega_0^2 s^4 + 8.60\omega_0^3 s^3 + 7.45\omega_0^4 s^2 + 3.95\omega_0^5 s + \omega_0^6$$
$$s^7 + 4.475\omega_0 s^6 + 10.42\omega_0^2 s^5 + 15.08\omega_0^3 s^4 + 15.54\omega_0^4 s^3 + 10.64\omega_0^5 s^2 + 4.58\omega_0^6 s + \omega_0^7$$
$$s^8 + 5.20\omega_0 s^7 + 12.80\omega_0^2 s^6 + 21.60\omega_0^3 s^5 + 25.75\omega_0^4 s^4 + 22.20\omega_0^5 s^3 + 13.30\omega_0^6 s^2 + 5.15\omega_0^7 s + \omega_0^8$$

Table 17-5.   Solution-time standard forms for zero step error systems

$$s + \omega_0$$
$$s^2 + 1.4\omega_0 s + \omega_0^2$$
$$s^3 + 1.55\omega_0 s^2 + 2.10\omega_0^2 s + \omega_0^3$$
$$s^4 + 1.60\omega_0 s^3 + 3.15\omega_0^2 s^2 + 2.45\omega_0^3 s + \omega_0^4$$
$$s^5 + 1.575\omega_0 s^4 + 4.05\omega_0^2 s^3 + 4.10\omega_0^3 s^2 + 3.025\omega_0^4 s + \omega_0^5$$
$$s^6 + 1.45\omega_0 s^5 + 5.10\omega_0^2 s^4 + 5.30\omega_0^3 s^3 + 6.25\omega_0^4 s^2 + 3.425\omega_0^5 s + \omega_0^6$$

The next criterion used was to minimize the solution time, which is the time to reach and remain within 5 per cent of the final value. Table 17-5 gives the standard forms using this criterion. The response to a step function using these standard forms is shown in Fig. 17-7 for systems of order 3 to 6.

A comparison of the response curves of Figs. 17-5 to 17-7 shows that, basically, all three criteria give good results. If a preference is to be given, the authors would select the ITAE criterion. However, it must be realized that, for a particular-order system, one of the other criteria may be better in some respect. For example, for the fifth-order system the time to first zero is 4.25 sec by using the solution time and is 5.5 sec for the ITAE criterion. But there must also be taken into account the fact that the maximum overshoot is 5 per cent for the solution time and only 2 per cent for the ITAE criterion. Further, the solution-time response is more oscillating than the ITAE response. The "optimum" response must be determined in terms of which is more important—rise time, peak overshoot, or frequency of oscillation.

17-7   Zero Steady-state Ramp Error Systems

The zero steady-state ramp error system is now studied. The control ratio of the closed-loop system is given by

$$\frac{C(s)}{R(s)} = \frac{b_1 s + b_0}{s^v + b_{v-1}s^{v-1} + \cdots + b_2 s^2 + b_1 s + b_0} \tag{17-20}$$

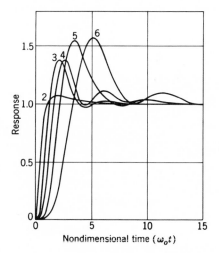

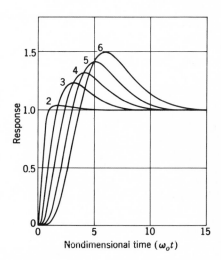

Fig. 17-8 *Step response using ITAE standard forms for zero ramp error systems.*

Fig. 17-9 *Step response using binomial standard forms for zero ramp error systems.*

Table 17-6. ITAE standard forms for zero ramp error systems

$$s^2 + 3.2\omega_0 s + \omega_0^2$$
$$s^3 + 1.75\omega_0 s^2 + 3.25\omega_0^2 s + \omega_0^3$$
$$s^4 + 2.41\omega_0 s^3 + 4.93\omega_0^2 s^2 + 5.14\omega_0^3 s + \omega_0^4$$
$$s^5 + 2.19\omega_0 s^4 + 6.50\omega_0^2 s^3 + 6.30\omega_0^3 s^2 + 5.24\omega_0^4 s + \omega_0^5$$
$$s^6 + 6.12\omega_0 s^5 + 13.42\omega_0^2 s^4 + 17.16\omega_0^3 s^3 + 14.14\omega_0^4 s^2 + 6.76\omega_0^5 s + \omega_0^6$$

Again the objective is to obtain a table of standard forms for the denominator of $C(s)/R(s)$ that produce the best response with a step input.

Applying the ITAE criterion results in the forms given by Table 17-6. In some cases the selection of the coefficients is arbitrary. For example, in the second-order system the response continues to improve without limit as b_1 is increased. Therefore an arbitrary selection of an overdamped system with $\zeta = 1.6$ is made on the basis that further increases result in negligible improvement in the response. Similarly, for the cubic system the optimum value of $b_2 = 1.75$, but the response continues to improve without limit as b_1 is increased. Therefore the value $b_1 = 3.25$ was chosen as marking the onset of diminishing returns. The response to a step input using the ITAE standard forms is plotted in Fig. 17-8. It is seen that there are large overshoots and rapid accelerations.

An alternative is to use the binomial standard forms with this system. The response to a step input is shown in Fig. 17-9. These responses show less overshoot and smaller accelerations than the ITAE responses but are appreciably slower. The designer must pick what he considers is the better response.

17-8 *Optimization Based on Open-loop Parameters*[11]

Previous sections have illustrated optimization criteria based on closed-loop parameters. It may be easier for the design engineer to have a table of standard forms based on the optimization of open-loop parameters. The following material is based on adjusting the open-loop gain and assuming that the open-loop time constants are fixed. *The ITAE criterion for optimum response is used.* The systems considered have unity feedback and are limited to those which have zero steady-state error with a step-function input.

A simple second-order system is considered first. Although this case is rather simple, it is presented to illustrate the procedure used. The open-loop

Table 17-7. *Optimum value of open-loop gain*

$G(s)$	α	$K' = KT$
$\dfrac{K}{s(1 + Ts)}$	—	1.25
$\dfrac{K}{s(1 + Ts)(1 + \alpha Ts)}$	1.000	0.30
	0.875	0.35
	0.750	0.375
	0.625	0.425
	0.500	0.50
	0.250	0.60

$$\frac{K}{s(1 + Ts)(1 + \alpha_1 Ts)(1 + \alpha_2 Ts)}$$

α_2 / α_1	1.00	0.875	0.75	0.625	0.50	0.25
1.00	0.2	0.2	0.2	0.2	0.2	0.3
0.875	0.2	0.2	0.225	0.225	0.225	0.3
0.75	0.2	0.225	0.225	0.25	0.25	0.3
0.625	0.2	0.225	0.25	0.3	0.3	0.3
0.50	0.2	0.225	0.25	0.3	0.3	0.3
0.25	0.3	0.3	0.3	0.3	0.3	0.3

Values of $K' = KT$ are read from the chart for sets of values of α_1 and α_2.

transfer function and control ratio are given by

$$\frac{C(s)}{E(s)} = \frac{K}{s(1 + Ts)} \tag{17-21}$$

$$\frac{C(s)}{R(s)} = \frac{K}{Ts^2 + s + K} \tag{17-22}$$

The time constant T is normally fixed, and the system is to be optimized by adjusting the gain K. To prevent T from appearing explicitly, it is advantageous to make a change of time scale. Multiplying both the numerator and the denominator of Eq. (17-22) by T, replacing KT by K', and making the time-scale change $t = \tau T$ gives

$$\frac{C(s')}{R(s')} = \frac{K'}{s'^2 + s' + K'} \tag{17-23}$$

The Laplace operator with the new time base is s'. The ITAE value becomes a minimum and remains constant for $K' > 1.25$. Since there is no advantage to larger gain, the optimum value chosen is $K' = 1.25$. The value of gain K can now be determined for any time constant T.

Using the same procedure results in the optimum values given in Table 17-7. This procedure can be extended to more complex systems.

17-9 Conclusions

There have been presented several criteria for optimizing the response of a system. The ITAE criterion appears to be the best over-all figure of merit for a zero steady-state step error system; it generally produces a smaller overshoot and less oscillation than the other criteria. However, if a faster rise time is required and the additional oscillation can be tolerated, the solution-time criterion may be preferable. For a zero steady-state ramp error system the designer must choose between rapid decay of the transient with its inherent high accelerations and oscillations as given by use of the ITAE criterion and the slower response but smaller accelerations given by the binomial criterion.

The procedure in system design is first to pick the response criterion that is considered optimum. Then the values of the design variables are solved algebraically to make the coefficients of either the control ratio or the forward transfer function numerically equal to those of the appropriate standard form.[13] Another approach is the use of the ISE criterion in conjunction with the root-locus technique.[16] This method also yields the cascade compensator $G_c(s)$ required to obtain optimum performance for a basic system.

In the study of linear or nonlinear systems for which no standard forms are available, the ITAE criterion may be used as a figure of merit for optimizing a particular system in terms of the adjustable parameters.

Bibliography

1. Graham, D., and R. C. Lathrop: The Synthesis of Optimum Response: Criteria and Standard Forms, *Trans. AIEE*, vol. 72, pt. II, pp. 273–288, Nov., 1953.
2. Seamons, Robert C., Jr.: "Automatic Control of Aircraft," Report of the Aeronautical Engineering Department, Massachusetts Institute of Technology, 1953, Appendix A, Summary of First and Second Order Responses.
3. Forgatt, John W., Jr.: "Investigation of Several Criteria for the Synthesis of Optimum Transient Response of Servomechanism Systems of Higher Orders," M.Sc. Thesis, Air Force Institute of Technology, Dayton, Ohio, 1954.
4. Oldenbourg, R. C., and H. Sartorius: "The Dynamics of Automatic Controls," translated and edited by H. L. Mason, ASME, 1948.
5. Stout, Thomas M.: A Note on Control Area, *J. Appl. Phys.*, vol. 21, pp. 1129–1131, November, 1950.
6. Nims, P. T.: Some Design Criteria for Automatic Controls, *Trans. AIEE*, vol. 70, pt. II, pp. 606–611, 1951.
7. Hall, A. C.: "The Analysis and Synthesis of Linear Servomechanisms," The Technology Press of the Massachusetts Institute of Technology, Cambridge, Mass., 1943.
8. Imlay, F. H.: "A Theoretical Study of Lateral Stability with an Automatic Pilot," TR 693, National Advisory Committee for Aeronautics, 1940.
9. Whitely, A. L.: The Theory of Servo Systems, with Particular Reference to Stabilization, *J. Inst. Elec. Engrs. (London)*, vol. 93, pp. 353–377, 1946.
10. Butterworth, S.: On the Theory of Filter Amplifiers, *Wireless Engr.*, vol. 7, pp. 536–541, 1930.
11. Davidson, K. A.: "Tabulation and Optimization of Some Commonly Encountered Servomechanisms Based upon Open-loop Parameters," M.Sc. Thesis, Air Force Institute of Technology, Dayton, Ohio, 1955.
12. Schultz, W. C., and V. C. Rideout: The Selection and Use of Servo Performance Criteria, *Trans. AIEE*, vol. 76, pt. II, pp. 383–388, January, 1958.
13. Bigelow, S. C.: "Design of Analog Computer Compensated Control Systems," Paper 58–800, American Institute of Electrical Engineers, Summer General Meeting, June, 1958.
14. Dunsmore, Chester L.: Computer Analogs for Common Nonlinearities, *Control Eng.*, vol. 6, pp. 109–111, October, 1959.
15. Gibson, J. E., and others: A Set of Standard Specifications for Linear Automatic Control Systems, *Trans. AIEE*, vol. 80, pt. II, pp. 65–74, May, 1961.
16. Chang, Sheldon S. L.: "Synthesis of Optimum Control Systems," chap. 2, McGraw-Hill Book Company, New York, 1961.
17. King, Leonard H.: Reduction of Forced Error in Closed-Loop Systems, *Proc. IRE*, vol. 41, pp. 1037–1042, August, 1953.

18

Nonlinearities and Describing Functions

18-1   Introduction[1-3]

The previous chapters have dealt only with linear systems, in order to establish firmly in the reader's mind the basic fundamentals and characteristics of feedback control systems. While nonlinearities did exist in the systems studied, they were considered negligible in order to use the simpler linear methods of analysis. However, the nonlinearities cannot always be neglected.

The systems considered were represented by linear differential equations of the form

$$r(t) = A_v\, D^v c + A_{v-1}\, D^{v-1} c + \cdots + A_1\, Dc + \cdots + A_{-w}\, D^{-w} c \quad (18\text{-}1)$$

where the A_i are constants. Generally, some of the coefficients vary as a function of the input. For example, the relationship between the applied force and the resulting deformation of a spring may be represented by Fig. 18-1 for a specific spring. This

597

Fig. 18-1 *Nonlinear spring characteristics.*

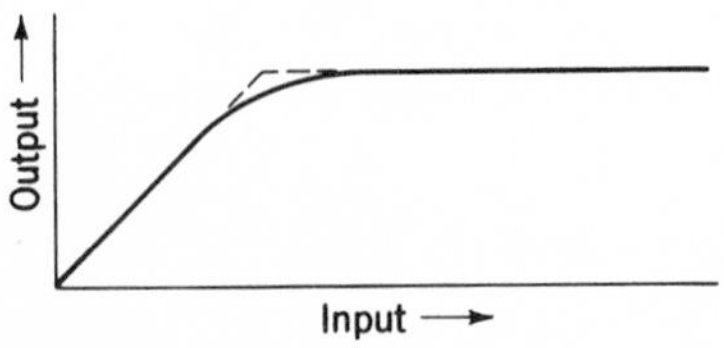

Fig. 18-2 *Saturation or limiting.*

characteristic represents a *continuous nonlinearity*. The dashed straight line can be used as a linear approximation to the true curve. A system is said to be "linearized" when the actual characteristics are approximated by a straight line.

Another type of nonlinearity is saturation or limiting. This characteristic is shown in Fig. 18-2. Up to a certain-size input the output is directly proportional to it, but for greater inputs the output remains constant. This type of characteristic is sometimes called a *discontinuous nonlinearity*.

When the coefficients A_i of the differential equation are continuous functions of the dependent variable $c(t)$ and its derivatives, the system is described as having a continuous nonlinearity. When the coefficients A_i are functions of $c(t)$ but have finite discontinuities, the system is described as having a discontinuous nonlinearity. The two types are shown in Fig. 18-3.

The nonlinearities may be intentional or unintentional. Intentional nonlinearities are introduced to improve the system response. In general, the use of nonlinear elements results in systems that are smaller and perform better than linear systems. In recent years control system engineers have recognized the burden placed on the design problem by restricting their thinking to the exclusive use of linear components. It requires components of high quality or large size to operate in a linear fashion for a variety of input values. It also limits the realizable system characteristics and the tasks that the system can be made to perform. The objection to nonlinear systems is the difficulty of design and the lack of clear-cut methods for predicting their behavior. However, it is usually too expensive to overdesign the equipment simply to make it linear in order to be able to analyze it more easily.

Direct methods of solving differential equations with some specific nonlinearities have been developed. However, there is no general method for treating all nonlinearities. One that has proved useful in the study and design of nonlinear feedback control systems is based on describing functions. This method provides information concerning system stability and

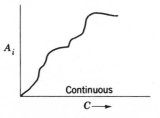 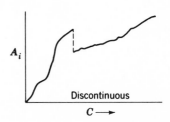

Fig. 18-3 *Nonlinearities.*

the possibility of sustained oscillations in the output, but it provides little information about transient behavior. The phase plane is useful for determining the time response of second-order systems with nonlinearities. In addition to stability, the transient response can also be determined by a phase-plane analysis.

This chapter is devoted to the describing-function method and its use in the determination of system stability. In a succeeding section some of the more commonly encountered nonlinearities are described.

18-2   Continuous Nonlinearities

Continuous nonlinearities are usually eliminated from the differential equations by assuming restricted ranges for the variables. Linear approximations can then be made to represent the nonlinear coefficients. It is recognized, of course, that the true response will not be exactly as predicted. However, the difference should be small as long as the disturbance is small. Of immediate concern is the stability of the system. In other words, is the stability determined from the linearized equations the same as for the actual nonlinear equations? The answer was determined by Liapunov, who investigated this problem. The nonlinear term in the differential equation is first expanded in a Taylor's series about the point of operation and then the linearized equation is formed. If the roots of this linear characteristic equation have negative real parts, the nonlinear system is stable for small disturbances around this operating point. The only ambiguity occurs if one or more of the roots are zero, in which case Liapunov's method cannot easily be used.

Liapunov's method is illustrated by the following example. The equation of motion of a pendulum subject to viscous damping is given by

$$J \ D^2\theta + B \ D\theta + Mgl \sin \theta = 0 \qquad (18\text{-}2)$$

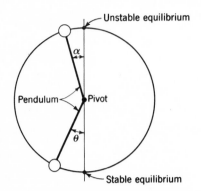

Fig. 18-4   Pendulum showing stable and unstable equilibrium points.

where g = acceleration of gravity
   l = length of pendulum
   θ = angular displacement measured from lowest position

The points of equilibrium occur when the velocity $D\theta$ and the acceleration $D^2\theta$ are zero. These points of equilibrium are $\theta = K\pi$, where K = any integer. The point of equilibrium $\theta = 0$ for stability is considered first. The linearized equation for small excursions is

$$J\,D^2\theta + B\,D\theta + Mgl\theta = 0 \qquad (18\text{-}3)$$

The roots of the characteristic equation have negative real parts. Therefore the point $\theta = 0$ is a stable point of equilibrium. Next, the stability of the point $\theta = \pi$ is investigated. The linearized equation, where α represents small excursions from this new equilibrium point, is given by

$$J\,D^2\alpha + B\,D\alpha - Mgl\alpha = 0 \qquad (18\text{-}4)$$

One of the roots of the characteristic equation is positive; therefore the point $\theta = \pi$ is an unstable point of equilibrium. This simple case is chosen because the natures of the stability of the pendulum at the two points of equilibrium are readily recognized. Figure 18-4 shows the stable and unstable equilibrium points for the pendulum.

It can be concluded that using linear approximations of continuous nonlinear equations gives the correct analysis of stability. However, this is valid only for small disturbances. The designer must be sure that the magnitude of the response remains within the linear region.

18-3   Types of Nonlinearities[1,4]

Several forms of discontinuous nonlinearities encountered fairly often are now described. These are frequently referred to as *fast nonlinearities*, which

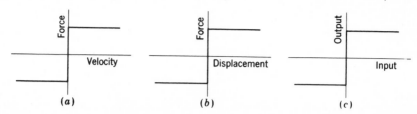

Fig. 18-5 Step nonlinearities. (a) Coulomb friction. (b) Preload. (c) Ideal relay.

means that the mode of operation changes at a fast rate compared with the response time of the system. For a fast nonlinearity the poles and zeros associated with the transfer function are moving rapidly about the complex plane. They cannot, therefore, be considered fixed over an interval of time that is of the order of the response time of the system. Many of the curves drawn show idealized nonlinearities; i.e., the nonlinearities are represented with straight-line sections.

Coulomb Friction, Preload, Ideal Relay (Fig. 18-5)

Coulomb friction exists in all physical systems and is caused by sliding motion. It is present between the brushes and commutator of a motor, between tight gears, and in bearings. The friction force is of constant magnitude and is always in a direction to resist the relative motion. A force equal to the friction is required to produce a displacement. A discontinuity occurs at the instant when reversal of motion takes place. Preload has the same characteristic as coulomb friction. The force can vary from plus to minus the preload value without causing any displacement. Preloading is often used to eliminate drift from the neutral position. An ideal two-position relay (with no "off" position) would deliver either a positive or a negative output, depending on the position of the relay. This relay characteristic would produce continuous oscillation in a control system. A system containing any of these nonlinearities may be represented by two linear equations. The equation that applies at any instant depends on the input at that time.

Dead Zone, Relay with Dead Zone, Backlash, and Hysteresis (Fig. 18-6)

A dead zone is a range of input for which there is no output. It is also called threshold, flat spot, or dead space. The addition of a dead zone to an ideal relay reduces the oscillations of the relay contacts. This reduces the arcing and therefore the wear of the relay contacts and also reduces

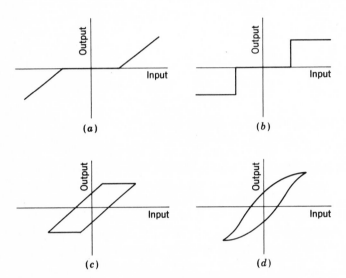

Fig. 18-6 *Dead-zone characteristics.* (*a*) *Dead zone.* (*b*) *Relay with dead zone.* (*c*) *Backlash with a cyclic input.* (*d*) *Hysteresis.*

the strain on the equipment.   For example, if the relay operates a motor, the motor is shut off, allowing the load to coast to a stop.   This means that an error is present in the output.   Backlash occurs in a system with dead zone.   However, the dead zone occurs every time there is a reversal of the input.   For a cyclical input the output has the closed-path variation shown in Fig. 18-6c.   Backlash is commonly associated with gearing and has a destabilizing influence.   To reduce backlash between mating gears requires careful adjustment of the meshes and minimizing eccentricities. Backlash is a form of hysteresis, which is usually associated with magnetic circuits.   Because of the residual magnetic characteristics the plot of output versus input is many-valued.   For cyclic inputs of constant amplitude the output is double-valued.

Saturation or Limiting (*Fig.* 18-7)

Many components are linear (the output is directly proportional to the input) up to a given input signal.   For larger signals the output may not increase proportionally and may actually reach a maximum value.   Although the change of slope of the characteristic may be gradual, an approximation is often made by assuming that the curve is made up of straight-line sections.

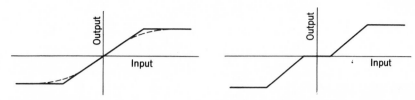

Fig. 18-7   Saturation.   Fig. 18-8   Dead zone and saturation.

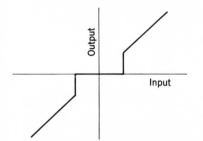

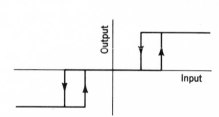

Fig. 18-9 *Dead zone with linear out-put outside the dead zone.*   **Fig. 18-10** *Relay with dead zone and hysteresis.*

The system is then said to be *piecewise linear*.   The solid curve in Fig. 18-7 shows the ideal saturation curve, and the dashed line shows the actual saturation characteristic.

Combined Nonlinearities

Several combinations of nonlinear characteristics may occur.   Figure 18-8 shows both dead zone and saturation.   Figure 18-9 shows a dead zone with output proportional to input outside the dead zone.   Figure 18-10 shows a typical relay characteristic with a dead zone and hysteresis.

18-4   Describing Functions[5,6]

The describing function is used to represent a nonlinearity by an approximately equivalent linear transfer function.   Its definition is based on applying a sinusoidal input to the nonlinearity.   In general, the output is periodic with the same period as the input, but it contains many harmonics in addition to the fundamental.   Most control systems, however, act as low-pass filters, with the result that the higher harmonics are attenuated compared

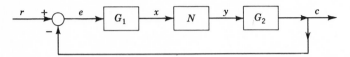

Fig. 18-11   *Feedback control system containing a nonlinear element N and linear elements G_1 and G_2.*

with the fundamental.   If the harmonics are sufficiently small, they can be neglected to obtain a good approximation to the performance of the non-linear system.   Figure 18-11 shows a block diagram with a nonlinear element.   All the nonlinearities of the system are lumped into the one block labeled N.   The output of this lumped nonlinear element is a function only of the present value and the past history of the input.   The characteristics of N are not a function of time.

In the system shown in Fig. 18-11 the higher harmonics appearing at the output of the nonlinear element N must pass through and are therefore attenuated by G_2.   They are further attenuated by G_1 so that the input to N is essentially a pure sine wave.   This reduction of the higher harmonics is produced when there are sufficient energy-storage elements in the system.   When this is true, the describing function gives a good first approximation to the system performance.

The conventional describing function for symmetric nonlinearities is defined as the ratio of the phasor representing the fundamental component of the output of the nonlinearity to the phasor representing the sinusoidal input.   In terms of the elements shown in Fig. 18-11, the describing function is

$$\mathbf{N}(j\omega) = \frac{\mathbf{Y}_1(j\omega)}{\mathbf{X}(j\omega)} \tag{18-5}$$

where $\mathbf{N}(j\omega)$ = describing function

$\qquad \mathbf{Y}_1(j\omega)$ = fundamental component of output, determined by Fourier analysis

$\qquad \mathbf{X}(j\omega)$ = sinusoidal input signal $x = X \sin \omega t$

$\mathbf{N}(j\omega)$ may be a function of the frequency as well as the amplitude of the input to the nonlinearity.   If no energy-storage elements are contained in the nonlinear element, then $\mathbf{N}(j\omega)$ is a function only of the amplitude X. In the case of simple saturation the value of $\mathbf{N}(j\omega)$ is real and single-valued for each value of X.

Calculation of a describing function requires a Fourier analysis to obtain the fundamental component of the output.   The input to the nonlinearity is given by

$$x(t) = X \sin \omega t$$

The output expressed as a Fourier series is

$$y(t) = \frac{A_0}{2} + A_1 \cos \omega t + A_2 \cos 2\omega t + \cdots$$

$$+ B_1 \sin \omega t + B_2 \sin 2\omega t + \cdots$$

$$= \frac{A_0}{2} + \sum_{n=1}^{\infty} (A_n \cos n\omega t + B_n \sin n\omega t) \tag{18-6}$$

where
$$A_n = \frac{1}{\pi} \int_0^{2\pi} y(t) \cos (n\omega t) \, d\omega t$$
$$\tag{18-7}$$
$$B_n = \frac{1}{\pi} \int_0^{2\pi} y(t) \sin (n\omega t) \, d\omega t$$

By combining the sine and cosine terms of the same frequency, an alternative form of the Fourier series is

$$y(t) = \frac{A_0}{2} + Y_1 \sin (\omega t + \phi_1) + Y_2 \sin (\omega t + \phi_2) + \cdots$$

$$= \frac{A_0}{2} + \sum_{n=1}^{\infty} Y_n \sin (n\omega t + \phi_n) \tag{18-8}$$

where
$$Y_n = \sqrt{A_n^2 + B_n^2}$$
$$\tag{18-9}$$
$$\phi_n = \tan^{-1} \frac{A_n}{B_n}$$

For a symmetric nonlinearity the term $A_0/2$ is zero. The phasor value of the first harmonic is $\mathbf{Y}_1(j\omega) = Y_1/\underline{\phi_1}$. This is used to obtain the conventional describing function, which is defined by Eq. (18-5).

For a number of common nonlinearities the output $y(t)$ is an odd function. This is the case if $y(t) = -y(-t)$. For odd functions all coefficients $A_n = 0$, and the Fourier series contains only sine terms. Evaluation of the coefficients B_n can be obtained by integrating Eqs. (18-7) from 0 to π only and multiplying the result by 2.

When the output $y(t)$ has half-wave symmetry, the Fourier series has ·only odd harmonics. Half-wave or mirror symmetry exists when $y(t) = -y(t + T/2)$, where T is the period. This means that the wave for half a period is the mirror image of the wave for the next half period. A wave which has half-wave symmetry and is also an odd function will have only odd sine harmonics in its Fourier series. For such a function, the evaluation of B_n can be obtained by integrating Eqs. (18-7) from 0 to $\pi/2$ only and multiplying the result by 4.

It is sometimes convenient to shift the axis of the output so that the curve of $y(t)$ will be an odd function. Then only B_n must be evaluated since $A_n = 0$. The phase shift necessary to accomplish this is then the phase angle of the fundamental component of the output.

The fundamental idea in this analysis is to represent a nonlinear device by a describing function, which is easier to handle. The assumption of a sinusoidal driving force simplifies the procedure but places some limitations on the validity of the analysis. Hence simplification is obtained only at the expense of a reduction in generality. Describing functions are applicable only when the input to the nonlinear element is nearly sinusoidal. Essentially they are still nonlinear functions, but they vary slowly compared with the original nonlinearities, which may change rapidly and over wide ranges during a cycle. The new system with slowly varying parameters can be visualized in terms of the poles and zeros of the response function, which move about the complex plane with changes in the frequency and amplitude of the input signal to the nonlinear element.

Systems which contain a nonlinearity may have a continuous oscillation (a limit cycle) in the output. The describing function is used to predict the amplitude and frequency of that oscillation. To obtain better prediction accuracy, a number of new or modified describing functions have been proposed. The two that seem to be consistently more accurate are the new rms describing function proposed by Gibson[14] and the corrected-conventional describing function proposed by Rankine.[15] The new rms describing function uses the rms value of the output of the nonlinearity. The phase of the output is taken as the shift in axis necessary to make the output wave odd periodic. It is defined by

$$\mathbf{N}_{\mathrm{rms}} = \frac{\left\{\dfrac{1}{\pi} \displaystyle\int_0^{2\pi} [y(t)]^2 \, d(\omega t)\right\}^{\frac{1}{2}} \underline{/-\phi}}{X\underline{/0°}} \tag{18-10}$$

The corrected-conventional describing function uses the rms value of the fundamental and third harmonic to represent the output. Only the phase shift of the fundamental is used. It is defined by

$$\mathbf{N}_{corr} = \frac{(A_1^2 + B_1^2 + A_3^2 + B_3^2)^{\frac{1}{2}}\underline{/\phi_1}}{X\underline{/0°}} \tag{18-11}$$

where the components are defined by Eqs. (18-7). A tabulation of these describing functions for a number of common nonlinearities is contained in reference 15. Only the conventional describing functions are derived in the next section.

18-5 Examples of Describing Functions[6-8]

This section covers the derivations for some of the more common nonlinearities. The conventional describing function is defined as the ratio of

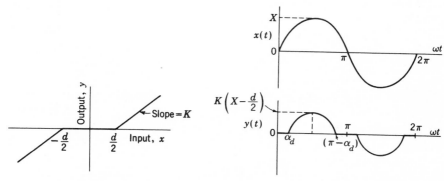

Fig. 18-12 *Dead zone.* **Fig. 18-13** *Input and output with dead zone.*

the complex number representing the fundamental output to the complex number representing the sinusoidal input. Use of the describing function is valid only when the higher-frequency harmonics are sufficiently small.

Example 1. Dead zone

Figure 18-12 shows the idealized representation of a dead-zone characteristic. Although the actual characteristic is more likely to change gradually from the dead zone to the "linear" characteristic, the idealized representation in terms of straight-line characteristics is simpler to analyze.

The sinusoidal input is $x(t) = X \sin \omega t$. If X is smaller than $d/2$, there will be no output. If X is larger than $d/2$, the output has the form shown in Fig. 18-13. The output is zero when x is smaller than $d/2$. Specifically, the output is zero for $0 < \omega t < \alpha_d$, where

$$\alpha_d = \sin^{-1} \frac{d}{2X} \tag{18-12}$$

Since the output is an odd function and there is symmetry of the half cycles, the Fourier series has only odd sine terms. Because of the symmetry of the output, the coefficients of the Fourier series can be evaluated by taking four times the integral over one-quarter of a cycle:

$$Y_k = \frac{4}{\pi} \int_0^{\pi/2} y(t) \sin k\omega t \, d(\omega t) \qquad k = 1, 3, 5, \ldots \tag{18-13}$$

When the input $x > d/2$, the output of the nonlinearity is

$$y(t) = K \left(X \sin \omega t - \frac{d}{2} \right) = KX(\sin \omega t - \sin \alpha_d) \tag{18-14}$$

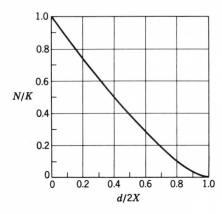

Fig. 18-14 *Nondimensionalized describing function for dead zone.*

The coefficient of the fundamental is therefore

$$Y_1 = \frac{4}{\pi} \int_{\alpha_d}^{\pi/2} K\left(X \sin \omega t - \frac{d}{2}\right) \sin \omega t \, d(\omega t)$$

$$= KX\left(1 - \frac{2\alpha_d}{\pi} - \frac{\sin 2\alpha_d}{\pi}\right) \tag{18-15}$$

To eliminate the slope K, the nondimensionalized ratio of describing function divided by K is used:

$$\frac{N}{K} = \frac{Y_1}{KX} = 1 - \frac{2\alpha_d}{\pi} - \frac{\sin 2\alpha_d}{\pi} \tag{18-16}$$

Equation (18-16) is plotted in Fig. 18-14 as a function of $d/2X$. For $d/2X > 1$ the output is zero and the value of the describing function is also zero.

Example 2.   Saturation

Figure 18-15 shows the idealized representation of saturation and the response to a sinusoidal input.   It should be realized that saturation does not occur as abruptly as shown here.

The input is $x = X \sin \omega t$.   If X is smaller than $S/2$, the output is directly proportional to the input.   When X is larger than $S/2$, the output reaches the value $KS/2$ and remains constant.   The output is proportional to the input for $0 < \omega t < \alpha_s$, where

$$\alpha_s = \sin^{-1} \frac{S}{2X} \tag{18-17}$$

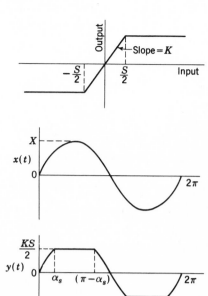

Fig. 18-15 *Input and output with satura-tion.*

The coefficients of the Fourier series are evaluated from the integral:

$$Y_k = \frac{4}{\pi} \int_0^{\pi/2} y(t) \sin k\omega t \, d(\omega t)$$

$$= \frac{4}{\pi} \int_0^{\alpha_s} KX \sin \omega t \sin k\omega t \, d(\omega t) + \frac{4}{\pi} \int_{\alpha_s}^{\pi/2} \frac{KS}{2} \sin k\omega t \, d(\omega t) \quad (18\text{-}18)$$

The fundamental is given by

$$Y_1 = KX \left(\frac{2\alpha_s}{\pi} + \frac{\sin 2\alpha_s}{\pi} \right) \quad (18\text{-}19)$$

The nondimensionalized describing function is

$$\frac{N}{K} = \frac{Y_1}{KX} = \frac{2\alpha_s}{\pi} + \frac{\sin 2\alpha_s}{\pi} \quad (18\text{-}20)$$

The curve of this function is plotted in Fig. 18-16 against $S/2X$. For $S/2X > 1$ the value of the describing function is unity.

Example 3.   Combined dead zone and saturation

The presence of both a dead zone and saturation is shown in Fig. 18-17. The output depends on the size of the input X.   For $X > S/2$ the output is shown in Fig. 18-18.

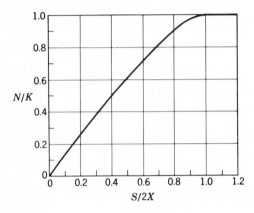

Fig. 18-16   *Nondimensionalized describing function for saturation.*

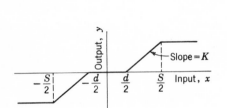

Fig. 18-17   *Dead zone and saturation.*

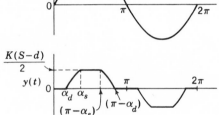

Fig. 18-18   *Input and output with both dead zone and saturation.*

The Fourier-series coefficients are given by

$$Y_k = \frac{4}{\pi} \int_0^{\pi/2} y(t) \sin k\omega t \, d(\omega t)$$

$$= \frac{4}{\pi} \left[\int_{\alpha_d}^{\alpha_s} K\left(X \sin \omega t - \frac{d}{2} \right) \sin k\omega t \, d(\omega t) + \int_{\alpha_s}^{\pi/2} \frac{K(S - d)}{2} \sin k\omega t \, d(\omega t) \right]$$

$$Y_k = \frac{2KX}{k\pi} \left[\frac{\sin (k - 1)\alpha_s}{k - 1} - \frac{\sin (k - 1)\alpha_d}{k - 1} + \frac{\sin (k + 1)\alpha_s}{k + 1} - \frac{\sin (k + 1)\alpha_d}{k + 1} \right] \quad (18\text{-}21)$$

The case of dead zone or saturation alone which were previously studied can be obtained from Eq. (18-21) by substituting $\alpha_s = \pi/2$ or $\alpha_d = 0$, respectively.

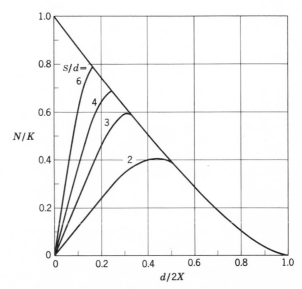

Fig. 18-19 *Nondimensionalized describing function for combined dead zone and saturation.*

The fundamental (using Lhopital's rule) is given by

$$Y_1 = \frac{2KX}{\pi}\left(\alpha_s - \alpha_d + \frac{\sin 2\alpha_s}{2} - \frac{\sin 2\alpha_d}{2}\right) \qquad (18\text{-}22)$$

The nondimensionalized describing function is

$$\frac{N}{K} = \frac{Y_1}{KX} = \frac{2\alpha_s}{\pi} - \frac{2\alpha_d}{\pi} + \frac{\sin 2\alpha_s}{\pi} - \frac{\sin 2\alpha_d}{\pi} \qquad (18\text{-}23)$$

The plot of this function depends on the relative values of α_d and α_s. The maximum value that may be used for α_d and α_s in this equation is $\pi/2$. A family of curves of this function is plotted in Fig. 18-19 against $d/2X$ for several values of S/d. For $d = S$ the value of the describing function is zero for all inputs. For $d/2X > 1$ the value of the describing function is zero. Also, as X becomes very large the describing function approaches zero in value.

Example 4. Friction-controlled backlash

For gear backlash that is friction-controlled, the characteristic is shown in Fig. 18-20. With this type of backlash the output member remains in contact with the input member until the velocity reaches zero. Then the output member stands still until the backlash is taken up, at which time it is assumed that the output member instantaneously takes on the velocity

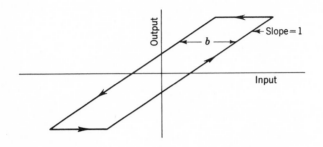

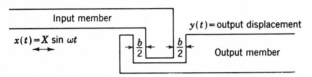

Fig. 18-20   *Backlash—physical model and characteristic.*

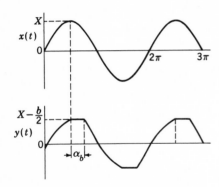

Fig. 18-21   *Input $x(t)$ and output $y(t)$ with friction-controlled backlash.*

of the input member.   The collision is assumed to take place without bouncing or deformation of the gears.   After contact, the output member follows the input member until the velocity again goes to zero.   Figure 18-21 shows the plot of output with a sinusoidal input.

The output y stands still for an angular change of the input equal to

$$\alpha_b = \sin^{-1} \frac{X - b}{X} \tag{18-24}$$

The nondimensionalized describing function as derived by Brearley[9] is given by

$$\mathbf{N}(\phi) = \frac{1}{\pi} [\sqrt{2\phi - \phi^2} \, (1 - \phi) + \pi - \cos^{-1} (1 - \phi)$$
$$+ j(\phi^2 - 2\phi)] \tag{18-25}$$

**Table 18-1.   Describing function for
friction-controlled backlash**

$\phi = b/X$	Re $[\mathbf{N}(\phi)]$	Im $[\mathbf{N}(\phi)]$	$|\mathbf{N}(\phi)|$	arg $\mathbf{N}(\phi)$, deg
0	1	0	1.000	0
0.1	0.976	−0.060	0.978	−3.5
0.2	0.948	−0.114	0.954	−6.7
0.3	0.907	−0.162	0.922	−10.2
0.4	0.856	−0.204	0.882	−13.4
0.5	0.802	−0.238	0.838	−16.5
0.6	0.748	−0.267	0.794	−19.7
0.7	0.688	−0.289	0.747	−22.8
0.8	0.627	−0.305	0.698	−26.0
0.9	0.563	−0.315	0.648	−29.3
1.0	0.500	−0.318	0.592	−32.5
1.1	0.436	−0.315	0.539	−35.9
1.2	0.372	−0.305	0.482	−39.5
1.3	0.309	−0.289	0.423	−43.0
1.4	0.252	−0.267	0.367	−46.6
1.5	0.197	−0.238	0.309	−50.4
1.6	0.142	−0.204	0.248	−55.2
1.7	0.092	−0.162	0.186	−60.4
1.8	0.051	−0.114	0.125	−66.0
1.9	0.022	−0.060	0.064	−69.8
2.0	0	0	0	−90.0

where $\phi = b/X$ varies over the range $0 \le \phi \le 2$.   The value $\phi = 2$ corresponds to a peak-to-peak amplitude of the input just equal to the backlash.   For input amplitude $x < b/2$ there is no output.   The describing function is a complex number because of the phase lag between the output and the input.   Table 18-1 gives values of the describing function in terms of ϕ.

Example 5.   Inertia-controlled backlash

When the friction on the output member of a system with backlash is very small compared with the inertia, the backlash is inertia-controlled.   In such a system the output member remains in contact with the input member as long as the acceleration is in the direction to keep the backlash spacing closed.   When the acceleration goes to zero, the output member leaves the input member and coasts at a constant velocity.   This separation of input and output members occurs when the input x is equal to zero and the velocity

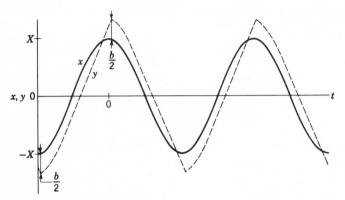

Fig. 18-22 *Input and output displacement waveforms of an inertia-controlled backlash element when $x(t) = X \cos \omega t$.*

is a maximum. The input member then slows down and is passed by the output member. When the output member has passed the input member by an amount equal to the backlash space, it comes in contact with the opposite face of the input member. It is assumed that the output member then instantaneously takes on the velocity of the input with no bouncing or deformation of the gears. This assumes that the output member does not load the input. The input and output are shown in Fig. 18-22.

The nondimensionalized describing function as derived from the Fourier expansion by Brearley is given by

$$\mathbf{N}(\phi) = \frac{1}{\pi} \{[(2 \sin \theta_1) \cos \theta_1 + (2\theta_1 + \pi - 2\phi) \sin \theta_1]$$

$$+ j[(2\theta_1 + \pi - 2\phi) \cos \theta_1 - 2 \sin \theta_1 - 3 + \sin^2 \theta_1]\} \quad (18\text{-}26)$$

where $\phi = b/X$ and $\theta_1 = \phi - \pi/2 + \cos \theta_1$. It is found for $\pi < \phi \le 3.721$ that the expression for $\mathbf{N}(\phi)$ can be simplified to

$$\mathbf{N}(\phi) = \frac{4}{\pi} \sin \theta_1 (\cos \theta_1 - j \sin \theta_1) \quad (18\text{-}27)$$

Table 18-2 gives values of the describing function in terms of ϕ. Subharmonics or nonperiodic waves appear for $\phi > 3.721$, rendering the describing-function approach inapplicable for such values of ϕ.

Example 6.   Backlash with inertia and viscous friction

The more general case consists of backlash with both inertia and friction acting on the output member. In this case the output member leaves the input member and begins coasting at some time after maximum velocity is

Table 18-2.   Describing function for
inertia-controlled backlash

$\phi = b/X$	Re $[\mathbf{N}(\phi)]$	Im $[\mathbf{N}(\phi)]$	$|\mathbf{N}(\phi)|$	arg $\mathbf{N}(\phi)$, deg
0	1.0	0	1.0	0
0.2	1.083	−0.0902	1.090	−4.8
0.4	1.125	−0.210	1.144	−10.6
0.6	1.135	−0.337	1.192	−16.4
0.8	1.121	−0.463	1.215	−22.4
1.0	1.085	−0.586	1.235	−28.4
1.2	1.031	−0.700	1.247	−34.2
1.4	0.963	−0.807	1.255	−40.0
1.6	0.881	−0.905	1.265	−45.7
2.0	0.686	−1.068	1.270	−57.2
2.4	0.462	−1.185	1.275	−68.7
2.8	0.216	−1.254	1.275	−80.2
3.0	0.0904	−1.270	1.274	−86.0
3.141	0	−1.273	1.273	−90.0
3.2	−0.042	−1.269	1.269	−91.8
3.4	−0.185	−1.245	1.268	−98.5
3.6	−0.369	−1.152	1.210	−107.8
3.721	−0.578	−0.906	1.075	−147.5

reached but before the maximum displacement is reached. *Coasting of the driven member starts when the deceleration of the driving member is greater than that of the driven member.* During the coasting stage the velocity of the driven member decreases exponentially. The case containing both inertia and friction falls between the friction-controlled and inertia-controlled cases.

The describing function is found to be a complex function of the two variables ϕ and α, which are defined by $\phi = b/X$ and $\alpha = \omega T = \omega J/B$, where ω is the frequency of the sinusoidal input and T is the ratio of inertia to friction of the output member. Figure 18-23 shows a complex plot of $N(\phi,\alpha)$ for several values of α. Friction-controlled backlash corresponds to $\alpha = 0$, and inertia-controlled backlash corresponds to $\alpha = \infty$. These limiting cases might have been predicted intuitively.

18-6   Effect of Nonlinearities on Stability[3,10]

Figure 18-24 shows the block diagram of a system which contains a nonlinearity. The describing function of the nonlinearity is N and can be used

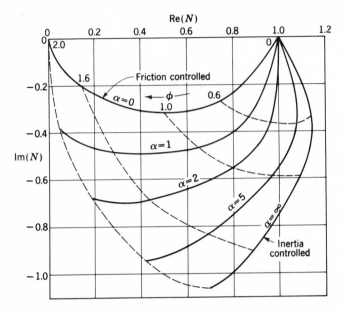

Fig. 18-23 *Plot of $N(\phi,\alpha)$, where N represents the describing function for a combination-type backlash in which the output member of the backlash element contains both inertia and viscous friction. $\phi = b/X$ and $\alpha = \omega T = \omega J/B$.*

to determine the system stability, provided that the harmonics are sufficiently attenuated. The control ratio of the closed-loop system is given by

$$\frac{\mathbf{C}(j\omega)}{\mathbf{R}(j\omega)} = \frac{\mathbf{N}G(j\omega)}{1 + \mathbf{N}G(j\omega)} \tag{18-28}$$

The stability can be determined from the zeros of the characteristic equation, which is

$$1 + \mathbf{N}G(j\omega) = 0 \tag{18-29}$$

Rewriting Eq. (18-29) in the form

$$G(j\omega) = -\frac{1}{\mathbf{N}} \tag{18-30}$$

describes the conditions that must be satisfied for continuous oscillations of the output. The function $\mathbf{N}$ may or may not be a function of frequency.

A linear system is considered stable if the transients die out. For nonlinear systems it may be found that there is stability in the same sense. Also it may be found that there are sustained oscillations in the output. If the amplitude of these sustained oscillations is sufficiently small, the system

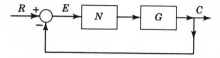

Fig. 18-24 *Feedback system with a nonlinearity.*

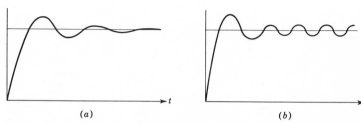

(a)

(b)

Fig. 18-25 (a) *Response of a stable system.* (b) *Limit cycle in the output of a nonlinear system.*

is still considered satisfactory. If it is excessive, the system is unsatisfactory. Figure 18-25 shows the time response of a stable linear system and the continuous oscillation (limit cycle) which is possible in a nonlinear system. The analysis that follows is used to determine the possibility of continuous oscillations and their amplitudes. It should be remembered that the use of describing functions is an approximation. The amplitudes and frequencies of oscillation determined experimentally may differ from those determined theoretically. However, the results should be qualitatively correct. The new rms and corrected-conventional describing functions usually predict the amplitude of the oscillations with greater accuracy than the conventional describing function.

Stability may be investigated by frequency-response analysis (using either the polar or log plot) or by root-locus analysis. A modification of the conventional Nyquist or Bode diagram must be made in order to apply Nyquist's stability criterion to the frequency-response plots. Based on Eq. (18-30), the locus of $-1/\mathbf{N}$ can be considered the locus of the critical point, which for linear systems is the -1 point in the complex frequency plane. When the critical point lies to the left of the $\mathbf{G}(j\omega)$ plot or is not enclosed by it, the poles of the control ratio have negative real parts. The system is then stable since any disturbances which appear tend to die out. Conversely, when the critical point lies to the right of the $\mathbf{G}(j\omega)$ locus and is therefore enclosed by it, the poles of the control ratio have positive real parts and the system is unstable. If the $\mathbf{G}(j\omega)$ plot passes through the critical point, the system may have a sustained oscillation, which may be either stable or unstable. If a slight disturbance in amplitude or frequency occurs and the oscillation returns to its original value, the oscillation is stable. If the oscillation continues to increase or decrease, it is unstable.

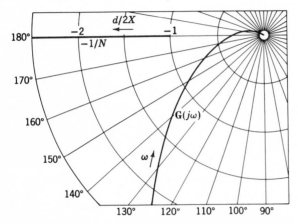

Fig. 18-26 Plot of $\mathbf{G}(j\omega) = 1/j\omega(1 + j0.5\omega)(1 + j\omega)$ and $-1/N$ representing dead zone.

18-7   Stability Using the Direct Polar Plot

As a first example, let the nonlinearity represent a dead zone, with the linear forward transfer function represented by

$$\mathbf{G}(j\omega) = \frac{1}{j\omega(1 + j\omega)(1 + j0.5\omega)} \tag{18-31}$$

Figure 18-26 shows a polar plot of $\mathbf{G}(j\omega)$ and $-1/\mathbf{N}$. The value of $\mathbf{N}$ is real and a function of the ratio of dead zone to amplitude of the input, as given in Fig. 18-14. The $-1/N$ locus is everywhere to the left of the $\mathbf{G}(j\omega)$ locus. Therefore this system is stable for all magnitudes of the input, and sustained oscillations do not occur. However, if the gain of $\mathbf{G}(j\omega)$ is increased to 3 or more, then the $\mathbf{G}(j\omega)$ and $-1/N$ curves will cross and oscillations will occur. But the resulting oscillations are unstable, that is, they either die out or increase in amplitude. This situation must therefore be avoided.

For a second example, let the nonlinearity represent inertia-controlled backlash, with the same linear forward transfer function represented by Eq. (18-31). Figure 18-27 shows a polar plot of $\mathbf{G}(j\omega)$ and $-1/\mathbf{N}$. The value of $\mathbf{N}$ as given in Table 18-2 is a complex quantity that is a function of the backlash spacing and the amplitude of the input X. Since the two curves cross at $\omega = 0.86$ with $\phi = 0.93$, a periodic solution is indicated. The stability of this solution can be tested by assuming that the system is operating at the amplitude and frequency corresponding to the intersection.

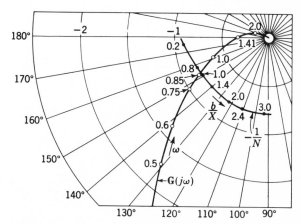

Fig. 18-27 *Plot of* $\mathbf{G}(j\omega) = 1/j\omega(1 + j0.5\omega)(1 + j\omega)$ *and* $-1/\mathbf{N}$ *representing inertia-controlled backlash.*

Assume that the amplitude is given a slight increase, so that the operating point $-1/\mathbf{N}$ shifts to the left, or stable, side of $\mathbf{G}(j\omega)$. The amplitude now tends to decrease, moving the operating point back to the original intersection value. In a similar manner, a slight decrease in the amplitude shifts the operating point to the right side of $\mathbf{G}(j\omega)$. The system now has poles with positive real parts and the amplitude tends to increase. This moves the operating point back to its original intersection value. The sustained oscillation indicated by the intersection of the curves in Fig. 18-27 is stable and describes a stable equilibrium point. This example illustrates how inertia-controlled backlash can produce sustained oscillations in a system which would otherwise be stable. Neglecting the backlash in the first linear approximation hides the presence of this sustained oscillation.

18-8   Stability Using the Log Plot

The same information regarding stability that is obtained on the polar plots can be obtained on the log plots. The log magnitude–angle diagram is the most convenient graph to use. The system with its $\mathbf{G}(j\omega)$ given by Eq. (18-31) is used again. For the case with dead zone the curves of $\mathbf{G}(j\omega)$ and $-1/\mathbf{N}$ are shown in Fig. 18-28. Since the $\mathbf{G}(j\omega)$ curve passes below the 0-db, $-180°$ point, the system is stable for all values of input signal. This is the same result as that obtained by use of the polar plot.

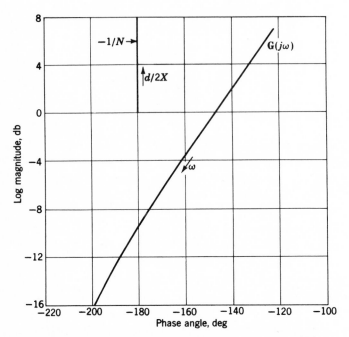

Fig. 18-28 *Plot of* $-1/N$ *representing dead zone and* $\mathbf{G}(j\omega) = 1/j\omega(1 + j0.5\omega)(1 + j\omega)$.

For the case of inertia-controlled backlash the curves of $\mathbf{G}(j\omega)$ and $-1/\mathbf{N}$ are shown in Fig. 18-29. The intersection of the two curves occurs at $\omega = 0.86$, $\phi = 0.93$. By following the analysis used in the preceding section with the polar plots, it is seen that this point of intersection represents a stable oscillation. Even if the amplitude of the oscillations is small enough to be considered unobjectionable, a sustained chatter occurs in the output.

A third example is presented next in which the describing function is also a function of frequency. The system contains $\mathbf{G}(j\omega)$ given by

$$\mathbf{G}(j\omega) = \frac{1}{j\omega(1 + j\omega)} \tag{18-32}$$

and $\mathbf{N}$ representing a backlash with both inertia and viscous friction. The plots of the functions $\mathbf{G}(j\omega)$ and $-1/\mathbf{N}$ are shown in Fig. 18-30. Since the describing function is a function of frequency, it is necessary to plot a family of curves $-1/\mathbf{N}$ for several frequencies. Oscillations can exist only if the curve $-1/\mathbf{N}(j\omega)$ and $\mathbf{G}(j\omega)$ cross at the same value of frequency. The possibility exists of sustained oscillations with a frequency ω_2 for this example. An analysis of this intersection shows that the oscillations are stable. This

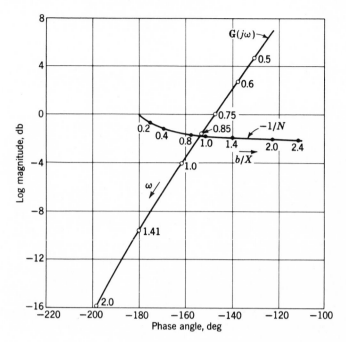

Fig. 18-29 *Plot of* $-1/\mathbf{N}$ *representing inertia-controlled backlash and of* $\mathbf{G}(j\omega) = 1/j\omega(1 + j0.5\omega)(1 + j\omega).$

case, in which the describing function is frequency-variant, is generally more difficult to analyze.

18-9 Stability Using the Root Locus[3,11]

Root-locus analysis can also be used to investigate the poles of $C(s)/R(s)$. For the system of Fig. 18-24 the poles of $C(s)/R(s)$ are determined by the condition

$$NG(s) = -1 \qquad (18\text{-}33)$$

The root-locus plot must be modified to satisfy the following conditions:

Magnitude Condition: $\quad |N| \cdot |G(s)| = 1 \qquad (18\text{-}34)$

Angle Condition: $\quad \underline{/N} + \underline{/G(s)} = (1 + 2m)180° \qquad (18\text{-}35)$

where m is any integer including zero. The inclusion of the magnitude and angle of N in these two conditions is the first difference from the linear

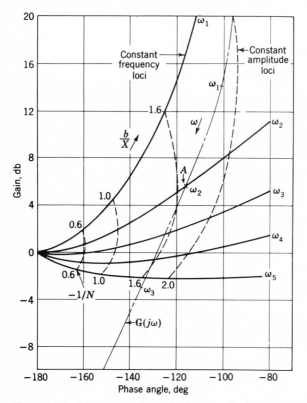

Fig. 18-30 *Log magnitude–angle diagram for* **N** *representing backlash with both inertia and friction and* **G**$(j\omega)$ = $1/j\omega(1 + j\omega)$.

system.   The locus is now a plot of the poles of $C(s)/R(s)$ as a function of the gain of the nonlinearity.   In other words, as the size of the signal input to the nonlinearity varies, the poles of $C(s)/R(s)$ move along the root locus. A correlation exists between this pole motion and the transient response of the system.   Such a correlation has been devised by Finnigan[3] but is not easily applied.

For a system in which the nonlinearity has a real value, such as in the case of saturation, the root locus of the system remains unchanged.   However, the poles of the closed-loop system are now a function of the product of the static loop sensitivity and the nonlinear describing function.   For example, with saturation and the transfer function given by

$$G(s) = \frac{K}{s(s + 1)} \tag{18-36}$$

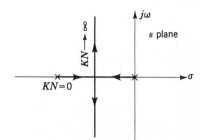

Fig. 18-31   *Root loci for KN/s(s + 1)* = *−1*.

the characteristic equation is given by

$$\frac{KN}{s(s + 1)} = -1 \tag{18-37}$$

The root locus obtained for the angle of $G(s) = (1 + 2m)180°$, as drawn in Fig. 18-31, is now a function of the gain KN. This means that if the input is small, so that $N = 1$, the poles of $C(s)/R(s)$ are determined by the value of K. As the size of the input increases, the value of N decreases, with the result that KN decreases and the system becomes less underdamped. This is typical of the results obtained with saturation. Saturation has not changed the stability of this system; i.e., it is stable for all values of gain.

If N is complex, the angle loci must be plotted for $G(s)$. The loci for angles of $\pm170°$, $\pm160°$, etc., can be plotted by using the method described in Appendix F. An example uses the transfer function $G(s)$ of Eq. (18-36) with $K = 1$ and N representing friction-controlled backlash. To illustrate the calculation of a point on the new locus, it is found from Table 18-1 that, at $\phi = 0.304$, $N = 0.9$, $/N = -10°$. A point on the locus is determined from the angle condition by use of Eq. (18-35),

$$\underline{/G(s)} = -180° - \underline{/N} = -170° \tag{18-38}$$

and the magnitude condition, by use of Eq. (18-34), is

$$|N| = \frac{1}{|G(s)|} = |s| \cdot |s + 1| = 0.9 \tag{18-39}$$

The point on the locus is on the $-170°$ locus of $G(s)$, which satisfies the magnitude condition of Eq. (18-39). A trial-and-error procedure is used to locate the point. This method yields a root-locus plot which is applicable for only one value of gain ($K = 1$, in this example). The root locus for any value of gain K in this system is obtained from

$$K \cdot |N| = |s| \cdot |s + 1| \tag{18-40}$$

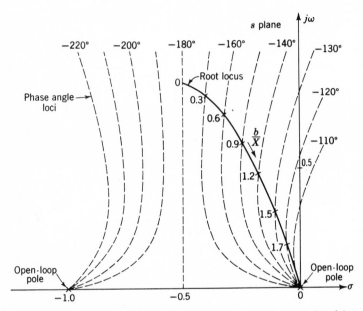

Fig. 18-32   *Root-locus plot corresponding to a system containing fric-tion-controlled backlash, where* $G(s) = 1/s(s + 1)$.

Figure 18-32 shows the root-locus plot in the upper-half s plane for this system with friction-controlled backlash.   The locus in the lower-half s plane is a reflection of the locus above the real axis.   Figure 18-33 shows the root-locus plot for this system with inertia-controlled backlash, and Fig. 18-34 shows the root locus for backlash with both friction and inertia.

The closed-loop system is unstable for values of the input which cause the poles of $C(s)/R(s)$ to lie in the right half of the s plane.   Periodic solutions correspond to intersections of the root locus and the real frequency axis. The case of friction-controlled backlash shows no possibility of sustained oscillation.   However, the inertia-controlled backlash shows that oscillation occurs for approximately $b/X = 1.65$.   The same information can be obtained from the polar or log plots.   The case of backlash with both friction and inertia shows the possibility of sustained oscillations at the frequency ω_2.   The same information is obtained from both the root-locus plot of Fig. 18-34 and the log plot of Fig. 18-30.

It must be emphasized that the describing function is applicable rigorously only when the input to the system is sinusoidal.   This limits the rigorous validity of the analysis to only those points on the imaginary axis of the s plane.

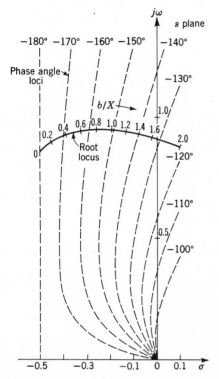

Fig. 18-33 *Root-locus plot for an oscillatory nonlinear system in which N, representing inertia-controlled backlash, is a function of input amplitude only and G(s) = 1/s(s + 1).*

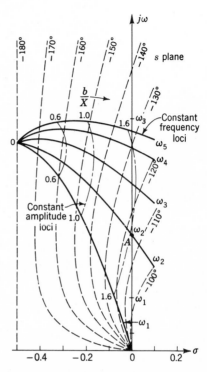

Fig. 18-34 *Root-locus plot for a nonlinear system in which N is a function of both input amplitude and frequency and where G(s) = 1/s(s + 1).*

An interesting case develops when the linear system is third-order, or in general is an odd order, and the nonlinearity is complex. This can be demonstrated with an inertia-controlled backlash and $G(s)$ given by

$$G(s) = \frac{5}{s(s^2 + 0.7s + 1.1)} \tag{18-41}$$

The root locus in the upper-half s plane is shown in Fig. 18-35. It is seen that the system has four branches. Therefore it behaves as a fourth- rather than a third-order system. A steady-state oscillation is indicated by the intersection of the dominant branch with the imaginary axis. The results compare closely to those obtained from the log plot, shown in Fig. 18-36.

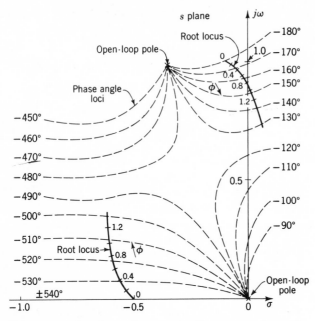

Fig. 18-35 Root-locus plot corresponding to a system with inertia-controlled backlash, where $G(s) = 5/s(s^2 + 0.7s + 1.1)$.

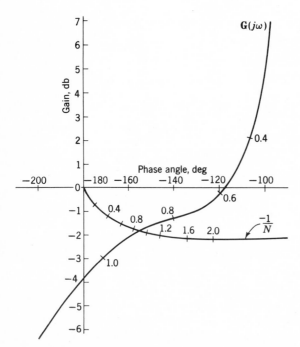

Fig. 18-36 Log magnitude–angle diagram with inertia-controlled backlash and $G(s) = 5/s(s^2 + 0.7s + 1.1)$.

18-10   Accuracy of Describing-function Predictions

The predicted amplitude and frequency of the limit cycle produced by a nonlinearity are known to be approximate values.   The accuracy depends on the correctness of the assumption made in deriving the describing function.   For the conventional describing function it is assumed that all the harmonics in the output of the nonlinearity are completely attenuated by the linear transfer function $G(j\omega)$.   Since harmonics do exist, they affect the actual frequency and amplitude of the limit cycle.   The *new rms*[14] and *corrected-conventional*[15] describing functions discussed in Sec. 18-4 include the effect of the harmonics and therefore give better accuracy.   As an example, consider the relay with dead zone whose characteristics are shown in Fig. 18-6b.   The three describing functions are:

Conventional:

$$N = \frac{4K}{\pi X} \sqrt{1 - \left(\frac{d}{2X}\right)^2}$$

New RMS:

$$N = \frac{K}{X} \sqrt{\frac{2}{\pi}\left(\pi - 2\sin^{-1}\frac{d}{2X}\right)}$$

Corrected Conventional:

$$N = \frac{4K}{\pi X} \sqrt{1 - \left(\frac{d}{2X}\right)^2} \sqrt{1 - \frac{1}{9}\left[1 - \left(\frac{d}{2X}\right)^2\right]^2}$$

The experimental and predicted amplitudes of the oscillation appearing in the output are listed in Table 18-3 for four different transfer functions $G(s)$.   These results show that the corrected-conventional describing function produces greater accuracy in predicting amplitude in most cases. All three describing functions predict the same frequency of oscillation. A comprehensive comparison with seven nonlinearities is contained in reference 15.   In most cases for which the degree of the denominator of $G(s)$ is three or more higher than the degree of the numerator, the corrected-conventional describing function seems to be the most accurate.

18-11   Multiplicative Systems

It is deemed proper to close this chapter by mentioning briefly a new type of control system which is gaining favor in the automatic control field

Table 18-3.   *Experimental and predicted oscillation for a relay with dead zone* $(d = 3)$

Linear transfer function $G(s)$	Experimental amplitude % error	Predicted amplitudes			Experimental frequency ω	Predicted frequency ω
		Conventional % error	New rms % error	Corrected-conventional % error		
$\dfrac{18}{s(s+1)(s+2)}$	3.537 / 0	3.45 / −2.460	3.615 / +2.205	3.45 / −2.460	1.391	1.414
$\dfrac{16.57(s+0.1)}{s^2(s+1)^2}$	13.49 / 0	13.05 / −3.262	14.145 / +4.855	13.71 / +1.631	0.868	0.894
$\dfrac{242.9(s+0.1)}{s^2(s+1)^2(s+5)^2}$	13.78 / 0	13.05 / −5.298	14.145 / +2.649	13.71 / −0.508	0.631	0.654
$\dfrac{301}{s(s+1)(s+2)(s+3)(s+4)}$	13.29 / 0	12.645 / −4.853	13.695 / +3.048	13.26 / −0.226	0.817	0.837

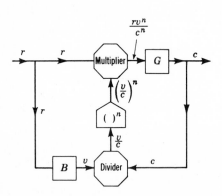

Fig. 18-37   *A multiplicative conditional feedback control system.*

(especially in process control systems). In Sec. 15-9 conditional feedback control systems are discussed. An extension of this type of system can be made by replacing the subtractor, $G_1(s)$, and the adder, in Fig. 15-23, by a divider, a power-raising device, and a multiplier, as shown in Fig. 18-37. This modification results in what is classified as a multiplicative conditional feedback control system.

Such nonlinearities as torque saturation in motors, backlash in gears, stiction on shafts, clipping in amplifiers, and nonlinear gain in synchros cause the designer great trouble. Most of these nonlinearities contribute to the instability of a feedback loop and especially to the instability of the classical feedback system, which does not permit any form of nonlinearity or any form of imperfect linearity to be fully tolerated; it strives for the ideal and commonly becomes unstable in so doing.

However, it is clear that in many practical applications certain residual nonlinearities in the relation of output to input are not undesirable. It is an unfortunate limitation of the classical feedback system that such admissible nonlinearities contribute to the instability of the feedback loop. The conditional feedback system, such as the one shown in Fig. 18-37, on the other hand, provides a direct means for accepting tolerable nonlinearities in the input-output response and at the same time largely prevents these nonlinearities from influencing the stability of the feedback loop.

The manner in which this is accomplished is described in detail in the literature. [12, 13]

18-12   *Conclusions*

This chapter has covered use of the describing function to represent a nonlinearity for the determination of system stability. When it is used in conjunction with the polar plots, the log plots, and the root locus, the stability of the system can be determined. It is found that stability may be a function of the magnitude of the input. The possibility of sustained oscillations and the magnitude of those oscillations can also be determined.

The main limitation in the use of the describing function is that it is based on a sinusoidal input. Determination of system response with other types of inputs is difficult. The current technical literature contains much material on treating and compensating nonlinear systems. This is an advanced topic which the reader can investigate further as the specific need arises. It is common to study nonlinear systems by the use of an analog computer. The computer produces the time response for many forms of the input, and the designer is therefore able to adjust the system to achieve satisfactory performance. The next chapter covers the use of analog computers.

Bibliography

1. "Fundamentals of Design of Piloted Aircraft Flight Control Systems," BuAer Report AE-61-4, vol. 1, chaps. 2 and 6, U.S. Navy, Bureau of Aeronautics, 1952.

2. Truxal, J. G. (ed.): "Control Engineers' Handbook," pp. 2-67 to 2-86, McGraw-Hill Book Company, New York, 1958.

3. Finnigan, R. E.: "Transient Analysis of Nonlinear Servomechanisms Using Describing Functions with Root-locus Techniques," Ph.D. Thesis, University of Illinois, 1957.

4. Tustin, A.: The Effects of Backlash and of Speed Dependent Friction on the Stability of Closed-cycle Control Systems, *J. Inst. Elec. Engrs.* (*London*), vol. 94, pp. 143–151, 1947.

5. Kochenburger, R. J.: A Frequency Response Method for Analyzing and Synthesizing Contactor Servomechanisms, *Trans. AIEE*, vol. 69, pt. I, pp. 270–284, 1950.

6. Johnson, E. C.: Sinusoidal Analysis of Feedback-control Systems Containing Nonlinear Elements, *Trans. AIEE*, vol. 71, pt. II, pp. 169–181, July, 1952.

7. Greif, H. D.: Describing Function Method of Servomechanisms Analysis Applied to Most Commonly Encountered Nonlinearities, *Trans. AIEE*, vol. 72, pt. II, pp. 243–248, September, 1953.

8. Satyendra, K. N.: Describing Functions Representing the Effect of Inertia, Backlash and Coulomb Friction on the Stability of an Automatic Control System, *Trans. AIEE*, vol. 75, pt. II, pp. 243–249, 1956.

9. Brearley, H. C., Jr.: "Prediction of Transient Response of Nonlinear Servomechanisms by Sinusoidal Analysis," Ph.D. Thesis, University of Illinois, 1953.

10. Kochenburger, R. J.: Limiting in Feedback Control Systems, *Trans. AIEE*, vol. 71, pt. II, pp. 180–194, 1953.

11. Truxal, J. G.: "Automatic Feedback Control System Synthesis," chap. 10, McGraw-Hill Book Company, New York, 1955.

12. Lang, G., and J. M. Ham: Conditional Feedback Systems—A New Approach to Feedback Control, *Trans. AIEE*, vol. 74, pt. II, pp. 152–161, July, 1955.

13. Aseltine, J. A., A. R. Mancini, and C. W. Sarture: Survey of Adaptive Systems, *IRE Trans. Auto. Control*, PGAC-6, pp. 102–108, December, 1958.

14. Gibson, J. E.: "Nonlinear Automatic Control," McGraw-Hill Book Company, New York, 1963.

15. Rankine, R. R.: "An Evaluation of Selected Describing Functions of Control System Nonlinearities," M.Sc. Thesis, Air Force Institute of Technology, Dayton, Ohio, 1964.

19

Analog
Computers

19-1 *Introduction*[1-4]

The preceding chapters have described the principal methods used in the analysis of feedback control systems. When the system is a simple single-loop linear device, the analysis may be performed in a straightforward manner by any of the methods described. However, when the system is multiloop, the effect on the over-all performance that is produced by variations of an inner-loop parameter is laborious and time-consuming to obtain. Aircraft control systems are an important example in which there are many degrees of freedom and many interconnected feedback loops. Some feedback loops are inherent in the dynamics of the aircraft itself, and others are added for proper stabilization in the design of the control system. In such a system the effect on the output response that is produced by the variation of any parameter becomes extremely difficult to determine. Even when the

631

preliminary design indicates the desirable ranges for several adjustable quantities, it is difficult to determine the best values for each of these quantities by analytical methods. This problem can be solved by the use of automatic computers which show directly the effect of parameter changes on the output response.

As pointed out earlier in the text, computers are utilized to obtain the frequency-response[6–8] and root-locus plots.[9,10] The digital computer can be used either as part of a control system in order to maintain, as closely as possible, a desired optimum performance or as a memory device in a sampled-data system.

Both analog and digital computers are used to solve complex problems. Each has its particular advantages, depending on the number of operations that must be performed by the computer and the accuracy desired. The material presented in this book is limited to the analog computer used to perform mathematical operations that include integration, multiplication, and addition. More specifically, the type of mathematical analog computer studied is the electronic differential analyzer, which is made up of "operational amplifiers" with assorted resistors, capacitors, and potentiometers.

The electronic analog computer is well suited to the solution of simultaneous differential equations. The presence of nonlinearities in these equations is usually handled directly by multipliers and some special components. The results are obtained in the form of voltages and, by use of a recorder, are obtained graphically as a function of time, which is very convenient for engineering use. For each value of an adjustable quantity the computer presents a graph of the response, and it is therefore possible to determine the best response and the corresponding value of each adjustable parameter.

The analog computer can operate in real time. This is an advantage since it permits the use of some physical components together with the mathematical representation of the remainder of the system on the analog computer. This is advantageous when the equations of a physical component are unknown. When the equations describing the performance of a piece of equipment are known but complex, fewer computer components are necessary. The accuracy of the results obtained from the analog computer depends on the accuracy of the components used. An accuracy of a few per cent is common and is usually acceptable. Acceptability of this accuracy must take into account that some parameters of control systems are not known precisely. Accuracy of the computer is a function of the uncertainty of the parameters.

In the following sections the components of the electronic analog computer are described and then used to solve for the performance of a system.

19-2   *Analog-computer Components*

D-C Amplifier

A high-gain d-c voltage amplifier is the basic component of the electronic analog computer. The d-c amplifier is used with feedback and input impedances to produce operational blocks which represent transfer functions on the computer. The gain of d-c amplifiers developed for use in computers is from 5×10^4 to 30×10^6. The output of d-c amplifiers is subject to drift, which is accentuated by the high gain. Drift may be produced by slow changes in the supply potentials used. It is therefore common to use well-regulated power supplies to furnish the potentials required. Changes in resistance of the resistors in the circuit due to changes in ambient temperature also contribute to drift. Therefore good-quality precision resistors that are temperature-compensated are frequently used. To further stabilize the d-c amplifier, it is common to use feedback which may be built into each d-c amplifier. Because of the large number of amplifiers and the expense involved, it is possible to use one stabilizing feedback circuit that operates on all the amplifiers in sequence. The short time interval between successive periods of stabilization minimizes the drift. The common symbol used for the high-gain d-c amplifier is the rectangle shown in Fig. 19-1.

Operational Amplifier[5]

The operational amplifier consists of a high-gain d-c voltage amplifier with an input impedance Z_i and a feedback impedance Z_f (see Fig. 19-2). The impedance Z_g represents the input impedance of the d-c amplifier

Fig. 19-1 *Symbolic representation of the d-c amplifier.*

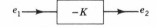

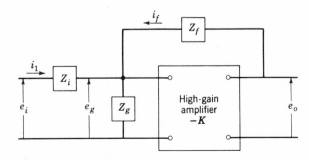

Fig. 19-2 *Operational amplifier.*

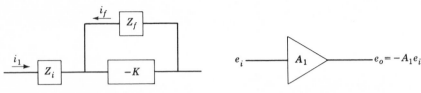

Fig. 19-3 *Simplified diagram of the operational amplifier.*

Fig. 19-4 *Operational amplifier.*

which has a gain $-K$. The over-all gain E_o/E_i of the operational amplifier is found from the equations

$$E_i - E_g = I_1 Z_i \tag{19-1}$$
$$E_o - E_g = I_f Z_f \tag{19-2}$$
$$E_g = (I_1 + I_f) Z_g \tag{19-3}$$
$$E_o = -K E_g \tag{19-4}$$

and is given by

$$\frac{E_o}{E_i} = -\frac{Z_f}{Z_i} \frac{1}{1 + (1/K)[1 + Z_f(Z_i + Z_g)/Z_i Z_g]} \tag{19-5}$$

For Z_g much larger than Z_i and Z_f and K very large, the gain can be approximated by the very useful relation

$$\frac{E_o}{E_i} = -\frac{Z_f}{Z_i} \tag{19-6}$$

The ratio $-Z_f/Z_i$ may be considered as an operator by which the input voltage is multiplied to produce the output voltage.   If the two impedances selected are resistances, the circuit has a known gain and a sign change.

Since Z_g is very large, it can be omitted from the circuit diagram. Figure 19-2 is redrawn in simplified form in Fig. 19-3.   When Z_f and Z_i are both resistors, the operational amplifier has a gain equal to $-R_f/R_i = -A_1$ and is represented by the symbol shown in Fig. 19-4.   Operational amplifiers in the REAC* use standard built-in resistances having the values $Z_f = 10^6$ ohms and $Z_i = 10^6$, 2.5×10^5, and 10^5 ohms so that standard gains of -1, -4, and -10 are readily available.   The GEDA† uses plug-in fixed and adjustable resistors to obtain the desired gains.

Adder or Summer

The operational amplifier tends to maintain the voltage e_g at approximately ground potential.   This can be shown by solving for the ratio E_g/E_i from

---

* Reeves electronic analog computer.
† Goodyear electronic differential analyzer.

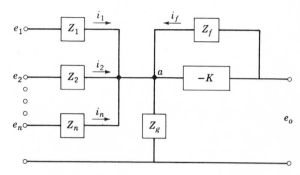

Fig. 19-5 *Adding circuit.*

Eqs. (19-1) to (19-4):

$$\frac{E_g}{E_i} = \frac{1}{K(Z_i/Z_f) + 1 + Z_i/Z_g + Z_i/Z_f} \qquad (19\text{-}7)$$

With large values of gain K the right side of this equation is approximately zero. This means that e_g remains essentially at zero potential.

Figure 19-5 shows an operational amplifier with several voltages e_1, e_2, $\ldots$, e_n applied to corresponding input impedances. The sum of the currents flowing into point a is given by

$$\frac{E_1 - E_g}{Z_1} + \frac{E_2 - E_g}{Z_2} + \cdots + \frac{E_n - E_g}{Z_n} + \frac{E_o - E_g}{Z_f} = \frac{E_g}{Z_g} \qquad (19\text{-}8)$$

As shown in the preceding paragraph, the voltage e_g is essentially zero. Therefore this equation can be rewritten in the form

$$E_o = -\frac{Z_f}{Z_1} E_1 - \frac{Z_f}{Z_2} E_2 - \cdots - \frac{Z_f}{Z_n} E_n = -\sum_{i=1}^{n} (A_i E_i) \qquad (19\text{-}9)$$

When all the input and feedback impedances are resistors, the operational amplifier is called an *adder* or *summer* and is represented by the symbol of Fig. 19-6. The gain associated with each input is marked on the amplifier.

Integrator

The operational amplifier of Fig. 19-2 acts as an integrator if a capacitor is used for the feedback element and a resistor is used for the input elements. In this case the Laplace transform of Eq. (19-6) gives

$$\frac{E_o(s)}{E_i(s)} = -\frac{1}{RCs} \qquad (19\text{-}10)$$

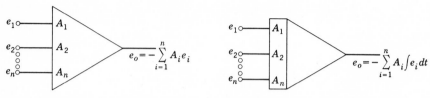

Fig. 19-6   *Symbol for an adder.*   **Fig. 19-7**   *Symbol for an integrator.*

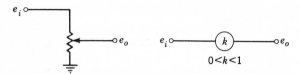

Fig. 19-8   *A schematic and a symbolic representation of a potentiometer.*

In the time domain this relation is

$$e_o(t) = -\frac{1}{RC} \int e_i(t) \, dt \tag{19-11}$$

The coefficient $1/RC$ represents a change of gain.   With $R = 10^6$ ohms and $C = 1$ μf the gain is equal to unity.   By the use of Eq. (19-9) it can be seen that several inputs can be connected through separate input resistances. The output is then equal to the sum of the integrals of each input.   By using the appropriate value for each resistor the gain can be adjusted separately for each integral.   The symbol for an integrator is shown in Fig. 19-7.

Potentiometer

When a differential equation is analyzed by the use of operational amplifiers, the gains of the amplifiers may represent the coefficients in the differential equation.   Rather than adjust the size of each input resistor to obtain the desired gain, it is common to use standard-size resistors and to use potentiometers at the input.   The potentiometer performs multiplication by an adjustable constant k, with a value between 0 and 1.   Figure 19-8 shows a schematic and a symbolic representation of a potentiometer.   When using a potentiometer it is necessary to make a correction for the loading effect of the input impedance of the operational amplifier to which it is connected.   This is frequently done by comparing the setting, with the load connected, with a reference standard.

19-3   Analog-computer Setup

The setup of the components described in the preceding section is best illustrated by examples.

Example 1

Consider the second-order differential equation given by

$$a_2 D^2 x + a_1 Dx + a_0 x = f(t) \tag{19-12}$$

The first step is to solve this equation for the highest-order derivative:

$$a_2 D^2 x = f(t) - a_1 Dx - a_0 x \tag{19-13}$$

There are three terms on the right side of the equation. Therefore three inputs are connected to an integrator, and each input is labeled to correspond to the terms on the right side of Eq. (19-13) (see Fig. 19-9, integrator 1). The output of this integrator is $-a_2 Dx$. This quantity is next used as the input to integrator 2, so that its output is $a_2 x$. This is the desired solution of the differential equation. However, to obtain this solution the specified inputs to integrator 1 must be inserted. This is done by connecting $f(t)$ from an external source. The quantity $-a_1 Dx$ is obtained by taking the output of integrator 1 and multiplying it by the proper factor k_1. Since the third input to integrator 1 is $-a_0 x$, this quantity can be obtained by applying the output of integrator 2 to amplifier 3. The function of amplifier 3 is to introduce the minus sign. The gain of amplifier 3 and the potentiometer setting k_2 are adjusted to furnish the input $-a_0 x$, which is required for integrator 1. It is noted that the second derivative $D^2 x$ does not appear explicitly anywhere in this circuit. The quantities x and Dx are connected to recorders and are obtained in graphical form. In an actual hookup the inputs to integrator 1 may be smaller than the indicated quantities.

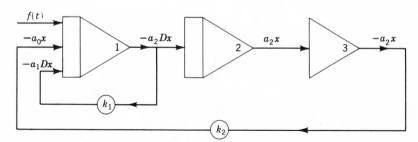

Fig. 19-9   *Computer setup for the differential equation* $a_2 D^2 x + a_1 Dx + a_0 x = f(t)$.

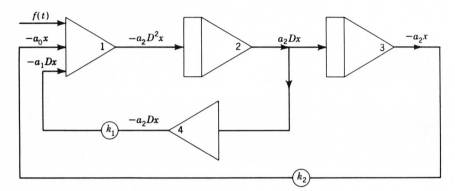

Fig. 19-10   *Alternative computer setup for $a_2 D^2 x + a_1 Dx + a_0 x = f(t)$.*

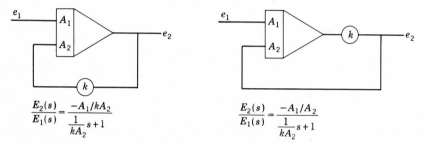

Fig. 19-11   *Computer setup for a lag component.*

The gain associated with each input is then adjusted to the proper value. If the second derivative $D^2 x$ is required, the alternative setup shown in Fig. 19-10 can be used. This setup requires one additional operational amplifier.

Example 2

The first-order lag unit given by Eq. (19-14) is a common factor in the transfer functions of feedback control systems. This factor corresponds to the differential equation (19-15). This characteristic can be obtained in several ways. Figure 19-11 shows two setups using an integrator:

$$\frac{E_2(s)}{E_1(s)} = \frac{A}{1 + Ts} \tag{19-14}$$

$$(TD + 1)e_2(t) = A e_1(t) \tag{19-15}$$

The same equation is obtained by an operational amplifier with a resistor as the input impedance and the parallel combination of a resistor and a

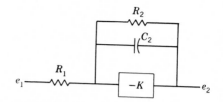

Fig. 19-12 Operational amplifier for the transfer function

$$\frac{E_2(s)}{E_1(s)} = - \frac{R_2}{R_1} \frac{1}{1 + R_2 C_2 s} = \frac{-A}{1 + Ts}$$

capacitor as the feedback impedance (see Fig. 19-12). The transfer function for this setup is given by

$$\frac{E_2(s)}{E_1(s)} = - \frac{R_2}{R_1} \frac{1}{1 + R_2 C_2 s} = \frac{-A}{1 + Ts} \tag{19-16}$$

Example 3

The transfer function given by Eq. (19-17), containing a zero and a pole, is often found in control systems. The lag and lead compensators are typical examples. There are several ways of obtaining this transfer function. Solving the differential equation for the highest-order derivative of the output $e_2(t)$ results in Eq. (19-18):

$$G(s) = \frac{E_2(s)}{E_1(s)} = \frac{1 + Ts}{1 + \alpha Ts} \tag{19-17}$$

$$\alpha T \, De_2 = T \, De_1 + e_1 - e_2 \tag{19-18}$$

Equation (19-18) is integrated to eliminate the need to differentiate the signal e_1. Upon rearranging terms, this gives

$$e_2 = \frac{e_1}{\alpha} + \frac{e_1 - e_2}{\alpha TD} \tag{19-19}$$

The first computer simulation is shown in Fig. 19-13. If $1/\alpha$ and $1/\alpha T$ are greater than unity, the gains of summer 3 can be increased so that the potentiometer settings are between 0 and 1.

Another computer setup can be obtained by performing the division in Eq. (19-17) to obtain

$$G(s) = \frac{E_2(s)}{E_1(s)} = \frac{1 + Ts}{1 + \alpha TS} = \frac{1}{\alpha}\left(1 + \frac{\alpha - 1}{1 + \alpha Ts}\right) \tag{19-20}$$

For values of α less than 1, the quantity $\alpha - 1$ is negative. The second fraction is a first-order lag and can be obtained from Fig. 19-11. The complete computer setup for Eq. (19-20) with $\alpha < 1$ (a lead compensator) is

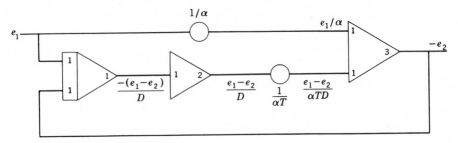

Fig. 19-13   *Computer setup for Eq. (19-19).*

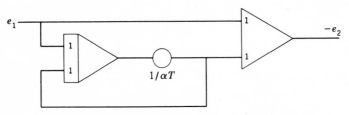

Fig. 19-14   *Computer setup for the lead compensator* $(\alpha < 1)$

$$G(s) = \frac{E_2(s)}{E_1(s)} = \alpha \frac{1 + Ts}{1 + \alpha Ts}$$

shown in Fig. 19-14.   Less equipment is required for the setup in Fig. 19-14 than for that in Fig. 19-13.

19-4   *Simulation of Nonlinearities*[11]

Several types of nonlinearities present in physical equipment are described in Chap. 18, where it is stated that the analytic solution of nonlinear equations is difficult and subject to various approximations.   A distinct advantage of the analog computer is the ability to simulate, in approximate form, many of the nonlinearities encountered.   The circuits for representing several of the more common nonlinearities are given in this section.   These circuits use two principal elements, the diode and the polarized relay.   The vacuum diode is frequently used and under ideal conditions is considered to have zero resistance for current flowing in one direction and infinite resistance for current flowing in the opposite direction.   The polarized relay is usually used with a high-gain amplifier so that it will switch with a voltage of 20 mv or less.   It has a normally "off" position and will switch in either direction, depending on the polarity of the applied voltage.   Switching time of 1 msec is usually satisfactory.

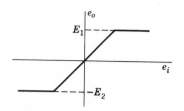

E_1

e_o

e_i

$-E_2$

Fig. 19-15 *Saturation character-istic.*

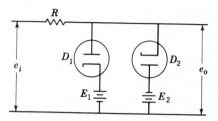

R

D_1 D_2

e_i e_o

E_1 E_2

Fig. 19-16 *Limiting circuit.*

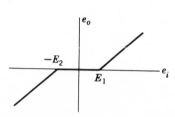

e_o

$-E_2$

E_1

e_i

Fig. 19-17 *Dead-zone character-istic.*

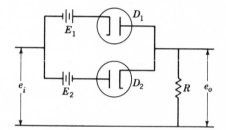

E_1 D_1

E_2 D_2

e_i R e_o

Fig. 19-18 *Dead-zone circuit.*

Limiting

The saturation characteristic shown in Fig. 19-15 can be accomplished by the shunt-type circuit shown in Fig. 19-16.   For values of the input $E_2 \leq e_i \leq E_1$ the output is equal to the input.   For values of $e_i < E_2$ the diode D_2 conducts and the voltage $-E_2$ appears across the output (neglecting the drop across the diode).   Similarly, for values $e_i > E_1$ the diode D_1 conducts and the voltage E_1 appears across the output.   The values of E_1 and E_2 need not be the same.

Dead Zone

The dead-zone characteristic shown in Fig. 19-17 can be achieved by the circuit of Fig. 19-18.   This figure is similar to Fig. 19-16 except that the reference and input terminals have been interchanged.   The dead zone need not be symmetrical in this circuit.

Coulomb Friction

The coulomb-friction characteristic of Fig. 19-19 can be obtained by the circuit of Fig. 19-20.   In this circuit the signal e_i is amplified and then

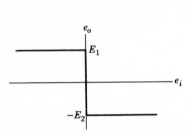

Fig. 19-19   Coulomb-friction characteristic.

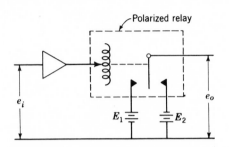

Fig. 19-20   Coulomb-friction circuit.

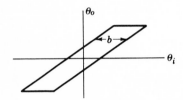

Fig. 19-21   Backlash characteristic.

energizes the coil of the polarized relay.   The polarity of the input determines whether the output is E_1 or $-E_2$.   The circuit of Fig. 19-16 can also be used if a very-high-gain amplifier is added at the input terminals.   This has the effect of increasing the slope of the curve in Fig. 19-15.   With an amplifier of sufficiently high gain at the input the circuit reaches saturation with very small inputs.

Backlash

The backlash characteristic shown in Fig. 19-21 can be produced by the circuit of Fig. 19-22.   If magnitude limiting is applied to the output of the integrator, an approximate representation of magnetic hysteresis is obtained.

Continuous Nonlinearities

There are many methods for producing a continuous nonlinear function such as that shown in Fig. 19-23.   A simple circuit using diodes is shown in Fig. 19-24.   The procedure is to draw a series of straight-line segments which represent a suitable approximation to the desired function.   Then the number of diodes needed is equal to the number of changes of slope.   For the example of Fig. 19-23 there are three changes of slope; therefore three diodes are needed.   The resistor between the voltages of -100 and $+100$ volts can be tapped at any desired point.   For values of $e_i < a$ the diode D_1 conducts, and the slope of the output voltage depends on the

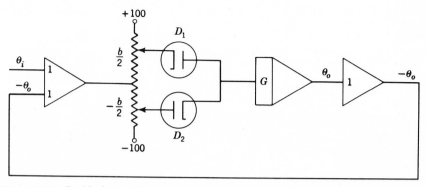

Fig. 19-22 *Backlash circuit.*

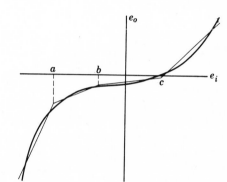

Fig. 19-23 *Continuous nonlinearity.*

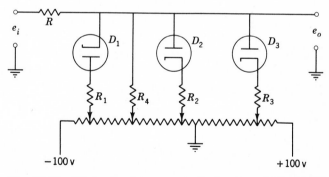

Fig. 19-24 *Diode circuit to generate the function of Fig. 19-23.*

relative sizes of R and R_1. When $e_i > a$ the diode D_1 cuts off. For $b > e_i > a$ none of the diodes conduct, and the slope of the curve from a to b is determined by the resistors R and R_4. For $e_i > b$ the diode D_2 conducts, and the slope of the output from b to c depends on the ratio of R to R_2. Similarly, for $e_i > c$ the diode D_3 conducts to produce the last portion of the desired function.   The position of the taps between the -100- and $+100$-volt sources is adjusted by trial and error.

19-5   *Time-scale Changes*

Often the actual solution time of a differential equation may be so fast that the recorder is unable to follow the response accurately.   Alternatively, the complete solution may take an excessively long time.   These problems are overcome by changing from real time to a machine time used by the computer.   This is accomplished by inserting $t = \tau/a$ in the differential equation.   If $a > 1$, the solution is slowed down.   If $a < 1$, the computer solution is speeded up.   When the change of time scale is inserted in the differential equation, the coefficients of all derivatives change.   This is expressed by the derivative operator relationship

$$\frac{d^n}{dt^n} = a^n \frac{d^n}{d\tau^n} \tag{19-21}$$

The change of time scale is often useful when the coefficients of the higher-order derivatives are much smaller than the coefficients of the lower-order derivatives.   This eliminates the need for very high gains in the operational amplifiers.   Examples 2 and 3 in the next section illustrate the use of a time-scale change.

19-6   *Use of the Analog Computer*

To show the usefulness of the analog computer, some problems are solved by this method.   These examples illustrate the complete computer setup. The problems are those previously solved in this text by the root-locus, polar-plot, and log-plot methods.   This permits a comparison of the results and a correlation of the various methods.

Example 1

The first problem is the unity feedback system (see Fig. 19-25) with the forward transfer function

$$\frac{C(s)}{E(s)} = G_x(s) = \frac{K_1}{s(1 + s)(1 + 0.2s)} \tag{19-22}$$

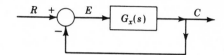

Fig. 19-25 *Basic unity feedback system.*

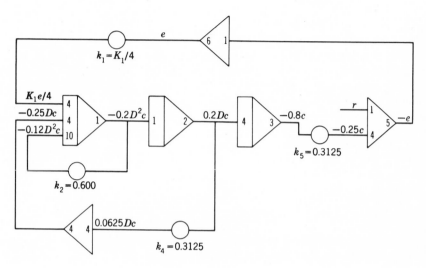

Fig. 19-26 *Detail analog-computer setup for Eqs. (19-24) and (19-25).*

The problem is to adjust the gain K_1 to achieve suitable performance. By the root-locus method the criterion established in Sec. 12-4 is $\zeta = 0.45$. This results in $K_1 = 0.83$, $\omega_n = 0.90$, $T_p = 4.25$ sec, $M_o = 0.20$, and $T_s = 10.0$ sec. In Sec. 13-5, with the log-plot method, the criterion used is $M_m = 1.25$, which corresponds to an effective $\zeta = 0.45$ for an equivalent second-order system. The results are $K_1 = 0.88$ and $\omega_m = 0.72$ radian/sec.

The equations for the unity feedback system obtained by use of Eq. (19-22) are

$$0.2D^3c + 1.2D^2c + Dc = K_1 e \qquad (19\text{-}23)$$
$$r - c = e \qquad (19\text{-}24)$$

Solving Eq. (19-23) for the highest-order derivative gives

$$0.2D^3c = K_1 e - 1.2D^2c - Dc \qquad (19\text{-}25)$$

By starting with Eq. (19-25), the three terms on the right are connected as inputs to integrator 1, as shown in Fig. 19-26. The output of this integrator $(-0.2D^2c)$ is fed to integrator 2 to generate $0.2Dc$. This in turn is the input to integrator 3 with a gain of 4 to generate $-0.8c$. Two of the inputs to integrator 1 are obtained from the outputs of integrators 1 and 2. A sign changer (4) is required to obtain $-0.25Dc$. Equation (19-24) is formed by means of adder 5, using the output of integrator 3 and r,

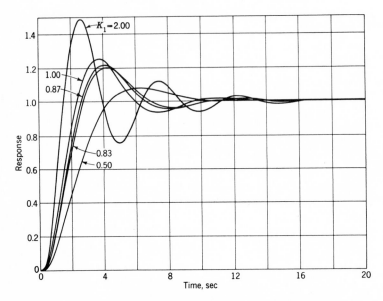

Fig. 19-27 *Response for several values of gain for the unity feedback system with $G(s) = K_1/s(1 + s)(1 + 0.2s)$.*

which is obtained from a separate d-c source. The output of adder 5 is $-e$, which is used to furnish the input to integrator 1. The complete computer-setup diagram showing the actual input and output of each unit is given in Fig. 19-26. The desired value of K_1 is inserted by adjusting the setting of potentiometer k_1. Figure 19-27 shows the response of the system with a step-function input and values of K_1 equal to 0.5, 0.83, 0.87, 1.00, and 2.00. These results show that the peak overshoot varies directly as the gain. A large gain is desired, as this reduces the steady-state error with a ramp input. The requirement for a large gain and a small overshoot are conflicting characteristics. The compromise value of $K_1 = 0.85$, which results in $M_p = 1.22$ and $T_s = 11.5$ sec, seems reasonable for this problem.

Example 2

The second problem uses the system of Example 1 with a cascade passive lead compensator (see Fig. 19-28a) which has the transfer function

$$\frac{E_1(s)}{E(s)} = G_c'(s) = \frac{1 + Ts}{1 + \alpha Ts} = \frac{1 + s}{1 + 0.1s} \qquad (19\text{-}26)$$

The main reason for adding the lead compensator is to improve the time response. This problem is worked by the root-locus method in Sec. 12-6,

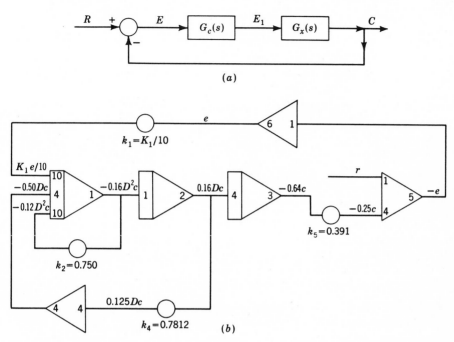

Fig. 19-28 (a) *Basic system with cascade compensator.* (b) *Computer setup based on Eqs. (19-29) and (19-24).*

with $\alpha = 0.1$. With a value $T = 1$ and with $\zeta = 0.45$ for the dominant complex roots, the results are $K_1 = 2.96$, $\omega_n = 3.52$, $T_p = 1.09$ sec, $M_o = 0.19$, and $T_s = 2.53$. The log-plot method of Sec. 13-7, with $M_m = 1.25$ specified, gives the results $K_1 = 3.1$ and $\omega_m = 2.95$ radians/sec.

The new forward transfer function with the lead compensator inserted is now given by

$$G(s) = G_c(s)G_x(s) = \frac{K_1}{s(1 + 0.2s)(1 + 0.1s)} \tag{19-27}$$

The differential equation obtained from this transfer function is

$$0.02D^3c + 0.3D^2c + Dc = K_1e \tag{19-28}$$

The setup of this equation requires the use of large gains in the amplifiers and integrators. This may necessitate the use of an excessive number of operational amplifiers. Therefore a time-scale change $t = \tau/2$ is inserted in Eq. (19-28). Solving for the highest-order derivative term after the insertion of the time-scale change results in

$$0.16D^3c = K_1e - 1.2D^2c - 2Dc \tag{19-29}$$

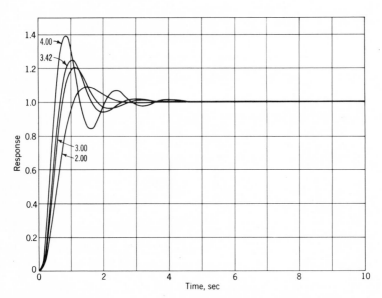

Fig. 19-29 *Response for several values of gain for the lead-compensated unity feedback system with* $G(s) = K_1/s(1 + 0.2s)(1 + 0.1s)$.

The computer setup is shown in Fig. 19-28b. The value of K_1 is inserted by adjusting the setting of the potentiometer k_1. The response to a step input is shown in Fig. 19-29 with several values of gain. Again it is seen that the peak overshoot and the settling time are a direct function of the gain. A gain $K_1 = 3$ results in a peak overshoot of 21 per cent and a settling time of 2.6 sec. A comparison with the results of Example 1 shows that the lead compensator has considerably reduced the settling time.

Example 3

The use of the lag-lead compensator has two effects. The lag portion produces an increase in the static error coefficient, and the lead portion reduces the settling time. The lag-lead compensator used has the transfer function

$$G_c(s) = \frac{(1 + 10s)(1 + s)}{(1 + 100s)(1 + 0.1s)} \tag{19-30}$$

The new forward transfer function using the basic system of Example 1 with a cascade lag-lead compensator is

$$G(s) = \frac{K_1(1 + 10s)}{s(1 + 0.2s)(1 + 100s)(1 + 0.1s)} \tag{19-31}$$

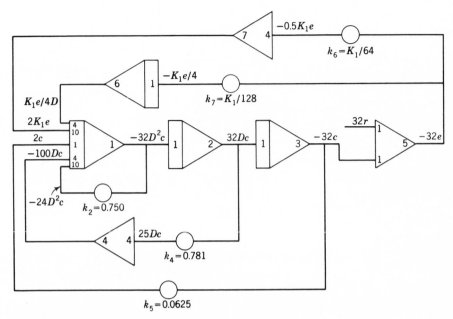

Fig. 19-30　*Computer setup for Eqs. (19-33) and (19-24).*

The differential equation obtained from this transfer function is

$$2D^4c + 30D^3c + 100D^2c + Dc = K_1e + 10K_1\,De \qquad (19\text{-}32)$$

To eliminate the differentiation of the actuating signal e, Eq. (19-32) is integrated. Also, because the coefficient of D^4c is much smaller than that of D^3c, it is appropriate to slow down the problem on the computer by using the time scale $t = \tau/2$. Solving for the highest-order derivative term now gives

$$32D^3c = \frac{K_1e}{D} + 20K_1e - 240D^2c - 400Dc - 2c \qquad (19\text{-}33)$$

The computer setup for this equation is shown in Fig. 19-30. The results with several values of gain are shown in Fig. 19-31. The uncompensated response is also shown for comparison purposes. The settling time is much faster than in the original system. The response time for the lead alone, as determined in Example 2, is essentially the same. But the lag component permits a much larger ramp error coefficient K_1.

Example 4

Chapter 14 covers the synthesis of a control system using a minor loop for compensation. The block diagram is shown in Fig. 19-32. The first

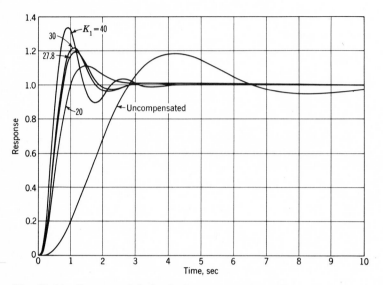

Fig. 19-31   *Response of the lag-lead compensated system.*

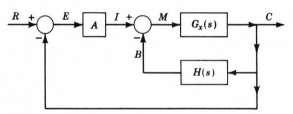

Fig. 19-32   *Basic system with a minor loop for compensation.*

case considered here uses simple tachometer feedback.   The transfer functions are

$$G_x(s) = \frac{K_x}{s(1 + s)(1 + 0.2s)} \tag{19-34}$$

$$H(s) = K_t s \tag{19-35}$$

The main difficulty is that there are three quantities that can be adjusted. These are K_x, K_t, and the amplifier A.   The root-locus, polar-plot, and log-plot methods require a series of trial solutions from which the best is chosen. This may be done with two of these quantities held constant and the third quantity adjusted.   Or the procedure may be based on solving in terms of the products of these quantities for a desired solution.   Of the analytical methods the log-plot method appears a little more versatile than the root-locus or polar-plot method.   In any event, much work is required.   The advantage of the computer is that each parameter may be adjusted inde-

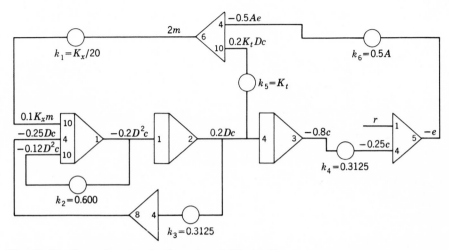

Fig. 19-33 *Computer setup for the system of Fig. 19-32.*

pendently and the result is observed immediately in the plot of the output. Once the system is set up on the computer, many combinations of parameter values may be examined in a much shorter time than can be done by any of the analytical methods. But it is usually desirable to start with an analysis based on one of the methods in Chap. 14. This provides a starting point for the adjustment of the system parameters and permits a check to determine the proper operation of the computer.

The equations set up on the computer to represent this system are

$$r - c = e \tag{19-36}$$
$$Ae = i \tag{19-37}$$
$$i - b = Ae - K_t Dc = m \tag{19-38}$$
$$0.2D^3c = K_x m - 1.2D^2c - Dc \tag{19-39}$$

The complete computer setup is shown in Fig. 19-33. The desired trial values of K_x, K_t, and A are inserted by adjusting potentiometers k_1, k_5, and k_6, respectively.

The use of a passive lead network in cascade with the tachometer in the minor feedback loop is suggested in Chap. 14 as a means of increasing the error coefficient. In this case the transfer function of $H(s)$ is

$$\frac{B(s)}{C(s)} = H(s) = \frac{K_t T s^2}{1 + Ts} \tag{19-40}$$

The differential equation to be set up on the computer to represent $H(s)$ is

$$0.2b = 0.2K_t Dc - \frac{0.2}{T} \frac{b}{D} \tag{19-41}$$

The computer setup for this equation is shown in Fig. 19-34. This setup is substituted for potentiometer k_5 in Fig. 19-33.

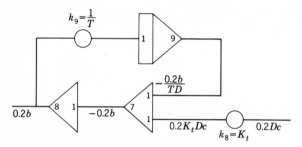

Fig. 19-34   *Computer setup for Eq. (19-41).*

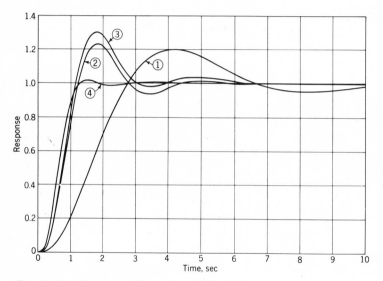

Fig. 19-35   *Response of the system of Fig. 19-32.*
1. *With no compensation*
2. *With tachometer compensation*
3. *Feedback compensation using tachometer and high-pass filter*
4. *Best time response using only tachometer compensator*

Figure 19-35 shows the response of the system with several sets of parameters for a step-function input.   Curve 1 is the response of the original system with no compensation and $K_1 = 0.85$.   Curve 2 is the response using tachometer feedback with the values obtained in the problem of Sec. 14-15 for Case 1.   The values used are $A = 1$, $K_t = 0.344$, $K_x = 10.8$, and $K_1 = 2.29$.   This curve shows the improvement possible with tachometer feedback.   The peak overshoot is essentially the same, but the settling time has been reduced to 4.2 sec.   Curve 3 is the response with both tachometric and lead compensation in the minor feedback loop.   The values used are those determined in Sec. 14-15, Case 3. They are $A = 1$, $K_t = 0.344$, $K_x = 12.1$, $T = 3.81$, and $K_1 = 12.0$. The computer response illustrates the shortcomings of the analytical frequency response method. While the value of K_1 has been increased considerably, the value of settling time has also increased to 5.3 sec. This does not show up in the frequency-response method since the value of ω_m was unchanged. Curve 4 shows the improvement that can be obtained by using tachometric compensation alone. The parameters were varied by trial to produce this result. The values are $A = 1.0$, $K_t = 0.525$, $K_x = 18$, and $K_1 = 1.47$. This is the best response since the peak overshoot is only 1.6 per cent and the settling time is reduced to 1.2 sec. However, the ramp error coefficient is only 1.47. A cascade lag compensator could be included with A to produce the desired value of K_1.

19-7   *Design Problem*

The block diagram of an airplane, representing motion in pitch, is shown in Fig. 19-36. The value θ_{comm} is the command pitch angle of the airplane. The elevator servo controls the position δ_e of the elevator control surfaces. The airplane responds to this elevator position to produce an airplane pitch angle θ. The aircraft motion and the definition of terms are given in Sec. 5-11. Typical transfer functions for an elevator servo and a high-performance aircraft are

$$G_a(s) = \frac{K_x}{s + 10} \tag{19-42}$$

$$G_b(s) = \frac{10(s + 0.4)}{s(s^2 + 0.7s + 5)} \tag{19-43}$$

The root locus for this system is shown in Fig. 19-37. This plot shows that the airplane becomes unstable for very low values of gain. The stabilization problem is to move the locus to the left. A common approach is to insert a rate gyro around the airframe and elevator servo. The block diagram illustrating this additional minor loop is shown in Fig. 19-38.

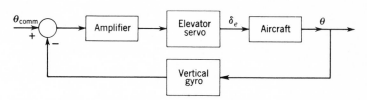

Fig. 19-36 *Pitch axis block diagram of an airplane.*

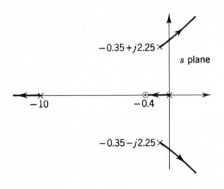

Fig. 19-37 *Root locus of the unmodified airplane (pitch motion).*

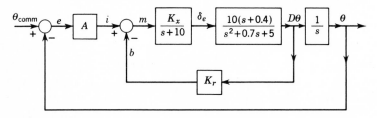

Fig. 19-38 *Pitch axis block diagram of an airplane with rate feedback for stabilization.*

Rate feedback has the effect of moving the poles of the inner loop to the left, as shown by the root locus of the inner loop, drawn in Fig. 19-39. The poles of the inner loop are a function of the product $K_x K_r$. The possible poles are those labeled s_1, s_2, and s_3. These poles show that the inner loop is oscillatory. There is a maximum value of damping ratio that is attainable because of the shape of the curves. The selection of a high gain would result in a low damping ratio. The root locus of the entire system is shown in Fig. 19-40. This root locus is similar in shape to that of the original system, for which the root locus is shown in Fig. 19-37. However, the complex poles have been moved farther to the left. Therefore the result is that larger values of gain are permissible and the oscillatory component of the transient dies out more rapidly. However, there is a real root s_a that produces a transient with a long settling time. The roots s_b and s_c produce a transient that is oscillatory. The root s_d is so far to the left that it has little effect on the performance. This is shown on the computer by setting up the following system equations:

$$\theta_{\text{comm}} - \theta = e \tag{19-44}$$

$$i - b = Ae - K_r\, D\theta = m \tag{19-45}$$

$$D\delta_e = K_x m - 10\delta_e \tag{19-46}$$

$$D^2\theta = 10\delta_e + \frac{4\delta_e}{D} - 0.7\, D\theta - 5\theta \tag{19-47}$$

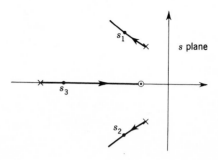

Fig. 19-39 *Root locus for the inner loop of Fig. 19-38.*

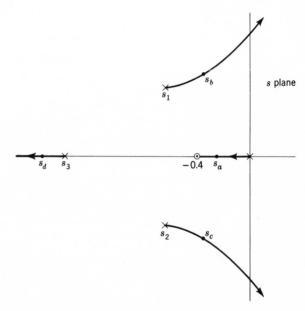

Fig. 19-40 *Over-all root locus of Fig. 19-38.*

The complete computer setup is shown in Fig. 19-41. The desired trial values of K_x, K_r, and A are inserted by adjusting the values of potentiometers k_{15}, k_{16}, and k_{17}. Figure 19-42 shows the results obtained on the computer with two sets of values for these quantities. Case 1 has $K_x = 2.8$, $K_r = 4.5$, and $A = 13$. Case 2 has $K_x = 9.0$, $K_r = 4.5$, and $A = 13$. Both cases show the presence of the long time constant. This performance is marginally acceptable and is typical of high-performance supersonic air-craft. More compensation must be added if the performance is to be improved. One possibility is to improve the elevator-servo performance. This example uses a servo time constant of 0.1 sec, but a time constant of 0.05 can be achieved. This moves the real root to the left to speed up the

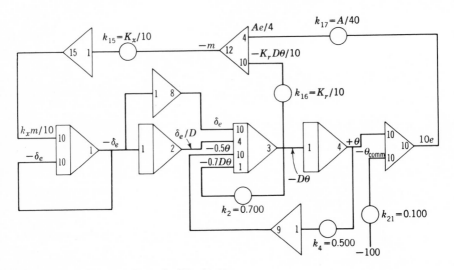

Fig. 19-41   *Computer setup for Fig. 19-38.*

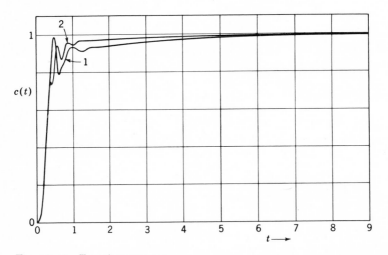

Fig. 19-42   *Transient response.*

response.   Also, the introduction of a pole on the negative real axis can improve the time response.   This is seen by referring to the manner in which the root locus is reshaped by the addition of a cascade compensator of the form

$$G_c(s) \;=\; \frac{1}{s+b} \tag{19-48}$$

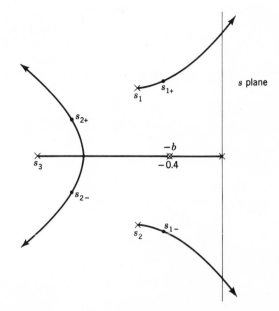

Fig. 19-43 *Over-all root locus with suggested compensation.*

Figure 19-43 shows the modified root locus with this compensator inserted in cascade with the amplifier A. The possible roots obtained from this figure are shown as s_{1+}, s_{1-}, s_{2+}, and s_{2-}. The real root s_a in Fig. 19-40 has been replaced by two roots which are much farther to the left. The transients should therefore die out faster, with a resulting improvement in the response. The details of adding this or other possible compensators are left as a problem for the reader.

19-8   *Conclusions*

This chapter presents the basic components of the electronic analog computer, which are used in several examples to demonstrate the computer setup representing a set of simultaneous equations. The response of the analog setup is the same as the response of the system. The adjustment of any individual parameter changes the response of the system. Since the adjustment of a parameter requires only the setting of a potentiometer, it is possible to make several tests in a relatively short time. Rather than set up the computer and make a large number of runs to determine an optimum response, it is preferable to make an analytical study first. Then a desirable range of values for the parameters is known.   This serves the double purpose of limiting the computer runs needed and providing a check on the correctness of the computer response.   Also, several nonlinearities can be approximated on the computer.   The response of the system with nonlinearities is therefore readily obtained, whereas the analytical solution may be tedious to secure.

Bibliography

1. Johnson, C. L.: "Analog Computer Techniques," 2d ed., McGraw-Hill Book Company, New York, 1963.
2. Karplus, W. J., and W. W. Soroka: "Analog Methods: Computation and Simulation," 2d ed., McGraw-Hill Book Company, New York, 1959.
3. Korn, G. A., and T. M. Korn: "Electronic Analog Computers," 2d ed., McGraw-Hill Book Company, New York, 1956.
4. "Fundamentals of Design of Piloted Aircraft Flight Control Systems," BuAer Report AE-61-4, vol. 1, U.S. Navy, Bureau of Aeronautics, 1952.
5. Seely, S.: "Electron-tube Circuits," chap. 8, McGraw-Hill Book Company, New York, 1958.
6. Vickers, D. B.: "Feedback System Analysis Using Frequency Response Curves from an Analog Computer," M.Sc. Thesis, Air Force Institute of Technology, Dayton, Ohio, 1960.
7. Ogar, G. W.: Obtaining the Frequency Response of Physical Systems by Analog Computer Techniques, *IRE Intern. Conv. Record*, vol. 9, 1961.
8. Dorrity, J. L.: "Frequency Response Program," Air Force Institute of Technology, Dayton, Ohio, 1964.
9. Paskin, H. M.: "Automatic Computation of Root Loci Using a Digital Computer," M.Sc. Thesis, Air Force Institute of Technology, Dayton, Ohio, March, 1962.
10. Liethen, F. E., C. H. Houpis, and J. J. D'Azzo: An Automatic Root Locus Plotter Using an Analog Computer, *Trans. AIEE*, vol. 79, pt. II, pp. 523–527, January, 1961.
11. Dunsmore, C. L.: Computer Analogs for Common Nonlinearities, *Control Eng.*, vol. 6, pp. 109–111, October, 1959.

20

Experimental and
Design Procedures

20-1   Introduction

This text presents various methods of analysis and synthesis of a basic feedback control system.   Compensation techniques that improve the system performance are presented for each of the methods.   In the presentation of this material the examples use the same particular basic system, which permits a comparison of both the work required by each method and the performance achieved.   It is the purpose of this chapter to unify this material by incorporating it all here in an analysis of the same basic system.

659

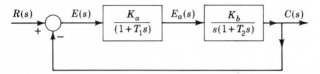

Fig. 20-1   *Form of Type 1 control system under consideration.*

20-2   *Statement of the Problem*

A basic unity feedback Type 1 control system has been designed analytically and constructed, as shown in Fig. 20-1. It is necessary to determine the constants and the performance of the basic system. The system, with compensation if required, must meet the following specifications:

$$K_1 = 30 \qquad T_s = 2.5 \text{ sec}$$
$$M_p = 1.20 \qquad T_p = 1.0 \text{ sec}$$

20-3   *Determination of System Constants by the Frequency-response Method*[1,5,6]

Two experimental methods that can be utilized to determine the constants are steady-state frequency-response and transient testing measurements. The frequency-response method requires the application of a steady-state sinusoidal signal to the input and the measurement of the magnitude of the input and output quantities and the phase relationship between them under open-loop conditions. This yields data for a plot of Lm $G(j\omega)$ versus ω. Instruments such as the Servoscope,* Solartron,† or a low-frequency oscillator may be used to obtain these data in conjunction with recorders or oscilloscopes. These instruments can be readily used in control systems where the input and output are electrical quantities. Both the Servoscope and the Solartron can be used for both d-c and a-c systems. Where either or both quantities are not electrical in nature, proper units must be designed and inserted at the input and the output to obtain a sinusoidal variation of the input and to measure these quantities. Under open-loop operation there may be a drift of the output quantity. If this occurs, the system must be operated closed-loop but still measure $C(j\omega)$ and $E(j\omega)$. The frequency-

* Servo Corporation of America: Servoscope.
† Solartron, Inc.: Transfer-function analyzer.

response method yields the parameters for a transfer function having the form given by Eq. (9-14).

The data secured by any of the means suggested in the last paragraph are used to plot the exact log magnitude and angle diagram. Asymptotes are drawn on the exact log magnitude curve by utilizing the fact that they must be multiples of ± 20 db/decade.

These asymptotes are determined by correlating the phase curve with the log magnitude curve. A suggested procedure follows:

1. From the low-frequency portion of the curves, determine the value of n for the $(j\omega)^n$ term of Eq. (9-14).
2. From the high-frequency portion of the curves, determine the value of $v - w$, the number of poles minus the number of zeros of $G(j\omega)$. This value yields the minimum number of poles that the transfer function must have.
3. From the mid-frequency portion of the curves, estimate the probable number of zeros present in the transfer function. With this value of w and with the information determined from step 2, estimate the number of poles $v = n + u$.

Steps 2 and 3 require a trial-and-error location of the poles and zeros, which must be located so that the synthesized transfer function fits the experimental curve with sufficient accuracy. From the final equation the system type and the approximate time constants are determined. In this manner the transfer function of the system is synthesized.

Care must be exercised in determining whether any poles or zeros of the transfer function are in the right-half s plane. As an example, consider the functions $1 + j\omega T$ and $1 - j\omega T$. The log magnitude plots of these functions are identical, but the angle diagram for the former goes from 0 to 90° whereas for the latter it goes from 0 to $-90°$. Most practical systems are in the minimum-phase category.

The experimental data utilized to obtain the log magnitude and phase diagrams can also be used to obtain a polar plot for the system whose transfer function is not known. Such a method is presented in the literature.[6]

The frequency response of a system may also be obtained from the system's transient response to an impulse signal by evaluating the Fourier integral, which has the form of Eq. (9-1).[1] Graphical or numerical methods must be applied to evaluate this integral because the equation representing the time response is not known. Either the log magnitude and phase angle plots or the polar-plot methods mentioned previously can then be applied to determine the system's transfer function.

20-4   *Determination of System Constants by Transient Testing Methods*[1,7,8]

The transient testing method requires a theoretical analysis of the system to determine the number of poles and zeros involved and possibly their anticipated general location in the s plane.   From this analysis it is possible to synthesize the form of the transfer function for each component, a group of components in cascade, and the control ratio of the system.   Whether transient testing is performed on each component, on a group of components in cascade, or on the entire closed-loop control system is dependent upon the following factors:

1.   The availability of input and output points for instrumentation.
2.   If the synthesized transfer function fits one of the known forms, the transient testing data can be utilized to determine its constants. If the synthesized function does not fit one of the known forms, one of the frequency-response methods of the last section must be used.
3.   If the control ratio fits one of the known forms, the effect of small-system nonlinearities may be taken into account.   That is, the transient testing data of the closed-loop system may be utilized to obtain an approximate linear transfer function over a limited range of system operation.

To each unit of the system, which has a transfer function of the form

$$\frac{K}{1 + Ts}$$

a step input is applied and the output response is recorded (see Fig. 20-2). From the definition of time constant given in Sec. 3-7, the point at which the output has reached 63.2 per cent of its final value yields the value of T. Another method is to extend the initial slope of the output response, $[dc/dt]_{t=0^+}$, until it intersects the line representing the steady-state value. The point of intersection yields the value of T.   Table 20-1 illustrates the manner in which the time constant can be determined for other forms of transfer functions containing a single-order time constant.

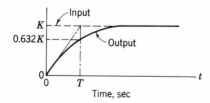

Fig. 20-2   Transient response to a step input of the transfer function $K/(1 + Ts)$.

Table 20-1. *Time response for a unit step input for typical transfer functions involving a single-order time constant*

Transfer function	Time response for a unit-step input
$\dfrac{K\,Ts}{1+Ts}$	
$\dfrac{K(1+Ts)}{(1+\alpha Ts)}$ $\alpha > 1.0$	
$K\,\dfrac{(1+Ts)}{(1+\alpha Ts)}$ $\alpha < 1.0$	
$\dfrac{K}{s(1+Ts)}$	

The time constants of a unit that has the transfer function

$$G(s) = \frac{1}{(s + 1/T_1)(s + 1/T_2)} \tag{20-1}$$

where $T_1 > T_2$, can be experimentally determined by utilizing either a step or an impulse input function. For a step input the output is given by

$$c(t) = A_0 - A_1 e^{-t/T_1} + A_2 e^{-t/T_2} \tag{20-2}$$

The experimental response of the above equation is shown in Fig. 20-3. The final value

$$c(t)_{ss} = A_0 \tag{20-3}$$

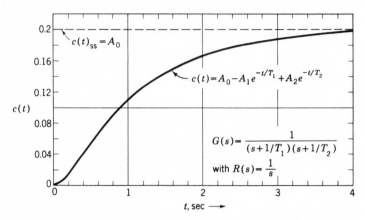

Fig. 20-3   *Plot of Eq. (20-2).*

can be graphically determined as shown in this figure. From this figure, data can then be obtained to plot the transient component

$$A_0 - c(t) = A_1 e^{-t/T_1} - A_2 e^{-t/T_2} \tag{20-4}$$

versus time on semilog paper, as shown in Fig. 20-4. The graphical determination of T_1 and T_2 requires that the two time constants be *appreciably different* so that the two exponential terms will each decay at significantly different rates. If this condition is satisfied, then for $t \gg T_2$ the component $A_2 e^{-t/T_2}$ becomes insignificant and Eq. (20-4) reduces to

$$A_0 - c(t) \approx A_1 e^{-t/T_1} \tag{20-5}$$

Since the plot of an exponential term on semilog paper is a straight line, this permits the graphical construction of the straight line that represents $A_1 e^{-t/T_1}$, as shown in Fig. 20-4. This straight line coincides with the actual plot $A_0 - c(t)$ for $t > t_1$, as shown on the graph.

Rearranging Eq. (20-4) yields

$$A_2 e^{-t/T_2} = A_1 e^{-t/T_1} - [A_0 - c(t)] \tag{20-6}$$

Data can now be obtained from Fig. 20-4 to plot $A_2 e^{-t/T_2}$ versus time, which is also a straight line. The vertical intercepts of these straight lines at $t = 0$ are the values A_1 and A_2. From the straight lines that represent Eqs. (20-5) and (20-6) the values T_1 and T_2 are readily determined. For the particular transfer function utilized for these curves the values are

$$\begin{array}{lll} A_0 = 0.2 & A_2 = 0.05 & T_2 = 0.2 \\ A_1 = 0.25 & T_1 = 1 \end{array}$$

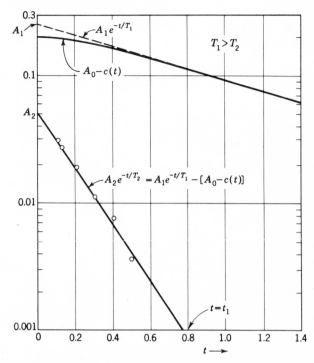

Fig. 20-4   *Evaluation of the transient parameters of Eq. (20-2).*

The same procedure can be used with an impulse input.   For the same transfer function the output is now given by

$$c(t) = Ae^{-t/T_1} - Ae^{-t/T_2} \tag{20-7}$$

The experimental response of Eq. (20-7) is plotted on semilog paper in Fig. 20-5.   If the time constants are appreciably different, the constants A, T_1, and T_2 can be determined graphically.   Note, as shown in the figure, that for the impulse function the two straight lines start at the same point for $t = 0$.   This is so since the coefficient A is the same for both exponential terms.   This was not the case for the step-input method.

By reference to Figs. 3-11 and 20-6 the parameters of the following quadratic transfer functions can be determined:[2,3]

$$\frac{1}{s^2/\omega_n^2 + (2\zeta/\omega_n)s + 1} \tag{20-8}$$

$$\frac{s}{s^2/\omega_n^2 + (2\zeta/\omega_n)s + 1} \tag{20-9}$$

$$\frac{1}{s[s^2/\omega_n^2 + (2\zeta/\omega_n)s + 1]} \tag{20-10}$$

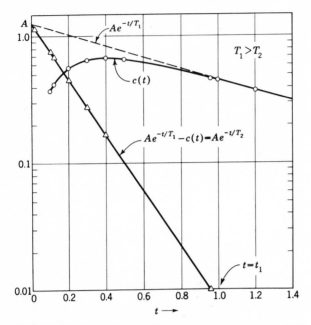

Fig. 20-5 *Evaluation of the transient parameters of Eq.*
(20-7).

To use Fig. 3-11 the amount of overshoot M_o must be determined. In determining the damping ratio ζ from these figures one must exercise caution, as these curves are approximations. From the actual time response the damped natural frequency ω_d is determined by measuring the period T of one oscillation ($\omega_d = 2\pi/T$). With these values, the undamped natural frequency ω_n is given by

$$\omega_n = \frac{\omega_d}{\sqrt{1 - \zeta^2}} \tag{20-11}$$

It is possible to determine, by transient testing, a transfer function that has the form

$$G(s) = \frac{K}{(s + b)(s^2 + 2\zeta\omega_n s + \omega_n^2)} \qquad \text{for } \zeta < 1 \tag{20-12}$$

if $b < \zeta\omega_n$. For a step input the time response is of the form

$$c(t) = A_a - A_b e^{-bt} + A_c e^{-\zeta\omega t} \sin(\omega_d t - \phi) \tag{20-13}$$

As a result of the condition imposed upon b, the damped sinusoid term dies out and becomes negligible before the exponential term, as shown in Fig. 20-7. Thus

$$c(t) \approx A_a - A_b e^{-bt} \qquad \text{for } t \geq t_1 \tag{20-14}$$

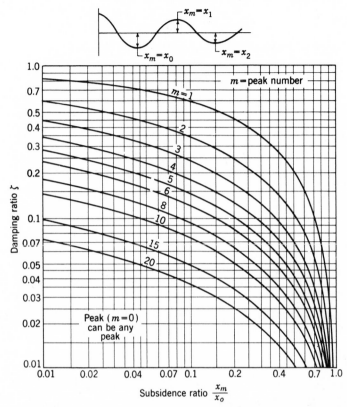

Fig. 20-6 *Damping ratio of oscillatory transients as a function of subsidence ratio for second-order systems. The first or reference peak x_0 ($m = 0$) can be any peak, and m represents the number of the peak that is being compared with the reference peak.*

$$\frac{s}{s^2/\omega_n^2 + (2\zeta/\omega_n)s + 1} \qquad \frac{1}{s^2/\omega_n^2 + (2\zeta/\omega_n)s + 1}$$

$$\frac{1}{s[s^2/\omega_n^2 + (2\zeta/\omega_n)s + 1]}$$

The graphical determination of A_b and b requires that for the time period $t_1 \leq t < t_2$ the exponential term still have an appreciable value. This permits a sufficient number of values to be obtained for $A_b e^{-bt}$, between t_1 and t_2 from Fig. 20-7, to obtain an accurate plot on semilog paper. Thus the values of A_b and b may be determined in the manner described previously. Once these values have been obtained, it is possible to obtain sufficient values of

$$A_c e^{-\zeta\omega_n t} \sin (\omega_d t - \phi) = c(t) - A_a + A_b e^{-bt} \qquad (20\text{-}15)$$

to plot it on regular graph paper. From this plot it is possible to determine

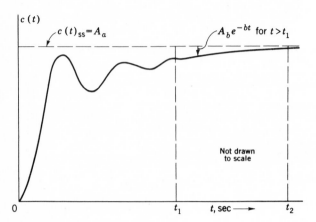

Fig. 20-7   *Plot of Eq. (20-13).*

the values of ζ and ω_n in the manner described previously in conjunction with Fig. 20-6.

In performing the transient measurements the following items must be considered:

1. The method for obtaining a step input signal.
2. The magnitude of the signal used must not cause the equipment to operate in the nonlinear region.
3. The signal-source and measuring-equipment impedances may modify the response.   Thus they must be taken into account.

20-5   *Evaluation of G(s) from the Closed-loop Transient Response*

If the transfer function derived by transient testing represents the control ratio $M(s)$ for a unity feedback control system, the system's forward transfer function $G(s)$ may be determined.   Since

$$M(s) = \frac{C(s)}{R(s)} = \frac{G(s)}{1 + G(s)} \qquad (20\text{-}16)$$

then

$$G(s) = \frac{M(s)}{1 - M(s)} \qquad (20\text{-}17)$$

As an example, if

$$M(s) = \frac{K}{s^2 + as + b}$$

where K, a, and b have been evaluated by the techniques of the previous sections, then

$$G(s) = \frac{K}{s^2 + as + (b - K)}$$

If $M(s)$ has the form of Eq. (20-12), the denominator of Eq. (20-17) is a third-degree polynomial. This polynomial can be factored to determine its roots.

Another approach to obtaining the forward transfer function $G(s)$ is to obtain the closed-loop transient response for a step input for various values of forward-loop gain. The $G(s)$ function of a unity feedback system may be determined by the root-locus method if it has one of the following forms:

$$G(s) = \frac{K}{s(s + a)} \tag{20-18}$$

$$G(s) = \frac{K}{s(s + a)(s + b)} \tag{20-19}$$

$$G(s) = \frac{K}{(s + a)(s^2 + 2\zeta\omega_n s + \omega_n^2)} \quad \text{for } \zeta < 1 \tag{20-20}$$

This method requires that the gain be varied between the values of $K_{\min}$ and $K_{\max}$, where

1. $K_{\min}$ is the value that just makes the response recognizably underdamped
2. $K_{\max}$ is the value that yields a peak value M_p. This peak value must not overstress the system and is dependent upon the components of the system.

Additional restrictions are:

3. For Eq. (20-19), where $a > b > 0$, the value of a must be sufficiently greater than the value of b that the root locus, for the lower portion of the gain range, is not appreciably affected by the pole at $s = -a$. This means that the corresponding root produced by the pole at $s = -a$ has no effect upon $c(t)$ for $t \geq t_0$ (the duplicating time).
4. For Eq. (20-20) the value of a must be sufficiently greater than $\zeta\omega_n$ so that, for $K_{\max} > K > 0$, it has no effect upon $c(t)$ for $t \geq t_0$.

This method is best described by the following example. A unity feedback system has a forward transfer function of the form of Eq. (20-19). Figure 20-8 shows the relative locations of the poles of $G(s)$. The value of a is such that it cannot be accurately determined by an open-loop test because it is much larger than b. For each time response obtained in the range of $K_{\min} < K < K_{\max}$, the dominant roots of the characteristic equation are determined by reference to Figs. 3-11 and 20-6. The dominant roots for

Fig. 20-8 *Pole-zero diagram of Eq. (20-19).*

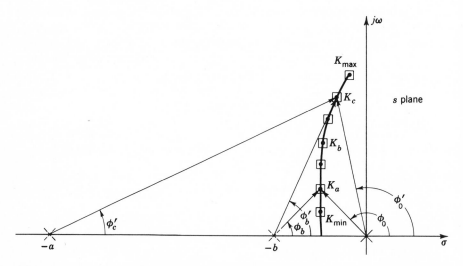

Fig. 20-9 *Experimentally determined root locus of a unity feedback system with a forward transfer function given by Eq. (20-19).*

each value of K are plotted in the s plane, as shown in Fig. 20-9.   These plotted roots yield portions of two branches of the root locus of the system. It is noted that the dominant roots for $K_{\min} < K < K_b$ all lie on a straight line perpendicular to the real axis.   This *straight line* implies that the angular contribution of the pole at $-a$ is negligible for this range of values for K. Thus for $K = K_a$, the angular condition is given by

$$\phi_0 + \phi_b = 180°$$

Since ϕ_0 can be measured from the graph, the angle ϕ_b is calculated from

$$\phi_b = 180° - \phi_0$$

Thus, by knowing ϕ_0, the pole $s = -b$ is located by use of the Spirule.   For $K_b < K < K_{\max}$, it is noted that the root locus bends to the right.   Thus for $K > K_b$ the angular contribution of the pole at $-a$ is now effective.   For $K = K_c$ the values of ϕ_0' and ϕ_b' are known.   Therefore the pole at $-a$ is located in order to satisfy the angular condition

$$\phi_0' + \phi_a' + \phi_b' = 180°$$

The transfer functions of Eqs. (20-18) and (20-20) can be determined in a similar manner.

The advantages of closed-loop transient testing are:

1. For systems that have small nonlinearities an equivalent linear forward transfer function may be derived for a specified range of values of K.
2. Consider those unity feedback systems whose forward transfer functions have a form more complex than the simple ones discussed in this section. It is possible that a complex $G(s)$ may produce a response which is essentially identical to a closed-loop output response of a system having a simple $G(s)$. In these situations the complex $G(s)$ may be replaced by an "equivalent" simple $G(s)$ for a specified range of values of K. These "equivalent" $G(s)$'s permit a simplified mathematical analysis to be performed on the system.

20-6   Analysis of the Basic System

By use of the methods indicated in the preceding section the forward transfer function of the basic system is experimentally determined to be

$$G_1(s) = \frac{C(s)}{E(s)} = \frac{0.83}{s(s+1)(0.2s+1)} \tag{20-21}$$

When the system parameters have been obtained, the next step is the determination of the performance of the basic control system. The choice that must be made at this point is whether to analyze the system by the root-locus or the logarithmic-plot approach. In general, the complexity of the control system dictates which of the methods is to be used. If the system is fairly simple (containing one or no minor loops, of the type discussed in Chap. 14) and if the characteristic equation is of the fifth order or less, one may be inclined to use the root-locus approach. This is so since it yields the roots of the characteristic equation and thus gives the time response of the output. For those systems that are more complicated than the above (of the type discussed in Chap. 15) one may be inclined to use the logarithmic approach. This is so because of the comparative ease of using the straight-line asymptotes or the templates in conjunction with the M-α contours (overlay) to obtain the M_m and ω_m of $C(j\omega)/R(j\omega)$. If overlays are not available, the polar plot can readily be obtained from the log plots.

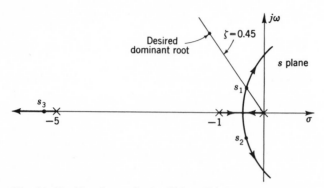

Fig. 20-10   *Root-locus plot for* $G(s) = K/s(s + 1)(s + 5)$.

The degree of correlation of the frequency response with the time response is less than that with the root locus. But this may not be objectionable when considering the time element involved in the analysis. Also, with the availability of a computer the results of either method can serve as a check in informing the designer that the programming and operation of the computer are correct. In other words, the logarithmic results permit the designer to know whether the results of the computer are in the "right ball park," and the computer provides the final adjustment of the system.

For the basic control system under consideration both methods of analysis are utilized for completeness' sake.

Since M_p is specified as 1.20, the *effective* damping ratio of the closed-loop response is determined by use of Eq. (3-87):

$$c_p = 1 + e^{-\zeta\pi/\sqrt{1-\zeta^2}} = 1.20 \tag{20-22}$$
$$\zeta = 0.45 \tag{20-23}$$

The root-locus plot of this control system is obtained by the method outlined in Chap. 7 and is shown in Fig. 20-10. For $\zeta = 0.45$ the following data are obtained:

> *Dominant Roots:* $s_{1,2} = -0.40 \pm j0.81$
> *Static Loop Sensitivity:* $K = 4.17$
> *Third Root:* $s_3 = -5.19$ (*Note:* The dominant roots having been found from the root locus, the third root can be determined mathematically without resorting to the use of the magnitude condition. Dividing the third-order characteristic polynomial by the dominant quadratic yields the third root.)

Undamped Natural
   Frequency: $\omega_n = 0.90$ radian/sec

Ramp Error Coefficient: $K_1 = \dfrac{K}{5} = 0.83$ sec^{-1}

Peak Time: $T_p = 4.25$ sec
Settling Time: $T_s = 10.0$ sec
Peak Overshoot: $M_p = 1.20$

By using the value $\zeta = 0.45$, the required M_m value is

$$M_m = \frac{1}{2\zeta\sqrt{1-\zeta^2}} = \frac{1}{0.9\sqrt{1-0.202}} = 1.25 \qquad (20\text{-}24)$$

Utilizing the specified settling time, the desired undamped frequency is

$$\omega_n = \frac{4.0}{\zeta T_s} = \frac{4.0}{(0.45)(2.5)} = 3.56 \text{ radians/sec} \qquad (20\text{-}25)$$

This yields a desired effective resonant frequency

$$\omega_m = \omega_n \sqrt{1 - 2\zeta^2} = 3.56\sqrt{1 - (2)(0.202)} = 2.74 \text{ radians/sec} \quad (20\text{-}26)$$

The log magnitude and phase diagrams of the basic system are shown in Fig. 20-11.   From this figure the log magnitude–angle diagram is plotted in Fig. 20-12.   By use of an M-α overlay the following values are obtained from the latter figure:

$$M_{m1} \approx 1.26 \qquad \omega_{m1} \approx 0.70 \text{ radian/sec}$$

From Fig. 20-12 the following values are obtained:

$$\gamma = +48° \qquad \omega_{\phi 1} = 0.66 \text{ radian/sec}$$

A comparison of the original system performance with the specifications shows the following:

1.   The ramp error coefficient K_1 must be increased by a factor of approximately 36.
2.   The peak overshoot M_p is satisfactory.
3.   The maximum frequency response M_m is satisfactory.   This is also a measure of the peak overshoot and confirms the fact that M_p is satisfactory.
4.   The peak time must be reduced by a factor of approximately 5.
5.   The settling time must be decreased by a factor of approximately 4. Therefore the poles of the closed-loop system must be moved to the left on the s plane.
6.   The value of the resonant frequency ω_m (or the phase-margin frequency ω_ϕ) is a measure of the settling time.   Therefore this value must be increased while maintaining the value of M_m.   Since the phase margin γ is indicative of M_m, it may also be used as a criterion in the design.

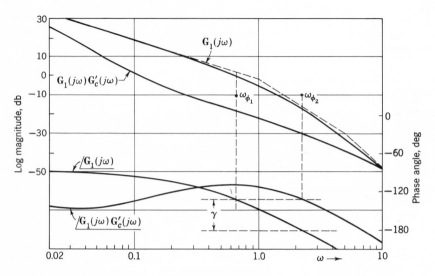

Fig. 20-11 *Log magnitude and phase diagram of*

$$G_1(j\omega) = \frac{0.83}{j\omega(1 + j\omega)(1 + j0.2\omega)}$$

$$G_1(j\omega)G_c'(j\omega) = \frac{0.83(1 + j10\omega)}{j\omega(1 + j0.2\omega)(1 + j100\omega)(1 + j0.1\omega)}$$

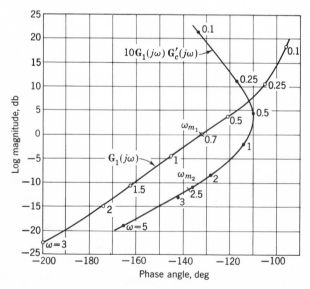

Fig. 20-12 *Log magnitude–angle diagrams for the original and the compensated systems.*

The results of the root-locus and frequency-response methods are complementary. They show the same performance in terms of different parameters.

20-7 Selection of a Cascade Compensator

Since both the ramp error coefficient and the settling time (or the resonant frequency) must be improved, this rules out immediately the use of a simple cascade lag compensator. Also, a simple cascade lead network alone cannot be used, although it results in an improvement of both of these factors. This is so since the lead network results in a smaller improvement in K_1 than can be obtained by the use of a lag network. The lag compensator results in an improvement of K_1 by a factor somewhat less than the value of $\alpha \leq 10$. Thus the next compensator to consider is a cascade lag-lead network, which both improves T_s and gives an increase in K_1 greater than that of a lag network. From the above analysis it seems that the use of a lag-lead compensator may yield the desired improvements. The lag-lead compensator has the form

$$G_c(s) = A \frac{(s + 1/T_1)(s + 1/T_2)}{(s + 1/\alpha T_1)(s + \alpha/T_2)} \tag{20-27}$$

Root-locus Method

To determine the location of the poles and zeros of the compensator, the following is known:

1. To obtain a settling time $T_s = 2.5$ sec, the real part of the desired dominant roots must be equal to

$$|\sigma| = \frac{4}{T_s} = \frac{4}{2.5} = 1.6$$

2. Since an M_o, and in turn the effective ζ, is specified, the above value of σ must lie on the radial line representing $\zeta = 0.45$. The desired dominant roots therefore are $s_{1,2} = -1.6 \pm j3.1$.

3. The lag portion of the compensator must be chosen so that it contributes a small angle at the location of the desired roots. The rules stated in Chap. 12 can be utilized to accomplish this. By using a value $\alpha = 10$, the values $1/T_1 = 0.1$ and $1/\alpha T_1 = 0.01$ are therefore selected.

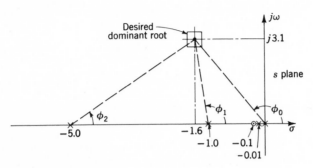

Fig. 20-13 *Plot of known data necessary to determine location of the pole and zero of the lead portion of the lag-lead compensation.*

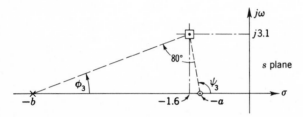

Fig. 20-14 *A pole-zero combination for the lead portion of a lag-lead compensation.*

Figure 20-13 shows the original poles of the system, the lag pole and zero, and the desired dominant root. For $(-1.6 + j3.1)$ to be a point on the root locus, the following angle condition must be satisfied:

$$(\phi_0 + \phi_1 + \phi_2) + (\phi_3 - \psi_3) = (1 + 2m)180° \qquad (20\text{-}28)$$

From Fig. 20-13 it is determined that the necessary angular contribution of the lead portion is

$$\phi_3 - \psi_3 = -80° \qquad (20\text{-}29)$$

Thus a pole-zero combination must be found such that Eq. (20-29) is satisfied. Figure 20-14 shows the necessary locations. The ratio of the zero and pole must satisfy the relationship

$$\frac{a}{b} = \frac{1/T_2}{\alpha/T_2} = \frac{1}{\alpha} = 0.1 \qquad (20\text{-}30)$$

A location satisfying this requirement is

$$T_2 = 1 \qquad \text{and} \qquad \frac{\alpha}{T_2} = 10$$

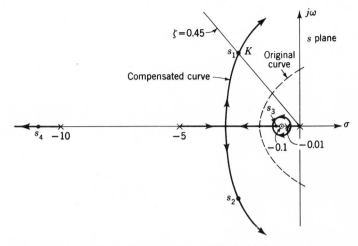

Fig. 20-15 *Root locus of* $G(s) = \dfrac{K(s + 0.10)}{s(s + 0.01)(s + 5)(s + 10)}$

The resulting compensator transfer function is

$$G_c(s) = A \frac{(s + 0.1)(s + 1.0)}{(s + 0.01)(s + 10)} \qquad (20\text{-}31)$$

The root locus of the compensated system is shown in Fig. 20-15 and yields the following information:

Dominant Roots:	$s_{1,2} = -1.55 \pm j3.08$
Other Roots:	$s_3 = -0.103$
	$s_4 = -11.81$
Ramp Error Coefficient:	$K_1 = 29.0 \text{ sec}^{-1}$
Peak Time:	$T_p = 1.08 \text{ sec}$
Settling Time:	$T_s = 2.58 \text{ sec}$
Peak Overshoot:	$M_p = 1.203$

When these results are compared with the desired specifications of $K_1 = 30$, $M_p = 1.20$, $T_p = 1.0$, and $T_s = 2.5$, it is seen that very good argeement exists. Thus no further adjustment is necessary. Generally, such close agreement may not result on the first trial. Further adjustments may be necessary to achieve the desired specifications. As a practical rule, it is not necessary to determine the complete root locus of the compensated system. This is true if only one pair of complex conjugate roots exists. Once the dominant roots have been located, the other roots, if they are all real, can be determined by applying the magnitude condition along the real axis until all roots have been accounted for. In searching the real axis, if not all

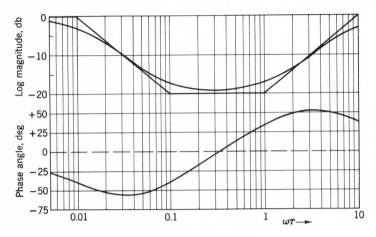

Fig. 20-16 *Log magnitude and phase diagram for the lag-lead compensa-*
tion $\mathbf{G}_c(j\omega) = \dfrac{(1 + j10\omega T)(1 + j\omega T)}{(1 + j100\omega T)(1 + j0.1\omega T)}$

the roots are accounted for, the remainder are complex.   Thus it is necessary to determine those portions of the root locus where they must exist.   In this problem, since the compensator is chosen to force the locus through the desired dominant roots $s_{1,2}$, it is not necessary, at this point in the design, to obtain the complete root locus.   The next step is to search the real axis for the remaining two roots by applying the magnitude conditions.   This results in locating two real roots s_3 and s_4.

Log-plot Method

The problem may also be solved by the log-plot method.   Since the desired M_m is approximately equal to M_{m1} the problem resolves to increasing ω_m (or ω_ϕ) by maintaining $\gamma = +48°$.   Remember that the phase-margin angle is indicative of the value of M_m.   An approximate value of ω_ϕ to strive for is

$$\frac{\omega_\phi}{\omega_{\phi 1}} = \frac{\omega_m}{\omega_{m 1}} \tag{20-32}$$

or $$\omega_\phi = \frac{\omega_m}{\omega_{m 1}} \omega_{\phi 1} = \frac{2.74}{0.70} 0.66 = 2.58 \text{ radians/sec}$$

For this value of frequency the angle of the original transfer function is approximately $-188°$ (see Fig. 20-11).   To maintain the same phase margin, the compensator must be chosen to give $\phi_c = +56°$ at this frequency.   By referring to Fig. 20-16, it is seen that this value of ϕ_c is slightly larger than the maximum positive angle of the lag-lead compensator having the form

$$G_c(j\omega) = A \frac{(1 + j10\omega T)(1 + j\omega T)}{(1 + j100\omega T)(1 + j0.1\omega T)} \tag{20-33}$$

From this figure the angle $\phi_{c,\max} \approx 52°$ occurs at

$$T = \frac{3.00}{\omega} = \frac{3.00}{2.58} \approx 1.16 \text{ sec}$$

For simplicity and for comparison purposes a value $T = 1$ sec is chosen. With this choice the Lm G_1G_c' and $\phi = \phi_1 + \phi_c$ versus ω plots are drawn in Fig. 20-11. To maintain the same phase margin of 48° (approximately the same M_m), a gain $A = 3.16$ can be inserted. This yields a new K_1 of approximately 26.2, which is a little less than the desired amount. To obtain more accurate results, the Lm–angle diagram of the compensated system is illustrated in Fig. 20-12. For ease in using the overlays a fixed gain of 20 db is added to Lm G_1G_c'. By using this curve, the following results are obtained for the compensated system:

$$M_m = 1.25$$
$$\omega_m = 2.6$$
$$K_1 = 29$$

Based upon the desired value of $K_1 = 30$ and the approximated value of $\omega_m = 2.74$, the above results are low. Taking into account the degree of accuracy one can expect with the frequency-response methods of analysis, these values are acceptable check points for a computer study.

Computer Study

The final decision on the suitability of the system performance should be based on a computer study if a computer is available. The computer setup is shown in Fig. 19-30. Figure 20-17 shows the plot of the response with a step-function input. The response is shown for both the original system and the compensated system, with the two values of gain determined by the root-locus and log-plot methods. They show a considerable improvement of the response of the original system.

Comparison of Results

In Table 20-2 a comparison is made of the values obtained by the root-locus method, the computer results, and the desired specifications. It is seen that a very favorable correlation exists between the root locus, computer values, and the desired specifications.

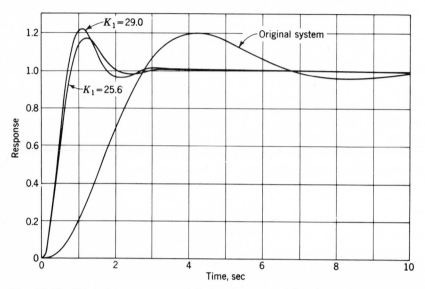

Fig. 20-17   *Computer responses—original and compensated systems.*

Table 20-2.   Results of the system utilizing the compensator
$G_c(s) = A[(1 + 10s)(1 + s)/(1 + 100s)(1 + 0.1s)]$

Performance	Desired specifications	Root-locus results	Computer results	Logarithmic results	Computer results
Peak overshoot	1.20	1.203	1.21	1.25	1.17
K_1	30	29.0	29.9	25.6	25.6
ω_m	—	—	—	2.4	—
T_s	2.5	2.58	2.45	—	1.9
T_p	1.0	1.08	1.1	—	1.2

In Table 20-2 a comparison is also made of values obtained by the logarithmic analysis and the computer with the desired specifications. Reviewing the manner in which the parameters of the compensator were chosen, the table indicates a satisfactory correlation. The actual computer values of M_p and T_s are better than those specified. If the peak time and error coefficient are not critical, this compensator is acceptable. With the problem already on the computer, further improvement may be sought by trial and error by altering the parameters of $G_c(s)$. If a computer is not available, the roots of the characteristic equation must be determined by utilizing any of the suggested methods in Appendix B to determine the response of the compensated system, if the logarithmic approach is used for the analysis.

20-8   Conclusions

In this chapter an attempt has been made to present to the reader an over-all picture of a control design problem. A simple control system is used in the problem in order to present clearly a philosophy of approach and an idea of the degree of correlation that can be expected between the two methods.

Table 20-2 presents a summary of results of the system utilizing the compensator

$$G_c(s) = A \frac{(1 + 10s)(1 + s)}{(1 + 100s)(1 + 0.1s)} \tag{20-34}$$

Note that the root-locus approach results in values that agree closely with the desired specifications. The logarithmic approach results in a poorer correlation, although the values of M_p and T_s are better. In the latter method, if the ramp error coefficient is too far below the desired value, additional trials must be made to achieve this value.

Root-locus Shortcuts

Below are listed some pointers that can be helpful in utilizing the root-locus method.

1. For the compensated system, it may not be necessary to determine the complete locus.
2. To determine break-in and breakaway points on the real axis it is easier to apply the magnitude condition to locate the "Hills and dales" (maxima and minima) in the value of the static loop sensitivity.
3. When the value of the static loop sensitivity for the dominant roots is known and point 2 above has been carried out, the number of existing real roots and their general locations can be determined. If real roots exist, further application of the magnitude condition will result in their exact location.
4. If complex roots other than the dominant pair exist, the branches that contain those roots must be determined.
5. If all roots are determined except one or two, they can be determined by factoring out the known roots from the characteristic polynomial or by Grant's method. This may be easier than locating them on the locus.

Selection of Method of Analysis

Selection of the root-locus or frequency-response method in analyzing and improving a system generally depends upon the following:

1. Specifications given
2. Degree of accuracy desired
3. Time available
4. Complexity of the control system

In some cases a combination of both approaches may be advantageous. In the end, it is the experience that an individual gains in design work that becomes the yardstick by which he makes his decisions.

Bibliography

1. Chestnut, H., and R. W. Mayer: "Servomechanisms and Regulating System Design," vol. 2, chap. 1, John Wiley & Sons, Inc., New York, 1955.
2. "Fundamentals of Design of Piloted Aircraft Flight Control Systems," BuAer Report AE-61-4, vol. 1, U.S. Navy, Bureau of Aeronautics, 1952.
3. Draper, C. S., W. McKay, and S. Lees: "Instrument Engineering," vol. 2, McGraw-Hill Book Company, New York, 1953.
4. Locke, A. S.: "Guidance," chap. 7, D. Van Nostrand Company, Inc., Princeton, N.J., 1955.
5. Bruns, R. A., and R. M. Saunders: "Analysis of Feedback Control Systems," chap. 14, McGraw-Hill Book Company, New York, 1955.
6. Ganapathy, S., and G. Krishna: A Method of Determining the Transfer Function of a Linear System in Terms of Its Poles and Zeros from the Frequency Response, *Trans. AIEE*, vol. 81, pt. II, p. 182, November, 1962.
7. Hoerner, G. M., Jr.: Finding System Characteristics from Initial Step Response, *Control Eng.*, vol. 9, pp. 103–104, May, 1962.
8. Hoerner, G. M., Jr.: Second Order System Characteristics from Initial Step Response, *Control Eng.*, vol. 9, pp. 93–95, December, 1962.

21

State of
the Art

The scope of this textbook is the analysis and synthesis of linear feedback control systems. It can best be summarized by saying that the coverage involves the following areas:

1. The analysis of linear feedback control systems by use of the root-locus and the steady-state frequency-response methods
2. The utilization of simple compensators for compensating a system to achieve the desired values of the figures of merit
3. The presentation of basic schemes for analyzing systems with multiple inputs and outputs
4. The brief introduction to optimum response and the analysis of common nonlinearities by use of the describing function

These techniques permit an individual to analyze a basic system.   By means of simple gain adjustment or in conjunction with a simple compensator, it is possible to achieve the values of the desired figures of merit.   That is, a satisfactory system design is obtained.

The procedures presented in this text yield a system design that is satisfactory.   They do not, however, indicate whether this particular system design is the one that yields the optimum performance obtainable.   To achieve optimum performance for a system, the following items must be taken into account:

1.   System environment
2.   Component limitations
3.   Noise
4.   Unwanted disturbances
5.   Space and weight requirements
6.   Nonlinearities
7.   Random inputs

Thus the frontiers of control theory present many challenges.

Extensive work is being done in the optimization of performance of control systems.   The techniques used include Wiener's least-squares optimization method, the maximum principle of Pontryagin, Bellman's principle of optimality, and various techniques in the classical calculus of variations and modifications thereof.   The utilization of computers for the optimization of both linear and nonlinear systems is finding extensive application.   Two highly developed advanced areas of automatic control theory are adaptive control and sampled-data theory.   An account of these areas is given in the next two sections.

21-2   Adaptive Systems[1,2]

Section 8-7 discusses the effect of the variation of a system parameter other than gain upon the system's performance.   The analysis of this effect is based upon the condition that the system is designed to satisfy the figures of merit, utilizing as a basis the nominal or reference value δ_0 of the variable parameter δ.   This variation is due to the changing environment, either external or internal, of the control system.   For those systems that have this characteristic, for which the changing environment can be measured and for which the effect upon a system parameter can be predicted in advance, an "adaptive" system may be designed to maintain optimum performance.

An early application of techniques in adaptive control systems was in

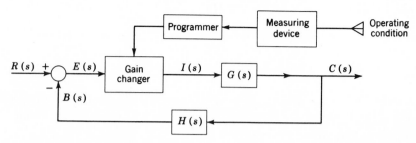

Fig. 21-1 *An adaptive control system.*

aircraft control systems. For the aircraft whose flight pattern is known in advance, the variations in speed (mach number) and altitude are also known or predicted in advance. These variations in turn cause a change in the plant (the aircraft—the controlled element) parameters. For particular types of aircraft, optimum performance may be obtained by varying the system gain in a prescribed manner according to the variation of the speed and altitude of flight. An adaptive system that accomplishes this is shown in Fig. 21-1. By proper instrumentation the operating conditions, in this example the mach number and altitude of the aircraft, are measured. These measured values are applied to the programmer. Since the variation of the operating condition is known in advance, the programmer is so designed that its output signal is a function of the mach number and altitude. The output of the programmer, in turn, causes the gain changer to have the prescribed value of gain for the system to maintain optimum performance. This type of adaptive system is sometimes referred to as *open-loop adaptation.*

The adaptive procedures discussed above cannot be utilized for situations in which the changing environment cannot be measured and its effect upon the system parameters cannot be predicted in advance. For high-speed aircraft and space vehicles the effect of the changing environment, even if measurement and prediction are possible (as required for the adaptive technique), affects the system parameters in such a manner that open-loop adaptation cannot be utilized or is not practicable. As an example, in a high-speed aircraft the changing environment not only can cause a change of system gain in a range as large as 240 to 1 but also affects the ζ and ω_n of the aircraft poles.

Where measurement and prediction are not possible or the effect of the changing environment is such that the application of the *open-loop* adaptive technique is not feasible, the self-adaptive technique may be applicable. A self-adaptive control system is defined[2] as one which *has the capability of changing its parameters through an internal process of measurement, evaluation, and adjustment to adapt to a changing environment,*

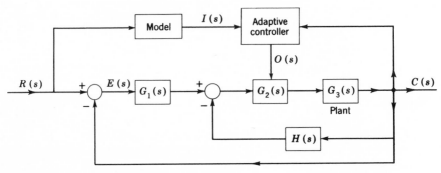

Fig. 21-2   *A minimum-error self-adaptive system.*

either external or internal, to the plant (controlled system) under control.
Figure 21-2 illustrates a self-adaptive system that utilizes the *model reference
principle.* The dynamics of the model, which are insensitive to the changing environment, are designed to yield an optimum response $I(s)$ for a given
input $R(s)$. The adaptive controller measures and compares the ideal
response $I(s)$ and actual response $C(s)$. The difference $I(s) - C(s)$ is
utilized by the controller computer to determine the required value of the
controller output $\acute{0}(s)$. The required value results in the necessary changes
in the dynamic characteristics of $G_2(s)$ (the adjustable parameter is usually
the forward gain) which cause the system output $C(s)$ to approximate, as
closely as possible, the ideal or optimum response. Note that this adaptive
system maintains the error between $I(s)$ and $C(s)$ at a minimum. Thus
this system is sometimes referred to as a *minimum-error self-adaptive system.*

Another form of self-adaptive system is shown in Fig. 21-3. The
control ratio for this system is

$$\frac{C(j\omega)}{I(j\omega)} = \frac{K_n G_1(j\omega) G_2(j\omega)}{1 + K_n G_1(j\omega) G_2(j\omega)} \tag{21-1}$$

If $K_n G_1(j\omega) G_2(j\omega) \gg 1$, then

$$\frac{C(j\omega)}{I(j\omega)} \approx 1 \tag{21-2}$$

Thus the output $c(t)$ is essentially identical to the input $i(t)$. Since $I(s)$
is the Laplace transform of the output of the model, which is identical to
the one in the previous example, $c(t)$ has the ideal or optimum response
characteristic. The requirement for successful operation is that $K_n \gg 1$
and that the system be stable. The procedure is to make the gain the
largest value possible while keeping the system stable. This maximum
value of gain produces roots of the characteristic equation on the imaginary
axis of the s plane. The *limit-cycle principle* is utilized to maintain the

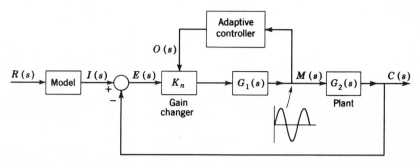

Fig. 21-3 *A high-gain, limit-cycle, self-adaptive system.*

condition $|K_n\mathbf{G}_1(j\omega)\mathbf{G}_2(j\omega)| \gg 1.0$.   The limit cycle is established by design-ing the system so that a pair of closed-loop poles are maintained on the $j\omega$ axis.   The limit-cycle oscillation appears in the signal $M(s)$ and is fed into the adaptive controller.   The controller utilizes the magnitude of this signal to determine the value $0(s)$.   In turn, $0(s)$ causes the gain K_n to change in order to maintain $|K_n\mathbf{G}_1(j\omega)\mathbf{G}_2(j\omega)| \gg 1$ under a changing environment.   This type of adaptive system is sometimes referred to as *closed-loop adaptation.*

Reference 1 gives a more detailed description, analysis, and synthesis of self-adaptive systems.

21-3   *Sampled-data Systems*[3-6]

The methods and systems discussed in this book deal with continuous signals.   That is, whether the system is linear or nonlinear, all variables are continuously present and are therefore known at all times.   Another category of control system is one in which one signal (or more) is *sampled* so that it appears as a pulse train.   Such sampling may be an inherent characteristic of the system.   For example, a radar tracking system supplies information on an airplane's position at discrete periods of time.   This information is therefore available as a succession of data points.   The sampling process can be performed at a constant rate or at a variable rate, or it may be random in nature.   The following discussion is based only on a constant-rate sampling.   Digital computers are available for perform-ing the computations necessary in a complex control system.   Since a digital computer must operate with discrete numbers, it is necessary to first convert a continuous signal to digital or sampled form.   Systems with

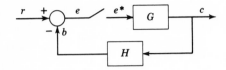

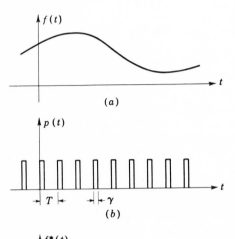

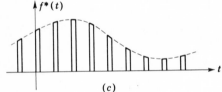

Fig. 21-4 *Block diagram of a system containing sampling of the actuating signal.*

Fig. 21-5 (a) *Continuous function f(t).* (b) *Sampling pulse train p(t).* (c) *Sampled function f*(t).*

digital computers can be analyzed in the same fashion as other sampled systems.

Sampling may occur in one or more places in a system. It is represented in a block diagram by the symbol for a switch. Figure 21-4 shows a system with sampling of the actuating signal. It should be noted that the output $c(t)$ is a continuous function of time.

The sampling process can be considered a modulation process in which a unit pulse train $p(t)$ multiplied by a continuous time function $f(t)$ produces the sampled function $f^*(t)$. This is represented by

$$f^*(t) = p(t)f(t) \tag{21-3}$$

These quantities are shown in Fig. 21-5. A Fourier-series expansion of $p(t)$ is

$$p(t) = \sum_{n=-\infty}^{+\infty} C_n e^{jn\omega_s t} \tag{21-4}$$

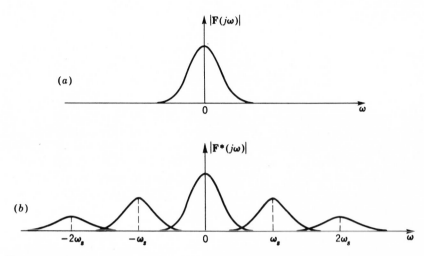

Fig. 21-6 *Frequency spectra for (a) a continuous function f(t); (b) a pulse sampled function f*(t).*

where the sampling frequency is $\omega_s = 2\pi/T$ and the Fourier coefficients C_n are given by

$$C_n = \frac{1}{T} \int_0^T p(t)e^{-jn\omega_s t}\, dt = \frac{\gamma}{T} \frac{\sin\,(n\omega_s\gamma/2)}{n\omega_s\gamma/2} e^{-jn\omega_s\gamma/2} \qquad (21\text{-}5)$$

The sampled function is therefore

$$f^*(t) = \sum_{n=-\infty}^{+\infty} C_n f(t)e^{jn\omega_s t} \qquad (21\text{-}6)$$

If the Fourier transform of the continuous function $f(t)$ is $\mathbf{F}(j\omega)$, the Fourier transform of the sampled function $f^*(t)$ is

$$\mathbf{F}^*(j\omega) = \sum_{n=-\infty}^{+\infty} C_n \mathbf{F}(j\omega + jn\omega_s) \qquad (21\text{-}7)$$

A comparison of the Fourier spectra of the continuous and sampled functions is shown in Fig. 21-6. It is seen that the sampling process produces a fundamental spectrum similar in shape to that of the continuous function. It also produces a succession of spurious complementary spectra which are shifted periodically by a frequency separation $n\omega_s$.

If the sampling frequency is sufficiently high, there is little overlap between the fundamental and complementary frequency spectra. In that case a low-pass filter could extract the spectrum of the continuous input signal by attenuating the spurious higher-frequency spectra. The forward

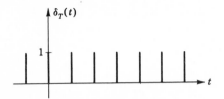

Fig. 21-7   *Impulse train.*

transfer function of a control system generally has a low-pass characteristic, so the system responds with more or less accuracy to the continuous signal.

If the duration γ of the sampling pulse is small compared with the sampling time T, the pulse train $p(t)$ can be represented by an ideal sampler followed by an ideal amplifier of gain γ. The ideal sampler produces the impulse train $\delta_T(t)$, which is shown in Fig. 21-7:

$$\delta_T(t) = \sum_{n=-\infty}^{+\infty} \delta(t - nT) \tag{21-8}$$

For convenience, only the ideal sampler is shown separately on the block diagram, and the amplifier gain is combined with the following "plant." The frequency spectrum of the function which is sampled by the ideal impulse train is similar to that shown in Fig. 21-6, except that the complementary spectra have the same amplitude as the fundamental spectrum. Since the forward transfer function of a control system attenuates the higher frequencies, the over-all system response is essentially the same with the idealized impulse sampling as with the actual pulse sampling. The use of impulse sampling simplifies the mathematical analysis of sampled systems and is therefore used extensively to represent the sampling process.

The Laplace transform of the impulse sequence

$$f^*(t) = \sum_{n=-\infty}^{+\infty} f(t)\, \delta_T(t - nT) \tag{21-9}$$

is given by

$$F^*(s) = \sum_{n=-\infty}^{+\infty} f(nT)e^{-nTs} \tag{21-10}$$

If the Laplace transform of $f(t)$ is a rational function, it is possible to write $F^*(s)$ in closed form. This closed form can also be obtained from

$$F^*(s) = \sum_{\substack{\text{At the poles} \\ \text{of } F(p)}} \text{residues of} \left[F(p)\, \frac{1}{1 - e^{-(s-p)T}} \right] \tag{21-11}$$

where $F(p)$ is the Laplace transform of $f(t)$ with s replaced by p. The expression $F^*(s)$ contains the term e^{Ts}, which means that it is not an algebraic expression but is a transcendental one.   Therefore a change of variable is made:

$$z = e^{Ts} \tag{21-12}$$

Equation (21-10) can now be written as

$$F^*(s)\Big]_{s = (1/T)\ln z} = F(z) = \sum_{n = -\infty}^{+\infty} f(nT)z^{-n} \tag{21-13}$$

Equation (21-11), which gives the transformed function in closed form, is rewritten as

$$F(z) = \sum_{\substack{\text{At the poles} \\ \text{of } F(p)}} \text{residues of} \left[F(p) \frac{1}{1 - e^{pT}z^{-1}} \right] \tag{21-14}$$

where $F(z)$ *is called the z transform of* $f^*(t)$.   The starred and z forms of the "pulsed transfer function" are easily obtained for the ordinary Laplace transfer function.   As an example, consider the transfer function

$$G(s) = \frac{K}{s(s + a)} \tag{21-15}$$

The corresponding pulsed transfer functions are as follows: in the s-domain,

$$G^*(s) = \frac{Ke^{-sT}(1 - e^{-aT})}{a(1 - e^{-sT})(1 - e^{-(s+a)T})} \tag{21-16}$$

and in the z domain,

$$G(z) = \frac{Kz^{-1}(1 - e^{-aT})}{a(1 - z^{-1})(1 - e^{-aT}z^{-1})} \tag{21-17}$$

The poles of $G^*(s)$ in Eq. (21-16) are infinite in number.   These poles exist at $s = jn\omega_s$ and $s = -a + jn\omega_s$ for all values of $-\infty < n < +\infty$.   Note that they are uniformly spaced with respect to their imaginary parts with a separation $j\omega_s$.   By contrast, there are just two poles of Eq. (21-17), located at $z = 1$ and $z = e^{-aT}$.   The root-locus method may therefore be applied in the z plane, whereas it is not very convenient to use in the s-plane because of the infinite number of poles.

The transformation of the s plane into the z plane can be investigated by inserting $s = \sigma + j\omega$ into

$$z = e^{Ts} = e^{\sigma T}e^{j\omega T} = e^{\sigma T}e^{j2\pi\omega/\omega_s} \tag{21-18}$$

1.   Lines of constant σ in the s plane map into circles of radius equal to $e^{\sigma T}$ in the z plane.   Specifically, the segment of the imaginary axis in the s plane of width ω_s maps into the circle of unit radius in the z plane; successive segments map into overlapping circles. This fact shows that the *proper* consideration of sampled-data

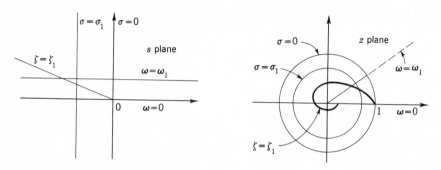

Fig. 21-8   *Transformation from the s plane to the z plane.*

systems in the z plane requires the use of a multiple-sheeted surface (i.e., a Riemann surface).   But by virtue of the uniform repetition of the roots of the characteristic equation, the condition for stability is that all roots of the characteristic equation contained in the principal branch lie within the unit circle in the principal sheet of the z plane.

2.   Lines of constant ω in the s plane map into radial rays drawn at the angle ωT in the z plane.   The portion of the constant ω line in the left half of the s plane becomes the radial ray within the unit circle in the z plane.   The negative part of the real axis $-\infty < \sigma \leq 0$ in the s plane is mapped on the segment of the real axis defined by $0 < z \leq 1$.

3.   The constant-damping-ratio ray in the s plane is defined by the equation

$$s = -\omega \cot \eta + j\omega$$

where $\eta = \cos^{-1} \zeta$.   Therefore

$$z = e^{sT} = e^{-\omega T \cot \eta + j\omega T} = e^{-\omega T \cot \eta} \underline{/\omega T} \qquad (21\text{-}19)$$

The corresponding map describes a logarithmic spiral in the z plane.

The corresponding paths, as discussed above, are shown in Fig. 21-8. For the block diagram of Fig. 21-4 the system equations are

$$C(s) = G(s)E^*(s) \qquad (21\text{-}20)$$
$$E(s) = R(s) - B(s) = R(s) - G(s)H(s)E^*(s) \qquad (21\text{-}21)$$

The starred transform of Eq. (21-21) is

$$E^*(s) = R^*(s) - GH^*(s)E^*(s) \qquad (21\text{-}22)$$

Solving for $C(s)$ from Eqs. (21-20) and (21-22) gives

$$C(s) = \frac{G(s)R^*(s)}{1 + GH^*(s)} \tag{21-23}$$

The starred transform $C^*(s)$ obtained from Eq. (21-23) is

$$C^*(s) = \frac{G^*(s)R^*(s)}{1 + GH^*(s)} \tag{21-24}$$

The z transform is obtained by replacing each starred transform by the corresponding function of z:

$$C(z) = \frac{G(z)R(z)}{1 + GH(z)} \tag{21-25}$$

Although the output $c(t)$ is continuous, the inverse of $C(z)$ in Eq. (21-25) yields only the set of values $c(nT)$, $n = 0, 1, 2, \ldots$, corresponding to the values at the sampling instants. Thus, one has the strength of the set of impulses from an ideal sampler located at the output which operates in synchronism with the sampler of the actuating signal.

Analysis of the performance of the sampled system, corresponding to the response given by Eq. (21-25), can be performed by the frequency-response or root-locus method. It should be kept in mind that the geometric boundary for stability in the z plane is the unit circle. As an example of the root-locus method, consider the sampled feedback system represented by Fig. 21-4 with $H(s) = 1$. Using $G(s)$ given by Eq. (21-15) with $a = 1$ and a sampling time $T = 1$, the equation of $G(z)$ is

$$G(z) = GH(z) = \frac{0.632Kz}{(z - 1)(z - 0.368)} \tag{21-26}$$

The characteristic equation is $1 + GH(z) = 0$ or

$$GH(z) = -1 \tag{21-27}$$

The usual root-locus techniques can be used to obtain a plot of the roots of Eq. (21-27) as a function of the sensitivity K. The root locus is drawn in Fig. 21-9. The maximum value of K for stability is obtained from the magnitude condition which yields $K_{max} = 2.73$. This occurs at the crossing of the root locus and the unit circle. The selection of the desired roots can be based on the damping ratio ζ desired. For the specified value of ζ the spiral given by Eq. (21-19) and shown in Fig. 21-8 must be drawn. The intersection of this curve with the root locus determines the roots. Alternatively, it is possible to specify the settling time. This determines the value of σ in the s plane, so the circle of radius $e^{\sigma T}$ can be drawn in the z plane. The intersection of this circle and the root locus determines the roots. For

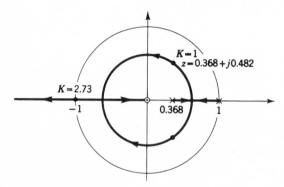

Fig. 21-9   *Root locus for Eqs. (21-26) and (21-27).*

the value $\zeta = 0.48$ the roots are $z = 0.368 \pm j0.482$ and the value of $K = 1$. The control ratio is therefore

$$\frac{C(z)}{R(z)} = \frac{0.632z}{z - 0.368 \pm j0.482} = \frac{0.632z}{z^2 - 0.736z + 0.368} \qquad (21\text{-}28)$$

For a step input the value of $R(z)$ is

$$R(z) = \frac{z}{z - 1} \qquad (21\text{-}29)$$

The expression for $C(z)$ is expanded by dividing its denominator into its numerator to get a power series in z^{-1}:

$$C(z) = 0.632z^{-1} + 1.096z^{-2} + 1.205z^{-3} + 1.120z^{-4} + 1.014z^{-5}$$
$$+ 0.98z^{-6} + \cdots \qquad (21\text{-}30)$$

The inverse transform of $C(z)$ is

$$c^*(t) = 0\delta(t) + 0.632\delta(t - T) + 1.096\delta(t - 2T) \cdots \qquad (21\text{-}31)$$

Hence the values $c(nT)$ at the sampling instants are recognizable as the coefficients of the terms in the series of Eq. (21-30) at the corresponding sampling instant. A plot of the values of $c(nT)$ is shown in Fig. 21-10. The curve of $c(t)$ is drawn as a smooth curve through these plotted points.

This section presents several basic points in the analysis of sampled-data systems. Additional topics of importance include the reconstruction of the continuous signal from the sampled signal by use of *hold* circuits, the entire problem of compensation to improve performance, and the effect of nonlinearities on system response. These topics are covered in some 10 books on sampled-data systems and in thousands of papers in the technical literature.

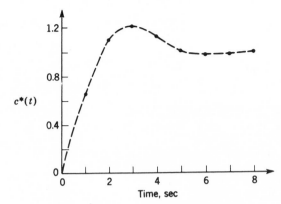

Fig. 21-10   *Plot of c*(t) for Eq. (21-30).*

21-4   *Conclusions*

This chapter presents an introduction to adaptive systems for optimization of system performance and to sampled-data systems for systems which have a sampled signal within the loop.   These are advanced topics in the field of feedback control systems.   It is intended that this account will bring to the reader's attention topics covered in detail in separate books on these areas.

Bibliography

1. Blakelock, J.: "Automatic Control of Aircraft and Missiles," John Wiley & Sons, Inc., New York, 1965.
2. "Proceedings of the Self Adaptive Flight Control Systems Symposium," WADC TR 59-49, Wright-Patterson Air Force Base, Dayton, Ohio, March, 1959.
3. Ragazzini, J. R., and G. F. Franklin: "Sampled-data Control Systems," McGraw-Hill Book Company, New York, 1958.
4. Tou, J. T.: "Digital and Sampled-data Control Systems," McGraw-Hill Book Company, New York, 1959.
5. Kuo, B. C.: "Analysis and Synthesis of Sampled-data Control Systems," Prentice-Hall, Inc., Englewood Cliffs, N.J., 1963.
6. Monroe, A. J.: "Digital Processes for Sampled Data Systems," John Wiley & Sons, Inc., New York, 1962.

Table of Laplace-transform Pairs

$F(s)$	$f(t) \qquad 0 \leq t$
1. 1	$u_1(t)$ unit impulse at $t = 0$
2. $\dfrac{1}{s}$	1 or $u(t)$ unit step at $t = 0$
3. $\dfrac{1}{s^2}$	$tu(t)$ ramp function
4. $\dfrac{1}{s^n}$	$\dfrac{1}{(n-1)!}\, t^{n-1}$ n is a positive integer
5. $\dfrac{1}{s}\, e^{-as}$	$u(t-a)$ unit step starting at $t = a$
6. $\dfrac{1}{s}\,(1 - e^{-as})$	$u(t) - u(t-a)$ rectangular pulse
7. $\dfrac{1}{s+a}$	e^{-at} exponential decay
8. $\dfrac{1}{(s+a)^n}$	$\dfrac{1}{(n-1)!}\, t^{n-1}e^{-at}$ n is a positive integer
9. $\dfrac{1}{s(s+a)}$	$\dfrac{1}{a}\,(1 - e^{-at})$
10. $\dfrac{1}{s(s+a)(s+b)}$	$\dfrac{1}{ab}\left(1 - \dfrac{b}{b-a}\,e^{-at} + \dfrac{a}{b-a}\,e^{-bt}\right)$
11. $\dfrac{s+\alpha}{s(s+a)(s+b)}$	$\dfrac{1}{ab}\left[\alpha - \dfrac{b(\alpha-a)}{b-a}\,e^{-at} + \dfrac{a(\alpha-b)}{b-a}\,e^{-bt}\right]$
12. $\dfrac{1}{(s+a)(s+b)}$	$\dfrac{1}{b-a}\,(e^{-at} - e^{-bt})$
13. $\dfrac{s}{(s+a)(s+b)}$	$\dfrac{1}{a-b}\,(ae^{-at} - be^{-bt})$
14. $\dfrac{s+\alpha}{(s+a)(s+b)}$	$\dfrac{1}{b-a}\,[(\alpha-a)e^{-at} - (\alpha-b)e^{-bt}]$

$F(s)$	$f(t)$ $0 \leq t$
15. $\dfrac{1}{(s+a)(s+b)(s+c)}$	$\dfrac{e^{-at}}{(b-a)(c-a)} + \dfrac{e^{-bt}}{(c-b)(a-b)} + \dfrac{e^{-ct}}{(a-c)(b-c)}$
16. $\dfrac{s+\alpha}{(s+a)(s+b)(s+c)}$	$\dfrac{(\alpha-a)e^{-at}}{(b-a)(c-a)} + \dfrac{(\alpha-b)e^{-bt}}{(c-b)(a-b)} + \dfrac{(\alpha-c)e^{-ct}}{(a-c)(b-c)}$
17. $\dfrac{\omega}{s^2+\omega^2}$	$\sin \omega t$
18. $\dfrac{s}{s^2+\omega^2}$	$\cos \omega t$
19. $\dfrac{s+\alpha}{s^2+\omega^2}$	$\dfrac{\sqrt{\alpha^2+\omega^2}}{\omega} \sin(\omega t + \phi) \qquad \phi = \tan^{-1}\dfrac{\omega}{\alpha}$
20. $\dfrac{s\sin\theta + \omega\cos\theta}{s^2+\omega^2}$	$\sin(\omega t + \theta)$
21. $\dfrac{1}{s(s^2+\omega^2)}$	$\dfrac{1}{\omega^2}(1 - \cos \omega t)$
22. $\dfrac{s+\alpha}{s(s^2+\omega^2)}$	$\dfrac{\alpha}{\omega^2} - \dfrac{\sqrt{\alpha^2+\omega^2}}{\omega^2}\cos(\omega t + \phi) \qquad \phi = \tan^{-1}\dfrac{\omega}{\alpha}$
23. $\dfrac{1}{(s+a)(s^2+\omega^2)}$	$\dfrac{e^{-at}}{a^2+\omega^2} + \dfrac{1}{\omega\sqrt{a^2+\omega^2}}\sin(\omega t - \phi) \qquad \phi = \tan^{-1}\dfrac{\omega}{a}$
24. $\dfrac{1}{(s+a)^2+b^2}$	$\dfrac{1}{b}e^{-at}\sin bt$
24a. $\dfrac{1}{s^2+2\zeta\omega_n s + \omega_n^2}$	$\dfrac{1}{\omega_n\sqrt{1-\zeta^2}}e^{-\zeta\omega_n t}\sin \omega_n\sqrt{1-\zeta^2}\,t$
25. $\dfrac{s+a}{(s+a)^2+b^2}$	$e^{-at}\cos bt$
26. $\dfrac{s+\alpha}{(s+a)^2+b^2}$	$\dfrac{\sqrt{(\alpha-a)^2+b^2}}{b}e^{-at}\sin(bt+\phi) \qquad \phi=\tan^{-1}\dfrac{b}{\alpha-a}$
27. $\dfrac{1}{s[(s+a)^2+b^2]}$	$\dfrac{1}{a^2+b^2} + \dfrac{1}{b\sqrt{a^2+b^2}}e^{-at}\sin(bt-\phi) \qquad \phi=\tan^{-1}\dfrac{b}{-a}$
27a. $\dfrac{1}{s(s^2+2\zeta\omega_n s+\omega_n^2)}$	$\dfrac{1}{\omega_n^2} - \dfrac{1}{\omega_n^2\sqrt{1-\zeta^2}}e^{-\zeta\omega_n t}\sin(\omega_n\sqrt{1-\zeta^2}\,t+\phi)$ $\phi = \cos^{-1}\zeta$
28. $\dfrac{s+\alpha}{s[(s+a)^2+b^2]}$	$\dfrac{\alpha}{a^2+b^2} + \dfrac{1}{b}\sqrt{\dfrac{(\alpha-a)^2+b^2}{a^2+b^2}}\,e^{-at}\sin(bt+\phi)$ $\phi = \tan^{-1}\dfrac{b}{\alpha-a} - \tan^{-1}\dfrac{b}{-a}$
29. $\dfrac{1}{(s+c)[(s+a)^2+b^2]}$	$\dfrac{e^{-ct}}{(c-a)^2+b^2} + \dfrac{e^{-at}\sin(bt-\phi)}{b\sqrt{(c-a)^2+b^2}} \qquad \phi=\tan^{-1}\dfrac{b}{c-a}$
30. $\dfrac{1}{s(s+c)[(s+a)^2+b^2]}$	$\dfrac{1}{c(a^2+b^2)} - \dfrac{e^{-ct}}{c[(c-a)^2+b^2]}$ $+ \dfrac{e^{-at}\sin(bt-\phi)}{b\sqrt{a^2+b^2}\sqrt{(c-a)^2+b^2}}$ $\phi = \tan^{-1}\dfrac{b}{-a} + \tan^{-1}\dfrac{b}{c-a}$

$F(s)$	$f(t)$ $\quad 0 \le t$
31. $\dfrac{s + \alpha}{s(s + c)[(s + a)^2 + b^2]}$	$\dfrac{\alpha}{c(a^2 + b^2)} - \dfrac{(c - \alpha)e^{-ct}}{c[(c - a)^2 + b^2]}$

$$+ \frac{\sqrt{(\alpha - a)^2 + b^2}}{b\sqrt{a^2 + b^2}\sqrt{(c - a)^2 + b^2}}\, e^{-at} \sin (bt + \phi)$$

$$\phi = \tan^{-1}\frac{b}{\alpha - a} - \tan^{-1}\frac{b}{-a} - \tan^{-1}\frac{b}{c - a}$$

32. $\dfrac{1}{s^2(s + a)}$	$\dfrac{1}{a^2}(at - 1 + e^{-at})$
33. $\dfrac{1}{s(s + a)^2}$	$\dfrac{1}{a^2}(1 - e^{-at} - ate^{-at})$
34. $\dfrac{s + \alpha}{s(s + a)^2}$	$\dfrac{1}{a^2}[\alpha - \alpha e^{-at} + a(a - \alpha)te^{-at}]$
35. $\dfrac{s^2 + \alpha_1 s + \alpha_0}{s(s + a)(s + b)}$	$\dfrac{\alpha_0}{ab} + \dfrac{a^2 - \alpha_1 a + \alpha_0}{a(a - b)}e^{-at} - \dfrac{b^2 - \alpha_1 b + \alpha_0}{b(a - b)}e^{-bt}$
36. $\dfrac{s^2 + \alpha_1 s + \alpha_0}{s[(s + a)^2 + b^2]}$	$\dfrac{\alpha_0}{c^2} + \dfrac{1}{bc}[(a^2 - b^2 - \alpha_1 a + \alpha_0)^2$

$$+ b^2(\alpha_1 - 2a)^2]^{1/2}e^{-at} \sin (bt + \phi)$$

$$\phi = \tan^{-1}\frac{b(\alpha_1 - 2a)}{a^2 - b^2 - \alpha_1 a + \alpha_0} - \tan^{-1}\frac{b}{-a}$$

$$c^2 = a^2 + b^2$$

| 37. $\dfrac{1}{(s^2 + \omega^2)[(s + a)^2 + b^2]}$ | $\dfrac{(1/\omega) \sin (\omega t + \phi_1) + (1/b)e^{-at} \sin (bt + \phi_2)}{[4a^2\omega^2 + (a^2 + b^2 - \omega^2)^2]^{1/2}}$ |

$$\phi_1 = \tan^{-1}\frac{-2a\omega}{a^2 + b^2 - \omega^2} \qquad \phi_2 = \tan^{-1}\frac{2ab}{a^2 - b^2 + \omega^2}$$

| 38. $\dfrac{s + \alpha}{(s^2 + \omega^2)[(s + a)^2 + b^2]}$ | $\dfrac{1}{\omega}\left[\dfrac{\alpha^2 + \omega^2}{c}\right]^{1/2} \sin (\omega t + \phi_1)$ |

$$+ \frac{1}{b}\left[\frac{(\alpha - a)^2 + b^2}{c}\right]^{1/2} e^{-at} \sin (bt + \phi_2)$$

$$c = (2a\omega)^2 + (a^2 + b^2 - \omega^2)^2$$

$$\phi_1 = \tan^{-1}\frac{\omega}{\alpha} - \tan^{-1}\frac{2a\omega}{a^2 + b^2 + \omega^2}$$

$$\phi_2 = \tan^{-1}\frac{b}{\alpha - a} + \tan^{-1}\frac{2ab}{a^2 - b^2 + \omega^2}$$

| 39. $\dfrac{s + \alpha}{s^2[(s + a)^2 + b^2]}$ | $\dfrac{1}{c}\left(\alpha t + 1 - \dfrac{2\alpha a}{c}\right) + \dfrac{[b^2 + (\alpha - a)^2]^{1/2}}{bc}e^{-at} \sin (bt + \phi)$ |

$$c = a^2 + b^2$$

$$\phi = 2\tan^{-1}\left(\frac{b}{a}\right) + \tan^{-1}\frac{b}{\alpha - a}$$

| 40. $\dfrac{s^2 + \alpha_1 s + \alpha_0}{s^2(s + a)(s - b)}$ | $\dfrac{\alpha_1 + \alpha_0 t}{ab} - \dfrac{\alpha_0(a + b)}{(ab)^2} - \dfrac{1}{a - b}\left(1 - \dfrac{\alpha_1}{a} + \dfrac{\alpha_0}{a^2}\right)e^{-at}$ |

$$- \frac{1}{a - b}\left(1 - \frac{\alpha_1}{b} + \frac{\alpha_0}{b^2}\right)e^{-bt}$$

Appendix B

Methods of Factoring Polynomials (Root Solving)

Finding the roots of an equation is often necessary. This is the case when the time response of a system is to be obtained by means of the inverse Laplace transform. The roots of the characteristic equation must be obtained before the inverse Laplace transformation can be performed. To systematize and simplify the process of obtaining the roots of an equation, the following rules, characteristics, and methods are outlined. Books on the theory of equations contain this material, often in more detail than is given here.

The problem is to find the roots of an equation of the form

$$f(s) = A_n s^n + A_{n-1} s^{n-1} + \cdots + A_i s^i + \cdots + A_2 s^2 + A_1 s + A_0 = 0 \tag{B-1}$$

The coefficients A_i of this equation are known real quantities. The following principles are useful in solving for the roots of the polynomial equation:

1. The number of roots is equal to the order of the equation, n.
2. Complex roots occur in pairs that are conjugates. Therefore, if n is odd, there must be at least one real root.
3. Descartes's rule of signs: The number of positive real roots is equal to, or an even number less than, the number of variations of sign of the coefficients of $f(s)$. In particular, there is exactly one positive real root if the coefficients present only one variation of sign. To investigate the possible number of negative real roots, apply Descartes's rule to the equation $f(-s) = 0$.
4. Routh's criterion: This criterion, described in Sec. 4-14, permits the determination of the exact number of roots with positive real parts.

The pure imaginary roots can also be determined.   If the equation being factored is the characteristic equation of a system, this method determines the system stability without actually solving for the roots.

5.   If the coefficients A_i are integers, then if any rational real roots exist they are given by r/m, where r is a divisor of A_0 and m is a divisor of A_n.

6.   Test for complex roots: The equation has complex roots if there are three consecutive coefficients A_{r+1}, A_r, A_{r-1} that satisfy the inequality

$$A_r^2 < A_{r+1}A_{r-1} \qquad \text{(B-2)}$$

This is a sufficient condition, but it is not necessary.   Therefore the converse is not necessarily true.

7.   Relations between roots and coefficients: In terms of the roots p_1, p_2, p_3, etc., which may be real or complex, the equation for $f(s)$ can be written

$$f(s) = A_n(s - p_1)(s - p_2)(s - p_3) \cdots (s - p_n) = 0 \qquad \text{(B-3)}$$

Multiplying the factors results in

$$f(s) = A_n s^n - A_n(p_1 + p_2 + \cdots + p_n)s^{n-1} + \cdots \\ + A_n(-1)^n(p_1 p_2 \cdots p_n) \qquad \text{(B-4)}$$

Equating the coefficients of like powers of s of Eqs. (B-1) and (B-4) results in the relationships

$$\frac{A_{n-1}}{A_n} = -(p_1 + p_2 + \cdots + p_n) \qquad \text{(B-5)}$$

$$\frac{A_0}{A_n} = (-1)^n(p_1 p_2 \cdots p_n) \qquad \text{(B-6)}$$

which are very useful in checking the values of the calculated roots.

8.   Change of variable: The work involved in solving for the roots of an equation can often be simplified.   The root-solving methods to be applied in this appendix are easier to use when the coefficient of s^n is made unity.   This is done by dividing the equation $f(s)$ by A_n. A further improvement may exist if the lowest-order term, the constant, is made unity.   This can be done by a change of variable by use of

$$s = VP$$

where V is a constant and P is the new variable.   After this substitution into $f(s)$, the equation is divided by $A_n V^n$ to make the coefficient of P^n equal to unity.   Then, the constant term $A_0/A_n V^n$ is equated to unity and the value of V is determined.

As an example, consider the equation

$$f(s) = s^4 + 1.4 \times 10^4 s^3 + 3.75 \times 10^8 s^2 + 1.65 \times 10^{12} s + 1.936 \times 10^{15} = 0$$

$$\text{(B-7)}$$

The reader may note from the equation that a substitution of $s = 10^4 P_1$ results in the equation

$$f(P_1) = P_1^4 + 1.4 P_1^3 + 3.75 P_1^2 + 1.65 P_1 + 0.1936 = 0 \qquad \text{(B-8)}$$

By use of the rule as outlined above,

$$\frac{1.936 \times 10^{15}}{V^4} = 1$$

resulting in $V = 0.663 \times 10^4$.   The new equation is

$$f(P) = P^4 + 2.11 P^3 + 8.52 P^2 + 5.66 P + 1 = 0 \qquad \text{(B-9)}$$

Either Eq. (B-8) or (B-9) is easier to work with than Eq. (B-7).

9.   The partition method of Sec. 8-5 and the shifting of the axis method described in Sec. 4-15 may also be used to help determine the location of the roots.

Some methods for finding the roots are now considered.

1.   Quadratic equation: direct solution.   For the quadratic equation

$$A_2 s^2 + A_1 s + A_0 = 0 \qquad \text{(B-10)}$$

the roots are obtained from the standard formula

$$s = -\frac{A_1}{2A_2} \pm \frac{1}{2A_2} \sqrt{A_1^2 - 4A_2 A_0} \qquad \text{(B-11)}$$

There are two roots which may be real but unequal, real and equal, or complex conjugates.   This is dependent on the quantity under the radical.

2.   Trial division: real roots.   When the order n of an equation is odd, there is one real root.   The factor containing the real root may be extracted by a series of trials.   As an example, consider the cubic equation

$$s^3 + 3.5 s^2 + 6.5 s + 10 = 0 \qquad \text{(B-12)}$$

The procedure is to assume a factor.   This factor is then divided into the original equation.   This is continued until the remainder is zero or sufficiently small.   The first trial divisor is chosen as the two lowest-order terms $(6.5s + 10)$ of the original equation.   Since this is known to be only a trial and is not the correct factor, it is best to round off the numbers of $s + 10/6.5$ to $s + 1$.   Performing tne division,

$$\begin{array}{r} s^2 + 2.5s + 4 \\ s + 1 \overline{\smash{\big)}\ s^3 + 3.5s^2 + 6.5s + 10} \\ \underline{s^3 +\ \ \ s^2} \\ 2.5s^2 + 6.5s \\ \underline{2.5s^2 + 2.5s} \\ 4s + 10 \leftarrow \\ \underline{4s +\ \ 4} \\ +\ \ 6 \end{array}$$

Since there is a remainder, the factor $s + 1$ is not correct. Before proceeding, the use of synthetic division to perform the division by $s + 1$ or $s = -1$ is indicated, as shown below.

$$\begin{array}{rrrr|r} 1 & 3.5 & 6.5 & 10 & -1 \\ & -1 & -2.5 & -4 & \\ \hline 1 & 2.5 & 4 & 6 \end{array}$$

It is seen that the last term is the remainder. Synthetic division is used hereafter. The next final factor is marked with an arrow in both the long division and synthetic division and is $4s + 10$ or $s + 2.5$. Using this trial divisor or $s = -2.5$,

$$\begin{array}{rrrr|r} 1 & 3.5 & 6.5 & 10 & -2.5 \\ & -2.5 & -2.5 & -10 & \\ \hline 1 & 1 & 4 & 0 \end{array}$$

shows a remainder of zero. Therefore the trial divisor $s + 2.5$ is correct, and the original equation can be factored into

$$s^3 + 3.5s^2 + 6.5s + 10 = (s + 2.5)(s^2 + s + 4) \qquad \text{(B-13)}$$

Although this case required two trials, it may be expected in general that more trials will be required.

3. Trial division of quadratic factors: Lin's method. The method outlined by Lin is basically one of successive trials using a quadratic factor. The process is continued until the remainder is considered sufficiently small. The first trial factor uses the three lowest-order terms of the original polynomial. This first trial is $s^2 + (A_1/A_2)s + A_0/A_2$. Performing the long division results in

$$s^2 + \frac{A_1}{A_2}s + \frac{A_0}{A_2} \overline{\smash{\big)}\begin{array}{l} A_n s^{n-2} + \cdots + B_2 \\ A_n s^n + A_{n-1}s^{n-1} + \cdots + A_2 s^2 + A_1 s + A_0 \\ \underline{A_n s^n + \cdots} \\ \qquad\qquad \cdots \\ \qquad\qquad B_2 s^2 + B_1 s + A_0 \leftarrow \\ \qquad\qquad \underline{B_2 s^2 + C_1 s + C_0} \\ \qquad\qquad\qquad D_1 s + D_0 \end{array}}$$

If the remainder $D_1s + D_0$ is too large, the next trial used is

$$s^2 + \frac{B_1}{B_2} s + \frac{A_0}{B_2}$$

This procedure is continued until the remainder is within an acceptable limit of approximation.   If the polynomial being factored has a pair of complex roots that are nearly equal in magnitude, the convergence is slow and the number of trials required may be very large.   In such a case it is well to investigate other methods of solution.

Lin's method is illustrated by means of an example.   Consider the polynomial

$$s^4 + 4.5s^3 + 12.5s^2 + 17.5s + 12.5 \tag{B-14}$$

The first trial divisor is

$$s^2 + \frac{17.5}{12.5} s + \frac{12.5}{12.5} = s^2 + 1.4s + 1$$

$$
\require{enclose}
\begin{array}{r}
s^2 + 3.1s + 7.1 \\[2pt]
s^2 + 1.4s + 1 \enclose{longdiv}{s^4 + 4.5s^3 + 12.5s^2 + 17.5s + 12.5} \\
\underline{s^4 + 1.4s^3 + s^2} \\
3.1s^3 + 11.5s^2 + 17.5s \\
\underline{3.1s^3 + 4.4s^2 + 3.1s} \\
7.1s^2 + 14.4s + 12.5 \\
\underline{7.1s^2 + 9.9s + 7.1} \\
4.5s + 5.4
\end{array}
$$

The remainder is large; therefore a second trial is made using

$$s^2 + \frac{14.4}{7.1} s + \frac{12.5}{7.1} = s^2 + 2s + 1.8$$

In these first trials it is common to use a small number of significant figures in order to reduce the work involved in the division.   When the remainder is sufficiently small, the number of significant figures is increased to obtain greater accuracy.   For the second trial division the quotient is $s^2 + 2.5s + 2.5$ and the remainder is $1.6s + 2.24$. The third trial divisor is $s^2 + 2.3s + 2.2$, the quotient is $s^2 + 2.2s + 5.25$, and the remainder is $0.6s + 1$.   The fourth trial divisor is $s^2 + 2.4s + 2.4$, the quotient is $s^2 + 2.1s + 5.05$, and the remainder is $0.33s + 0.38$.   The fifth trial divisor used is $s^2 + 2.47s + 2.48$, the quotient is $s^2 + 2.03s + 5$, and the remainder is $0.11s + 0.1$.   Considering this remainder as sufficiently small gives the factors

$$(s^2 + 2.47s + 2.48)(s^2 + 2.03s + 5) \tag{B-15}$$

The exact factors are

$$(s^2 + 2.5s + 2.5)(s^2 + 2s + 5) \tag{B-16}$$

It is seen that each successive approximation came closer to the exact solution. This trial procedure can be carried out by additional trials to the accuracy desired.

4. Method of Luke and Ufford.[1] The method devised by Lin has been extended by Luke and Ufford to obtain all the roots of the polynomial equation simultaneously. This has resulted in the root-solving tables* given below. The notation in the upper right-hand corner, such as (1,2,1), means that the quartic equation is to be factored into a quadratic and two linear factors. Usually it is best to factor the real root from an equation of odd order. Then the table for an even-order equation can be used with the remaining factor. Since equations of higher order are cumbersome for the iterative process used, the tables shown are up to the fifth order only. The procedure is to evaluate the coefficients of the factors by using the formulas given in each table and using the set of values of the other coefficients previously calculated. This is continued until these coefficients repeat themselves. On the first trial, when little is known regarding the roots, the choice would be the table having only quadratics. If the table does not converge, then the table showing two real factors can be used. Since the method is not infallible, a lack of convergence may indicate that another method should be used.

The example solved by Lin's method uses the (2,2) chart shown at the top of page 709. Note that the starting point is the assumption that $b_0 = b_1 = 0$. Then the value a_1 can be calculated. With this value of a_1 the value of a_0 can be calculated. Then b_0 is calculated, using the previous value of b_1, a_1, and a_0. The table has been carried out to 10 calculations of each coefficient, showing that it converges slowly. Since the coefficients change continuously in the same direction, convergence could have been speeded up by anticipating the correct value.

The choice between Lin's method and the Luke and Ufford charts is a matter of personal preference.

Advanced techniques that permit the evaluation of the roots of polynomials with both real coefficients and complex coefficients are available in the current mathematical literature.[2,3]

---

* Charts reproduced by courtesy of Yudell L. Luke and Dolores Ufford.

$$(s + b_0)(s^2 + a_1 s + a_0) = s^3 + A_2 s^2 + A_1 s + A_0$$

$$b_0 = \frac{A_0}{a_0}$$

$$a_1 = A_2 - b_0$$

$$a_0 = A_1 - a_1 b_0$$

Cubic (1,2)

b_0	a_1	a_0
0		

$$(s^2 + b_1 s + b_0)(s + a_0) = s^3 + A_2 s^2 + A_1 s + A_0$$

$$b_0 = \frac{A_0}{a_0}$$

$$b_1 = \frac{A_1 - b_0}{a_0}$$

$$a_0 = A_2 - b_1$$

Cubic (2,1)

b_0	b_1	a_0
0	0	

$$(s + x_0)(s^2 + b_1 s + b_0)(s + x_1) = s^4 + A_3 s^3 + A_2 s^2 + A_1 s + A_0$$

Quartic
(1,2,1)

$$b_0 = \frac{A_1 - x_0 A_2 + x_0^2 A_3 - x_0^3}{x_1}$$

$$b_1 = \frac{A_2 - x_0 A_3 + x_0^2 - b_0}{x_1}$$

$$x_0 = \frac{A_0}{x_1 b_0}$$

$$x_1 = A_3 - x_0 - b_1$$

b_0	b_1	x_0	x_1	$x_1 b_0$	x_0^2
0	0	0			

$$(s^2 + b_1 s + b_0)(s^2 + a_1 s + a_0) = s^4 + A_3 s^3 + A_2 s^2 + A_1 s + A_0$$

Quartic
(2,2)

$$b_0 = \frac{A_0}{a_0}$$

$$b_1 = \frac{A_1 - a_1 b_0}{a_0}$$

$$a_1 = A_3 - b_1$$

$$a_0 = A_2 - b_0 - a_1 b_1$$

b_0	b_1	a_1	a_0
0	0		

$$(s^2 + c_1 s + c_0)(s^2 + b_1 s + b_0)(s + x_1)$$

Quintic
(2,2,1)

$$= s^5 + A_4 s^4 + A_3 s^3 + A_2 s^2 + A_1 s + A_0$$

$$c_0 = \frac{A_2 - b_1 A_3 - b_0 A_4 + b_1^2 A_4 + 2b_1 b_0 - b_1^3}{x_1}$$

$$c_1 = \frac{A_3 - b_1 A_4 - b_0 + b_1^2 - c_0}{x_1}$$

$$b_0 = \frac{A_0}{c_0 x_1}$$

$$b_1 = \frac{A_1 - b_0(c_1 x_1 + c_0)}{c_0 x_1}$$

$$x_1 = A_4 - b_1 - c_1$$

c_0	c_1	b_0	b_1	x_1	$c_0 x_1$	$c_1 x_1 + c_0$	b_1^2	$2b_1 b_0$
0	0	0	0		0	0	0	0

$$(s + x_0)(s^2 + b_1 s + b_0)(s^2 + c_1 s + c_0) = s^5 + A_4 s^4 + A_3 s^3 + A_2 s^2 + A_1 s + A_0$$

Quintic
(1,2,2)

$$b_0 = \frac{A_1 - x_0 A_2 + x_0^2 A_3 - x_0^3 A_4 + x_0^4}{c_0}$$

$$b_1 = \frac{A_2 - x_0 A_3 + x_0^2 A_4 - x_0^3 - c_1 b_0}{c_0}$$

$$x_0 = \frac{A_0}{b_0 c_0}$$

$$c_1 = A_4 - x_0 - b_1$$

$$c_0 = A_3 - x_0 A_4 + x_0^2 - b_0 - c_1 b_1$$

b_0	b_1	x_0	c_1	c_0	$b_0 c_0$	x_0^2	x_0^3
0	0	0			0	0	0

$$(s^2 + b_1 s + b_0)(s^2 + a_1 s + a_0) = \begin{matrix} s^4 + 4.5s^3 + 12.5s^2 + 17.5s + 12.5 \quad \text{Quartic} \\ s^4 + A_3 s^3 + A_2 s^2 + A_1 s + A_0 \quad \quad \text{(2,2)} \end{matrix}$$

$$b_0 = \frac{A_0}{a_0} = \frac{12.5}{a_0}$$

$$b_1 = \frac{A_1 - a_1 b_0}{a_0} = \frac{17.5 - a_1 b_0}{a_0}$$

$$a_1 = A_3 - b_1 = 4.5 - b_1$$

$$a_0 = A_2 - b_0 - a_1 b_1 = 12.5 - b_0 - a_1 b_1$$

	b_0	b_1	a_1	a_0
1	0	0	4.5	12.5
2	1	1.1	3.4	7.8
3	1.6	1.5	3.0	6.4
4	2.0	1.8	2.7	5.6
5	2.2	2.1	2.4	5.3
6	2.4	2.2	2.3	5.0
7	2.5	2.4	2.1	4.96
8	2.52	2.46	2.04	4.96
9	2.52	2.49	2.01	4.97
10	2.51	2.49	2.01	4.99

Bibliography

1. Luke, Y. L., and D. Ufford: On the Roots of Algebraic Equations, *J. Math. Phys.*, vol. 30, no. 2, July, 1951.
2. Wall, H. S.: Polynomials Whose Zeros Have Negative Real Parts, *Am. Mathematical Monthly*, vol. 52, 1945.
3. Frank, E.: On the Zeros of Polynomials with Complex Coefficients, *Bull. Am. Mathematical Soc.*, vol. 52, 1946.

Appendix C

Standard
Block-diagram
Terminology

C-1   Nomenclature

The terminology described here is based on the proposed standards issued by the IEEE Subcommittee on Terminology and Nomenclature of the Feedback Control Systems Committee. These standards were prepared in cooperation with the ASA Committee on Letter Symbols for Feedback Control Systems.[1]

Figure C-1 shows a block-diagram representation of a feedback control system containing the basic elements. Figure C-2 shows the block diagram and symbols of a more complicated system with multiple paths.

Numerical subscripts are used to distinguish between blocks of similar functions in the circuit. For example, in Fig. C-1 the control elements are designated by G_1 and the controlled system by G_2. In Fig. C-2 the control elements are divided into two blocks, G_1 and G_2, to aid in representing relations between parts of the system. Also, in Fig. C-2 the primary feedback is represented by H_1. Additional feedback loops might be designated by H_2, H_3, and so forth.

The idealized system represented by the block enclosed with dashed lines in Fig. C-1 can be understood to show the relation between the basic input to the system and the performance of the system in terms of the desired output. This would be the system agreed upon to establish the ideal value for the output of the system. In systems where the command is actually the desired value or ideal value, the idealized system would be represented by unity. The arrows and block associated with the idealized system, the ideal value, and the system error are shown in dashed lines on

710

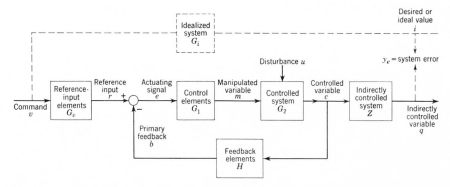

Fig. C-1 *Block diagram of feedback control system containing all basic elements.*

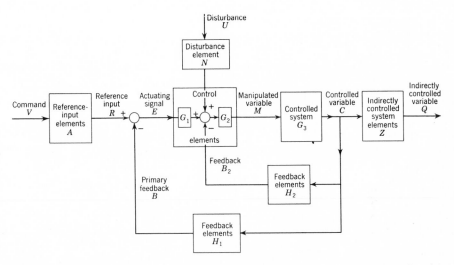

Fig. C-2 *Block diagram of representative feedback control system showing use of multiple feedback loops (all uppercase letters to denote transformation).*

the block diagram because they do not exist physically in any feedback control system. For any specific problem, it represents the system (conceived in the mind of the designer) that will give the nearest approach, when considered as a perfect system, to the desired output or ideal value.

C-2 *Definitions: Variables in the System*

The *command* (v) is the input which is established by some means external to and independent of the feedback control system.

The *reference input* (r) is derived from the command and is the actua signal input to the system.

The *controlled variable* (c) is the quantity that is directly measured and controlled. It is the output of the controlled system.

The *primary feedback* (b) is a signal which is a function of the controlled variable and which is compared with the reference input to obtain the actuating signal.

The *actuating signal* (e) is obtained from a comparison measuring device and is the reference input minus the primary feedback. This signal, usually at a low energy level, is the input to the control elements that produce the manipulated variable.

The *manipulated variable* (m) is that quantity obtained from the control elements which is applied to the controlled system. The manipulated variable is generally at a higher energy level than the actuating signal and may also be modified in form.

The *indirectly controlled variable* (q) is the output quantity and is related through the indirectly controlled system to the controlled variable. It is outside the closed loop and is not directly measured for control.

The *ultimately controlled variable* is a general term that refers to the indirectly controlled variable. In the absence of the indirectly controlled variable, it refers to the controlled variable.

The *ideal value* (i) is the value of the ultimately controlled variable that would result from an idealized system operating from the same command as the actual system.

The *system error* (y_e) is the ideal value minus the value of the ultimately controlled variable.

The *disturbance* (u) is the unwanted signal that tends to affect the controlled variable. The disturbance may be introduced into the system at many places.

C-3   Definitions: System Components

The *reference input elements* (G_v) produce a signal proportional to the command.

The *control elements* (G) produce the manipulated variable from the actuating signal.

The *controlled system* (G) is the device that is to be controlled. This is frequently a high-power element.

The *feedback elements* (H) produce the primary feedback from the controlled variable. This is generally a proportionality device but may also modify the characteristics of the controlled variable.

The *indirectly controlled system* (*Z*) relates the indirectly controlled variable to the controlled quantity. This component is outside the feedback loop.

The *idealized system* (*G_i*) is one whose performance is agreed upon to define the relationship between the ideal value and the command. In the adaptive systems this is called the *model* system.

The *disturbance element* (*N*) denotes the functional relationship between the variable representing the disturbance and its effect on the control system.

Bibliography

1. ASA–C85.1 Terminology for Automatic Control 1963.

Appendix D

The Spirule

D-1   Description

The Spirule is a device used chiefly by feedback control system engineers to multiply and divide complex quantities in the application of the root-locus method of analysis. Complex quantities are multiplied by multiplying their magnitudes and adding the angles between each directed line segment and a reference axis. Division of complex quantities is the inverse operation. The word *Spirule* is the trade name given to this device by W. R. Evans and is a combination of the words spiral and rule. The device consists of a protractor and an arm which are made of transparent plastic and have a common pivot point. A commercial model made by The Spirule Company of Whittier, California, is shown in Fig. D-1.

The protractor is 4.5 in. in diameter and is graduated in 1° increments. Cross-hair lines are drawn on the 0 to 180° axis and the 90 to 270° axis. The arm extends 9.25 in. from the pivot point. One side of the arm is aligned with the diameter of the protractor. The scale printed along the edge of the arm is marked in tenths (which are ½ in. long) as a scale factor. Each half-inch is further subdivided into 10 parts. A logarithmic spiral S is drawn on the arm and is used for multiplication and division of numbers. The arm has a solid reference line marked R which starts at the pivot and is approximately diagonal on the arm. The angle between the reference line on the arm and a radial line from the pivot to any point on the spiral curve is proportional to the logarithm of the distance from the pivot to the point. The intersection of the spiral with the reference line on the arm corresponds to the unit distance 5 in. A factor of 10 for the numbers being multiplied or divided corresponds to a rotation of 90°. Therefore three of the index arrows are labeled with the factors X–0.1, X–1, and X–10. The spiral beyond 7 in. from the pivot point is reflected about the edge of the arm and is drawn as a dashed curve. The pivot permits the protractor

714

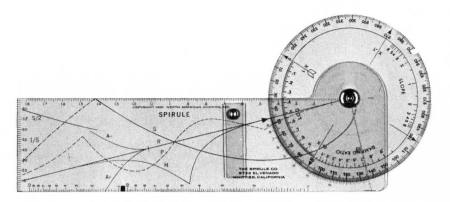

Fig. D-1   *A Spirule.*

and ₃he arm to be moved separately.   There is also enough friction between the arm and the protractor to cause them to rotate together.

Section 4-11 describes the procedures for using the Spirule to obtain the magnitudes and angles of the coefficients of the partial-fraction expansion of a function $F(s)$.   The same basic procedures are applied in this appendix to the root locus.

D-2   Angle Measurement

The angle condition for points on the root locus states that the sum of the angles from the poles of $G(s)$ minus the sum of the angles from the zeros of $G(s)$ must be an odd multiple of 180° for positive values of gain:

$$\Sigma\underline{/poles\ of\ G(s)} - \Sigma\underline{/zeros\ of\ G(s)} = (1 + 2m)180° \qquad (D\text{-}1)$$

Note that in the root locus it is customary to take the sum of the angles of the poles minus the angles of the zeros of $G(s)$, whereas for the partial-fraction coefficients (see Sec. 4-11) the sign of the angles is reversed.   The points on the complex portion of the s plane are obtained by trial and error by applying the angle condition of Eq. (D-1).   For the transfer function

$$G(s) = \frac{K(s - z)}{s(s - p)} \qquad (D\text{-}2)$$

the pole-zero diagram is shown in Fig. D-2.   The angle condition for this problem requires that

$$\phi_0 + \phi_1 - \psi = (1 + 2m)180° \qquad (D\text{-}3)$$

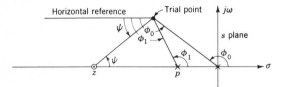

Fig. D-2 *Trial point for root locus of*

$$G(s) = \frac{K(s - z)}{s(s - p)}$$

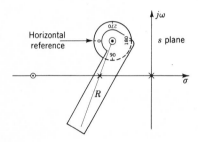

Fig. D-3 *Operation of Spirule.*

Figure D-2 shows that these angles can also be measured from a horizontal reference line drawn to the left from the trial point.   The use of the Spirule to measure these angles is now described.

The first step is to set the reference line R on the arm in line with the $0°$ index on the protractor.   The center of the protractor is placed on a trial point with the $0°$ index aligned with a horizontal reference line of the graph paper and pointed to the left.   A finger of the right hand is placed on the pivot point and the protractor is restrained from rotating by pressing on it with another finger of the same hand.   The arm is rotated counterclockwise, by using the left hand, until a pole is aligned with the reference line on the arm (see Fig. D-3).   The reading on the protractor is now equal to ϕ_1.   The protractor is now released, but a finger is kept on the pivot point.   Next, the arm and protractor are rotated together in a clockwise direction until the reference line R is again horizontal.   The procedure is repeated to add the angles from the other poles.

Angles from the zeros are subtracted by rotating the arm and pro-tractor together until the line R of the arm is aligned with a zero.   The protractor is then restrained, and the arm is rotated back to the horizontal.   When this procedure has included all the poles and zeros, the reading on the protractor is equal to the sum of the angles from the poles minus the sum of the angles from the zeros.   If the sum of these angles is $180°$, the trial point is a point on the root locus.   If the angle is not $180°$, then another trial point must be chosen.   A systematic procedure should be used for

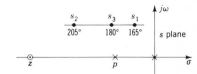

Fig. D-4 *Successive trial points for locating the root locus.*

successive trial points. For example, successive trial points may be taken along a horizontal line. Figure D-4 shows a trial point s_1 for which the sum of the angles is 165°. The sum of the angles increases if the second trial point is taken to the left of s_1. The second trial point s_2 shows an angle of 205°. The point on the root locus must lie between s_1 and s_2. The point s_3 is then located for which the sum of the angles is 180°.

D-3   *Shortcuts in Angle Measurements*

With experience, shortcut methods can be derived and used. For example, with the protractor and arm aligned with the zero, as shown in Fig. D-5, the protractor is held and the arm is rotated until the R line is over with the pole. The net angle indicated on the protractor is $\phi - \psi$. This operation measures simultaneously, in effect, the angle contributed by both the pole and the zero. The procedure can be applied repeatedly to reduce the work of determining the sum of the angles at a point.

The *angle of departure* from a complex pole can be determined by use of the Spirule. To do this, the center of the Spirule is placed on the complex pole. Then the angles from all the other poles and zeros are measured, using the procedure outlined in this appendix. With the relative position of the arm and protractor maintained, the Spirule is rotated until the line R is aligned with the horizontal reference and points to the left. The 0° index now points along the departure angle from the complex pole. An example will illustrate the procedure.

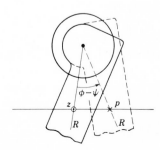

Fig. D-5 *Net angle due to a pole and a zero.*

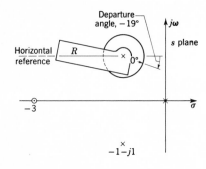

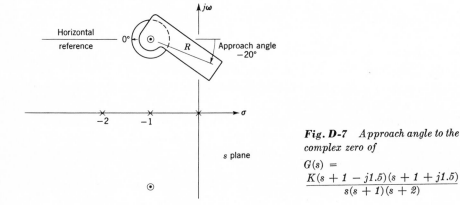

Fig. D-6 *Departure angle from complex poles determined by use of the Spirule.*

Fig. D-7 *Approach angle to the complex zero of*

$$G(s) = \frac{K(s + 1 - j1.5)(s + 1 + j1.5)}{s(s + 1)(s + 2)}$$

Consider the transfer function

$$G(s) = \frac{K(s + 3)}{s(s^2 + 2s + 2)} \tag{D-4}$$

The center of the Spirule is placed on the complex pole $s = -1 + j1$, and the angles due to the other two poles and the zero are measured. The Spirule reads 199°. With the arm and protractor in position, the line R on the arm is aligned with the horizontal reference, as shown in Fig. D-6. The arrow on the 0° index points in the direction of the departure angle and is $-19°$.

The *approach angle* to a complex zero can also be obtained by use of the Spirule. The procedure is to place the center of the Spirule on the complex zero. The sum of the angles from all the poles minus the angles from the other zeros is determined. Then, with the relative position of the arm and protractor maintained, the Spirule is rotated until the 0° index is directed to the left and is aligned with the horizontal reference. The reference line R on the arm now points in the direction of the approach angle. Figure D-7 illustrates this procedure.

D-4 Damping Ratio

The Spirule is calibrated to read damping ratio directly. To use the damping-ratio scale the protractor is placed with its center at the origin of the *s* plane and with the 0° index pointing along the horizontal line to the right. The arm is then turned until its intersection with the damping-ratio scale occurs at the desired value of ζ. The edge of the arm now describes the constant ζ line in the *s* plane.

D-5 Multiplication of Lengths

Multiplication and division of lengths are performed by using the logarithmic spiral S drawn on the arm of the Spirule. The need to perform these operations occurs after the exact shape of the root locus has been determined by applying the angle condition. Based on the specifications for the desired system performance, the roots of the characteristic equation are selected. The next step is to evaluate the required magnitude of the static loop sensitivity K. This is done by use of the magnitude condition which requires that

$$|G(s)| = 1 \tag{D-5}$$

Solving Eq. (D-5) for the static loop sensitivity gives

$$K = \frac{\displaystyle\prod_{c=1}^{v} |s - p_c|}{\displaystyle\prod_{m=1}^{w} |s - z_m|} \tag{D-6}$$

For the system described by Eq. (D-2) the magnitude condition requires that

$$K = \frac{|s| \cdot |s - p|}{|s - z|} \tag{D-7}$$

The root locus for this system is drawn in Fig. D-8, and it is assumed that the roots s_1 and s_2 have been selected. The static loop sensitivity K for these roots is determined by inserting the magnitudes of the lengths shown in Fig. D-8 into Eq. (D-7).

The value of K is determined by use of the Spirule in the following manner: First, line R on the arm is set in line with the X—1 index arrow.

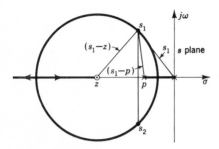

Fig. D-8 *Root locus for* $G(s) = \dfrac{K(s-z)}{s(s-p)}$

The pivot of the Spirule is placed over the point s_1, to which the lengths are to be measured.   To measure the length to a pole such as p, both the arm and the disk are rotated until the line R of the arm is aligned with the pole. The disk is held stationary and the arm is rotated until the spiral curve S crosses over the pole.   The numerical value of the magnitude $s - p$, to the given scale, can be read from the index arrow and the spiral calibration scale on the arm.   The next directed line segment is multiplied by releasing the protractor so that the assembly can rotate and aligning the arm with the next pole.   The protractor is then restrained, and the arm is rotated until the spiral lies over the second pole.   The index arrow now points to a reading on the spiral calibration scale that represents the product of the two magnitudes.   Division by the length from a zero $(s - z)$ is accomplished by releasing the protractor, so that the assembly can rotate, and aligning the Spirule so that the spiral curve crosses over the zero.   The protractor is then held while the arm is rotated until the R line of the arm is aligned with the zero.   The index arrow now points to a reading on the spiral calibration scale that is equal to the previous number divided by the length $s - z$.   This procedure is continued until all the lengths have been included.   When lengths greater than 1.42 units on the arm scale are involved, the reflected spiral curve which is drawn as a dashed curve may be used.   However, since this is a reflected curve, the procedure for poles and zero must be modified.   The curve $S/2$ can also be used but the angle associated with it must be introduced twice.   The correct value of K is obtained by multiplying the reading on the spiral scale by the index factor and the scale factor:

$$K = (\text{Spirule reading})(\text{index factor})(\text{scale factor})^x \qquad \text{(D-8)}$$

The index factor is read from the Spirule and is 0.1, 1, or 10, depending on whether the X–0.1, X–1, or X–10 index arrow points to the Spirule reading on the spiral calibration scale.   The scale factor is the numerical value of the scale of the plot corresponding to 5 in. (the unit length on the Spirule arm) and x is equal to the number of poles minus the number of zeros of $G(s)$.

As an example, consider the transfer function

$$G(s) = \frac{K(s + 4)}{s(s + 1)} \tag{D-9}$$

The pole-zero diagram is drawn to the scale 1 unit equals 1 in.; the scale factor is therefore equal to 5. The point $s_1 = 1 + j1.7$ is a point on the root locus. The gain sensitivity at this point is found in the following manner: The arm of the Spirule is placed at the index X–1 and the arm is aligned with the pole at the origin. Holding the protractor and rotating the line R on the arm until the curve crosses the pole results in a reading of 0.39. The procedure is repeated for the pole at $s = -1$, and the Spirule reading on the spiral curve is now 0.133. The Spirule is then rotated until the spiral curve crosses the zero at $s = -4$. The protractor is held while the line R on the arm is aligned with the zero. The X–1 index points to the reading 0.193 on the spiral curve. The value of gain sensitivity K is

$$K = 0.193 \times 1 \times (5^{2-1}) = 0.965 \tag{D-10}$$

Familiarity with the use of the Spirule expedites the work of plotting the root locus and the determination of static loop sensitivities.

Appendix **E**

Angle to a Search Point near the Real Axis Which Is Produced by Complex Poles and Zeros

Calculation of breakaway and break-in points of the root locus on the real axis is covered in Sec. 7-8. The change in the angle due to the complex roots at a small distance δ off the real axis is given by

$$\Delta\phi = \Delta\phi_1 + \Delta\phi_2 = \frac{2\delta b}{a^2 + b^2} \tag{E-1}$$

The dimensions are shown in Fig. E-1.

Equation (E-1) can be derived from the expressions for the areas of the two triangles having the common side δ. The area of each triangle is calculated by using both of the following: (1) one-half the base times the altitude and (2) one-half the product of two sides times the sine of the included

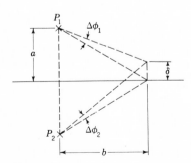

Fig. E-1 *Incremental angles produced by complex poles.*

angle. Thus

$$A_1 = \tfrac{1}{2}\delta b = \tfrac{1}{2}\sqrt{a^2 + b^2}\sqrt{(a - \delta)^2 + b^2}\sin \Delta\phi_1 \qquad \text{(E-2)}$$
$$A_2 = \tfrac{1}{2}\delta b = \tfrac{1}{2}\sqrt{a^2 + b^2}\sqrt{(a + \delta)^2 + b^2}\sin \Delta\phi_2 \qquad \text{(E-3)}$$

When the angle is small, the angle is approximately equal to the sine of the angle. Thus $\sin \Delta\phi_1 \approx \Delta\phi_1$ and $\sin \Delta\phi_2 \approx \Delta\phi_2$. This approximation yields, from Eqs. (E-2) and (E-3), the following:

$$\Delta\phi = \Delta\phi_1 + \Delta\phi_2 \approx \frac{\delta b}{\sqrt{a^2 + b^2}\sqrt{(a - \delta)^2 + b^2}}$$
$$+ \frac{\delta b}{\sqrt{a^2 + b^2}\sqrt{(a + \delta)^2 + b^2}} \qquad \text{(E-4)}$$

Taking the limit of this equation as δ becomes very small results in the approximate equation

$$\Delta\phi = \Delta\phi_1 + \Delta\phi_2 = \frac{2\delta b}{a^2 + b^2} \qquad \text{(E-5)}$$

Appendix F

Phase Angle Loci

A straightforward and logical method for the determination of the root locus of a system is the method of phase angle loci which was introduced by Chu. This permits the plotting of the root locus in a systematic manner and provides a convenient means for reshaping the locus to obtain more desirable performance characteristics.

For a given transfer function $G(s)$, the plot on the complex s plane of the family of curves satisfying the relationship

$$\text{angle of } G(s) = \phi \qquad \text{(F-1)}$$

is called the phase angle loci. The variable $s = \sigma + j\omega$ is the Laplace operator. A separate locus is drawn for each angle ϕ. The root locus is that particular phase angle locus for which $\phi = (1 + 2m)180°$, where m takes on integer values.

The most basic form of the phase angle loci is obtained for the transfer function

$$G'(s) = (s + s_1)^{\pm 1} \qquad \text{(F-2)}$$

The value of s_1 may be zero, real, or complex. If the exponent is negative the transfer function has a pole, and if the exponent is positive the transfer function has a zero. The phase angle loci for this transfer function are a family of radial straight lines that start at $-s_1$ and have a slope equal to $\tan \phi$. Figure F-1 shows these angle loci for a real pole.* The angle loci for a zero are the same, but the signs of the angles are changed.*

Another basic form of the phase angle loci is obtained for a transfer function which has a pole and a zero on the real axis of the s plane:

$$G'(s) = \frac{s + \sigma_1}{s + \sigma_2} \qquad \text{(F-3)}$$

---

* In this appendix the angle due to a pole is shown as negative and the angle due to a zero is shown as positive. Note that in Chap. 7 the signs are reversed.

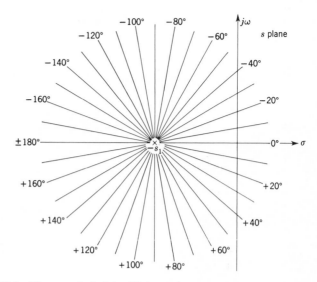

Fig. F-1   *Phase angle loci for $G'(s) = 1/(s + s_1)$.*

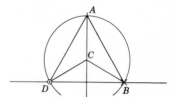

Fig. F-2   *Phase-angle-locus construction.*

The phase angle loci are circles that go through the pole and zero and have their centers along the perpendicular bisector of the line connecting the pole and zero. The coordinates of the center of the circle are

$$-\frac{\sigma_1 + \sigma_2}{2} + j\frac{\sigma_2 - \sigma_1}{2}\cot\phi \qquad\qquad \text{(F-4)}$$

The centers of the phase-angle-loci circles can also be determined graphically. Figure F-2 shows one circle representing a phase angle locus for the angle $\phi = \underline{/DAB}$. From the geometry of the figure, where the point C is the center of the circle, $\underline{/DBC}$ is equal to

$$\underline{/DBC} = 90° - \underline{/\phi} \qquad\qquad \text{(F-5)}$$

Now the center of the circle can be determined by first drawing the perpendicular bisector of the line joining the pole and zero. Then, from B, a line is drawn at $\underline{/DBC}$ determined from Eq. (F-5) to intersect the perpendicular. This intersection point C is the center of the circle.

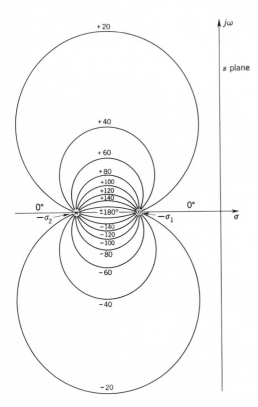

Fig. F-3 *Phase angle loci for* $G'(s) = (s + \sigma_1)/(s + \sigma_2)$.

Figure F-3 shows the family of phase angle loci for the case where the pole is to the left of the zero.

The method of obtaining the phase angle loci for a complex transfer function makes use of the principle of superposition. The transfer function is divided into sufficiently simple factors for which the phase angle loci are known. The loci for all the factors are then combined to form the phase angle loci of the complete system. This method is illustrated in the following examples.

Example 1

$$G'(s) = \frac{1}{(s + \sigma_1)(s + \sigma_2)} \tag{F-6}$$

The phase angle of $G'(s)$ in terms of its factors is

$$\underline{/G'(s)} = \underline{/\frac{1}{s + \sigma_1}} + \underline{/\frac{1}{s + \sigma_2}} \tag{F-7}$$

$$\phi = \phi_1 + \phi_2 \tag{F-8}$$

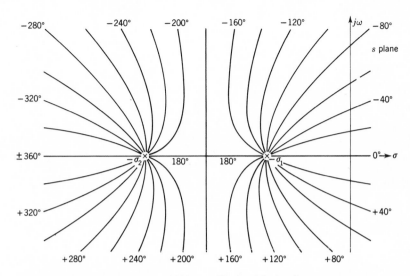

Fig. F-4 *Phase angle loci for* $G'(s) = 1/(s + \sigma_1)(s + \sigma_2)$.

The phase angle loci for each factor on the right side of Eq. (F-7) are the same as those shown in Fig. F-1 except that the poles are located at $-\sigma_1$ and $-\sigma_2$. These curves are superimposed on the same s plane at the proper locations. A large number of points of intersection result between these two superimposed loci. Only intersections of the two families of curves at which the sum of ϕ_1 and ϕ_2 is equal to a desired value of ϕ are of particular interest. These points are marked, and then the phase angle locus of $G'(s)$ for that angle ϕ is obtained by drawing a curve connecting these points. Figure F-4 shows the composite phase angle curve for various values of ϕ, for the transfer function which has two real poles. The procedure outlined above would also be used if the poles were complex.

Example 2

$$G'(s) = \frac{1}{s(s + \sigma_1)(s + \sigma_2)} \tag{F-9}$$

The phase angle of $G'(s)$ is given by

$$\underline{/G'(s)} = \underline{/\frac{1}{s}} + \underline{/\frac{1}{(s + \sigma_1)(s + \sigma_2)}} \tag{F-10}$$

The phase angle loci of $1/s$ are the same as those drawn in Fig. F-1 except that they are drawn from the origin, and the phase angle loci of $1/[(s + \sigma_1)(s + \sigma_2)]$ are the same as those drawn in Fig. F-4. The method of super-

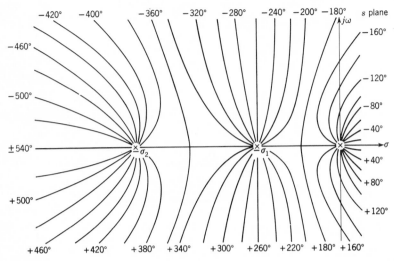

Fig. F-5 *Phase angle loci for* $G'(s) = 1/s(s + \sigma_1)(s + \sigma_2)$.

position can now be used with these curves to obtain the composite curves shown in Fig. F-5.

Example 3

$$G'(s) = \frac{(s + \sigma_1)(s + \sigma_2)}{(s + \sigma_3)(s + \sigma_4)} \qquad (F\text{-}11)$$

This is the transfer function of the lag-lead compensator. The phase angle $G'(s)$ in terms of its factors is

$$\underline{/G'(s)} = \underline{/\frac{s + \sigma_1}{s + \sigma_4}} + \underline{/\frac{s + \sigma_2}{s + \sigma_3}} \qquad (F\text{-}12)$$

$$\phi = \phi_1 + \phi_2 \qquad (F\text{-}13)$$

The phase angle loci of the basic factor $(s + \sigma_2)/(s + \sigma_3)$ have the form shown in Fig. F-3. The phase angle loci of the factor $(s + \sigma_1)/(s + \sigma_4)$ have the same form except that the signs of the angle are changed. Figure F-6 shows a family of phase-angle-locus circles 10° apart for each of the two factors. The intersection points of the two loci locate points on the phase angle locus of $G'(s)$ which have the magnitude $\phi = \phi_1 + \phi_2$. Additional points for which the sum of the two component angles yields the same value are located in the same manner. The composite phase angle loci obtained by this procedure are drawn in Fig. F-7 for a lag-lead compensator.

When the use of a lag-lead compensator is warranted, the phase angle loci of the compensator can be superimposed on those of the original open-

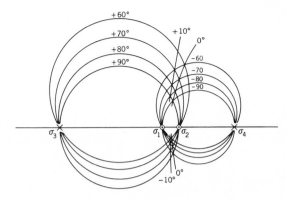

Fig. F-6 *Phase-angle-loci construction for two poles and two zeros.*

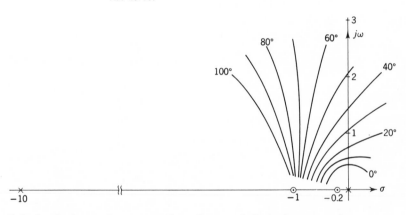

Fig. F-7 *Lag-lead cascade-compensation phase angle loci of compensator* $G'(s) = (s + 1)(s + 0.2)/(s + 10)(s + 0.02)$.

loop transfer function. The dominant roots can then be determined fairly easily since they fall on the line representing the desired damping ratio ζ. The point on this line for which the sum of all the angles is equal to 180° is a dominant complex root. The phase angle loci of the compensator can be placed at several trial positions to obtain the desired results.

There are several important properties of the phase angle loci which are an aid in their construction.

1. Phase angle loci are symmetrical to the real axis of the s plane.
2. The shape of the phase angle loci depends on the relative positions of the poles and zeros and is independent of the axes of the s plane.
3. The locus for each angle ϕ approaches the straight-line asymptote that makes the angle $-\phi/(v - w)$ with the real axis, where v is the number of poles and w is the number of zeros of $G(s)$.

4.   The asymptotes all intersect the real axis at the point

$$\sigma_o = \frac{\displaystyle\sum_{c=1}^{v} (p_c) - \sum_{m=1}^{w} (z_m)}{v - w} \tag{F-14}$$

The angle loci can be obtained from the root-locus digital-computer program of reference 1.   Introducing a hypothetical zero at a large magnitude and appropriate angle will convert the program to compute the angle loci.

Bibliography

1. Paskin, H. M., and C. W. Richard, Jr.: Computation of Root Loci, available from IBM 1620 Users Library, P. O. Box 790, White Plains, New York.

Conversion of Decibels
to Magnitude

To convert decibels to the corresponding magnitude, locate the decibel value in the proper row in Fig. G-1. Then read off the value $|A|$. The magnitude corresponding to this value of decibels is expressed by

$$B = |A|(10)^n \qquad \text{(G-1)}$$

where n represents the row in which the decibel value is located.

Example

Given:

$$\text{Lm } A(10)^n = 46 \text{ db}$$

From Fig. G-1, for 46 db,

$$n = 2 \quad \text{and} \quad |A| = 2$$

Therefore

$$B = |A|(10)^n = 2 \times 10^2 = 200$$

For negative values of decibels, the corresponding magnitude can be found by determining $|A|(10)^n$ for $|-\text{db}|$ from Fig. G-1. When this value has been found, the desired magnitude is

$$B = \frac{1}{|A|(10)^n}$$

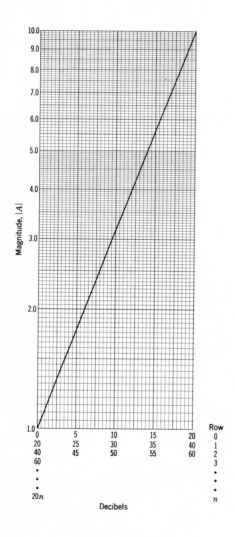

Fig. G-1 *Conversion of decibels to magnitude.*

Design of a Bridged-T Compensator*

In this appendix, the application of the elementary bridged-T network is developed for d-c system compensation. The locus of roots of the transfer function of the bridged-T network is presented in dimensionless form. The transfer function of the bridged-T possesses two zeros, one or the other of which may be adjusted to fall upon any desired point in the left half-plane, including any complex conjugate pair. The bridged-T also produces a pair of poles which fall upon the negative real axis on either side of the real component of the pair of zeros.

A set of curves is presented which allows very rapid synthesis of the *unloaded* bridged-T network with its compensating complex conjugate zeros falling at any desired left-half-plane points. Both types of simple bridged-T network shown in Fig. H-1 can be synthesized from the same graphs. The effects of loading upon the bridged-T network are demonstrated by a general method which shows the locus of the poles and zeros of any four-terminal passive network with varying load termination. The load does not affect the location of the zeros but can be employed to change materially the location of the compensating poles upon the negative real axis.

H-1   Analysis of the Bridged-T Network

Bridged-T RC networks fall in two configurations (see Fig. H-1). The network in Fig. H-1a is referred to as bridged-T Type Ia and the other as Type Ib. The transfer functions of these networks have two poles and two zeros. The poles are always negative and real, but the zeros can be

---

* This appendix is based on reference 1.

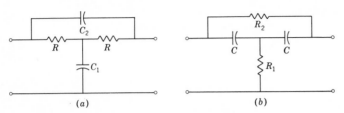

Fig. H-1   *Common types of bridged-T configuration.*

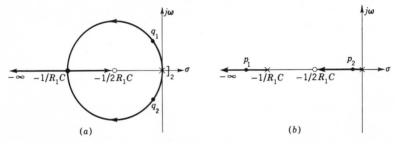

Fig. H-2   *Root loci of zeros and of poles of transfer function versus loop gain* $1/R_2$.   (a) *Of zeros.*   (b) *Of poles.*

made negative real or complex, with negative real parts depending on the ratio of capacitances. The bridged-T Type I*b* is selected for illustration. With zero input source impedance and infinite output load impedance, the voltage transfer function of the network I*b* is

$$G_c(s) = \frac{E_{\text{out}}}{E_{\text{in}}} = \frac{1 + 2R_1Cs + R_1R_2C^2s^2}{1 + (2R_1 + R_2)Cs + R_1R_2C^2s^2} \tag{H-1}$$

The equation for the zeros of Eq. (H-1) is

$$1 + 2R_1Cs + R_1R_2C^2s^2 = 0 \tag{H-2}$$

which can be rearranged as

$$\frac{1 + 2R_1Cs}{R_1R_2C^2s^2} = -1 \tag{H-3}$$

The pole equation can also be arranged as

$$\frac{1 + 2R_1Cs}{R_2Cs(1 + R_1Cs)} = -1 \tag{H-4}$$

If $1/R_2$ is assumed to be the loop gain and allowed to vary from zero to infinity, the root-locus plot of Eq. (H-3) gives the zero configuration for the network, while the root locus of Eq. (H-4) yields the pole characteristics, as shown in Fig. H-2. From these plots, it may be seen that the zeros of the transfer function can be made complex or real in the left half-plane, while the poles are always negative real when R_2 is varied.

The transfer function of the network Type Ia with zero input impedance and infinite output load impedance is found to be

$$G(s) = \frac{E_{\text{out}}}{E_{\text{in}}} = \frac{1 + 2RC_2 s + R^2 C_1 C_2 s^2}{1 + R(C_1 + 2C_2)s + R^2 C_1 C_2 s^2} \qquad \text{(H-5)}$$

By comparing the transfer functions of both networks, Eqs. (H-1) and (H-5), it is obvious that they have similar characteristics. If R, C_1, and C_2 (Type Ia) in Eq. (H-5) are replaced by C, R_2, and R_1, respectively, Eq. (H-5) becomes the transfer function for Type Ib as shown in Eq. (H-1). A single graph giving the pole-zero laws for both types of network in normalized coordinates is presented in this appendix. It allows the rapid location of a pair of complex conjugate zeros anywhere in the left-half plane.

H-2   *Normalized Characteristics*

For design convenience, a graphical method is presented. It should be made clear at the beginning that all derivations are given with *no loading*. It is desired to design the bridged-T networks with specified conjugate complex zeros. The poles are dependent on the zero locations. However, loading influences the pole-zero location of the transfer function. This effect is taken into account under the synthesis procedure presented later.

As shown, Type Ia and Ib networks are equivalent to each other. The same transfer function can be obtained just by switching their resistances and capacitances. Therefore, for the development of the design charts, only network Ib is analyzed. As shown in Fig. H-2a, the locus of complex zeros is along the circle centered at $-1/2R_1 C$. If Eq. (H-2) is rewritten in terms of its damping ratio ζ and undamped natural frequency ω_n, it becomes

$$\frac{1}{R_1 R_2 C^2} + \frac{2}{R_2 C}s + s^2 = \omega_n^2 + 2\zeta\omega_n s + s^2 = 0 \qquad \text{(H-6}a\text{)}$$

where

$$\text{Undamped frequency} = \omega_n = \frac{1}{C\sqrt{R_1 R_2}}$$

$$\text{Damping ratio} = \zeta = \sqrt{\frac{R_1}{R_2}} \qquad \text{(H-6}b\text{)}$$

$$\frac{\omega_n}{\zeta} = \frac{1}{R_1 C} \qquad \text{(H-6}c\text{)}$$

The damping ratio ζ is seen to depend upon the ratio R_2/R_1. To normalize the locus of zeros in Fig. H-2a, a scale change is introduced to cause the circle diameter to be equal to unity or, in other words, to cause $1/R_1C = 1$. With $R_1C = 1$, Eq. (H-2) is normalized and can be rewritten as

$$\frac{1}{R_1R_2C^2} + \frac{2}{R_2C}s + s^2 = \frac{R_1}{R_2} + 2\frac{R_1}{R_2}s + s^2 = \omega_{nz}^2 + 2\zeta_{nz}\omega_{nz}s + s^2 = 0$$

$$(\text{H-7})$$

from which it may be seen that the normalized natural frequency and the normalized damping ratio are, respectively,

$$\omega_{nz} = \sqrt{\frac{R_1}{R_2}} = \frac{1}{\sqrt{a}} \qquad (\text{H-8}a)$$

$$\zeta_{nz} = \sqrt{\frac{R_1}{R_2}} = \frac{1}{\sqrt{a}} \qquad (\text{H-8}b)$$

where $a = R_2/R_1$.

But by comparing Eqs. (H-6b), (H-8a), and (H-8b), it is seen that the original damping ratio, the normalized damping ratio, and the normalized natural frequency of oscillation are all equal:

$$\zeta = \zeta_{nz} = \omega_{nz} = \sqrt{\frac{R_1}{R_2}} = \frac{1}{\sqrt{a}} \qquad (\text{H-9})$$

For complex conjugate zeros it is necessary for R_2 to be greater than R_1. By normalizing Eq. (H-2) into Eq. (H-7) by setting $1/R_1C = 1$, the locus of zeros is plotted in Fig. H-3 as the ratio R_2/R_1 is varied from unity to infinity. Similarly, Eq. (H-4) can also be normalized with the use of the scale factor R_1C and the locus of poles plotted in Fig. H-3 as the ratio R_2/R_1 is varied from unity to infinity. Note that the plots could be presented in terms of R_1/R_2, but it is desirable to use large numbers rather than small decimals.

The ability of the bridged-T network to synthesize any left-half-plane complex zero can be realized from Fig. H-3. The complex zeros can be readily placed with any desired damping ratio between zero and unity. Further, since the damping ratio is invariant with scale change, the designer has only to change scale by selecting the proper value of $1/R_1C$ to adjust the complex zeros to any desired values on the required ζ line.

The chart of Fig. H-3 can be used for networks of both Type Ia and Type Ib. The numbers in the bracket indicate the ratio C_1/C_2 for network Ia, or R_2/R_1 for network Ib. Direct readings of poles and zeros can be obtained only under the condition that $1/RC_2$ (network Ia) or $1/R_1C$ (network Ib) equal unity. The exact location of the poles and zeros can be obtained by multiplying the chart reading by $1/RC_2$ or $1/R_1C$. These

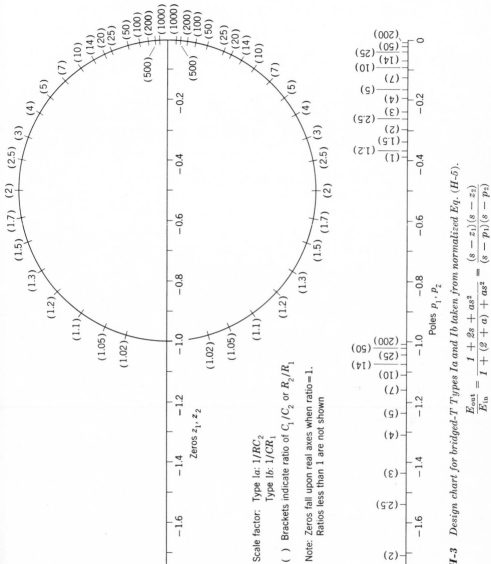

Scale factor: Type Ia: $1/RC_2$
Type Ib: $1/CR_1$

( ) Brackets indicate ratio of C_1/C_2 or R_2/R_1

Note: Zeros fall upon real axes when ratio=1.
Ratios less than 1 are not shown

Zeros z_1, z_2

Poles p_1, p_2

Fig. H-3 *Design chart for bridged-T Types Ia and Ib taken from normalized Eq.* (H-5).

$$\frac{E_{\text{out}}}{E_{\text{in}}} = \frac{1 + 2s + as^2}{1 + (2 + a)s + as^2} = \frac{(s - z_1)(s - z_2)}{(s - p_1)(s - p_2)}$$

737

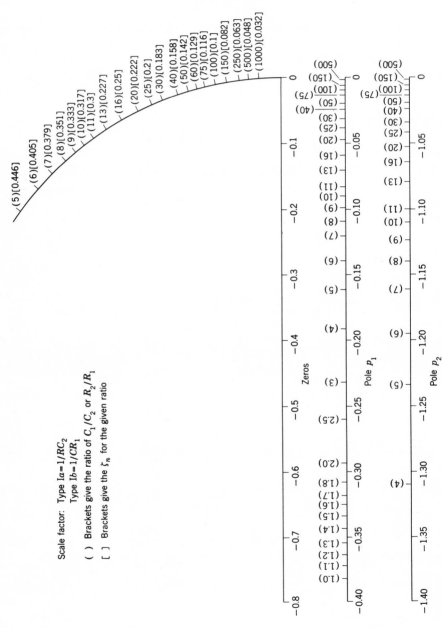

Scale factor: Type $Ia=1/RC_2$
Type $Ib=1/CR_1$

( ) Brackets give the ratio of C_1/C_2 or R_2/R_1

[ ] Brackets give the ζ_n for the given ratio

Fig. H-4 Expanded chart for small damping ratio taken from normalized Eq. (H-5).

factors are therefore the scale factors for the determination of the actual pole-zero locations. On the other hand, the desired locations of the zeros might be given and the problem would then be one of determining the required network. In this case, the given value of ζ would produce R_2/R_1 or C_1/C_2 from the chart. The ratio between the actual coordinate of the zeros and the normalized coordinate would produce the scale factor which gives $1/RC_2$ or $1/R_1C$.

For compensation in many practical applications, one must place the complex zeros close to the imaginary axis. To obtain more accuracy, an expanded chart of the region in the vicinity of the imaginary axis is provided in Fig. H-4. Two examples of the application of the bridged-T charts are given.

Example 1

The problem is to determine the voltage transfer function of the network shown in Fig. H-1a with $R = 1$ megohm, $C_1 = 1$ μf, and $C_2 = 0.1$ μf. This is a Type Ia network with $1/RC_2 = 10$ and $C_1/C_2 = 10$. $1/RC_2$ becomes the scale factor. From the chart in Fig. H-4, the zero and pole locations are:

Zeros:	$-0.1 \pm j0.3$
Poles:	$-0.09, -1.11$

Multiply the singularities by $1/RC_2$ to obtain the correct transfer function

$$\frac{E_{\text{out}}}{E_{\text{in}}} = \frac{(s + 1 + j3)(s + 1 - j3)}{(s + 0.9)(s + 11.1)} \tag{H-10}$$

Example 2

The problem is to design a bridged-T Type Ia network with a pair of complex zeros at $\zeta = 0.2$ and $\omega_n = 2$ and to find its voltage transfer function. From Eq. (H-6b), if $\zeta = 0.2$, then R_2/R_1 or $C_1/C_2 = 25$. From the expanded chart, Fig. H-4, for $C_1/C_2 = 25$, the zero and pole locations are:

Zeros:	$-0.04 \pm j0.195$
Poles:	$-0.0384, -1.042$

From Eq. (H-6c), $1/RC_2 = \omega_n/\zeta = 2/0.2 = 10$. The actual pole-zero locations are:

Zeros:	$-0.4 \pm j1.95$
Poles:	$-0.384, -10.42$

The transfer function is

$$G_c(s) = \frac{(s + 0.4 + j1.95)(s + 0.4 - j1.95)}{(s + 0.384)(s + 10.42)}$$

Network Type Ia can be obtained by selecting its components under the condition that $C_1/C_2 = 25$ and $1/RC_2 = 10$.   If one selects $R = 1$ megohm, then $C_2 = 0.1$ μf and $C_1 = 2.5$ μf.

H-3   Summary of Properties of Unloaded Bridged-T

From the previous analysis, the properties of both types of networks without loading may be summarized as follows.   Refer to Fig. H-1a and b.

1. Product of poles = product of zeros [see Eqs. (H-1) and (H-5)], that is, the coefficients of s^2 in both the numerator and the denominator are the same.
2. Poles are always on the negative real axis.
3. Zeros can be made negative real or complex according to the ratio of C_1/C_2 (Type Ia) or R_2/R_1 (Type Ib).

Network Ia	Network Ib	Zeros
$C_1/C_2 > 1$	$R_2/R_1 > 1$	Complex
$C_1/C_2 = 1$	$R_2/R_1 = 1$	Double, negative real
$C_1/C_2 < 1$	$R_2/R_1 < 1$	Different, negative real

4. One pole is always closer to the origin than the real part of the zeros.
5. On the complex plane, the locus of zeros is a cricle.
6. There is no d-c attenuation.
7. For very large C_1/C_2 or R_2/R_1, one pole and two zeros approach the origin; another pole approaches $-1/RC_2$ for Type Ia and $-1/R_1C$ for Type Ib.

H-4   Effects of Loading Impedance

If a four-terminal network is terminated by load impedance, the pole-zero configuration of its original unloaded transfer function is affected and a new configuration results.   A technique that takes into account loading effects

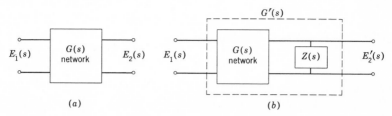

Fig. H-5   *Transfer function G(s) of four-terminal passive networks.*

is used as part of the synthesis procedure in the bridged-T design.   Generally, this method can be applied to any four-terminal network to observe the locus of the pole-zero change of the original transfer function when a variable load impedance is connected across its output terminals.   It can be used, as well, for adjustment of the configuration.   The pole-zero change of the network transfer function due to loading can be obtained by the following steps:

1.   Having determined the original unloaded system transfer function, use Thévenin's theorem to find a new transfer function with load impedance $Z(s)$ placed across the output terminal.
2.   Convert this function into an equivalent feedback control system.
3.   Apply the root-locus technique to the equivalent control system in step 2.   The roots of the locus plot now show the changing poles and zeros under the variable load conditions.

The three steps are analyzed as follows: Suppose $G(s)$ is the original unloaded voltage transfer function, $Z(s)$ is the load impedance function, and $G'(s)$ is the new transfer function including load impedance $Z(s)$.

From Fig. H-5a and b, respectively,

$$\frac{E_2(s)}{E_1(s)} = G(s) \tag{H-11}$$

$$\frac{E_2'(s)}{E_1(s)} = G'(s) \tag{H-12}$$

By Thévenin's theorem, Fig. H-5b can be transformed to Fig. H-6.   The new transfer function is

$$G'(s) = \frac{E_2'(s)}{E_1(s)} = \frac{G(s)Z(s)}{Z'(s) + Z(s)} = \frac{G(s)}{1 + Z'(s)/Z(s)} \tag{H-13}$$

where $Z'(s)$ = impedance looking back into the output of network $G(s)$ with the input short-circuited.

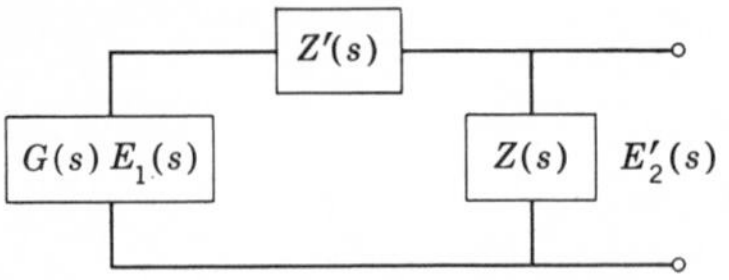

Fig. H-6 *Transformed circuit by Thévenin's theorem.*

Suppose $G(s)$, $Z(s)$, and $Z'(s)$ are rewritten as the quotients of polynomials as follows:

$$G(s) = \frac{G_a(s)}{G_b(s)} \qquad Z(s) = \frac{Z_a(s)}{Z_b(s)} \qquad Z'(s) = \frac{Z'_a(s)}{Z'_b(s)}$$

Note that $Z'_b(s) = G_b(s)$ by the property of all four-terminal passive networks. From Eq. (H-13) the poles and zeros of the function $G'(s)$ can be easily obtained as follows:

1. Zeros of $G'(s)$ = combined zeros of $G(s)$ and $Z(s)$.
2. Poles of $G'(s)$ are the roots of $Z'(s)/Z(s) = -1$, which can be found by the root-locus technique.
3. D-c attenuation = $G'(0)$ can be obtained from Eq. (H-13) with $s = 0$.
4. Finally, $G'(s)$ can be written as

$$G'(s) = G'(0) \frac{(1 - s/a_1)(1 - s/a_2) \cdots (1 - s/a_n)}{(1 - s/r_1)(1 - s/r_2) \cdots (1 - s/r_n)} \qquad \text{(H-14)}$$

where $a_1, a_2, \ldots, a_n$ = zeros of $G(s)$ and $Z(s)$
$r_1, r_2, \ldots, r_n$ = roots of $Z'(s)/Z(s) = -1$

H-5   *Loading Effect on Type Ia*

Case 1

$Z(s)$ is pure resistance (see Fig. H-5). Suppose that

$$G(s) = \frac{1 + as + bs^2}{(1 + \alpha_1 s)(1 + \alpha_2 s)}$$
$$Z(s) = R_t$$

$Z'(s)$ is found to be

$$Z'(s) = \frac{2R(1 + RC_1 s/2)}{1 + R(C_1 + 2C_2)s + R^2 C_1 C_2 s^2} = \frac{2R(1 + RC_1 s/2)}{(1 + \alpha_1 s)(1 + \alpha_2 s)} \qquad \text{(H-15)}$$

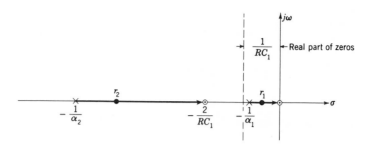

Fig. H-7 *Root locus of* $2RC_t \dfrac{s(1 + RC_1s/2)}{(1 + \alpha_1s)(1 + \alpha_2s)} = -1$

Thus

$$\frac{Z'(s)}{Z(s)} = \frac{2R}{R_t} \frac{(1 + RC_1s/2)}{(1 + \alpha_1s)(1 + \alpha_2s)} = -1 \qquad \text{(H-16)}$$

The poles of $G'(s)$ may now be obtained from the root-locus plot of Eq. (H-16). Thus

$$G'(s) = \frac{R_t}{2R + R_t} \frac{1 + as + bs^2}{(1 - s/r_1)(1 - s/r_2)} \qquad \text{(H-17)}$$

Analysis of bridged-T Type Ia with load resistance R_t shows the following properties for $G'(s)$:

1. D-c attenuation $= R_t/(2R + R_t)$.
2. Zeros are the same as the original function $G(s)$.
3. Both poles of $G'(s)$ are real and move to the left if R_t is decreased.
4. Limit of movement for the pole near the origin is $-2/RC_1$ or twice the real part of the zeros. The other pole moves toward infinity along the negative real axis.

It is of interest to note that, with the introduction of d-c attenuation, the poles of the original bridged-T transfer function can be pulled farther into the left half-plane and are even capable of moving to the left of the real part of the zeros.

Case 2

$Z(s)$ is pure capacitance (see Fig. H-5). Assume that $G(s)$ and $Z'(s)$ are the same as in Case 1 but that $Z(s) = 1/C_ts$. Then

$$\frac{Z'(s)}{Z(s)} = 2RC_t \frac{s(1 + RC_1s/2)}{(1 + \alpha_1s)(1 + \alpha_2s)} = -1 \qquad \text{(H-18)}$$

The root locus of $Z'(s)/Z(s) = -1$ gives the locations of poles in $G'(s)$. The locations of the new poles of $G'(s)$ depend on C_t if $G(s)$ is fixed. The pole movement versus loop gain C_t is shown in Fig. H-7. If $G'(0) = 1$, then

$$G'(s) = \frac{1 + as + bs^2}{(1 - s/r_1)(1 - s/r_2)} \tag{H-19}$$

The analysis of bridged-T Type Ia with pure capacitive load C_t shows the following properties for $G'(s)$:

1. It contains no d-c attenuation.
2. Zeros are the same as the original.
3. Both poles of $G'(s)$ are real and move to the right if C_t is increased.
4. When C_t approaches infinity, one pole approaches the origin and the other pole approaches $-2/RC_1$, which is twice the real part of the zero.

Case 3

$Z(s)$ is a series combination of resistance and capacitance (see Fig. H-5). $G(s)$ and $Z'(s)$ are the same as in the previous two cases.

$$Z(s) = \frac{1 + R_t C_t s}{C_t s}$$

Zeros of $G'(s) = (1 + as + bs^2)(1 + R_t C_t S)$
Poles of $G'(s) =$ roots of equation

$$2RC_t \frac{s(1 + RC_1 s/2)}{(1 + \alpha_1 s)(1 + \alpha_2 s)(1 + R_t C_t s)} = -1 \tag{H-20}$$

The root-locus plot of Eq. (H-20) yields the poles of $G'(s)$. If R_t and C_t are arbitrarily selected, one zero in $G'(s)$ will be fixed at $-1/R_t C_t$ and the poles of $G'(s)$ are fixed. Obviously, this network has no d-c attenuation. Suppose the roots of the system are fixed at r_1, r_2, and r_3; then

$$G'(s) = \frac{(1 + as + bs^2)(1 + R_t C_t s)}{(1 - s/r_1)(1 - s/r_2)(1 - s/r_3)} \tag{H-21}$$

To make this analysis applicable, it should be emphasized that one negative real zero may be added at any point, depending on the selection of R_t and C_t. Poles can be adjusted by varying C_t with the condition that $R_t C_t$ be kept unchanged. Under this condition, C_t [see Eq. (H-20)] can be considered as the open-loop gain of the system.

Analysis of $G'(s)$ of bridged-T Type Ia with series resistance R_t and capacitance C_t as load impedance shows the following properties:

1. There is no d-c attenuation.
2. It contains three poles and three zeros (two original/zeros and one negative real zero at $-1/R_tC_t$).
3. For a specified position of the new zero, the poles can be adjusted by varying C_t with R_tC_t maintained constant.

A similar analysis of loading effects can be made on the Type Ib network.

H-6   Use of Loading Effects

The behavior of the voltage transfer functions of the Type Ia bridged-T network under three types of loading conditions has been presented. It should be pointed out that under the same loading condition the two types of network behave in a similar manner. The only slight difference is the position of one zero in the equivalent open-loop transfer function $Z'(s)/Z(s)$ for the determination of poles in $G'(s)$. This zero gives a different termination of one root. Even when $G(s)$ for networks Ia and Ib are the same, different pole locations result in $G'(s)$ for equivalent load resistance R_t. Also, if ζ is large, the Type Ia network is preferred.

With a knowledge of the root-locus technique and the results of the analysis given in this appendix, the proper type of bridged-T can be selected to meet system requirements. Generally, the effect of loading can be used to adjust the pole positions of the transfer function of the bridged-T. While it is impossible to fix both poles at will, one can fix the pole which is most significant.

In actual applications, the bridged-T without load should be designed first by consultation of the design charts. Zeros of this network are fixed at the desired positions; then the complete pole-zero configuration of $G(s)$ is obtained. Modifications of $G(s)$ by loading effects are made by selecting the proper type of loading. The exact pole-zero location of $G'(s)$ is obtained by applying the root-locus technique to the equivalent control system $Z'(s)/Z(s)$.

If all specifications of the system were not entirely met, additional measures would, of course, be used. It should also be recognized that in actual systems, the exact pole-zero locations may not be exactly known and will often vary with change in environmental conditions. Exact cancellation of terms is, therefore, seldom if ever achieved.

Reference 1 presents numerical examples of the technique presented in this Appendix. Also, references 2 to 4 cover the design procedures for twin-T compensation networks.

Bibliography

1. Chandaket, P., and A. B. Rosenstein: Notes on Bridged-T Complex Conjugate Compensation and 4-Terminal Network Loading, *Trans. AIEE*, vol. 78, pt. II, pp. 148–163, July, 1959.
2. Slaughter, J. B., and A. B. Rosenstein: Twin-C Compensation Using Root Locus Methods, *Trans. AIEE*, vol. 81, pt. II, pp. 339–349, January, 1963.
3. Lazear, T. J., and A. B. Rosenstein: "On-Pole-Zero Synthesis and the General Twin-T," *Trans. IEEE*, vol. 83, pt. II, pp. 389–393, November, 1964.
4. Barker, A. C., and A. B. Rosenstein: "s-Plane Synthesis of the Symmetrical Twin-T Network," *Trans. IEEE*, vol. 83, pt. II, pp. 382–388, November, 1964.

Appendix I

Nyquist Stability Criterion

The qualitative approach of Sec. 10-2 for the understanding of the Nyquist stability criterion is now presented in more rigorous detail. This criterion was developed by H. Nyquist[1] in 1932 as an application to feedback amplifier design. Since then, it has been applied to other related engineering areas such as feedback control system design. This criterion utilizes a graphical method of determining the number of roots of the characteristic equation $1 + G(s)H(s) = 0$ that lie in the right-half s plane and therefore produce the instability of the control system. Two mathematical theorems are presented first, as they provide the basis for the Nyquist stability criterion.

Theorem 1. Cauchy's theorem

The integral of $W(s)$ between the points s_1 and s_2 in the s plane is the same for either of two paths (see Fig. I-1), provided that $W(s)$ is analytic on both paths and everywhere in the region between these two paths. The function $W(s)$ is analytic if it can be expressed in terms of a power series in the neighborhood of a point lying in a region R.

$$\int_{s_1}^{s_2} W(s)_a \, ds = \int_{s_1}^{s_2} W(s)_b \, ds \tag{I-1}$$

As a consequence, the integral around the closed path or contour in the s plane, for which $W(s)$ is analytic within the enclosed region, is always zero. This characteristic is expressed as

$$\oint W(s) \, ds = 0 \tag{I-2}$$

which is known as Cauchy's Theorem.

In the s plane, consider a region R bounded by a simple closed path C such that $W(s)$ is not analytic at some isolated singular point s_d lying in R (see Fig. I-2). In a circular neighborhood of s_d, lying in R, the function

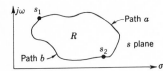

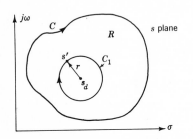

Fig. I-1   An analytic region R in the s plane.

Fig. I-2   A singular point s_d in the s plane.

can be represented by the Laurent series

$$W(s) = \frac{b_1}{s - s_d} + \frac{b_2}{(s - s_d)^2} + \cdots + \frac{b_i}{(s - s_d)_i} + \sum_{v=0}^{\infty} [a_v(s - s_d)^v] \quad \text{(I-3)}$$

for which it is assumed that an integer i exists such that all the b coefficients beyond the ith term vanish and that $b_i \neq 0$. Under these assumptions, the isolated singular point s_d is defined as a pole of order i of the function $W(s)$. A simple pole is defined when the order of the pole is 1, that is, $i = 1$. The function $W(s)$ is then said to have a pole if the value of the function becomes arbitrarily large as s approaches s_d. The coefficients of Eq. (I-3) are given by

$$a_v = \frac{-1}{2\pi j} \int_C \frac{W(s')\,ds'}{(s' - s_d)^{v+1}} \qquad \text{for } v = 0, 1, 2, 3, \ldots \quad \text{(I-4)}$$

$$b_i = \frac{-1}{2\pi j} \int_{C_1} \frac{W(s')\,ds'}{(s' - s_d)^{-i+1}} \qquad \text{for } i = 1, 2, 3, \ldots \quad \text{(I-5)}$$

where each integral is taken clockwise.

The values of s that will make the function $W(s)$ equal zero are called zeros of $W(s)$. Let

$$W(s) = a(s - s_0)^{a_w} U(s) = 0$$

Then s_0 is a zero of order a_w, where s_0 is not a zero of $U(s)$.

Theorem 2.   Residue theorem

Let $W(s)$ be a function that has a simple pole and whose Laurent expansion is given by

$$W(s) = \frac{b_1}{s - s_d} + a_0 + a_1(s - s_d) + a_2(s - s_d)^2 + \cdots \quad \text{(I-6)}$$

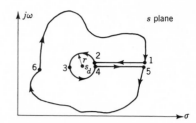

Fig. I-3   An integration path in the s plane.

In Fig. I-3 the path of integration in the s plane about the simple pole s_d is indicated. Since the chosen path does not enclose the isolated point s_d and since the function $W(s)$ is analytic in the region encircled by this path, then, by Cauchy's theorem,

$$\oint W(s) \, ds = 0 \tag{I-7}$$

The portions of the path 1 to 2 and 4 to 5 are parallel straight lines of equal length that are arbitrarily close together. Under these conditions,

$$\int_{C_{4,5}} W(s) \, ds = - \int_{C_{1,2}} W(s) \, ds \tag{I-8}$$

Thus the contributions of these two portions cancel out in the integration over the complete path. To satisfy Eq. (I-7),

$$\int_{C_{5,6,1}} W(s) \, ds = - \int_{C_{2,3,4}} W(s) \, ds \tag{I-9}$$

If both paths had been taken in the clockwise direction, the integral around each loop would have been equal. From this it can be seen that the integral around a closed path enclosing the pole is independent of the path as long as the path encircles the pole in the manner described in Fig. I-3. The integral $\oint W(s) \, ds$ around the path 4,3,2 can be evaluated by letting $s - s_d = re^{j\phi}$. If r is arbitrarily small, then all terms, except the first one, in Eq. (I-6) can be neglected since they become negligible compared with the first term. Thus

$$\int_{C_{4,3,2}} W(s) \, ds = \int_{2\pi}^{0} \frac{b_1}{s - s_d} \, ds = b_1 \int_{2\pi}^{0} \left(\frac{1}{r} e^{-j\phi} \right) (jre^{j\phi} \, d\phi) \tag{I-10}$$

Then

$$\int_{C_{4,3,2}} W(s) \, ds = jb_1 \int_{2\pi}^{0} d\phi = -2\pi jb_1 \tag{I-11}$$

Thus for the simple pole s_d the value of the coefficient b_1 can be expressed as

$$b_1 = \frac{-1}{2\pi j} \int_{C_{4,3,2}} W(s) \, ds \tag{I-12}$$

The value b_1, expressed by Eq. (I-12), is defined to be the *residue* of the function $W(s)$ in the neighborhood of s_d. By integrating the Laurent expansion of $W(s)$ it can be shown that poles of higher order than 1 do not

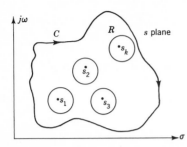

Fig. I-4 Singular points in the s plane.

contribute to the value of the integral of the function about a simple closed path containing the pole.

Derivation of Nyquist's Stability Criterion

Assume now that a finite number of simple poles of the function $W(s)$ lie in a region R and are bounded by a simple closed path C (see Fig. I-4). By utilizing the same technique as for the single simple pole, the general formula for determining the integral of the $W(s)$ around the path C in the s plane is

$$\oint W(s)\, ds = -2\pi j(b_1 + b_2 + b_3 + \cdots + b_k) \qquad \text{(I-13)}$$

where $b_1, b_2, \ldots, b_k$ are the residues at the simple poles $s_1, s_2, \ldots, s_k$ of the function $W(s)$ in R. The above is known as the residue theorem.

The two theorems that have been presented are utilized to derive an expression relating the number of poles and zeros of $B(s)$ lying within a region R bounded by a simple closed path C. The path to be considered is the one that encloses the entire right-half s plane (see Fig. I-5). Consider the function $B(s)$ which has a zero of a_1th order at $s = Z_1$ in the region R. Assume that there are no other poles or zeros in this region. The function $B(s)$ is written as the product of two factors as given below, where $U(s)$ is analytic in R:

$$B(s) = (s - Z_1)^{a_1} U(s) \qquad \text{(I-14)}$$

The derivative of $B(s)$ with respect to s is

$$B'(s) = a_1(s - Z_1)^{a_1-1} U(s) + (s - Z_1)^{a_1} U'(s) \qquad \text{(I-15)}$$

Next, take the ratio of Eq. (I-15) to Eq. (I-14), which is

$$\frac{B'(s)}{B(s)} = \frac{a_1}{s - Z_1} + \frac{U'(s)}{U(s)} \qquad \text{(I-16)}$$

Let $\qquad W(s) = \dfrac{B'(s)}{B(s)} \qquad Y(s) = \dfrac{a_1}{s - Z_1} \qquad V(s) = \dfrac{U'(s)}{U(s)}$

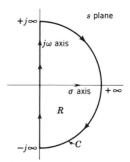

Fig. I-5   *A contour enclosing the entire right-half s plane.*

The term $V(s)$ is analytic since it does not contain any poles or zeros. The function $Y(s)$ has a simple pole at $s = Z_1$. Thus

$$\int_C W(s)\ ds = \int_C Y(s)\ ds + \int_C V(s)\ ds \qquad (\text{I-17})$$

Since the function $V(s)$ is analytic, then

$$\int_C V(s)\ ds = 0 \qquad (\text{I-18})$$

Thus Eq. (I-17) reduces to

$$\int_C W(s)\ ds = \int_C Y(s)\ ds = \int_C \frac{a_1}{s - Z_1}\ ds \qquad (\text{I-19})$$

When this equation is compared with Eqs. (I-10) and (I-11), it follows that

$$\int_C W(s)\ ds = -2\pi j a_1 \qquad (\text{I-20})$$

Thus the order of the zero of $B(s)$ is the residue of the function $W(s)$.   Note that by taking the ratio $B'(s)/B(s)$ the zero of order a_1 of $B(s)$ becomes a simple pole of $W(s)$.   It is this property that allows the use of the residue theorem in the derivation of the Nyquist stability criterion.

Now consider the function $B(s)$ to be of the following general form:

$$B(s) = \frac{(s - Z_1)^{a_1}(s - Z_2)^{a_2} \cdots (s - Z_w)^{a_w}}{(s - P_1)^{b_1}(s - P_2)^{b_2} \cdots (s - P_v)^{b_v}}\ U(s) \qquad (\text{I-21})$$

where all the poles P and zeros Z in the region R are separated from the remainder of the function.   The exponents $a_1, a_2, \ldots, a_w$ and $b_1, b_2, \ldots, b_v$ are the orders of the zeros and poles, respectively.   Taking the derivative of Eq. (I-21) and then using the ratio of $B'(s)/B(s)$, as in the previous example, yields

$$W(s) = \frac{B'(s)}{B(s)} = \left(\frac{a_1}{s - Z_1} + \frac{a_2}{s - Z_2} + \cdots + \frac{a_w}{s - Z_w} \right)$$
$$- \left(\frac{b_1}{s - P_1} + \frac{b_2}{s - P_2} + \cdots + \frac{b_v}{s - P_v} \right) + \frac{U'(s)}{U(s)} \qquad (\text{I-22})$$

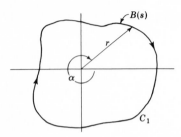

Fig. I-6   *A plot of $B(s)$.*

Note that all the terms in both parentheses are functions each having a simple pole. Since the function $W(s)$ consists only of simple poles, by the residue theorem,

$$\int_C W(s)\,ds = -2\pi j[(a_1 + a_2 + \cdots + a_w) - (b_1 + b_2 + \cdots + b_v)]$$

$$(I\text{-}23)$$

Let
$$P_R = b_1 + b_2 + \cdots + b_v$$
$$Z_R = a_1 + a_2 + \cdots + a_w$$

where both P_R and Z_R take into account the multiplicity of each pole and zero. For example, if

$$B(s) = \frac{(s - Z_1)^1(s - Z_2)^3}{(s - P_1)^2(s - P_2)^1}\,U(s)$$

$$(I\text{-}24)$$

then $P_R = 2 + 1 = 3$ and $Z_R = 1 + 3 = 4$.

Equation (I-23) can be rewritten as

$$\int_C W(s)\,ds = 2\pi j(P_R - Z_R)$$

$$(I\text{-}25)$$

The above result can be interpreted by considering first the contour C of Fig. I-5, which maps into the contour C_1, shown in Fig. I-6, under $B(s)$, as expressed by Eq. (I-21). Secondly, the following properties of complex numbers are also utilized:

(1) $$B(s) = re^{j\alpha}$$ $$(I\text{-}26)$$
(2) $$\ln B(s) = \ln r + j\alpha$$ $$(I\text{-}27)$$
(3) $$\frac{d[\ln B(s)]}{ds} = \frac{d(\ln r)}{ds} + \frac{j\,d\alpha}{ds}$$ $$(I\text{-}28)$$
(4) $$\frac{d[\ln B(s)]}{ds} = \frac{B'(s)}{B(s)} = W(s)$$ $$(I\text{-}29)$$

Substituting Eqs. (I-28) and (I-29) into Eq. (I-25) yields

$$\int_C W(s)\,ds = \int_C \frac{d[\ln B(s)]}{ds}\,ds = \int_C d(\ln r) + j\int_C d\alpha = 2\pi j(P_R - Z_R)$$

$$(I\text{-}30)$$

The integral $\int_C d(\ln r)$ is zero since the magnitude of $\ln B(s)$ is the same at the start α_2 and end α_1 of the contour C_1. Therefore

$$j \int_C d\alpha = 2\pi j(P_R - Z_R) \qquad (I\text{-}31)$$

or

$$\frac{1}{2\pi}(\alpha_2 - \alpha_1) = P_R - Z_R \qquad (I\text{-}32)$$

Let

$$N = \frac{1}{2\pi}(\alpha_2 - \alpha_1) \qquad (I\text{-}33)$$

Thus

$$N = P_R - Z_R \qquad (I\text{-}34)$$

The difference between the final and initial values of α, as expressed by Eq. (I-33), represents the total change in phase angle of $B(s)$ as s takes on all values lying on the path of integration C. Since there are 2π radians/revolution, then N, as expressed by Eq. (I-33), represents the number of encirclements that the $B(s)$ directed line segment makes about the origin (see Fig. I-6) as the s point traverses once the contour C in the s plane (see Fig. I-5).

Assume that counterclockwise rotation of α is positive and that clockwise rotation of α is negative. In other words, N is positive if the net encirclement $\alpha_2 - \alpha_1$ of the origin by $B(s)$ is counterclockwise as s makes one complete clockwise rotation in the s plane. Thus, if any two factors in Eq. (I-34) are given, the other can be found. As stated in Chap. 10, P_R can be readily determined for the $B(s) = 1 + G(s)H(s)$ function from the $G(s)H(s)$ term. Also, N can be readily determined from the $B(s)$ plot. The method presented indicates the existence of zeros Z_R of $B(s)$ in the right-half s plane. Since $B(s)$ represents the characteristic polynomial of the closed-loop system, the stability of the system can now be determined. In other words, the presence of zeros of $B(s)$ in the right-half s plane indicates an unstable system.

Bibliography

1. Nyquist, H.: Regeneration Theory, *Bell System Tech. J.*, vol. 11, pp. 126–147, January, 1932.

Appendix **J**

Block-diagram
*Manipulations**

Manipulation	Original diagram	Equivalent diagram	Equation
1. Interchange of blocks			$b = aG_1G_2$
2. Rearrangement of summing points			$d = a - b - c$
3. Interchange of summing points			$d = a - b + c$
4. Moving a summing point ahead of an element			$d = aG - c$
5. Moving a summing point beyond an element			$c = (a - b)G$
6. Moving a take-off point ahead of an element			$b = aG$
7. Moving a take-off point beyond an element			$b = aG$ $a = b/G$

Manipulation	Original diagram	Equivalent diagram	Equation
8. Moving a take-off point ahead of a summing point			$c = a - b$
9. Moving a take-off point beyond a summing point			$c = a - b$ $a = c + b$
10. Removing an element from a forward path			$d = a(G_1 - G_2)$
11. Inserting an element in a forward path			$d = aG_1 - a$
12. Removing an element from a feedback path			$d = aG_1/(1 + G_1 H)$
13. Inserting an element in a feedback path			$d = aG_1/(1 + G_1)$
14. Inserting a feedback path			$d = aG_1$
15. Inserting a feedback path			$d = aG_1$
16. Combining cascade elements			$b = aG_1G_2$
17. Eliminating a forward loop			$d = a(G_1 - G_2)$

Manipulation	Original diagram	Equivalent diagram	Equation
18. Eliminating a feedback loop			$d = aG_1/(1 + G_1H)$
19. Eliminating a feedback loop			$d = a/(1 + H)$
20. Consolidation of summing points			$e = a - (b + d)$
21. Expansion of summing points			$d = a + b + c$
22. Converting a nonunity feedback system to an equivalent unity feedback system			$H = 1 - H_x$
23. Elimination of a minor loop			$I = G_yR - (G_y + H_x)C = G_yE - H_xC$

* Adapted from T. D. Graybeal, Block Diagram Network Transformation, *Elec. Eng.*, vol. 70, no. 11, 1951.

Problems

2-1. *a.* Draw the mechanical network. *b.* From the mechanical network write the differential equations of performance.

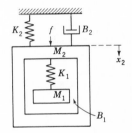

2-2. *a.* Draw the mechanical network. *b.* Write the differential equations of performance. *c.* Draw the analogous electric circuit.

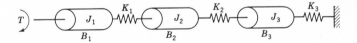

2-3. *a.* Derive the differential equation relating the position $y(t)$ and the force $f(t)$. *b.* Draw an analogous electric circuit. List all the analogous quantities.

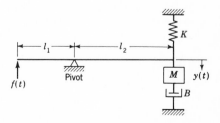

2-4. For the hydraulic preamplifier write the differential equations of performance relating x_1 to y_1. *a.* Neglect the load reaction. *b.* Do not neglect load reaction.

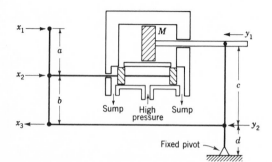

2-5. The polarized solenoid produces a magnetic force proportional to the current in the coil, $f = K_i i$. The coil has resistance and inductance. Write the differential equations of performance.

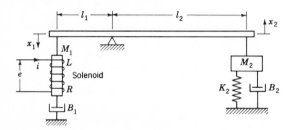

2-6. *a.* Draw the mechanical network for the mechanical system shown. *b.* Draw the analogous electric circuit in which force is analogous to current.

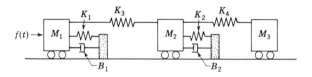

2-7. In the mechanical system shown, r_2 is the radius of the drum. *a.* Write the necessary differential equations of performance for this system. *b.* Obtain a differential equation expressing the relationship of the output x_3 in terms of the input θ_1.

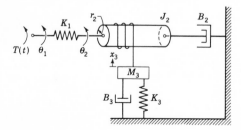

2-8. Write all the necessary equations to determine v_o. *a.* Use nodal equations. *b.* Use loop equations.

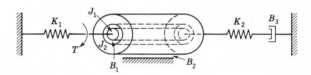

2-9. The figure represents a cylinder of inertia J_1 inside a sleeve of inertia J_2. There is viscous damping B_1 between the cylinder and the sleeve. *a.* Draw the mechanical network. *b.* Write the system equations. *c.* Draw the analogous electric circuit.

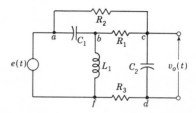

2-10. Write the loop and node equations for the circuit shown after the switch is closed.

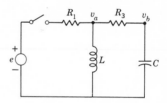

2-11. A sketch of a moving-coil microphone is shown. The diaphragm has the spring elastance K, mass M, and damping B. Fastened to the diaphragm is a coil which moves in the magnetic field produced by the permanent magnets. *a.* Derive the differential equations of the system considering changes from the equilibrium conditions. *b.* Draw an analogous electric circuit.

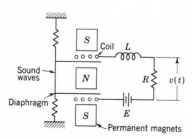

2-12. The two gear trains shown have an identical net reduction, have identical inertias at each stage, and are driven by the same motor. The number of teeth on each gear is indicated on the figures. At the instant of starting, the motor develops a torque T. Which system has the higher initial load acceleration?

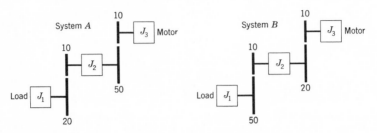

2-13. Write the differential equations describing the motion of the following system, assuming small displacements.

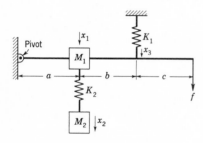

2-14. Most control systems require some type of motive power. One of the most commonly used units is the electric motor. Write the differential equation for the angular displacement of an inertia connected directly to a d-c motor shaft when a d-c voltage is suddenly applied to the armature terminals with the field separately energized.

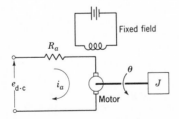

2-15. Write the loop and node equations for the circuit shown after the switch S is closed.

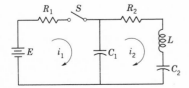

2-16. The circuit shown is in the steady state with the switch S closed. At time $t = 0$, S is opened. Write the necessary differential equations for determining $i_1(t)$.

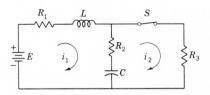

3-1. *a.* What is the initial value of current in R_1 and R_2 when the switch is closed?
b. Solve for the voltage across C_2 as a function of time.

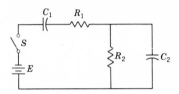

3-2. In Prob. 2-2, the parameters have the following values:

$J_1 = 74,150$ oz-in.^{2} $B_1 = 0.5$ lb-ft/(radian/sec) $K_1 = 0.5$ lb-ft/radian
$J_2 = 1.0$ lb-ft-sec^2 $B_2 = 12.8$ oz-ft/(radian/sec) $K_2 = 8.0$ oz-ft/radian
$J_3 = 1.0$ slug-ft^2 $B_3 = 3.35$ oz-in./(deg/sec) $K_3 = 96.0$ oz-in./radian

Solve for $\theta_3(t)$ if $T(t) = tu(t)$.

3-3. In Prob. 2-5, the parameters have the following values:

$M_1 = M_2 = 0.05$ slug $L = 1$ henry $l_1 = 10$ in.
$K_2 = 1.2$ lb/in. $R = 10$ ohms $l_2 = 20$ in.
$B_1 = B_2 = 15$ oz/(in./sec) $K_i = 20$ oz/amp

Solve for $x_2(t)$ if $e(t) = u(t)$.

3-4. In Prob. 2-4a, the parameters have the following values:

$a = b = 6$ in. $C_1 = 6.0$ (in./sec)/in.
$c = 10$ in.
$d = 4$ in.

Solve for $y_1(t)$ with $x_1(t) = 0.2u(t)$ inches and zero initial conditions.

3-5. Given: (a) $r(t) = [(D + 1)(D^2 + 4D + 5)]c(t)$; (b) $r(t) = tu(t)$; (c) all initial conditions are zero. Determine the complete solution with all constants evaluated.

3-6. The rotational hydraulic transmission described in Sec. 2-8 has an inertia load coupled through a spring. The system equation is

$$4(D^2 + 1)x = (3D^2 + 3D + 1) D\theta_m$$

The system is originally stationary. With $x(t) = 5 \cos (5t + 30°)$, find the motor velocity $\omega_m(t)$. Note: $\theta_m(0) = \theta_L(0) = \omega_L(0) = 0$, but $\omega_m(0) \neq 0$. Its value must be determined by use of Eqs. (2-75) and (2-79).

3-7. A hydraulic motor with inertia load is driven by a variable displacement pump. Determine (a) the undamped natural frequency ω_n; (b) the damping ratio ζ;

(c) the damped natural frequency of the system, ω_d.   The parameters have the following values:

$$S_p = 100 \text{ in.}^3/\text{sec} = d_p\omega_p \qquad L = 0.005 \text{ (in.}^3/\text{sec})/(\text{lb/in.}^2)$$
$$d_m = 1.0 \text{ in.}^3/\text{radian} \qquad C = 1 \text{ in.}^3$$
$$V = 20 \text{ in.}^3 \qquad\qquad K_B = 2.5 \times 10^5 \text{ lb/in.}^2$$
$$J = 200 \text{ lb*-in.}^2$$

3-8.   The system shown is initially at rest.   At time $t = 0$ the string connecting M to W is severed at X.   Find $x(t)$.

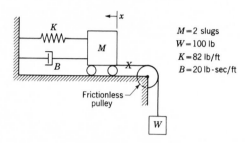

M = 2 slugs
W = 100 lb
K = 82 lb/ft
B = 20 lb-sec/ft

3-9.   Switch S_1 is open, S_2 is closed, and there is no energy stored in the circuit.   *a.* Find $v_o(t)$ after switch S_1 is closed.   *b.* Switch S_2 is opened 0.001 sec after switch S_1 is closed.   Find the current through the inductor.   *c.* Switch S_2 is closed 0.002 sec after it has been opened.   Find $v_o(t)$.

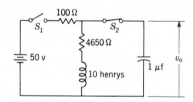

3-10.   Solve the following differential equations.   Assume zero initial conditions.   Sketch the solutions.   *a.* $D^2x + 9x = 1$.   *b.* $D^2x + 5Dx + 4x = 8$.   *c.* $D^2x + Dx + 4.25x = t + 1$.   *d.* $D^3x + D^2x + 4Dx + 4x = 10 \sin 10t$.

3-11.   For Prob. 2-15, solve for $i_2(t)$, where

$$E = 10 \text{ volts} \qquad L = 1 \text{ henry} \qquad C_1 = C_2 = 0.01 \text{ farad}$$
$$R_1 = 10 \text{ ohms} \qquad R_2 = 15 \text{ ohms}$$

3-12.   Repeat Prob. 3-11 with the switch relocated between R_2 and L.

3-13.   In Prob. 2-16, the parameters have the values $R_1 = R_2 = R_3 = 10$ ohms, $C = 1\mu\text{f}$, and $L = 100$ henrys.   If $E = 100$ volts, (*a*) find $i_1(t)$; (*b*) sketch $i_1(t)$ and label significant points; (*c*) determine the value of T_s (± 2 per cent).

4-1.   Take the inverse Laplace transform of

$$(a) \ F(s) = \frac{10}{(s+1)^2(s^2+2s+2)} \qquad (b) \ F(s) = \frac{13}{s(s^2+4s+13)}$$

* Pound of mass.

4-2. Find $x(t)$ for

(a) $X(s) = \dfrac{1}{s^3 + 5s^2 + 12s + 8}$

(b) $X(s) = \dfrac{10}{s^5 + 4s^4 + 14s^3 + 66s^2 + 157s + 130}$

(c) $X(s) = \dfrac{2(s^2 + 2s + 2.2)}{s(s + 1)(s^2 + 2s + 2)}$

(d) $X(s) = \dfrac{10(s + 1.1)}{s(s + 1)(s^2 + 2s + 2)}$

(e) Sketch $x(t)$ for parts a to d.

4-3. Given an a-c servomotor with inertia load, find $\omega(t)$ by (a) the classical method; (b) the Laplace-transform method. At $t = 0$,

$$\omega_m = 10^4 \text{ radians/sec} \qquad K_\omega = -1.5 \times 10^{-4} \text{ oz-in./(radian/sec)}$$

$$K_c = 1.2 \frac{\text{oz-in.}}{\text{volt}} \qquad e(t) = 5u(t) \text{ volts}$$

$$J = 38.8 \text{ oz-in.}^2$$

4-4. For each of the following cases, determine the range of values of K for which the response $x(t)$ is stable, where the driving function is a step function. Determine the roots that lie on the imaginary axis that yield sustained oscillations.

(a) $X(s) = \dfrac{K}{s[s(s + 2)(s^2 + s + 10) + K]}$

(b) $X(s) = \dfrac{K}{s[s(0.02s + 1)(0.01s + 1) + K]}$

(c) $X(s) = \dfrac{K(s + 5)}{s[(s + 5)(s^2 + 8s + 20) + K(s + 5)]}$

(d) $X(s) = \dfrac{K}{s[(s + 1)(s + 2)(s + 5) + K]}$

(e) $X(s) = \dfrac{K}{s[s^3 + 6s^2 + 11s + (6 + K)]}$

(f) $X(s) = \dfrac{K(s + 4)}{s[s(s + 1)(s + 2)(s + 3) + K(s + 4)]}$

4-5. Repeat the following problems by use of the Laplace transform:

(a) Prob. 3-1   (d) Prob. 3-5   (f) Prob. 3-8
(b) Prob. 3-2   (e) Prob. 3-6   (g) Prob. 3-13
(c) Prob. 3-3

4-6. The sketches show $f(t)$. Find $F(s)$.

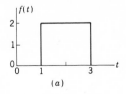

(a)

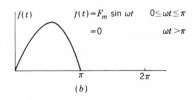

$f(t) = F_m \sin \omega t \qquad 0 \le \omega t \le \pi$
$= 0 \qquad \omega t > \pi$

(b)

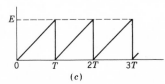

(c)

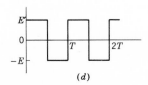

(d)

4-7. Find the partial-fraction expansions of the following:

(a) $F(s) = \dfrac{1}{(s+1)(s+4)}$

(g) $F(s) = \dfrac{10(s+1.01)}{s(s+1)(s^2+2s+10)}$

(b) $F(s) = \dfrac{1}{s(s+2)(s+5)}$

(h) $F(s) = \dfrac{10(s^2+2.2s+10.21)}{s(s+1)(s^2+2s+10)}$

(c) $F(s) = \dfrac{1}{s(s^2+2s+10)}$

(i) $F(s) = \dfrac{10}{s(s^2+2s+2)(s^2+6s+10)}$

(d) $F(s) = \dfrac{s+2}{s^2(s+1)(s+6)}$

(j) $F(s) = \dfrac{10}{s(s^2+2s+2)(s^2+10s+26)}$

(e) $F(s) = \dfrac{1}{s(s+2)(s^2+6s+10)}$

(k) $F(s) = \dfrac{10(s^2+2s+2.21)}{s(s^2+6s+10)(s^2+2s+2)}$

(f) $F(s) = \dfrac{s+1}{s(s+2)(s^2+6s+10)}$

(l) $F(s) = \dfrac{1}{s^2(s+15)(s^2+6s+10)}$

4-8. Solve the differential equations of Prob. 3-10 by means of the Laplace transform.

4-9. Write the Laplace transforms of the following equations:

$$\qquad\qquad\qquad\qquad\qquad\qquad\qquad\qquad\textit{Initial conditions}$$

(a) $\qquad\qquad Dx + 7x = 0 \qquad\qquad x(0) = -1$

(b) $\qquad D^2x + 2Dx + 5x = 10 \qquad x(0) = 2,\ Dx(0) = 0$

(c) $\qquad D^2x + 3Dx + x = t \qquad\quad x(0) = 0,\ Dx(0) = -2$

(d) $D^3x + 4D^2x + 8Dx + 4x = \sin 5t \qquad x(0) = -4,\ Dx(0) = 1,\ D^2x(0) = 0$

4-10. Factor the following equations:

(a) $\quad s^3 + 13s^2 + 33s + 30 = 0$

(e) $\qquad\qquad s^3 + 5s^2 + 8s + 6 = 0$

(b) $\qquad s^3 + 4s^2 + 6s + 4 = 0$

(f) $\qquad\qquad s^3 + 7s^2 + 16s + 10 = 0$

(c) $s^4 + 2s^3 + 2s^2 + 3s + 6 = 0$

(g) $s^4 + 12s^3 + 54s^2 + 108s + 80 = 0$

(d) $\ s^4 + s^3 + 2s^2 + 9s + 5 = 0$

(h) $\qquad s^4 + 7s^3 + 12.2s^2 + 11.05s = 0$

(i) $s^5 + 3s^4 + 28s^3 + 226s^2 + 600s + 400 = 0$

4-11. Use Routh's criterion to determine the number of roots in the right-half s plane for the equations of Prob. 4-10.

4-12. Determine the final value for each of the following: (a) Prob. 4-1; (b) Prob. 4-4; (c) Prob. 4-7.

4-13. Determine the initial value for each of the following: (a) Prob. 4-1; (b) Prob. 4-4; (c) Prob. 4-7.

4-14. For the functions of Prob. 4-7, plot M versus ω and α versus ω.   Use the Spirule.

4-15. Determine the location of the roots, by shifting the origin of the s plane and applying the Routh criterion to

(a) $s^2 + 4s + 4.25 = 0$

(c) $s^4 + 4s^3 + 6s^2 + 4s + 2 = 0$

(b) $s^3 + 6s^2 + 10s + 12.4 = 0$

(d) $s^4 + 4.9s^3 + 9.44s^2 + 8.5s + 2.44 = 0$

4-16. Find the complete solution of $x(t)$ for the following equation:

$$(D^2 + D + 2)(D + 4)x = (D + 2)f(t)$$

Use the Spirule to evaluate all residues.   The forcing function $f(t)$ is

(a) $\delta(t)$ $\qquad\qquad\qquad$ (b) $10u(t)$ $\qquad\qquad\qquad$ (c) $tu(t)$

4-17. The output of a control system is related to its input $r(t)$ by

$$[s^4 + 2s^3 + 2s^2 + (3 + K)s + K]C(s) = K(s + 1)R(s)$$

K represents the positive gain of an amplifier. *a.* With $K = 6$ and a step input, will the output response be stable? *b.* Determine the limiting positive value that K can have for a stable output response.

4-18. The equation relating the output $y(t)$ of a control system to its input is

$$[s^5 + s^4 + 2s^3 + s^2 + (K + 1)s + (K + 2)]Y(s) = 5X(s)$$

Determine the range of K for stable operation of the system. Consider both positive and negative values of K in your analysis.

4-19.

$$F(s) = \frac{1}{s[(s^4 + 2s^3 + 3s^2 + s + 1) + K(s + 1)]}$$

K is real, but may be positive or negative. *a.* Find the range of values of K for which the time response is stable. *b.* Select a value of K which will produce imaginary poles for $F(s)$. Find these poles. What is the physical significance of imaginary poles as far as the time response is concerned?

5-1. An industrial process requires that the temperature θ of materials being mixed in a tank be held constant. The complete system in the accompanying diagram shows a voltage r as the input calibrated in terms of the desired temperature. The actual temperature is measured by means of a thermocouple immersed in the tank. The voltage produced in the thermocouple is amplified to produce a voltage $b = K_b\theta$.

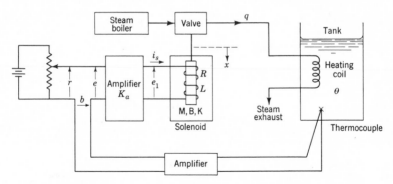

The actuating signal e is amplified to produce the voltage b_1. This voltage energizes the solenoid, which has resistance R and inductance L, producing a force proportional to the current in the solenoid, $f_s = K_s i_s$. This force acts on the mass M, damping B, and restraining spring K of the solenoid armature and valve to control the valve position x. The valve position in turn controls the flow q of hot steam into the heating coil, $q = K_q x$. The temperature of the tank is directly proportional to the steam flow with a time delay, $\theta = K_c q/(D + a)$. *a.* Draw a block diagram showing each component of this system. Label the input and output of each block explicitly. *b.* Determine the transfer function of each block. *c.* Determine the forward transfer function.

5-2. Find an example of a practical closed-loop control system not covered in this book. Briefly describe the system and show a block diagram.

5-3. Find the transfer function for the circuit shown.

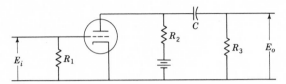

$$r_p = 5,000 \text{ ohms} \qquad R_1 = 1 \text{ megohm}$$
$$\mu = 20 \qquad R_2 = 5,000 \text{ ohms}$$
$$g_m = 2 \times 10^{-3} \text{ mho} \qquad R_3 = 0.5 \text{ megohm}$$
$$C = 1 \ \mu f$$

5-4. A linear rotational position servomechanism with unity feedback has a moment of inertia of 4 slug-ft^2 and a viscous friction coefficient of 2 lb-ft/(radian/sec). If the developed torque of the controller in pound-feet is

$$T(t) = 100e(t) + K \int_0^t e(t) \, dt$$

what would be the value of K that would just make the servo system stable? $e(t)$ is the actuating signal.

5-5. *a.* Find the open-loop transfer function. *b.* Find the closed-loop transfer function (control ratio) where the units of A_1 are volts per radian and those of A_2 are torque per volt.

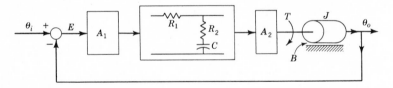

5-6. Use the hydraulic valve and power piston shown in a closed-loop position control system. Derive the transfer function of the system.

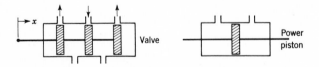

5-7. Draw a diagram of a control system for the elevators on an airplane. Use a hydraulic actuator.

5-8. Find the transfer function of the amplidyne shown in Sec. 2-9.

5-9. In the circuit shown, both armature and field voltages are varied to control the output θ_m. *a.* Draw a block diagram that relates both inputs, v_1 and v_2, to the output θ_m. This is a nonlinear system containing a multiplication. The block diagram should be labeled in the D-operator form. *b.* Linearize the system and use Laplace transforms. Draw the new block diagram. Multiplication is replaced by addition for small excursions of the variables by using

$$df(x,y) = \frac{\partial f}{\partial x}\, dx + \frac{\partial f}{\partial y}\, dy = K_x\, dx + K_y d_y$$

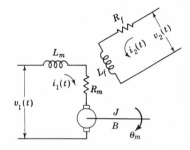

5-10. In the diagram $[q_c]$ represents compressibility flow, and the pressure p is the same in both cylinders. Assume that there is no leakage flow around the pistons. Draw a block diagram that relates the output $Y(s)$ to the input $X(s)$.

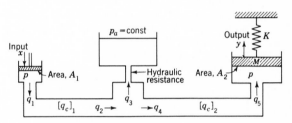

5-11. The diagram represents a gyroscope that is often used in autopilots, automatic gun sights, etc. Assume that the speed of the rotor is constant and that the total developed torque about the output axis is

$$T_0 = H \frac{d\theta_i}{dt}$$

where H is a constant. The inner gimbal exerts a moment of inertia J about the output axis. Draw a block diagram that relates the output $\theta_o(s)$ to the input $\theta_i(s)$.

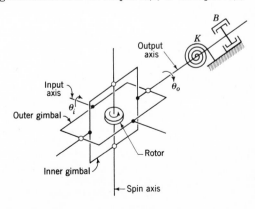

5-12. The figure shows a system for controlling the output movement $y(t)$ of a hydraulic linear actuator. The load reaction on the actuator can be considered negligible. The solenoid constant is K_c (pounds per ampere), the output potentiometer constant is K_p (volts per inch), and the output of the tachometer is given by

$$e_t = K_t \frac{dy}{dt} \qquad \text{volts}$$

where K_t has the units of volts per (inch per second). *a.* Draw a detailed block diagram that shows all variables of the system explicitly. *b.* Derive the transfer functions in terms of the Laplace operator for each block in the block diagram. *Note:* The lever at the input to the hydraulic unit has no fixed pivot. Consider that x, x_1, and y have only horizontal motion.

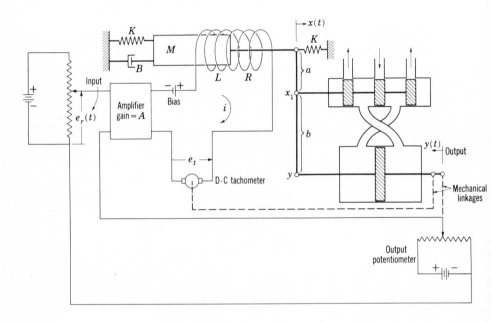

5-13. For the control system shown: (1) The force of attraction on the solenoid is given by $f_c = K_c i_c$, where K_c has the units pounds per ampere. (2) The voltage appearing across the generator field is given by $e_f = K_x x$, where K_x has the units volts per inch and x is in inches. (3) When the voltage across the solenoid coil is zero, the spring K_s is unstretched and $x = 0$. *a.* Derive all necessary equations relating all the variables in the system. *b.* Draw a block diagram for the control system. The diagram should include enough blocks to indicate specifically the variables $I_c(s)$, $X(s)$, $I_f(s)$, $E_g(s)$, and $T(s)$. Give the transfer function for each block in the diagram. *c.* Draw a signal flow graph for this system.

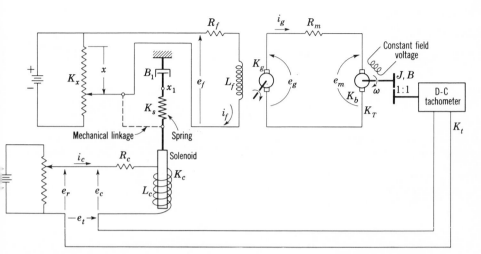

5-14. The figure indicates a possible arrangement for using an error signal to control the movement of a potentiometer arm. The solenoid produces a magnetic force proportional to the current in the coil; that is, $f_c = K_c i$, where i is the change of current from the equilibrium condition. *a.* Draw a detailed block diagram showing all variables explicitly. *b.* Determine the transfer function for each block. *c.* Determine the transfer function $X(s)/E(s)$.

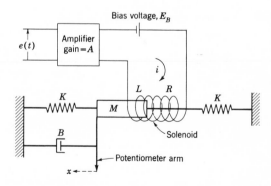

6-1. A system has

$$G(s) = \frac{N}{s(as + 1)(bs^2 + cs + d)}$$

Find (*a*) step coefficient K_0; (*b*) ramp coefficient K_1; (*c*) parabolic coefficient K_2.

For the unity feedback closed-loop system find the steady-state error for (*d*) step input $R_0 u(t)$; (*e*) ramp input $R_1 t u(t)$; (*f*) parabolic input $R_2 t^2 u(t)$.

6-2.   *a.* Find K_0, K_1, K_2,   *b.* State the type of the system.   *c.* This system is required to follow a parabolic input signal $r(t)$.   Will the system perform satisfactorily with this input?   Give the *reason* for your answer.

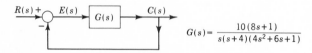

$$G(s) = \frac{10(8s+1)}{s(s+4)(4s^2+6s+1)}$$

6-3.   For the system shown, find (a) $C(s)/R(s)$ and (b) the steady-state value of $c(t)$ if $r(t) = 10u(t)$ radians. (c) What type of systems do $C(s)/M(s)$ and $C(s)/E(s)$ represent?

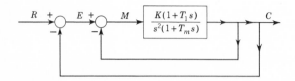

6-4.   For the system shown, draw a block diagram.   Indicate a block for each main component of the system.   The output appears at the motor shaft.   The generator is driven at constant speed so that its generated voltage is $100i_f$ (volts).   The motor is separately excited so that its counter emf is $1.0\omega_0$ (volts).   The tachometer gives 6 volts at $\omega_0 = 100$ radians/sec.   The torque of the motor in pound-feet is $0.704i_a$.   The motor and its load have a moment of inertia of J slug-ft^2 and negligible friction. e_i is the reference voltage.   *a.* Determine the transfer function for each block and the type of this control system.   *b.* Determine $C(s)/R(s)$.   *c.* If $e_i = 6$ volts, what is the value of the steady-state actuating signal?   *d.* Change the block diagram to the form shown in Appendix J, manipulation 12.   Indicate the units of all variables in both diagrams.   *e.* What are the significance and the value of the actuating signal in the equivalent diagram?

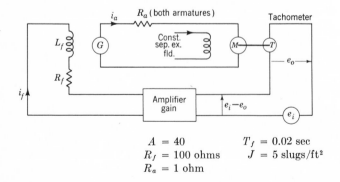

$$A = 40 \qquad T_f = 0.02 \text{ sec}$$
$$R_f = 100 \text{ ohms} \qquad J = 5 \text{ slugs/ft}^2$$
$$R_a = 1 \text{ ohm}$$

6-5.   *a.* Determine the open-loop transfer function of Fig. (a).   *b.* Determine the over-all transfer function.   *c.* Figure (b) is an equivalent block diagram of Fig. (a).   What must be the transfer function of $H_x(s)$ in order for Fig. (b) to be equivalent to Fig. (a)?   *d.* Figure (b) represents what type of system?   *e.* Determine the static error coefficients of Fig. (b).   What is the significance of the minus sign for

the ramp error coefficient?   *f.* If $r(t) = u(t)$, determine the final value of $c(t)$.
g. What are the values of $e_x(t)_{ss}$ and $e(t)_{ss}$?

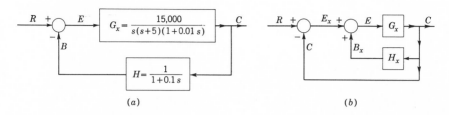

(a)   (b)

6-6. Find the step, ramp, and parabolic error coefficients for unity feedback systems
which have the following forward transfer functions:

(a) $G(s) = \dfrac{10}{(0.2s + 1)(0.5s + 1)}$ (c) $G(s) = \dfrac{12}{s(s + 2)(0.4s + 1)}$

(b) $G(s) = \dfrac{10}{s^2 + 5s + 25}$ (d) $G(s) = \dfrac{14(s + 3)}{s(s + 5)(s^2 + 2s + 2)}$

(e) $G(s) = \dfrac{8}{s^2(s^2 + 4s + 8)(s^2 + 3s + 12)}$

(f) $G(s) = \dfrac{11}{s^3(s + 1)(0.2s + 1)(s^2 + 5s + 15)}$

(g) $G(s) = \dfrac{15}{(s + 3)(0.2s + 1)}$

(h) $G(s) = \dfrac{10(s + 3)}{(s + 1)(s^2 + 2s + 2)}$

(i) $G(s) = \dfrac{-6(1 + 0.04s)}{s(1 + 0.1s)(1 + 0.1s + 0.01s^2)}$

(j) $G(s) = \dfrac{4(s^2 + 10s + 100)}{s^2(s + 3)(s^2 + 6s + 10)}$

(k) $G(s) = \dfrac{100(s - 1)}{s(s + 2)(1 + 0.25s)}$

(l) $G(s) = \dfrac{26}{s^2(s - 1)(s^2 + 6s + 13)}$

6-7. For Prob. 6-6, find $e(t)_{ss}$, by use of the definitions of the error coefficients, with
the following inputs: (a) $r(t) = 5$; (b) $r(t) = 2t$; (c) $r(t) = t^2$.

6-8. Repeat Prob. 6-5 with $H(s)$ replaced in Fig. (a) by

$$H(s) = \frac{s + 10}{s + 1}$$

6-9. A unity feedback control system has

$$G(s) = \frac{K_1}{s(s + 1)(0.5s + 1)}$$

and $r(t) = 5t$.   *a.* If $K_1 = 1.5$ sec^{-1}, determine $e(t)_{ss}$.   *b.* It is desired that, for a
ramp input, $e(t)_{ss} \leq 0.1$.   What minimum value must K_1 have for this condition
to be satisfied?   *c.* For the value of K_1 determined in part *b*, is the system stable
or unstable?

6-10. *a.* Derive the ratio $G(s) = C(s)/E(s)$. *b.* Based upon $G(s)$, the figure has the characteristic of what type of system? *c.* Derive the control ratio for this control system.

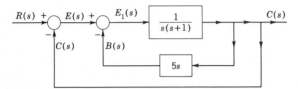

6-11. A unity feedback system has the forward transfer function

$$G(s) = \frac{K_1(2s + 1)}{s(4s + 1)(s + 1)^2}$$

The input $r(t) = 1 + 5t$ is applied to the system. *a.* It is desired that the steady-state value of the error be equal to or less than 0.1 for the given input function. Determine the minimum value that K_1 must have to satisfy this requirement. *b.* By use of Routh's stability criterion, determine whether the system is stable for the minimum value of K_1 determined in part *a.*

6-12. A unity feedback system has

$$G(s) = \frac{20(s + 2)}{s(s + 3)(s + 4)}$$

a. Find the gain. *b.* Find the step, ramp, and parabolic error coefficients. *c.* With $r(t) = 3u(t) + 5tu(t)$, find $e(t)_{ss}$.

6-13. Assume that the control system shown is stable and

$$G(s) = \frac{10(5s + 1)}{s^2(2s + 1)(s + 1)}$$

$$H(s) = \frac{3s + 1}{4s + 1}$$

a. Determine the value of $c(t)_{ss}$ for $r(t) = 10u(t)$. *b.* Manipulate the block diagram to obtain a unity feedback system containing a minor loop (see manipulation 22, Appendix J). *c.* Determine the value of $e(t)_{ss}$. *d.* Is the system stable?

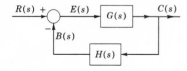

6-14. The RC circuit shown can be represented by an equivalent feedback system form with appropriate forward and feedback transfer functions which are not equal to unity. Determine the expressions for $G_1(s)$, $G_2(s)$, and $H(s)$. Assume that output load is negligible.

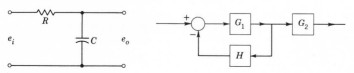

7-1. For the transfer function

$$G(s)H(s) = \frac{K(s + 12)}{s(s^2 + 16s + 100)} = \frac{K(s + 12)}{s(s + 8 - j6)(s + 8 + j6)}$$

sketch the root locus and determine all pertinent data.

7-2. For the transfer function

$$G(s)H(s) = \frac{K_0}{(1 + 0.5s)(1 + 0.2s)(1 + s)^2}$$

sketch the locus of roots and give all pertinent characteristics of the curves that are useful in establishing the locus. Indicate at least two calculated points not on the real axis and far enough apart to give a fairly accurate plot. Use a Spirule. Label completely.

7-3. Draw the root locus of the following:

(a) $G(s)H(s) = \dfrac{K}{(s + 5)(s^2 + 4s + 7)} = \dfrac{K}{(s + 5)(s + 2 - j\sqrt{3})(s + 2 + j\sqrt{3})}$

(b) $G(s)H(s) = \dfrac{K}{s(s^2 + 6s + 12)} = \dfrac{K}{s(s + 3 - j\sqrt{3})(s + 3 + j\sqrt{3})}$

(c) $G(s)H(s) = \dfrac{K}{(s^2 + 2s + 2)(s^2 + 6s + 10)}$

$$= \frac{K}{(s + 1 + j1)(s + 1 - j1)(s + 3 + j1)(s + 3 - j1)}$$

(d) $G(s)H(s) = \dfrac{K}{s(s^2 + 2s + 2)(s^2 + 6s + 10)}$

Does the root locus cross any of the asymptotes?

7-4. A system has the following transfer functions:

$$G(s) = \frac{K}{s(s^2 + 8s + 20)} \qquad H(s) = 1$$

a. Plot the root locus. *b.* A damping ratio of 0.5 is required for the dominant roots. Find $C(s)/R(s)$. The denominator should be in factored form. *c.* With a unit step input, find $c(t)$. *d.* Evaluate graphically $\mathbf{C}(j\omega)/\mathbf{R}(j\omega)$ versus ω. Plot the magnitude and angle of $\mathbf{C}(j\omega)/\mathbf{R}(j\omega)$ versus ω.

7-5. A certain feedback control system with unity feedback has a transfer function

$$G(s) = \frac{K_1}{s(1 + 0.02s)(1 + 0.01s)}$$

a. Plot the locus of the roots of $1 + G(s) = 0$ as the static loop sensitivity is varied. *b.* Determine the value of K_1 that just makes the system unstable. Compare this with the value obtained in Prob. 4-4b. *c.* From the root-locus plot determine the value of K_1 for a $\zeta = 0.388$. *d.* For the value of K_1 found in part *c*, determine $e(t)$ for $r(t) = u(t)$. *e.* Determine graphically from the root-locus plot the data for plotting the curve of M versus ω for the closed-loop system. Plot this curve.

7-6. Sketch the root locus for the control system having the following open-loop transfer function. Calculate the value of K_1 that causes instability.

$$G(s) = \frac{K_1(1 + 0.04s)}{s(1 + 0.1s)(1 + 0.1s + 0.0125s^2)}$$

7-7. *a.* Sketch the root locus for a control system having the following forward and feedback transfer functions:

$$G(s) = \frac{K_2(1 + s/5)}{s^2(1 + s/12)} \qquad H(s) = 1 + \frac{s}{10}$$

b. Choose closed-loop pole locations which produce a time constant $T = \frac{1}{3}$ sec for the complex roots and indicate these locations on the root locus. Using these locations, write the factored form of the closed-loop transfer function.

7-8. For

$$G(s)H(s) = \frac{K}{(0.5s + 3)(0.4s + 1.2)(0.5s^2 + s + 1)}$$

determine the value of K from the root locus that makes the closed-loop system a perfect oscillator.

7-9. A unity feedback control system has the following transfer function:

$$G(s) = \frac{K_1}{s(1 + 0.25s)(1 + 0.5s + 0.125s^2)}$$

a. Sketch the root locus for positive and negative values of K_1. *b.* For what value of K_1 does the system become unstable? *c.* When the ramp error coefficient has a value of 0.5, all the poles of the closed-loop system are located at $s = -2$. What is the impulse response of the control system for this case?

7-10. A unity feedback control system has the following transfer function:

$$G(s) = \frac{K_1(1 - 0.5s)}{s(1 + s)(1 + 0.5s)(1 + 0.25s)}$$

a. Sketch the root locus for positive and negative values of K_1. *b.* What range of values of K_1 makes the system unstable?

7-11. For positive values of gain, sketch the root locus for unity feedback control systems having the following open-loop transfer functions. For what value or values of gain does the system become unstable in each case?

(a) $G(s) = \dfrac{K_1}{s(1 + 0.1s)(1 + 0.02s)}$

(b) $G(s) = \dfrac{K_0(1 + 10s)}{(1 + 100s)(1 + 0.001s)(1 + s + s^2)}$

(c) $G(s) = \dfrac{K_0(1 - 100s)}{(1 + 10s)(1 + 0.002s)(1 + s + s^2)}$

(d) $G(s) = \dfrac{K_0}{1 - 0.1s}$

(e) $G(s) = \dfrac{K(s^2 + 4s + 5)}{s^2(s + 1)(s + 3)}$ (g) $G(s) = \dfrac{K(s + 2)^2}{s(s^2 - 2s + 2)}$

(f) $G(s) = \dfrac{K(s^2 + 8s + 20)}{s(s + 8)(s^2 + 2s + 10)}$ (h) $G(s) = \dfrac{K(s + 2)}{(s + 18)(s^2 + 2.5s + 12)}$

7-12. Without drawing the entire root locus of Prob. 7-3, determine the value of the static loop sensitivity for a $\zeta = 0.5$.

7-13. A nonunity feedback control system has the following transfer functions:

$$G(s) = \frac{K_A(1 + s/3.9)}{(1 + s/10)[1 + 2(0.7)s/23 + (s/23)^2][1 + 2(0.49)s/7.6 + (s/7.6)^2]}$$

$$H(s) = \frac{K_B(1 + s/10)}{1 + 2(0.89)s/42.7 + (s/42.7)^2}$$

a. Sketch the root locus for the system, using as few trial points as possible. Determine the angles and locations of the asymptotes, and the angles of departure of the branches from the open-loop poles. b. Estimate the gain $K_A K_B$ at which the system becomes unstable, and determine the approximate locations of all the closed-loop poles for this same value of gain. c. Using the closed-loop configuration determined in part b, write the expression for the closed-loop transfer function of the system.

7-14. A nonunity feedback control system has the following transfer functions:

$$G(s) = \frac{10A(s^2 + 8s + 20)}{(s + 1)(s + 4)} \qquad H(s) = \frac{0.2}{s + 2}$$

a. Determine the value of the amplifier gain A that will produce complex roots having the *minimum* possible value of ζ. b. Express $C(s)/R(s)$ in terms of its poles, zeros, and constant term.

7-15. a. Show that, when $G(s)H(s)$ contains two real poles to the right of a real zero, the complex portion of the root locus is a circle with its center at the zero and the radius equal to the square root of the product of the distances from the zero to each pole (see Fig. 7.4b). b. Repeat for two complex poles located anywhere (see Fig. 7-5b).

8-1. For the root locus of Prob. 7-1, do the following: a. Assuming the dominancy of the complex pair of roots and the effect of the real root being negligible, determine M_o, T_p, T_s, and N for K equal to 30, 50, 85, and 175. b. Modify the values determined in (a) to take into account the effect of the real root. c. For each value of K in (a), plot M versus ω.

8-2. For Prob. 7-4, do the following: a. By neglecting the contribution of the real root, determine M_o, T_p, T_s, and N. b. Modify the values determined in (a) to take into account the effect of the real root. c. Plot $c(t)$ versus t. Select sufficient values of time in the vicinity of the values of T_p and T_s obtained in part a. Use an analog computer if possible. d. Compare the actual values of M_o, t_p, t_s, and N with those determined in (a) and (b).

8-3. Repeat Prob. 8-2 but for Probs. 4-2c and d.

8-4. Repeat Prob. 8-2 but for Prob. 4-7e to h.

8-5. a. For Prob. 4-7i, neglecting the contribution of the nondominant complex roots, determine M_o, T_p, T_s, and N. b. Plot $|j\omega F(j\omega)|$ versus ω.

8-6. Repeat Prob. 4-10, using the partition method.

8-7. Determine the roots of the following polynomial by the partition method:

$$F(s) = s^6 + 13s^5 + 58s^4 + 506s^3 + 2,860s^2 + 6,400s + 4,000 = 0$$

8-8. By the partition method, determine the breakaway and break-in points of one of the Probs. 7-1 through 7-12 and Prob. 7-14.

8-9. A system has the transfer functions

$$G(s) = \frac{3.88}{s(s/2 + 1)} \quad \text{and} \quad H(s) = \frac{T_2 s + 1}{T_1 s + 1}$$

where $T_2 = \frac{1}{3}$ and $T_1 = \frac{1}{8}$. a. Sketch the root locus and determine the roots of the characteristic equation. b. Consider that T_2 is fixed and T_1 varies. Determine the root locus of this system as a function of the variation in T_1. c. Consider that T_1 is fixed and T_2 varies. Determine the root locus of this system as a function of the variation in T_2. d. For parts b and c, express $C(s)$ in factored form for a unit step input, with $\delta' = 0.1\delta_0$ and compare with the case when $\delta' = 0$.

8-10. A system has the transfer function

$$G(s) = \frac{3.88(T_2 s + 1)}{s(0.5s + 1)(T_1 s + 1)}$$

where $T_2 = \frac{1}{3}$ and $T_1 = \frac{1}{8}$. *a.* Consider that T_2 is fixed and T_1 varies. (See Prob. 8-9a.) Express $C(s)$ in factored form for a unit step input with $\delta' = 0.1\delta_0$ and $\delta' = 0$. *b.* Compare (a) above with the corresponding part of Prob. 8-9c.

8-11. A system has the transfer function

$$G(s) = \frac{K}{s(s^2 + 4\zeta_x s + 4)}$$

For $K = 2.19$ and $\zeta_x = 0.5$ the closed-loop poles are $s = -0.75$ and $s = -0.625 \pm j1.63$. Investigate the effect of a variation of ζ_x on the system performance, with K constant at the value 2.19. Determine the mathematical function that permits plotting a root locus with the variation in ζ_x as the variable parameter. This function must be in factored form, and the values of all its parameters must be determined. Plot this root locus.

8-12. The block diagram shown below represents the longitudinal or pitch node of an aerospace vehicle where

$$G_x(s) = \frac{26.01(s + 1.4)(s + 0.01)}{[s^2 + 2(0.07)(0.06)s + (0.06)^2][s^2 + 2(\zeta_x)(4.3)s + (4.3)^2]}$$
$$\underbrace{}_{\text{phugoid node}} \underbrace{}_{\text{short-period node}}$$

and
$$0.1 \leq \zeta_x \leq 0.7$$

a. Obtain the root locus for each of the following values of K_H, as a function of δ' where $\delta_0 = 0.1$. Take values of K_H equal to 1, 10, 30, 50, and 65. For simplicity,

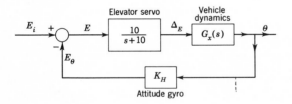

replace the two complex poles and the zero in the vicinity of the origin by an equivalent real pole located at their geometric center. *b.* From the results of (a), determine the optimum value of K_H, which is that value for which the roots vary the least as ζ_x varies and the system is always stable.

8-13.
$$s^3 + 9s^2 + 33s + 25 = 0$$

Use partitions one and two and draw *only* the real-axis portions of the resulting root loci. Identify a region of the real axis where a real root of the polynomial equation *must* exist.

9-1. Plot to scale the angle versus log ω curve for $\mathbf{G}(j\omega)$. Is it an integral or a derivative compensating network?

$$\mathbf{G}(j\omega) = \frac{10(1 + j2\omega)}{1 + j6\omega}$$

9-2. A system has

$$G(s) = \frac{2(1 + T_2 s)(1 + T_3 s)}{s^2(1 + T_1 s)^2(1 + T_4 s)}$$

where $T_1 = 4$, $T_2 = 1$, $T_3 = \frac{1}{2}$, $T_4 = \frac{1}{4}$. *a.* Draw the asymptotes of $\mathbf{G}(j\omega)$ on a decibel versus log ω plot. Label the corner frequencies on the graph. *b.* What is the *total* correction from the asymptotes at $\omega = 1$?

9-3. *a.* What characteristic must the plot of magnitude in decibels versus log ω possess in order that a velocity servo system (ramp input) may have no *steady-state velocity error* for a constant velocity input $[dr(t)/dt]$? *b.* What is true of the corresponding phase angle characteristic?

9-4. Explain why the phase angle curve cannot be calculated from the plot of $|\mathbf{G}(j\omega)|$ in decibels versus log ω if some of the factors are not minimum phase.

9-5. For each plot shown, (*a*) evaluate the transfer function; (*b*) find the correction that should be applied to the straight-line curve at $\omega = 8$.

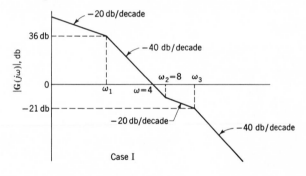

Case I

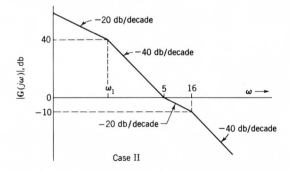

Case II

9-6. Determine the value of the error coefficient from Figs. (a) and (b).

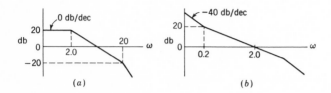

(a) (b)

9-7. For each of the transfer functions

(1) $$G(s) = \frac{K_1}{s(1 + 0.1s)(1 + 0.5s)}$$ $\qquad K_1 = 10$

(2) $$G(s) = \frac{K_1}{s(1 + 0.04s)(1 + s/50 + s^2/2,500)}$$ $\qquad K_1 = 1$

(a) draw the log magnitude (exact and asymptotic) and phase diagrams; (b) from the curves of part a, obtain the data for plotting the direct polar plots; (c) from the curves of part a, obtain the data for plotting the inverse polar plots.

9-8. For each of the transfer functions

(1) $G(s) = \dfrac{20}{(1 + 0.2s)(1 + 0.4s)(1 + s)}$ $\qquad$ (4) $G(s) = \dfrac{10}{(1 + s)(1 + 0.1s)}$

(2) $G(s) = \dfrac{5}{s(s^2 + 16s + 400)}$ $\qquad$ (5) $G(s) = \dfrac{2}{s^2(1 + 0.1s)(1 + 0.4s)}$

(3) $G(s) = \dfrac{2(1 + 0.3s)}{s^2(1 + 0.1s)(1 + 0.05s)}$ $\qquad$ (6) $G(s) = \dfrac{2(1 + 0.3s)}{s(1 + 0.1s)(1 + 0.4s)}$

(a) draw the log magnitude (exact and asymptotic) and phase diagrams; (b) from the curves of part a, obtain the data for plotting the direct polar plots; (c) from the curves of part a, obtain the data for plotting the inverse polar plots.

9-9. A control system with unity feedback has the forward transfer function

$$G(s) = \frac{K_1}{s(1 + 0.1s)(1 + 30s/625 + s^2/625)}$$

a. Plot the polar plot $G'(j\omega)$. Determine all key points of the curve. b. Plot the polar plot of $G'(j\omega)^{-1}$. Determine all key points of the curve.

9-10. An experimental transfer function gave the following results:

ω	$G(j\omega)$		ω	$G(j\omega)$	
0.04	-0^+	$- j25$	2.0	-0.431	$- j0.431$
0.05	-0^+	$- j40$	3.0	-0.121	$- j0.091$
0.1	-2.51	$- j19.8$	4.0	-0.165	$- j0.1002$
0.2	-1.53	$- j9.86$	6.0	-0.044	$- j0.01995$
0.5	-1.225	$- j3.44$	8.0	-0.0528	$- j0.0202$
1.0	-0.832	$- j1.283$	16.0	-0.0075	$- j0.0014$
1.5	-0.278	$- j0.334$	20.0	-0.00958	$- j0.00287$

a. Determine the transfer function represented by the above data. b. What type of system does it represent?

9-11. Repeat Prob. 9-10 for each of the following:

$a.$	ω	$\lvert G(j\omega)\rvert$	Angle, deg		$b.$	ω	Lm $\mathbf{G}(j\omega)$, db	Ang $\mathbf{G}(j\omega)$, deg
	0.1	9.95	−96.9			1	0.08	−92.9
	0.3	3.19	−110.1			2	−5.71	−95.9
	0.6	1.42	−127.8			4	−10.77	−103.4
	0.8	0.963	−137.8			6	−12.55	−115.1
	1.0	0.693	−146.3			8	−12.68	−138.0
	2.0	0.21	−175.2			10	−13.98	−180.0
	3.0	0.092	−192.5			15	−26.80	−239.0
	5.0	0.028	−213.7			20	−36.00	−251.6
	8.0	0.0082	−230.9			30	−47.75	−259.4
	12.0	0.0027	−242.6			40	−55.64	−262.4
	16.0	0.0012	−249.0			50	−61.63	−264.1
	20.0	0.0006	−253.1			55	−64.17	−264.6

9-12. Determine the transfer function by using the straight-line asymptotic log plot shown and the fact that the correct angle is −129.7° at $\omega = 0.8$. Assume a minimum-phase system.

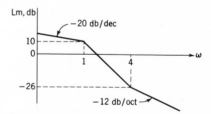

10-1. Determine whether each system shown is stable or unstable in the absolute sense by sketching the *complete* Nyquist diagrams. [$H(s) = 1$.]

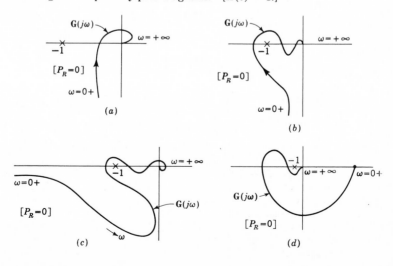

10-2. Shown are plots of the open-loop transfer functions for a number of control systems. Only the curves for positive frequencies are given. *Using Nyquist's criterion,* determine the closed-loop system stability. $G(s)$ has no poles or zeros in the right-half s plane.

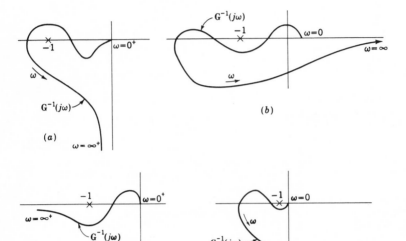

10-3. For the following transfer functions, sketch a direct and an inverse Nyquist locus to determine the closed-loop stability. Determine the values of K that correspond to stable closed-loop operation and those that correspond to unstable closed-loop operation. $[H(s) = 1.]$

(a) $G(s) = \dfrac{K}{1 + s}$

(b) $G(s) = K\dfrac{1 + s}{1 - s}$

(c) $G(s) = \dfrac{K}{s^2}$

(d) $G(s) = \dfrac{K}{s(1 + s)(2 + s)}$

(e) $G(s) = \dfrac{K(s + 2)}{(s + 3)(s - 1)}$

(f) $G(s) = \dfrac{K}{(s - 1)(s + 2)(s + 4)}$

(g) $G(s) = \dfrac{K_1(1 + 2s)}{s(1 + s)(1 + s + s^2)}$

(h) $G(s) = \dfrac{K_3}{s^3(1 + s)^2}$

10-4. A unity feedback control system has the following forward transfer function:

$$G(s) = \frac{K_0(1 - s)}{(1 + 2s/3)(1 + s)}$$

a. Draw the polar plot of $\mathbf{G}'(j\omega)$. *b.* Use the Nyquist stability criterion to determine the range of positive and negative values of K_0 for which the system is stable. *c.* By use of the Nyquist stability criterion determine whether $C(s)/R(s)$ has any time constants greater than 1.0 sec for $K_0 = 1.0$. Repeat for $K_0 = 0.5$. (Remember that only roots of the characteristic equation that lie in the left half-plane are of concern for stable operation.) *d.* By use of the Nyquist stability criterion, determine whether $C(s)/R(s)$ has any roots with a value ζ less than 0.5 for $K_0 = 1.0$.

10-5. For the control systems having the transfer functions

(1) $G(s) = \dfrac{K_1}{s(1 + 0.001s)(1 + 0.025s)(1 + 0.10s)}$

(2) $G(s) = \dfrac{K(1 + 0.2s)}{(1 + 0.1s)(2 + 3s + s^2)}$

(a) determine, from the logarithmic curves, the required value of K_n so that each system will have a positive phase margin of $45°$; (b) from these same curves, determine the maximum permissible value of K_n for stability.

10-6. Repeat Prob. 10-5a for a positive phase margin of $60°$.

10-7. A system has the transfer functions

$$G(s) = \frac{K}{s(s + 10)(s^2 + 16s + 1,600)} \qquad H(s) = 1$$

a. Draw the log magnitude and phase diagram of $G'(j\omega)$. Draw both the straight-line and the corrected log magnitude curves. b. Draw the log magnitude–angle diagram. Determine the maximum value of K_1 for stability.

10-8. What value of gain would just make the system in Prob. 9-10 unstable?

10-9. By use of the Nyquist stability criterion, determine whether the system having the following inverse polar plot is stable or unstable.

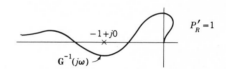

10-10. By use of the Nyquist stability criterion, determine whether a system having the following transfer function is stable or unstable (a) if $K_x = 10$ and (b) if $K_x = -10$. If either case is unstable, determine the number of poles of $C(s)/R(s)$ in the right-half s plane.

$$G(s)H(s) = \frac{K_x}{s^2(1 - s/2)}$$

10-11. A system has the transfer function

$$G(s) = \frac{K_0}{(1 + s)^3}$$

a. Draw the log magnitude and phase diagrams of $G'(s)$. Draw both the straight-line and the corrected log magnitude curves. b. Determine the phase margin γ for $K_0 = 4$. c. What is the maximum value of K_0 for stability?

10-12. For

$$G(s)^{-1} = \frac{s^2(0.5s + 1)(3s + 1)}{K_2(s - 1)}$$

sketch the inverse Nyquist locus and determine the closed-loop stability for $K_2 = 2$.

10-13. Use the Nyquist criterion to determine the maximum value of K_n for stability of the closed-loop systems having the following transfer functions:

(a) $G(s)H(s) = \dfrac{K_1 e^{-s}}{s(1 + s)(1 + 0.5s)}$ \qquad (b) $G(s)H(s) = \dfrac{K_0 e^{-2s}}{s^2 + 2s + 2}$

(c) The transfer functions of Prob. 10-5 with the transport lag $e^{-1.5s}$ included in the numerator

10-14. Use the Nyquist criterion to determine the maximum value of T for stability of the closed-loop system which has the open-loop transfer function

$$G(s)H(s) = \frac{1.25e^{-Ts}}{s(s^2 + 6s + 10)}$$

Hint: Determine the frequency ω_x for which $|\mathbf{G}(j\omega)\mathbf{H}(j\omega)| = 1$. What additional angle ωT, due to the transport lag, will make the angle of $\mathbf{G}(j\omega_x)\mathbf{H}(j\omega_x)$ equal to $-180°$?

11-1. For Prob. 9-7, find K_1 and ω_m for an $M_m = 1.2$ by use of the direct-polar-plot method.

11-2. Repeat Prob. 11-1, using the inverse-polar-plot method.

11-3. For the feedback control system of Prob. 7-4, (a) determine, by use of the polar-plot method, the value of K_1 that just makes the system unstable; (b) determine the value of K_1 that makes $M_m = 1.15$; (c) for the value of K_1 found in part b, find $e(t)$ if $r(t) = u(t)$; (d) determine graphically from the polar plot of $\mathbf{G}(j\omega)$ the data for plotting the curve of M versus ω for the closed-loop system and plot this curve; (e) compare the results of parts a to d with the results obtained in Prob. 7-4.

11-4. Determine the value K_1 must have for an $M_m = 1.3$.

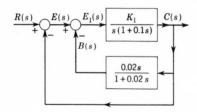

11-5. For the $1/G'(j\omega)$ curve shown it is found that two values of gain, K_a and K_b, will produce a desired M_m. The corresponding resonant frequencies are ω_a and ω_b. Which value of gain gives the better performance for the system? Give the reasons for your choice.

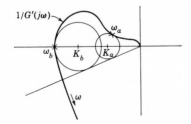

11-6. Using the plot of $\mathbf{G}'(j\omega)$ shown, determine the number of values of gain K_n which produce the same value of M_m. Which of these values yields the best system performance? Give the reasons for your answer.

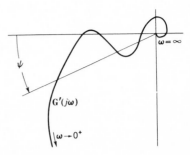

11-7. *a.* Specify the value of K_n that will make the peak in the frequency response as small as possible. *b.* At what frequency does this peak occur? *c.* What value does $\mathbf{C}(j\omega)/\mathbf{R}(j\omega)$ have at the peak?

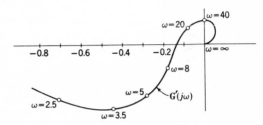

11-8. A Type 1 servo with unity feedback has an $M_m = 1.3$ and an $\omega_m = 8.2$ radians/ sec. The closed-loop response is a second-order system. The system shown must have the same form of transient response, ζ and ω_d, as the system described above. Find the necessary values of K and T.

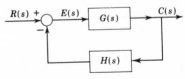

$$G(s) = \frac{100}{1 + 10s} \qquad H(s) = \frac{K}{1 + Ts}$$

11-9. *a.* If $K_n = 1$ in Prob. 10-5, determine the values of M_m and ω_m. *b.* What are the values of M_m and ω_m for Prob. 10-5a? *c.* What are the values of M_m and ω_m for Prob. 10-6?

11-10. *a.* In Prob. 10-7, adjust the gain for $M_m = 2$ db and determine ω_m. For this value of gain find the phase margin, and plot M versus ω and α versus ω. *b.* Repeat for Prob. 9-9.

11-11. For Prob. 9-7, determine the values of M_m and ω_m. How much gain must be added to achieve an $M_m = 1.3$? What is the value of ω_m for this value of M_m for each case? *a.* Use log plots. *b.* Use polar plots.

11-12. For Prob. 9-8, determine the values of M_m and ω_m. What is the phase margin for each case? How much gain must be added to achieve an $M_m = 1.25$? What is the value of ω_m for this value of M_m for each case?

11-13. Refer to Prob. 9-7. *a.* Determine the values of K_1 and ω_m corresponding to the following values of M_m: 1.05, 1.1, 1.2, 1.4, 1.6, 1.8, and 2.0. For each value of M_m determine ζ_{eff} from Eq. (11-18) and then plot M_m versus ζ_{eff}. *b.* For each value of K_1 determined in part *a*, calculate the value of M_p from the root-locus plot. For each value of M_p determine ζ_{eff} from Fig. 3-11 and plot M_p versus ζ_{eff}. (See the last paragraph of Sec. 7-12.) *c.* What is the degree of correlation between M_m versus ζ_{eff} and M_p versus ζ_{eff}?

11-14. For Prob. 10-13, determine the value of K_n for an $M_m = 3$ db.

12-1. It is desired that the control system of Prob. 9-7 (2)* have a damping ratio of 0.425 for the dominant complex roots. Using the root-locus method, (*a*) add a lag compensator, with $\alpha = 10$, so that this value of ζ can be obtained; (*b*) add a lead compensator with $\alpha = 0.1$; (*c*) add a lag-lead compensator with $\alpha = 10$. Indicate the time constants of the compensator in each case. Compare the results obtained by the use of each type of compensator with respect to the error coefficient, ω_d, T_s, and M_o.

12-2. A control system has the forward transfer function

$$G_x(s) = \frac{K_x}{s^2(1 + 0.1s)}$$

The closed-loop system is to be made stable by adding a compensator $G_c(s)$ and an amplifier A in cascade with $G_x(s)$. A ζ of 0.357 is desired with a value of $\omega_n \approx 1.6$ radians/sec. By use of the root-locus method, determine the following. *a.* What kind of compensator is needed? *b.* Select an appropriate α and T for the compensator. *c.* Select the value of static error coefficient K_2. *d.* Plot the compensated locus. *e.* Plot M versus ω for the compensated system. *f.* From the plot of part *e* determine the M_m and ω_m. With this value of M_m determine the effective ζ of the system by the use of $M_m = (2\zeta \sqrt{1 - \zeta^2})^{-1}$. Compare the effective ζ and ω_m with the values obtained from the dominant pair of complex roots. *Note:* The effective $\omega_m = \omega_n \sqrt{1 - 2\zeta^2}$.

12-3. *a.* Write the transfer function $V_o(s)/V_i(s)$. *b.* Under what conditions would this approximate an ideal integrator?

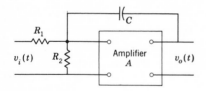

12-4. With

$$G_x(s) = \frac{K}{s(s + 10)(s^2 + 30s + 625)} \qquad H(s) = 1$$

find K_1, ω_n, M_o, T_p, T_s, N for each of the following cases: (*a*) original system; (*b*) lag compensator added, $\alpha = 10$; (*c*) lead compensator added, $\alpha = 0.1$; (*d*) lag-lead compensator added, $\alpha = 10$. Use $\zeta = 0.5$ for the dominant roots.

Note: Except for the original system it is not necessary to obtain the complete root locus for each type of compensation. When the lead and lag-lead compensators are added to the original system, there may be other dominant roots in addition to the complex pair.

---

* The transfer function $G_x(s)$ given on page 804 may be used instead of the one given by Prob. 9-7 (2).

12-5. Using the root-locus plot of Prob. 7-9, adjust the damping ratio to $\zeta = 0.5$ for the dominant roots of the system. Find K_1, ω_n, M_o, T_p, T_s, N, $C(s)/R(s)$ for (a) the original system and (b) the original system with a cascade lag compensation using $\alpha = 10$; c. Suggest a cascade compensator that will improve the response time, i.e., will move the dominant branch to the left in the s plane.

12-6. A control system has

$$G_x(s)H(s) = \frac{K}{(s+1)(s+2)(s+5)(s+6)} \qquad H(s) = 1$$

a. Determine the roots of the characteristic equation of this system for $\zeta = 0.5$.
b. Determine the ratio $\sigma_{3,4}/\sigma_{1,2}$.   c. It is desired that this ratio be increased to at least 10 with no increase in T_s for $\zeta = 0.5$.   A network that may accomplish this has the following transfer function:

$$G_c = A \frac{s+a}{s+b}$$

Determine appropriate values of a and b that can be achieved with practical values of α.

12-7. A unity feedback system has the transfer function $G_x(s)$.   The closed-loop roots must satisfy the specifications $\zeta = 0.5$ and $T_s = 2$ sec.   A suggested compensator $G_c(s)$ must maintain the same degree as the characteristic equation for the basic system:

$$G_x(s) = \frac{K_x(s+6)}{s(s+5)} \qquad G_c(s) = \frac{A(s+a)}{s+b}$$

a. Graphically determine the values of a and b.   b. Determine the value of α. c. Is this a lag or a lead network?

12-8. A unity feedback system has the transfer function $G_x(s)$, where $K = 4.46$ for $\zeta = 0.7$.   System performance to be obtained with the addition of a simple lead compensator is: $T_s = 1.6$ sec (± 2 per cent overshoot), $\zeta = 0.7$, $K_0 \geq 2$, and the characteristic equation must be reduced to second order.   Determine the necessary values A, α, and T of the compensator.

$$G_x(s) = \frac{K(s+3)}{(s+1)(s+2)(s+4)}$$

12-9. A unity feedback system has

$$G_x(s) = \frac{K}{(s+1)(s+2)(s+6)}$$

a. For $\zeta = 0.5$, determine the roots and the value of K.   b. Add a lead compensator which cancels the pole at $s = -1$.   c. Add a lead compensator which cancels the pole at $s = -2$.   d. Compare the results of parts b and c.   Establish a "rule" for adding a lead compensator to a Type 0 system.

12-10. a. Determine $c(t)$ for a unity feedback control system having $G_x(s) = 1/(s+2)$ with $r(t) = u(t)$ b. Add the simple cascade lag compensator $G_c(s) = 8/(s+8)$. Determine $c(t)$ for $r(t) = u(t)$.   c. Compare the responses and the settling times for the original and compensated systems.

12-11. For the system of Prob. 7-5, the desired roots are $s = -10 \pm j35$ and the desired gain is $K_1 = 100$.   a. Design a compensator which will achieve these characteristics.   b. Determine $c(t)$ with a unit step input.   c. Are the desired complex roots dominant?

12-12. A unity feedback system has the transfer function

$$G_x(s) = \frac{K}{s^2(1 + 0.2s)}$$

The poles of the closed-loop system must be $s = -0.75 \pm j2$.   *a.* Design a lead compensator with the maximum possible value of α which will produce these roots.   *b.* Determine the control ratio for the compensated system.   *c.* What kind of compensation should be added to increase the gain without increasing the settling time?

12-13. A system has the transfer function

$$G_x(s) = \frac{K}{(s^2 - 2s + 2)(s + 10)}$$

The roots of the closed-loop system must have $\zeta = 0.5$.   Add a lead compensator, with two zeros and two poles, to produce a stable system.   Determine $C(s)/R(s)$.

13-1. It is desired that the control system of Prob. 9-9 have an $M_m = 1.26$.   *a.* Add a lag compensator, with $\alpha = 10$, so that this value of M_m can be obtained.   *b.* Add a lead compensator with $\alpha = 0.1$.   *c.* Add a lag-lead compensator with $\alpha = 10$. Indicate the time constants of the compensator in each case.   Compare the results obtained by the use of each type of compensator.   Also, compare the results of this problem with those of Prob. 12-4.

13-2. A control system has the forward transfer function

$$G_x(s) = \frac{K_2}{s^2(1 + 0.1s)} \qquad K_2 = 1$$

The closed-loop system is to be made stable by adding a compensator G_c and an amplifier A in cascade with $G_x(s)$.   An $M_m = 1.5$ is desired with $\omega_m \approx 1.4$ radians/sec.   *a.* What kind of compensator is needed?   *b.* Select an appropriate α and T for the compensator.   *c.* Select the necessary value of amplifier gain A.   *d.* Plot the compensated curve.   *e.* Compare the results of this problem with those obtained in Prob. 12-2.

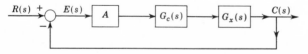

13-3. A control system with unity feedback has a forward open-loop transfer function

$$G_x(s) = \frac{K_1}{s(1 + 0.5s)(1 + 0.1s)^2}$$

a. Find the gain K_1 for 45° phase margin, and determine the corresponding phase-margin frequency ω_ϕ.   *b.* For the same phase margin as in *(a)*, it is desired to increase the phase-margin frequency to a value of $\omega_\phi = 3.0$ with the maximum possible improvement in gain.   To accomplish this, a lead compensator is to be used.   Determine the values of α and T that will satisfy these requirements. For these values of α and T, determine the new value of gain.   *c.* Repeat part *b* with the lag-lead compensator of Fig. 20-16.   Select an appropriate value for T. With $G_c(s)$ inserted in cascade with $G_x(s)$, find the gain needed for 45° phase margin.   *d.* Show how the compensator has improved the system performance.

13-4. The mechanical system shown has been suggested for use as a compensating component in a mechanical system.   *a.* Determine whether it will function as a lead or a lag compensator by finding $X_2(s)/F(s)$.   *b.* Sketch $\mathbf{X}_2(j\omega)/\mathbf{F}(j\omega)$.

Label the key points on the curve.   The force f is applied at the point shown. x_1 and x_2 are displacements.   Neglect all masses.

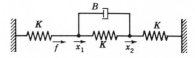

13-5.   A unity feedback control system has the following transfer function:

$$G(s) = \frac{K}{(1+s)^2} G_c(s)$$

a. What is the value of the gain for an $M_m = 1.4$ if $G_c(s) = 1.0$?   *b.* Design a compensator $G_c(s)$ that will increase the step error coefficient by a factor of 8 while maintaining the same M_m.   *c.* What effect does the compensation have on the closed-loop response of the system?

13-6.   A unity feedback control system has the transfer function

$$G(s) = G_c(s)G_1(s)$$

where $G_1(s) = 1/s^2$.   It is desired to have an $M_m = 1.4$ with a damped natural frequency of about 10.0 radians/sec.   Design a compensator $G_c(s)$ that will help to meet these specifications.

13-7.   For the lag compensator, plot T versus ω for $\alpha = 10$ and $\phi_c = -5°$ [see Eq. (13-4)].

13-8.   A system has

$$G_1(s) = 10 \qquad G_2(s) = \frac{1}{(1+0.1s)(1+0.01s)} \qquad G_3(s) = \frac{1}{s}$$

For the control system shown in (a), determine the cascade compensator $G_c(s)$ required to make the system meet the desired open-loop frequency-response characteristic shown in (b).   What is the value of M_m for the compensated system?

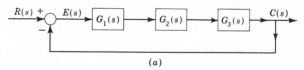

(a)

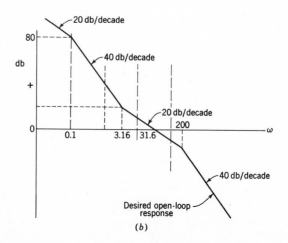

(b)

13-9. A system has the transfer function

$$G_x(s) = \frac{K_1(1 + 4s)}{s(1 + 0.5s)(1 + 0.5s + s^2)}$$

a. For an $M_m = 1.41$, determine K_1 and ω_m for the original system.   *b.* Add a lag compensator with $\alpha = 10$ and determine T, K_1, and ω_m for the same M_m. *c.* Add a lead compensator with $\alpha = 0.1$ and determine T, K_1, and ω_m for the same M_m.

13-10. Repeat Prob. 12-1, using $M_m = 1.3$ as the basis of design.   Compare the results.

14-1. In the figure the servomotor has inertia but no viscous friction.   The feedback through the accelerometer is proportional to the acceleration of the output shaft. *a.* When $H(s) = 1$, is the servo system stable?   *b.* Determine an $H(s)$ that causes the system to be stable.   Show that the system is stable with the value of $H(s)$ selected by sketching the transfer function $\mathbf{C}(j\omega)/\mathbf{E}(j\omega)$.

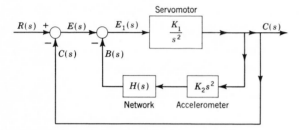

14-2. For the system shown, find (*a*) the transfer function $C(s)/E(s)$; (*b*) ω_n and ζ for the open-loop system.

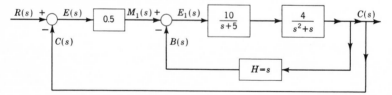

14-3. A servo with tachometric feedback can be represented as shown in (*a*).   Note that $E(s) = R(s) - C(s)$ and $E_1(s) = E(s) - B(s)$. $G_x(s)$ represents an ideal servomotor which develops a torque $T = K_T E_1$.   The total inertia on the output shaft is J_0, and there is negligible viscous friction.   The torque-speed characteristics are shown in (*b*) for several values of E_1. $H_1(s)$ represents a tachometric generator which develops voltage $B = K_g\, Dc$ and $G_3(s) = C(s)/E(s)$.   *a.* Find, in terms of system constants, $G_x(s)$, $H_1(s)$, and $G_3(s)$.   *b.* Sketch the locus of $\mathbf{G}_x(j\omega)$, $\mathbf{H}_1(j\omega)$, and $\mathbf{G}_3(j\omega)$.

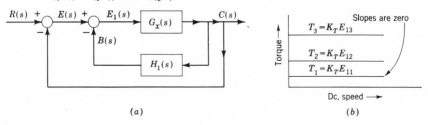

(*a*)   (*b*)

14-4. The accompanying schematic shows a method for maintaining a constant rate of discharge from a water tank by regulating the level of the water in the tank. The dynamics of the flow in and out of the tank may be taken into account by the relationships

$$Q_1 - Q_2 = 20\frac{dh}{dt}$$
$$Q_2 = 4.0h$$
$$Q_1 = 7.0\theta$$

where Q_1 = volumetric flow into tank
Q_2 = volumetric flow out of tank
h = pressure head in tank
θ = angular rotation of control valve

The motor damping is much smaller than the inertia. The system shown is unstable for small values of system gain. Therefore, feedback from the control valve to the amplifier is proposed. Show conclusively which of the following feedback functions would produce the best results: (*a*) feedback signal proportional to control-valve position; (*b*) feedback signal proportional to rate of change of control-valve position; (*c*) feedback signal comprising a component proportional to valve position and a component proportional to rate of change of valve position.

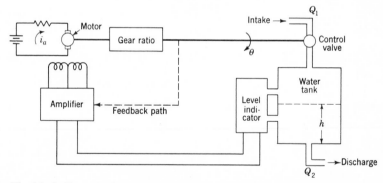

14-5–14-9. The block diagram shows a simplified form of roll control for an airplane. Over-all system specifications are t_s = 1.0 sec and M_p = 1.3. A_h is the gain of an amplifier in the H_2 feedback loop. G_2 is an amplifier of adjustable gain with a maximum value of 100. Restrict b between 15 and 50.

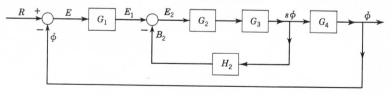

$$G_3(s) = \frac{1.7}{(1 + 0.25s)(1 + s)}$$
$$H_2(s) = 0.2A_h\frac{s + a}{s + b}$$
$$G_4(s) = \frac{1}{s}$$

14-5.   *a.* By use of the root locus, Method 1, design the RC network to be used in $H_2(s)$ to meet the over-all system specifications with $G_1(s) = 1$.   *b.* Design $G_1(s)$ to improve the value of K_1 of part a by a factor of 5 while maintaining the desired over-all system specifications.   Determine all roots of the characteristic equation for Φ/R after compensation.   *c.* For the values obtained in part a, determine T_s and M_p for the inner loop.   Compare these values with those given in Prob. 14-6a.

14-6.   *a.* By use of the root locus, Method 2, design the RC network to be used in $H_2(s)$ to meet the inner-loop specifications $M_p = 1.05$ and $T_s = 1.0$ sec.   Determine the necessary values of A_h and G_2.   *b.* Design $G_1(s)$ to improve the value of K_1, resulting from part a, by a factor of 5 and also to meet the over-all system specifications.

14-7.   Repeat Prob. 14-5.   Use the polar-plot method, Case 3.   *Note:* Incorporate G_4 into the minor loop.   It is not necessary to initially adjust the gain of $G_2G_3G_4$ to make the curve of $1/G_2G_3G_4$ tangent to the required M_m circle.   It is not necessary to plot $1/G_3G_4$.   All that is needed is $1/\mathbf{G}_3(j\omega_a)\mathbf{G}_4(j\omega_a)$, which can be evaluated analytically.   The gain G_2 can then be adjusted in accordance with your design procedure.

14-8.   Repeat Prob. 14-7; but use Case 4.

14-9.   Repeat Prob. 14-5, but solve by the use of the logarithmic plots.   *Note:* Incorporate G_4 into the minor loop.   Limit G_1 and G_2 to less than 100.

14-10.   A system has

$$G_1(s) = 3 \qquad G_3(s) = \frac{1}{1+s} \qquad H_1(s) = \frac{s}{1+s}$$

$$G_2(s) = 10 \qquad G_4(s) = \frac{1}{s} \qquad U(s) = 0$$

System specifications are

$$K_1 = 30 \text{ sec}^{-1} \qquad \gamma \geqq +50°$$
$$\omega_\phi = 3 \text{ radians/sec}$$

Determine, by using approximate techniques, the following: (a) whether the feedback-compensated system satisfies all the specifications; (b) a cascade compensator to be added between $G_1(s)$ and $G_2(s)$, with the feedback loop omitted, to produce the same $\mathbf{C}(j\omega)/\mathbf{E}(j\omega)$ as in part a; (c) whether the system of part b satisfies all the specifications.

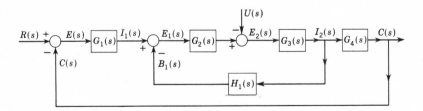

14-11.   In the figure of Prob. 14-10,

$$G_1(s) = A_1 \qquad G_3(s) = \frac{1}{(1+0.1s)(1+0.01s)}$$

$$G_2(s) = A_2 \qquad G_4(s) = \frac{1}{s}$$

$H_1(s)$ is a passive network.   Choose A_1, A_2, and $H_1(s)$ to meet the desired open-loop response characteristic shown in the figure of Prob. 13-8 for $\omega < 120$.

14-12.   For the feedback-compensated system of Fig. 14-8, $A = 1$,

$$G_x(s) = \frac{K_x}{s(s+2)} \qquad H(s) = \frac{0.1s^2}{s+6}$$

a. Use Method 1 to determine $C(s)/R(s)$ using $T_s = 2$ sec (2 per cent criterion) for the dominant roots.   *b.* Design a cascade compensator which yields the same dominant roots as part *a.*   *c.* Compare the gains of the two systems.

14-13.   For the feedback-compensated system of Fig. 14-31,

$$G_x(s) = \frac{1}{s(1+s)} \qquad H(s) = \frac{2s}{1+1.25s}$$

a. For the original system, without feedback compensation, determine the values of A, K_1, and ω_ϕ for a phase margin of 45°.   *b.* Add the minor-loop compensator $H(s)$ and determine, for a phase margin of 45°, the corresponding values of A, K_1, and ω_ϕ.   *c.* Compare the results.

14-14.   A unity feedback system has

$$G_x(s) = \frac{K_x}{s(s+7)}$$

a. For $\zeta = 0.5$, determine $C(s)/R(s)$.   *b.* Minor-loop and cascade compensation are to be added (see Fig. 14-8) with the amplifier replaced by $G_c(s)$:

$$G_c(s) = \frac{As}{s+b} \qquad H(s) = \frac{K_h s}{s+a}$$

Use the value of K_x determined in part *a.*   By Method 1 of the root-locus design procedure, determine values of a, b, K_h, and A to produce dominant roots having $\zeta = 0.5$ and $\sigma = -8$.   Only positive values of K_h and A are acceptable.   To minimize the effect of the third real root, it should be located near a zero of $C(s)/R(s)$.

14-15. K_a is an amplifier,

$$G_x(s) = \frac{K_x}{s(T_1s+1)(T_2s+1)} \qquad H(s) = \frac{K_h T^2 s^3}{T^2 s^2 + ATs + 1}$$

Describe the procedure to be used to adjust K_a, K_x, K_h, T, and A to improve the response of this control system (*a*) by using the root-locus plots (Method 1) and (*b*) by using the log plots.   Show qualitatively that the system performance after H is inserted as shown is better than the performance possible without H. Sketch typical curves where necessary to illustrate your descriptions.

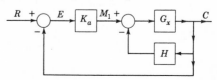

14-16. For $G(s)$ given in Prob. 12-4, add feedback compensation containing a tachometer. Use root locus Method 1 or 2 as assigned. Design the tachometer so that an improvement in system performance is achieved while maintaining a $\zeta = 0.5$ for the dominant roots.

14-17. Repeat Prob. 14-16 using the log plots for $M_m = 1.26$.

15-1. A system has

$$G_x(s) = \frac{K_x}{1 + T_1 s} \qquad G_y = \frac{K_y}{s(1 + T_2 s)}$$

The K_1 of the system is just satisfactory, and T_1 is larger than desired. *a.* Mathematically, show how the effective gain and time constant of $G_x(s)$ can be changed by use of direct feedback around $G_x(s)$ only. *b.* State in what manner the values of gain and time constant are changed. *c.* State whether the changes in both parameters are desired changes.

15-2. A system has

$$G_1(s) = 2 \qquad\qquad G_3(s) = \frac{1}{s}$$

$$G_2(s) = \frac{10}{(1 + s)(1 + 0.2s)} \qquad H_2(s) = \frac{s}{1 + s}$$

By use of the straight-line asymptotes, draw the approximate plot of $\mathrm{Lm}\ [\mathbf{C}(j\omega)/\mathbf{R}(j\omega)]$.

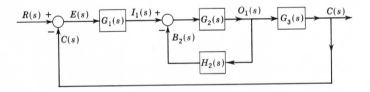

15-3. Sketch the approximate closed-loop magnitude response for the open-loop functions of (*a*) Prob. 9-5; (*b*) Prob. 9-7; (*c*) Prob. 9-8. Superimpose these sketches on the log magnitude diagram of the open-loop functions.

15-4. Determine $G_c(s)$ in the figure to achieve a zero final error to a step-function input.

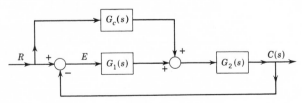

15-5.   Draw the signal flow graph and find C/R for the system shown.

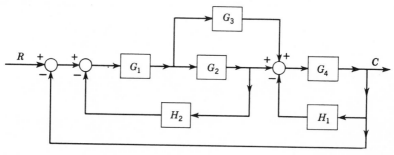

15-6.   A system has

$$G_x(s) = \frac{10}{1+s} \qquad G_y(s) = \frac{1}{s(1+s)} \qquad N_1(s) = 0.01\,\frac{1+0.1s}{1+0.2s}$$

a. What must be added to the system shown so that $R_2(s)$ has the same steady-state effect on $C(s)$ as $R_1(s)$ does?   *b.* Apply the principle of invariance to reduce the effect of $r_2(t)$ on the output to zero.

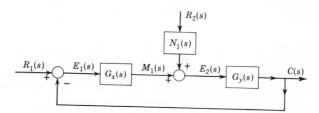

15-7.   *a.* Find C/R.   *b.* Find C/T_L.   *c.* Which block or blocks would you work on to reduce the effect of T_L $\left[\text{as } \lim_{s \to 0} (C/T_L)\right]$?   State the relative magnitude each must have to accomplish this.   *d.* Which block or blocks would you work on to improve the response to R $\left[\text{as } \lim_{s \to 0} (C/R)\right]$?   State the relative magnitude each block or blocks must have to accomplish this.

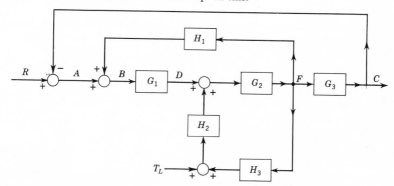

15-8. Where, in the servo loop shown, must an integral device be inserted to eliminate the steady-state error due to a constant load torque?   Show the reason for your answer.

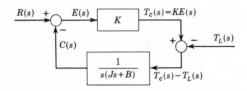

15-9. Refer to Prob. 14-10.   An additional specification is $\mathbf{E}(j\omega)/\mathbf{U}(j\omega) \leq -30$ db for all values of ω.   *a.* Does the feedback-compensated system of Prob. 14-10*a* meet this specification?   *b.* Does the cascade-compensated system of Prob. 14-10*b* meet this specification?

15-10. Draw a signal flow graph for the block diagram of Prob. 14-10.   From the signal flow graph determine $C(s)/U(s)$ and $C(s)/R(s)$.

15-11. Draw the signal flow graph and determine $C(s)/R(s)$ for the block diagrams shown.

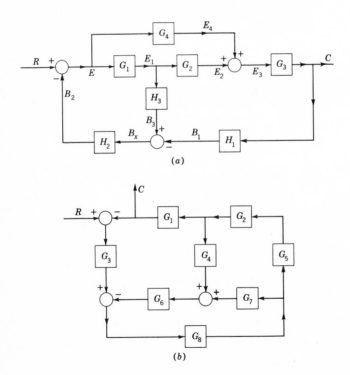

15-12. In the block diagram shown, it is required that (*a*) with $y(t) = 0$, the response to the input $r(t) = u(t)$ be $c(t)_{ss} = 1$ and (*b*) with $r(t) = 0$, the response to the input $y(t) = u(t)$ be $c(t)_{ss} = 0$.   Three feedback units $H(s)$ are suggested.   Select the one which satisfies the requirements.

$$G(s) = \frac{1}{(1 + T_1 s)(1 + T_2 s)}$$

$$H_a(s) = \frac{K_a(1 + T_3 s)}{1 + T_4 s} \qquad H_b(s) = K_b s \qquad H_c(s) = \frac{K_c(1 + T_5 s)}{s(1 + T_6 s)}$$

15-13. By use of the straight-line log magnitude approximations, determine the approximate equation for $C(s)/R(s)$.

$$G_1(s) = \frac{10(1 + 0.1s)}{1 + s} \qquad G_2(s) = \frac{4}{s(1 + s)(1 + 0.1s)}$$

15-14. Refer to the example of Sec. 15-7. (*a*) By means of the root locus, determine the poles and zeros of $[\Omega(s)/E_r(s)]_{T_L=0}$ and $[\Omega(s)/T_L(s)]_{E_r=0}$ for the lag compensator. (*b*) Perform a correlation analysis between the frequency response given in Sec. 15-7 and the pole-zero diagram obtained in part *a*. Use the approach of Secs. 8-3 and 8-4.

15-15. Determine the function $G_c(s)$ which produces steady-state correspondence between $c(t)$ and $r(t)$ for the system shown in the figure of Prob. 15-4 for $r(t) = u(t)$

(*a*) $G_1(s) = 5 \qquad G_2(s) = \dfrac{1}{(s + 1)(s + 3)}$

(*b*) $G_1(s) = \dfrac{-5}{s + 3} \qquad G_2(s) = \dfrac{1}{(s + 1)(s + 2)}$

15-16. Apply the method of Sec. 15-13 to the example of Sec. 15-12.

16-1. The circuit shown is intended to be used as a lead compensator in an a-c (carrier) servo. The carrier frequency is ω_c. The signal frequency is ω_s. Derive the transfer function $\mathbf{E}_o(j\omega)/\mathbf{E}_i(j\omega)$. Put it in the standard form $\alpha(1 + j\omega_s T)/(1 + j\omega_s \alpha T)$ by evaluating T and α.

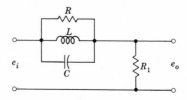

16-2. If the initial $G(s_s)_{a\text{-}c}$ is identical with $G(s_s)_{d\text{-}c}$, prove that the compensation $G_c(s)$ needed for the a-c servo is identical with the compensation $G_c(s_s)$ needed for the d-c servo.

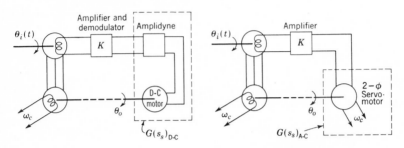

16-3. It is desired to build a lead compensator whose transfer function is

$$\frac{E_o(j\omega_s)}{E_i(j\omega_s)} = \frac{1 + j0.01\omega_s}{10 + j0.01\omega_s} = 0.1\,\frac{1 + j0.01\omega_s}{1 + j0.001\omega_s}$$

where ω_s is the (angular) signal frequency. The device is to be a static network to operate into an open circuit. The input impedance of the network is to be 10^5 ohms when $\omega_s = 0$.

a. Design the network of a figure (*a*) as a dc compensator ($\omega_c = 0$) to meet the above specifications.

b. Repeat (*a*) for figure (*b*).

c. Repeat (*a*) for figure (*c*) but for a carrier frequency of 60 cycles.

d. Design a bridged T consisting of resistances and capacitances only which has the same transfer function as part *c* and an input impedance whose magnitude is 10^5 ohms at the carrier frequency. *e.* Design (*c*) and (*d*) for a 400-cycle carrier if possible. *f.* What is the highest carrier frequency for which a network of type *d* is realizable?

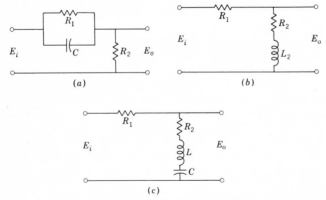

16-4. The network shown has been suggested as a compensating network for a carrier-type servo system. *a.* Find $E_o(j\omega)/E_i(j\omega)$. *b.* The frequency seen by the network is $\omega = \omega_c \pm \omega_s$, where ω_c is the carrier frequency and ω is the signal frequency. Let $LC = 1/\omega_c^2$ and $x = \omega/\omega_c - \omega_c/\omega$. Find $E_o(jx)/E_i(jx)$. *c.* Under what conditions is x approximately proportional to the servo frequency ω_s? Show that this is so.   *d.* Sketch the transfer function $\mathbf{E}_o(j\omega_s)/\mathbf{E}_i(j\omega_s)$.   Is this an integral or derivative network?

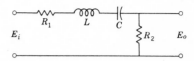

16-5.   Calculate the transfer function for the signal frequency ω_s for the bridged-T network of Fig. 16-28a.

16-6.   The following networks have been suggested as compensators for an a-c (carrier-type) system.   The carrier frequency is ω_c and the signal frequency is ω_s, where $\omega_c \gg \omega_s$.   *a.* For each network, determine the transfer function $\mathbf{G}_c(j\omega_s)$.   *b.* State whether each network is a lag or a lead compensator.

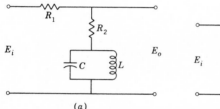

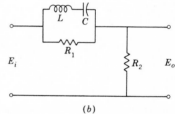

(a)   (b)

16-7.   Derive the transfer function $\mathbf{E}_2'(j\omega)/\mathbf{E}_1'(j\omega)$ for the network of Fig. 16-28b when an inductance L is placed in series with R_A.   Rationalize this function and simplify by using the condition $\omega_c{}^2 R_B L C_A' C_B' = R_B(C_A' + C_B')$ and $\mathbf{G}(j\omega_c) = 0$.   Show that this network can be employed as an a-c compensator when used in the block diagram of Fig. 16-30.   Obtain $\mathbf{G}_c(j\omega_s)$.

16-8.   Plot the transfer function of the compensator $\mathbf{G}_c(j\omega_s) = \alpha[(1 + j\omega_s T)/(1 + j\omega_s \alpha T)]$ on a log magnitude–angle diagram.   Show that it is a semicircle by determining its center and radius.

18-1.   Plot the describing function for (a) Eq. (18-16); (b) Eq. (18-20); (c) Eq. (18-23).

18-2.   A certain relay without dead band or hysteresis may be represented by the idealized characteristic shown in (a).   When a control system with the $\mathbf{G}(j\omega)$ shown in (b) is driven by the given relay, how many periodic solutions for zero reference input are obtained from the describing-function analysis?   Indicate which of these are stable.

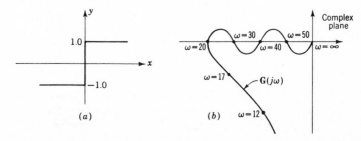

(a)   (b)

18-3. The control system shown has a nonlinear element whose output $y(t)$ is the following function of its input $x(t)$ and the derivative of its input dx/dt:

$$y(t) = \left(\frac{dx}{dt}\right)^3 + x^2 \frac{dx}{dt}$$

a. Derive the describing function for the above nonlinear element.   *b.* Determine
the possibility of a limit cycle in the system shown.

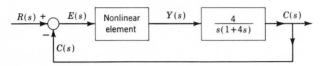

18-4. For the system shown in (*a*), determine the frequency and amplitude of oscillation
of $x(t)$ that are predicted by the describing-function method.   Open-loop measurements indicate that the response of $G(s)$ when $y(t)$ is a unit impulse at $t = 0$ is as
shown in (*b*).   The plot of the log magnitude of the describing function $N(x)$ is
shown in (*c*).

(*a*)

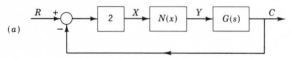

(*b*)

(*c*)

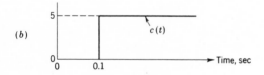

18-5. A system has

$$G(j\omega) = \frac{10\sqrt{2}}{j\omega(1 + j0.5\omega)} \qquad \text{and} \qquad N = \frac{1}{X} \underline{/-45°}$$

where X is the amplitude of the sinusoidal input to the nonlinear element.   Find
the amplitude and frequency of the possible periodic solution.   Is it a stable
oscillation?

18-6. A relay has the dead zone shown below.   Determine the describing function.
Sketch the plot of N versus $d/2X$.   Determine the amplitude, frequency, and
stability of the possible periodic solution for $G(s) = 10/s(1 + s)^2$.

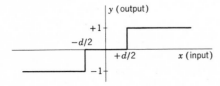

18-7. A dead zone $d = 1$ is in cascade with a transfer function $G(j\omega)$. The Lm–angle diagram of $G(j\omega)$ is shown on the sketch below. Find the frequencies and amplitudes of the stable oscillations, if any exist.

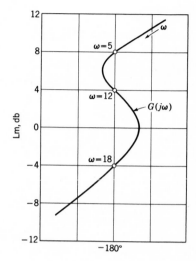

Answers to Selected Problems

Chapter 2

2-2(b) (1) $T = (J_1D^2 + B_1D + K_1)\theta_1 - K_1\theta_2$

(2) $0 = -K_1\theta_1 + [J_2D^2 + B_2D + (K_1 + K_2)]\theta_2 - K_2\theta_3$

(3) $0 = -K_2\theta_2 + [J_3D^2 + B_3D + (K_2 + K_3)]\theta_3$

2-3(a) $f(t) = \dfrac{l_2}{l_1}(MD^2 + BD + K)y$

2-4(a) $\left[\dfrac{C_1ad}{(a+b)(c+d)} + D\right]y_1 = \dfrac{C_1b}{a+b}x_1$

(b) $\dfrac{VM}{K_BC}D^3y_1 + \left[\dfrac{VB}{K_BC} + \dfrac{M}{C}(L + C_p)\right]D^2y_1 + \left[C_b + \dfrac{B}{C}(L + C_p)\right]Dy_1$

$$+ \dfrac{C_xad}{(a+b)(c+d)}y_1 = \dfrac{C_xb}{a+b}x_1$$

2-11 $f(t) = \left(MD + B + \dfrac{K}{D}\right)Dx + \dfrac{U(U\,Dx - v)}{LD}$

where $f(t)$ is the force exerted by the sound waves on the diaphragm, x is the distance the diaphragm moves, and $U = 2\pi r N\beta$, $2\pi r$ is the length of one turn of coil, N is the number of turns in the coil, and β is the flux density of permanent magnets.

Chapter 3

3-2 $\theta_3(t) = -16.4 + 2t + 21.7e^{-0.22t} - 4.81e^{-0.48t}$
$\qquad\qquad + 0.852e^{-0.389t}\sin(0.744t - 1.03) + 0.113e^{-0.411t}\sin(1.19t - 4.83)$

3-6 $\omega(t) = -6.35\sin(5t - 48.5°) - 16.82e^{-0.5t}\sin(0.289t - 81.5°)$

3-10(b) $x(t) = 2 + 0.667e^{-4t} - 2.667e^{-t}$

Chapter 4

4-1(a) $f(t) = 10e^{-t}(t - \sin t)$

4-2(b) $x(t) = [77e^{-2t} + 7.42e^{-2.035t}\sin(0.785t - 87.2°)$
$\qquad\qquad\qquad\qquad + 2.67e^{+1.035t}\sin(3.54t - 136.6°)]10^{-2}$

4-4(a)	$0 < K < 35.5$	**(c)**	$-20 < K < \infty$
(b)	$0 < K < 150$	**(d)**	$-10 < K < 126$

4-5(b) See Prob. 3-2.

4-6(a) $F(s) = \dfrac{2}{s}(e^{-s} - e^{-3s})$

(c) $F(s) = E\left[\dfrac{1}{Ts^2} - \dfrac{e^{-Ts}}{s(1 - e^{-Ts})}\right]$

4-7(d) $F(s) = -\dfrac{2}{9s} + \dfrac{1}{3s^2} + \dfrac{1}{5(s+1)} + \dfrac{1}{45(s+6)}$

4-8(b) See Prob. 3-10b.

4-9(b) $(s^2 + 2s + 5)X(s) - 2s - 4 = \dfrac{10}{s}$

(d) $(s^3 + 4s^2 + 8s + 4)X(s) + 4s^2 + 15s + 28 = \dfrac{5}{s^2 + 25}$

4-10(b) $(s + 2)(s^2 + 2s + 2)$
(c) $(s^2 + 3s + 3)(s^2 - s + 2)$
(i) $(s^2 + 6s + 10)(s^2 - 4s + 40)(s + 1)$

4-11(b) None
(c) 2
(i) 2

4-12(a) Prob. 4-1a: 0
(b) Prob. 4-4a: 1
            4-4b: 1

4-13(a) Prob. 4-1a: 0
(b) Prob. 4-4a: 0
            Prob. 4-4b: 0

Chapter 5

5-4 $K < 50$

5-5(a) $G(s) = \dfrac{K_1(1 + T_1 s)}{s(1 + T_2 s)(1 + T_3 s)}$

(b) $\dfrac{C(s)}{R(s)} = \dfrac{K_1(1 + T_1 s)}{s(1 + T_2 s)(1 + T_3 s) + K_1(1 + T_1 s)}$

5-8 $\dfrac{E_d(s)}{E_c(s)} = \dfrac{K_q K_d / L_c L_q}{(s + R_c/L_c)(s + R_q/L_q)}$

5-11 $(Js^2 + Bs + K)\theta_0(s) = Hs\theta_i(s)$

Chapter 6

6-1 (a) ∞; (b) N/d; (c) 0; (d) 0; (e) $R_1 d/N$; (f) ∞
6-6(e) ∞, ∞, $\frac{1}{2}$
6-7(e) (a) 0; (b) 0; (c) 24

Chapter 7

7-3(c)   Real-axis branches: none; 4 branches; $\gamma = \pm 45°$, $\pm 135°$; $\sigma_0 = -2$; angles of departure: $\pm 45°$, $\pm 135°$; imaginary-axis crossing: $\pm j2$ with $K = 60$; and the branches are straight lines.

(d)   Real-axis branch: 0 to $-\infty$; 5 branches; $\gamma = \pm 36°$, $\pm 108°$, $180°$; $\sigma_0 = -\frac{8}{5}$; no breakaway or break-in points; angles of departure: $-90°$, $-26°$; and the imaginary-axis crossing: $\pm j0.9$ with $K = 20.6$. Locus crosses asymptotes at $\pm 36°$.

7-5(a)   Real-axis branches: 0 to -50 and -100 to $-\infty$; 3 branches; $\gamma = \pm 60°$, $180°$; $\sigma_0 = -50$; breakaway point: -21.2 $(K_1 = 9.5)$; and imaginary-axis crossing is $\pm j70.7$ $(K_1 = 150)$.

(b) $K_1 = 150$

(c) $K_1 = 36.1$

(d) $e(t) = 0.12e^{-121t} - 1.171e^{-15.t} \sin (35.t - 132°)$

(e)

ω	0	10	20	30	40	60
M	1.00	1.03	1.18	1.32	1.16	0.48
α	0°	$-18°$	$-35°$	$-73°$	$-115°$	$-166°$

Chapter 9

9-3(a)   It must have an initial slope of -40 db/decade.   **(b)** The phase-angle curve must approach an angle of $180°$ as $\omega \to 0$.

9-5(a) $$G(s) = \frac{31.6(1 + 0.125s)}{s(1 + 2s)(1 + 0.0445s)}$$
Total correction of $+2.4$ db at $\omega = 8$.

9-6(a) $K_0 = 10$

9-9(a) $\lim_{\omega \to 0} [\mathbf{G}(j\omega)] = \infty \underline{/-90°}$ $\omega_y = \pm 30.4$

$\lim_{\omega \to \infty} [\mathbf{G}(j\omega)] = 0 \underline{/-360°}$ $\omega_x = \pm 12.5$

$V_x = \lim_{\omega \to 0} [\text{Re } \mathbf{G}(j\omega)] = -0.148K_1$

Chapter 10

10-1(a) $N = 0$, $Z_R = 0$; stable

(b) $N = -2$, $Z_R = 2$; unstable

10-2(a) $N' = 0$, $Z_R = 0$; stable

(b) $N' = -2$, $Z_R = 2$; unstable

10-3(a)   Stable for $K > -1$

(b)   Unstable for $-1 < K < 1$

10-5(a)   (1) a. For $\gamma = +45°$, $K_1 = 8.4$
    b. $K_1 < 50.2$ for stability

Chapter 11

11-4 $K_1 \approx 100$ and $\omega_m \approx 37$

11-7(a) $K_2 \approx 3.57$

   (b) $\omega_m \approx 5$

   (c) $M_m \approx 1.52$

11-8 $T = 0.116$ and $K = 1.21$

11-12   (4) $M_m \approx 0.908$, $\omega_m = 0$, and $\gamma \approx +91°$; for $M_m = 1.25$, $A \approx 1.78$, $\omega_m \approx$ 10.45, and $\gamma \approx +48°$

       (5) A completely unstable system and $\gamma \approx -27°$

       (6) $M_m \approx 1.0$, $\omega_m \approx 0$, and $\gamma \approx +71°$; for $M_m = 1.25$, $A \approx 5.8$, $\omega_m \approx 6.4$, and $\gamma \approx +48°$

Chapter 12

12-1

System	Dominant roots	Other roots	K_1, sec^{-1}	T_p, sec	T_s, sec	M_o
Basic	$-5.9 \pm j12.25$	$-56.4 \pm j18.3$	10	0.291	0.679	0.207
Lag compensated $\alpha = 10, T = 1$	$-5.6 \pm j12.2$	$-56.4 \pm j18.3$ -1.1	104	0.292$^+$	0.679$^+$	0.226
Lead compensated $\alpha = 0.1, T = \tfrac{1}{25}$	$-11.8 \pm j24.5$	-80.5 -248.5	22.1	0.147	0.339	0.194
Lag-lead compensated $\alpha = 10, T_1 = 1,$ $T_2 = \tfrac{1}{25}$	$-11.8 \pm j24.5$	-1.045 -80.5 -248.5	226	0.147$^+$	0.339$^+$	0.212

Note: This solution is based on using the transfer function

$$G_x(s) = \frac{K_1}{s(1 + 0.04s)(1 + s/26 + s^2/2,600)}$$

for the basic system instead of the one given in the statement of the problem.

Chapter 13

13-1

	K_1	ω_m
Basic	6.1	6.1
Lag $\alpha = 10,$ $T = 2$	52.1	4.75
Lead $\alpha = 0.1$ $T = 0.1$	10.2	15
L-L $\alpha = 10,$ $T_1 = 2,$ $T_2 = 0.1$	102.3	15

Chapter 14

14-2(a) $\quad G(s) = \dfrac{20}{s(s^2 + 6s + 45)}$

14-3(a) $\quad G_x(s) = \dfrac{K_T}{Js^2} \qquad H_1(s) = K_g s \qquad G_3(s) = \dfrac{AK_T}{s(Js + K_T K_g)}$

Chapter 15

15-4 $\quad$ $G_c(s)$ may have any form as long as it satisfies the condition
$$\lim_{s \to 0} [G_c(s)] = \lim_{s \to 0} \frac{1}{G_2(s)}$$

15-5(a) $\quad \dfrac{C}{R} = \dfrac{G_1 G_2 G_4 + G_1 G_3 G_4}{1 + G_4 H_1 + G_1 G_2 H_2 + G_1 G_2 G_4 H_1 H_2 + G_1 G_2 G_4 + G_1 G_3 G_4}$

15-8 $\quad$ Insert an integrating element, $1/s$, in parallel with K or one having the form of $(1 + Ts)/T_s$ in cascade with K, where $T > J/B$.

15-9 $\quad$ (a) Yes; (b) Yes

Chapter 16

16-3(a) $\quad R_2 = 10^4$ ohms $\qquad R_1 = 9 \times 10^4$ ohms $\qquad$ and $\qquad C = 0.111\ \mu$f

(b) $\quad R_2 = 10^4$ ohms $\qquad R_1 = 9 \times 10^4$ ohms $\qquad$ and $\qquad L_2 = 100$ henrys

(c) $\quad R_2 = 10^4$ ohms $\qquad R_1 = 9 \times 10^4$ ohms
$\qquad L = 50$ henrys $\qquad C = 0.141\ \mu$f

(d) $\quad R_1 = 2.13 \times 10^4$ ohms $\qquad C_2 = 0.0399\ \mu$f,
$\qquad R_2 = 31.65 \times 10^4$ ohms $\qquad C_1 = 0.0259\ \mu$f

(e)   Cannot be designed for 400-cycle carrier.

(f) $f_c < 96$ cps

Chapter 18

18-2 $\omega = 30$ and $\omega = 50$ are unstable, and $\omega = 20$ and $\omega = 40$ are stable.

18-4   Stable oscillation at $\omega = 10$ with $X = 6.37$.

Index